Studies in
Natural Products Chemistry

Volume 1
Stereoselective Synthesis (Part A)

Studies in Natural Products Chemistry

Volume 1

Stereoselective Synthesis (Part A)

Edited by

Atta-ur-Rahman
H.E.J. Research Institute of Chemistry,
University of Karachi, Karachi 32, Pakistan

ELSEVIER

Amsterdam — Oxford — New York — Tokyo   1988

ELSEVIER SCIENCE PUBLISHERS B.V.
Sara Burgerhartstraat 25
P.O. Box 211, 1000 AE Amsterdam, The Netherlands

Distributors for the United States and Canada:

ELSEVIER SCIENCE PUBLISHING COMPANY INC.
52, Vanderbilt Avenue
New York, NY 10017, U.S.A.

ISBN 0-444-42970-0 (Vol. 1A)
ISBN 0-444-42971-9 (Series)

Printed in The Netherlands

FOREWORD

The explosive growth in the natural sciences in this century has created an
ever-growing need for books and journals which provide state-of-the-art over-
views in specific fields of research. With the advances made in spectroscopy,
as well as in chromatographic techniques, it has become possible to purify and
unravel the structures of complex primary and secondary metabolites from the
plant and animal kingdoms of an increasing order of complexity. While this has
promoted a deeper understanding of some of the underlying chemistry which controls living processes, it has also provided organic chemists with complex synthetic targets, and posed new challenges to their genius for developing synthetic
approaches to these substances. The study of the chemistry of natural products
has therefore had a profound impact on the development of organic chemistry,
having attracted the efforts of such giants as Woodward, Robinson, Todd and
Perkin, to mention but a few. The Herculean efforts of Woodward and Eschenmoser
which finally resulted in the first synthesis of vitamin B_{12} two decades ago,
constituted an important landmark, and heralded a new era in organic synthesis.
Since the 1960's the emphasis has shifted to asymmetric synthesis, efforts
having been directed by a number of leading groups towards the development of
new synthetic methods which would afford the desired products with a high enantiomeric excess.

In view of these developments, it was felt that there was a strong need for a
series of volumes which would provide comprehensive accounts by leading scientists in each area, covering the broad developments as well as highlighting the
research contributions of the authors. The present volume is the first of a
series of volumes which will be devoted to advances made in stereoselective
synthesis of natural products. Other volumes in the series will be devoted to
structure elucidation techniques and other selected areas of natural products
chemistry. It is hoped that the series will provide a platform on which the
major developments in the field can be presented by renowned experts and that it
will prove to be an important and useful addition to the current literature.

This volume covers synthetic approaches to a wide variety of natural products
including indole alkaloids, nucleoside antibiotics, anthracyclines and a number
of other classes with diverse structures. The approaches adopted by the authors
either highlight the various synthetic strategies used for a particular class
of natural products or focus attention on the versatility of a certain approach
to synthesising a wide diversity of natural products. All the contributors are
eminent scientists who have made significant contributions to the progress of
natural products chemistry. It is hoped that the articles will provide stimulat-

ing and enjoyable accounts of the work accomplished in each field, and will prove to be useful to a large community of synthetic organic chemists.

I wish to express my deep gratitude to Miss Khurshid Zaman for her assistance in the preparation of the manuscript.

March 1988   Atta-ur-Rahman, Editor

CONTENTS

CONTRIBUTORS

Jan Bergman	Department of Organic Chemistry, Royal Institute of Technology, S-100 44 Stockholm, Sweden
Josep Bonjoch	Laboratory of Organic Chemistry, Faculty of Pharmacy, University of Barcelona, 08028 Barcelona, Spain
Joan Bosch	Laboratory of Organic Chemistry, Faculty of Pharmacy, University of Barcelona, 08028 Barcelona, Spain
Michael T. Crimmins	Department of Chemistry, University of North Carolina, Chapel Hill, NC 27514, U.S.A.
Philip Paul Garner	Department of Chemistry, Case Western Reserve University, Cleveland, OH 44106, U.S.A.
Gordon W. Gribble	Department of Chemistry, Dartmouth College, Hanover, NH 03755, U.S.A.
Miyoji Hanaoka	Faculty of Pharmaceutical Sciences, Kanazawa University, Takara-machi, Kanazawa, 920 Japan
W. Gary Hollis, Jr.	Department of Chemistry, University of North Carolina, Chapel Hill, NC 27514, U.S.A.
Chihiro Kibayashi	Tokyo College of Pharmacy, 1432-1 Horinouchi, Hachioji, Tokyo 192-03, Japan
Mauri Lounasmaa	Laboratory for Organic and Bioorganic Chemistry, Technical University of Helsinki, SF-02150 Espoo, Finland
Eugene A. Mash	Department of Chemistry, University of Arizona, Tucson, AZ 85721, U.S.A.
Fuyuhiko Matsuda	Sagami Chemical Research Centre, Nishi-Ohnuma 4-4-1, Sagamihara, Kanagawa 229, Japan
C.J. Moody	Department of Chemistry, Imperial College of Science and Technology, London SW7 2AY, United Kingdom

Kenji Mori — Department of Agricultural Chemistry, The University of Tokyo, Yayoi 1-1-1, Bunkyo-ku, Tokyo 113, Japan

Yoshio Nishimura — Institute of Microbial Chemistry, 14-23 Kamiosaki 3-chome, Shinagawa-ku, Tokyo 141, Japan

Mugio Nishizawa — Faculty of Pharmaceutical Science, Tokushima Bunri University, Yamashiro-cho, Tokushima 770, Japan

Rosemary O'Mahony — Department of Chemistry, University of North Carolina, Chapel Hill, NC 27514, U.S.A.

William H. Pearson — Department of Chemistry, The University of Michigan, Ann Arbor, MI 48109, U.S.A.

A.V. Rama Rao — Regional Research Laboratory, Hyderabad - 500 007, India

James H. Rigby — Department of Chemistry, Wayne State University, Detroit, MI 48202, U.S.A.

Shiro Terashima — Sagami Chemical Research Centre, Nishi-Ohnuma 4-4-1, Sagamihara, Kanagawa 229, Japan

Stereoselective Synthesis

INDOLOCARBAZOLE ALKALOIDS

JAN BERGMAN

INTRODUCTION

At present only a few natural products possessing an indolocarbazole unit are known. They include the antibiotics staurosporine (1) (from <u>Streptomyces staruosporeus</u>)[1-4], rebeccamycin (2) (a new antitumor antibiotic produced by <u>Nocardia</u> <u>aeroligenes</u>)[5] and the pigments[6] arcyriaflavin B (3b) and C (3c) from the slime mould <u>Arcyria</u> <u>denudata</u>. Arcyriarubin B (4b) and C (4c) as well as arcyroxepin A (5) are non-indolocarbazolic congeners to 3.

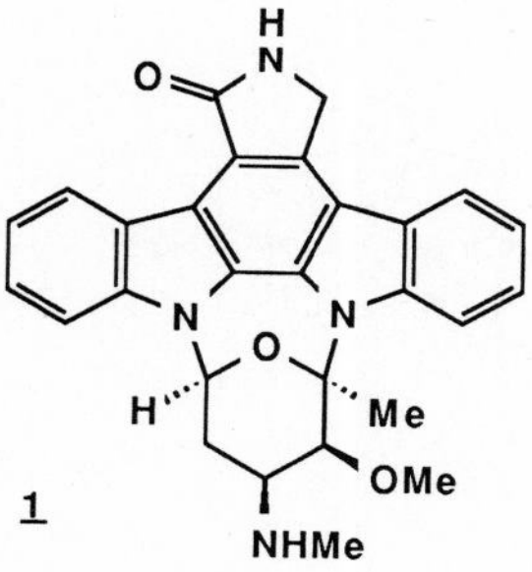

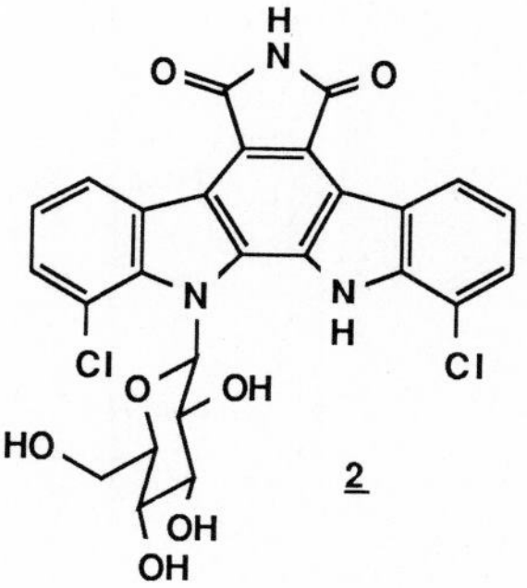

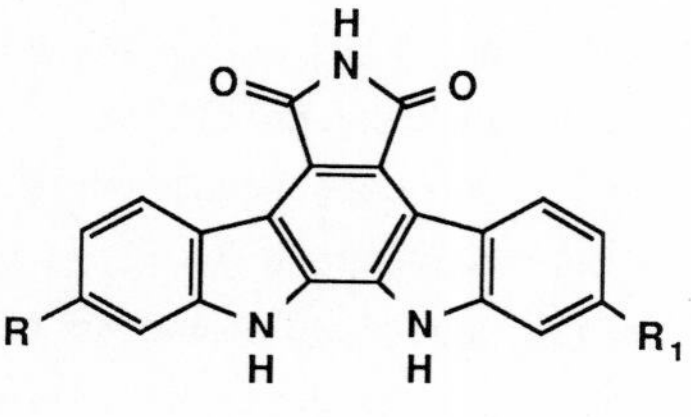

<u>3</u> a R=R$_1$=H
    b R=H, R$_1$=OH
    c R=R$_1$=OH

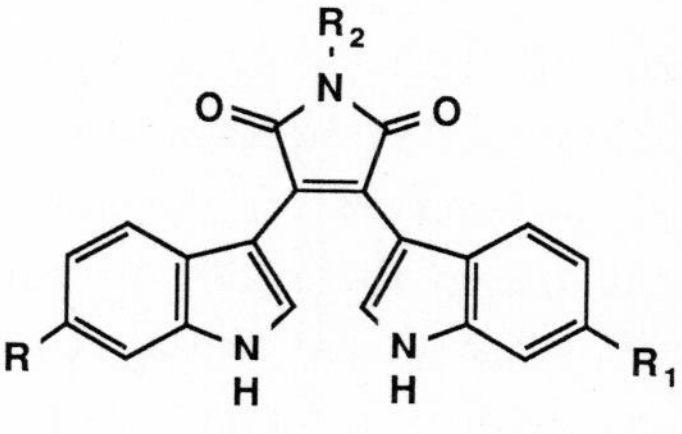

<u>4</u> a R=R$_1$=R$_2$=H
    b R=R$_2$=H, R$_1$=OH
    c R=R$_1$=OH, R$_2$=H
    d R=R$_1$=H, R$_2$=CH$_3$
    e R=R$_1$=H, R$_2$=Bn

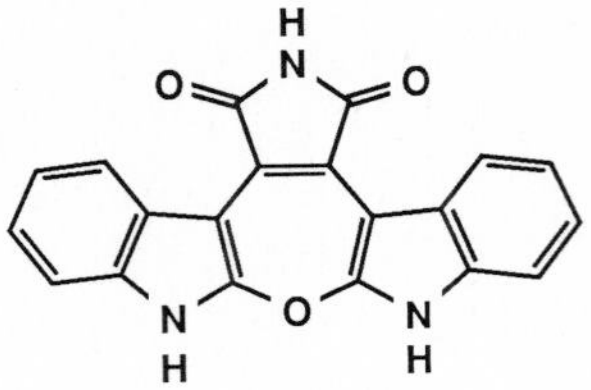

5

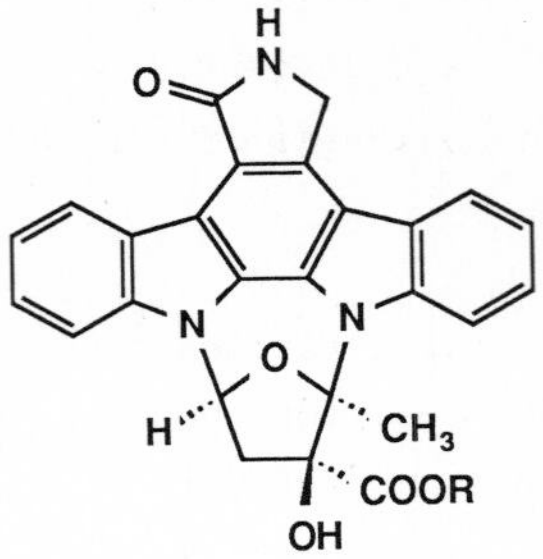

6 a R=CH₃
   b R=H

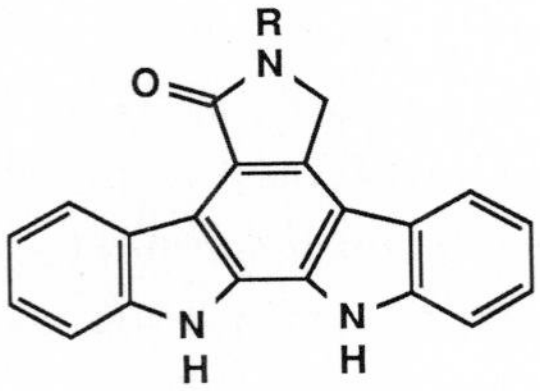

7 a R=H
   b R=Bn

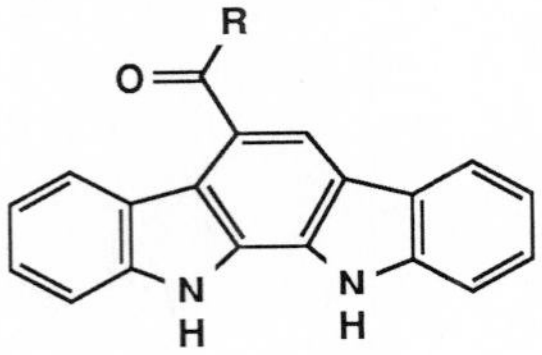

8 a R=NH₂
   b R=OCH₃

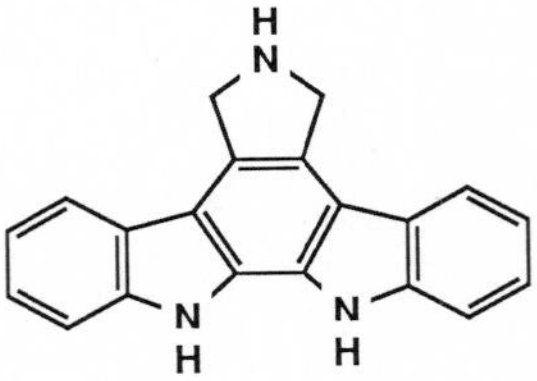

9

Staurosporine (1) is a potent platelet aggregation inhibitor[3] and likewise a potent inhibitor[4] of protein Kinase C. Subsequently similar inhibiting activity was reported for the new antibiotic 6 (and some congeners) isolated from *Nocardiopsis* sp. K-252. In this study, the indolocarbazole 7a (*i.e.* the aglycon of 1 and 6) was isolated as a natural product and also found to be a potent inhibitor of protein Kinase C. Compounds 1, 6 and 7a seriously affect the function of platelets, mast cells and several other cells and tissues. The availability of these indolocarbazoles should facilitate studies on the physiological role of protein Kinase C and calmodulin in the Ca^{2+}-messenger system. In this connection it appears to be of importance to evaluate the activity of "simplified" synthetic analogues of 7a such as 8a and 9.

The first synthetic efforts in the field were reported 1980 by Steglich[6] who prepared the arcyriarubin analogue 4d from N-methyl-3,4-dibromomaleimide and the indole Grignard reagent. More recently several other routes, more or less related to the synthesis of staurosporine (1) or its aglycon 7a, have been reported[8-17].

All the indolocarbazoles isolated so far are indolo[2,1-b]carbazoles; no members of the other four possible systems have been reported as a unit in natural products. Recently it has been suggested[18,19] that an acid-induced selfcondensation (cf. ref. 20) of indole-3-carbinol (10) is responsible for the antitumor effect associated with 10, since indolo[3,2-b]carbazole(11) binds to the TCDD (2,3,7,8-tetrachlorodibenzo-p-dioxin) receptor almost as efficiently as TCDD itself, whereas 10 seems not to bind at all. However, it should be stressed that 11 (or a derivative) has yet to be isolated from a natural source.

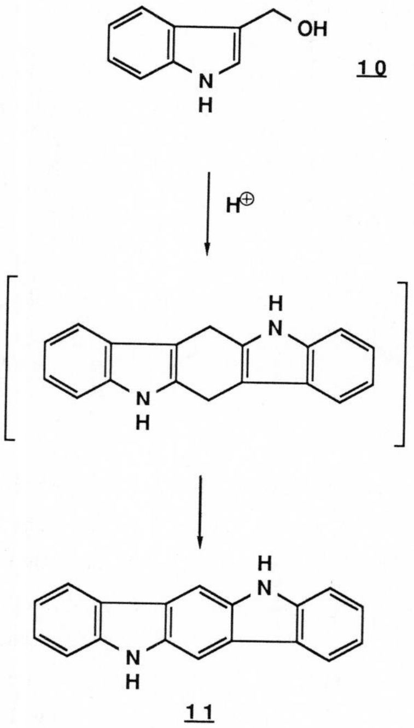

Scheme 1

Nothing is known about the biosynthesis of the indolocarbazole alkaloids but an obvious candidate for involvement, as speculated in Scheme 2, is tryptophan or indole-3-acetic acid (IAA). An un-confirmed report[21] which states that 12 is responsible for the plant hormone effect of IAA is of interest in this connection.

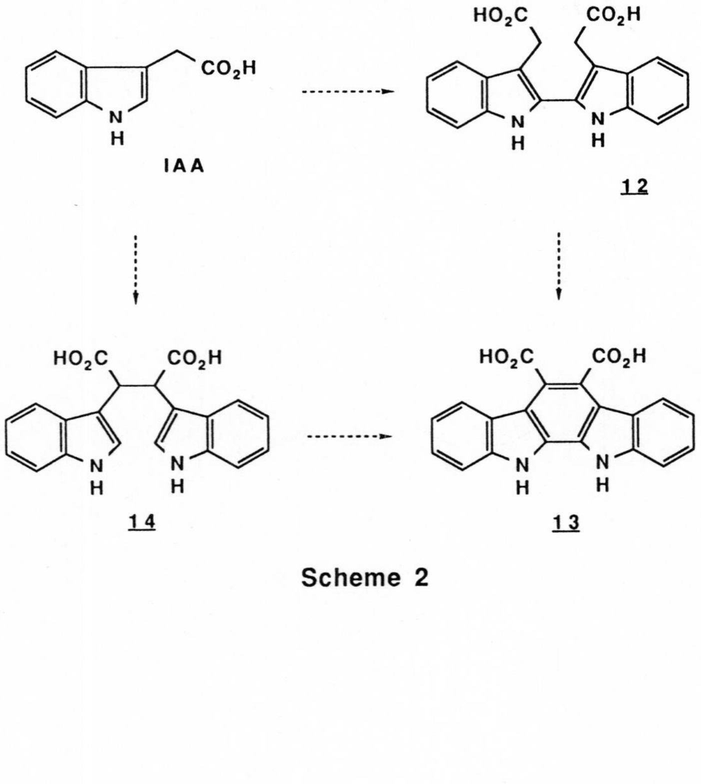

Scheme 2

Synthetic Studies

In connection with the isolation work of the pigments from <u>Arcyria denudata</u> Steglich[6] synthesized 4d in a straightforward manner by reacting indolylmagnesium iodide with <u>N</u>-methyl-3,4-dibromomaleimide in benzene in the presence of HMPA (Scheme 3). This efficient coupling methodology was later adopted by Weinreb[10] in a slightly modified form for the synthesis of 4e, as outlined in Scheme 4. Attempts to reduce the imide from the 2,2'-coupling reaction with LiAlH$_4$ and related reagents were unsuccessful because only partial reduction to the hydroxy lactam occurred, even under forcing conditions.

The conditions for the Clemmensen reduction step (15→7b) was originally worked out by Raphael[9] in connection with his approach to arcyriaflavin B, which is outlined in Scheme 5.

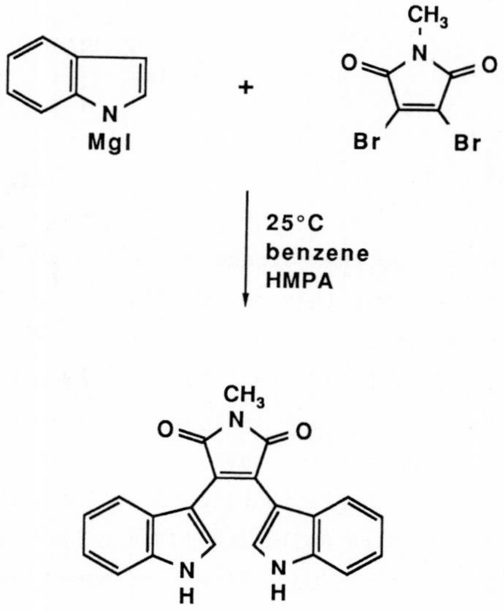

Scheme 3

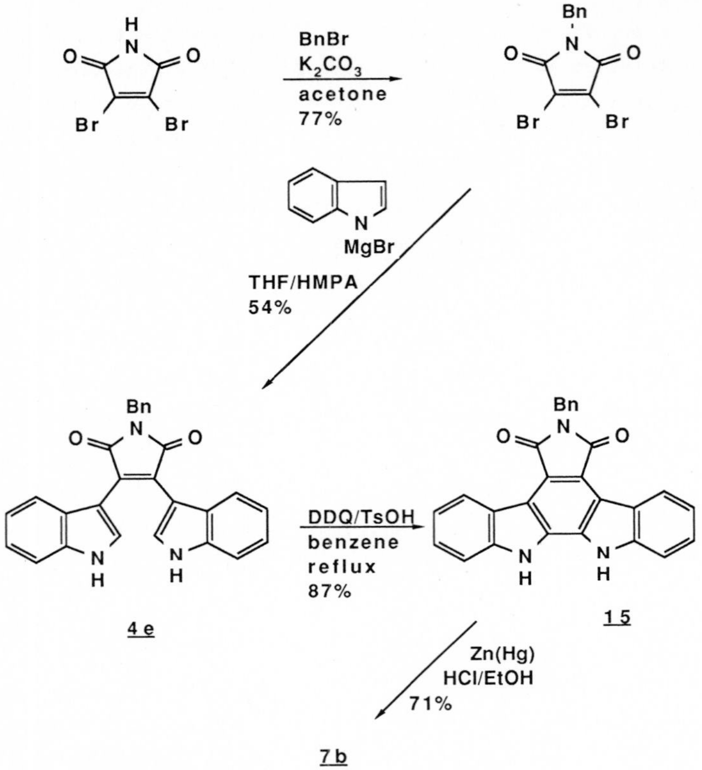

Scheme 4

The ylid was generated from the corresponding phosphonium bromide (K_2CO_3/18-crown-6/CH_2Cl_2) and was then condensed with o-nitrocinnamaldehyde to yield a (Z,E)-diene mixture which was quantitatively converted to the pure (E,E)-diene 17. Reaction with maleimide subsequently gave the desired Diels-Alder adduct in excellent yield. Dehydrogenation of this adduct (18) with DDQ in t-butylbenzene gave the required substituted terphenyl 19(75-80% yield). The crucial double indolization by nitrene insertion was effected (65% yield) with PPh_3 in refluxing (40h) collidine. Minor byproducts in this step included two isomeric mononitrene carbonyl insertion products. The final demethylation required some experimentation but was finally effected by Prey's method (heating with neat pyridine-HCl).

N-benzylmaleimide reacted similarly with the parent compound of 17 (i.e. 1,4-di(o-nitrophenyl)butadiene) to give a Diels-Alder adduct which could in steps be converted to 7b (cf Scheme 4).

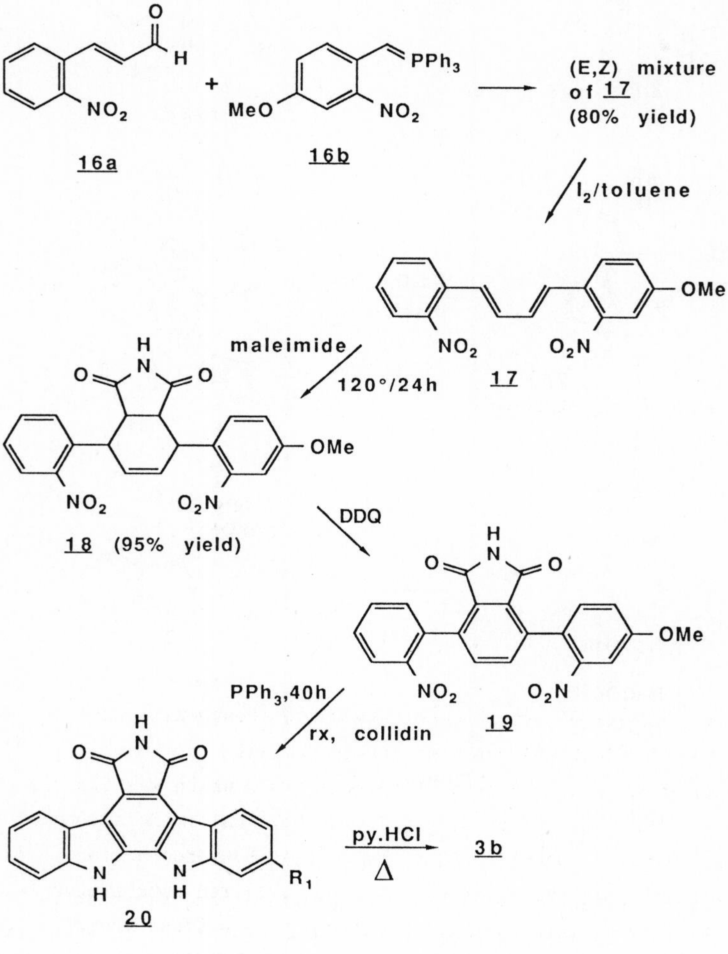
O
H
NO2
16a
+
PPh3
MeO
NO2
16b
(E,Z) mixture
of 17
(80% yield)
I2/toluene
OMe
NO2
O2N
17
maleimide
120°/24h
H
N
O
O
OMe
NO2
O2N
18 (95% yield)
DDQ
H
N
O
O
OMe
NO2
O2N
19
PPh3,40h
rx, collidin
H
N
O
O
N
H
N
H
R1
20
py.HCl
Δ
3b

Scheme 5

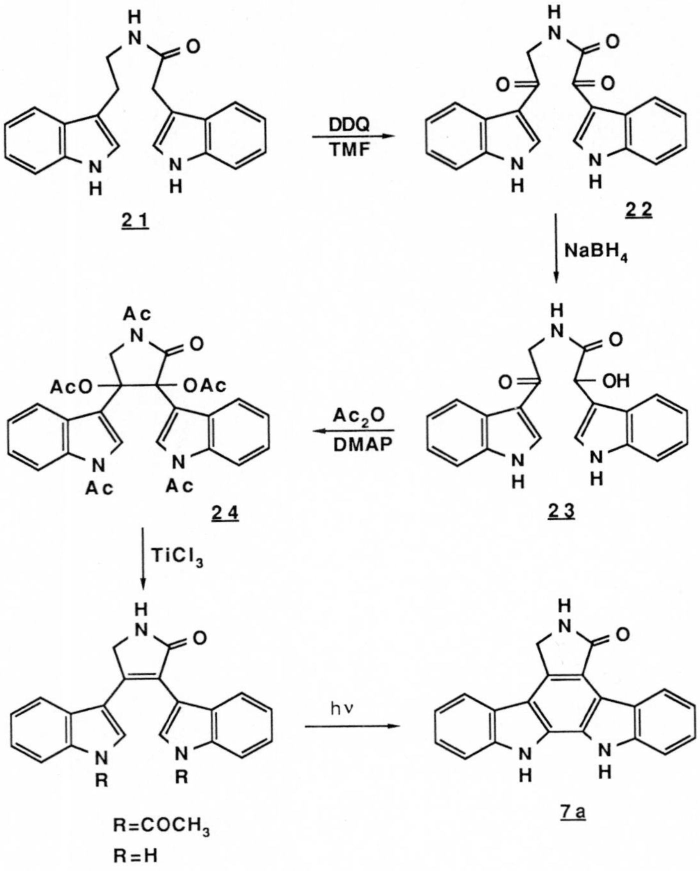

Scheme 6

The synthetic approach[8] (Scheme 6) of Winterfeldt and Sarstedt is interesting as it illustrates a biosynthetic model reaction. In the first step the amide is oxidized with DDQ containing small amounts of water in THF (cf. refs 22 and 23). The diketone 22 obtained is then selectively reduced to the hydroxy-ketone 23, which upon treatment with Ac_2O-DMAP was directly cyclized to the pentacetate 24. This latter compound could, in acceptable yields, be reduced only with $TiCl_3$ in aqueous acetone. All efforts to photocyclize the diacetate were unsuccessful, though smooth deacetylation with sodium bicarbonate in aqueous methanol followed by irradiation in methanol did indeed produce the aglycon of staurosporin (7a).

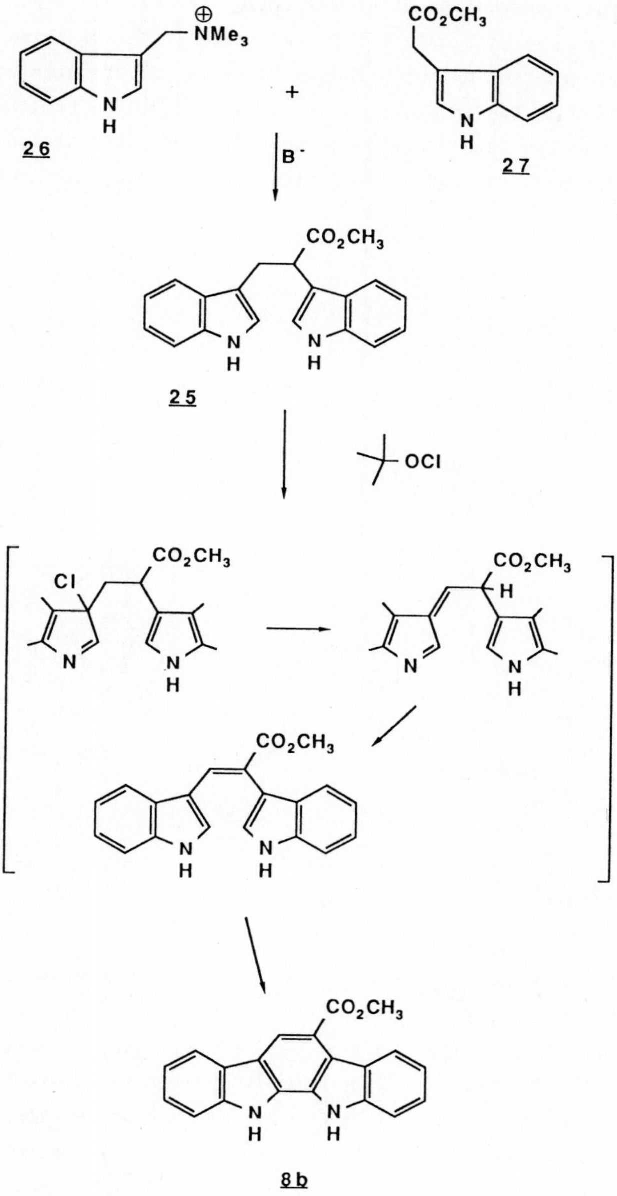

Scheme 7

12

As two different building blocks are involved (Winterfeldt and Sarstedt used tryptamine and indolyl-3-acetyl chloride)[24] the method in Scheme 6 is of interest for the construction of unsymmetrical systems. Sarstedt and Winterfeldt also reported the cyclization outlined in Scheme 7. The starting diindolylethane 24, easily obtained from methyl 3-indolyl acetate (27) by gramine alkylation, directly cyclized to the indolo [2,3-a] carbazole (8b) on treatment with t-butyl hypochlorite. Clearly this useful sequence merits further studies not the least as a source for model compounds for evaluation of biological mechanisms (cf compound 8a).

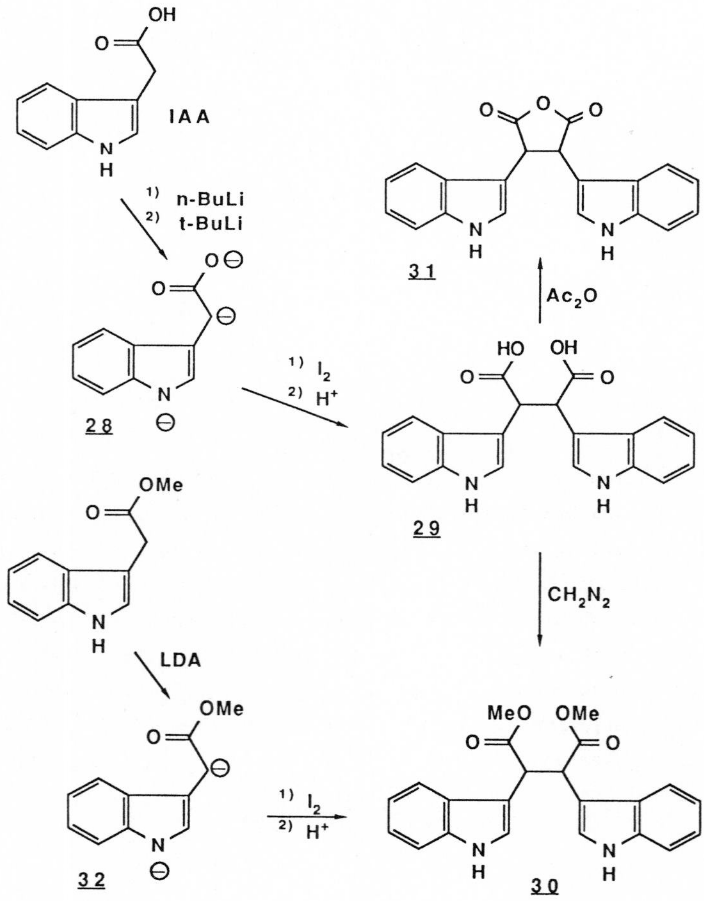

Scheme 8

As already discussed it is reasonable to assume that the skeleton in the indolocarbazole alkaloids is derived from indole-3-acetic acid or tryptophan moieties. Also, it is likely that the a or the b bond (Fig.1) is the first to be biosynthetically formed, as suggested by the coexistence of 3c, 4c and 5 in the slime mould <u>Arcyria denudata</u>. As shown in Scheme 6, formation of the a bond was used by Sarstedt and Winterfeldt as the first connective step in their biomimetic synthesis of 7.

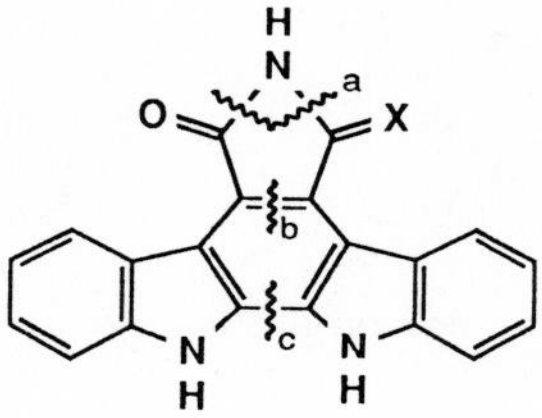

Fig. 1

Bergman and Pelcman have reported a biomimetic synthesis of the indolo-[2,3-α]pyrrolo[3,4-c]carbazole system where the b bond is formed in the first step by oxidative coupling of the indole-3-acetic acid trianion or the methyl indole-3-acetate dianion. The trianion 28 was formed by the sequential addition of n-BuLi(2 eq.) and t-BuLi(1 eq.) to indole-3-acetate acid (IAA) in THF at -78 °C. Coupling with iodine (0.5 eq.) followed by an acidic workup, afforded the bisindolesuccinic acid 29 which was isolated as the corresponding dimethyl ester 30 or as the anhydride 31 (Scheme 8.) A higher yield (85%) of 30 was obtained by iodine promoted coupling of the dianion 32 prepared from methyl 3-indolylacetate and LDA.

14

The diester 30 was formed as a mixture of dl and meso forms which could easily be separated by crystallization and chromatography. Heating 30 or 31 with DDQ afforded the maleimide 4e which previously has been transformed to 15 with DDQ/p-TsOH. (Scheme 4) By this method it was also possible to convert 32 directly to 15.

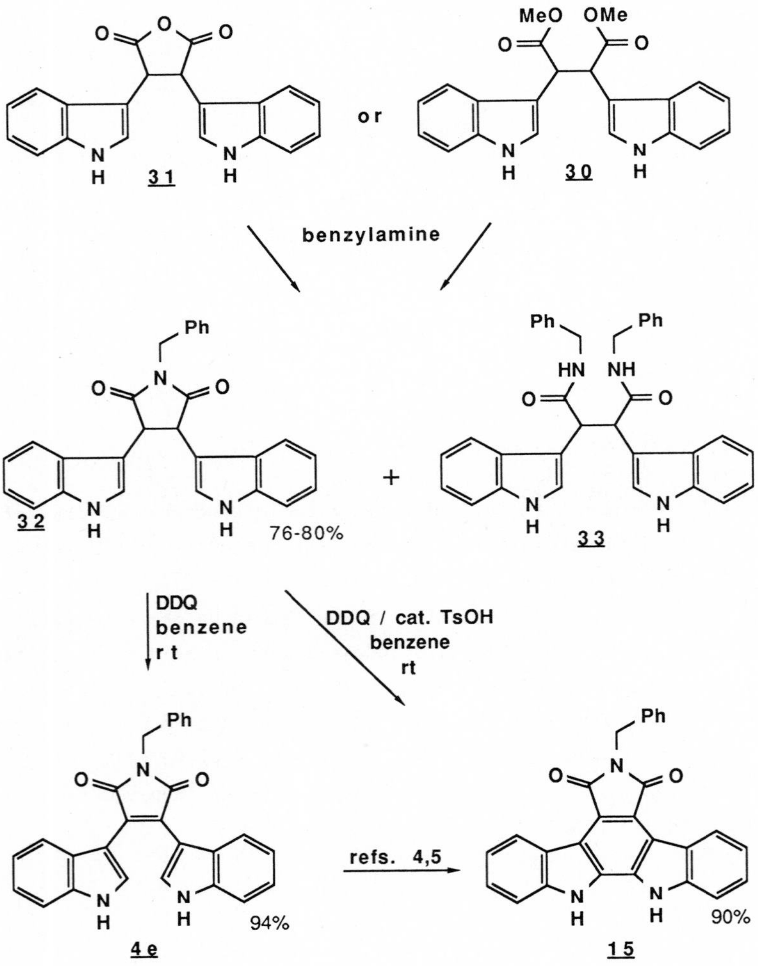

Scheme 9

Recently Bergman and Pelcman have developed[14] the method outlined in Scheme 10. The readily available Diels-Alder adduct 36 gives with e.g. phenylhydrazine a bis-phenylhydrazone which can be forced to undergo a two-fold Fischer indolization[25] in the crucial step. This step requires PPSE[26] in sulfolane as the cyclization agent. Conventional agents such as polyphosphoric acid, $ZnCl_2$, HCl, and H_2SO_4 failed or gave complex mixtures. Utilization of o-chlorophenylhydrazine in this method gives a fast and convenient entry to the aglycon of rebeccamycin (2). Scheme 10 could readily be modified for preparation of unsymmetrical indolo[2,3-a]pyrrolo[3,4-c]carbazoles by using two different phenylhydrazines in two distinct derivatization steps.

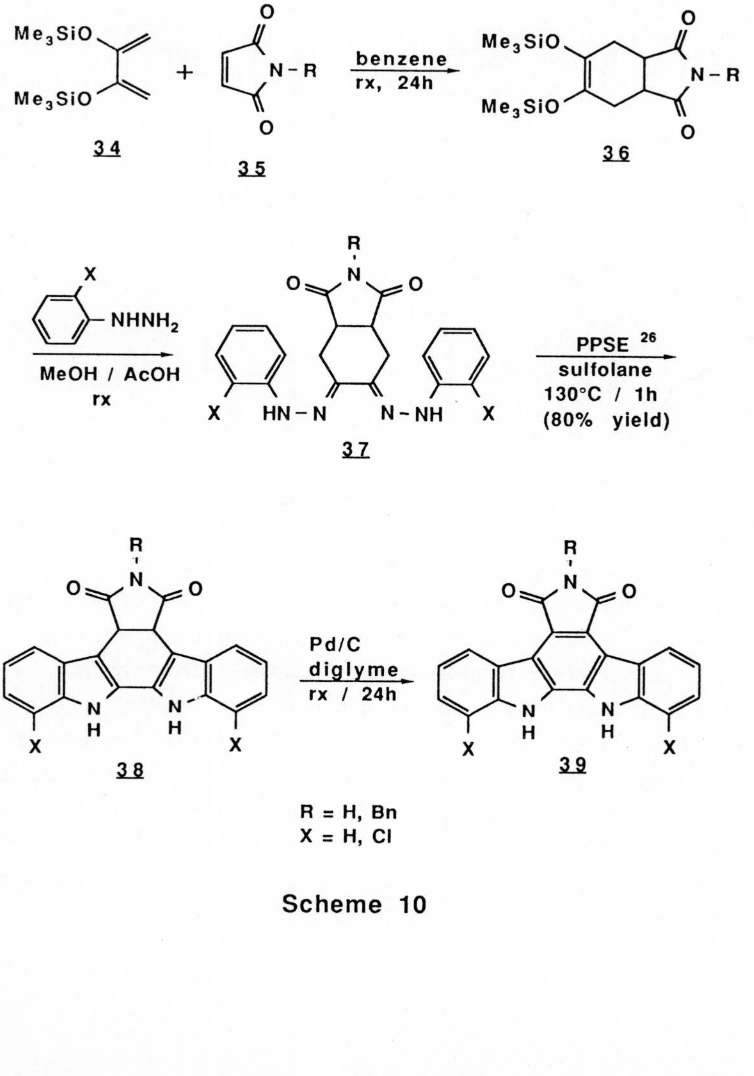

Scheme 10

Another variant of Scheme 10 is outlined in Scheme 11. In this method fewer transformations are required and only inexpensive starting materials are employed.

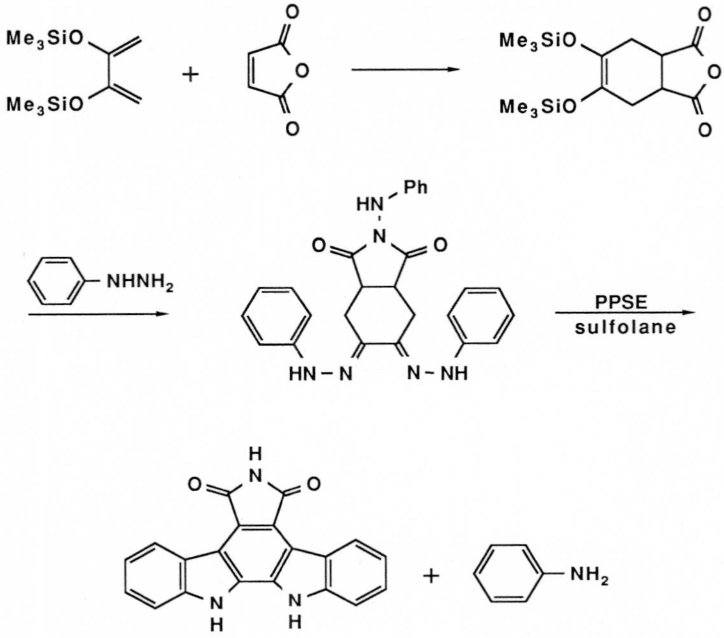

Scheme 11

Magnus and Sear[15-17] have described the use of indole-2,3-quinodimethane intermediates for the synthesis of indolocarbazoles. Thus, both nitrogen atoms of tryptamine were protected and the resulting product, 40, was formylated in 2-position with Cl$_2$CHOCH$_3$/ TiCl$_4$ at -35° to 41; condensation was then effected with 2-aminostyrene to give imine 42 (Scheme 12). This imine was then subjected to the standard conditions for generating an indole-2,3-quinodimethane intermediate (methyl chloroformate-chlorobenzene/140 °C), which gave the pentacyclic carbamate 43 as a 4:1 mixture of epimers. Dehydrogenation, followed by treatment with hydrazine hydrate in THF at 20 °C removed the phthalimido protecting group. (Somewhat surprisingly, the carbamate group survived this transformation). Treatment of amine 45 with phosgene in CH$_2$Cl$_2$, followed by TiCl$_4$ (at 0°C) gave the hexacyclic indolocarbazole 46.

Although this approach involves several interesting operations its overall complexity and low total yields renders it noncompetitive with the methods described in the previous Schemes.

In the connection with the development of the chemistry shown in Scheme 12, Magnus and Sear also studied simpler variants as outlined in Scheme 13. All attempts to cyclize compound 49 failed.

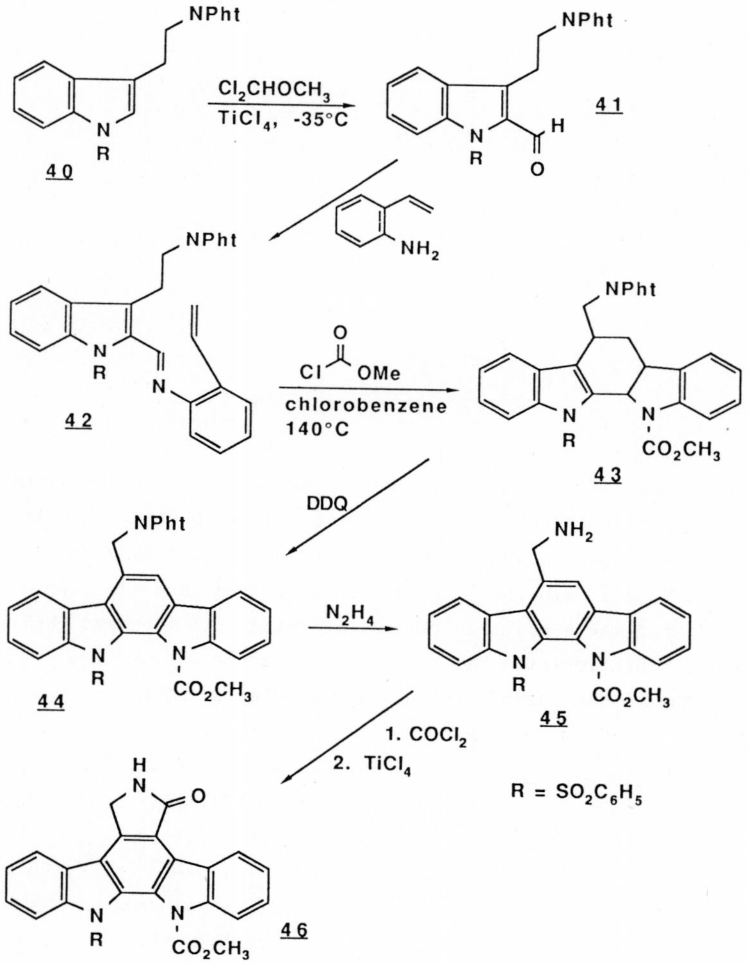

Scheme 12

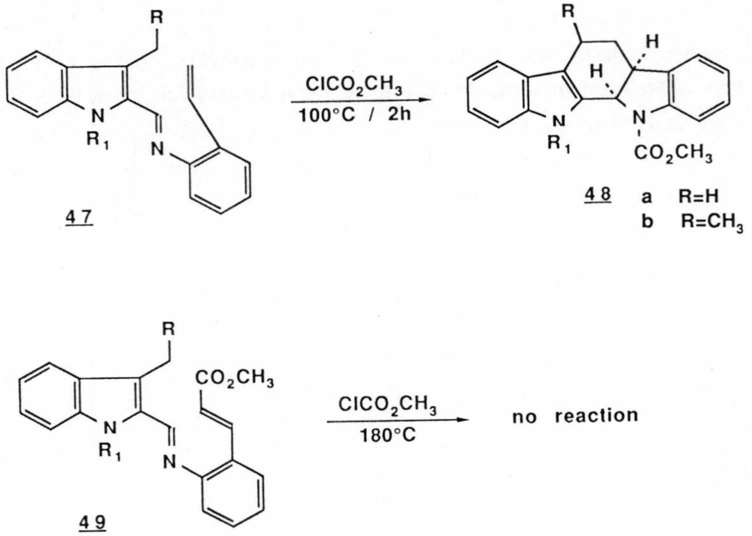

Scheme 13

Dehydrogenation of 48a with DDQ, followed by hydrolysis gave the parent compound, which is also readily available [27] from 1,2-cyclohexanedione and phenylhydrazine as outlined in Scheme 14. Compound 48b similarly gave 6-methyl-11H-indolo[2,3-a]carbazole. All attempts to formylate 11H-indolo [2,3-a]carbazole (50) in the 6-position failed. Likewise, all attempts (NBS, CrO3, KMnO4 etc) to functionalize the methyl group in 6-methyl-11H-indolo[2,3-a]carbazole failed (N-protected or not).

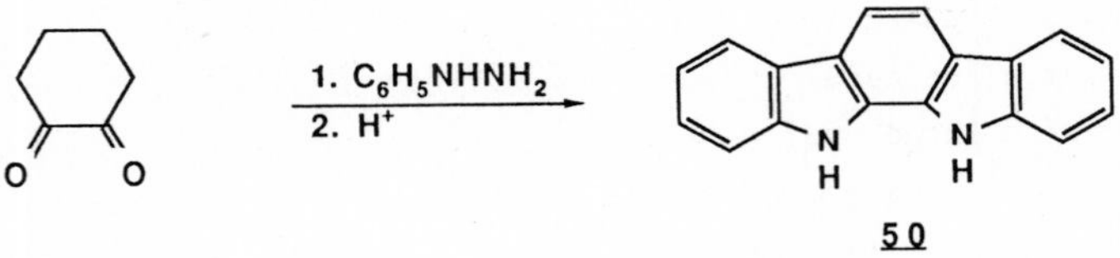

Scheme 14

Cycloadditions based on 2,2'-biindolyls should be an attractive approach to indolocarbazoles. A new synthesis[28] of 2,2'-biindolyls (Scheme 15) is based on conversion of indoles to 1,1'-carbonylindoles followed by 2,2'-coupling induced by Pd(OAc)$_2$ in acetic acid. This method can be used also for the preparation of unsymmetrical 2,2'-biindolyls if an indole-1-carbonyl chloride is condensed with a different indole nucleus in the first step. For symmetrical indoles 1,1-carbonylimidazole is the preferred reagent. The method shown in Scheme 15 has recently,[29] been used for the preparation of a number of alkoxy-2,2'-biindolyls.

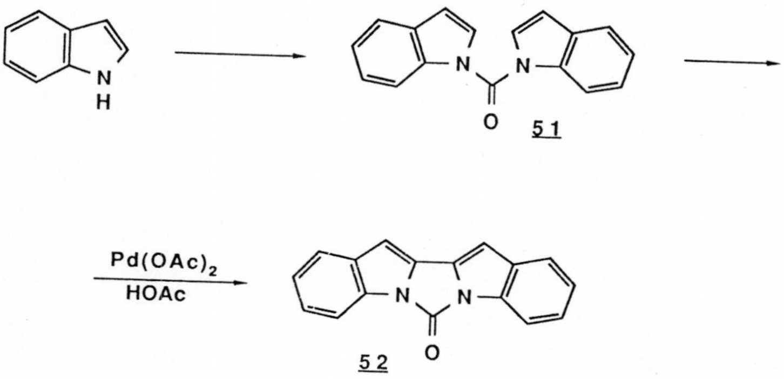

Scheme 15

The coupled product 52 seemed, with its forced cisoid arrangement, to be an ideal candidate for cycloadditions such as the one outlined in Scheme 16 Cycloadditions of related systems had previously been studied by Zander[30] (2,2'-Bibenzo[b]thienyl and maleic anhydride) and by Clardy and Kaneko who prepared[5] the aglycon of rebeccamycin in a low yield. The approach in Scheme 16 was however found to be inferior to the methods outlined in Scheme 10 and 11, particularly for the synthesis of the aglycon of rebeccamycin.

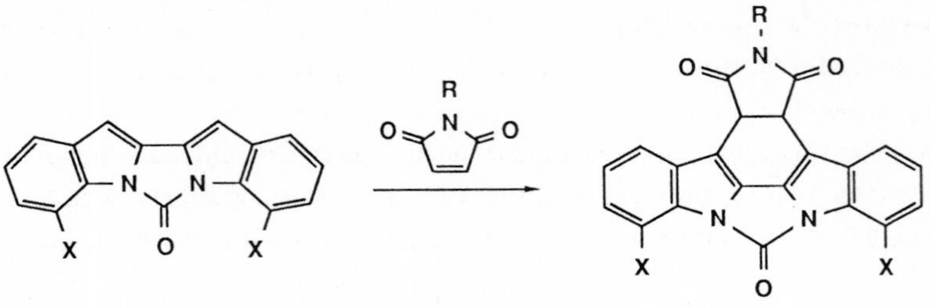

Scheme 16

A minor modification of Scheme 16 is outlined in Scheme 17. The key compound 55 had previously been prepared by Bergman[31] by reduction of indigo (53) with hydrazine.

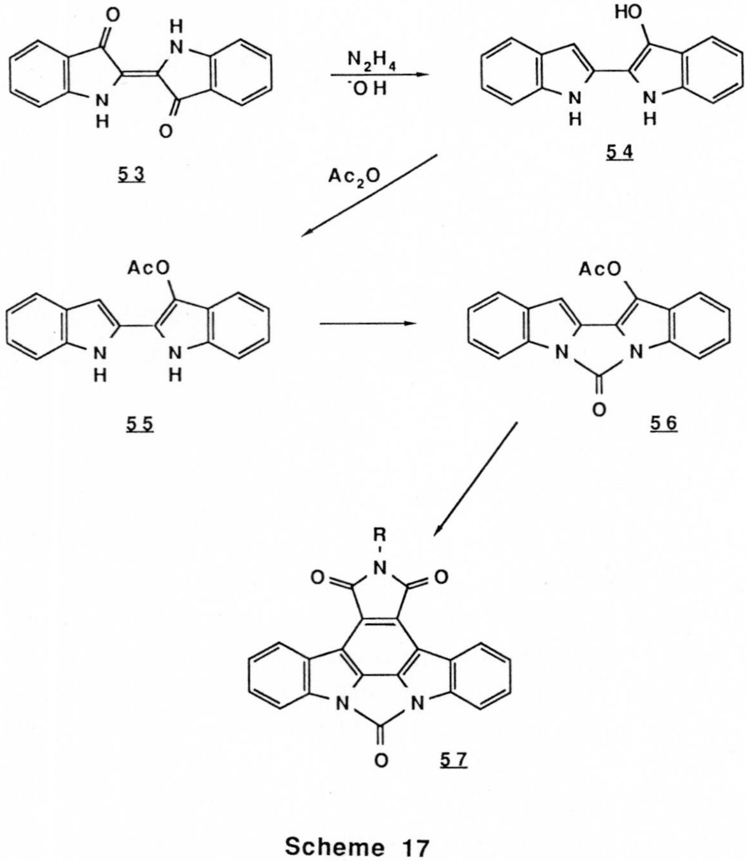

Scheme 17

Clardy and Kaneko[5] have recently utilized the 7,7'-dichloro derivative of 55 in a low yield cycloaddition reaction key-step towards rebeccamycin (2), which is, so far, the only sugar-containing indolocarbazole alkaloid that has been synthezised (Scheme 18).

The synthetic material was identical to the natural product by NMR, IR, TLC and optical rotation. Since the sugar moiety was prepared from D-glucose, the absolute configuration of rebeccamycin (2) was determined to be the same as D-glucose.

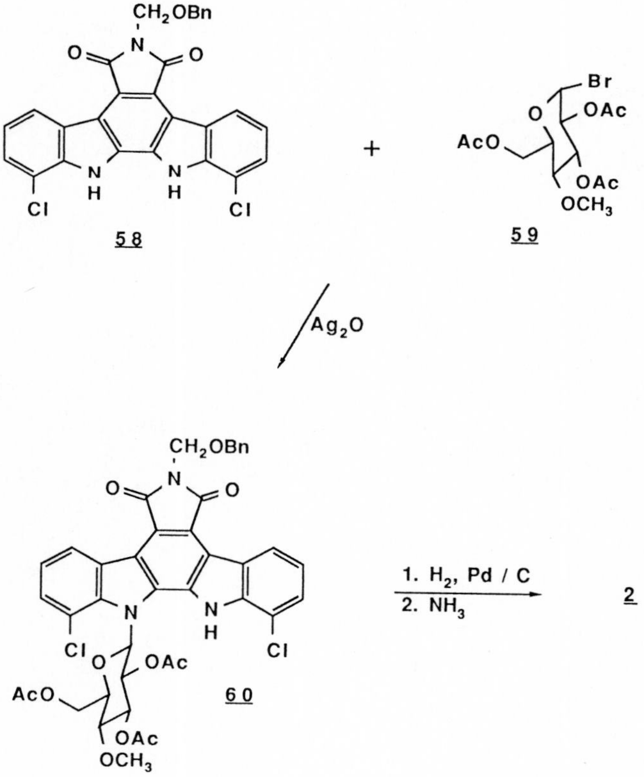

Scheme 18

A total synthesis of staurosporin (1) will pose considerable problems due to the complexity of the amino sugar component and the unsymmetrical nature of the whole molecule. As discussed above, several approaches are available for the aglycon whereas Weinreb[10] has reported the only known approach to the aminosugar component. Suitable aminohexose precursors both in a pyranose form and as protected acyclic ketoaldehyde derivatives were prepared by using a strategy based upon N-sulfinyl Diels-Alder cycloadditions. The required dienes were prepared as outlined in Scheme 19.

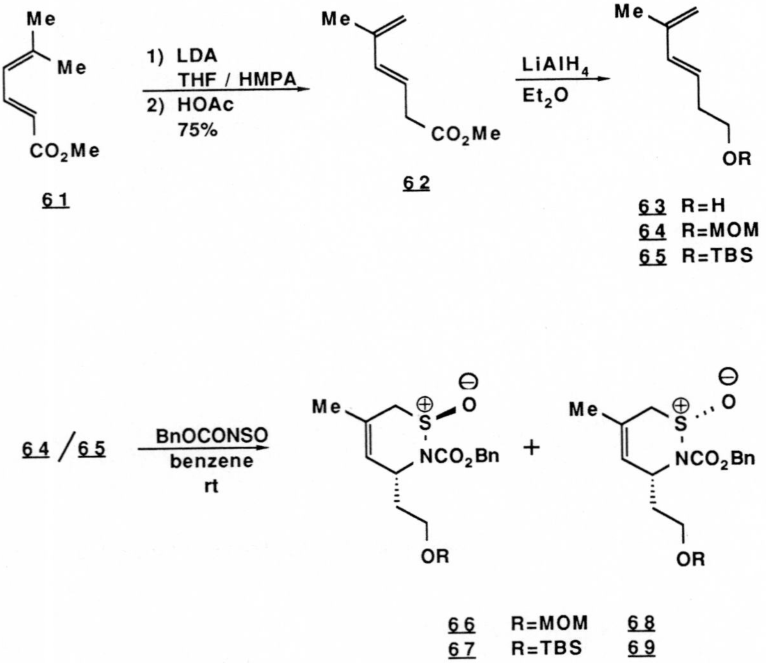

Scheme 19

Cycloadditions of both dienes (64/65) with benzyl sulfinylcarbamate
proceeded in a completely regioselective manner in high yields in toluene at
room temperature. Diene 64 afforded a chromatographically separable 2:1 mixture
of epimeric 3,6-dihydrothiazine oxides 66 and 68, respectively. Similarly, diene
65 gave a 2.2:1 mixture of adducts 67 and 69. These additions are in accord with
the known orientational preferences of N-sulfinyl Diels-Alder reactions. No
regioisomers were detected in either series.

Epoxidation of the major isomer in the MOM series gave exclusively the
desired β-epoxy sultam 70. (Scheme 20). The stereochemistry of 70 was
established by subsequent transformations.

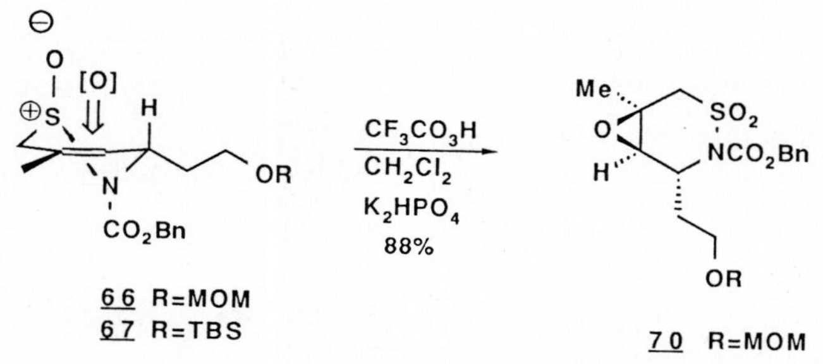

66 R=MOM
67 R=TBS

CF₃CO₃H / CH₂Cl₂ / K₂HPO₄ / 88%

70 R=MOM

Scheme 20

Epoxidation of the minor cis adduct 68 under the same conditions took a totally different stereochemical course. In this case, the reaction was more complex, giving a mixture of α-epoxide 71, α-epoxy sultam 72, and some unepoxidized sultam 73. Further oxidation of 71 at sulfur gave sultam 72, indicating that both compounds were in the same α-epoxide stereochemical series. None of the β-epoxide was detected in this oxidation.

Treatment of 73 with trifluoroperacetic acid under similar conditions gave only β-epoxide 70. The stereoselectivity of this epoxidation can be explained if one assumes that sultam 73 has conformation 73a (Scheme 22).

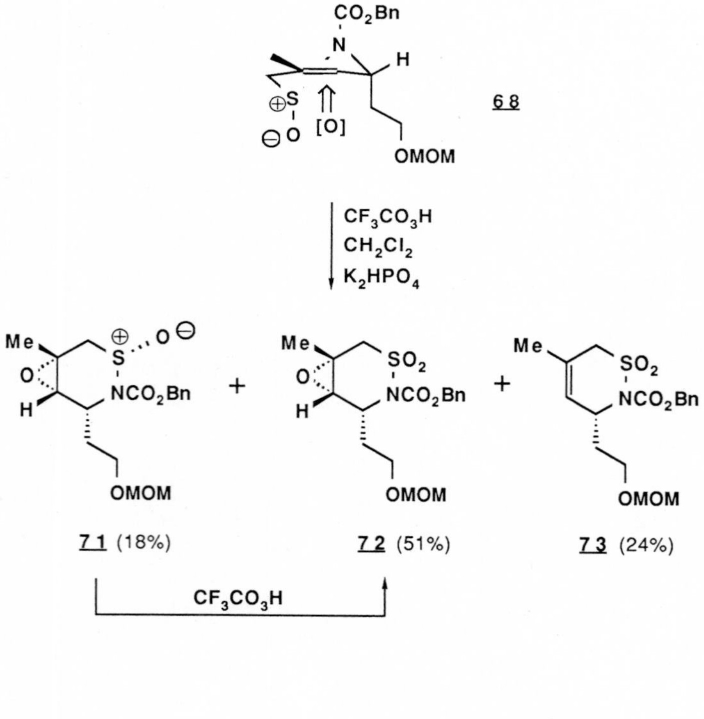

Scheme 21

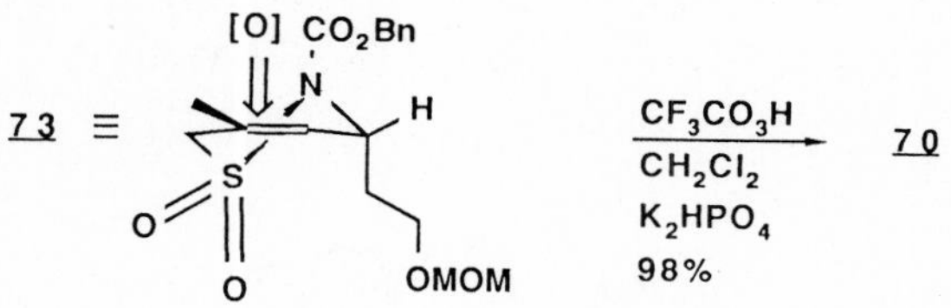

73 ≡ [structure: sultam with [O], CO₂Bn, N, H, S, O, O, OMOM]

$$\xrightarrow[\substack{CH_2Cl_2 \\ K_2HPO_4 \\ 98\%}]{CF_3CO_3H}}$$ 70

Scheme 22

In order to prepare aminohexose derivatives it was necessary to adjust the oxidation level of the terminal side-chain carbon of sultam 73 to that of an aldehyde.

However, deprotection of 73 with methanolic HCl gave the undesired (but interesting) ring expansion product 74 (Scheme 23).

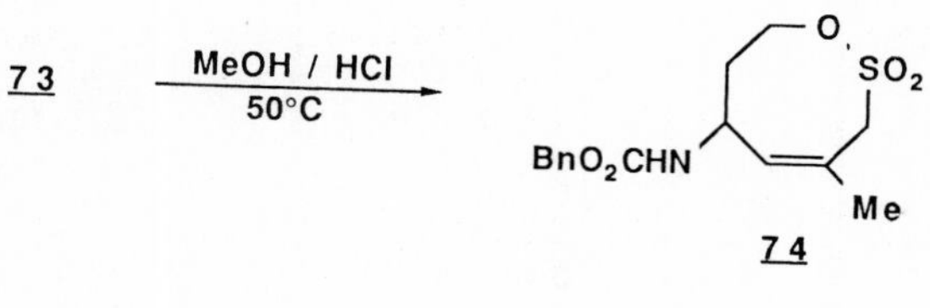

73 $\xrightarrow[50°C]{MeOH / HCl}$ [structure: ring expansion product with O, SO₂, BnO₂CHN, Me] 74

Scheme 23

To avoid this problem TBS-protected adducts were used instead (Scheme 24). The epoxide 77 was then transformed to an acetate and oxidatively cleaved with ozone to give the sugar models 80 and 81 (Scheme 25).

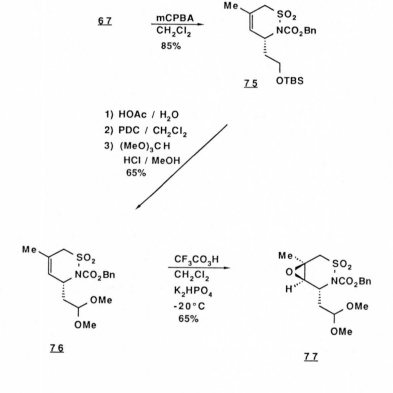

Scheme 24

Since the staurosporin sugar has O-and N-methyl substituents, the series of compounds in Scheme 26 was also prepared. The transformation (83→84) could be neatly effected with RuO$_4$, whereas attempted O$_3$ treatment gave intractable tars.

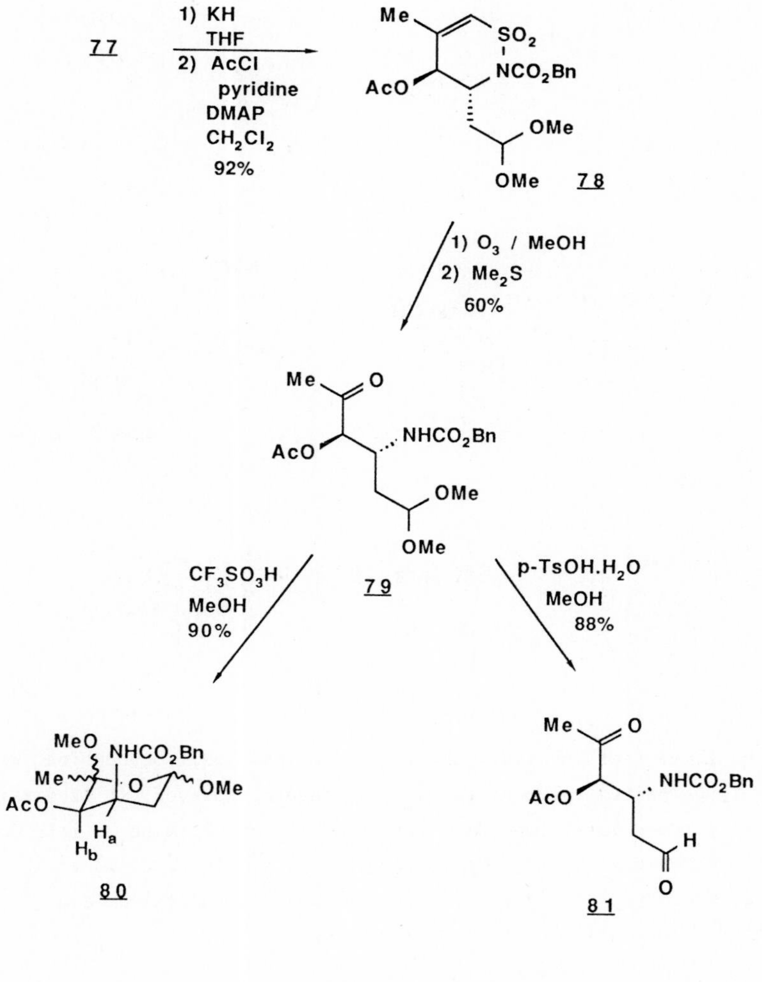

Scheme 25

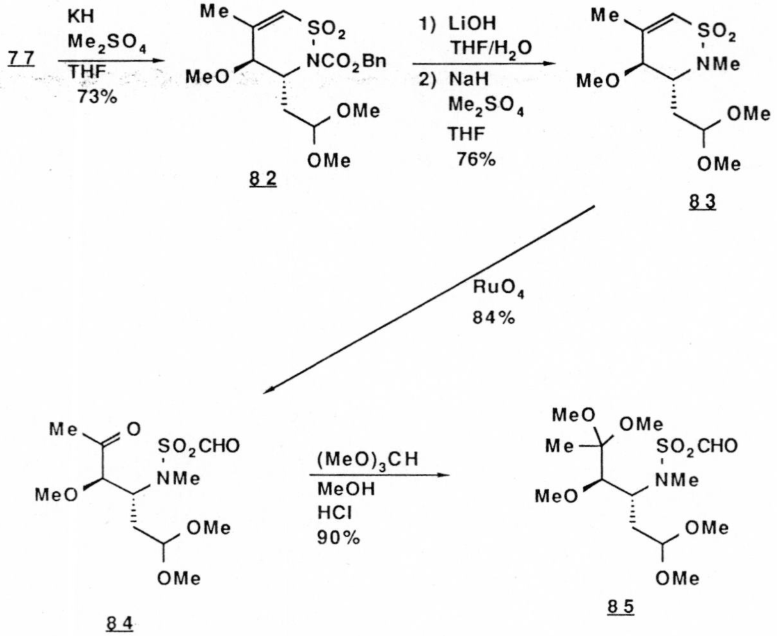

Scheme 26

It is expected that the sugar units now available will be combined with the appropriately protected aglycons in the near future. However, it is first necessary to solve several important problems. The development of methodology to generate the bis-N-glycosidic bonds between the aromatic unit and an aminohexose moiety is a priority, and additionally, nontrivial regiochemical and stereochemical considerations must be addressed.

REFERENCES

1 A. Furusaki, N. Hashiba, T. Matsumoto, A. Hirano, Y. Iwai, S. Omura, Chem. Comm., 800 (1978)..

2 A. Furusaki, N. Hashida, T. Matsumoto, A.Hirano, Y.Iwai and S.Omura, Bull.Chem.Soc. Japan 55, 3681 (1982).

3 S.Oka, M. Kodama, H. Takeda, N. Tomizuka and H. Suzuki, Agric.Biol.Chem. 50, 2723 (1986).

4 T. Tamaoki, H. Nomoto, I. Takahashi, Y. Kato M. Morimoto and F. Tomita, Biochem.Biophys. Res.Commun., 135 397 (1986).

5 T. Kaneko, H. Wong, K.T. Okamoto and J Clardy, Tetrahedron Letters, 4015, (1985).

6 W. Steglich, B. Steffon, L. Kopanski and G. Eckhardt, Angew. Chem. Int. Ed., 19, 459 (1980).

7 S. Nakanishi, Y Matsuda, K. Iwahashi and H. Kase, J. Antibiotics, 39, 1066 (1986) and refs therein.

8 B. Sarstedt and E. Winterfeldt, Heterocycles, 20, 469 (1983).

9 I. Hughes and R.A. Raphael, Tetrahedron Letters, 1441 (1983)

10 S.M. Weinreb, R.S. Garigipati and J.A. Gainor, Heterocycles, 21 309 (1984).

11 J.A. Gainor, Diss. Penn. State Univ. (1983).

12 R.P. Joyce, J.A. Gainor, and S.M. Weinreb, J.Org.Chem. 52, 1177 (1987).

13 J. Bergman and B. Pelcman, Tetrahedron Letters, 4441 (1987)

14 J. Bergman and B. Pelcman. To be published.

15 P.D. Magnus, C. Exon and N.L. Sear, Tetrahedron Letters, 3725 (1983)

16 P.D. Magnus and N.L. Sear, Tetrahedron, 40, 2795 (1984).

17 N.L. Sear, Diss. Indiana University (1985).

18 G. Gillner, J. Bergman, Chr. Cambillau, B Fernström and J.-A. Gustafsson, Mol.Pharmacol. 28, 357 (1985).

19 C.A. Bradfield and L.F. Bjeldanes, J. Tox.Environ.Health, 21, 311 (1987).

20 J. Bergman, Tetrahedron, 26, 3353 (1970).

21 A.A. Bitancourt, Nature, 200, 548 (1963).

22 J. Bergman, R. Carlsson and B. Sjöberg. J.Het.Chem. 14, 1123(1977).

23 Y. Oikawa, T. Yoshioka, K. Mohri, and O. Yonemitsu, Heterocycles, 12, 1457 (1979).

24a Tryptamine and methyl indolyl-3-acetate gives better yields of the precursor when heated together at 170° under N_2.[24b]

24b J. Bergman, unpublished experiments.

25a Two-fold Fischer cyclizations are frequently problematic.[25b,28]

25b J. Fränkel Diss. Aachen (1965).

26a This reagent is prepared [26b] from P_2O_5 and $(CH_3)_3 SiOSi(CH_3)_3$.

  b S. Ogata, A Mochizuki, M. Kakimoto and Y. Imai, <u>Bull.Chem.Soc.Japan</u> <u>59</u>, 2171 (1986)

27a F.G. Mann and T.J. Willcox, <u>J.Chem.Soc.</u> 1525 (1958).

  b W. Moldenhauer and H. Simon, <u>Chem.Ber.</u> <u>102</u>, 1198 (1969).

28   J. Bergman, N. Eklund, <u>Tetrahedron</u>, <u>36</u>, 1439 (1980).

29   D. Black, N. Kumar and L.C.H. Wong, <u>Chem. Comm.</u>, 1174 (1985)

30   M. Zander, <u>Chemiker-Zeitung</u>, <u>101</u>, 507 (1977).

31   J. Bergman and N. Eklund, <u>Chemica Scripta</u>, <u>19</u>, 193 (1982).

PENTACYCLIC *STRYCHNOS* INDOLE ALKALOIDS

JOAN BOSCH and JOSEP BONJOCH

1 INTRODUCTION

1.1 Biosynthesis and Structural Types

Among the numerous indole alkaloids derived from secologanin and tryptophan, the *Strychnos* type includes those alkaloids in which an unrearranged secologanin skeleton may be identified as being attached to the indole nucleus by C_7-C_3 (or C_{21}) and C_2-C_{16} bonds (1). These alkaloids are found only in species of the plant families Apocynaceae and Loganiaceae (2,3).

From a biogenetic standpoint, the *Strychnos* alkaloids are formed from geissoschizine, an early common intermediate in the biosynthesis of the monoterpenoid class of indole alkaloids. Although several mechanisms have been postulated to interconnect the *Corynanthe* alkaloid geissoschizine with those of the *Strychnos* type (4), the details of the rearrangement to dehydropreakuammicine and pre-

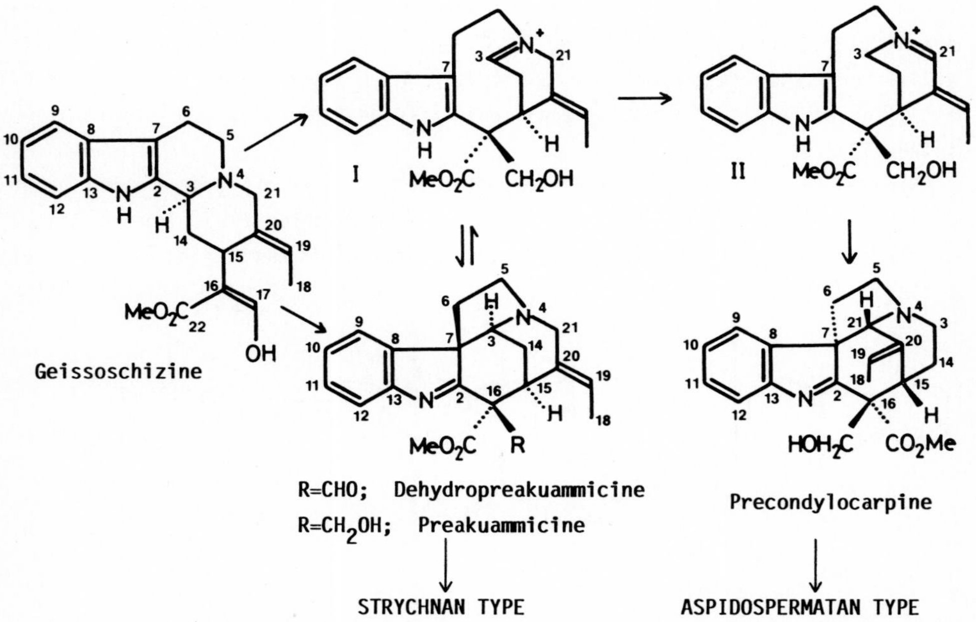

R=CHO; Dehydropreakuammicine

R=CH₂OH; Preakuammicine

Scheme 1. Biosynthesis and biogenetic numbering of *Strychnos* indole alkaloids.

akuammicine remain still unknown. These bases are the precursors of the *Strych-nos* alkaloids having the Strychnan skeleton (C_3-C_7 bond). The intermediate imi-nium salt I can undergo a 1,3 hydrogen shift, with migration of the double bond to the N_4-C_{21} position, to afford II. This conjugated iminium salt may cyclize to precondylocarpine, which is the precursor of the *Strychnos* alkaloids having the Aspidospermatan skeleton (C_7-C_{21} bond). For clarity, the numbering system based on the biogenetic interrelationship of indole alkaloids, as proposed by Le Men and Taylor (5), will be used throughout this chapter.

Hence, two main skeletal types of *Strychnos* alkaloids, depending on the lo-cation of the two-carbon (C_{18}-C_{19}) chain, may be identified. According to the Hesse's classification (2), these skeletal types can be subdivided depending on their chemical complexity. The Strychnan type is composed of 11 skeletal varieties whereas the alkaloids of the Aspidospermatan type are distributed among six skeletal variations.

The absolute configuration of pentacyclic *Strychnos* indole alkaloids follows from their biogenetic origin. All these alkaloids contain an unrearranged seco-loganin moiety and, consequently, have the same absolute configuration at C-15 as natural (−)-secologanin at the corresponding position (C-7). Three exceptions (see also Addendum) are the alkaloids of the Strychnan type pseudoakuammicine [(±)-akuammicine] [(±)-**2**] (6), (+)-20-epilochneridine (**15**) (7), and (±)-schol-arine [(±)-**18**] (8). They are probably formed from the corresponding alkaloids having the usual C-15 configuration by inversion at C_3, C_7, and C_{15} through a mechanism that involves rupture of C_3-C_7 and C_{15}-C_{16} bonds followed by double cyclization (2b,4) (see Scheme 19, Section 2.5). This interpretation is in agree-ment with the result of a model experiment effected from (−)-19,20-dihydroakuam-micine (**12**) (9).

Also as a consequence of its biogenetic origin, the ethylidene substituent present in some *Strychnos* alkaloids has an E-configuration (10). This substitu-ent is formed, after hydrolysis of strictosidine, by isomerization of the vinyl double bond and further reduction of the resulting conjugated iminium cation

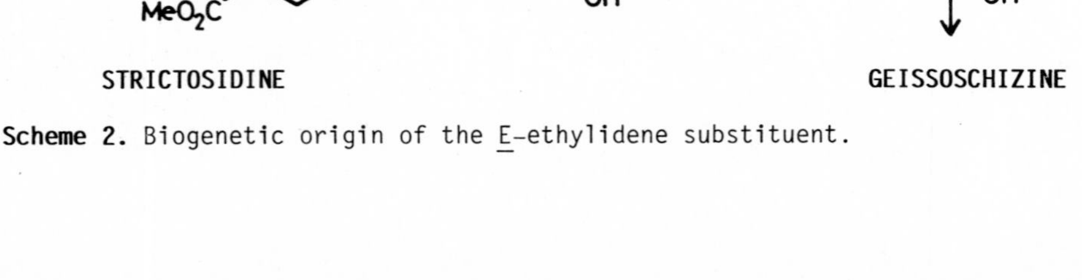

STRICTOSIDINE   GEISSOSCHIZINE

Scheme 2. Biogenetic origin of the E-ethylidene substituent.

(Scheme 2). The latter exists as the more stable $\underline{E}$-isomer since the steric interactions between C_{21}-H and C_{19}-H in the iminium salt with an $\underline{E}$ configuration
are lower than those between C_{21}-H and C_{19}-CH$_3$ in the $\underline{Z}$ isomer (11). The same
reason applies to the Aspidospermatan series, in which precondylocarpine is
formed by cyclization of a conjugated iminium salt (see Scheme 1).

A compilation of the information about the *Strychnos* alkaloids until 1964 can
be found in *The Alkaloids* (N.Y.), edited by Manske at that time (12).The literature in this area has been covered annually up to 1982 by Joule (13a) and Saxton
(13b) in *The Alkaloids* (London) as alkaloids of the strychnine-akuammicine-con-
dylocarpine group and during the last years by Saxton (14) in *Natural Product
Reports* as alkaloids of the strychnine group. A review about the *Strychnos* alkaloids, specially those studied during the 1973-1983 period, has been published
by Husson (15). Although it includes all the Strychnan and Aspidospermatan types,
the *Strychnos* alkaloids are classified as α-methyleneindoline, indole, and indoline alkaloids. A comprehensive review of the *Strychnos* alkaloids of the Aspidospermatan type has been recently published by Lounasmaa (16).

This chapter deals with the synthesis of *Strychnos* indole alkaloids having
the pentacyclic 3,5-ethano-3H-pyrrolo[2,3-d]carbazole framework. These pentacyclic *Strychnos* indole alkaloids are representatives of the Strychnan (S) and
Aspidospermatan (A) types (S4 and A4 skeletal variations, respectively, according
to the Hesse's classification) (2) and are indexed as derivatives of curan (or
17-norcuran) and condyfolan stereoparents by the *Chemical Abstracts* from 1972 (17).
The latter are frequently known as alkaloids of the aspidospermatidine group.

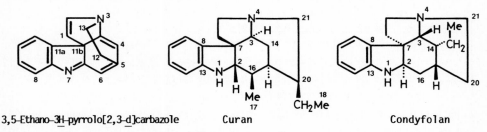

3,5-Ethano-3H-pyrrolo[2,3-d]carbazole   Curan   Condyfolan

Curan: [3aS-(3aα,5ß,6ß,6aß,11bR*,12R*)]-12-ethyl-1,2,3a,4,5,6,6a,7-octahydro-
6-methyl-3,5-ethano-3H-pyrrolo[2,3-d]carbazole

Condyfolan: [3aR-(3aα,4ß,5ß,6aß,11bR*)]-4-ethyl-1,2,3a,4,5,6,6a,7-octahydro-
3,5-ethano-3H-pyrrolo[2,3-d]carbazole

Scheme 3. *Chemical Abstracts* nomenclature and numbering of pentacyclic *Strychnos*
indole alkaloids.

Two points about the above condyfolan stereoparent are worthy of comment.
Firstly, the numbering adopted for carbons 3, 14, 20, and 21 is not the usual
biogenetic numbering (5). Secondly, the absolute $\underline{R}$-configuration at C-14 (C-20
in the biogenetic numbering) is the opposite of that found (20$\underline{S}$) in tubotaiwine

(=19,20-dihydrocondylocarpine) (16,18-20) and in related natural and synthetic (condyfoline) (18) bases with chirality at that carbon. In these bases the ethyl substituent is equatorial with regard to the piperidine ring and lies near the aromatic nucleus. However, in *Chemical Abstracts* they are listed as having the same C-14 absolute configuration as the condyfolan stereoparent (17). This 20<u>S</u> stereochemistry is the same that the one of the tetracyclic alkaloids uleine and dasycarpidone, which incorporate a 1,2,3,4,5,6-hexahydro-1,5-methanoazocino [4,3-<u>b</u>]indole skeleton containing four of the five rings of pentacyclic *Strychnos* indole alkaloids.

Tridimensional representations of geissoschizoline and tubotaiwine, characteristic examples of pentacyclic alkaloids of the Strychnan and Aspidospermatan types, respectively, are depicted in the following figure.

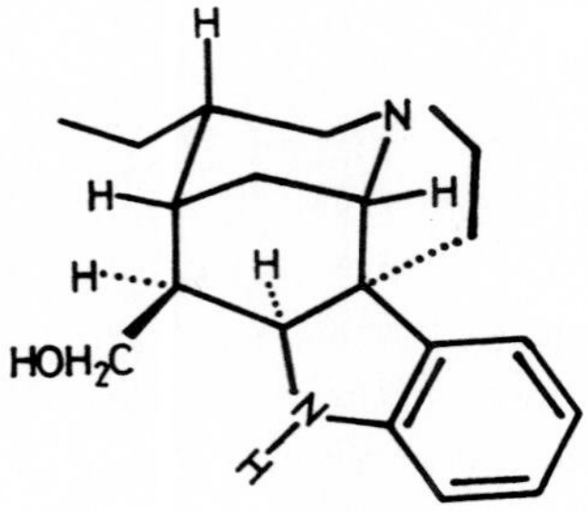

Geissoschizoline
[(16α)-Curan-17-ol]

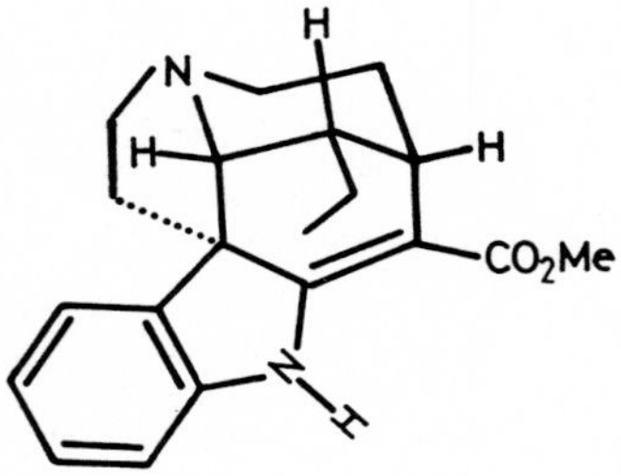

Tubotaiwine [Methyl (14ß)-2,16-Didehydrocondyfolan-16-carboxylate]

1.2 <u>Survey of Pentacyclic *Strychnos* Alkaloids</u>

The *Strychnos* alkaloids of the Strychnan and Aspidospermatan types that possess a pentacyclic 3,5-ethanopyrrolo[2,3-<u>d</u>]carbazole ring system are listed in Tables 1 and 2, respectively, along with pertinent references about their structural elucidation, partial or total synthesis, and NMR data. We have also included as partial syntheses simple synthetic transformations between alkaloids used for chemical correlation, even if they had been reported before the alkaloids were known as natural products. The structures of the alkaloids are given in the associated figures. For each skeletal type, the alkaloids have been assembled according to the C-16 substituent and the C-20 two-carbon chain.

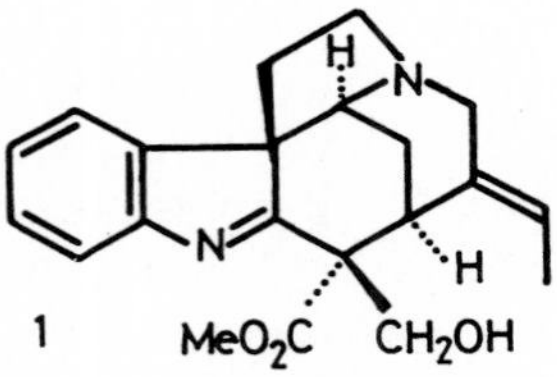

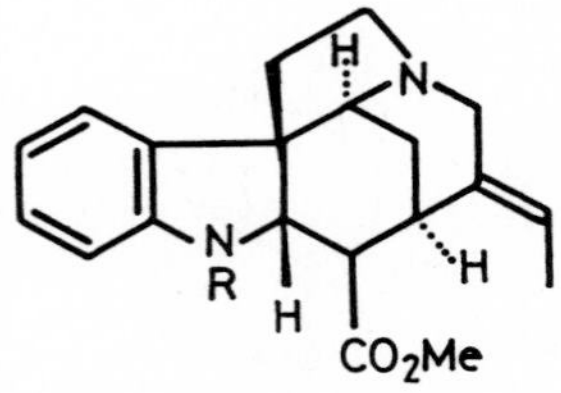

Table 1.   **Pentacyclic _Strychnos_ Alkaloids with the Strychnan Skeleton**

Name	Structure	Elucidation (ref.)	Synthesis[a-c] (ref.)	NMR data[d,e] (ref.)
A. With a Methoxycarbonyl Substituent at C-16				
Ethylidene side chain at C-20				
Preakuammicine	1	21,22		
Akuammicine[f,g]	2	23-27	21,22,28-30	31[d],32-34[e]
Akuammicine $\underline{N}_b$-metho salt	3	35	25,30,35	
Sewarine	4	36		
11-Methoxyakuammicine	5	37		
Vinervine	6	38		39[d]
Vinervinine	7	38	40	34[e]
2,16-Dihydroakuammicine	8	41	22[h],42[h],26-28[i],43[i]	
$\underline{N}_a$-Methyl-2,16-dihydroakuammicine	9	41,44		
$\underline{N}_a$-Methyl-2β,16β-dihydroakuammicine $\underline{N}_b$-methochloride	10	45		
Mossambine	11	46		

a.Partial synthesis unless otherwise indicated; b.Total synthesis; c.Formal synthesis; d. ^{1}H-NMR data (≥200 MHz); e.^{13}C-NMR data; f.(±)-akuammicine is also known (ref.6); g.The corresponding $\underline{N}$-oxide is also known; h.16-H_αEpimer; i.16-H_βEpimer;

	R¹	R²	R³	
2	H	H	H	
3	H	H	H	N$\bar{b}$ metho salt
4	OH	H	H	
5	H	OMe	H	
6	H	H	OH	
7	H	H	OMe	

Table 1 (Continued). Pentacyclic *Strychnos* Alkaloids with the Strychnan Skeleton

Name	Structure	Elucidation (ref.)	Synthesis[a-c] (ref.)	NMR data[d,e] (ref.)
Modified side chain at C-20				
19,20-Dihydroakuammicine	12	47	24,26,48	
Angustimicine	13	49		
Lochneridine	14	28,50	33[j]	
20-Epilochneridine	15	7		
Echitamidine	16	51,52	53	52[d,e],53[e]
20-Epiechitamidine	17	52		52[d,e]
Scholarine[k]	18	54	53	53[e],54[e]
Scholaricine	19	55		55[e]
N_a-Formylechitamidine	20	53		53[e]
N_a-Formyl-12-methoxy echitamidine	21	53		53[e]
Compactinervine	22	50,56		57[e]
Alstovine	23	58		40[e]
12-Methoxycompactinervine	24	40		40[d,e]
12-Methoxy-19α,20α-epoxy- akuammicine	25	40	40	40[d,e]

B. With a Formyl Substituent at C-16

Name	Structure	Elucidation (ref.)	Synthesis[a-c] (ref.)	NMR data[d,e] (ref.)
Ethylidene side chain at C-20				
Nor-C-fluorocurarine[g]	26	59	29,60[b],61,62	31[d],34[e]
C-Fluorocurarine	27	61,63,64	60[b],61	
10-Methoxy-nor-C-fluorocurarine	28	65		
Vincanicine	29	38		
Vincanidine	30	38		34[e]
18-Desoxy-Wieland-Gumlich aldehyde	31	66	61,67	
Isoretulinal	32	68		69[d],70[e]
12-Hydroxyisoretulinal	33	69		69[d]
16-Hydroxyisoretulinal	34	71		71[d]
Retulinal	35	69		69[d]
12-Hydroxyretulinal	36	69		69[d]

j.20α-OH Epimer; k.($\pm$)-Scholarine is also known (ref.8);

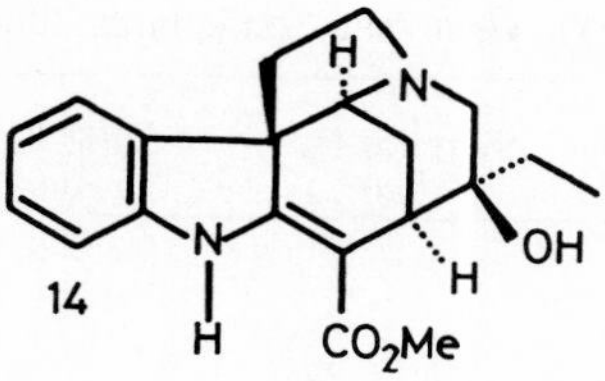

14 CO₂Me

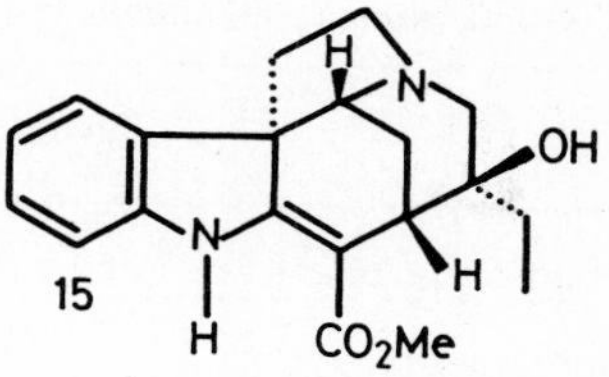

15 CO₂Me

16–21 (R¹, R²)

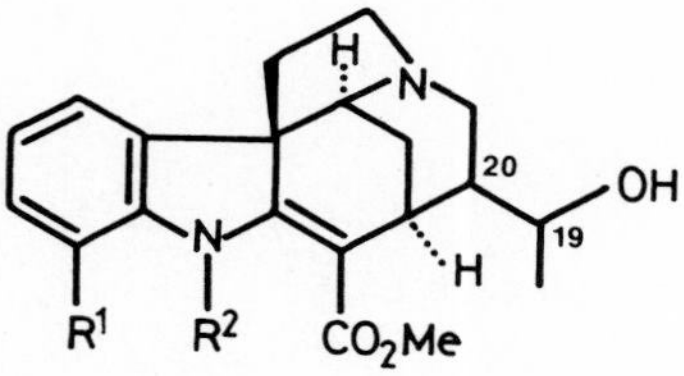

	R¹	R²	
16	H	H	(19S,20S)
17	H	H	(20R)
18	OMe	H	
19	OH	H	
20	H	CHO	(19S,20S)
21	OMe	CHO	(20S)

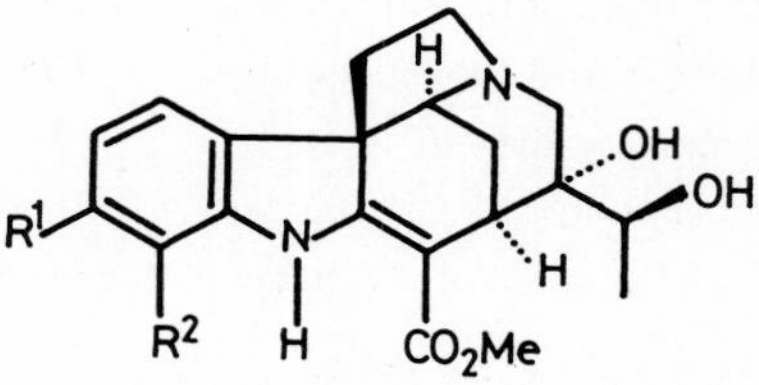

22–24

	R¹	R²
22	H	H
23	OMe	H
24	H	OMe

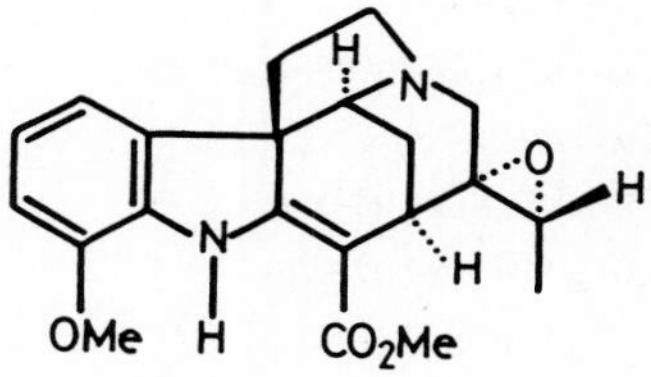

25

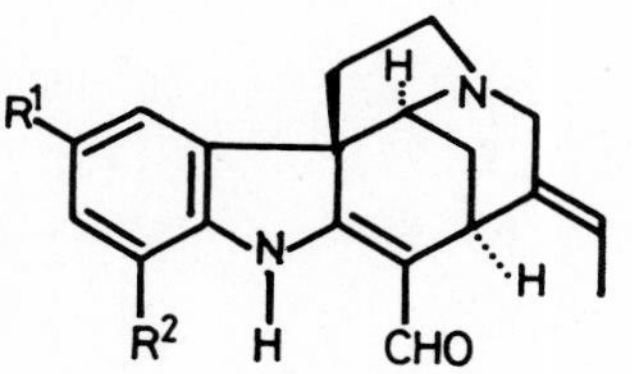

26–30

	R¹	R²	
26	H	H	
27	H	H	N_b-metho salt
28	OMe	H	
29	H	OMe	
30	H	OH	

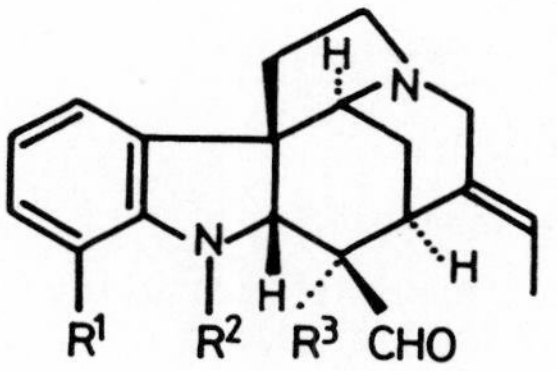

31–34

	R¹	R²	R³
31	H	H	H
32	H	Ac	H
33	OH	Ac	H
34	H	Ac	OH

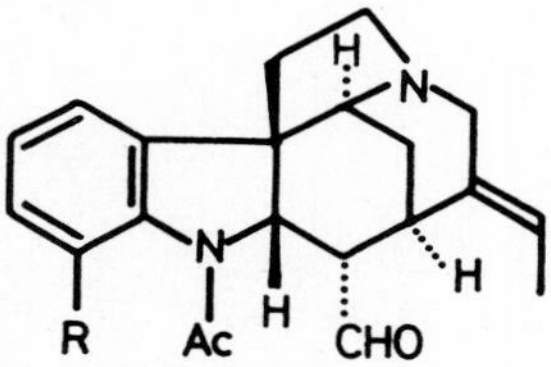

35–36

	R
35	H
36	OH

37 CH₃ CHO

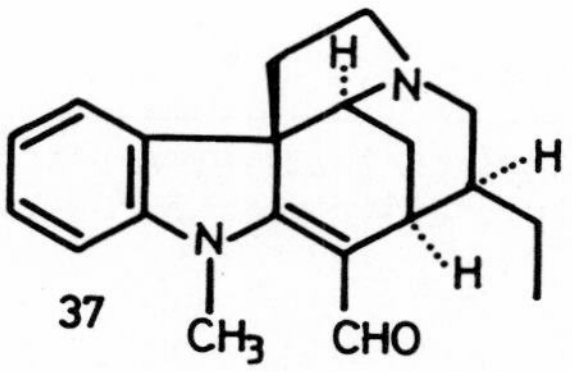

Table 1 (Continued). Pentacyclic *Strychnos* Alkaloids with the Strychnan Skeleton

Name	Structure	Elucidation (ref.)	Synthesis[a-c] (ref.)	NMR data[d,e] (ref.)
Modified side chain at C-20				
Strychnofluorine	37	72	48[1],73[1]	
18-Hydroxynorfluorocurarine	38	74	74	
18-Acetoxynorfluorocurarine	39	74		
Strychnozairine	40	75		75[d,e]

C. With a Hydroxymethyl Substituent at C-16

Ethylidene side chain at C-20				
Retuline[g]	41	76-78	68,69,77-79	80[d]
Deacetylretuline	42	68,81	71,77,78,82	80[d]
O-Acetylretuline	43	83	68,77,78	
11-Methoxyretuline	44	70	65[m]	
1,2-Dehydrodeacetylretuline	45	74		74[d]
23-Hydroxy-2,16-dehydroretuline	46	74		74[d]
Isoretuline	47	84	68,69,77	80[d],85[e]
11-Methoxyisoretuline	48	70		70[e]
N-Deacetylisoretuline	49	86	26,61,67	80[d],85[e]
O-Acetylisoretuline	50	87	68,77	
Tsilanimbine	51	86	65[n],86	
Modified side chain at C-20				
Geissoschizoline	52	81,88,89	27,73[b],78,82,90[b]	91[o]
18-Hydroxyisoretuline	53	86	77	85[e]
N$_a$-Deacetyl-18-hydroxyisoretuline (Wieland-Gumlich diol)	54	86	27,61,67,74, 86,92,93	85[d,e]
N$_a$-Deacetyl-17-O-acetyl-18-hydroxy- isoretuline	55	86		
18-Acetoxy-N$_a$-deacetylisoretuline	56	74		74[d]

D. Lacking Substituent at C-16

Strychnopivotine	57	71		71[d]
Tubifoline	58	47	18,25,48,94-95[b],96-97[c]	
Tubifolidine	59	47	24,47,94-95[b],98[b]	85[p],98[e]

1.N$_a$-Demethyl derivative; m.Deacetyl derivative; n.16-H$_\beta$ Epimer; o.^{13}C-NMR of N,O$_a$-Diacetyl derivative; p.^{13}C-NMR of 16ß-Methyltubifolidine (curan).

	R
38	H
39	Ac

| 40 |

	R¹	R²	R³
41	H	Ac	H
42	H	H	H
43	H	Ac	Ac
44	OMe	Ac	H

	R¹	R²	R³	R⁴
47	H	H	Ac	H
48	H	OMe	Ac	H
49	H	H	H	H
50	H	H	Ac	Ac
51	OMe	H	H	H

| 45 |

| 46 |

| 52 |

	R¹	R²	R³
53	Ac	H	H
54	H	H	H
55	H	Ac	H
56	H	H	Ac

| 57 |

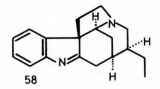

58   59

Table 2.   Pentacyclic *Strychnos* **Alkaloids with the Aspidospermatan Skeleton**

Name	Structure	Elucidation (ref.)	Synthesis[a,b] (ref.)	NMR data[c,d] (ref.)
A. With a Methoxycarbonyl Substituent at C–16				
Precondylocarpine	**60**	22,99	100[e]	
Condylocarpine[f]	**61**	101	99,100,102–104,105[b]	106[d]
Tubotaiwine[f]	**62**	19,20,47	18,104 107[b],108,109	19–20[c,d],57[d] 74[d],106[d],110[c,d]
11–Methoxytubotaiwine	**63**	56		57[d]
19,20–Dihydro–19–hydroxy–[g] condylocarpine	**64**	52		52[c]
B. With a Formyl Substituent at C–16				
Tubotaiwinal	**65**	74		74[d]
C. Lacking Substituent at C–16				
Aspidospermatidine	**66**	111	99,101	
N_a–Acetylaspidospermatidine	**67**	111,112		
N_a–Methylaspidospermatidine	**68**	111		
N_a–Acetyl–11–hydroxyaspidos- permatidine	**69**	56		
Aspidospermatine	**70**	111	113	
N_a–Deacetylaspidospermatine	**71**	111	111	
Limatine	**72**	112,114		
11–Methoxylimatine	**73**	115		
Limatinine	**74**	113,115		
11–Methoxylimatinine	**75**	115		
N_a–Acetyl–11,12–dihydroxy- aspidospermatidine	**76**	116		
19,20–Dihydroaspidospermatine	**77**	111	111	

a.Partial synthesis unless otherwise indicated; b.Total synthesis; c. ^{1}H–NMR data (≥200 MHz); d.^{13}C–NMR data; e.O–Acetyl derivative; f.The corresponding N–oxide is also known; g.Two epimers.

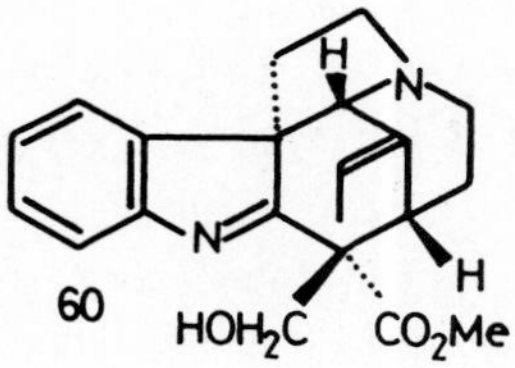

60   HOH₂C   CO₂Me

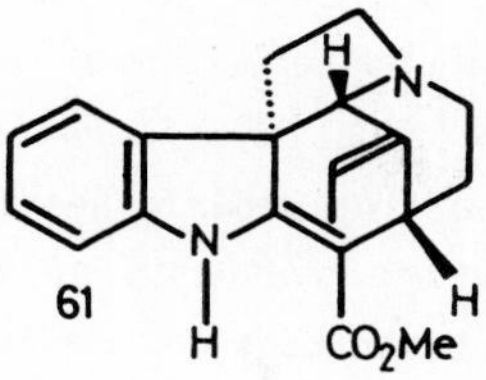

61

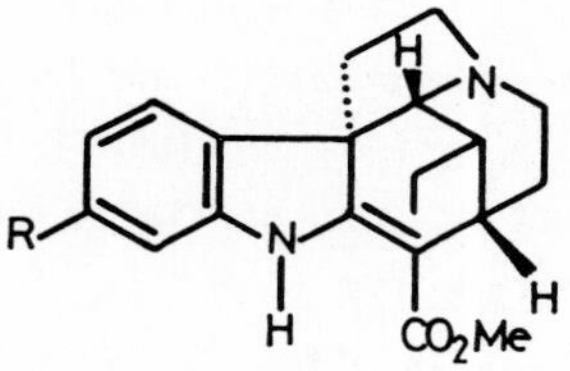

62   H
63   OMe

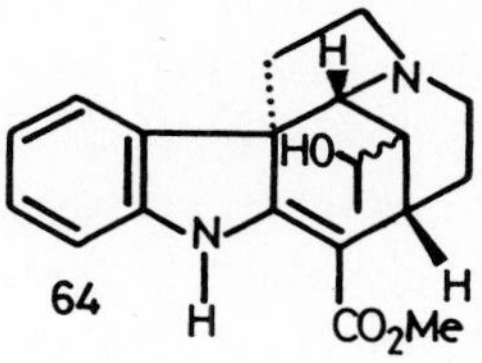

64

65   CHO

66   67   68   69   70   71   72   73   74   75   76

77

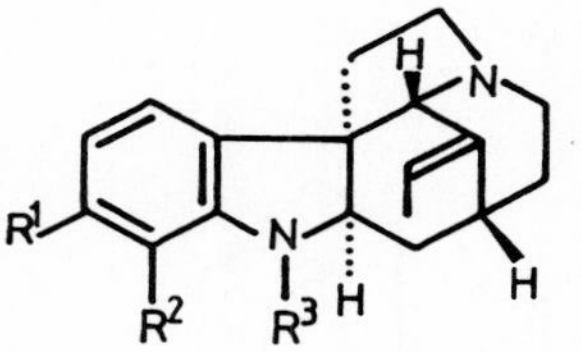

	R¹	R²	R³
66	H	H	H
67	H	H	Ac
68	H	H	CH₃
69	OH	H	Ac
70	H	OMe	Ac
71	H	OMe	H
72	H	OH	COEt
73	OMe	OH	COEt
74	H	OH	Ac
75	OMe	OH	Ac
76	OH	OH	Ac

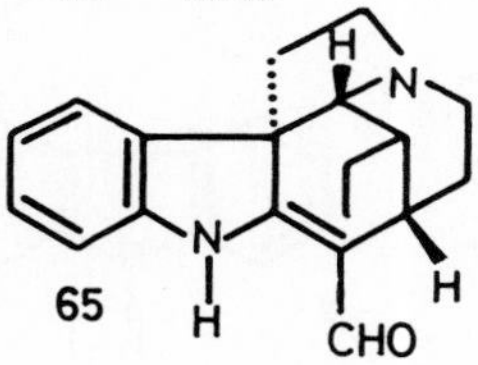

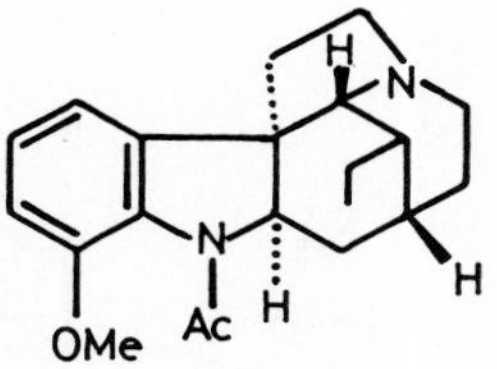

1.3 Synthesis: Overview

The pentacyclic *Strychnos* alkaloids constitute one of the groups of indole alkaloids that has received comparatively less attention from a synthetic stand-point. Only a few alkaloids of this type have been obtained so far by total syn-thesis, all of them in the racemic series: tubifoline (94,95), tubifolidine (94, 95,98), geissoschizoline (73,90), fluorocurarine (60), and nor-C-fluorocurarine (60) in the Strychnan series, and condylocarpine (105) and tubotaiwine (107) in the Aspidospermatan series. In addition, two formal syntheses of tubifoline have been reported (96,97).

Most of these syntheses converge to tetracyclic structures having the ring skeleton of stemmadenine, which are further subjected to transannular cyclization by formation of the bond between the indole 3-position and one of the α-carbons of the piperidine ring (Section 2).

Only one synthesis of tubifolidine, effected in our laboratory, emploies a different strategy to complete the pentacyclic ring system of the alkaloid (98). The crucial step of our synthesis is the closure of the five membered E ring by cyclization upon the indole 3-position using an appropriately N-substituted 1,2,3,4,5,6-hexahydro-1,5-methanoazocino[4,3-b]indole system (Section 3).

All synthetic strategies adopted for constructing the partially reduced 3,5-ethanopyrrolo[2,3-d]carbazole ring system of pentacyclic *Strychnos* alkaloids are outlined in the following scheme. For clarity, C-16 and C-20 substituents have been omitted.

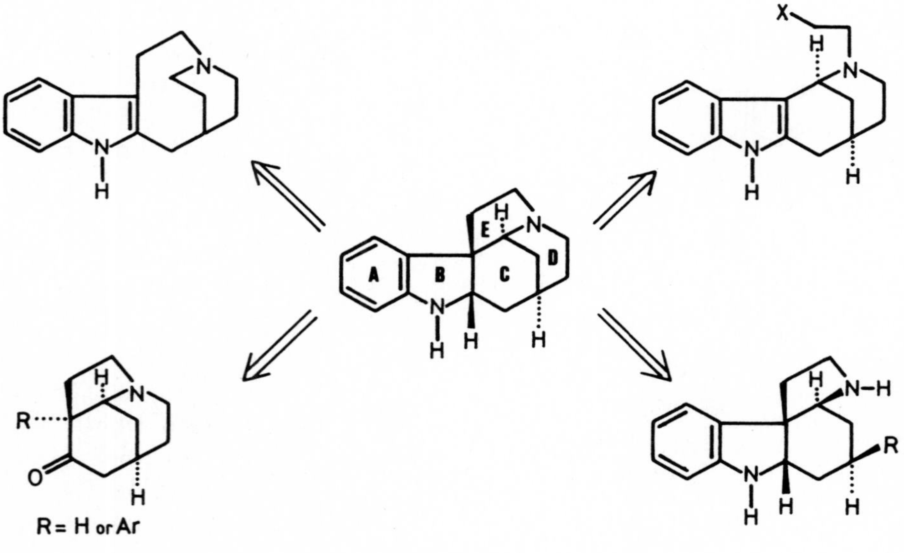

Scheme 4

The approaches based on the elaboration of the piperidine ring (117-120) or the modified indole nucleus (121,122) in the last steps have not succeeded so far in obtaining natural products (Section 4).

At this point, it is worth mentioning that the pioneering synthetic studies in this field were done by van Tamelen (123), who in 1960 reported a straightforward construction of the pentacyclic curan skeleton (79) by closure of rings C, D, and E in a single synthetic step (Scheme 5). After two successive reductions, a compound (80) having the same constitution than geissoschizoline (52) but undefined relative configuration was obtained. The intermediate spiro indolenine **78** does not rearrange, as usual (124), to an indolo[2,3-g]quinolizidine system but undergoes a new cyclization promoted by the enolizable aldehyde group. The presence of the amide carbonyl group could account for the feasibility of the process (125).

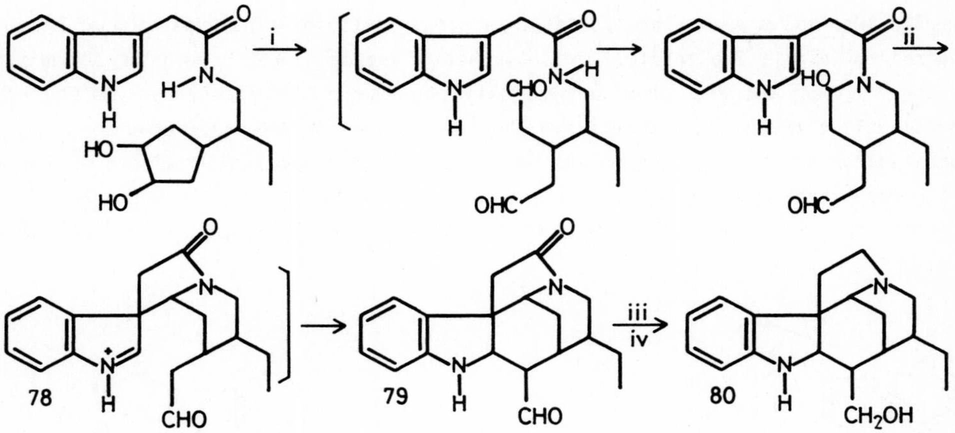

Scheme 5. Reagents: (i) NaIO$_4$; (ii) AcOH, NaOAc; (iii) NaBH$_4$; (iv) LiAlH$_4$.

Attempts to induce a similar C-ring closure from imino ether **81** (mixture of epimers at C-7) failed to yield products of the *Strychnos* type and led only to pentacyclic spiro oxindole **82** (126) (Scheme 6).

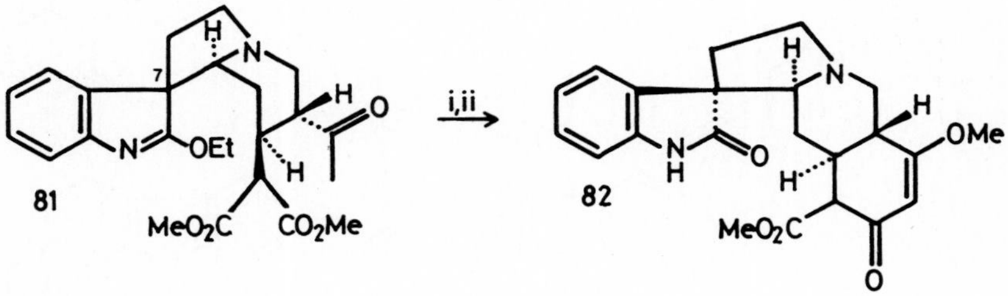

Scheme 6. Reagents: (i) t-BuOK, THF; (ii) HCl, MeOH.

2 TRANSANNULAR CYCLIZATION OF STEMMADENINE-TYPE SYSTEMS

This approach to pentacyclic *Strychnos* alkaloids can be considered as biomimetic since the crucial step consists in a transannular cyclization that imitates the biogenetic pathway to these alkaloids. Four different strategies have

44

been developed to construct the required tetracyclic stemmadenine-type system. They are summarized in Scheme 7. In order to better visualize bond disconnections, the biogenetic numbering has been maintained for the synthetic precursors.

The Harley-Mason's strategy (disconnection A) is based on the construction of the piperidine ring by cyclization from an appropriately substituted azonino [5,4-b]indole system, which, in turn, is prepared (B) by ring expansion of hexahydroindolizino[8,7-b]indole. This synthetic approach has proved to be quite general (105) and has been successfully employed to synthesize a variety of pentacyclic *Strychnos* alkaloids, of both Strychnan and Aspidospermatan types, possessing an ethyl or ethylidene C-20 side chain, and having or not a functionalized one-carbon C-16 substituent.

The Ban's approach (disconnection C) differs from the preceding in that the

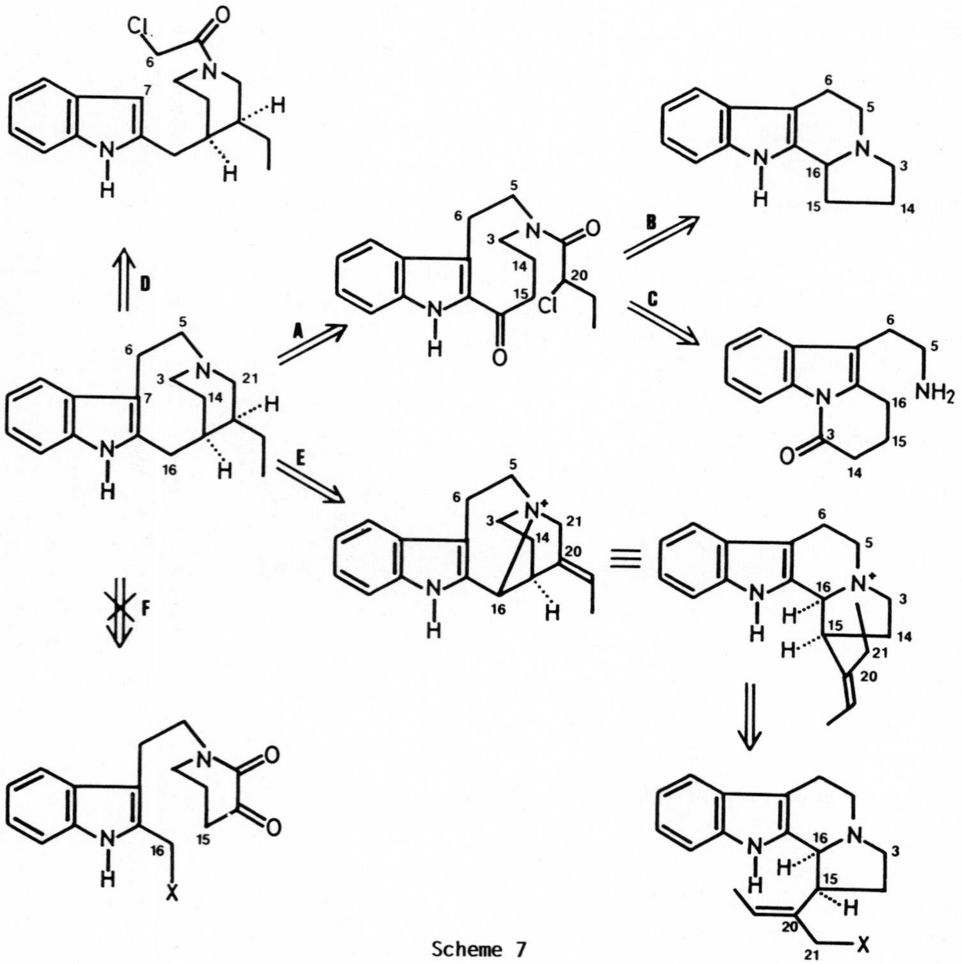

Scheme 7

required nine-membered ring system is formed by photoisomerization of an appropriate 1-acylindole (95).

The strategy employed by Wu and Snieckus (disconnection D) to produce the stemmadenine ring skeleton implies the closure of the nine-membered ring by formation of C_6-C_7 bond. This was satisfactorily achieved by photocyclization of a 2-(N-chloroacetylpiperidylmethyl)indole (96).

The synthesis of Takano *et al.* (disconnection E) takes advantage again of the easy ring enlargement reaction of hexahydroindolizino[8,7-b]indoles, although in this case the piperidine ring has been built in a previous step and, consequently, the key tetracyclic intermediate is obtained from a pentacyclic precursor (97).

Finally, the strategy (F) adopted by Lounasmaa and Somersalo is also based on the elaboration of the medium-sized ring in the key step. It implies the formation of C_{15}-C_{16} bond through intramolecular C-4 alkylation of a suitable 1-(3-indolylethyl)-3-hydrazono-2-piperidone system (127).

2.1 The Harley-Mason´s Strategy

The synthesis of (±)-tubifoline and (±)-tubifolidine, published by Harley-Mason in 1968, constituted the first total synthesis of pentacyclic *Strychnos* indole alkaloids (94). The easy cleavage of N-C_{11b} bond of hexahydroindolizino-[8,7-b]indole (**83**) with acid anhydrides to give N-acyl-7-acyloxyazonino[5,4-b]-indoles had been previously observed by the author (128). The use of bis(2-chloro-butyric) anhydride afforded amido ester **84**, which, on successive hydrolysis, oxidation, and base-catalyzed cyclization, was elaborated in a stereoselective manner into keto lactam **85**, having the appropriate relative configuration at C-20 (Scheme 8). After two consecutive reductions, **85** was converted to the tetracyclic indole **86**, a product that had been obtained earlier from akuammicine (24) and condylocarpine (18).

Transannular cyclization of the synthetic racemic base **86** was conducted by catalytic aerial oxidation over platinum, as described previously by Schumann and Schmid (18) from the same product obtained from natural sources by degradation. The process is not wholly regioselective and a mixture (4:1) of (±)-tubifoline and (±)-condyfoline, arising from cyclization of the regioisomeric iminium salts **87** and **88**, respectively, was obtained. Reduction of (±)-tubifoline gave (±)-tubifolidine. These two alkaloids had been obtained by partial synthesis (see Table I) from more complex *Strychnos* alkaloids before they were isolated. Condyfoline was a known product arising from degradation of condylocarpine (18).

The synthetic route to the 17-norcuran skeleton developed by Harley-Mason opened a general entry to pentacyclic *Strychnos* alkaloids. The ketone carbonyl group of the above intermediate **85** allows the introduction of the functionalized one-

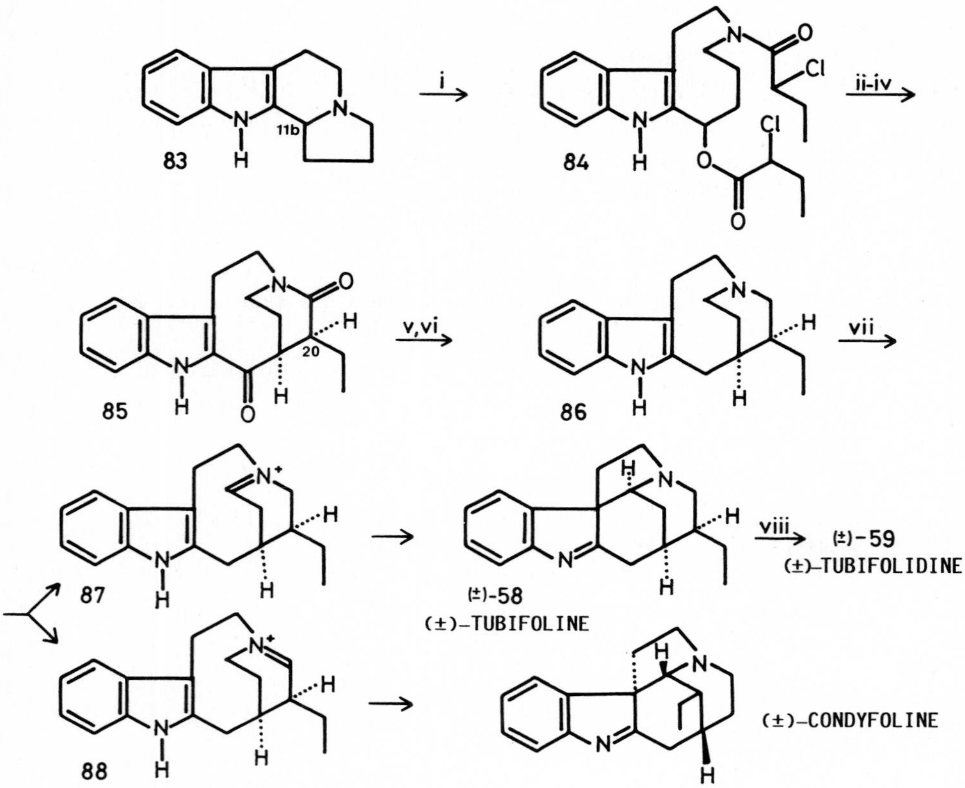

Scheme 8. Harley–Mason's synthesis of (±)–tubifoline and (±)–tubifolidine. Reagents: (i) (EtCHClCO)$_2$O; (ii) OH$^-$, H$_2$O; (iii) MnO$_2$ (ref.94) or Pb(OAc)$_4$ (ref.
105); (iv) t–AmONa; (v) Wolff–Kishner reduction; (vi) LiAlH$_4$; (vii) Pt,O$_2$; (viii)
H$_2$, cat. (ref.94) or LiAlH$_4$ (ref.105).

carbon substituent at C–16 present in the alkaloids having the curan skeleton.
Two alternative stereoselective total syntheses (73,90) of (±)–geissoschizoline
and a synthesis (73) of (±)–dihydronorfluorocurarine (N$_a$–demethylstrychnofluo-
rine), a degradation product of strychnine (48), illustrate the usefulness of
this synthetic approach (Scheme 9). It is noteworthy that in these cases, when a
substituent is present at the 16–position (as in **90** and **91**), the oxidative cyclization step is regioselective and leads only to the Strychnan skeletal type.

A conceptually similar approach was employed in the total synthesis of the
E–ethylidene bearing alkaloids nor–C–fluorocurarine and fluorocurarine (60)
(Scheme 10). The use of bis(2–bromo–3–methoxybutyric) anhydride in the first
step of the synthesis resulted, after hydrolysis and oxidation, in the formation
of keto amide **92**. The ethylidene substituent was formed simultaneously to the

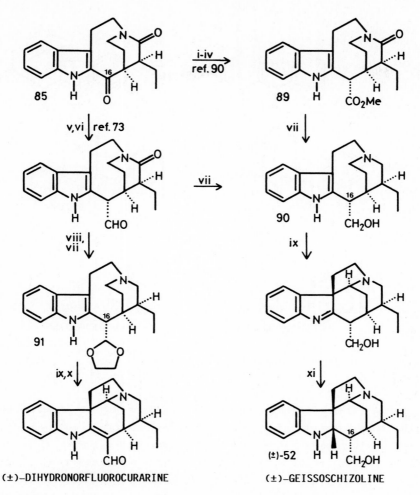

Scheme 9. Introduction of the C-16 substituent: synthesis of (±)-geissoschizoline and (±)-dihydronorfluorocurarine. Reagents: (i) NaBH$_4$; (ii) Ac$_2$O; (iii) NaCN, DMSO; (iv) CH$_3$OH,H$_2$SO$_4$; (v) (CH$_3$)$_2$S=CH$_2$,DMSO; (vi) MgBr$_2$.Et$_2$O or Δ; (vii) LiAlH$_4$; (viii) (CH$_2$OH)$_2$; (ix) Pt, O$_2$; (x) H$_3$O$^+$; (xi) B$_2$H$_6$ (ref.90) or LiAlH$_4$ (ref.73).

piperidine ring closure by base-catalyzed elimination of methanol. Thus, on treatment with excess of sodium <u>tert</u>-amyloxide, **92** underwent a cyclization-elimination process to give a 6:4 mixture of <u>E</u> keto lactam **93** and the unnatural <u>Z</u>-isomer. The latter could be equilibrated with the natural one on treatment with methoxide ion (105). The functionalized substituent at C-16 was introduced again by taking advantage of the ketone carbonyl group of keto amide **93**. However, in this case the application of dimethylsulfonium methylide was not fruitful since the resulting epoxide failed to rearrange to the desired aldehyde, so integration was effected with (methoxymethylene)triphenylphosphorane. Reductive removal

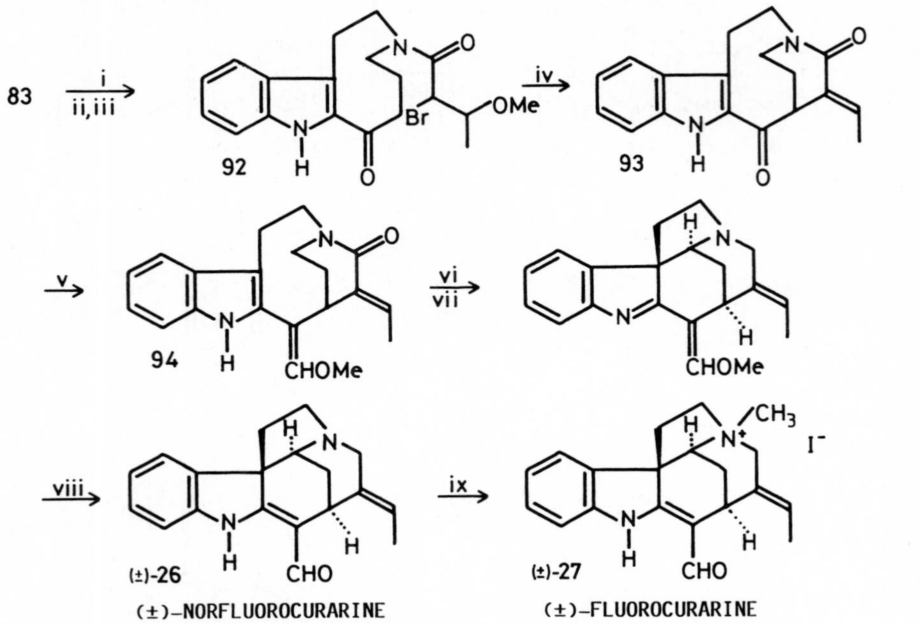

Scheme 10. Elaboration of the E-ethylidene substituent: synthesis of (±)-fluoro-curarine. Reagents: (i) (CH$_3$CHOCH$_3$CHBrCO)$_2$O,CH$_3$CN; (ii) OH$^-$,H$_2$O; (iii) Pb(OAc)$_4$; (iv) t-AmONa,THF; (v) Ph$_3$P=CHOCH$_3$; (vi) AlH$_3$; (vii) Pt,O$_2$; (viii) H$_3$O$^+$; (ix) CH$_3$I.

of the lactam carbonyl group of **94** occurred with concomitant reduction of the ethylidene double bond to give, after oxidative cyclization and hydrolysis, a mixture of (±)-norfluorocurarine and (±)-dihydronorfluorocurarine. In a similar manner, the Z-isomer of **93** was converted to isonorfluorocurarine (60), although the above phosphorane was ineffective to integrate the required functionalized one-carbon atom upon the 2-acylindole carbonyl group and a Horner-Wittig reaction with methoxymethyldiphenylphosphine oxide ought to be used (129).

As in the above synthesis of (±)-geissoschizoline, transannular cyclization proceeded only in the direction indicated to give the Strychnan skeleton. This regioselectivity was also observed in the platinum mediated oxidative cyclization

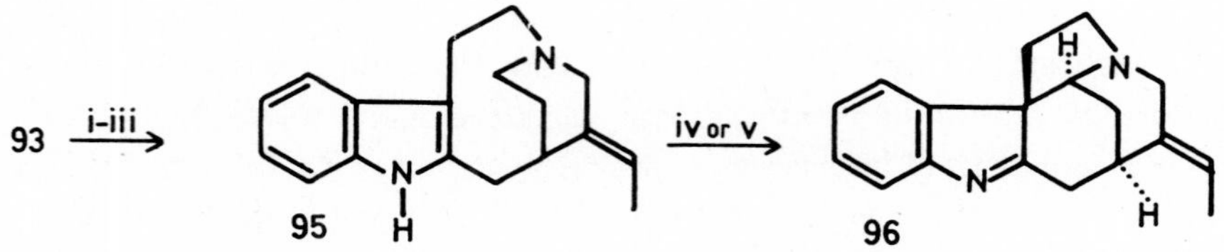

Scheme 11. Reagents: (i) (CH$_2$SH)$_2$; (ii) Raney Ni; (iii) LiAlH$_4$; (iv) Pt,O$_2$; (v) m-CPBA; then (CF$_3$CO)$_2$O.

of the C-16 unsubstituted tetracyclic base **95**, which was prepared from **93** as outlined in Scheme 11 (60). Several years later, Takano *et al.* (97) corroborated the above result and reported again the exclusive formation of pentacyclic amine **96**, having the Strychnan skeletal type, when the iminium salt required for cyclization was generated either by catalytic oxidation of **95** over platinum or by Polonovski-Potier reaction (130) from the corresponding $\underline{N}_b$-oxide.

Interestingly, oxidative cyclization of stemmadenine (102,103) and the corresponding acetate **97** (100) gives condylocarpine derivatives (Aspidospermatan type), as illustrated in Scheme 12. On the contrary, oxidation of 19,20-dihydrostemmadenine acetate (**98**) occurs regioselectively to give 19,20-dihydropreakuammicine acetate (**99**) (Strychnan type) (100).

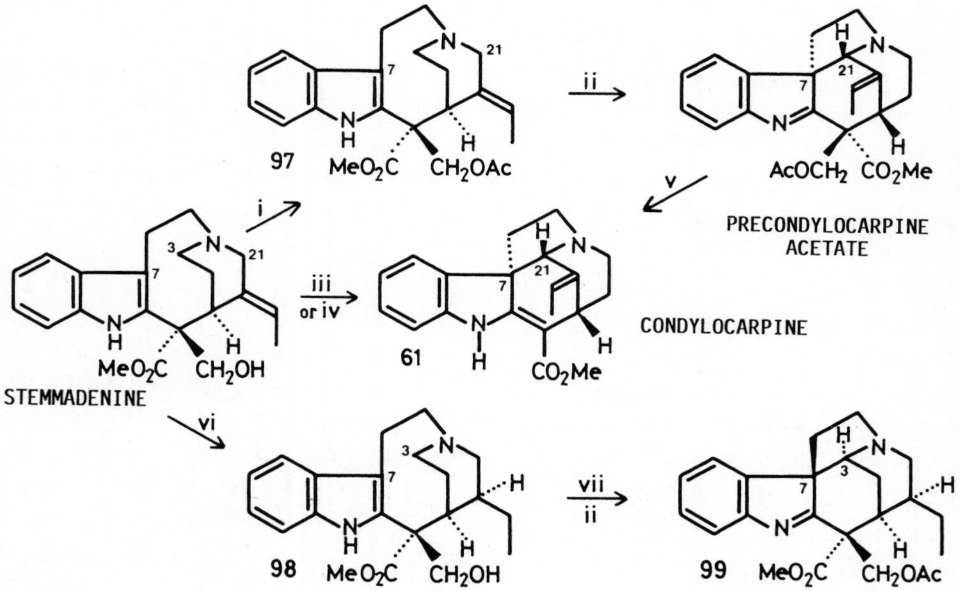

Scheme 12. Oxidative cyclization of stemmadenine derivatives. Reagents: (i) Ac$_2$O; (ii) Pt,O$_2$; (iii) from the hydrochloride, aq. KMnO$_4$ and then heating (ref.102); (iv) Hg(OAc)$_2$, then NaBH$_4$; isolated as borine adduct (ref.103); (v) NaOMe; (vi) H$_2$,Pt; (vii) Ac$_2$O,py.

The extension of the Harley-Mason´s strategy to the synthesis of *Strychnos* alkaloids of the Aspidospermatan type implies the formation of a bond between C-7 and C-21. This was satisfactorily attained by generating the iminium salt required for cyclization by means of a non-oxidative process taking advantage of the functionality at C-21 in the tetracyclic lactams **89** and **94**. Thus, treatment of **89** with phosphorus oxychloride in boiling benzene, followed by

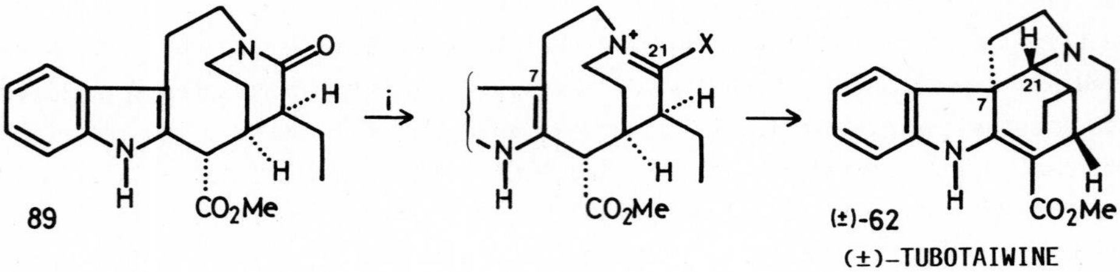

50

Scheme 13. Synthesis of *Strychnos* alkaloids of the Aspidospermatan type: synthesis of (±)-tubotaiwine. Reagents: (i) $POCl_3, C_6H_6$; then base.

addition of base, led to the isolation of (±)-tubotaiwine (107), probably via a Vilsmeier-type intermediate (Scheme 13). This is an unusual reaction since it involves a change in the oxidation level.

In the synthesis of (±)-condylocarpine (105) the regioselective formation of the required iminium salt ($N-C_{21}$) was accomplished by boron trifluoride treatment of amino ether 101. This was prepared in six steps from keto lactam 94, as illustrated in Scheme 14. The enol ether function of 94 was converted to a methoxycarbonyl substituent (see 100) and the lactam carbonyl was reduced to an amino ether function (see 101), via the corresponding imidate salt.

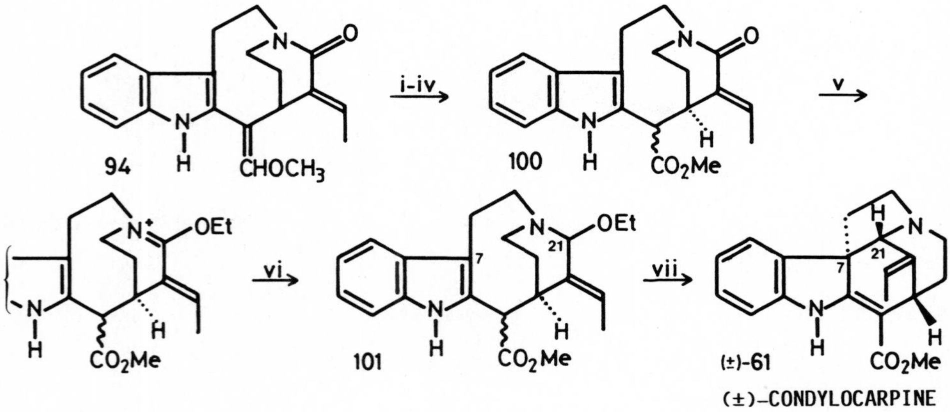

Scheme 14. Synthesis of *Strychnos* alkaloids of the Aspidospermatan type: synthesis of (±)-condylocarpine. Reagents: (i) H_3O^+; (ii) NH_2OH; (iii) $TiCl_4$; (iv) CH_3OH, H^+; (v) $Et_3O^+BF_4^-$; (vi) $NaBH_4$; (vii) BF_3.

In summary, the Harley-Mason's strategy provides a highly general and flexible route for the synthesis of *Strychnos* indole alkaloids.

2.2 Photoisomerization of 1-Acylindoles

Ban *et al.* had observed (95a) the formation of 3-acylindolenines by photo-isomerization of 1-acyl-3-alkylindoles. The intramolecular version of this process was applied (95) to the total synthesis of the *Strychnos* alkaloids tubifoline and tubifolidine through a versatile nine-membered lactam intermediate (**106**), prepared in a one pot reaction by photolysis and simultaneous ring enlargement (Scheme 15). Lactam **106** was also employed by Ban to achieve the total synthesis of pentacyclic *Aspidosperma* alkaloids (131).

The required 1-acylindole **103**, carrying a nucleophilic aminoethyl chain at the 3-position, was prepared from keto diacid **102** by Fischer indolization, followed by Curtius rearrangement and hydrogenolysis of the resulting benzyl carbamate. Irradiation of **103** with a 300 W high-pressure mercury lamp afforded the key intermediate lactam **106** as the sole product in an excellent yield of 89%. This remarkable ring expansion probably proceeds by a photochemical[1,3]-acyl migration to give the labile carbazolenine **104**, which spontaneously rearranges, via carbinol amine **105**, to the stable lactam **106**. Elaboration of **106** to the Harley-Mason tubifoline intermediate **107** involved reduction, acylation with 2-chlorobutyryl chloride, and, finally, oxidation with iodine pentoxide. The later stages of the synthesis are similar to those previously reported by Harley-Mason (94), although the authors include experimental specifications.

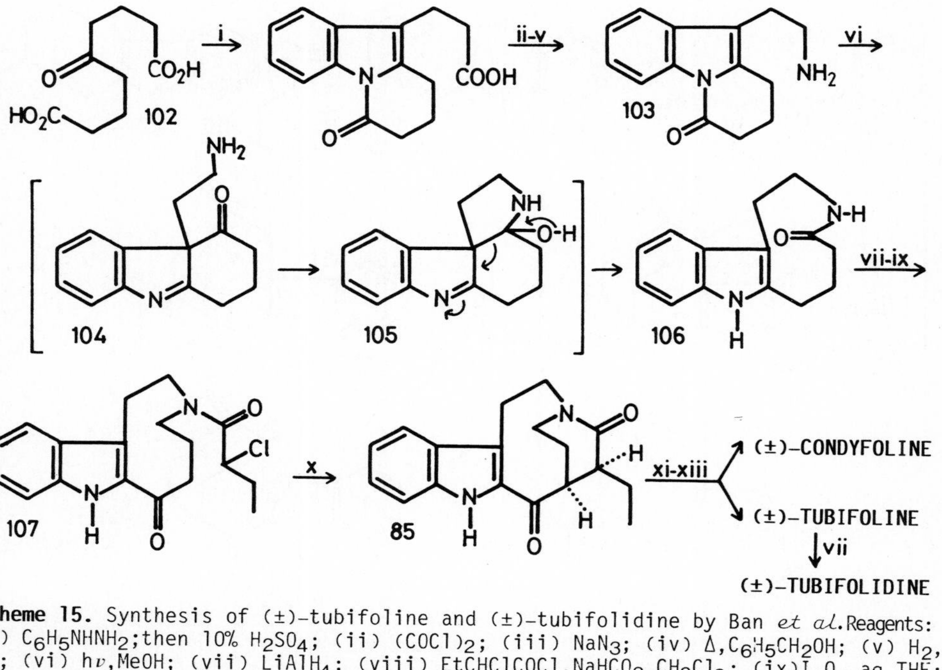

Scheme 15. Synthesis of (±)-tubifoline and (±)-tubifolidine by Ban *et al.* Reagents: (i) $C_6H_5NHNH_2$; then 10% H_2SO_4; (ii) $(COCl)_2$; (iii) NaN_3; (iv) Δ, $C_6H_5CH_2OH$; (v) H_2, Pd; (vi) $h\nu$, MeOH; (vii) $LiAlH_4$; (viii) $EtCHClCOCl$, $NaHCO_3$, CH_2Cl_2; (ix) I_2O_5, aq. THF; (x) $\underline{t}$-AmONa, THF; (xi) NH_2NH_2, KOH; (xii) B_2H_6, THF; (xiii) Pt, O_2, EtOAc.

2.3 Photocyclization of Chloroacetamides

The key step in the Snieckus synthesis (96) of the tetracyclic intermediate **86** consists in the formation of a nine-membered lactam by Witkop photocycliza- tion (132) from a N_b-chloroacetyl derivative of a 2-(4-piperidylmethyl)indole. The synthesis is outlined in Scheme 16 and begins with Wadsworth-Emmons conden- sation between amido phosphonate **108** and 1-methyl-4-piperidone. The ethyl side chain was introduced by regioselective γ-alkylation of the resulting α,β-unsatu- rated amide **109**, via a dianion reaction. Subsequent stereoselective hydrogena- tion and Madelung reaction gave piperidylmethylindole **110**, having the appropri- ate piperidine H-3/H-4 cis-relationship. After replacing the N-methyl substitu- ent by the required chloroacetyl group, the resulting chloroacetamide **111** was photocyclized to the tetracyclic lactam **112** in 20% yield. A similar low-yield- ing cyclization in the 20-deethyl series had been reported before by the author (133). Subsequent hydride reduction provided the target compound **86**. Since this tetracyclic base had previously been transformed (94) into (±)-tubifoline and (±)-tubifolidine, this route constitutes a formal synthesis of these alkaloids.

This synthetic approach has not received further application and seems not to be quite general since no photocyclization products were observed from the related N_a-methyl derivatives **113** (134).

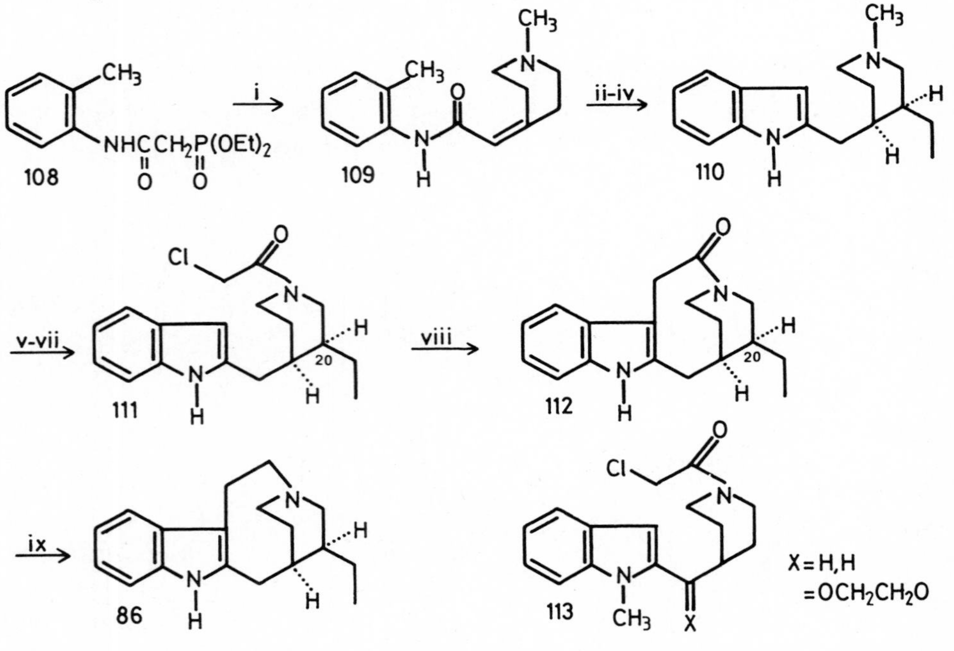

Scheme 16. Formal synthesis of (±)-tubifoline by Wu and Snieckus. Reagents: (i) 1-methyl-4-piperidone, NaOEt, EtOH; (ii) 2 eq.n-BuLi, TMEDA, Et₂O; then EtBr; (iii) H₂, Pd; (iv) t-BuOK, 340°C; (v) ClCO₂CH₂CCl₃⁻, C₆H₆; (vi) Zn, AcOH, MeOH; (vii) ClCOCH₂Cl, NaOH; (viii) hν, aq.EtOH; (ix) LiAlH₄, THF.

2.4 Thio-Claisen Rearrangement

One of the major difficulties in the synthesis of pentacyclic *Strychnos* alkaloids lies in the introduction of the exocyclic C-20 ethylidene substituent
with the requisite E-stereochemistry. Takano *et al*. have developed (97) a simple
and selective route to the 20E-ethylidene substituted stemmadenine-type tetracyclic base **95** (Scheme 17).

The nine-membered ring of **95** was formed by reductive cleavage of the quaternary indolizidinoindole derivative **118**, whereas the 3E-ethylidene piperidine
moiety was built (see **118**) by intramolecular N-alkylation of the unstable mesylate
obtained by reduction of E-ester **117** and further methanesulfonation.

The natural E configuration of 19,20-double bond was gained (see **116**) employing the thio-Claisen rearrangement in the first steps of the synthesis. Thus,
thio lactam **114** was converted to sulfonium salt **115**, which was treated with sodium methoxide in order to induce proton abstraction, [3,3]-sigmatropic rearrangement, and stereoselective double bond migration; the α,β-unsaturated E-ester **116** was obtained in 83% yield. Cyclization of thio lactam **116** followed by
stereoselective reduction of the resulting iminium salt provided the cis tertiary amine **117** (54%). Under the reaction conditions, the partial isomerization
of the double bond occurred and the Z isomer was obtained in 8% yield.

Cyclization of the tetracyclic base **95** to the pentacyclic Strychnan system
96 has already been discussed (Scheme 11). On the other hand, since **95** had been
converted into the *Strychnos* alkaloids tubifoline and tubifolidine, via catalytic hydrogenation (24) followed by oxidative cyclization (18,94), the above route
constitutes an alternative formal synthesis of these alkaloids.

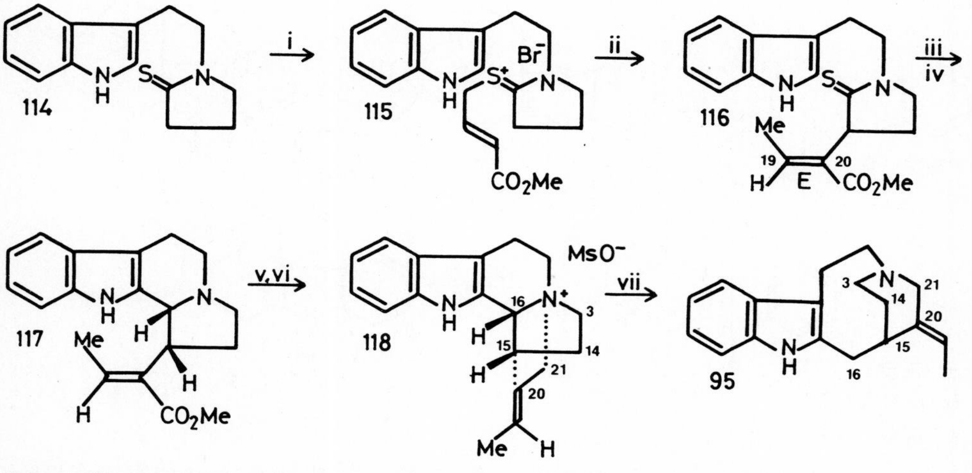

Scheme 17. Synthetic route to the *Strychnos* framework by Takano *et al*. Reagents;
(i) BrCH$_2$CH=CHCO$_2$Me, THF; (ii) NaOMe; (iii) POCl$_3$; (iv) NaBH$_4$; (v) DIBAH, C$_6$H$_6$;
(vi) MsCl; (vii) Na, NH$_3$.

2.5 Formation of C_{15}-C_{16} Bond by Intramolecular Alkylation

With the aim of developing a general method for the synthesis of *Strychnos* alkaloids of the Aspidospermatan type, Lounasmaa and Somersalo (127) have explored an alternative synthetic entry to tetracyclic stemmadenine-type systems. The authors have recently reported the unsuccessful attempts of constructing the medium-sized ring by formation of C_{15}-C_{16} bond using the intramolecular alkylation of 1-(indolylethyl)-3-hydrazono-2-piperidones **122** in the crucial step (Scheme 18).

Reaction between 2-hydroxymethyltryptamine **119** and bromo ester **120**, followed by several selective protection-deprotection steps, gave 1-(indolylethyl)-2,3-piperidinedione **121**. After the corresponding dimethylhydrazone **122a** was formed, the hydroxyl group was converted to a leaving group, either by benzoylation to **122b** or by conversion to the chloride **122c**. Although a simple 3-hydrazono-2-piperidone was satisfactorily alkylated at C-4 with ethyl bromide, neither **122b** nor **122c** could be intramolecularly alkylated to yield the desired tetracyclic compound **123**.

Formation of C_{15}-C_{16} bond from a related intermediate system **124**, in which the carbon atoms at positions 3, 7, and 15 are not chiral, has been proposed (9) as the crucial step in the isomerization of (−)-19,20-dihydroakuammicine (**12**) to its diastereoisomer **125** (Scheme 19). A similar mechanism could also explain the formation in nature of alkaloids possessing the opposite absolute configuration at C-15 than (−)-secologanine. However, at variance with the strategy adopted by Lounasmaa, this process implies nucleophilic attack of a C-16 enolate anion to an electrophilic center at C-15.

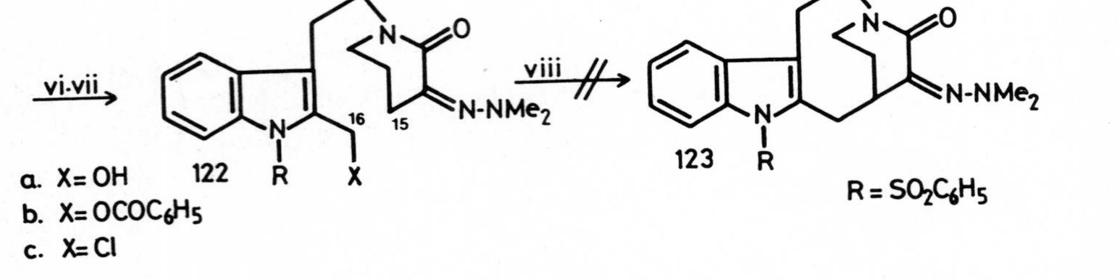

Scheme 18. Reagents: (i) Br(CH$_2$)$_3$C(OMe)$_2$CO$_2$Me (**120**),NaHCO$_3$,KI; (ii) THPCl,$\underline{i}$-Pr$_2$EtN; (iii) LDA, then C$_6$H$_5$SO$_2$Cl; (iv) TsOH, MeOH; (v) TsOH, acetone; (vi) NH$_2$-NMe$_2$; (vii) C$_6$H$_5$COCl, py or MsCl, 2,6-diMepy; (viii) LDA.

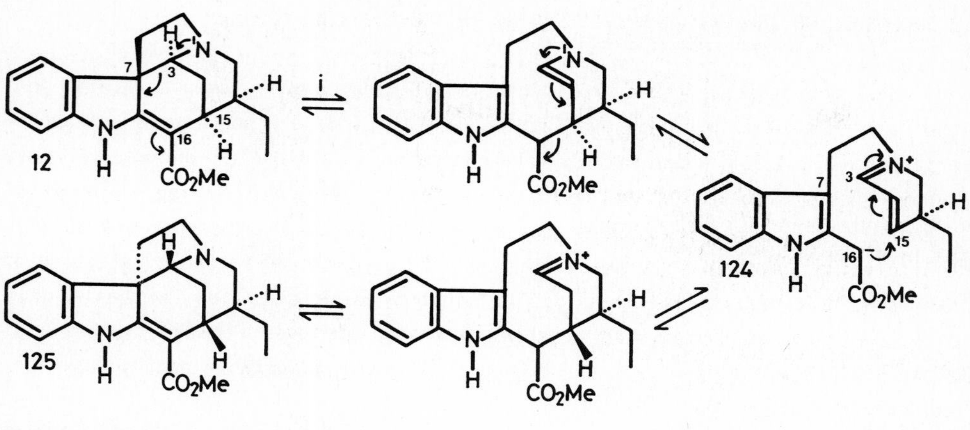

Scheme 19. Proposed mechanism of isomerization of (−)-19,20-dihydroakuammicine (**12**). Reagents: (i) degassed absolute MeOH, reflux, 50 h.

In contrast, it is worth commenting that base-promoted cyclization of N-(2-cyanomethyltryptophyl)pyridinium salts **126a–e** occurs by nucleophilic attack at the α-position of the pyridine ring (formation of C_3–C_{16} bond) with closure of a seven-membered ring, to give (135) in excellent yields the corresponding tetracyclic pyridoazepinoindole system (**127a–e**) characteristic of the indole alkaloid ngouniensine (Scheme 20). Formation of stemmadenine-type nine-membered ring products, such as **128a**, coming from nucleophilic attack at the γ-position of the pyridinium salt, i.e., in which a C_{15}–C_{16} bond had been formed, was never detected.

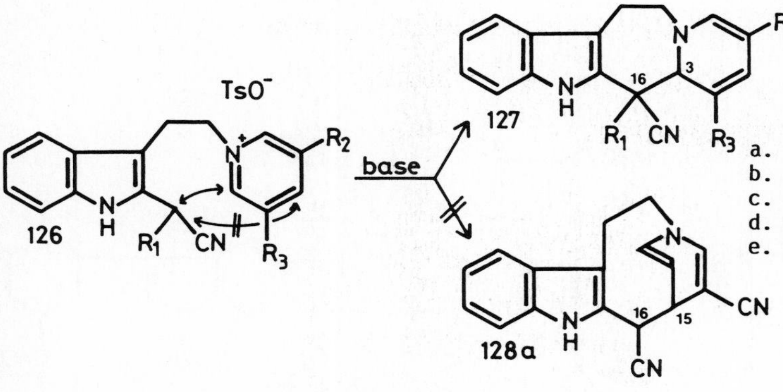

	R_1	R_2	R_3
a.	H	CN	H
b.	H	COOMe	COOMe
c.	CH_3	COOMe	COOMe
d.	CH_3	COOMe	H
e.	CH_3	CN	H

Scheme 20

The synthetic approach to pentacyclic *Strychnos* indole alkaloids based on the formation of C_{15}–C_{16} bond is worthy of further study.

3   CLOSURE OF THE FIVE-MEMBERED E RING IN THE KEY STEP

The elaboration of the pyrrolidine ring (E ring) of *Strychnos* alkaloids in the last synthetic steps by cyclization upon the indole 3-position of an appropriate 1,2,3,4,5,6-hexahydro-1,5-methanoazocino[4,3-b]indole derivative (see Scheme 4) constitutes a synthetic entry to the pentacyclic ring skeleton of these alkaloids, that has been successfully applied to the synthesis of (±)-tubifolidine (98). This is the only synthetic alternative to the cyclization of stemmadenine-type systems that has culminated to date in the synthesis of a *Strychnos* alkaloid. A similar strategy has been employed for the synthesis of pentacyclic *Aspidosperma*-type structures starting from suitable octahydropyrido-[3,2-c]carbazoles (136).

The application of this strategy to the synthesis of *Strychnos* alkaloids of the Strychnan type requires two well-definite phases. Firstly, construction of the tetracyclic hexahydro-1,5-methanoazocino[4,3-b]indole system, which embodies rings ABCD of these alkaloids, through a route that allows the introduction of the required ring substituents, i.e. a) ethyl or ethylidene at C-4 (C-20 in the biogenetic numbering), b) a functionalized one-carbon at C-6 (C-16 in the biogenetic numbering), and c) a suitable two-carbon unit (the tryptamine bridge) on the piperidine nitrogen atom. Secondly, closure of E ring, with formation of the crucial quaternary C-7 center, by intramolecular alkylation upon the 3-position of a 2,3-disubstituted indole ring.

3.1 Construction of the Tetracyclic ABCD Ring System

The hexahydro-1,5-methanoazocino[4,3-b]indole system has received considerable attention from a synthetic standpoint because it constitutes the fundamental tetracyclic framework of indole alkaloids uleine and dasycarpidone. Previous

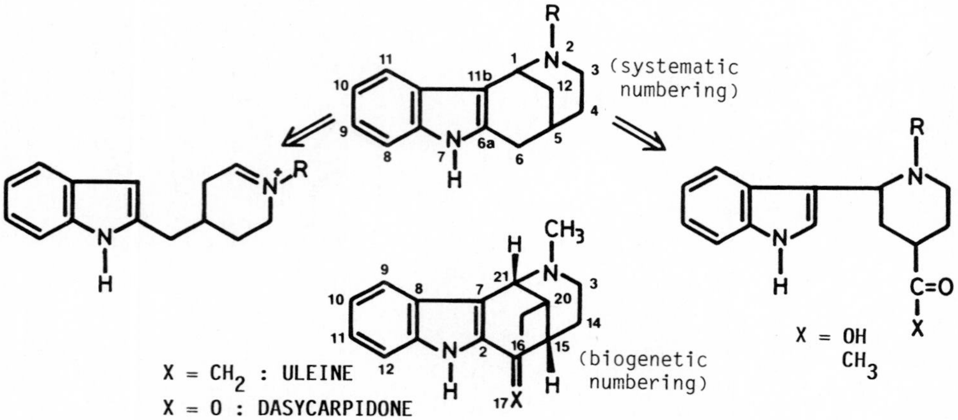

X = CH$_2$: ULEINE
X = O : DASYCARPIDONE

X = OH
CH$_3$

Scheme 21

synthetic approaches to this ring system had been accomplished (137) in the context of the synthesis of these alkaloids and structural analogues by closure of carbocyclic C ring according to one of the two bond disconnections depicted in Scheme 21: a) formation of C_1-C_{11b} bond by cyclization of an iminium salt upon the indole 3-position (138), and b) formation of C_6-C_{6a} bond by cyclization upon the indole 2-position of either 2-(indolyl)piperidine-4-carboxylic acids (139) or 2-(indolyl)-4-acetylpiperidines (140). Although one of these syntheses (Scheme 22) had been applied (139c) to the preparation of a 4-ethyl derivative (iso- and epiisodasycarpidone), its very low overall yield made evident that new approaches had to be developed to accomplish the first phase of our strategy.

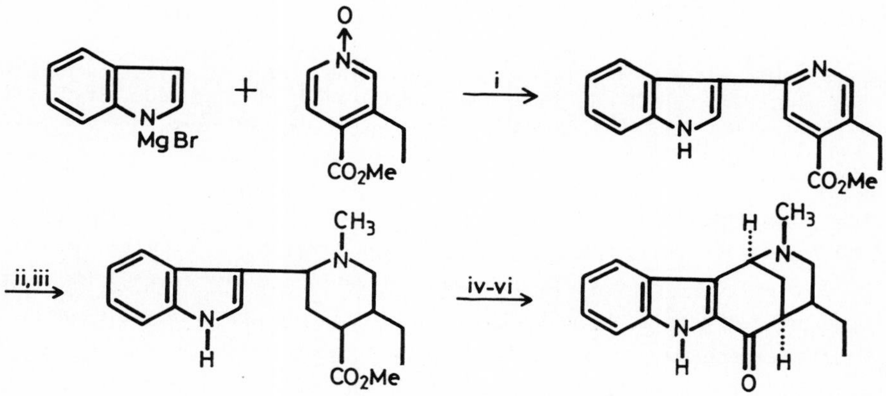

Scheme 22. Synthesis of (±)-iso- and (±)-epiisodasycarpidone. Reagents: (i) C_6H_5COCl, CH_2Cl_2, THF; (ii) CH_3I; (iii) H_2, Pt; (iv) KOH, aq EtOH; (v) PPA; (vi) separation of diastereoisomers.

A more convenient route to the required tetracyclic ABCD ring system, involving the intramolecular acylation of a 3-(2-piperidyl)indole in the last step, is outlined in Scheme 23 (141). This synthesis illustrates the usefulness of 2-cyanotetrahydropyridines as synthetic intermediates. These compounds can be considered as latent forms of dihydropyridinium salts that are able to react with indole itself (140c) or with indole activated either as Grignard reagent (142) or alkali metal salt (143), thus allowing the introduction of a tetrahydropyridine fragment onto the indole 3-position.

Thus, condensation of indole with 2-cyano-1,2,3,6-tetrahydropyridine **129**, which was easily accessible by reductive cyanation of the corresponding pyridinium salt, gave tetrahydropyridylindole **130** in nearly quantitative yield. Catalytic hydrogenation of **130** brough about both tetrahydropyridine double bond reduction and hydrogenolysis of N-benzyl group to give piperidylindole **131**, lacking N_b-substituent as required for further introduction of an appropriate functionalized two-carbon chain on the nitrogen atom.

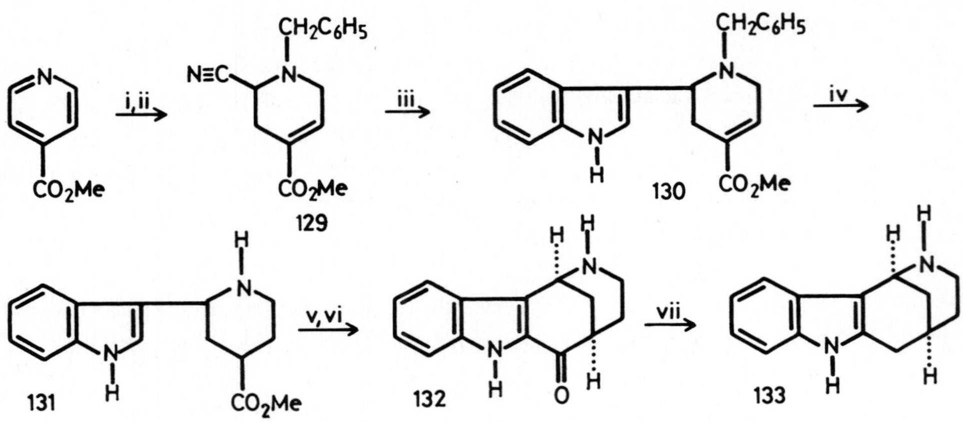

Scheme 23. Elaboration of the tetracyclic ABCD ring system by acylation of indole ring. Reagents: (i) $C_6H_5CH_2Cl$, MeOH; (ii) $NaBH_4$, NaCN, HCl, aq MeOH, Et_2O; (iii) indole, AcOH; (iv) from the hydrochloride, H_2,Pd; (v) $Ba(OH)_2$; then CO_2; (vi) PPA; (vii) $LiAlH_4$, dioxane.

Piperidine **131** was obtained as a 5:1 mixture of cis–trans isomers, which could be separated. Obviously, only the diastereoisomer in which the substituents at C-2 and C-4 are cis can undergo cyclization to the tetracyclic system **132**. However, tetracyclic ketone **132** was obtained in comparable yields when, in a duplicate experiment, pure **cis**-**131** and pure **trans**-**131** were separately subjected to alkaline hydrolysis and then to polyphosphoric acid cyclization. It was demonstrated that isomerization at the piperidine 4-position in the trans isomer to give the most stable cis configuration occurred under the alkaline conditions of the saponification step (144). Finally, reduction of the 2-acyl-indole carbonyl group was accomplished by lithium aluminium hydride (145). Attempted Wolff-Kishner reduction caused cleavage of C_1-N_b bond and further cyclization upon the carbonyl group to give a dihydropyrrolo[2,3-a]carbazole (146).

The above route allows to incorporate the required ethyl substituent by means of a conjugate addition on the α,β-unsaturated ester moiety of **130** (Scheme 24). Piperidylindole **134** was obtained in excellent yield as a mixture of stereoisomers, which could partially be separated. However, this operation is unnecessary from the synthetic point of view since separation of isomers can be more efficiently accomplished after the cyclization step. The mixture **134** was debenzylated and then converted, as in the above deethyl series, to a nearly equimolecular mixture of tetracyclic ketone **135** and its C-4 epimer (**epi**-**135**), thus pointing out the lack of stereoselectivity of the conjugate addition step. In a similar way, starting from the N-methyl analogue of **129**, an improved synthesis of (±)-iso- and (±)-epiisodasycarpidone was reported (144). Reduction of **135** gave the key tetracyclic intermediate **136**, having the same relative configuration

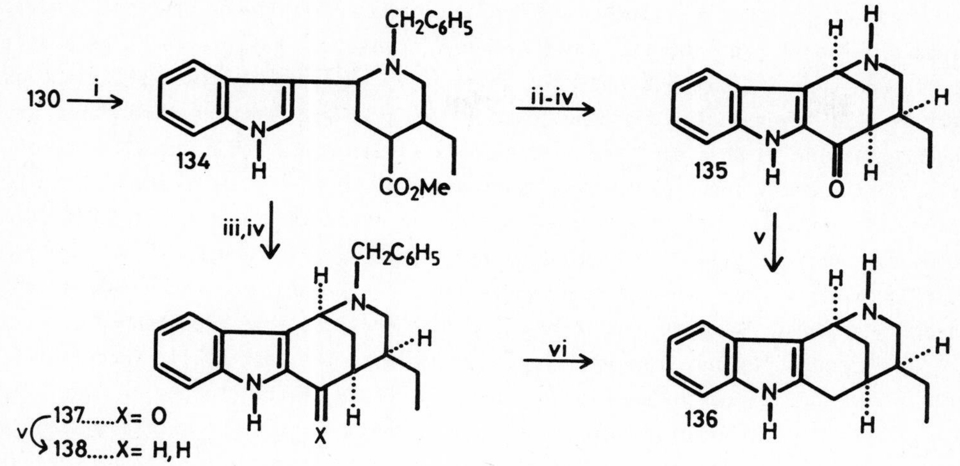

Scheme 24. Introduction of the ethyl substituent. Reagents: (i) EtMgBr/CuI, THF; (ii) from the hydrochloride, H_2, Pd; (iii) Ba(OH)$_2$; then CO_2; (iv) PPA; (v) LiAlH$_4$, dioxane; (vi) H_2, Pd(OH)$_2$.

at C–4 as the *Strychnos* alkaloid tubifolidine at the corresponding position (C–20). Tetracycle **136** was also obtained by reversing the above synthetic steps, as illustrated in Scheme 24. As could be expected, cyclization of **134** afforded again a mixture of C–4 epimeric ketones, **137** and **epi–137**. Ketones **epi–135** and **epi–137** were converted, as in the natural series, to the tetracyclic system **epi–136** (98).

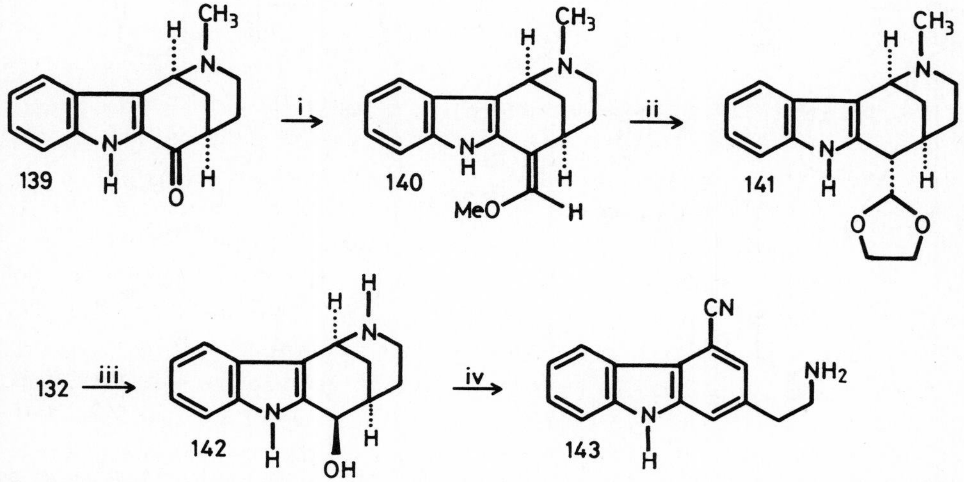

Scheme 25. Reagents: (i) (C$_6$H$_5$)$_2$POCH$_2$OCH$_3$, LDA, THF, 85% yield; (ii) (CH$_2$OH)$_2$, TsOH, C$_6$H$_6$, 38% yield; (iii) NaBH$_4$; (iv) NaCN, DMSO.

The introduction of a functionalized one-carbon substituent upon the 2-acyl-indole carbonyl group of the above tetracyclic systems has proved to be diffi-cult and, for the present, has not been fully explored (Scheme 25). Although ketone **139** (deethyldasycarpidone), which was prepared through a reaction se-quence similar to that depicted in Scheme 23 (141), failed to react with dimeth-ylsulfonium methylide, tosylmethyl isocyanide, and methyl methylthiomethyl sulf-oxide, it was satisfactorily converted to enol ether **140**. However, neither this compound nor the acetal **141** could be hydrolyzed to the corresponding aldehyde (147). On the other hand, attempted nucleophilic substitution by cyano of the hydroxy group of **142** gave carbazole **143** as the only isolable product (148).

An alternative but less efficient synthesis of the tetracyclic intermediate **138** (149) and its deethyl analogue **148** (150), based on the closure of ring C by cyclization of an iminium salt generated by mercuric acetate oxidation from a suitable 2-(4-piperidylmethyl)indole, is outlined in Scheme 26. Although several synthetic routes to 2-(4-piperidylmethyl)indoles had been previously reported, two of them in the more interesting 3-ethylpiperidine series (96, see Scheme 16) (151), the required compounds **146a** and **146b** were prepared by a new procedure involving Wadsworth-Emmons condensation from an appropriate 4-piperidone **144** as the first step and Fischer indolization from a 4-acetonylpiperidine **145** as the last step. From 4-piperidone **144a** the acetonyl chain was directly introduced with diethyl 2-oxopropylphosphonate followed by catalytic hydrogenation, whereas in the 3-ethyl series it was more conveniently accomplished, in a stereoselective

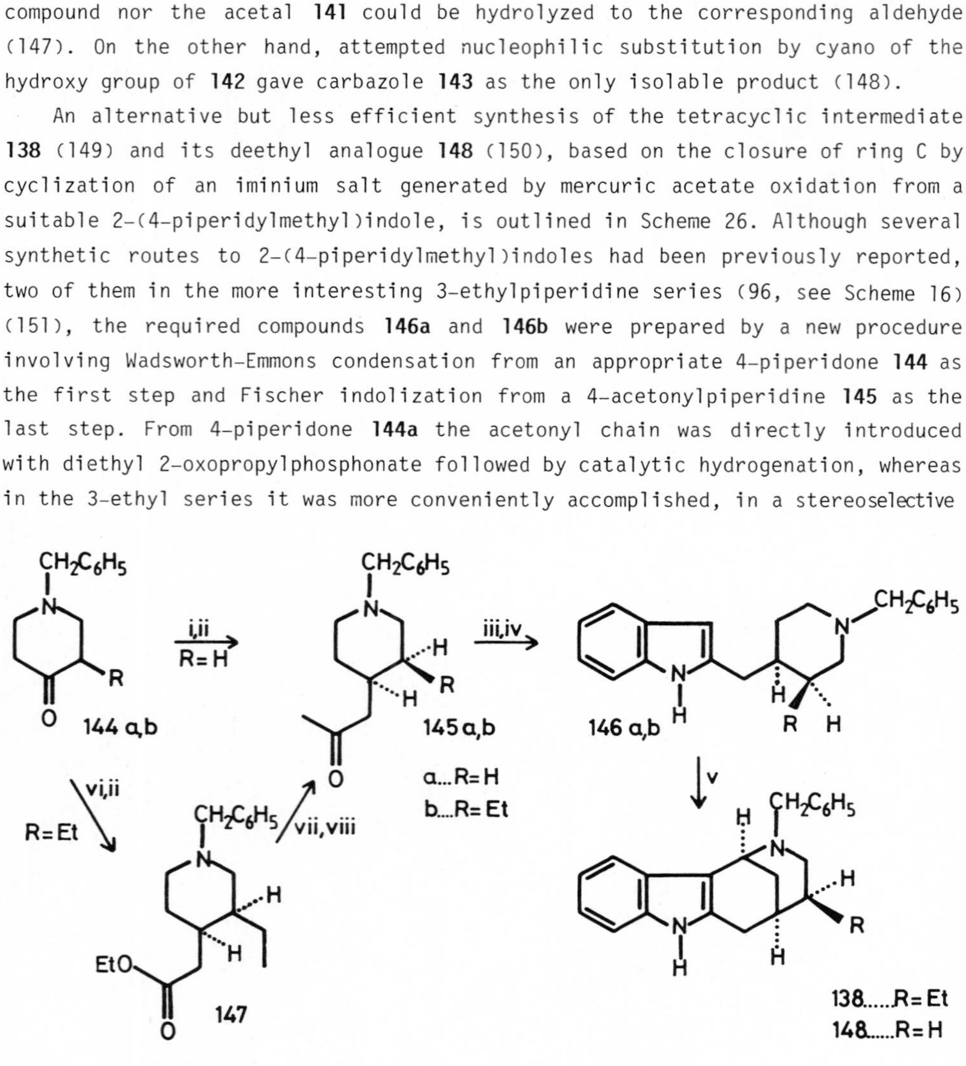

Scheme 26. Elaboration of the tetracyclic ABCD ring system by cyclization of an iminium salt. Reagents: (i) $(EtO)_2POCH_2COCH_3$, KOH, aq EtOH; (ii) H_2, Pd; (iii) $C_6H_5NHNH_2$; (iv) PPA; (v) $Hg(OAc)_2$, $EDTA.2Na.2H_2O$; then $NaBH_4$; (vi) $(EtO)_2POCH_2CO_2Et$, NaH, glyme; (vii) $MeSOCH_2Na$, DMSO; (viii) Zn, AcOH.

manner, via 4-piperidineacetate **147**, having the natural cis relationship between C-3 and C-4 substituents (152). The main limitation of this route is the lack of regioselectivity of the indolization step, since mixtures (approximately 1:1 ratio) of the desired (piperidylmethyl)indoles **146** and the corresponding 2,3-disubstituted indoles coming from cyclization upon the methylene group were obtained (however, see Scheme 46, Section 4.2.1).

The strategy based on the oxidative cyclization of (piperidylmethyl)indoles has been applied to the synthesis of tetracyclic systems bearing a functionalized one-carbon substituent at C-6. The synthesis of the model tetracyclic derivative **152**, which can be envisaged as a synthetic precursor of 20-deethyl-strychnofluorine, exemplifies the applicability of this methodology (153) (Scheme 27). A similar reaction sequence had been previously employed in the synthesis of 19,20-dihydro-16-epivinoxine (154).

The methoxycarbonyl group was introduced from 2-(4-pyridylmethyl)indole **149** in an early stage of the synthesis by taking advantage of the acidic character of the interannular methylene protons. Alkylation of the pyridine nitrogen of **150** followed by catalytic hydrogenation and mercuric acetate oxidation of the resultant piperidine **151** afforded tetracycle **152** as a mixture of epimers at C-6. The 2-(benzyloxy)ethyl substituent on the piperidine nitrogen incorporates the functionalized two-carbon unit required for subsequent closure of the tryptamine bridge.

A straightforward entry to hexahydro-1,5-methanoazocino[4,3-b]indoles, that allows to incorporate the characteristic C_{16}-methoxycarbonyl and C_{20}-ethylidene appendages present in some *Strychnos* alkaloids of the Strychnan type, e.g. akuammicine, has been recently reported (155). This synthesis also involves closure of ring C by attack of an iminium salt upon the indole 3-position and consists of only two separate steps. First, intermolecular addition of the lithium enolate

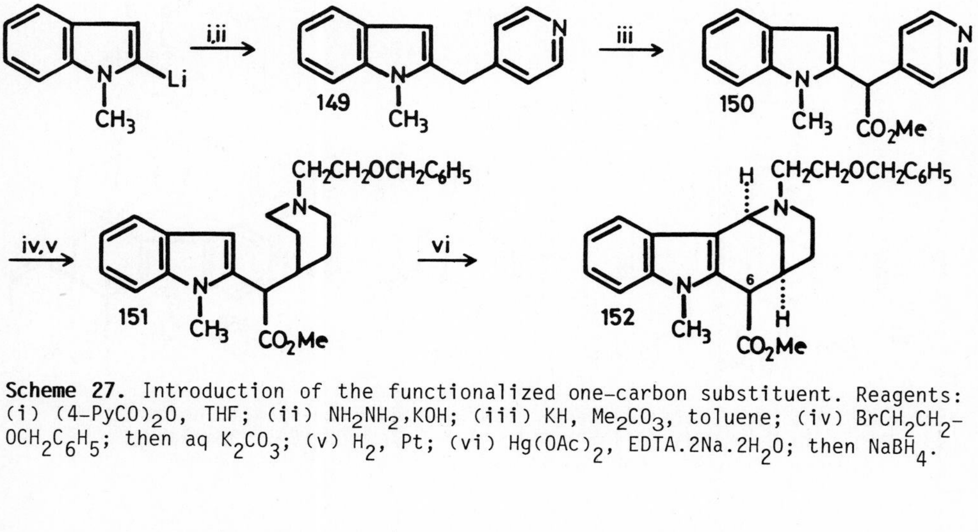

Scheme 27. Introduction of the functionalized one-carbon substituent. Reagents: (i) (4-PyCO)$_2$O, THF; (ii) NH$_2$NH$_2$,KOH; (iii) KH, Me$_2$CO$_3$, toluene; (iv) BrCH$_2$CH$_2$-OCH$_2$C$_6$H$_5$; then aq K$_2$CO$_3$; (v) H$_2$, Pt; (vi) Hg(OAc)$_2$, EDTA.2Na.2H$_2$O; then NaBH$_4$.

of ester **153** to the γ-position of pyridinium salt **154** (156) followed by regiospecific acid-promoted cyclization of the intermediate 1,4-dihydropyridine **155**,
which occurs in 75% overall yield. Second, stereoselective elaboration of the
E-ethylidene substituent (10); this transformation was effected in 32% yield
taking advantage of the doubly vinylogous urethane moiety present in tetracycle
156, by treatment with hydrochloric acid, further reesterification of the C-6
carboxy group in the intermediate **157**, and subsequent sodium borohydride reduction of the iminium cation. The natural E-configuration of the ethylidene group
results as a consequence of its formation by reduction of a conjugated iminium
salt (see Scheme 2). A similar reaction sequence had been used in the first
total synthesis of the bridged indole alkaloid vinoxine (157). The N-hydroxy-
ethyl substituent of **158** can allow further elaboration of ring E of pentacyclic
Strychnos indole alkaloids.

   This short route offers the potential to provide a synthetic entry to diver-
sely substituted hexahydro-1,5-methanoazocino[4,3-b]indoles and, if further closure of ring E could be accomplished, to a variety of pentacyclic alkaloids having the Strychnan skeleton. In fact, the above two-step synthetic sequence also
works in the N-unsubstituted indole series (155) and, when forcing the conditions of the hydrolytic step, decarboxylation of the intermediate iminium salt
occurs to give C-6 unsubstituted analogues of **158** (158). On the other hand, the
ethylidene substituent can be considered as a precursor of an ethyl group, and
the required reduction would presumably lead to the natural stereochemistry
(24).

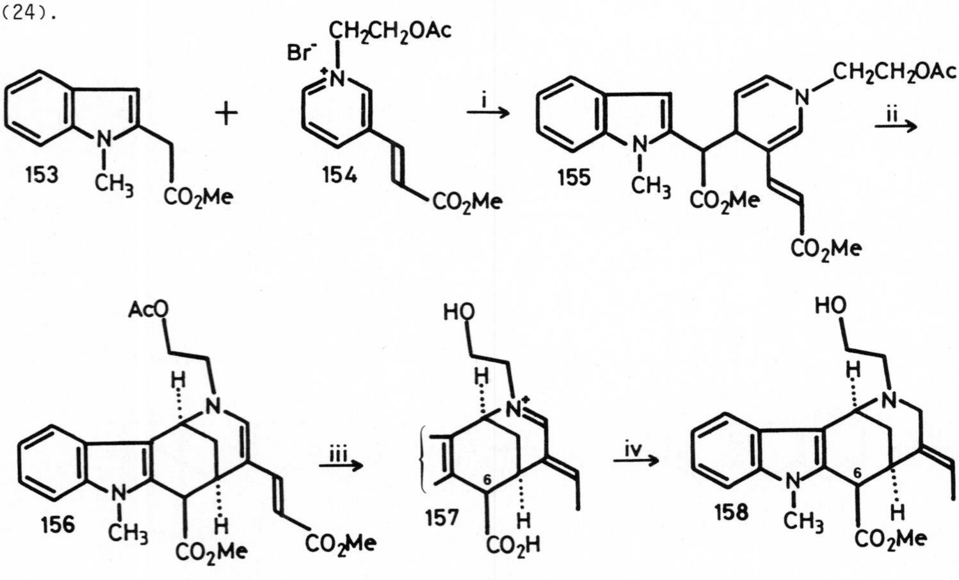

Scheme 27. A direct entry to the tetracyclic ABCD ring system. Reagents: (i)
LDA, THF; (ii) C_6H_6-HCl; (iii) 4N HCl; (iv) 1.2N MeOH-HCl; then NaBH$_4$.

3.2 Closure of Ring E

With viable methods in hand for the construction of the tetracyclic ABCD ring substructure of *Strychnos* alkaloids and for the introduction of the required ring substituents, the second phase of our strategy could be undertaken.

The more readily available tetracycle **133**, lacking the two-carbon substituent at C-4, was selected as a model starting material to explore what method was suitable for the closure of the tryptamine bridge.

The first method studied was the photocyclization of chloroacetamide **159** (145) (Scheme 28). However, when compound **159** was subjected to photolysis, the pentacycle **160a** coming from photocyclization at the indole 4-position, instead of the desired *Strychnos*-type system, was obtained in 40% yield. 2-Acylindole **160b**, presumably formed by photooxidation of **160a**, was formed as a by-product (10%).

Some factors could account for the failure to construct the pentacyclic *Strychnos* skeleton: a) Photocyclization of chloroacetamides upon activated aromatic rings is a good method of forming medium-sized lactams; in fact, pentacycles **160** have a seven-membered lactam while photocyclization at the indole 3-position would imply closure of a five-membered ring. b) The indole ring is already substituted at the 3-position, a structural feature normally not present in this type of photocyclizations (however, see reference 159). c) As can be observed from stereomodels, the chloroacetyl group lies nearer the C-4 of the indole ring than the C-3 and cyclization upon the latter position would imply to disturb the planarity, and hence the resonance of the amide bond. This reason could also explain the failure of the base-promoted cyclization of chloroacetamide **159** under a variety of reaction conditions (160). In this context, it is worth mentioning that attempts to obtain a pentacyclic *Aspidosperma* structure by cyclization of a chloroacetamide, either light- or base-induced, on the substituted indole 3-position of an octahydropyrido[3,2-c]carbazole also failed (161).

In contrast, cyclization of 2-(phenylsulfinyl)acetyl derivatives of this tetracyclic ring system by intramolecular Pummerer-type reaction has proved to be a good method of constructing the five-membered E ring of *Aspidosperma* alka-

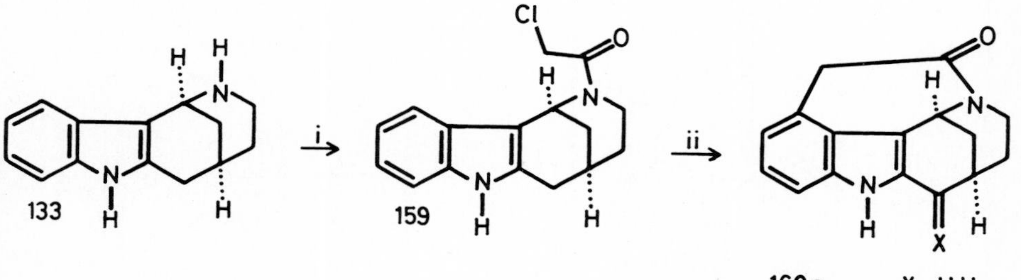

160a..........X= H,H
160b..........X= O

Scheme 28. Chloroacetamide cyclization. Reagents: (i) ClCH$_2$COCl, aq NaOH; (ii) hν, MeOH, K$_2$CO$_3$, 45 min.

loids (161). However, the extension of this methodology to the *Strychnos* series did not result in the desired cyclization (Scheme 29). When the required (phenyl-sulfinyl)acetamide **161a** was treated with trifluoroacetic anhydride and the re-sulting α-(trifluoroacetoxy)sulfide **162a** (detected by NMR) was heated at 135°C, the dithioacetal **163a** was obtained in 70% yield. No cyclization products were detected (162).

Formation of dithioacetals from sulfoxides under Pummerer-type conditions had been previously reported and involves intermolecular attack of the sulfur atom of the initially formed α-acyloxysulfide at the methine carbon of a second molecule, in which the acyloxy group acts as a leaving group (163). This means that sulfur favorably compites with indole as nucleophile. In accordance with the lower leaving group character of acetoxy group as compared with trifluoro-acetoxy, sulfoxide **161a** was cleanly converted (95%) to α-acetoxysulfide **164a**, which was stable to heating. However, in the presence of a Lewis acid, which enhances the ability as leaving group of the acetoxy group, **164a** gave again dithioacetal **163a** (160).

Similar results were obtained from (methylsulfinyl)acetamide **161b**: dithio-acetal **163b** or α-acetoxysulfide **164b** were obtained depending on the anhydride used, thereby indicating that the failure in the cyclization step (formation of a quaternary center) can not be attributed to steric interactions due to the bulky phenyl group. The reluctance of sulfoxides **161** towards cyclization can not be attributed either to the inductive deactivation exerted by the tosyl substit-

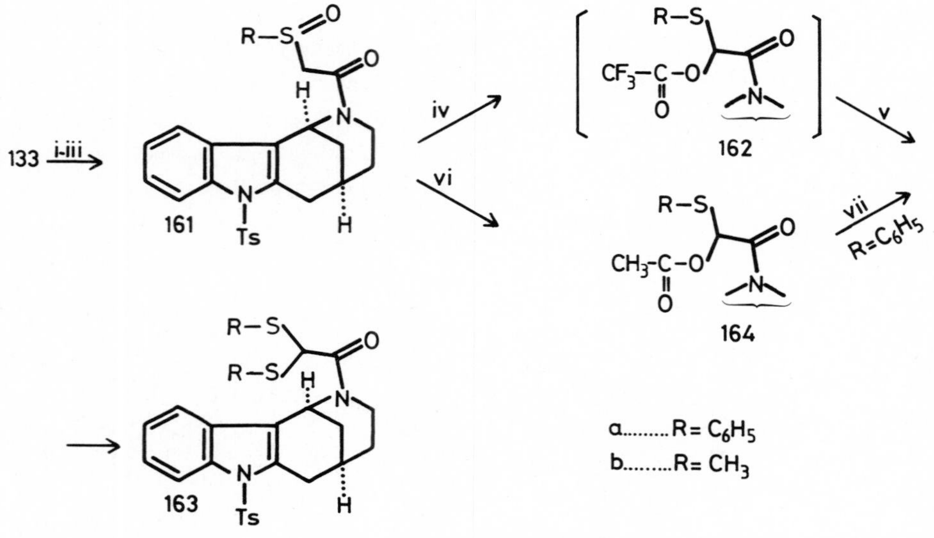

Scheme 29. Reagents: (i) $C_6H_5SCH_2COCl$ or CH_3SCH_2COCl, aq NaOH, CH_2Cl_2; (ii) TsCl, $Bu_4N^+HSO_4^-$, aq NaOH, C_6H_6; (iii) m-CPBA; (iv) TFAA, CH_2Cl_2, 0°C; (v) C_6H_5Cl, Δ; (vi) Ac_2O, Δ; (vii) $BF_3 \cdot Et_2O$, CH_2Cl_2, Δ, 40 h.

uent attached to the indole nitrogen since the desulfonylated analogue of **161a** also gave the corresponding dithioacetal under Pummerer reaction conditions (162). In this case the reaction was performed in the presence of an equimolecular amount of triethylamine since in absence of base an intractable polymeric mixture was obtained.

A common feature of substrates used to study both the chloroacetamide photo-cyclization and the Pummerer reaction is that the exocyclic carbon atom linked to the piperidine nitrogen is sp^2 hybridized. Changing the hybridation of this carbon from sp^2 to sp^3 not only reduces the distance between the indole 3-position and the potential electrophilic carbon atom but also would avoid the possible unfavorable effect due to the loss of planarity of the amide bond.

Therefore, attention was focused to substrates in which the exocyclic carbon attached to the piperidine nitrogen was sp^3 hybridized (Scheme 30). However,

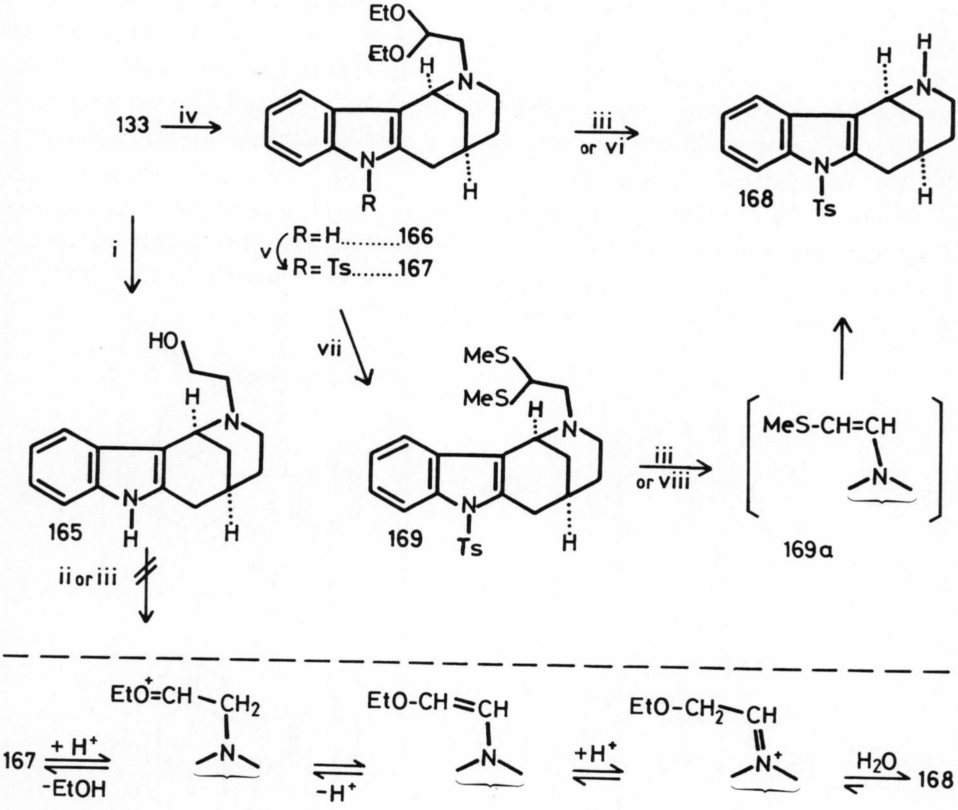

Scheme 30. Reagents: (i) oxirane; (ii) MsCl, Et$_3$N, THF; then t-BuOK; (iii) BF$_3$.Et$_2$O; (iv) (EtO)$_2$CHCH$_2$Br, K$_2$CO$_3$, dioxane; (v) TsCl, Bu$_4$N$^+$HSO$_4^-$, aq NaOH,C$_6$H$_6$; (vi) TsOH,C$_6$H$_6$; (vii) CH$_3$SH, BF$_3$.Et$_2$O,CH$_2$Cl$_2$,r.t.; (viii) [Me$_2$S-SMe]$^+$BF$_4^-$,CH$_2$Cl$_2$,rt.

alcohol **165** could not be cyclized either directly in the presence of a Lewis acid or under basic conditions by way of the corresponding mesylate. On the other hand, attempted cyclization of amino acetal **167** under acidic conditions unexpectedly gave secondary amine **168** as the only isolable product. A similar result was obtained from dithioacetal **169**, although in this case the intermediate enamine **169a** could be detected from the reaction mixture (160).

These discouraging results can be rationalized by considering the equilibria depicted in Scheme 30 and raised the question regarding whether the deactivating effect of the tosyl substituent was the responsible of the failure of the above cyclizations involving trigonally hybridized oxonium or thionium intermediates. This question could not be solved in the first case since the desulfonylated amino acetal **166** gave only polymeric materials when it was treated with p-toluenesulfonic acid in boiling benzene. However, removal of the indole protecting group was crucial for the success of the cyclization in the dithioacetal series, and treatment of the unprotected indole dithioacetal **170** with dimethyl (methylthio)sulfonium fluoroborate (DMTSF) afforded pentacycle **171** in 50% yield (162) (Scheme 31). A great part of the success of this reaction probably stems from the reagent used to generate the thionium ion, since DMTSF is an excellent initiator to generate thionium ions in very mild conditions (164), compatible with the unprotected indole ring.

Further reduction of the imine double bond of **171** followed by desulfurization gave 20-deethyltubifolidine (**172**), the fundamental ring system of pentacyclic *Strychnos* indole alkaloids. This transformation could be more efficiently done in a single step by Raney nickel treatment (160).

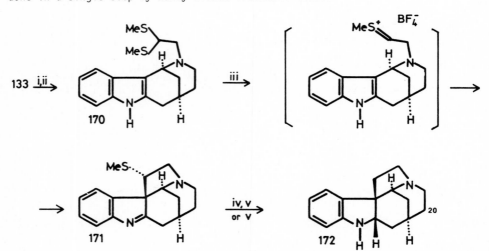

Scheme 31. Closure of E ring: synthesis of 20-deethyltubifolidine. Reagents: (i) $(EtO)_2CHCH_2Br$, K_2CO_3, dioxane; (ii) CH_3SH, $BF_3.Et_2O$, CH_2Cl_2, 0°C; (iii) $[Me_2S-SMe]^+BF_4^-$, CH_2Cl_2, 0°C, 3 h; (iv) $LiAlH_4$; (v) Raney Ni W-2, EtOH.

The problem of the closure of ring E having been solved, the same methodology was applied to the synthesis of (±)-tubifolidine from the ethyl substituted tetracycle **136**. Alkylation of secondary amine **136** with bromoacetaldehyde diethyl acetal followed by exchange of ethoxy groups by methylthio gave dithioacetal **173**, which was then cyclized with DMTSF to pentacycle **174**. Hydrogenolysis of the carbon-sulfur bond with simultaneous reduction of the indolenine ring gave (±)-tubifolidine (98) (Scheme 32).

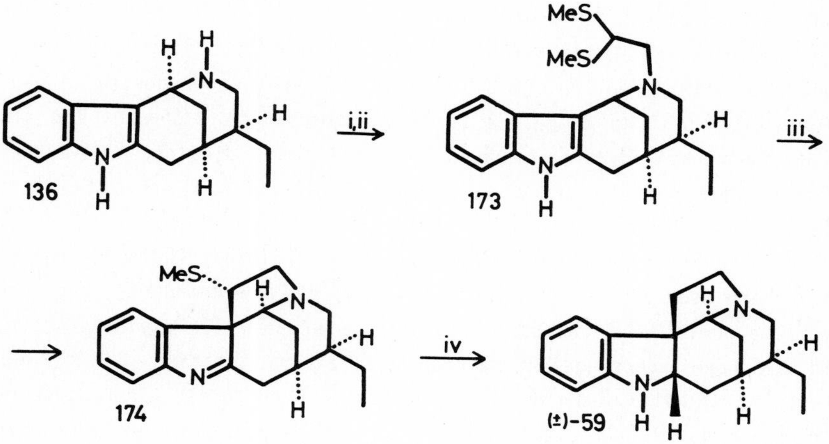

(±)-TUBIFOLIDINE

Scheme 32. Total synthesis of (±)-tubifolidine by Bosch *et al.* Reagents: (i) $(EtO)_2CHCH_2Br$, Na_2CO_3, dioxane; (ii) CH_3SH, $BF_3.Et_2O$, CH_2Cl_2, 0°C; (iii) $[Me_2S\text{-}SMe]^+BF_4^-$, CH_2Cl_2, 0°C, 3 h; (iv) Raney Ni W-2, EtOH.

Through a similar four-step reaction sequence, tetracycle **epi-136** was converted to (±)-20-epitubifolidine (Scheme 33).

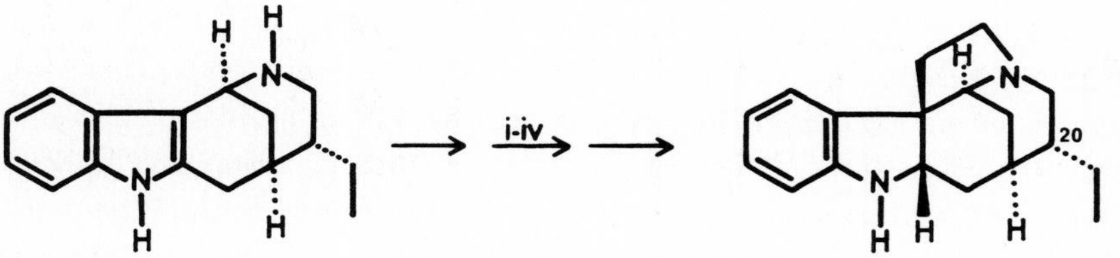

epi-136 (±)-20-EPITUBIFOLIDINE

Scheme 33. Synthesis of (±)-20-epitubifolidine. Reagents: (i-iv) as in Scheme 32.

The extension of this work to the synthesis of more complex pentacyclic *Strychnos* alkaloids is in progress in our laboratory. With available methods for the introduction of the functionalized one-carbon substituent at C-16 and the ethylidene group at C-20, the strategy discussed in this Section can provide a general synthetic entry to most of pentacyclic *Strychnos* alkaloids.

4 OTHER APPROACHES

4.1 Closure of the Piperidine Ring

Although the approaches to pentacyclic *Strychnos* alkaloids based on the closure of the piperidine ring (D ring) in the last synthetic steps by formation of
N_4-C_{21} (or $-C_3$ in the Aspidospermatan series) bond from partially reduced pyrrolo[2,3-$\underline{d}$]carbazoles have received some attention in recent years. (117-120),
the efforts in this field have not culminated to date in the total synthesis of
individual alkaloids. In contrast, several total (165) and formal (166) syntheses of *Aspidosperma* alkaloids (167) by elaboration of the piperidine ring from
functionalized octahydropyrrolo[2,3-$\underline{d}$]carbazoles have been reported. These pyr-
rolocarbazoles, e.g. the so-called Büchi´s intermediate (165a,c), can also be
envisaged as potential synthetic precursors of *Strychnos* alkaloids. Nevertheless,
inasmuch as they have not received further application in this field, their
syntheses will not be discussed in this report.

In the context of the synthesis of *Strychnos* alkaloids, two different strategies have been used for constructing, with complete stereocontrol, the required
tetracyclic pyrrolo[2,3-$\underline{d}$]carbazole skeleton. Overman (117) emploies the tandem
cationic aza-Cope rearrangement-Mannich cyclization (Section 4.1.1), whereas
Vercauteren *et al.* (118) make use of an intramolecular ˝Diels-Alder˝ type cyclization (Section 4.1.2). The latter approach has made possible the first enantioselective entry to the pentacyclic Aspidospermatan skeleton.

4.1.1 Tandem Aza-Cope-Mannich Approach

The ˝Mannich-directed˝ cationic aza-Cope (2-azonia-[3,3]-sigmatropic) rearrangement constitutes a useful synthetic transformation that has been extensively developed by Overman and co-workers since 1979 (168) and successfully applied
to the stereoselective synthesis of alkaloids (169), including the *Aspidosperma*
indole alkaloid 16-methoxytabersonine (170). In 1985, with the final aim of
synthesizing akuammicine, Overman reported (117) the stereocontrolled synthesis
of a series of hexahydropyrrolo[2,3-$\underline{d}$]carbazoles **181** using his aza-Cope-Mannich
strategy. The route adopted for the synthesis of **181b** is summarized in Scheme 34
(R=Si$\underline{t}$BuMe$_2$) (the biogenetic numbering has been maintained for all synthetic intermediates). It implies the formation of rings B, C, and E in a single synthetic step, with complete stereocontrol, from a cyclopentoxazolidine derivative **177**
having the appropriate stereochemistry, i.e. : i) natural $\underline{E}$-configuration for
the C-20 ethylidene substituent; ii) cis-relationship between C_3-amino and C_{15}-
methylpropenyl substituents, as necessary for further closure of piperidine ring;
and iii) trans-relationship between C_2-vinyl and C_3-amino substituents in order
to allow the stereospecific formation of cis-fused octahydroindole **180**. As a result of this trans orientation, the intermediate iminium ion **178** can undergo an

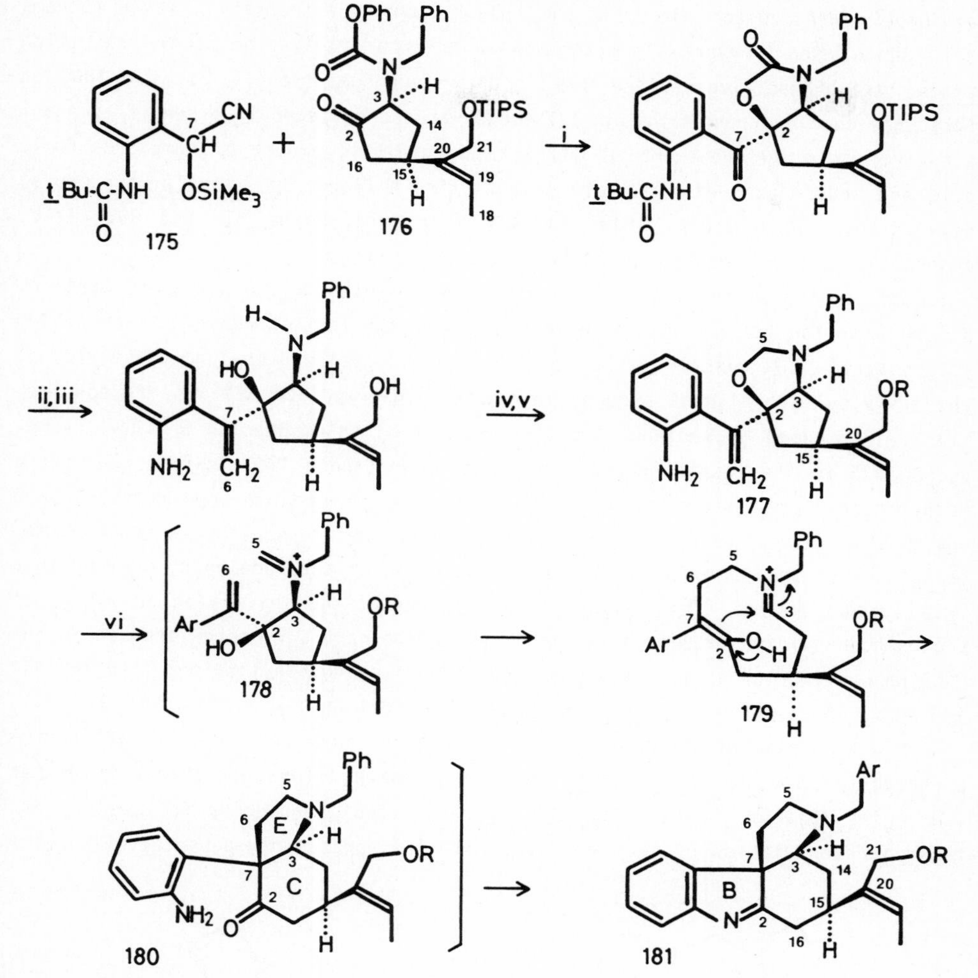

181a (Ar=C_6H_5; R=H); 181b (Ar=C_6H_5; R=Si($\underline{t}$-Bu)Me$_2$; 181c (Ar=$\underline{p}$-MeOC$_6$H$_4$; R=CH$_2$C$_6$H$_5$)

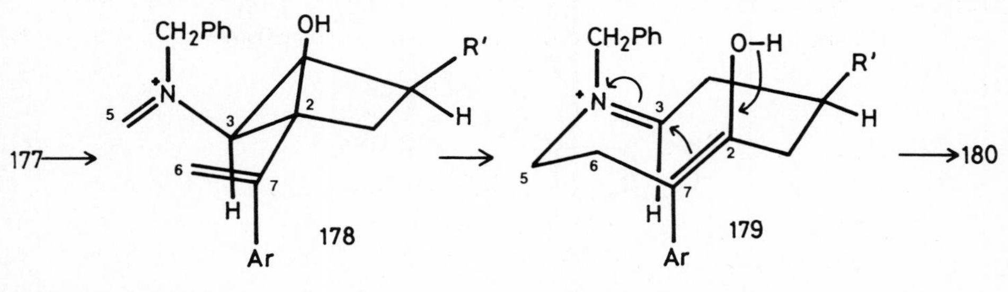

Scheme 34. Synthesis of hexahydropyrrolo[2,3-d]carbazoles by Overman and Angle. Reagents: (i) 2 eq $\underline{n}$-BuLi; (ii) Ph$_3$P=CH$_2$; (iii) KOH, MeOH; (iv) (CH$_2$O)$_n$, THF; (v) $\underline{t}$-BuMe$_2$SiCl; (vi) camphorsulfonic acid, Na$_2$SO$_4$, toluene.

aza-Cope rearrangement via only a single "chairlike" transition state (Scheme 34). Subsequent irreversible transannular Mannich cyclization of the resulting trans, trans-azacyclononadiene **179** occurs stereospecifically to give **180**. In fact, the complex transformation **177**→**181b** was accomplished in essentially quantitative yield by treatment of **177** with a catalytic amount of camphorsulfonic acid and anhydrous sodium sulfate in refluxing toluene. By a similar procedure, hexahydropyrrolocarbazoles **181a** and **181c** were prepared in 51% and 99% yield, respectively, from the appropriate cyclopentoxazolines.

Four remarkable features of this convergent synthesis are worthy of mention: a) the starting ketone **176** was prepared in an efficient stereoselective fashion (eight steps, 20% overall yield) from ethyl α-(3-cyclopentenyl)acetoacetate; b) the arylvinyl substituent present in the key intermediate **177** was stereoselectively introduced by nucleophilic attack of the lithium dianion of silyl cyanohydrin **175** to the less hindered face of cyclopentanone **176**, followed by Wittig methylenation; c) the iminium salt required to promote aza-Cope rearrangement was generated by opening of an oxazolidine ring, and d) non-indole starting materials were used, which required the elaboration of the indolenine nucleus in a late synthetic step. This was efficiently accomplished by taking advantage of the aminophenyl substituent attached to the quaternary C-7 center created in the Mannich cyclization step, since the resulting α-(aminophenyl)ketone intermediate **180** undergoes cyclization under the reaction conditions.

Unfortunately, all attempts to convert hexahydropyrrolocarbazoles **181** to the pentacyclic Strychnan ring system were unsuccessful. Thus, treatment of either indolenine **181a** or indoline **182** with methanesulfonic anhydride followed by attempted debenzylation of the resulting material (presumably a pentacyclic salt)

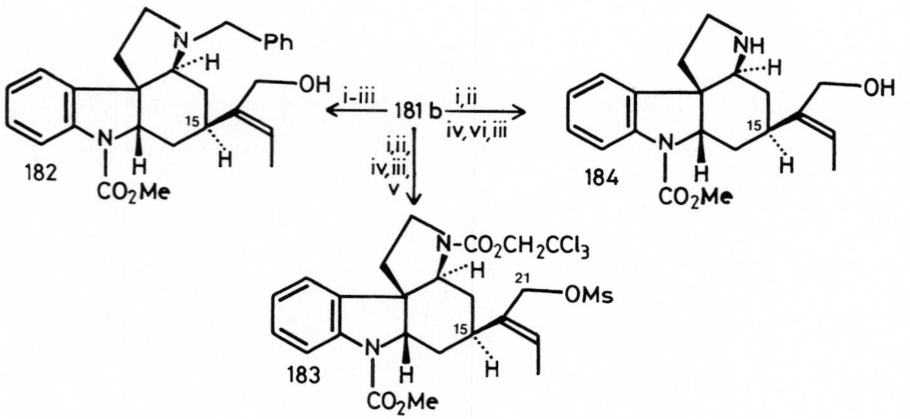

Scheme 35. Reagents: (i) LiAlH$_4$; (ii) ClCO$_2$Me; (iii) Bu$_4$N$^+$F$^-$, THF; (iv) ClCO$_2$CH$_2$CCl$_3$, NaHCO$_3$, CHCl$_3$; (v) Ms$_2$O, Et$_3$N; (vi) Zn, KH$_2$PO$_4$, THF.

under a variety of experimental conditions led only to intractable product mixtures. Similarly, all attempts to induce cyclization by removing the trichloro-
ethyl carbamate group of **183**, in the presence of the allylic mesylate [or in similar carbamates in which C-21 was $CH(OMe)_2$ or CO_2Me], failed. The desired cyclization could not be accomplished from amino alcohol **184** either, since the hydroxy
group could not be selectively activated in front of the secondary amino group.

The failure of the above cyclizations has been explained (117) by considering the severe steric interactions exerted by the C-15 substituent when it
adopts an axial orientation, as required for ring closure.

4.1.2 Intramolecular "Diels-Alder" Approach

Cyclization of secodine-type intermediates constitutes a biomimetic entry to
pentacyclic *Aspidosperma* alkaloids, that has been successfully exploited by the
Kuehne group (167,171). When a simple enamine instead of a 1,4,5,6-tetrahydro-
pyridine is used as the dienophile moiety, the intramolecular "Diels-Alder"
cycloaddition provides hexahydropyrrolo[2,3-d]carbazoles (172). Recently, Vercauteren and co-workers (118) have applied this route to the synthesis of hexahydropyrrolocarbazole **187**, a potential intermediate towards *Strychnos* alkaloids
of the Aspidospermatan type. However, at variance with the Kuehne's approach,
the required enamine precursor **186** is easily generated from a 1,1-disubstituted
tetrahydro-β-carboline derivative, as outlined in Scheme 36. The methoxycarbonyl
substituent at C-15 allows further elaboration of the piperidine ring (see below,
Scheme 38).

Thus, treatment of tetrahydro-β-carboline **185**, which was prepared by a modi-

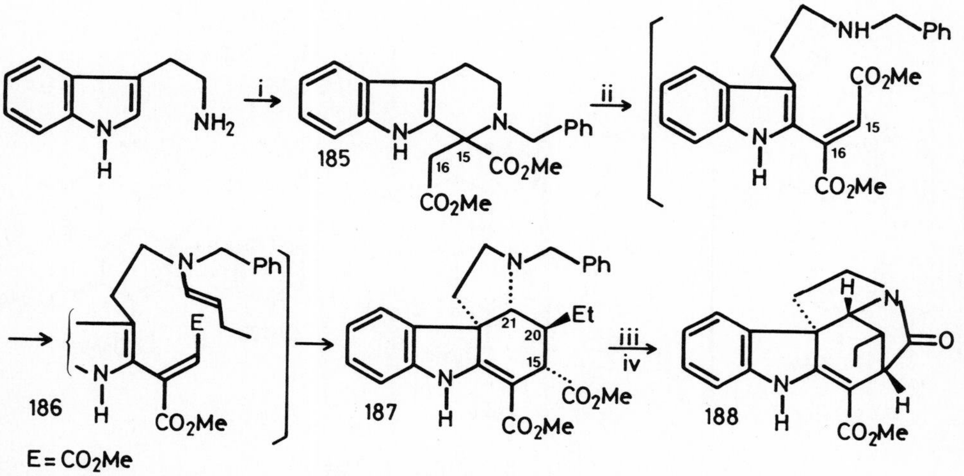

E = CO_2Me

Scheme 36. Reagents: (i) $MeO_2C\equiv C-CO_2Me$; then TFA; (ii) $CH_3CH_2CH_2CHO$, toluene,
AcOH; (iii) H_2,Pd, AcOH; (iv) NaOMe. (Racemic compounds **187** and **188** are shown in
the same absolute configuration as the alkaloids of the Aspidospermatan type).

fied Pictet-Spengler condensation between tryptamine and dimethyl acetylenedicarboxylate (173), with butyraldehyde in the presence of acetic acid afforded tetracycle **187** (59% yield) in a single synthetic step. The crucial cis relationship
between C_{15}-ester and C_{21}-amino substituents was unambiguously proved by conversion of **187** into pentacyclic lactam **188**, having a D-nor Aspidospermatan skeleton. On the other hand, the relative configuration at C-20, which implies a
trans relationship between C_{20}-ethyl and C_{21}-amino substituents, as in tubotaiwine and related alkaloids, follows from the expected more stable E-configuration of the intermediate enamine **186**.

The above straightforward synthetic entry to hexahydropyrrolo[2,3-d]carba-
zoles has been extended to the obtention of the corresponding optically active
series (119,120). Thus, the use of the chiral (S)-tryptophan derivative **189**
instead of tryptamine allowed the preparation of chiral tetrahydro-ß-carbolines
190 as an equimolecular mixture of diastereomers at C-15 (Scheme 37). Reaction
of this mixture with butyraldehyde in acetic acid afforded (68% yield) a mixture
of chiral hexahydropyrrolocarbazoles (+)-**192** and (-)-**193** in a 85:15 ratio, respectively. The same ratio was obtained when the reaction was effected from
the separate ß-carboline isomers, thus providing evidence of an open fumarate
intermediate **191**, without chirality at C-15, in the process.

By X-ray analysis it was ascertained that the new asymmetric centers of the
major isomer (+)-**192** possessed the same absolute configuration as the *Strychnos*
alkaloids of the Aspidospermatan type. On the contrary, the minor isomer (-)-**193**

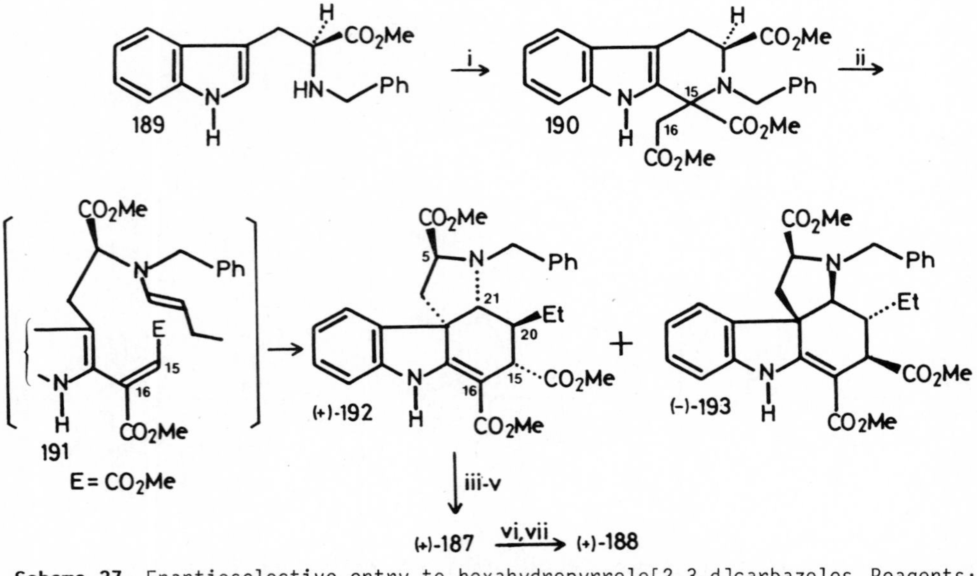

Scheme 37. Enantioselective entry to hexahydropyrrolo[2,3-d]carbazoles. Reagents:
(i) $MeO_2C-C \equiv C-CO_2Me$, MeOH; then TFA; (ii) $CH_3CH_2CH_2CHO$, toluene, AcOH, Δ, 72 h;
(iii) NH_3, 110°C; (iv) TFAA, THF, py; (v) $NaBH_4$; (vi) H_2, Pd; (vii) NaOMe.

belongs to the non—natural enantiomeric series of these alkaloids.

The chiral auxiliary substituent at C—5 was removed in three steps to give an enantiomerically pure form [(+)—187] of the previously prepared hexahydro—pyrrolocarbazole 187, which was further converted into lactam (+)—188.

The same methodology, although with better diastereoselectivity, was used from the chiral N—(phenylethyl)tryptamine 194 (120). This was converted into a C—15 diastereoisomeric mixture of chiral tetrahydro—ß—carbolines (—)—195 and (—)—196 (55:45 ratio) and then into a single hexahydropyrrolocarbazole (—)—197 (Scheme 38).

To establish the absolute and relative configurations of the new asymmetric centers, the chiral inducer was removed in a single step and the resultant sec—ondary amine was transformed into lactam (—)—188, which was the enantiomer of lactam (+)—188 obtained before (119) from L—tryptophan derivative 189. In ac—cordance with a mechanism involving an open intermediate without chirality at C—15, when the diastereoisomeric tetrahydro—ß—carbolines (—)—195 and (—)—196 were separately debenzylated and the resulting enantiomers were independently treated with butyraldehyde, the same racemic lactam (±)—188 was obtained in both cases.

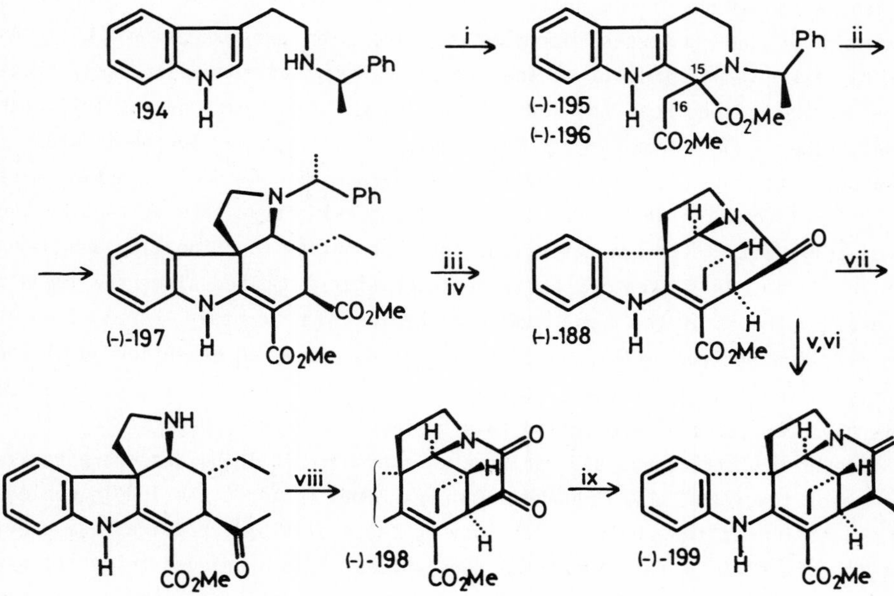

Scheme 38. Enantioselective entry to the pentacyclic Aspidospermatan ring system. Reagents: (i) MeO$_2$C≡C—C—CO$_2$Me; then TFAA; (ii) CH$_3$CH$_2$CH$_2$CHO, toluene, AcOH, Δ, 12 h; (iii) H$_2$,Pd; (iv) NaOMe; (v) C$_6$H$_5$SOCH$_2^-$; (vi) I$_2$, MeOH; (vii) CH$_3$Li; (viii) SeO$_2$; (ix) NaBH$_4$.

With enantioselective methods in hand for constructing both enantiomeric series of hexahydropyrrolocarbazoles, which allow to incorporate the C_{16}-methoxy-
carbonyl and C_{20}-ethyl substituents present in tubotaiwine, the piperidine ring
was built by enlargement of γ-lactam ring of pentacycle **188**. However, rather
surprisingly, the enantiomer chosen by the authors was (-)-**188** (Scheme 38), i.e.
the one possessing the absolute configuration opposite to that of the alkaloids
of the Aspidospermatan type. Expansion of ring D was brough about by two alternative procedures. Whereas the keto sulfoxide route led to alcohol (-)-**199** in a
very poor overall yield, methyl-lithium addition to (-)-**188** followed by selenium
dioxide oxidation afforded keto lactam (-)-**198** in 71% overall yield.

It is worth mentioning that the procedure used for the closure of N_4-C_3 bond
of the pentacyclic Aspidospermatan skeleton is similar to that employed by
Woodward (174) in his synthesis of strychnine and that this successful result
contrasts with the failure (117) to construct the Strychnan ring system from
intermediates such as **181a**, **182-184** (see Section 4.1.1).

The authors (120) have announced further work in order to extend the "Diels-
Alder" approach to the synthesis of natural products of both the Strychnan and
Aspidospermatan types.

4.2 Elaboration of the Indole Nucleus

In all of the syntheses discussed so far, with one exception (117), the
starting material was indolic or the indole nucleus was formed in early stages
of the synthetic sequence; the indole-containing target compound was built taking advantage of the reactivity of the indole ring in the key synthetic steps. A
conceptually different strategy consists in the use of non-indole starting materials with elaboration of the indole, or related, ring at a late synthetic stage
from azapolycyclic intermediates having the appropriate functionality and stereochemistry (175). This strategy has been successfully applied since the 1960s to
the total synthesis of the *Aspidosperma* alkaloids that contain a 3,3-disubsti-
tuted indoline nucleus (176). However, it has not received attention until very
recently in the context of the synthesis of *Strychnos* alkaloids and no total synthesis has been accomplished yet in this series.

Since from a structural standpoint the *Strychnos* alkaloids are characterized
by a 4-azatricyclo[5.2.2.0^{4,8}]undecane ring system fused to the indole nucleus,
the application of the above methodology to these alkaloids requires the previous synthesis of C-11 functionalized derivatives of this bridged tricyclic system, so that later synthetic steps may allow the elaboration of the indole ring.

The only precedent (177) about the formation of the required azatricycloun-
decane ring system is showed in Scheme 39. However, neither the substitution
pattern of **200** nor the low yields of the process seem to be suitable for its
further use in the synthesis of *Strychnos* alkaloids.

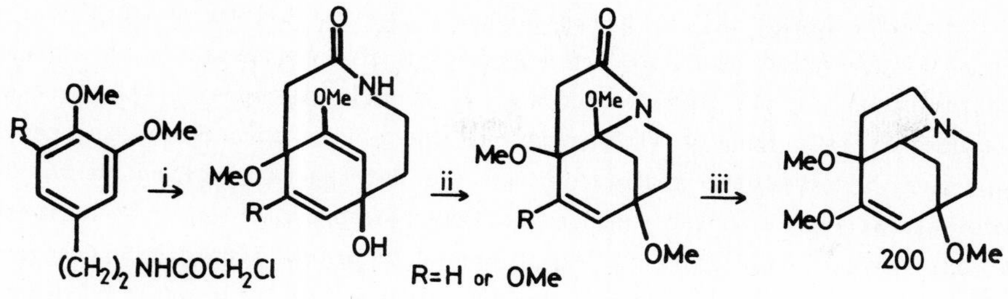

Scheme 39. Reagents: (i) hν, aq EtOH; (ii) HCl–MeOH; (iii) LiAlH$_4$.

In the synthetic approaches developed in our laboratory (Section 4.2.1), the synthesis of the requisite azatricyclic ketone that incorporates rings C, D, and E of *Strychnos* alkaloids was accomplished by closure of the five-membered E ring through intramolecular alkylation of appropriate 2-azabicyclo[3.3.1]nonan-7-ones (Scheme 40).

A route to a 10-oxo derivative, developed with the final objective of the synthesis of strychnine (122), involves closure of the piperidine ring from a 5-oxooctahydroindole in the last step (Section 4.2.2).

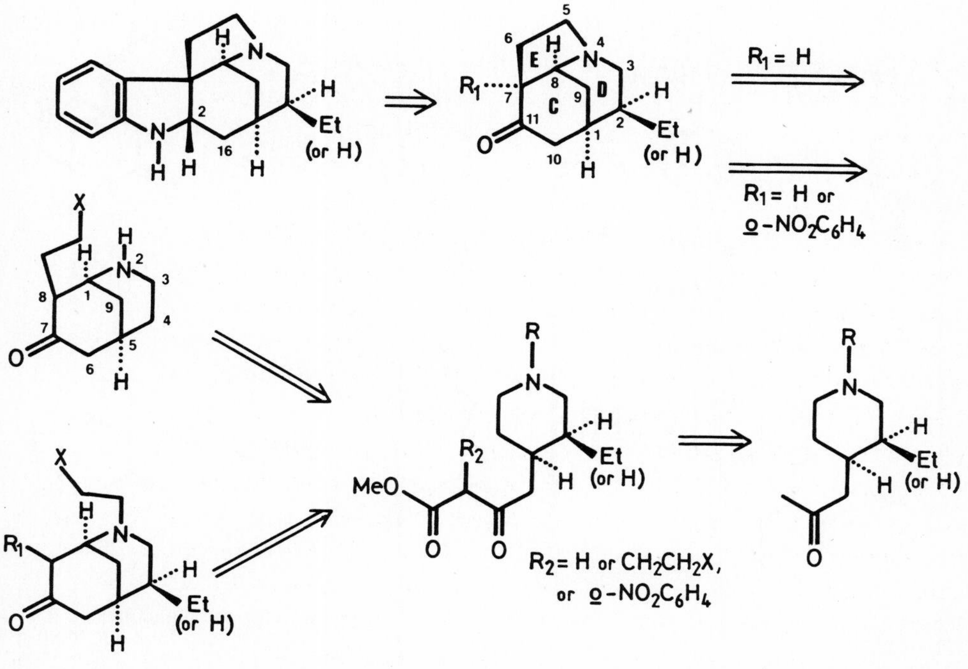

Scheme 40

4.2.1 Use of 2-Azabicyclo[3.3.1]nonan-7-ones as Synthetic Intermediates

Some years ago we dealt with the synthesis of heteroaromatic analogues of 6,7-benzomorphans, a class of compounds with potential analgesic activity due to its structural relationship with morphine. One of the strategies explored for this purpose involved the elaboration of the heteroaromatic ring in the last synthetic steps from a suitably functionalized 2-azabicyclo[3.3.1]nonane (morphan) (178), a structural unit common to morphine and *Strychnos* alkaloids. Continuing our interest in the use of functionalized 2-azabicyclo[3.3.1]nonanes as synthetic intermediates (179), we planned to explore a new synthetic entry to pentacyclic *Strychnos* alkaloids, according to bond disconnections depicted in Scheme 40. In order to evaluate the effectiveness of this proposal we initially selected as synthetic goal the simplified model structure **204**, lacking the C-20 two-carbon substituent present in the natural products, from which the indolenine nucleus would be formed by Fischer indolization.

The required azatricyclic ketone **204** was synthesized from 1-benzyl-4-piperidone by two alternative routes (121,180), both of them using a 2-azabicyclo-[3.3.1]nonan-7-one derivative as the key intermediate (Scheme 41).

The problems associated to these syntheses can be subdivided as follows. First, introduction and elaboration of the requisite piperidine substituents, either on the nitrogen (2-hydroxyethyl) or at the 4-position (acetonyl, aceto-

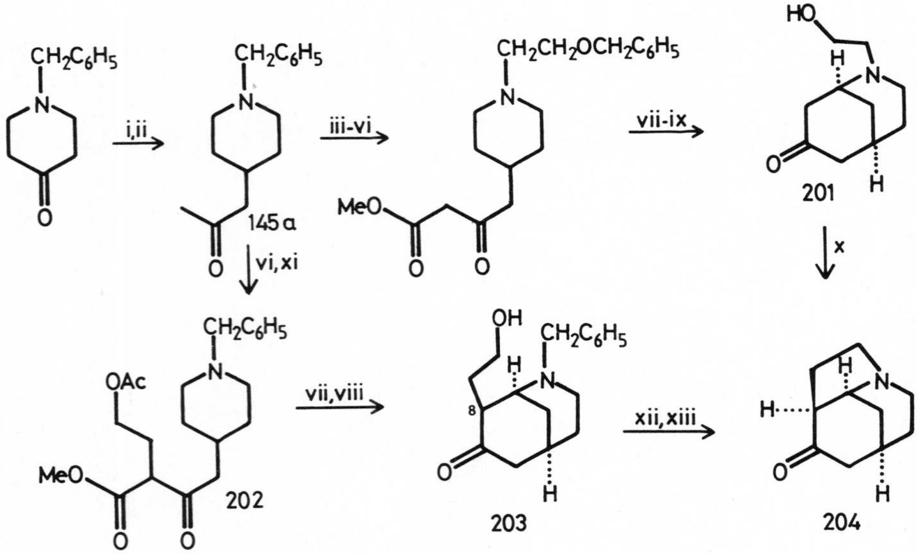

Scheme 41. Syntheses of 4-azatricyclo[5.2.2.0^{4,8}]undecan-11-one. Reagents: (i) (EtO)$_2$POCH$_2$COCH$_3$, KOH, aq EtOH; (ii) H$_2$, Pd; (iii) ClCO$_2$Et, C$_6$H$_6$; (iv) 5N H$_2$SO$_4$; (v)BrCH$_2$CH$_2$OCH$_2$C$_6$H$_5$, K$_2$CO$_3$, C$_6$H$_6$; (vi) Me$_2$CO$_3$, NaH, THF; (vii) Hg(OAc)$_2$,H$_2$O;then H$_2$S; (viii) aq HCl, Δ; (ix) BF$_3$.Et$_2$O, Me$_2$S, CH$_2$Cl$_2$; (x) MsCl, Et$_3$N,THF; then t-BuOK (xi) BrCH$_2$CH$_2$OAc, NaH; (xii) H$_2$, Pd, MeOH-HCl; (xiii) HBr-AcOH; then aq Na$_2$CO$_3$.

acetate chain). Activation of the methyl ketone group as a ß-keto ester allows the incorporation of the 2-acetoxyethyl substituent present in **202** (180) and has beneficial effects in the cyclization step to the bicyclo derivatives **201** (121) and **203** (180).

Second, construction of the carbocyclic ring. The synthesis of 2-azabicyclo-[3.3.1]nonan-7-ones had previously been accomplished by two similar routes (181). Both of them start with diethyl 3-oxoglutarate and utilize the intramolecular Michael-type addition of an amino group upon an α,ß-unsaturated ketone as the last step. Our more direct synthesis involves cyclization between an iminium salt generated by mercuric acetate oxidation and the α-position of a ketone group. The use of 4-piperidineacetoacetates, in which the nucleophilic carbon is activated by two electron-withdrawing substituents, appears to be better than the direct cyclization of 4-acetonylpiperidines (121) (see also Scheme 43) due to the slightly higher overall yields and, more important, to the higher purity of the azabicyclo derivatives obtained after acid hydrolysis of the initially formed bicyclic ß-keto esters. A point worthy of mention is that, as a conse-quence of the type of cyclization, construction of the bridged bicyclic systems **201** and **203** does not offer any stereochemical problem.

Third, closure of the five-membered E ring. This was successfully achieved either by intramolecular <u>C</u>-alkylation of a ketone α-position or by <u>N</u>-alkylation, taking advantage of the functionalized two-carbon chain present at N_b or C-8, respectively. Cyclization of **201** to all-cis tricyclo **204** (121) is unambiguous from a stereochemical standpoint since the relative configuration of **204** follows from steric grounds: (i) substituents at bridgehead carbons 1 and 8 must neces-sarily adopt a cis relationship and (ii) the fused octahydroindole moiety cannot be trans as a consequence of the axial disposition of the nitrogen substituent with regard to the carbocyclic ring (see a tridimensional drawing of **204** in Scheme 42). On the other hand, although azabicyclo **203** was obtained (180) as a C-8 epimeric mixture, in which the <u>exo</u> stereochemistry of the major isomer was unappropriate for further cyclization, this fact did not constitute a serious trouble since carbon 8, in α-position with respect to the carbonyl group, is epimerizable (Scheme 42). This stereochemistry presumably reflects the thermo-dynamic bias of the hydroxyethyl substituent to release the interactions with the bulky <u>N</u>-benzyl group. Once debenzylation was done, an epimeric mixture of secondary amino ketones **205**, in which the most stable <u>endo</u> isomer was prepon-derant, was obtained. This isomer has the C-8 substituent in equatorial disposi-tion as required for subsequent cyclization.

In contrast to the above successful cyclizations to azatricyclic ketone **204**, all attempts to prepare the related tricyclic keto lactam **211** by base-catalyzed cyclization of haloacetamides **210** failed (Scheme 43) and only polymeric materials were obtained (121). As in cyclizations discussed in Section 3.2, the sp^2-hy-

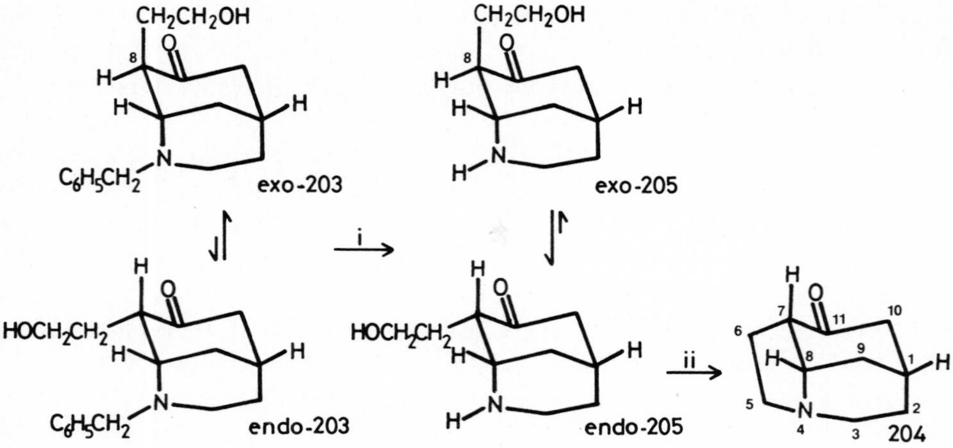

Scheme 42. Reagents: (i) H$_2$, Pd, HCl–MeOH; (ii) HBr–AcOH; then aq Na$_2$CO$_3$.

bridation of the exocyclic carbon atom linked to the piperidine nitrogen could account for this failure.

Haloacetamides **210** were prepared from acetonylpiperidine **145a** through the reaction sequence depicted in Scheme 43 (121). The synthesis of bicyclic ketones **207** and **208** by mercuric acetate oxidation of acetonylpiperidines **206** and **145a**,

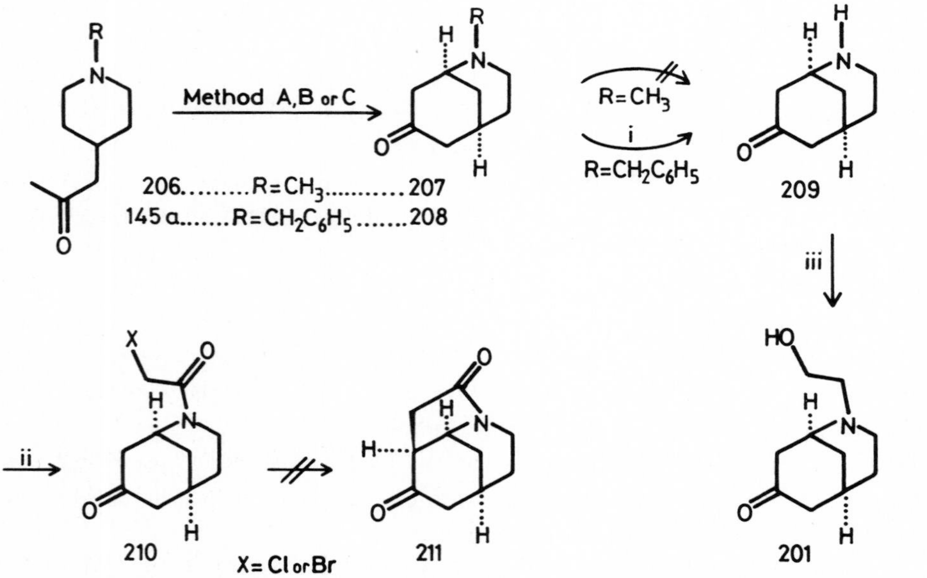

Scheme 43. Synthesis of 2-azabicyclo[3.3.1]nonan-7-ones. Method A: Hg(OAc)$_2$ (ref.182); Method B: (a) Me$_2$CO$_3$, NaH; (b) Hg(OAc)$_2$; (c) aq HCl (ref.121); Method C: (a) m–CPBA; then TFAA; then aq KCN pH4, 90% yield; (b) HCl,MeOH,Δ, 63% yield (ref.183). Reagents: (i) H$_2$,Pd, HCl–EtOH; (ii) XCOCH$_2$X,Et$_3$N; (iii) oxirane.

respectively, either directly (method A) (182) or, more efficiently, via the corresponding acetoacetates (method B) (121), further illustrates the usefulness of this synthetic route to 2-azabicyclo[3.3.1]nonan-7-one derivatives. In these cases cyclization was also accomplished (183) by generating the required inter- mediate iminium salt by Polonovski-Potier-Husson reaction (184) (method C). The secondary amino ketone **209**, obtained by debenzylation of bicyclo **208**, could also be converted, although in low yield, to the useful N-(hydroxyethyl) derivative **201**.

With the requisite tricyclic ketone **204** in hand, the last and crucial step of the synthesis was the Fischer indolization. Although it is well-known that, in the Fischer indole synthesis of phenylhydrazones derived from unsymmetrical ketones, the use of weak (i.e., carboxylic) acids promotes cyclization towards the more branched α-carbon atom (185), when **204** phenylhydrazone was treated with acetic or formic acid under a variety of experimental conditions, only the un- natural regioisomer **213** coming from cyclization upon the methylene group was obtained (121). The desired indolenine **212**, having the pentacyclic ring system of *Strychnos* alkaloids, was not detected.

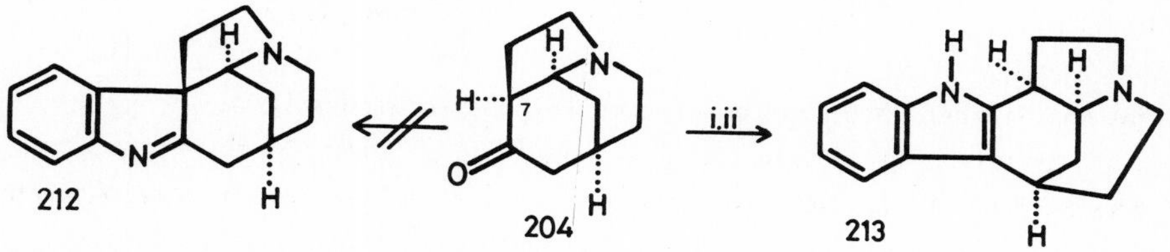

Scheme 44. Reagents: (i) $C_6H_5NHNH_2$; (ii) AcOH or HCOOH.

This result contrasts with those obtained in the synthesis of *Aspidosperma* alkaloids. In such cases, Fischer indolization from a fused hydrolilolidin-9-one ring system (e.g.**214**) leads to the corresponding indolenine (176) (Scheme 45). Two reasons could account for this different behavior: a) the instability of the de- sired indolenine **212** under the acidic reaction conditions; and b) the greater stability of the intermediate enehydrazine leading to the indole **213**, as compared with the more substituted double-bond isomer required for cyclization to the indolenine **212**, owing to the steric strain of the bridged tricyclic system when carbon 7 is sp^2 hybridized. The formation, albeit as a minor product, of hexa- hydropyrrolo[2,3-d]carbazole **216** by Fischer indolization of **215** (186), a bicyclic ketone lacking the ethano bridge that performs the piperidine ring of **204**, is in accordance with the latter explanation.

As a last example about the regioselectivity of the Fischer indole synthesis in systems related to **204**, it is worthy of mention that Fischer indolization of ketone **207** affords indole **217** as the sole product (182). The formation of the *Strychnos*-type regioisomer **218** was not detected.

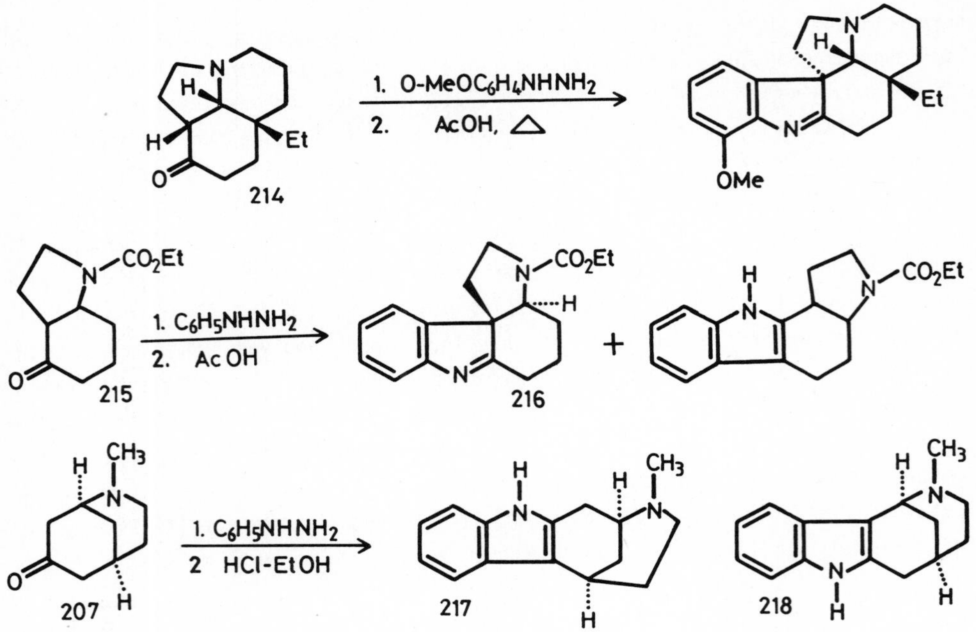

Scheme 45. Fischer indolization of some ketones related to 204.

The failure in the formation of the C_7-aryl bond by Fischer indole synthesis made evident that further efforts on the synthesis of pentacyclic *Strychnos* alkaloids by construction of the modified indole ring in the last synthetic steps had to be directed to the use of a tricyclic ketone similar to 204, bearing a suitable aryl substituent at C-7 (see Scheme 40), in which the bond between the quaternary C-7 center and the aryl group had been previously formed. Recent work by Overman on the synthesis of *Aspidosperma* alkaloids (170b) (see also section 4.1.1) has established the usefulness of this strategy.

Synthetic routes to 7-aryl-4-azatricycloundecan-11-ones, according to the strategy depicted in Scheme 40 ($R_1 = \underline{o}$-$NO_2C_6H_4$), are currently being explored in our laboratory in order to elaborate the indoline ring by reductive cyclization at a late stage of the synthesis. Scheme 46 illustrates the state of our studies in this respect. The required o-nitrophenyl group was incorporated by arylation of acetonylpiperidine 145a (activated again as a β-keto ester), whereas cyclization to 8-aryl-2-azabicyclo[3.3.1]nonan-7-one system 221a was efficiently accomplished by Polonovski-Potier-Husson reaction (184), via an iminium salt generated from 2-cyanopiperidine 220a. The accessibility of cis-3-ethyl-4-acetonylpiperidine 145b (see Section 3.1) has allowed so far the extension (187) of the same reaction sequence to the synthesis of the derivative 219b, having the ethyl substituent present in many *Strychnos* alkaloids. As could be expected, reductive cyclization of nitrophenyl ketones 219 gave in excellent yields the useful (see Scheme 26) piperidylmethylindoles **146**. Further work in this field is in progress.

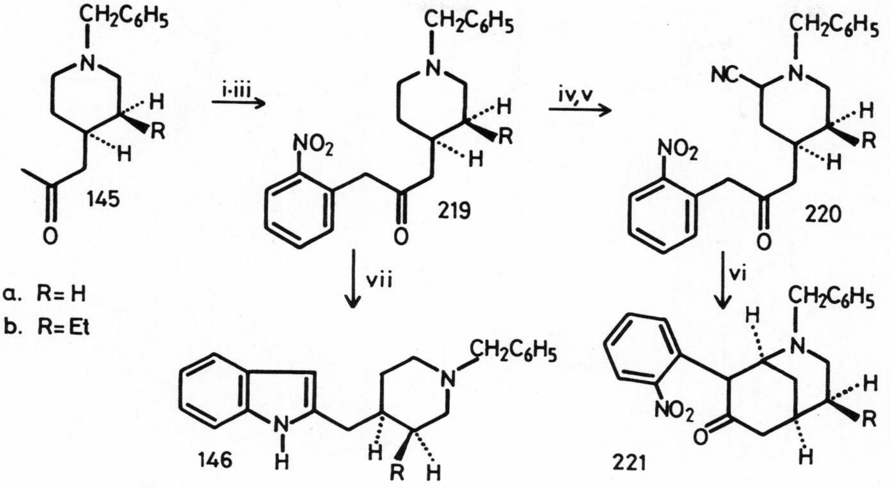

a. R= H
b. R= Et

Scheme 46. Synthesis of the potential intermediate 8-aryl-2-azabicyclo[3.3.1]-nonan-7-one. Reagents: (i) Me_2CO_3, NaH; (ii) o-$FC_6H_4NO_2$, t-BuOK, HMPA (a,44%, b, 56%); (iii) 3N HCl (a,95%; b,85%); (iv) m-CPBA; (v) TFAA, then aq KCN pH4, 48% overall yield; (vi) TsOH, toluene, 40% yield; (vii) H_2, Pt (a,95%; b,70%).

4.2.2 Synthesis via a 3a-Aryl-*cis*-octahydroindole

Recently, in the context of studies towards the total synthesis of strychnine, the stereoselective synthesis of a potential synthetic intermediate **227**, having a 7-phenyl-4-azatricycloundecane skeleton, has been reported (122). The synthesis begins with N-phenacyl amide **222** and implies successive construction of carbocyclic (C), pyrrolidine (E), and piperidine (D) rings, as outlined in Scheme 47 (to better visualize bond formations, the biogenetic numbering has been employed for all synthetic intermediates).

The most remarkable features of this approach are the following: (i) efficient one-step elaboration of cyclohexenone **223** (ring C) by Robinson annelation; (b) stereoselective construction of the cis-fused bicyclic lactam **225** by Michael cyclization of enone **224**, which occurs in quantitative yield; it is at this step when the important C-7 quaternary center is created; and (c) closure of the piperidine ring by formation of the $C_{15}-C_{20}$ bond through intramolecular conjugate addition to the acetylenic Michael acceptor in **226**. The authors have indicated (122) that transformation of tricyclic ketone **227** to strychnine is under study in their laboratories.

It seems evident that approaches to pentacyclic *Strychnos* alkaloids discussed in Section 4 await further developments.

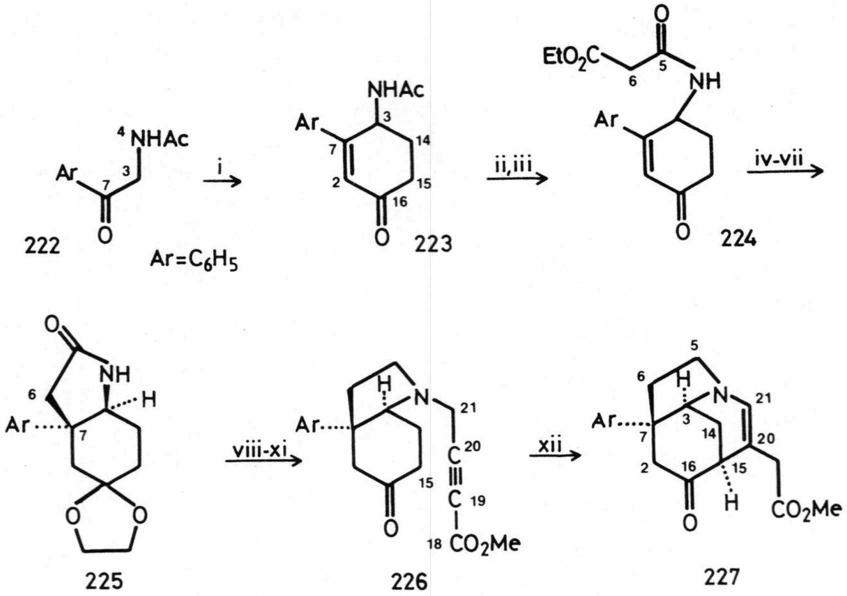

Scheme 47. Reagents: (i) methyl vinyl ketone, NaOEt; (ii) $Et_3O^+BF_4^-$, $NaHCO_3$; then AcOH; (iii) $ClCOCH_2CO_2Et$, $NaHCO_3$; (iv) NaH, THF; (v) $(CH_2OH)_2$, TsOH; (vi) NaOH, aq MeOH; (vii) LiI, diglyme; (viii) $LiAlH_4$; (ix) $BrCH_2C \equiv CH$, Na_2CO_3; (x) LDA; then $ClCO_2Me$; (xi) TsOH, acetone; (xii) Triton B, DME.

Addendum

Recently, the isolation of (+)-stricticine, a new alkaloid having the same absolute configuration at C-15 as (+)-20-epilochneridine (**15**), has been reported (188).

Acknowledgment

The work conducted in our laboratory has been supported by the Comisión Asesora de Investigación Científica y Técnica, Spain.

REFERENCES

1. K. S. J. Stapleford, The Indole Alkaloids, in: S. Coffey (Ed.), Rodd's Chemistry of Carbon Compounds 2nd ed. Vol. IVB, Elsevier, Amsterdam, **1977**, pp 111-116.
2. a) M. V. Kisakürek and M. Hesse, Chemotaxonomic Studies of the Apocynaceae, Loganiaceae, and Rubiaceae, with Reference to Indole Alkaloids, in: J. D. Phillipson and M. H. Zenk (Eds.), Indole and Biogenetically Related Alkaloids, Academic Press, London, **1980**, chapter 2; b) M. V. Kisakürek, A. J. M. Leeuwenberg and M. Hesse, A Chemotaxonomic Investigation of the Plant Families of Apocynaceae, Loganiaceae, and Rubiaceae by Their Indole Alkaloid Content, in: S. W. Pelletier (Ed.), Alkaloids: Chemical and Biological Perspectives, Vol. 1, John Wiley and Sons, New York, **1983**, pp. 211-376.
3. N. G. Bisset, Alkaloids of the Loganiaceae, in J. D. Phillipson and M. H. Zenk (Eds.), Indole and Biogenetically Related Alkaloids, Academic Press, London, **1980**, chapter 3.

4. Atta–ur–Rahman and A. Basha, Biosynthesis of Indole Alkaloids, Clarendon Press, Oxford, **1983**.
5. J. Le Men and W. I. Taylor, *Experientia* **1965**, *21*, 508–510.
6. P. N. Edwards and G. F. Smith, *Proc. Chem. Soc. (London)* **1960**, 215.
7. P. Lathuillière, L. Olivier, J. Lévy, and J. Le Men, *Ann. Pharm. Fr.* **1966**, *24*, 547–549.
8. A. Banerji and A. K. Siddhanta, *Phytochemistry* **1981**, *20*, 540–542.
9. A. I. Scott and C. L. Yeh, *J. Am. Chem. Soc.* **1974**, *96*, 2273–2274.
10. For a review: J. Bosch and M.–L. Bennasar, *Heterocycles* **1983**, *20*, 2471–2511.
11. M. R. Uskoković, R. L. Lewis, J. J. Partridge, C. W. Despreaux, and D. L. Pruess, *J. Am. Chem. Soc.* **1979**, *101*, 6742–6744.
12. a) J. E. Saxton, Alkaloids of *Picralima Nitida*, in: R. H. F. Manske (Ed.), *The Alkaloids*, vol. 8, Academic Press, New York, **1965**, chapter 7; b) J. E. Saxton, Alkaloids of *Alstonia* Species, *ibid.*, chapter 8; c) B. Gilbert, The Alkaloids of *Aspidosperma, Diplorrhyncus, Kopsia, Ochrosia, Pleiocarpa*, and Related Genera, *ibid.*, chapter 14; d) A. R. Battersby and H. F. Hodson, Alkaloids of Calabash Curare and *Strychnos* Species, *ibid.*, chapter 15; e) G. F. Smith, *Strychnos* Alkaloids, *ibid.*, chapter 17; f) R. H. F. Manske and W. A. Harrison, The Alkaloids of *Geissospermum* Species, *ibid.*, chapter 19.
13. a) J. A. Joule, Indole Alkaloids, in: J. E. Saxton (Ed.), *The Alkaloids*, Vols. 1–5, The Chemical Society, London, **1971–1975**; b) J. E. Saxton, Indole Alkaloids, in: M. F. Grundon (Ed.), *The Alkaloids*, Vols. 6–13, The Chemical Society, London, **1976–1983**.
14. J. E. Saxton, *Nat. Prod. Rep.* **1984**, *1*, 21–51; *ibid.* **1985**, *2*, 49–80; *ibid.* **1986**, *3*, 353–394.
15. H.–P. Husson, The *Strychnos* Alkaloids, in: J. E. Saxton (Ed.), Indoles, part four, The Monoterpenoid Indole Alkaloids, in A. Weissberger and E. C. Taylor (Eds.), The Chemistry of Heterocyclic Compounds, Vol. 25, John Wiley and Sons, New York, **1983**, pp. 293–330.
16. M. Lounasmaa and P. Somersalo, The Condylocarpine Group of Indole Alkaloids, in: W. Herz, H. Grisebach, G. W. Kirby, and Ch. Tamm (Eds.), Progress in the Chemistry of Organic Natural Products, Vol. 50, Springer–Verlag, Wien, **1986**, pp. 28–56.
17. Chemical Abstracts Service, Selection of Index Names for Chemical Substances, in Index Guide, Appendix IV, Ninth Collective Index, **1972–1976**, Columbus.
18. D. Schumann and H. Schmid, *Helv. Chim. Acta* **1963**, *46*, 1996–2003.
19. M. Lounasmaa, A. Koskinen, and J. O´Connell, *Helv. Chim. Acta* **1986**, *69*, 1343–1348.
20. J. Schripsema, T. A. van Beek, R. Verpoorte, C. Erkelens, P. Perera, and C. Tibell, *J. Nat. Prod.* **1987**, *50*, 89–101.
21. A. I. Scott and A. A. Qureshi, *J. Am. Chem. Soc.* **1969**, *91*, 5874–5876.
22. J. P. Kutney and G. B. Fuller, *Heterocycles* **1975**, *3*, 197–204.
23. K. Aghoramurthy and R. Robinson, *Tetrahedron* **1957**, *1*, 172–173.
24. G. F. Smith and J. T. Wróbel, *J. Chem. Soc.* **1960**, 792–795.
25. K. Bernauer, W. Arnold, Ch. Weissmann, H. Schmid, and P. Karrer, *Helv. Chim. Acta* **1960**, *43*, 717–726.
26. J. Lévy, J. Le Men, and M.–M. Janot, *Bull. Soc. Chim. Fr.* **1960**, 979–981.
27. P. N. Edwards and G. F. Smith, *J. Chem. Soc.* **1961**, 152–156.
28. Y. Nakagawa, J. M. Wilson, H. Budzikiewicz, and C. Djerassi, *Chem. Ind.* **1962**, 1986.
29. L. Olivier, J. Lévy, J. Le Men, M.–M. Janot, H. Budzikiewicz, and C. Djerassi, *Bull. Soc. Chim. Fr.* **1965**, 868–876.
30. W. Boonchuay and W. E. Court, *Planta Medica* **1976**, *29*, 380–390.
31. J. Schripsema, A. Hermans–Lokkerbol, R. van der Heijden, R. Verpoorte, A. B. Svendsen, and T. A. van Beek, *J. Nat. Prod.* **1986**, *49*, 733–735.
32. S. Mukhopadhyay and G. A. Cordell, *J. Nat. Prod.* **1981**, *44*, 335–339.
33. C. Mirand, G. Massiot, L. Le Men–Olivier, and J. Lévy, *Tetrahedron Lett.* **1982**, *23*, 1257–1258.
34. M. R. Yagudaev, *Khim. Prir. Soedin.* **1983**, 210–212; *Chem. Abstr.* **1983**, *99*, 140209u.

84

35. M. F. Bartlett, B. Korzun, R. Sklar, A. F. Smith, and W. I. Taylor, *J. Org. Chem.* **1963**, *28*, 1445–1449.
36. a) Y. Ahmad, P. W. Le Quesne, and N. Neuss, *J. Chem. Soc. Chem. Commun.* **1970**, 538–539; b) Y. Ahmad, P. W. Le Quesne, and N. Neuss, *J. Pharm. Sci.* **1971**, *60*, 1581–1583.
37. J. M. Cook and P. W. Le Quesne, *J. Org. Chem.* **1975**, *40*, 1367–1368.
38. M. R. Yaqudaev, V. M. Malikov, and S. Yu. Yunusov, *Khim. Prir. Soedin.* **1974**, 260–261; *Chem. Abstr.* **1974**, *81*, 63833s.
39. T. A. van Beek, R. Verpoorte, and A. B. Svendsen, *J. Nat. Prod.* **1985**, *48*, 400–423.
40. B. Legseir, A. Cherif, B. Richard, J. Pusset, S. Labarre, G. Massiot, and L. Le Men–Olivier, *Phytochemistry* **1986**, *25*, 1735–1738.
41. P. Rasoanaivó , N. Langlois, and P. Potier, *Phytochemistry* **1972**, *11*, 2616–2617.
42. P. N. Edwards and G. F. Smith, *J. Chem. Soc.* **1961**, 1458–1462.
43. W. B. Hinshaw Jr., J. Lévy, and J. Le Men, *Tetrahedron Lett.* **1971**, 995–998.
44. Z. M. Khan, M. Hesse, and H. Schmid, *Helv. Chim. Acta* **1967**, *50*, 1002–1010.
45. Z. Votický, L. Dolejš, and E. Grossmann, *Coll. Czech. Chem. Commun.* **1979**, *44*, 123–127.
46. X. Monseur, R. Goutarel, J. Le Men, J. M. Wilson, H. Budzikiewicz, and C. Djerassi, *Bull. Soc. Chim. Fr.* **1962**, 1088–1092.
47. W. G. Kump, M. B. Patel, J. M. Rowson, and H. Schmid, *Helv. Chim. Acta* **1964**, *47*, 1497–1503.
48. Ch. Weissmann, H. Schmid, and P. Karrer, *Helv. Chim. Acta* **1961**, *44*, 1877–1880.
49. K. Bojthe–Horvath, A. Kocsis, I. Mathe, J. Tamas, and O. Clauder, *Acta Pharm. Hung.* **1974**, *44*, 66–69; *Chem. Abstr.* **1974**, *81*, 136347y.
50. C. Djerassi, Y. Nakagawa, J. M. Wilson, H. Budzikiewicz, B. Gilbert, and L. D. Antonaccio, *Experientia* **1963**, *19*, 467–469.
51. C. Djerassi, Y. Nakagawa, H. Budzikiewicz, J. M. Wilson, J. Le Men, J. Poisson, and M.–M. Janot, *Tetrahedron Lett.* **1962**, 653–659.
52. M. Zèches, T. Ravao, B. Richard, G. Massiot, L. Le Men–Olivier, J. Guilhem, and C. Pascard, *Tetrahedron Lett.* **1984**, *25*, 659–662.
53. J. U. Oguakwa, C. Galeffi, I. Messana, M. Patamia, M. Nicoletti, and G. B. Marini–Bettòlo, *Gazz. Chim. Ital.* **1983**, *113*, 533–535.
54. J. Banerji, R. Mustafi, and D. J. Roy, *Indian J. Chem.* **1984**, *23B*, 455.
55. Atta–ur–Rahman, M. Asif, M. Ghazala, J. Fatima, and K. A. Alvi, *Phytochemistry* **1985**, *24*, 2771–2773.
56. B. Gilbert, A. P. Duarte, Y. Nakagawa, J. A. Joule, S. E. Flores, J. A. Brissolese, J. Campello, E. P. Carrazzoni, R. J. Owellen, E. C. Blossey, K. S. Brown, Jr., and C. Djerassi, *Tetrahedron* **1965**, *21*, 1141–1166.
57. R. Verpoorte, E. Kos–Kuyck, A. T. A. Tsoi, C. L. M. Ruigrok, G. de Jong, and A. B. Svendsen, *Planta Medica* **1983**, *48*, 283–289.
58. S. Mamatas–Kalamaras, T. Sévenet, C. Thal, and P. Potier, *Phytochemistry* **1975**, *14*, 1637–1639.
59. D. Stauffacher, *Helv. Chim. Acta* **1961**, *44*, 2006–2015.
60. G. C. Crawley and J. Harley–Mason, *J. Chem. Soc. Chem. Commun.* **1971**, 685–686.
61. a) H. Fritz, E. Besch, and T. Wieland, *Angew. Chem.* **1959**, *71*, 126; b) *Liebigs Ann. Chem.* **1963**, *663*, 150–156.
62. Y. Ahmad, K. Fatima, Atta–ur–Rahman, J. L. Occolowitz, B. A. Solheim, J. Clardy, R. L. Garnick, and P. W. Le Quesne, *J. Am. Chem. Soc.* **1977**, *99*, 1943–1946.
63. a) W. von Philipsborn, H. Meyer, H. Schmid, and P. Karrer, *Helv. Chim. Acta* **1958**, *41*, 1257–1273; b) W. von Philipsborn, K. Bernauer, H. Schmid, and P. Karrer, *Helv. Chim. Acta* **1959**, *42*, 461–463.
64. V. Boekelheide, O. Ceder, M. Natsume, and A. Zürcher, *J. Am. Chem. Soc.* **1959**, *81*, 2256–2259.
65. T. Ravao, B. Richard, T. Sévenet, G. Massiot, and L. Le Men–Olivier, *Phytochemistry* **1982**, *21*, 2160–2161.
66. R. Marini–Bettòlo and F. D. Monache, *Gazz. Chim. Ital.* **1973**, *103*, 543–549.
67. K. Bernauer, F. Berlage, W. von Philipsborn, H. Schmid, and P. Karrer, *Helv. Chim. Acta* **1958**, *41*, 2293–2308.

68. C. Richard, C. Delaude , L. Le Men-Olivier, J. Lévy, and J. Le Men, *Phyto-chemistry* **1976**, *15*, 1805-1806.
69. M. Tits, L. Angenot, and D. Tavernier, *Tetrahedron Lett.* **1980**, *21*, 2439-2442.
70. P. Thepenier, M.-J. Jacquier, G. Massiot, L. Le Men-Olivier, and C. Delaude, *Phytochemistry* **1984**, *23*, 2659-2663.
71. M. Tits, D. Tavernier, and L. Angenot, *Phytochemistry* **1980**, *19*, 1531-1534.
72. C. A. Coune and L. J. G. Angenot, *Herba Hung.* **1980**, *19*, 189-193; *Chem. Abstr.* **1980**, *93*, 217906e.
73. J. Harley-Mason and C. G. Taylor, *J. Chem. Soc. Chem. Commun.* **1970**, 812.
74. G. Massiot, P. Thépenier, M.-J. Jacquier, J. Lounkokobi, C. Mirand, M. Zèches, L. Le Men-Olivier, and C. Delaude, *Tetrahedron* **1983**, *39*, 3645-3656.
75. M. Tits, D. Tavernier, and L. Angenot, *Phytochemistry* **1985**, *24*, 205-207.
76. a) N. G. Bisset, *Chem. Ind.* **1965**, 1036-1037; b) J. L. Occolowitz, K. Biemann, and J. Bosly, *Il Farmaco Ed. Sci.* **1965**, *20*, 751-756.
77. E. Wenkert and R. Sklar, *J. Org. Chem.* **1966**, *31*, 2689-2691.
78. J. R. Hymon and H. Schmid, *Helv. Chim. Acta* **1966**, *49*, 2067-2071.
79. M. Koch, J. Garnier, and M. Plat, *Ann. Pharm. Fr.* **1972**, *30*, 299-306.
80. D. Tavernier, M. J. O. Anteunis, M. Tits, and L. Angenot, *Bull. Soc. Chim. Belg.* **1978**, *87*, 595-607.
81. L. Angenot, N. G. Bisset, and M. Franz, *Phytochemistry* **1975**, *14*, 2519-2520.
82. M.-M. Janot, *Tetrahedron* **1961**, *14*, 113-125.
83. L. Angenot and M. Tits, *Planta Medica* **1981**, *41*, 240-243.
84. M. Tits and D. Tavernier, *Plant. Med. Phytother.* **1978**, *12*, 92-95; *Chem. Abstr.* **1979**, *90*, 69097n.
85. E. Wenkert, H. T. A. Cheung, H. E. Gottlieb, M. C. Koch, A. Rabaron, and M. M. Plat, *J. Org. Chem.* **1978**, *43*, 1099-1105.
86. M. Koch, E. Fellion, and M. Plat, *Phytochemistry* **1976**, *15*, 321-324.
87. M. Tits and L. Angenot, *Plant. Med. Phytother.* **1980**, *14*, 213-217; *Chem. Abstr.* **1981**, *95*, 58049a.
88. H. Rapoport, T. P. Onak, N. A. Hugues, and M. G. Reinecke, *J. Am. Chem. Soc.* **1958**, *80*, 1601-1609.
89. A. Bertho and M. Koll, *Chem. Ber.* **1961**, *94*, 2737-2746.
90. B. A. Dadson and J. Harley-Mason, *J. Chem. Soc. Chem. Commun.* **1969**, 665.
91. R. Goutarel, M. Païs, H. E. Gottlieb, and E. Wenkert, *Tetrahedron Lett.* **1978**, 1235-1238.
92. F. A. L. Anet and R. Robinson, *J. Chem. Soc.* **1955**, 2253-2262.
93. J. R. Hymon, H. Schmid, P. Karrer, A. Boller, H. Els, P. Fahrni, and A. Fürst, *Helv. Chim. Acta* **1969**, *52*, 1564-1602.
94. B. A. Dadson, J. Harley-Mason, and G. H. Foster, *J. Chem. Soc. Chem. Commun.*, **1968**, 1233.
95. a) Y. Ban, K. Yoshida, J. Goto, and T. Oishi, *J. Am. Chem. Soc.* **1981**, *103*, 6990-6992; b) Y. Ban, K. Yoshida, J. Goto, T. Oishi, and E. Takeda, *Tetra-hedron* **1983**, *39*, 3657-3668.
96. A. Wu and V. Snieckus, *Tetrahedron Lett.* **1975**, 2057-2060.
97. S. Takano, M. Hirama, and K. Ogasawara, *Tetrahedron Lett.* **1982**, *23*, 881-884.
98. M. Amat, A. Linares, M.-L. Salas, M. Alvarez, and J. Bosch, *J. Chem. Soc. Chem. Commun.*, submitted for publication.
99. A. Walser and C. Djerassi, *Helv. Chim. Acta* **1965**, *48*, 391-404.
100. A. I. Scott and C. C. Wei, *Tetrahedron* **1974**, *30*, 3003-3011.
101. K. Biemann, A. L. Burlingame, and D. Stauffacher, *Tetrahedron Lett.* **1962**, 527-532.
102. A. Sandoval, F. Walls, J. N. Shoolery, J. M. Wilson, H. Budzikiewicz, and C. Djerassi, *Tetrahedron Lett.* **1962**, 409-414.
103. A. H.-J. Wang and I. C. Paul, *Acta Crystallogr. Sect. B* **1977**, *33*, 2977-2979.
104. H. Achenbach and B. Raffelsberger, *Z. Naturforsch.* **1980**, *35B*, 885-891.
105. J. Harley-Mason, *Pure Appl. Chem.* **1975**, *41*, 167-174.
106. M. Urrea, A. Ahond, H. Jacquemin, S.-K. Kan, C. Poupat, P. Potier, and M.-M. Janot, *C. R. Hebd. Seances Acad. Sci.*, *Ser. C* **1978**, *287*, 63-67.
107. B. A. Dadson and J. Harley-Mason, *J. Chem. Soc. Chem. Commun.* **1969**, 665.
108. S. Baassou, H. Mehri, and M. Plat, *Phytochemistry* **1978**, *17*, 1449-1450.

109. M. Pinar, U. Renner, M. Hesse, and H. Schmid, *Helv. Chim. Acta* **1972**, *55*, 2972-2974.
110. Atta-ur-Rahman, K. A. Alvi, and A. Muzaffar, *Planta Medica* **1986**, 325-326.
111. K. Biemann, M. Spiteller-Friedmann and G. Spiteller, *J. Am. Chem. Soc.* **1963**, *85*, 631-638.
112. W. Klyne, R. J. Swan, B. W. Bycroft, and H. Schmid, *Helv. Chim. Acta* **1966**, *49*, 833-841.
113. R. R. Arndt, S. H. Brown, N. C. Ling, P. Roller, C. Djerassi, J. M. Ferreira, B. Gilbert, E. C. Miranda, S. E. Flores, A. P. Duarte, and E. P. Carrazoni, *Phytochemistry* **1967**, *6*, 1653-1658.
114. M. Pinar, B. W. Bycroft, J. Seibl, and H. Schmid, *Helv. Chim. Acta* **1965**, *48*, 822-825.
115. M. Pinar and H. Schmid, *Helv. Chim. Acta* **1967**, *50*, 89-93.
116. M. Hesse, Indolalkaloide in Tabellen, Springer, Berlin, **1964**, p. 41.
117. L. E. Overman and S. R. Angle, *J. Org. Chem.* **1985**, *50*, 4021-4028.
118. J. Vercauteren, A. Bideau, and G. Massiot, *Tetrahedron Lett.* **1987**, *28*, 1267-1270.
119. J. Henin, G. Massiot, J. Vercauteren, and J. Guilhem, *Tetrahedron Lett.* **1987**, *28*, 1271-1274.
120. B. Legseir, J. Henin, G. Massiot, and J. Vercauteren, *Tetrahedron Lett.* **1987**, *28*, 3573-3576.
121. J. Bonjoch, N. Casamitjana, J. Quirante, M. Rodríguez, and J. Bosch, *J. Org. Chem.* **1987**, *52*, 267-275.
122. M. L. Quesada, D. Kim, S. K. Ahn, N. S. Jeong, Y. Hwang, M. Y. Kim, and J. W. Kim, *Heterocycles* **1987**, *25*, 283-286.
123. E. E. van Tamelen , L. J. Dolby, and R. G. Lawton, *Tetrahedron Lett.* **1960**, 30-35.
124. F. Ungemach and J. M. Cook, *Heterocycles* **1978**, *9*, 1089-1119.
125. E. Wenkert, K. Orito, D. P. Simmons, N. Kunesch, J. Ardisson, and J.Poisson, *Tetrahedron* **1983**, *39*, 3719-3724.
126. A. J. Gaskell, H.-E. Radunz, and E. Winterfeldt, *Tetrahedron* **1970**, *26*, 5353-5360.
127. M. Lounasmaa and P. Somersalo, *Tetrahedron* **1986**, *42*, 1501-1509.
128. G. H. Foster, J. Harley-Mason, and W. R. Waterfield, *J. Chem. Soc. Chem. Commun.* **1967**, 21.
129. C. Earnshaw, C. J. Wallis, and S. Warren, *J. Chem. Soc. Perkin Trans 1*, **1979**, 3099-3106.
130. a) P. Potier, *Rev. Latinoamer. Quím.* **1978**, *9*, 47-54; b) P. Potier, Is the Modified Polonovski Reaction Biomimetic?, in: J. D. Phillipson and M. H. Zenk (eds.), Indole and Biogenetically Related Alkaloids, Academic Press, London, **1980**, chapter 8; c) M. Lounasmaa and A. Koskinen, *Heterocycles* **1984**, *22*, 1591-1612.
131. K. Yoshida, Y. Sakuma, and Y. Ban, *Heterocycles* **1987**, *25*, 47-50 and references cited therein.
132. O. Yonemitsu, T. Tokuyama, M. Chaykovsky, and B. Witkop, *J. Am. Chem. Soc.* **1968**, *90*, 776-784.
133. K. S. Bhandari, J. A. Eenkhoorn, A. Wu, and V. Snieckus, *Synth. Commun.* **1975**, *3*, 79-86.
134. R. J. Sundberg and F. X. Smith, *J. Org. Chem.* **1975**, *40*, 2613-2621.
135. a) F. DiNinno, Jr., W. L. Heckle, Jr., D. K. Rehse, and R. M. Wilson, *Tetrahedron Lett.* **1972**, 2639-2642; b) R. M. Wilson, R. A. Farr, and D. J.Burlett, *J. Org. Chem.* **1981**, *46*, 3293-3302.
136. a) H.-P. Husson, C. Thal, P. Potier, and E. Wenkert, *J. Chem. Soc. Chem. Commun.* **1970**, 480-481; b) F. E. Ziegler and E. B. Spitzner, *J. Am. Chem. Soc.* **1973**, *95*, 7146-7149; c) M. Natsume and I. Utsunomiya, *Heterocycles* **1982**, *17*, 111-115; d) P. Magnus and P. A. Pappalardo, *J. Am. Chem. Soc.* **1986**, *108*, 212-217 and references cited therein.
137. J. A. Joule, The Uleine-Ellipticine-Vallesamine Group, in J. E. Saxton (Ed.), Indoles, part four, The Monoterpenoid Indole Alkaloids, in A. Weissberger and E. C. Taylor (Eds.), The Chemistry of Heterocyclic Compounds, Vol. 25, John Wiley and Sons, New York, 1983, pp. 265-292.

138. a) A. Jackson, N. D. V. Wilson, A. J. Gaskell, and J. A. Joule, *J. Chem. Soc. (C)* **1969**, 2738–2747; b) R. Besselièvre and H.-P. Husson, *Tetrahedron* **1981**, *37, Suppl n° 1*, 241–246.

139. a) L. J. Dolby and H. Biere, *J. Org. Chem.* **1970**, *35*, 3843–3845; b) T. Kametani and T. Suzuki, *J. Chem. Soc. (C)* **1971**, 1053–1054; c) T. Kametani and T. Suzuki, *J. Org. Chem.* **1971**, *36*, 1291–1293; d) T. Kametani and T. Suzuki, *Chem. Pharm. Bull.* **1971**, *19*, 1424–1425.

140. a) G. Büchi, S. J. Gould, and F. Näf, *J. Am. Chem. Soc.* **1971**, *93*, 2492–2501; b) M. Natsume and Y. Kitagawa, *Tetrahedron Lett.* **1980**, *21*, 839–840; c) M. Harris, R. Besselièvre, D. S. Grierson, and H.-P. Husson, *Tetrahedron Lett.* **1981**, *22*, 331–334; d) D. S. Grierson, M. Harris, and H.-P. Husson, *Tetrahedron* **1983**, *39*, 3683–3694.

141. M. Feliz, J. Bosch, D. Mauleón, M. Amat, and A. Domingo, *J. Org. Chem.* **1982**, *47*, 2435–2440.

142. J. Bosch, M. Alvarez, R. Llobera, and M. Feliz, *An. Quím.* **1979**, *75*, 712–717.

143. J. Bosch and M. Feliz, *An. Quím.* **1982**, *78C*, 240–243.

144. J. Bosch, M. Rubiralta, A. Domingo, J. Bolós, A. Linares, C. Minguillón, M. Amat, and J. Bonjoch, *J. Org. Chem.* **1985**, *50*, 1516–1522.

145. J. Bosch, M. Amat, E. Sanfeliu, and M.-A. Miranda, *Tetrahedron* **1985**, *41*, 2557–2566.

146. J. Bosch, M. Amat, and A. Domingo, *Heterocycles* **1984**, *22*, 561–564.

147. J. Bonjoch, A. Linares, M.-L. Pérez, and J. Bosch, unpublished results.

148. J. Bosch and M. Amat, *An. Quím.* **1985**, *81C*, 277–279.

149. J. Bonjoch, A. Linares, and J. Bosch, unpublished results.

150. J. Bosch, J. Bonjoch, A. Díez, A. Linares, M. Moral, and M. Rubiralta, *Tetrahedron* **1985**, *41*, 1753–1762.

151. R. L. Augustine, A. J. Gustavsen, S. F. Wanat, I. C. Pattison, K. S. Houghton, and G. Koletar, *J. Org. Chem.* **1973**, *38*, 3004–3011.

152. J. Bonjoch, A. Linares, M. Guardià, and J. Bosch, *Heterocycles* **1987**, *26*, 2165–2174.

153. M. Alvarez, R. Lavilla, C. Roure, E. Cabot, and J. Bosch, *Tetrahedron* **1987**, *43*, 2513–2522.

154. J. Bosch, M.-L. Bennasar, and E. Zulaica, *J. Org. Chem.* **1986**, *51*, 2289–2297.

155. M. Alvarez, R. Lavilla, and J. Bosch, *Tetrahedron Lett.* **1987**, *28*, 0000.

156. For a review: M.-L. Bennasar, R. Lavilla, M. Alvarez, and J. Bosch, *Heterocycles* in press.

157. J. Bosch, M.-L. Bennasar, E. Zulaica, and M. Feliz, *Tetrahedron Lett.* **1984**, *25*, 3119–3122.

158. M. Alvarez, R. Lavilla, and J. Bosch, unpublished results.

159. O. Schindler, P. Niklaus, U. Stauss, and H. P. Härter, *Helv. Chim. Acta* **1976**, *59*, 2704–2710.

160. M. Amat and J. Bosch, unpublished results.

161. T. Gallagher, P. Magnus, and J. C. Huffman, *J. Am. Chem. Soc.* **1983**, *105*, 4750–4757.

162. J. Bosch and M. Amat, *Tetrahedron Lett.* **1985**, *26*, 4951–4954.

163. T. D. Harris and V. Boekelheide, *J. Org. Chem.* **1976**, *41*, 2770–2772.

164. a) B. M. Trost and E. Murayama, *J. Am. Chem. Soc.* **1981**, *103*, 6529–6530. b) B. M. Trost and J. Sato, *J. Am. Chem. Soc.* **1985**, *107*, 719–721.

165. a) G. Büchi, K. E. Matsumoto, and H. Nishimura, *J. Am. Chem. Soc.* **1971**, *93*, 3299–3301; b) Y. Ban, Y. Sendo, M. Nagai, and T. Oishi, *Tetrahedron Lett.* **1972**, 5027–5030; c) M. Ando, G. Büchi, and T. Ohnuma, *J. Am. Chem. Soc.* **1975**, *97*, 6880–6881; d) J. P. Brennan and J. E. Saxton, *Tetrahedron* **1987**, *43*, 191–205.

166. a) Y. Ban, T. Sekine, and T. Oishi, *Tetrahedron Lett.* **1978**, 151–154; b) S. Takano, K. Shishido, J. Matsuzaka, M. Sato, and K. Ogasawara, *Heterocycles* **1979**, *13*, 307–320; c) S. J. Veenstra and W. N. Speckamp, *J. Am. Chem. Soc.* **1981**, *103*, 4645–4646.

167. L. E. Overman and M. Sworin, Recent Advances in the Total Synthesis of Pentacyclic *Aspidosperma* Alkaloids, in: S. W. Pelletier (ed.), Alkaloids: Chemical and Biological Perspectives, Vol. 3, John Wiley, New York, **1985**, pp. 275–307.

168. a) L. E. Overman and M. Kakimoto, *J. Am. Chem. Soc.* **1979**, *101*, 1310-1312; b) L. E. Overman , M. Kakimoto, M. E. okazeki, and G. P. Meier, *J. Am. Chem. Soc.* **1983**, *105*, 6622-6629.

169. L. E. Overman, L. T. Mendelson, and E. J. Jacobsen, *J. Am. Chem. Soc.* **1983**, *105*, 6629-6637.

170. a) L. E. Overman, M. Sworin, L. S. Bass, and J. Clardy, *Tetrahedron* **1981**, *37*, 4041-4045; b) L. E. Overman, M. Sworin, and R. M. Burk, *J. Org. Chem.* **1983**, *48*, 2685-2690.

171. M. E. Kuehne and P. J. Seaton, *J. Org. Chem.* **1985**, *50*, 4790-4796 and previous papers in this series.

172. a) M. E. Kuehne, T. H. Matsko, J. C. Bohnert, L. Motyka, and D. Oliver-Smith, *J. Org. Chem.* **1981**, *46*, 2002-2009; b) M. E. Kuehne and J. C. Bohnert, *J. Org. Chem.* **1981**, *46*, 3443-3447; c) M. E. Kuehne and W. G. Earley, *Tetrahedron* **1983**, *39*, 3707-3714.

173. J. Vercauteren, C. Lavaud, J. Lévy, and G. Massiot, *J. Org. Chem.* **1984**, *49*, 2278-2279.

174. R. B. Woodward, M. P. Cava, W. D. Ollis, A. Hunger, H. U. Daeniker, and K. Schenker, *Tetrahedron* **1963**, *19*, 247-288.

175. J. P. Kutney, The Synthesis of Indole Alkaloids, in: J. ApSimon (Ed.), The Total Synthesis of Natural Products, John Wiley and Sons, New York, **1977**, pp. 273-438.

176. a) G. Stork and J. E. Dolfini, *J. Am. Chem. Soc.* **1963**, *85*, 2872-2873; b) Y. Ban, Y. Sato, I. Inoue, M. Nagai, T. Oishi, M. Terashima, O. Yonemitsu, and Y. Kanaoka, *Tetrahedron Lett.* **1965**, 2261-2268; c) I. Inoue and Y. Ban, *J. Chem. Soc. (C)* **1970**, 602-610; d) S. S. Klioze and F. P. Darmory, *J. Org. Chem.* **1975**, *40*, 1588-1592; e) G. Lawton, J. E. Saxton, and A. J. Smith, *Tetrahedron* **1977**, *33*, 1641-1653; f) A. J. Pearson and D. C. Rees, *J. Chem. Soc. Perkin Trans. 1* **1982**, 2467-2476.

177. a) H. Nakai, K. Hemmi, T. Iwakuma, and O. Yonemitsu, *Chem. Pharm. Bull.* **1972**, *20*, 998-1005; b) For the synthesis of a benzo fused derivative, see: S. Shiotani and T. Kometani, *Chem. Pharm. Bull.* **1980**, *28*, 1928-1931.

178. For a review: J. Bosch and J. Bonjoch, *Heterocycles* **1980**, *14*, 505-529.

179. a) J. Bosch, J. Bonjoch, and I. Serret, *Heterocycles* **1980**, *14*, 1983-1988; b) J. Bosch and J. Bonjoch, *J. Org. Chem.* **1981**, *46*, 1538-1543; c) J. Bosch, J. Bonjoch , and I. Serret, *Tetrahedron Lett.* **1982**, *23*, 1297-1298.

180. J. Bonjoch, N. Casamitjana, J. Quirante, A. Torrens, A. Paniello, and J. Bosch, *Tetrahedron* **1987**, *43*, 377-381.

181. a) M. Mokotoff and R. C. Cavestri, *J. Org. Chem.* **1974**, *39*, 409-411; b) J. Adachi, K. Nomura, and K. Mitsuhashi, *Chem. Pharm. Bull.* **1976**, *24*, 85-91.

182. J. Bonjoch, N. Casamitjana, and J. Bosch, *Tetrahedron* **1982**, *38*, 2883-2888.

183. J. Bonjoch, N. Casamitjana, and J. Bosch, 5th European Symposium on Organic Chemistry, Jerusalem, **1987**.

184. D. S. Grierson, M. Harris, and H.-P. Husson, *J. Am. Chem. Soc.* **1980**, *102*, 1064-1082.

185. a) F. M. Miller and W. N. Schinske, *J. Org. Chem.* **1978**, *43*, 3384-3388; b) B. Robinson, The Fischer Indole Synthesis, John Wiley, Chichester, **1982**.

186. H. Fritz and G. Rubach, *Liebigs Ann. Chem.* **1968**, *715*, 135-145.

187. J. Bonjoch, J. Quirante, and J. Bosch, unpublished results.

188. Atta-ur-Rahman, S. Khanum, and T. Fatima, *Tetrahedron Lett.* **1987**, *28*, 3609-3612.

SYNTHETIC STUDIES IN THE FIELD OF INDOLE ALKALOIDS

MAURI LOUNASMAA

1. INTRODUCTION
    For more than a decade, research efforts in our laboratory have been
directed toward the development of synthetic methods in the field of the
alkaloids, in particular the indole alkaloids. This article reviews some of
our achievements.

2. VALLESIACHOTAMINE STRUCTURE
    I shall begin by describing our synthetic efforts toward vallesiachotamine
1 (Fig. 1) and some of its derivatives.

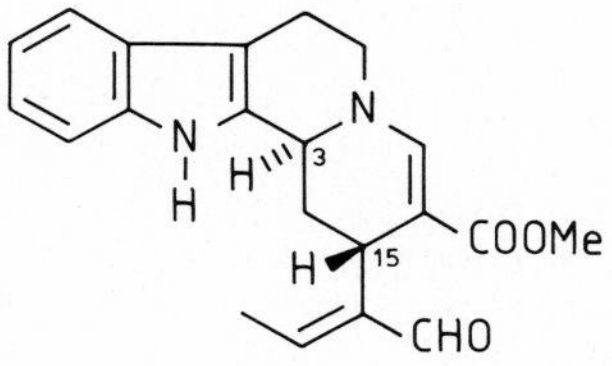

Fig. 1. Structural formulae of vallesiachotamine 1 and strictosidine 2.

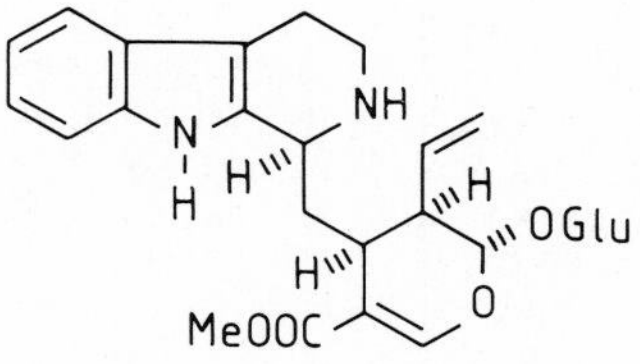

There were two main reasons for our interest in the vallesiachotamine 1
synthesis:
    First, vallesiachotamine 1 with its C(15)Hβ stereochemistry is highly
unusual among the indole alkaloids and this peculiarity makes it attractive
from the biogenetical point of view (ref. 1).
    Second, and perhaps more importantly, vallesiachotamine 1 and some of its
derivatives have played a central role in the determination of the stereostructure of strictosidine 2 (Fig. 1) (ref. 2), which is the crucial intermediate in the whole indole alkaloid biosynthesis (refs. 3-5). It is worth
mentioning that earlier there was considerable controversy concerning the
structure and role of strictosidine 2 in the alkaloid biosynthesis (refs. 6-7).

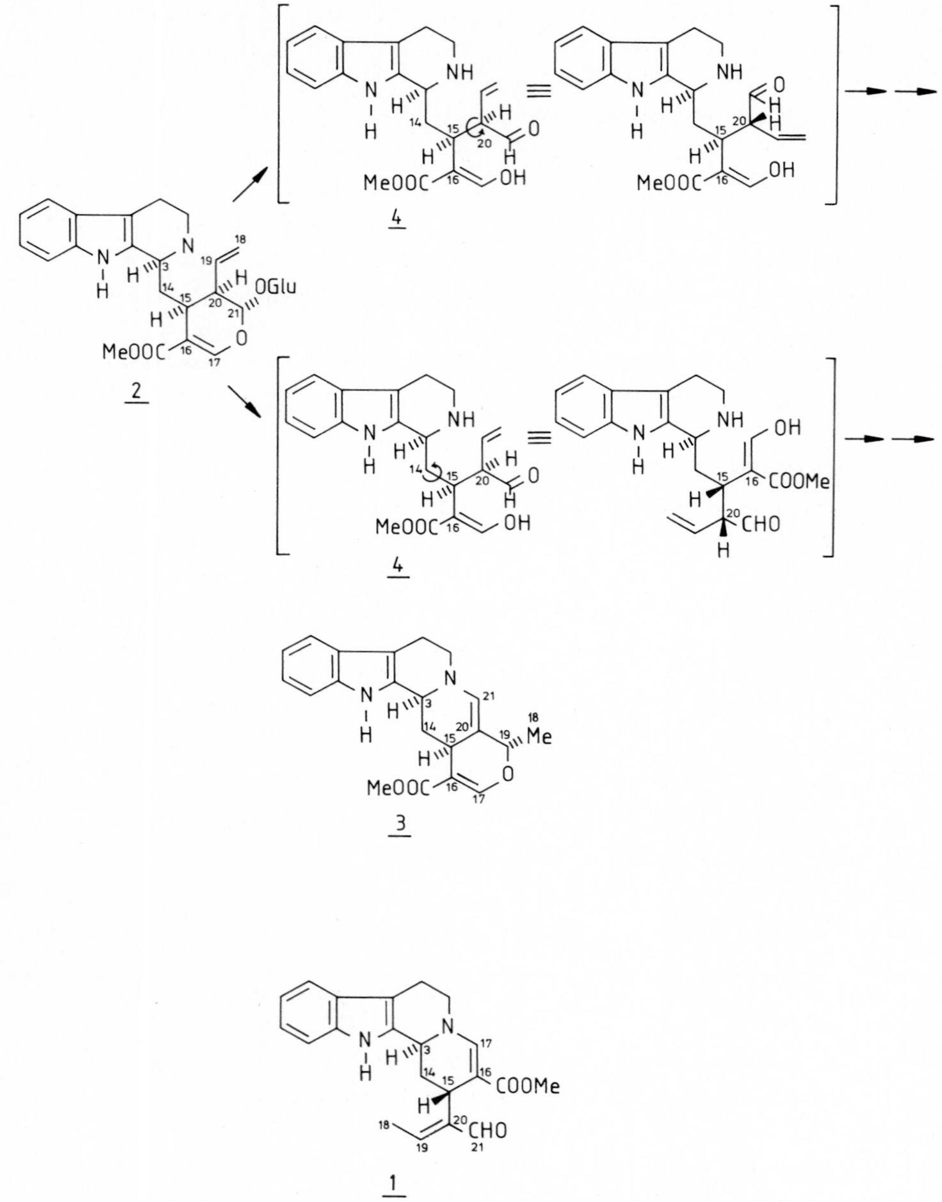

Scheme 1. Simplified representation of the biosynthetic formation of cathenamine 3 and vallesiachotamine 1 from strictosidine 2.

Scheme 1 presents a succinct summary of the biosynthetic formation of
indole alkaloids from strictosidine 2. The normal series, characterized by the

C(15)Hα stereochemistry, is represented by cathenamine 3 and the "non-normal" series, characterized by the C(15)Hβ stereochemistry, is represented by vallesiachotamine 1.

Cleavage of the mixed acetal group of strictosidine 2 leads to the aldehyde 4. If the molecule is now turned 180 degrees around the C(15)-C(20) bond, the secondary amino group and the aldehyde group are brought near to each other and can form a new bond between N_b and C(21). Normal biochemical transformations then lead to cathenamine 3. However, if the initially formed aldehyde 4 is turned 180 degrees around the C(14)-C(15) bond, it is the other aldehyde function, here presented in the enol form, which comes to lie near the secondary amino group, and the bond that forms this time is between N_b and C(17). This leads after isomerization to vallesiachotamine 1.

3. SYNTHETIC VALLESIACHOTAMINE DERIVATIVES

I now return to our synthetic strategy in the vallesiachotamine 1 field.

Consideration of the structure of vallesiachotamine 1 led us to focus our attention on two structural details. The vinylogous urethane structure >N-CH=CH-COOMe and trans stereochemical relationship between C(3)H and C(15)H.

The sodium dithionite reduction of appropriate pyridinium salts seemed to be the ideal starting point for the synthesis of the vinylogous urethane structure. It was known (ref. 8) that if a pyridinium salt possessing an electron withdrawing group (e.g. -COOMe) at C(3) is reduced with sodium dithionite, the corresponding 1,4-dihydropyridine is normally formed (Scheme 2). In certain special cases (vide infra) the sodium dithionite reduction can also lead to 1,2-dihydropyridines.

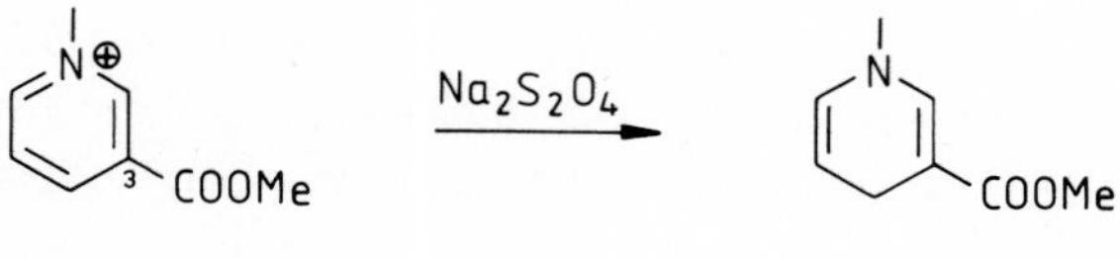

Scheme 2. Sodium dithionite reduction of a pyridinium salt to the corresponding 1,4-dihydropyridine.

In the course of investigating the C(3)H-C(15)H stereochemical relationship, we prepared the three simplified vallesiachotamine analogues 5, 6 and 7 (Scheme 3).

In all three cases we were able to introduce the desired C(3)H-C(15)H trans relationship (ref. 9).

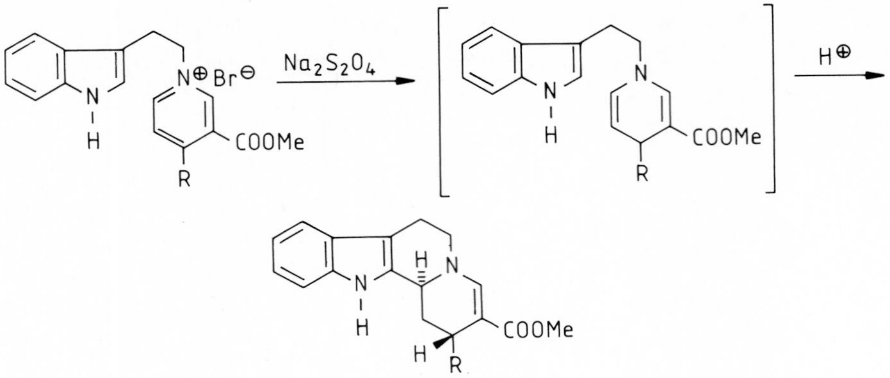

5 R = CH$_3$

6 R = CH$_2$ - CH$_2$ - CH$_3$

7 R = CH$_2$ - COOMe

Scheme 3. Preparation of simplified vallesiachotamine analogues 5, 6 and 7.

With these favourable results in hand we moved on to apply the developed procedure to the preparation of more sophisticated vallesiachotamine analogues.

For the next step we needed the pyridine derivative 8, which was prepared according to Scheme 4 (ref. 10).

Success in the synthesis of derivative 8 permitted us to prepare the pyridinium salt 9, which was then reduced with sodium dithionite. The obtained 1,4-dihydropyridine derivative 10 was cyclized by acid treatment to the vallesiachotamine analogue 11 (Scheme 5).

This is already a relatively good vallesiachotamine derivative. However, the C(19)-C(20) bond is unsaturated in vallesiachotamine 1 and saturated in 11, and instead of the aldehyde group in 1 there is a methoxycarbonyl group.

Encouraged by the above results we took as our next goal the preparation of the vallesiachotamine analogue containing the C(19)-C(20) double bond (ref. 11). For that purpose, the pyridine derivative 12 was prepared (Scheme 6).

Unfortunately, this approach did not permit the preparation of the desired product 13 (Scheme 7) (ref. 11). When the reaction was carried out without buffering, it led directly, owing to the acidity of the reaction mixture, to the earlier described vallesiachotamine derivative 11, where the C(19)-C(20) double bond is reduced. Further, the over-reduced tetrahydropyridine derivative 14, where, however, the C(19)-C(20) double bond was intact, could be isolated. Buffering the reaction mixture with NaHCO$_3$ led to the 1,4-dihydropyridine derivative 15, where again the exocyclic double bond was reduced. The desired 1,4-dihydropyridine 16 with the exocyclic double bond was not detected.

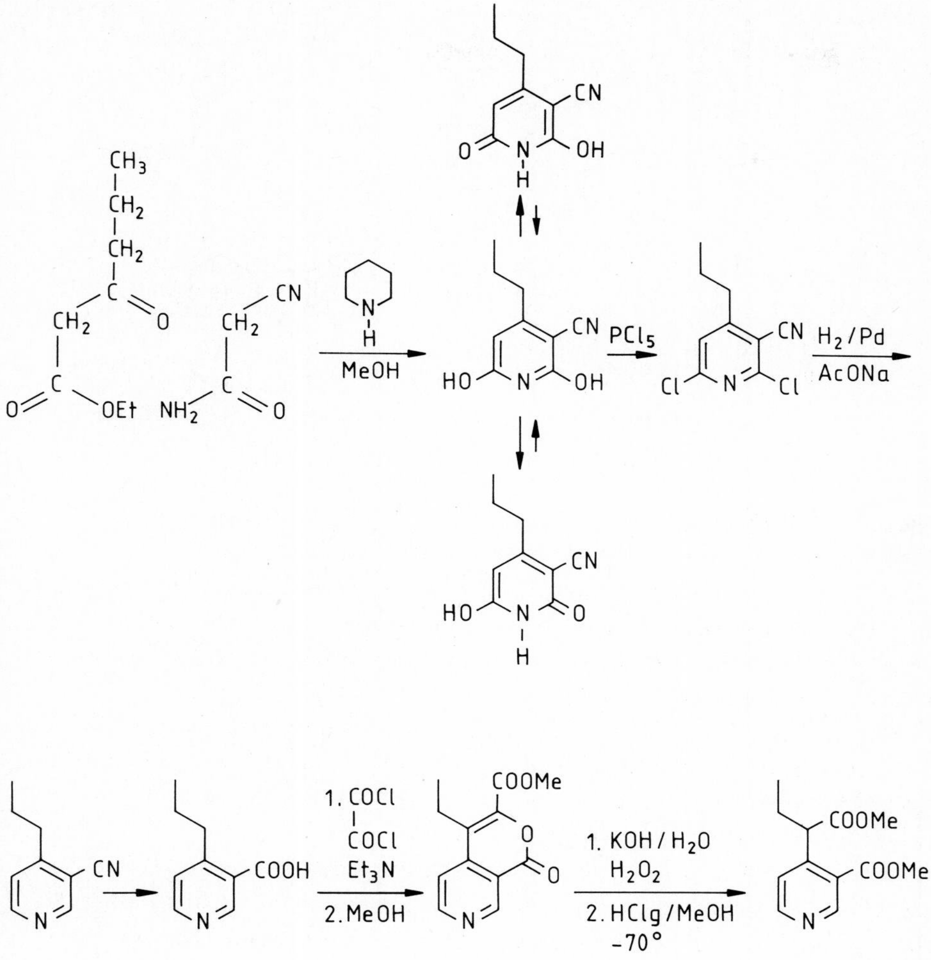

Scheme 4. Preparation of the pyridine derivative 8.

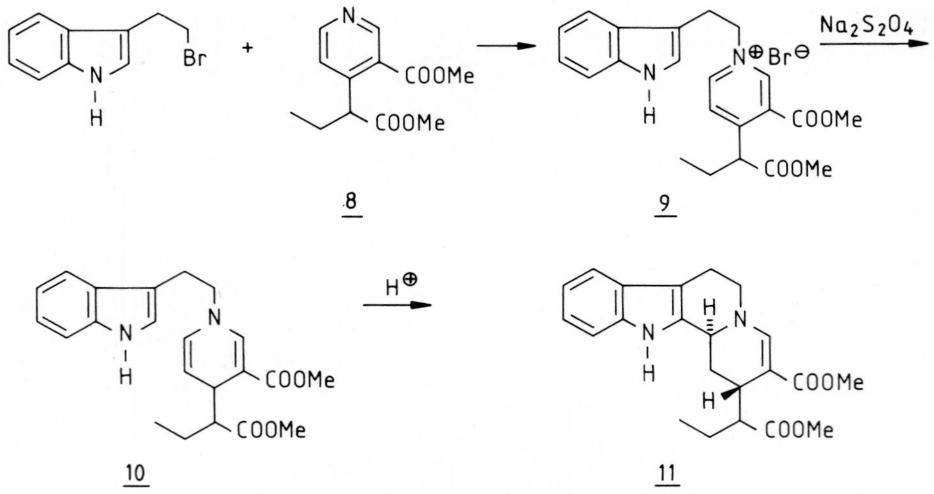

94

Scheme 5. Preparation of the vallesiachotamine analogue 11.

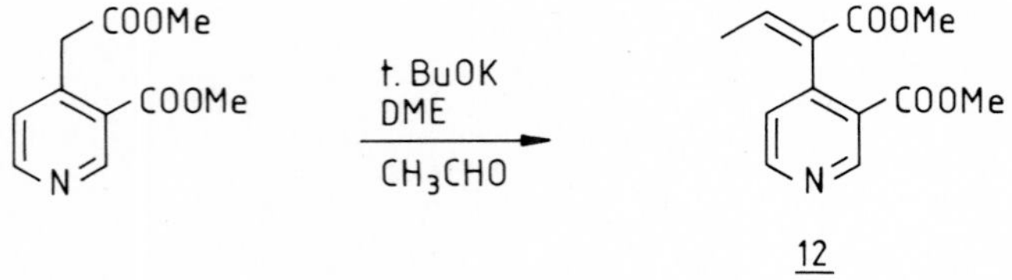

Scheme 6. Preparation of the pyridine derivative 12.

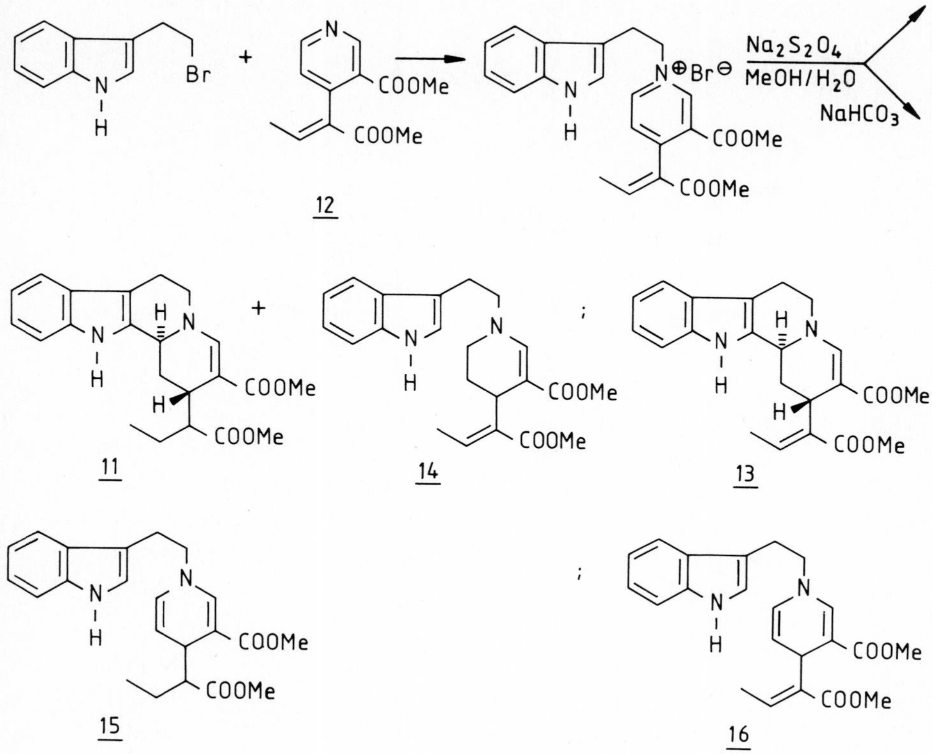

Scheme 7. Attempt to prepare the vallesiachotamine analogue 13 or its precursor 16.

4. VALLESIACHOTAMINE LACTONES

The disappointing results forced us to modify our synthetic strategy and the vallesiachotamine derivatives 17 and 18 were chosen as the new target molecules (Fig. 2).

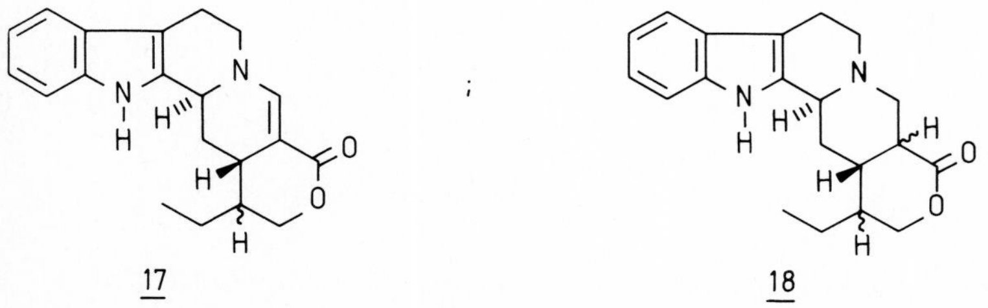

Fig. 2. Structural formulae of vallesiachotamine derivatives 17 and 18.

The pyridine derivative 19 (Scheme 8) and thereby the pyridinium salt 20 were prepared (ref. 12).

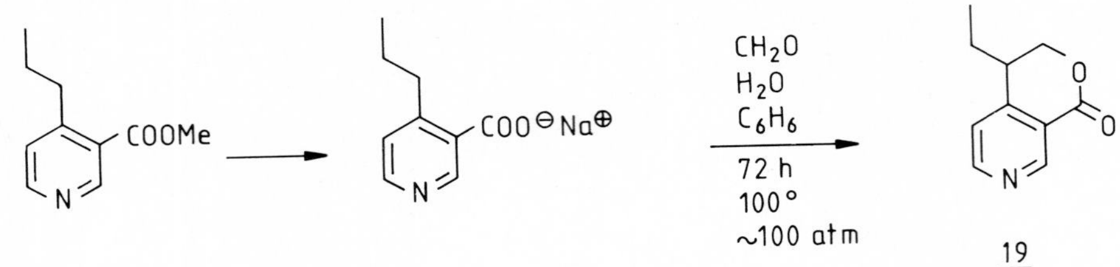

Scheme 8. Preparation of the pyridine derivative 19.

Unfortunately, when the sodium dithionite reduction of compound 20 was carried out in the usual way, the over-reduction leading to compound 21 could not be avoided (ref. 12), and the desired 1,4-dihydropyridine derivative 22 was not found (Scheme 9).

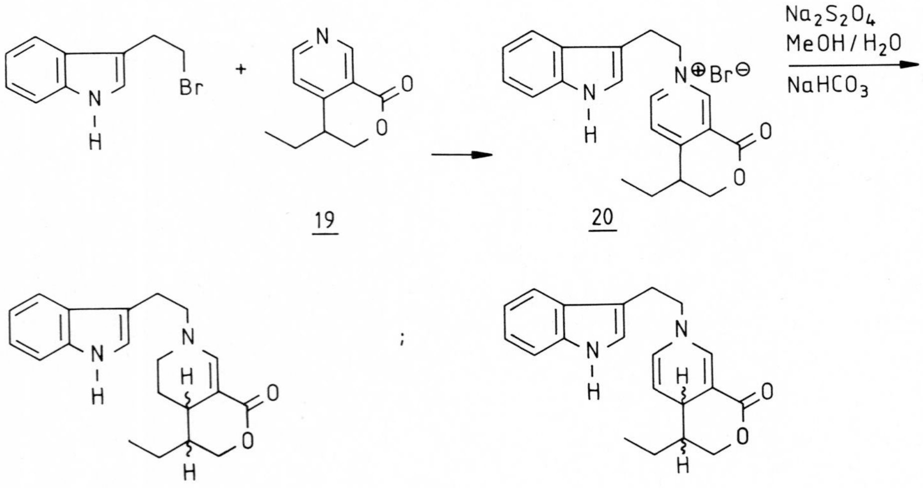

Scheme 9. Attempt to prepare the 1,4-dihydropyridine derivative 22.

However, when the sodium dithionite reduction of 20 was carried out in a two-phase system (H_2O/CH_2Cl_2), the corresponding 1,2-dihydropyridine derivative 23 could be isolated (ref. 13). An acid-induced cyclization led to the C(20) epimers 24a and 24b, which could be separated (Scheme 10).

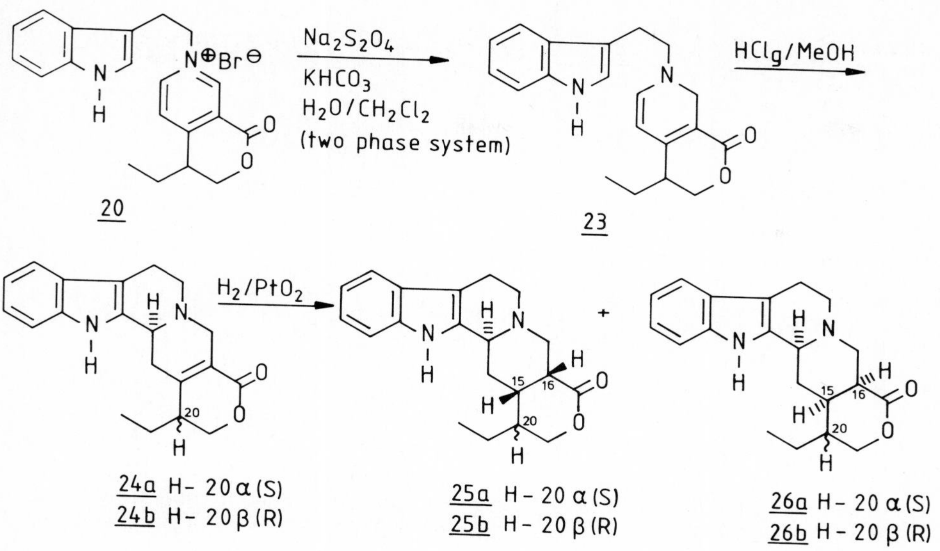

Na$_2$S$_2$O$_4$
KHCO$_3$
H$_2$O/CH$_2$Cl$_2$
(two phase system)

HCl$_g$/MeOH

20

23

H$_2$/PtO$_2$

+

15 16

20

24a H- 20 α (S)
24b H- 20 β (R)

25a H- 20 α (S)
25b H- 20 β (R)

26a H- 20 α (S)
26b H- 20 β (R)

Scheme 10. Preparation of vallesiachotamine lactones 25a, 25b, 26a and 26b.

Proof of the stereochemical designations used for C(20) will be provided below.

Catalytic hydrogenation of the C(20) epimers 24a and 24b carried out separately produced in both series two compounds (25a and 26a, and 25b and 26b, respectively), where C(15)H and C(16)H are cis to each other (Scheme 10).

We then turned our attention back to the authentic vallesiachotamine 1 (ref. 13) where the C(3) and C(15) stereochemistry was known (ref. 1).

Sodium borohydride reduction of the aldehyde function of vallesiachotamine 1 produced the corresponding alcohol 27. The exocyclic double bond of 27 was catalytically reduced, and the obtained C(20) epimers 28a and 28b were separated (Scheme 11).

Sodium borohydride reduction in the presence of acetic acid (protonation and iminium formation) afforded in both cases two C(16) epimers (25a and 29a, and 25b and 29b, respectively) (Scheme 11), one of each of which proved to be identical with the hydrogenation products described above. In this, we had the first case where totally synthetic compounds were proven identical with real vallesiachotamine derivatives (ref. 13).

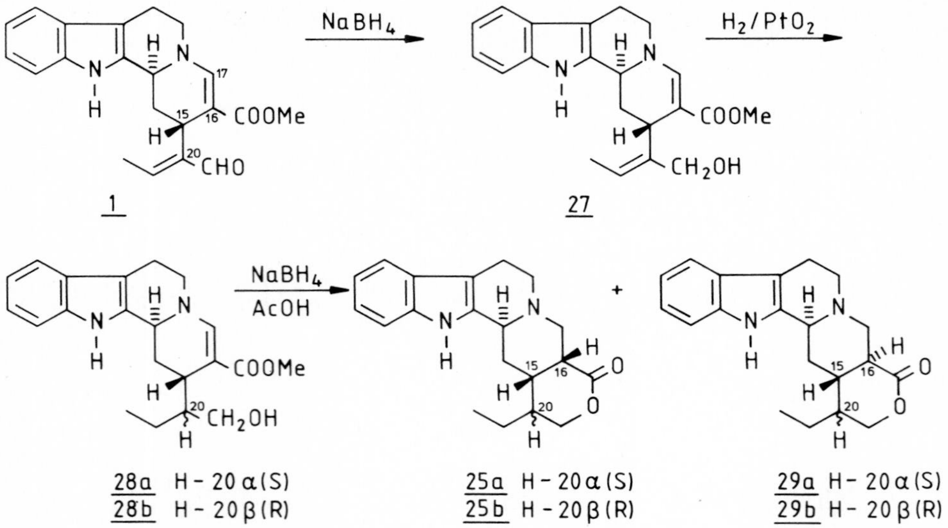

Scheme 11. Preparation of vallesiachotamine lactones 25a, 25b, 29a and 29b.

5. STEREOCHEMICAL CONSIDERATIONS

I now return to our stereochemical determination of the above compounds (Schemes 10 and 11).

Before proceeding, however, a word needs to be said about the conformational behaviour of indoloquinolizidines (Fig. 3) in general.

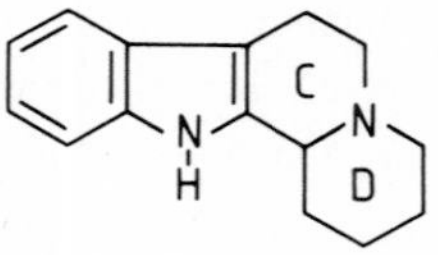

Fig. 3. The ring nomenclature for the indoloquinolizidine skeleton.

In indoloquinolizidines of the present type, the conformational equilibrium between the C and D rings as shown in Scheme 12 must be taken into account (refs. 9, 14-19).

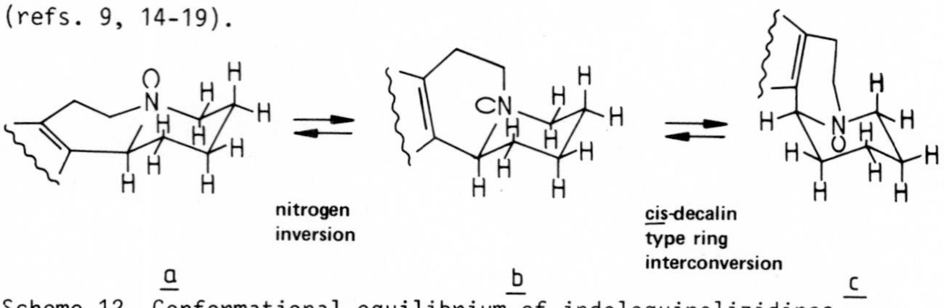

Scheme 12. Conformational equilibrium of indoloquinolizidines.

Analogously, the a, b and c conformers of compound 25a are the following (Scheme 13):

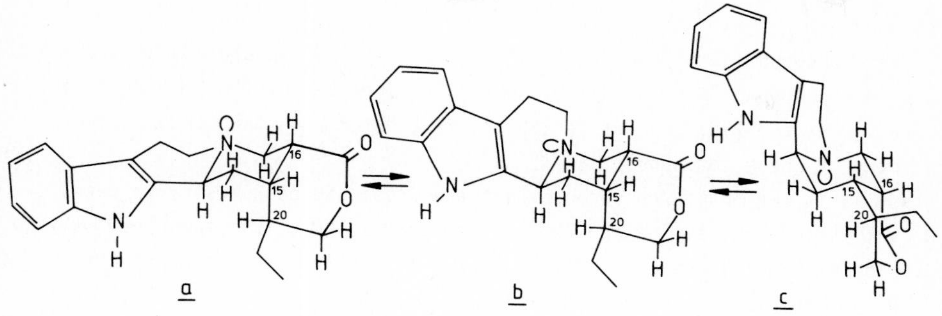

Scheme 13. Conformational equilibrium of the vallesiachotamine lactone 25a.

Note that the ring D substituents that are in axial positions in conformers a and b are in equatorial positions in conformer c, and correspondingly the equatorial substituents in conformers a and b are in axial positions in conformer c. What is seen in NMR measurements carried out at room temperature is the average of the contribution of these three conformers to the conformational equilibrium. Frequently this kind of situation complicates the interpretation of the spectra.

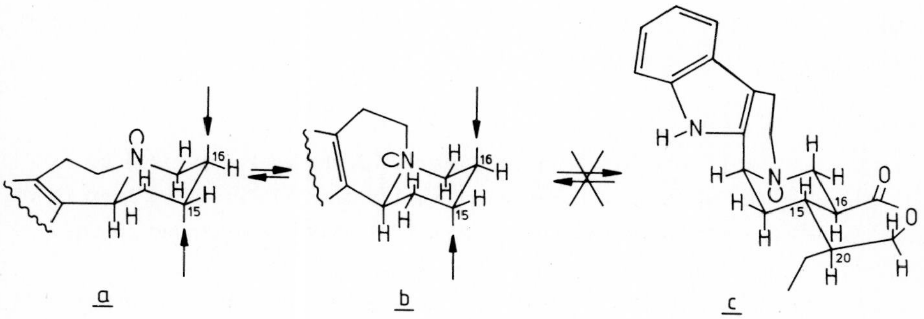

Scheme 14. Conformational rigidity of the vallesiachotamine lactone 29a.

However, in the case of compound 29a (or 29b), where the C(15)H-C(16)H relationship is trans, the above-described conformational equilibrium cannot exist (Scheme 14): the compound is totally in conformation c, because in conformations a and b the bonds from C(15) and C(16), by which ring E is connected to ring D, would be in trans diaxial positions (marked by arrows) and this is not possible for a six-membered ring.

The conformational rigidity of compound 29a (and 29b) makes it ideally suited for NMR based determination of the C(20) stereochemistry, and this was easily done.

The determination of the C(20) stereochemistry in compound 29a as C(20)Hα (S) means that compound 29b has the C(20)Hβ (R) stereochemistry. The transformation of 28a to 25a and 29a, and 28b to 25b and 29b, determines the C(20) stereochemistry of these compounds.

Now it is reasonable to suppose that the C(15) stereochemistry in tetra-hydrovallesiachotamines 28a and 28b is the same as the C(15) stereochemistry in vallesiachotamine 1, which is known. Since then the transformation of 28a to 25a and 29a and 28b to 25b and 29b determines a like C(15) stereochemistry for these four compounds, the difference between 25a and 29a and between 25b and 29b must be at C(16). The identity of the fully synthetic and vallesiachotamine derived 25a and 25b means that both C(15)H and C(16)H must be β. As a consequence, the stereochemistries of 25a and 25b, and as a corollary those of 24a, 24b, 26a, 26b, 28a, 28b, 29a and 29b, are completely determined.

6. GAMBIRTANNINE SYNTHESIS

One further application of the dithionite reduction bears notice (Scheme 15). This one permits the preparation of several indole alkaloids of gambirtannine-type (dihydrogambirtannine 30a and demethoxycarbonyldihydro-gambirtannine 30b, ourouparine 31a and gambirtannine 32a) and their congeners (demethoxycarbonylourouparine 31b and demethoxycarbonylgambirtannine 32b) from easily available starting materials (ref. 20).

7. DITHIONITE REDUCTION (SUMMARY)

Schemes 16 and 17 present a concise summary of the work described to this point. Clearly, the most tedious and conceptually least interesting task is the production of the 3,4-disubstituted pyridine derivatives to be used in the crucial, two-step, general alkaloid synthesis.

It thus became highly desirable to devise alternative routes, which would permit the use of simple pyridines as starting compounds.

8. KRÖHNKE PROCEDURE

For that purpose we turned our attention to the Kröhnke-procedure, which consists of the γ-addition of ketones (or similar compounds) to N-alkyl-pyridinium salts (Scheme 18) (refs. 21-23). The method has also been applied by Wenkert et al. to some of their alkaloid syntheses (refs. 24, 25).

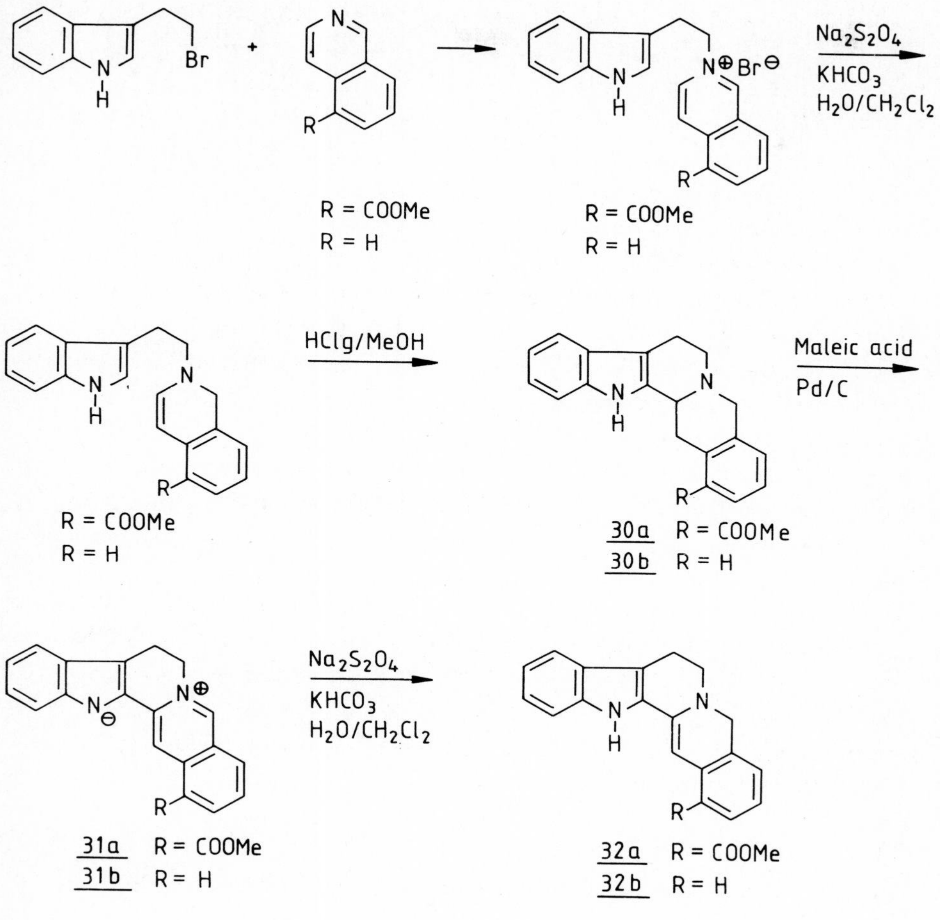
Br
Na₂S₂O₄
KHCO₃
H₂O/CH₂Cl₂
R = COOMe
R = H
R = COOMe
R = H
HClg/MeOH
Maleic acid
Pd/C
R = COOMe
R = H
30a R = COOMe
30b R = H
Na₂S₂O₄
KHCO₃
H₂O/CH₂Cl₂
31a R = COOMe
31b R = H
32a R = COOMe
32b R = H

Scheme 15. Preparation of indole alkaloids of gambirtannine type.

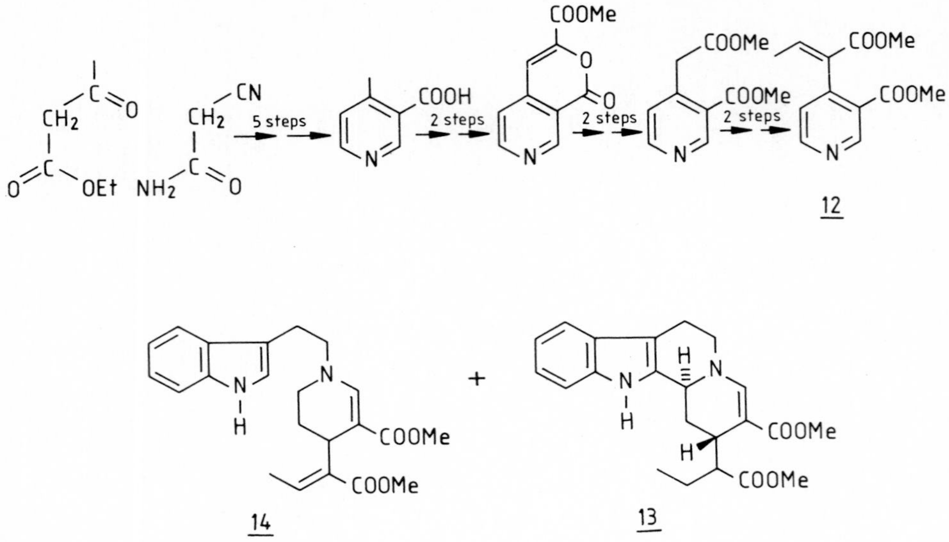

Scheme 16. Short summary of the preparation of compounds 13 and 14.

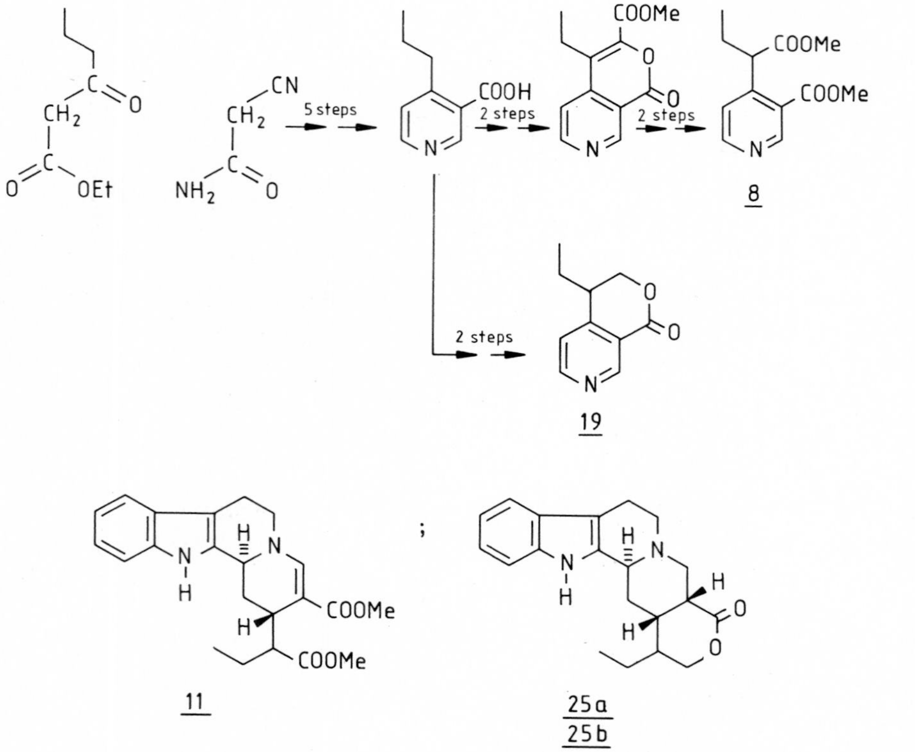

Scheme 17. Short summary of the preparation of compounds 11 and 25a (and 25b).

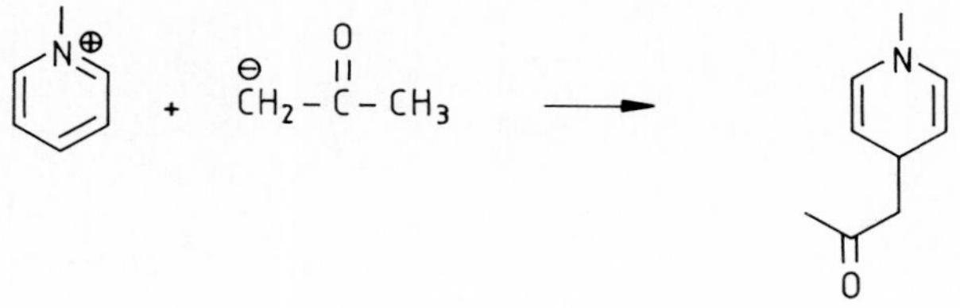

Scheme 18. The γ-addition to an N-alkylpyridinium salt.

Following are two applications. In the first (Scheme 19) the method is applied for the preparation (via 33 and 34) of intermediate 35 for our quinu-clidine synthesis (vide infra) (ref. 26).

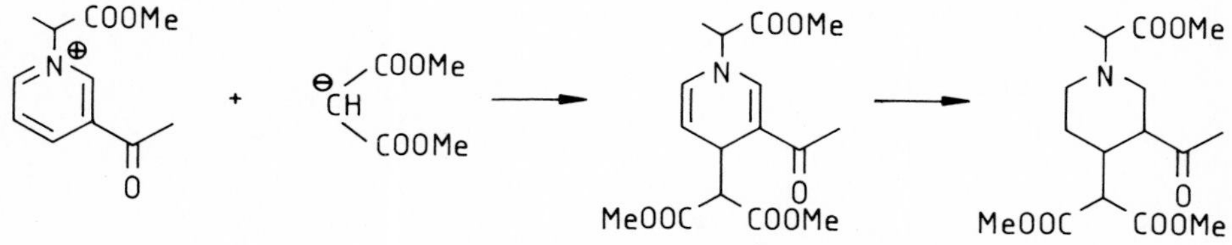

Scheme 19. Preparation of the intermediate 35.

The second (Scheme 20) presents our stereoselective synthesis of (+)-3-iso-19-epiajmalicine 36 (Scheme 20) (ref. 27). The intermediate 37, prepared by the Kröhnke procedure followed by acid induced cyclization, was first treated with NaBH$_4$ in the presence of glacial acetic acid, and then with an excess of methanol (to eliminate the acetic acid through ester formation) and more NaBH$_4$. Four successive reactions take place in one pot:

1. Pronation of the enamine (= vinylogous amide) and hydride reduction of the formed iminium ion.
2. Reduction of the created ketonic carbonyl group.
3. Lactonization between the formed alcohol function and one of the ester functions.
4. Reduction of the lactonic carbonyl group.

Polyphosphoric acid treatment of the formed intermediate 38 completed the preparation of (+)-3-iso-19-epiajmalicine 36.

However, there is one serious limitation on the application of the Kröhnke-procedure: namely, the pyridinium salt in question must have an electron with-drawing group at the 3-position (Scheme 21).

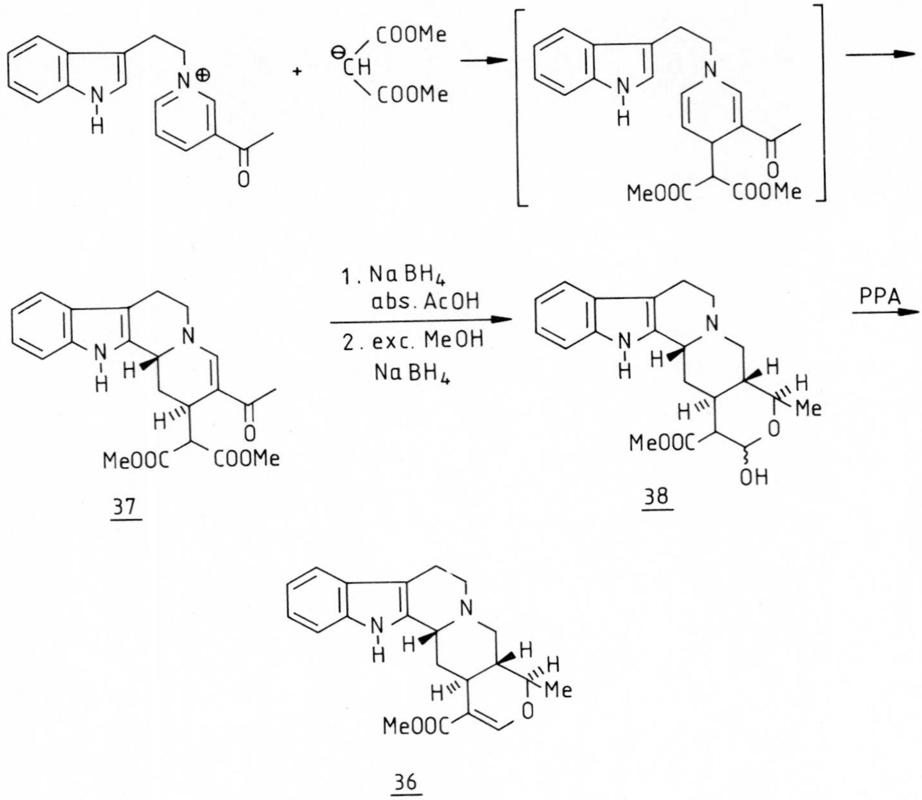

Scheme 20. Stereoselective total synthesis of (±)-3-iso-19-epiajmalicine 36.

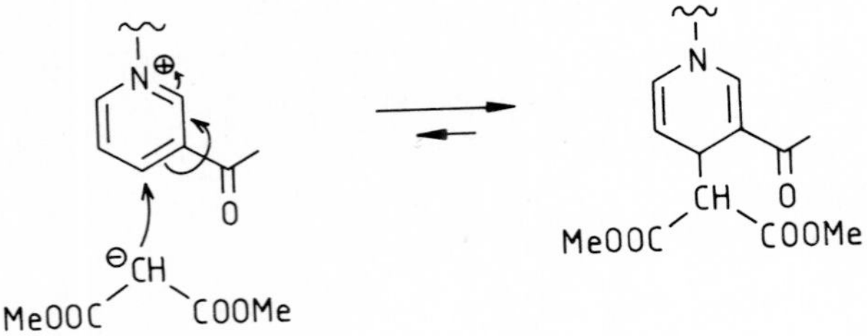

Scheme 21. Equilibrium between pyridinium salts and 4-substituted 1,4-dihydro-pyridine derivatives containing an electron withdrawing group at the 3-position

If this group is lacking, the equilibrium between the pyridinium salt and the 1,4-dihydropyridine derivative will be very unfavourable for the desired product (Scheme 22).

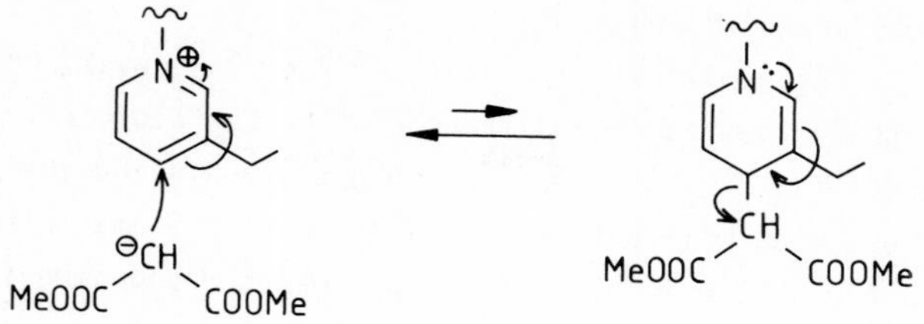

Scheme 22. Equilibrium between pyridinium salts and 4-substituted 1,4-dihydro-pyridine derivatives not containing an electron withdrawing group at the 3-position.

9. MODIFIED POLONOVSKI REACTION

I move on now to another topic, namely to the application of the modified Polonovski reaction to alkaloid syntheses (refs. 28-30). The method was initially developed by Prof. Pierre Potier and his group in France (ref. 31).

In the modified Polonovski reaction a tertiary amine oxide is treated with trifluoroacetic anhydride, producing an iminium ion. This ion is relatively stable and can be attacked by external nucleophiles. If the cyanide ion is used as the external nucleophile, the procedure, leading to an α-aminonitrile, is called the cyanotrapping method (Scheme 23).

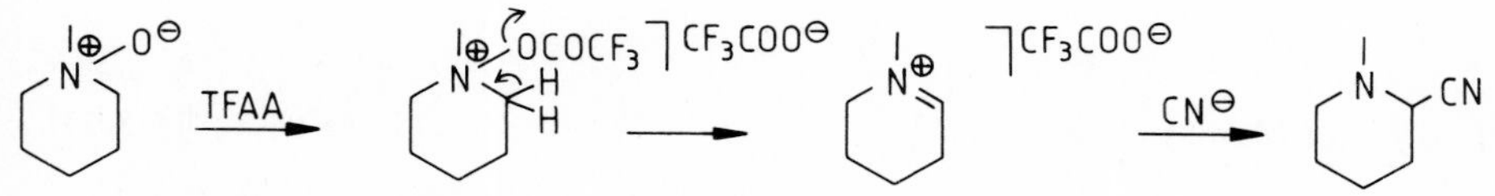

Scheme 23. The modified Polonovski reaction followed by cyanotrapping.

The α-aminonitriles are stable, versatile synthetic intermediates which can be considered synthetic equivalents of the corresponding iminium species (Fig. 4).

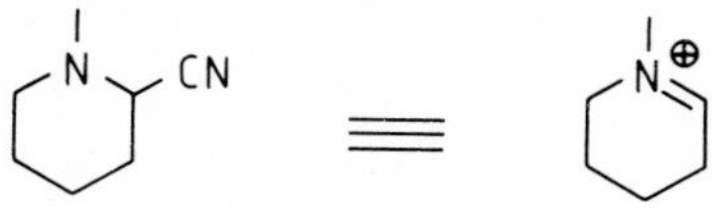

Fig. 4. α-Aminonitriles as synthetic equivalents of the corresponding iminium species.

10. REGIOSELECTIVE FUNCTIONALIZATION

We have developed novel applications of the modified Polonovski reaction, which permit the regioselective functionalization of carbon atoms α to the heterocyclic nitrogen (refs. 32, 33). Thus, by slightly modifying the reaction conditions, we can prepare at will both exo- and endocyclic α-aminonitriles, which, as just noted, are synthetic equivalents of the corresponding iminium species (Scheme 24).

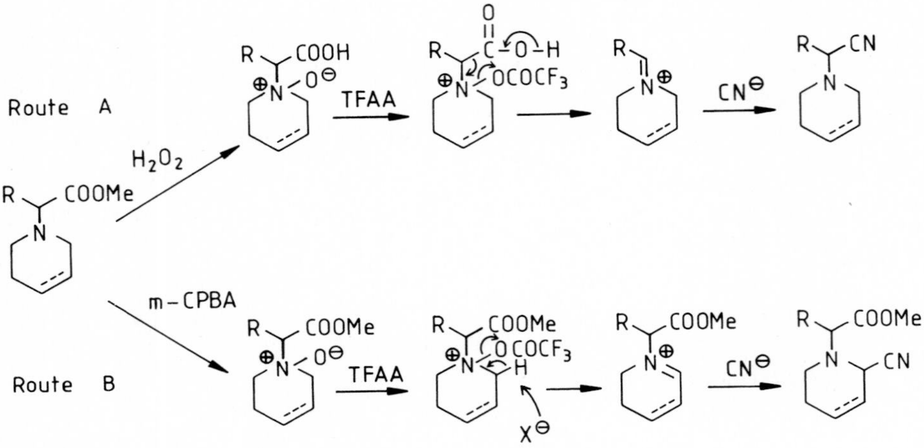

Route A   1) Oxidation with H_2O_2 ; N-Oxide formation and ester hydrolysis

2) Treatment with TFAA ; Generation of the iminium species

3) Addition of KCN   ; Trapping of the iminium species

Route B   1) Oxidation with m-CPBA; N-Oxide formation

2) Treatment with TFAA ; Generation of the iminium species

3) Addition of KCN   ; Trapping of the iminium species

Scheme 24. Regioselective functionalization of carbon atoms α to the heterocyclic nitrogen.

Shown below are some compounds prepared by the developed procedure (Scheme 25).

To summarize, our procedure, combined with the reductive cyanation method developed by Fry (refs. 34, 35), permits selective preparation of any of the

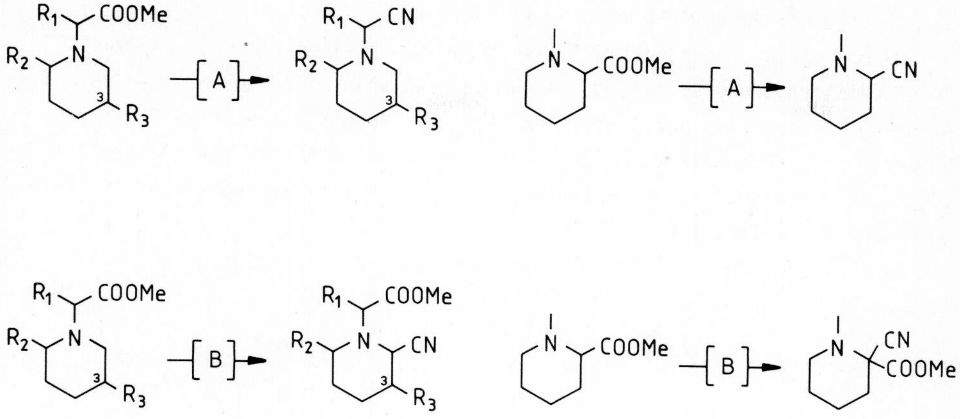

Scheme 25. Some compounds prepared by regioselective functionalization.

	R_1	R_2	R_3	
a	Me	H	H	
b	Et	H	H	
c	Ph	H	H	
d	Me	Me	H	
e	Me	H	Et	Δ^3

following three iminium salt equivalents in the presence of other
functionalities (Scheme 26).

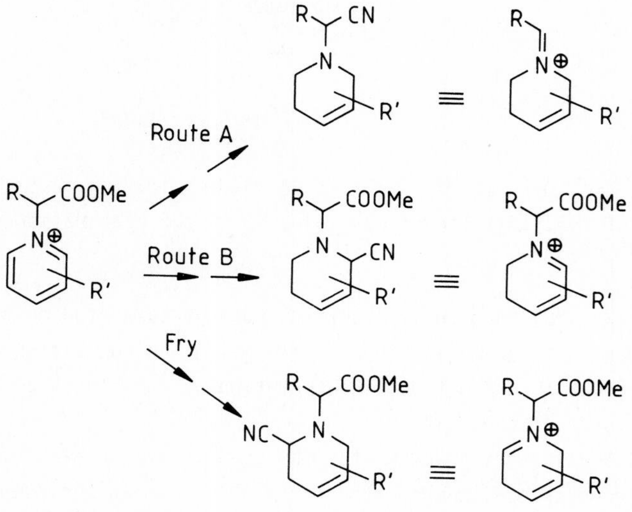

Scheme 26. Regioselective preparation of iminium salt equivalents.

11. 1,2,5,6-TETRAHYDROPYRIDINES

Husson _et al_. (refs. 36, 37) have described the formation of 2-cyano-
1,2,5,6-tetrahydropyridine 42 from the corresponding 1,2,5,6-tetrahydropyridine
43 using the modified Polonovski reaction. The formed 2-cyano-1,2,5,6-tetra-
hydropyridine 42, which is a synthetic equivalent of 5,6-dihydropyridinium salt
44, can then be used for the preparation of different synthons (e.g. 45 and 46)
for alkaloid synthesis (Scheme 27).

However, in the case of methyl acetoacetate, the reaction goes further,
leading to the bicyclic compound 47.

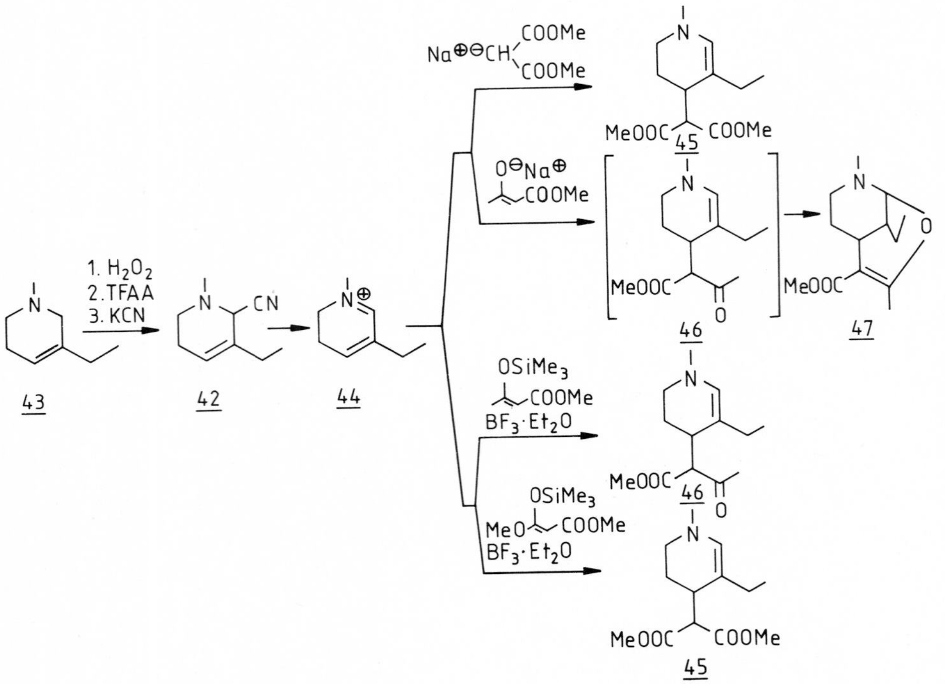

Scheme 27. The use of 1,2,5,6-tetrahydropyridine 43 in the preparation of indole
alkaloid synthons 45 and 46.

We found (refs. 38, 39) that these reactions could be executed more
efficiently and products obtained in higher yield using the corresponding silyl
enol ethers. Moreover, with our method, the formation of the bicyclic compound
47 can be totally avoided (Scheme 27).

Although the method is very powerful for the preparation of alkaloid
synthons of the type described, complications may arise during subsequent
synthetic steps, because the equilibrium tends to be unfavourable for the
desired products (Scheme 28) (see also Scheme 22).

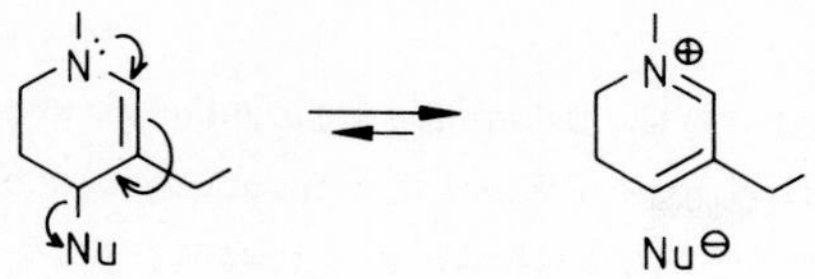

Scheme 28. Equilibrium between 4-substituted 1,4,5,6-tetrahydropyridines and 5,6-dihydropyridinium salts not containing an electron withdrawing group at the 3-position.

By contrast, the following structures (Fig. 5) representing vinylogous amides are more stable under similar conditions.

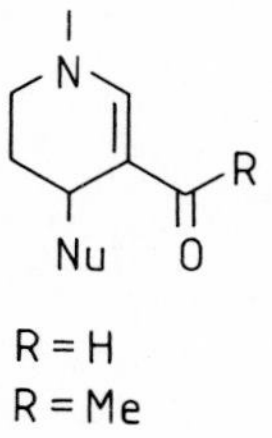

R = H
R = Me

Fig. 5.   4-Substituted vinylogous amide structures.

Nevertheless, it was clear that, owing to competitive reactions (cf. Baeyer-Villiger oxidation), tetrahydropyridines of the present type were not particularly suitable for direct preparation of the corresponding 5,6-dihydro-pyridinium salts or their equivalents by the modified Polonovski reaction (Scheme 29).

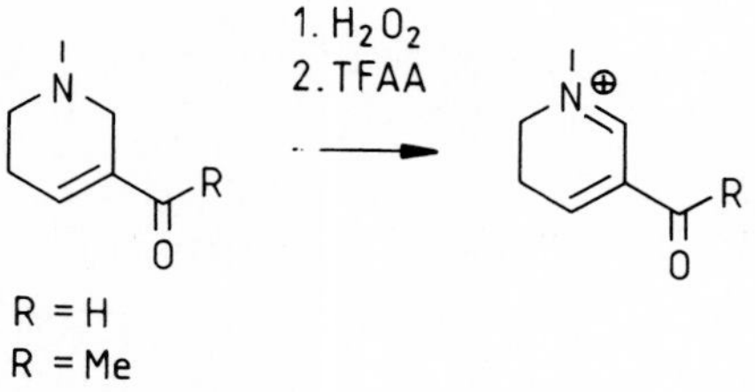

R = H
R = Me

Scheme 29. Direct preparation of a 5,6-dihydropyridinium salt from the corresponding 1,2,5,6-tetrahydropyridine by the modified Polonovski reaction in the presence of an aldehydic or ketonic carbonyl group.

12. NITROGEN ASSISTED RING CLEAVAGE

If the modified Polonovski reaction is to be used, the carbonyl group ought to be protected. And it is desirable that the protecting group should be easily cleavable, under mild conditions, if possible without supplementary steps in the reaction path.

For the purpose, we developed a method (refs. 40-42) which we call nitrogen assisted ring cleavage. In this method, an appropriate 5,6-dihydropyridine acetal derivative, prepared by the modified Polonovski reaction from the corresponding 1,2,5,6-tetrahydropyridine, is attacked by an external nucleophile at the 4-position. This creates a 4-substituted 1,4,5,6-tetrahydropyridine acetal derivative, where the nitrogen assisted acetal ring cleavage can take place leading to the desired vinylogous amide (Scheme 30).

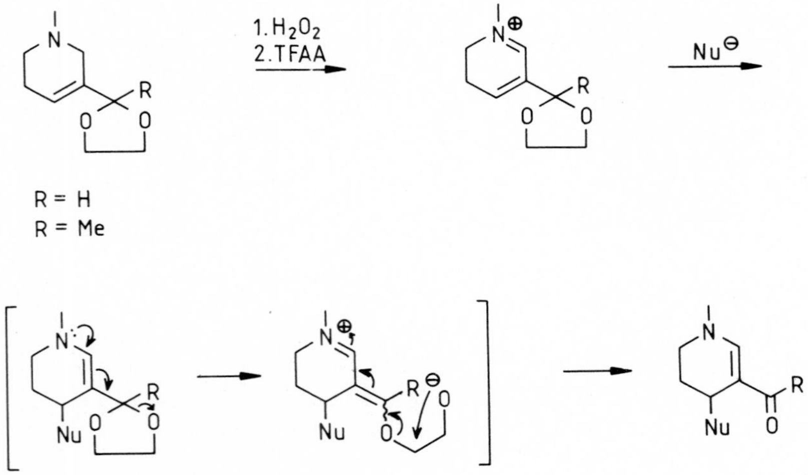

R = H
R = Me

Scheme 30. Nitrogen assisted acetal ring cleavage.

13. INDOLOQUINOLIZIDINE ALDEHYDES

Scheme 31 shows one application of the method (ref. 42). The BOC protected indole derivative 48 was converted by the modified Polonovski reaction to the iminium species 49. Attack by an external reagent (acting as a nucleophile) at the 4-position led to the intermediate 50, where the nitrogen assisted acetal ring cleavage took place, yielding compound 51. Acid treatment of 51 led to the intermediate 52, and cyclization of 52 afforded compound 53.

Alternatively, if the iminium species 49 is attacked by an external reagent (acting as a base) at the 5-position, the intermediate 54 is formed. This time nitrogen assisted acetal ring cleavage leads to compound 55, and acid treatment yields, via compound 56, compound 57.

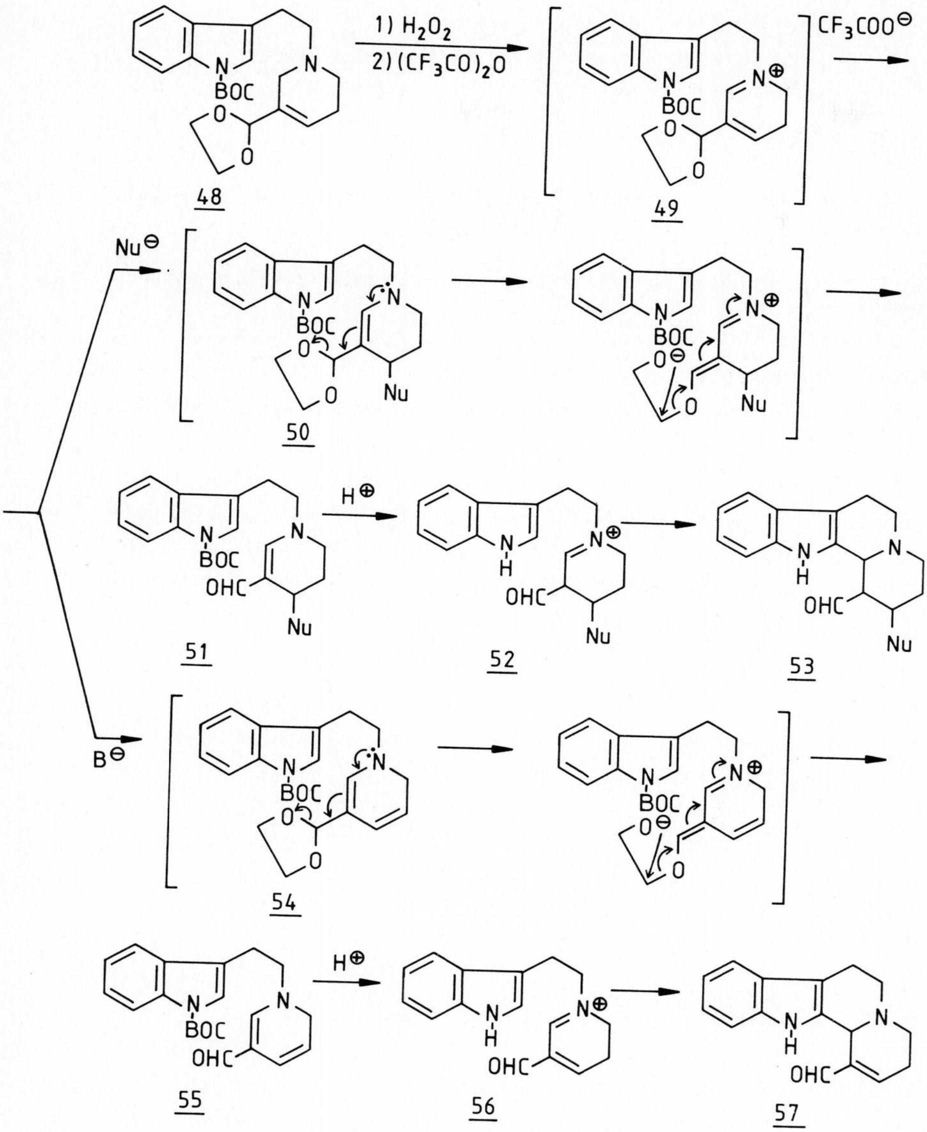

Scheme 31. Preparation of indoloquinolizidine aldehydes 53 and 57.

Thus, the method permits the preparation of a compound such as indoloquinolizidine aldehyde 57, which is a useful intermediate for the preparation
of desethyl eburnamine-vincamine alkaloid synthons (e.g. 58a, 58b and 59),
and which is difficult and tedious to prepare by other means (Scheme 32)
(ref. 38).

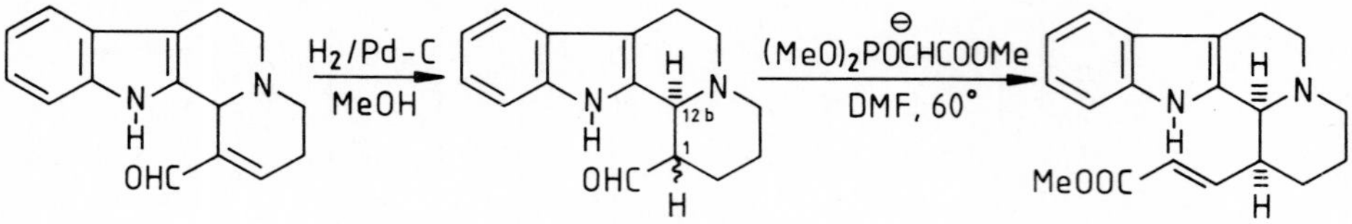

57 58a (C(1)H-α) 58b (C(1)H-β) 59

Scheme 32. Preparation of desethyl eburnamine-vincamine alkaloid synthons 58a, 58b and 59.

14. FURTHER APPLICATIONS

A few further novel applications of the modified Polonovski reaction are noted below.

Using our method, which permits the preparation of the exocyclic iminium species from 35 (*vide supra*), we were able to synthesize the quinuclidine derivative 60 (ref. 26), albeit in low yield (partially due to the oxidation of the ketonic carbonyl group present). Compound 60 contains the crucial part present in the sarpagine alkaloids (Scheme 33).

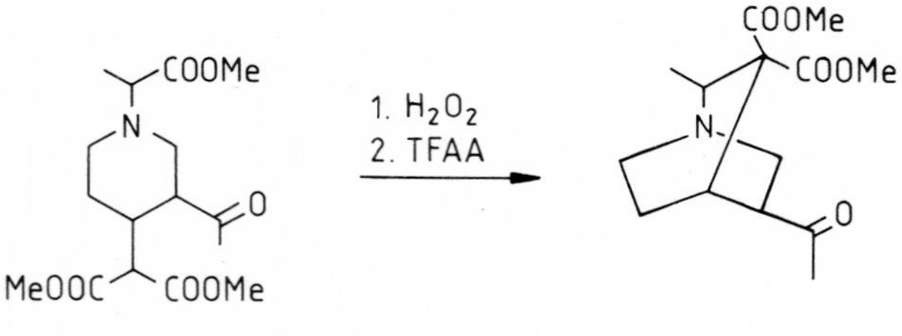

35 60

Scheme 33. Preparation of the quinuclidine derivative 60.

Similarly, the starting material 61 permitted the preparation of compound 62 (ref. 43), which contains the isoquinuclidine skeleton present in the *Iboga* alkaloids (Scheme 34).

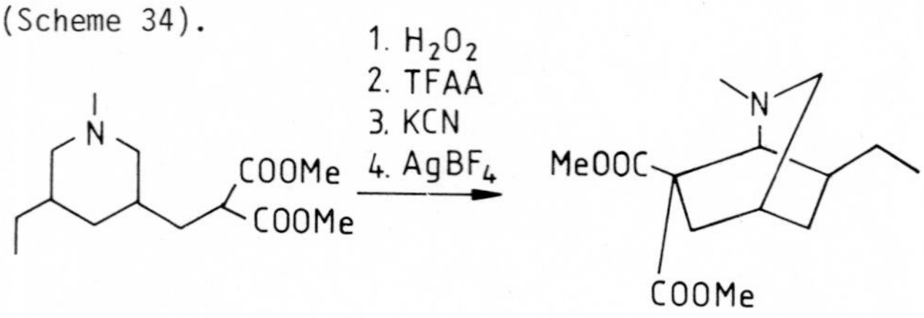

61 62

Scheme 34. Preparation of the isoquinuclidine derivative 62.

15. WENKERT'S ENAMINE

Using the modified Polonovski reaction, we have also developed (ref. 44) a new route to Wenkert's enamine 63 (Scheme 35), which has proven to be a key intermediate in the preparation of several eburnamine-vincamine alkaloids. The easily accessible salt 64 was NaBH$_4$-reduced to the enol ether 65. The indolic N was BOC-protected, leading to compound 66, which was subjected to the modified Polonovski reaction conditions, yielding after cyanide trapping compound 67. The protecting group (BOC) was cleaved to give the α-aminonitrile 68, which was then cyclized to the enol ether 69.

As all our attempts to convert the enol ether 69 directly to the corresponding ketone failed, we transformed it (by hydrogenolysis + reduction) first to alcohols 70a and 70b and then by Swern oxidation and Grignard reaction to the alcohols 71a and 71b. Finally, TFA treatment of the mixture of alcohols 71a and 71b, afforded the enamine 63 in almost quantitative yield.

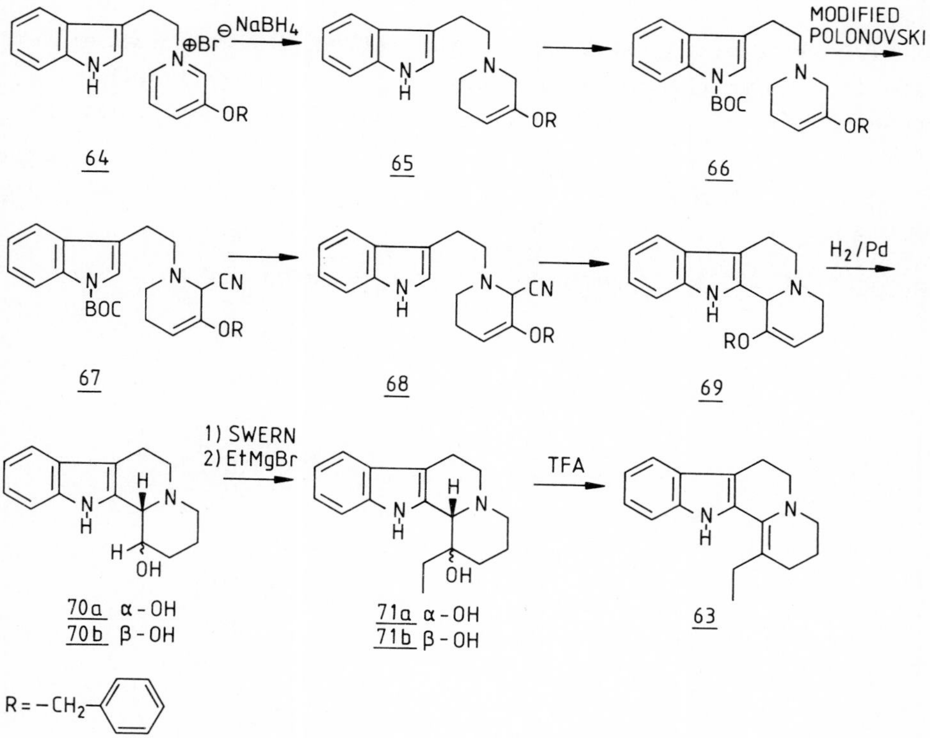

Scheme 35. Preparation of the Wenkert's enamine 63.

16. PSEUDOVINCAMINES

A few years ago the Dutch group under Prof. Anders Baerheim Svendsen isolated the first pseudovincamine alkaloids (tacamines) (ref. 45, 46), here represented by the basic compound of the group, tacamine 72 (Fig. 6). The existence of this new group of indole alkaloids had been predicted earlier by the late Prof. Jean Le Men (ref. 47) and the first synthetic derivatives were prepared by Massiot et al. (ref. 48).

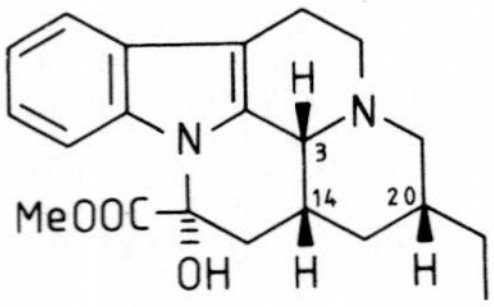

72

Fig. 6. Structural formula of tacamine 72.

Our interest was aroused in applying the modified Polonovski reaction to the preparation of compounds of pseudovincamine type. We proceeded as follows (Scheme 36) (ref. 49). In the piperidine derivative 73 the indolic N was BOC-

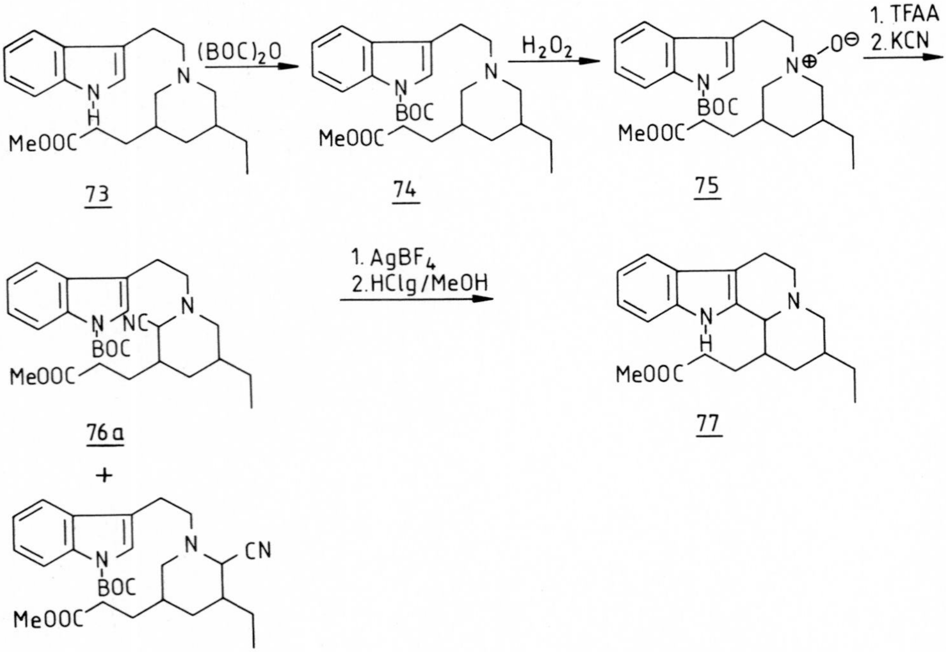

Scheme 36. Preparation of the pseudovincamine analogue 77.

protected leading to compound 74. The corresponding N-oxide 75 was subjected to the modified Polonovski reaction conditions, followed by cyanide trapping, and the two α-aminonitriles obtained (76a and 76b) were separated by preparative TLC. Thereafter, the α-aminonitrile 76a was cyclized to indoloquinolizidine 77.

As we have seen before, the following three forms (Scheme 37), have to be taken into account in considering the conformational equilibrium of indolo-quinolizidines of the present type. To avoid confusion, please note that in Scheme 37 I have used the optical antipodes to the earlier drawings (Schemes 12, 13 and 14). This is because of the absolute stereochemistry of tacamines.

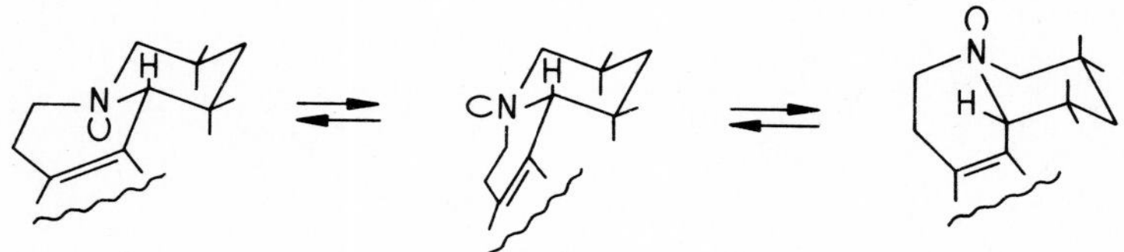

a   b   c

Scheme 37. Conformational equilibrium of indoloquinolizidines using optical antipodes to those of Schemes 12, 13 and 14.

In view of the considerable confusion in the literature concerning the conformational equilibrium of certain indoloquinolizidines, we preceded our study of the conformational behaviour of compound 77 with a look at the ethyl-indoloquinolizidines 78, 79, 80 and 81 (Fig. 7) (ref. 50).

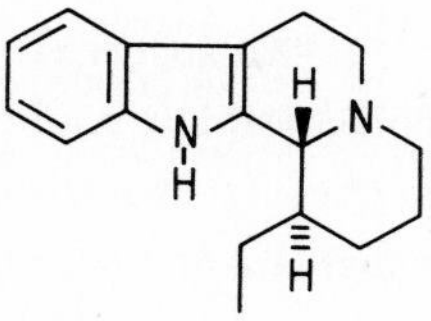

78 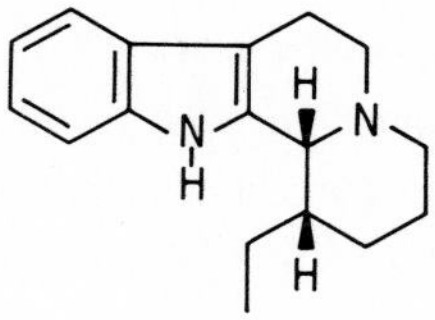   80

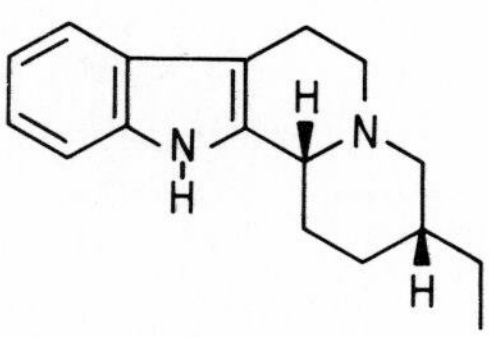

79 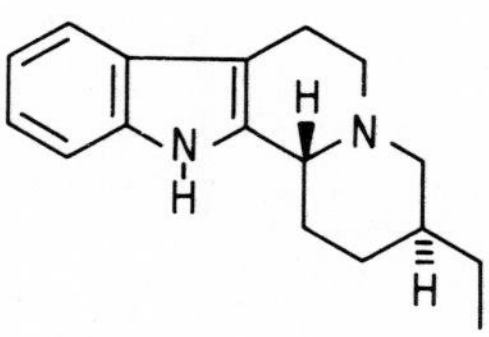   81

Fig. 7. Structural formulae of ethylindoloquinolizidines 78, 79, 80 and 81.

We were able to show, mainly by ^{13}C NMR spectral analysis, that the contribution of conformer c to the conformational equilibrium between a and c (the contribution of conformer b is considered negligible) is about 48% and 56% for compounds 78 and 79, respectively, whereas compounds 80 and 81 exist predominantly in conformation a (Fig. 8).

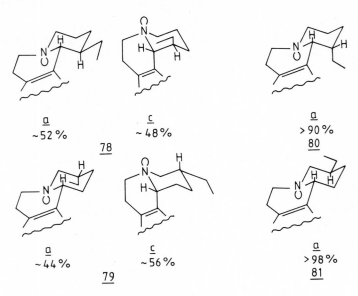

a
~52%

c
~48%

78

a
>90%

80

a
~44%

79

c
~56%

a
>98%

81

Fig. 8. The contribution of different conformations in the conformational equilibrium of ethylindoloquinolizidines 78, 79, 80 and 81.

I now return to the pseudovincamine analogue 77. The ^{13}C NMR spectral analysis showed the compound to exist almost totally in conformation c (Scheme 38). The only stereostructure for compound 77 that allows this is stereostructure 82 (Fig. 9), where in conformation c the C(1) substituent is axially oriented (to avoid steric interactions with the indolic part) and the C(3)

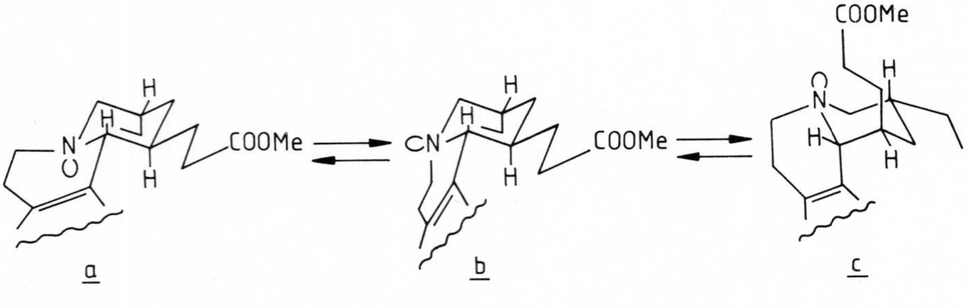

a

b

c

Scheme 38. Conformational equilibrium of the pseudovincamine analogue 77.

substituent is equatorially oriented. This means that compound <u>82</u> and tacamine <u>72</u> have different stereostructures at C(1). A little later, Prof. Csaba Szántay and his group in Hungary published (ref. 51) the preparation of the three other possible stereoisomers of compound <u>77</u> (as ethyl esters).

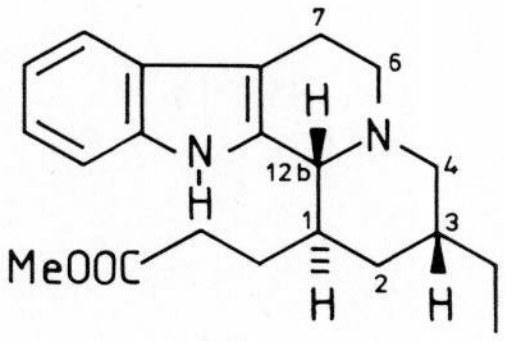

<u>82</u>

Fig. 9. The stereostructure <u>82</u> of the pseudovincamine analogue <u>77</u>.

17. VINCAMINE SYNTHESIS

We have developed a new formal total synthesis for (±)-vincamine <u>83</u> (Scheme 39) (ref. 52). Compound <u>84</u> was treated with $AgBF_4$ leading to enamine <u>85</u>, which was alkylated with methyl acrylate to afford the intermediate <u>86</u>. The protecting group (BOC) was cleaved by acid treatment. This permitted the nucleophilic attack to take place, leading to the cyclized products <u>87</u> and <u>88</u>, of which the former has been transformed earlier to (±)-vincamine <u>83</u>.

18. OTHER CASES

Finally, I would like to mention two other cases where we have exploited the nitrogen assistance in the cleavage.

The first represents short routes to the synthesis of the indole alkaloid (±)-deplancheine <u>89</u> (Scheme 40). $LiAlH_4$ treatment of the easily accessible indoloquinolizidine derivative <u>90</u> leads to intermediate <u>91</u>, where the nitrogen assisted cleavage takes place yielding the iminium compound <u>92</u>. This is attacked by a hydride ion leading to (±)-deplancheine <u>89</u> (ref. 53). Further, when the dihydropyridine derivative <u>93</u> is treated with $NaBH_4$, the hydride ion attacks the 2-position affording the intermediate <u>94</u>, where again the nitrogen assisted cleavage takes place. The formed iminium compound <u>92</u> is further attacked by a hydride ion, as above, to yield (±)-deplancheine <u>89</u> (ref. 54).

In the second example nitrogen assisted cleavage is applied to the transformation of the earlier described indoloquinolizidine derivative <u>37</u>, <u>via</u> <u>85</u> and <u>96</u>, to the heteroyohimbine analogue <u>97</u> (Scheme 41) (ref. 55).

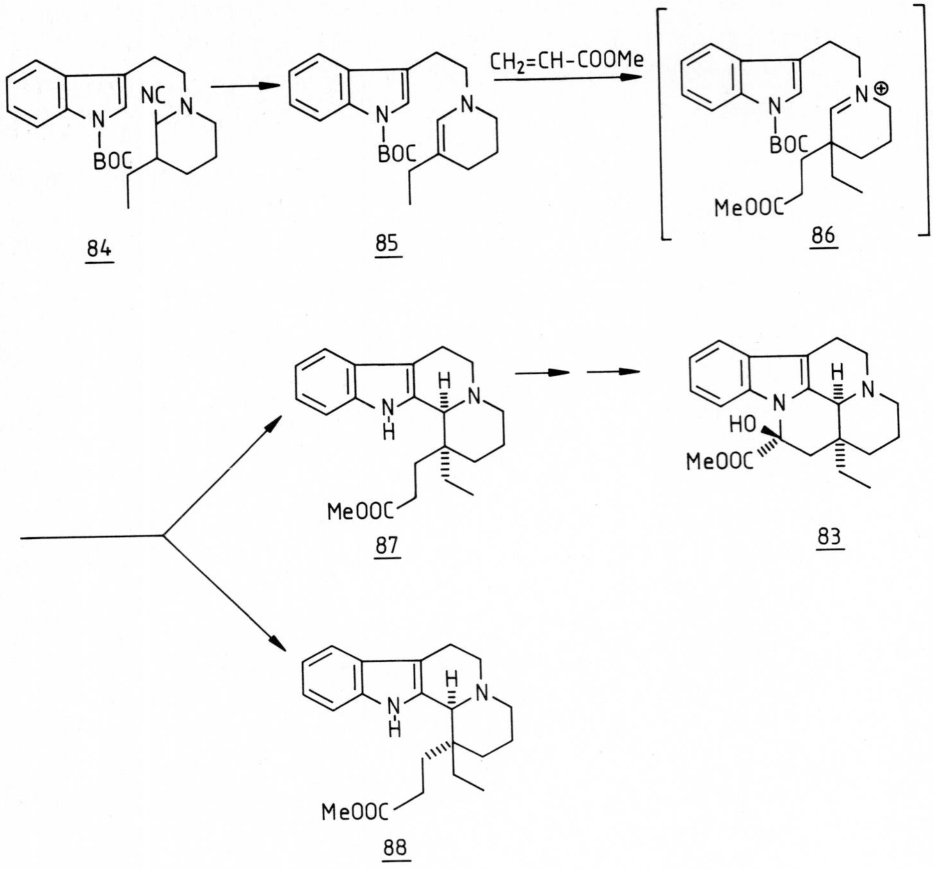

Scheme 39. Formal total synthesis of (±)-vincamine 83.

19. CONCLUSIONS

The results described illustrate some of our synthetic achievements related to the indole alkaloids. Much work remains to be done in this fascinating field. Our recent results (ref. 56), showing that certain indoloquinolizidines exhibit selective affinity for benzodiazepine, tryptamine and serotonin binding sites in rat brain, are providing extra stimulus for our efforts.

ACKNOWLEDGEMENTS

I am pleased to have this opportunity to thank all the past and present members of my research group for their valued and enthusiastic contributions. The financial support from the Finnish Academy is gratefully acknowledged.

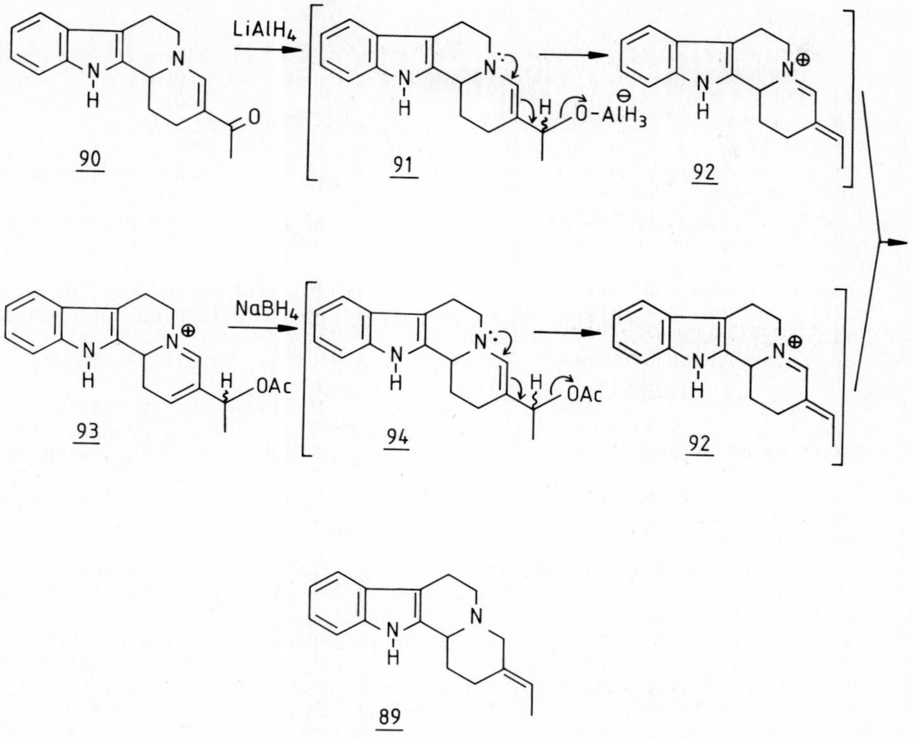

Scheme 40. Two total syntheses of (+)-deplancheine 89.

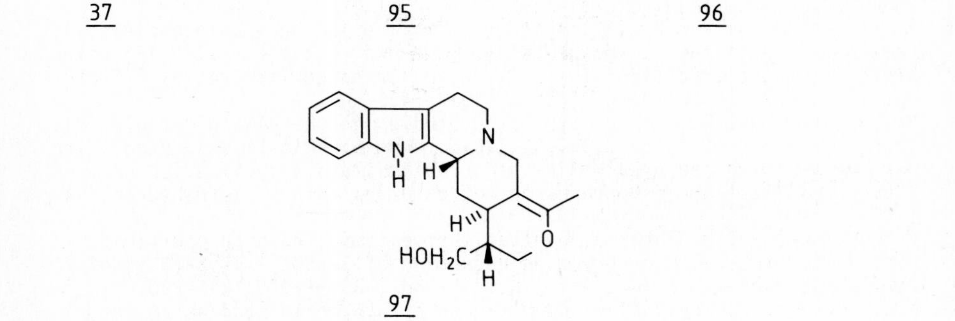

Scheme 41. Preparation of the heteroyohimbine analogue 97.

REFERENCES

1. C. Djerassi, H.J. Monteiro, A. Walser and L.J. Durham, Alkaloid studies, LVI. The constitution of vallesiachotamine, J. Am. Chem. Soc., 88 (1966) 1792-1798.
2. K.T.D. De Silva, G.N. Smith and K.E.H. Warren, Stereochemistry of strictosidine, J. Chem. Soc., Chem. Commun. (1971) 905-907.
3. G.N. Smith, Strictosidine: A key intermediate in the biogenesis of indole alkaloids, Chem. Commun. (1968) 912-914.
4. A.I. Scott, S.L. Lee, P. le Capite, M.G. Gulver and C.R. Hutchinson, The role of isovincoside (strictosidine) in the biosynthesis of the indole alkaloids, Heterocycles, 7 (1977) 979-984.
5. J. Stöckigt and M.H. Zenk, Strictosidine (isovincoside): the key intermediate in the biosynthesis of monoterpenoid indole alkaloids, J. Chem. Soc., Chem. Commun. (1977) 646-648.
6. G.A. Cordell, The biosynthesis of indole alkaloids, Lloydia 37 (1974) 219-298.
7. A. Koskinen and M. Lounasmaa, The sarpagine-ajmaline group of indole alkaloids, in: W. Herz, H. Grisebach and G.W. Kirby (Eds.), Progress in the Chemistry of Organic Natural Products, Vol. 43, Springer-Verlag, Wien - New York, 1983, pp. 267-346.
8. J.H. Supple, D.A. Nelson and R.E. Lyle, The synthesis of partially reduced indolo and benzoquinolizines via 1,4-dihydropyridine intermediates, Tetrahedron Lett. (1963) 1645-1649.
9. M. Lounasmaa and C.-J. Johansson, Synthetic studies in the alkaloid field - IV. The sodium dithionite reduction of 1-[2-(3-indolyl)-ethyl]-3-methoxycarbonyl pyridinium bromides, Tetrahedron, 33 (1977) 113-117.
10. M. Lounasmaa, P. Juutinen and P. Kairisalo, Synthetic studies in the alkaloid field - IX. Stereospecific total synthesis of (+)-19,20-dihydro-20-desformyl-20-methoxycarbonylvallesiachotamine and its 20-desethyl analogue, Tetrahedron, 34 (1978) 2529-2532.
11. M. Lounasmaa and R. Jokela, Synthetic studies in the alkaloid field - XII. Vallesiachotamine models, Tetrahedron Lett. (1978) 3609-3612.
12. M. Lounasmaa and H.-P. Husson, Anomalous sodium dithionite reduction of a 1-[2-(3-indolyl)-ethyl]-pyridinium bromide, Acta Chem. Scand., B33 (1979) 466-467.
13. R. Jokela and M. Lounasmaa, Total synthesis of (+)-desmethylhexahydro-vallesiachotaminelactones, Tetrahedron, 38 (1982) 1015-1018.
14. M. Lounasmaa and C.-J. Johansson, Synthetic studies in the alkaloid field. Part II. On the preparation of N-(β-indolylethyl)-dihydro- and (N-(β-indolylethyl)-tetrahydropyridines by catalytic hydrogenation, and their acid-induced cyclization, Acta Chem. Scand., B29 (1975) 655-661.
15. M. Lounasmaa, C.-J. Johansson and J. Svensson, Synthetic studies in the alkaloid field. Part III. Selective alkaline decarboalkoxylative cyclization of some N-alkyldihydro- and N-alkyltetrahydropyridines, Acta Chem. Scand., B30 (1976), 251-254.
16. M. Lounasmaa and M. Hämeilä, Synthetic studies in the alkaloid field. V. Determination of the stereochemistry of several 1,2,3,4,6,7,12,12b-octahydro-3-methoxycarbonylindolo-[2,3-a]quinolizine derivatives by ^{13}C NMR spectral analysis, Tetrahedron, 34 (1978) 437-442.
17. M. Lounasmaa and R. Jokela, Synthetic studies in the alkaloid field. VII. Stereochemistry of 1,2,3,4,6,7,12,12b-octahydroindolo-[2,3-a]quinolizine derivatives prepared by alkaline decarboalkoxylative cyclization or acid-induced cyclization of appropriate tetrahydropyridines, Tetrahedron, 34 (1978) 1841-1844.
18. M. Lounasmaa and M. Puhakka, Synthetic studies in the alkaloid field. Part VIII. The sodium borohydride reduction of 3-acetyl-1-[2-(3-indolyl)-ethyl]-pyridinium bromide, Acta Chem. Scand., B32 (1978) 216-220.
19. M. Lounasmaa, H. Merikallio and M. Puhakka, Synthetic studies in the alkaloid field. X. Preparation and stereostructure determination of several indolo-[2,3-a]quinolizine derivatives, Tetrahedron, 34 (1978) 2995-2999.

20. E. Frostell, R. Jokela and M. Lounasmaa, Sodium dithionite reduction in the preparation of indole alkaloids of gambirtannine-type, Acta Chem. Scand., B35 (1981) 671-672.

21. F. Kröhnke and K. Ellegast, Anlagerungen von Methyl- und Methylen-Ketonen and Pyridiniumbasen. Über Pseudobasen I. Liebigs Ann. Chem., 600 (1956) 176-197.

22. F. Kröhnke and K. Ellegast, Spaltung 4-substituierter Pyridiniumsalze. Ein ergiebiger Weg zu 4-Acalkylpyridinen und zum 2,7-Naphthyridin-Ringsystem. Über Pseudobasen II. Liebigs Ann. Chem. (600) (1956) 198-210.

23. F. Kröhnke and J. Vogt, Methylketon-Addukte der Chinolinium-Reihe. Über Pseudobasen III. Leibigs Ann. Chem., 600 (1956) 211-228.

24. E. Wenkert and G.D. Reynolds, Vallesiachotamine models, Synth. Commun., 3 (1973) 241-243.

25. E. Wenkert, C.-J. Chang, H.P.S. Chawla, D.W. Cochran, E.W. Hagaman, J.C. King and K. Orito, General methods of synthesis of indole alkaloids. 14. Short routes of construction of yohimboid and ajmalicinoid alkaloid systems and their ^{13}C nuclear magnetic resonance spectral analysis, J. Am. Chem. Soc., 98 (1976) 3645-3655.

26. M. Lounasmaa and A. Koskinen, Novel applications of the modified Polonovski reaction. A biomimetic synthesis of quinuclidines, Tetrahedron Lett., 23 (1982) 349-352.

27. M. Lounasmaa and R. Jokela, Stereoselective total synthesis of (±)-3-iso-19-epiajmalicine, Tetrahedron Lett., 27 (1986) 2043-2044.

28. M. Lounasmaa and A. Koskinen, Modified Polonovski reaction, a versatile synthetic tool, Heterocycles, 22 (1984) 1591-1612.

29. H. Volz, Die Bedeutung der Polonovski-Reaktion für die Organische Synthese, Kontakte (Darmstadt) (1984) (3) 14-28.

30. P. Potier, La reaccion de Polonovski modificada, Rev. Latinoamer. Quim., 9 (1978) 47-54.

31. Ad. Cavé, C. Kan-Fan, P. Potier and J. Le Men, Modification de la réaction de Polonovski. Action de l'anhydride trifluoroacétique sur un aminoxyde, Tetrahedron, 23 (1967) 4681-4689.

32. M. Lounasmaa and A. Koskinen, Novel applications of the modified Polonovski reaction - II. Regiocontrolled iminium ion formation, Heterocycles, 19 (1982) 2115-2117.

33. A. Koskinen and M. Lounasmaa, Regiospecific functionalization of carbon atoms α to heterocyclic nitrogen, Tetrahedron, 39 (1983) 1627-1633.

34. E.M. Fry, Stereospecific tautomerism in a 1,2-dihydropyridine. A β-benzomorphan synthesis, J. Org. Chem., 28 (1963) 1869-1874.

35. E.M. Fry, 6-Cyano-1,2,5,6-tetrahydropyridines in the preparation of 1,2-dihydropyridines. Tautomerism of the dienes, J. Org. Chem., 29 (1964) 1647-1650.

36. D.S. Grierson, M. Harris and H.-P. Husson, Synthesis and chemistry of 5,6-dihydropyridinium salt adducts. Synthons for general electrophilic and nucleophilic substitution of the piperidine ring system. J. Am. Chem. Soc., 102 (1980) 1064-1082.

37. H.-P. Husson, Dihydropyridine equivalents as intermediates for the synthesis of alkaloids, Bull. Soc. Chim. Belg., 91 (1982) 985-995.

38. A. Koskinen and M. Lounasmaa, Novel applications of the modified Polonovski reaction - V. Silyl enolates as nucleophilic alkylating reagents, Tetrahedron Lett., 24 (1983) 1951-1952.

39. A. Koskinen and M. Lounasmaa, Mild general synthesis of 4-substituted piperidines. J. Chem. Soc., Chem. Commun. (1983) 821-822.

40. M. Lounasmaa and A. Tolvanen, Nitrogen assisted enol ether formation, Heterocycles, 24 (1986) 651-654.

41. M. Lounasmaa and A. Tolvanen, Nitrogen assisted acetal ring cleavage. Part II. Preparation of synthons for eburnamine-vincamine alkaloids, Heterocycles, 24 (1986) 1279-1284.

42. A. Tolvanen and M. Lounasmaa, Nitrogen assisted ring cleavage. Part III. Synthesis and reaction of 1-formyl-3,4,6,7,12,12b-hexahydroindolo[2,3-a]-quinolizine, Tetrahedron 43 (1987) 1123-1127.

122

43. M. Lounasmaa, R. Jokela and T. Tamminen, Novel applications of the modified Polonovski reaction - VII. Preparation of isoquinuclidines, Tetrahedron Lett., 26 (1985) 801-802.
44. M. Lounasmaa, E. Karvinen, A. Koskinen and R. Jokela, Novel applications of the modified Polonovski reaction - IX. A new route to Wenkert's enamine, Tetrahedron, 43 (1987) 2135-2146.
45. T.A. van Beek, P.P. Lankhorst, R. Verpoorte and A. Baerheim Svendsen, Tacamine, the first example of a new class of indole alkaloids, Tetrahedron Lett., 23 (1982) 4837-4840.
46. T.A. van Beek, R. Verpoorte and A. Baerheim Svendsen, Alkaloids of Tabernaemontana eglandulosa, Tetrahedron, 40 (1984) 737-748.
47. J. Le Men, C. Caron-Sigaut, G. Hugel, L. Le Men-Olivier and J. Lévy, Synthèse partielle de la (20 S)-Ψ-vincamine et de la (20 S)-épi-16Ψ-vincamine, Helv. Chim. Acta, 61 (1978) 566-570.
48. G. Massiot, F. Sousa Oliveira and J. Lévy, Synthèses en série indolique. IX. Synthèse totale des pseudovincamines, Bull. Soc. Chim. Fr. II, (1982) 185-190.
49. R. Jokela, S. Schüller and M. Lounasmaa, Novel applications of the modified Polonovski reaction - VIII. Synthetic studies in the pseudovincamine series, Heterocycles, 23 (1985) 1751-1757.
50. M. Lounasmaa, R. Jokela and T. Tamminen, Conformational studies of 1- and 3-ethyl-1,2,3,4,6,7,12,12b-octahydroindolo[2,3-a]quinolizines, Heterocycles, 23 (1985) 1367-1371.
51. L. Szabó, E. Márványos, G. Tóth, Cs. Szántay Jr., G. Kalaus and Cs. Szántay, Synthesis of Vinca alkaloids and related compounds XXX. Total synthesis of (+)-tacamine, (+)-apotacamine and their 20-epimers, Heterocycles, 24 (1986) 1517-1525.
52. M. Lounasmaa and R. Jokela, A new formal total synthesis of (+)-vincamine, Heterocycles, 24 (1986) 1663-1665.
53. M. Hämeilä and M. Lounasmaa, A short stereospecific synthesis of (+)-deplancheine, Acta Chem. Scand., B35 (1981) 217-218. See also M. Hämeilä and M. Lounasmaa, Formation of exocyclic terminal methylene groups in piperidine derivatives, Heterocycles, 19 (1982) 1517-1522.
54. R. Jokela, A. Juntunen and M. Lounasmaa, Elaboration of the ethylidene side chain in the synthesis of indole alkaloids: Preparation of (+)-deplancheine and its analogues, Planta Med., 53 (1987) 386-388.
55. R. Jokela, T. Taipale, K. Ala-Kaila and M. Lounasmaa, Heteroyohimbine alkaloid synthesis, Heterocycles, 24 (1986) 2265-2271.
56. M. Lounasmaa, V. Saano, M.M. Airaksinen, R. Jokela and A. Huhtikangas, Indoloquinolizidines, formal derivatives of tetrahydro-β-carbolines, show selective affinity for benzodiazepine, tryptamine and serotonin binding sites in rat brain, Neuropharmacology, 25 (1986) 915-918.

SYNTHESIS OF ZWITTERIONIC INDOLO[2,3-a]QUINOLIZINE ALKALOIDS

G. W. GRIBBLE

1. INTRODUCTION

The relatively small collection of biogenetically and theoretically interesting zwitterionic indolo -[2,3-a]quinolizine alkaloids has received less attention from synthetic chemists than have most other classes of alkaloids. Perhaps for this reason, syntheses of these alkaloids *per se* do not appear to have been reviewed previously (ref. 1). Moreover, the newly discovered antitumor activity of some of these alkaloids (*vide infra*) is certain to pique the interest of the synthetic community. For these reasons and because of our own research in this area, we have surveyed herein the known synthetic approaches to indolo[2,3-a]quinolizine alkaloids.

The fact that this ring system can be represented by the two resonance structures **A** and **B** is evidenced by the colored nature and the high dipole moments of compounds of this type, as well as by their pH dependent ultraviolet-visible spectra (ref. 2).

In addition to reviewing alkaloids of the indolo[2,3-a]quinolizine type, we will cover the related 6,7-dihydro (**C**) and 1,2,3,4-tetrahydro (**D**) derivatives, which, like the parent indolo[2,3-a]quinoli -zine system, may or may not exist in a zwitterionic form (pH dependent). But we will not include the 1,2,3,4,6,7-hexahydro alkaloids (**E**), which at high pH will exist as an enamine (**F**) rather than as the zwitterionic form.

Schemes 1-2 summarize the known examples of indolo[2,3-*a*]quinolizine and related alka - loids. It should be noted that at least some of these alkaloids may exist in the protonated form in the plant. In fact, as will be seen, these compounds are normally isolated and purified via their acid salts. Tetracyclic examples (**Scheme 1**) include the parent alkaloid indolo[2,3-*a*]quinolizine (**1**) and a dihy - dro derivative **2** (conceivably the Δ-2,3 isomer) (both **1** and **2** isolated from *Gonioma kamassi* (ref. 3)), flavopereirine (**3**) (also known as melinonine G) (*Geissospermum laeve* (ref. 4), *Geissospermum vellosii* (ref. 5a), *Strychnos melinoniana* (ref. 5b), *Strychnos longicaudata* (ref. 6), and *Strychnos ngouniensis* (ref. 6)), 6,7-dihydroflavopereirine (**4**) (alkaloid numbering gives this as 5,6-dihydro - flavopereirine) (*Strychnos usambarensis* (ref. 7)), flavocarpine (**5**) (*Pleiocarpa mutica* (ref. 8)), vin - carpine (**6**) and dihydrovincarpine (**7**) (both from *Vinca major elegantissima* (ref. 9)). In addition, several tetracyclic alkaloids have been recently isolated and identified as being close relatives to their more well known indolo[2,3-*a*]quinolizidine counterparts. These (alkaloid numbering shown) are 3,4,5,6-tetradehydro-18,19-dihydrocorynantheol (**8**) (*Aspidosperma marcgravianum* (ref. 10)), 10-methoxy-3,4,5,6-tetradehydro-18,19-dihydrocorynantheol (**9**), 3,4,5,6-tetradehydroochropposinine (**10**), 10-methoxy-3,4,5,6-tetradehydrocorynantheol (**11**), and 3,4,5,6,18,19-hexadehydroochrop - posinine (**12**) (all from *Neisosperma glomerata* (ref. 11)), 3,4,5,6-tetradehydrositsirikine (**13**) (*Aspidosperma oblongum* (ref. 12)).

Scheme 1

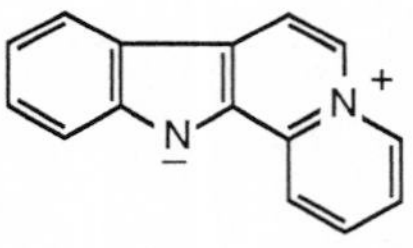

1

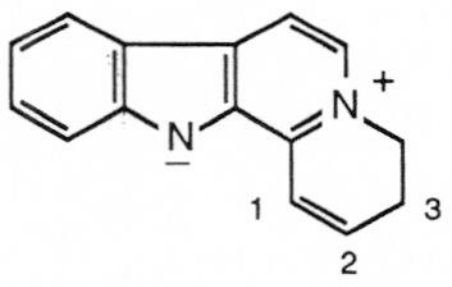

2

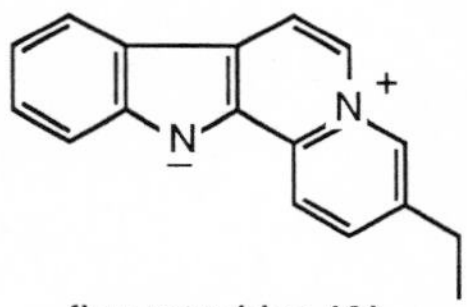

flavopereirine (**3**)

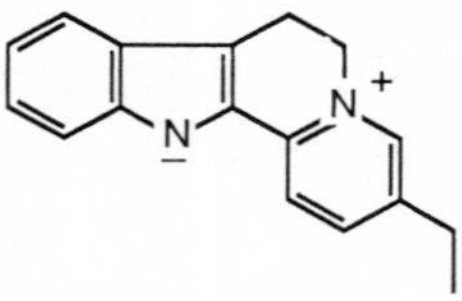

dihydroflavopereirine (**4**)

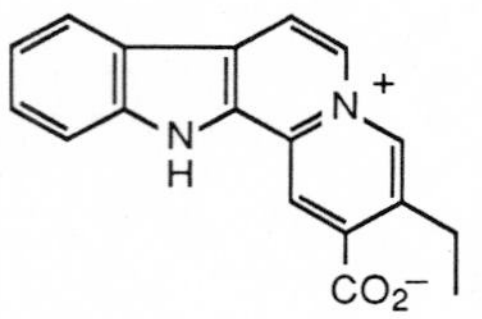

flavocarpine (**5**)

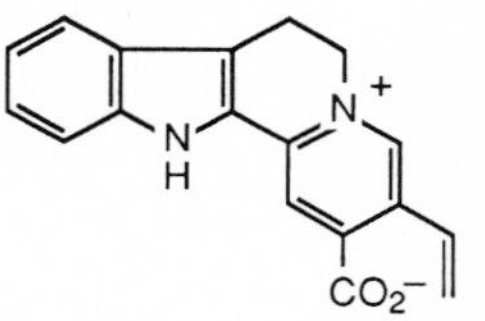

vincarpine (**6**)

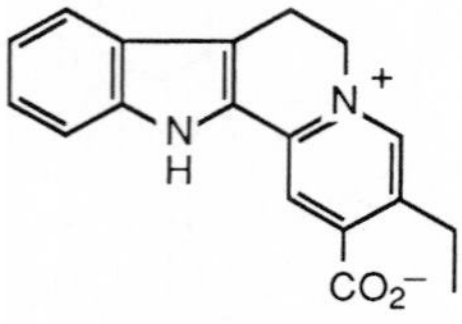

dihydrovincarpine (**7**)

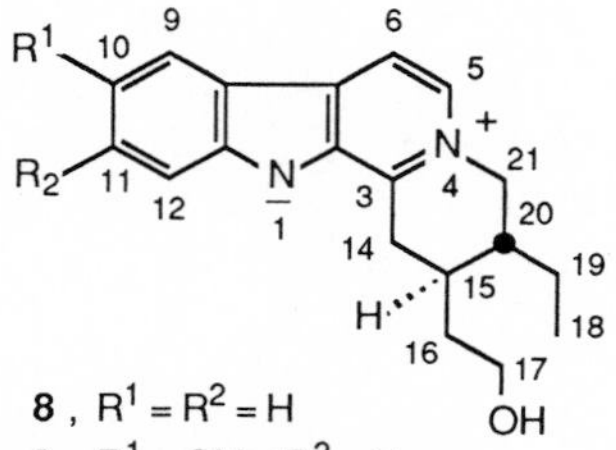

8 , $R^1 = R^2 = H$
9 , $R^1 = OMe, R^2 = H$
10 , $R^1 = R^2 = OMe$

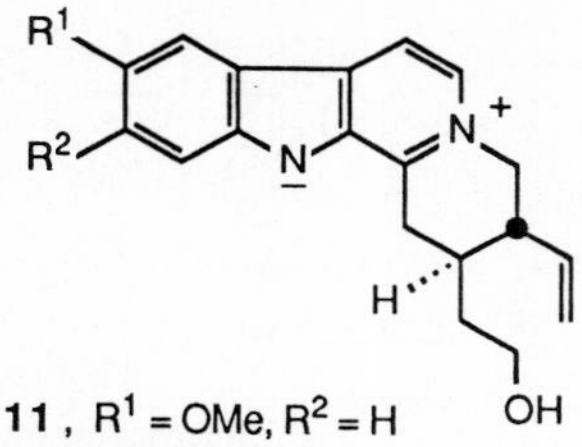

11, R¹ = OMe, R² = H
12, R¹ = R² = OMe

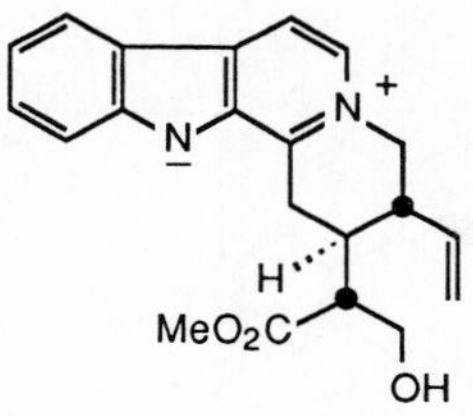

13

Pentacyclic indole alkaloids in the indolo[2,3-*a*]quinolizine (ref. 13) class (**Scheme 2**) are sempervirine (**14**) (*Gelsemium sempervirens* (ref. 14)), serpentine (**15**) (*Rauwolfia serpentina* (ref. 15), *Rauwolfia canescens* (ref. 16), *Rauwolfia heterophylla* (ref. 17), *Lochnera (Vinca) rosea* (ref. 18), *Rauwolfia sellowii* (ref. 19), *Rauwolfia nitida* (ref. 20)), alstonine (**16**) (*Alstonia constricta* (ref. 21), *Rauwolfia vomitora* (ref. 22), *Rauwolfia obscura* (ref. 22), *Rauwolfia hirsuta* (ref. 23), *Rau-wolfia nitida* (ref. 20)), serpenticine (**17**) (*Rauwolfia vomitoria* (ref. 24)), bleekerine (**18**) (*Bleekeria vitiensis* (ref. 25)), alstoniline (**19**) (*Alstonia constricta* (ref. 26)), ourouparine (**20**) (*Ourouparia gambir* (ref. 27)), neooxygambirtannine (**21**) (*Aspidosperma oblongum* (ref. 12)), anhydroalstona-tine (**22**) (*Alstonia venenata* (ref. 28)), 3,4,5,6-tetradehydro-β-yohimbine (**23**) (*Amsonia elliptica* (ref. 29)), melinonine E (**24**) (*Strychnos melinoniana* (refs. 5, 30)), strychnoxanthine (**25**) (*Strych-nos gossweileri* (ref. 31), and the unusual pentacyclic alkaloids schoberidine (**26**) (*Nitraria schoberi* (ref. 32)), and ophiorines A (**27a**) and B (**27b**) (*Ophiorrhiza japonica* (ref. 33)).

Scheme 2

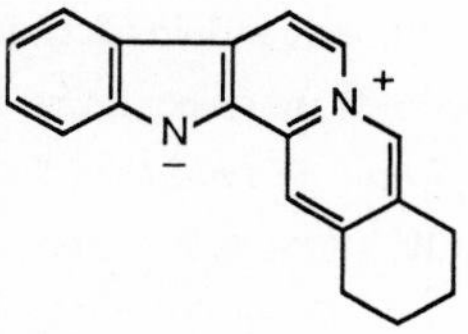

sempervirine (**14**)

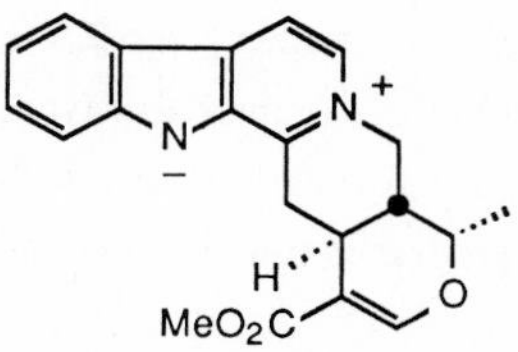

serpentine (**15**)

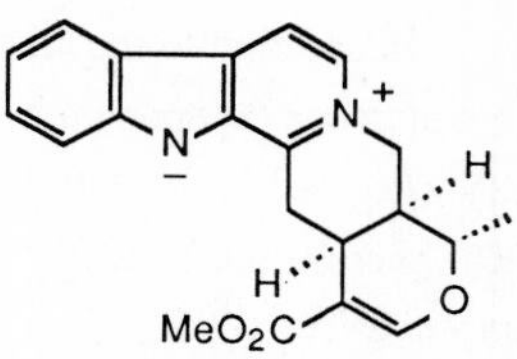

alstonine (**16**)

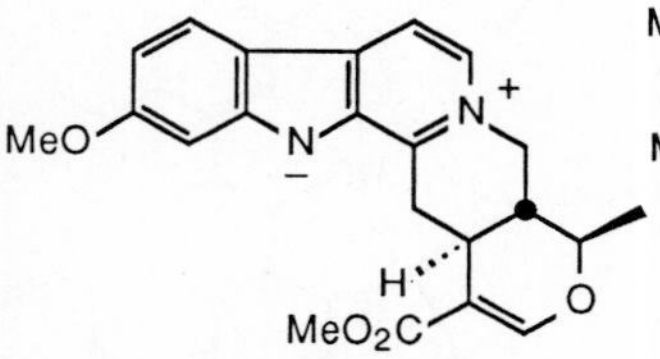

serpenticine (**17**)

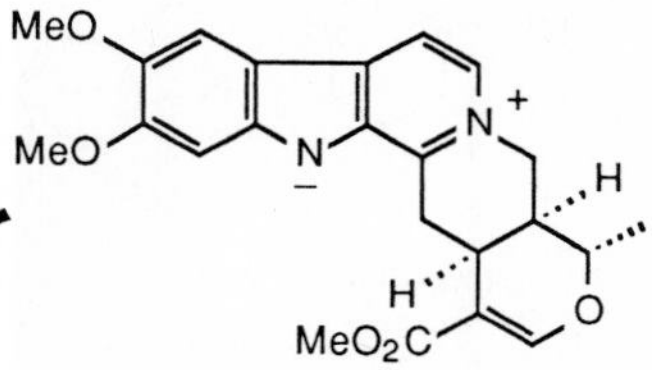

bleekerine (**18**)

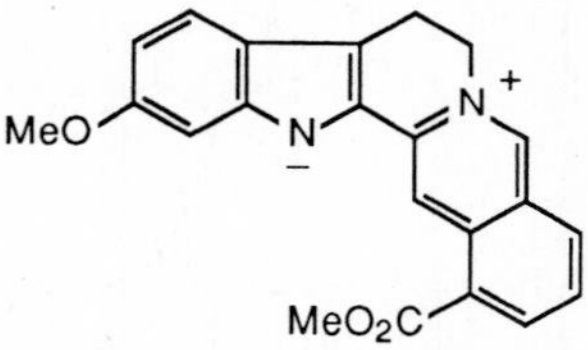

alstoniline (**19**)

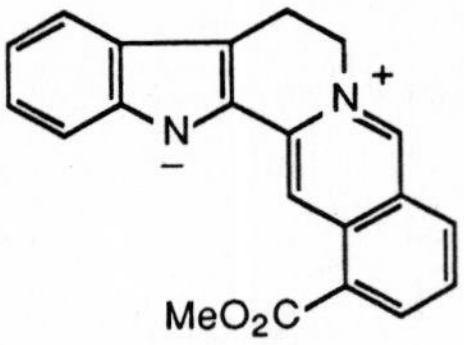

ourouparine (20)

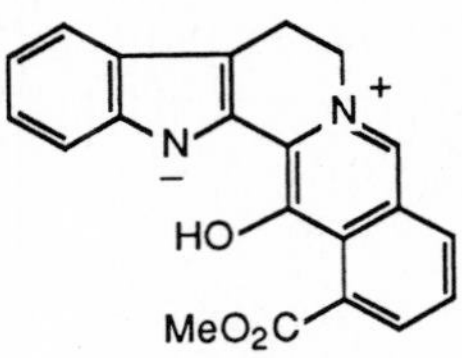

neooxygambirtannine (21)

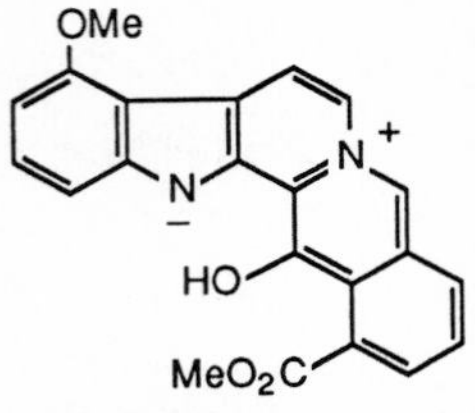

anhydroalstonatine (22)

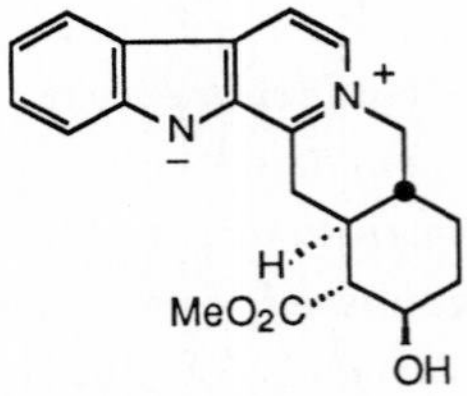

3,4,5,6-tetradehydro-
ß-yohimbine (23)

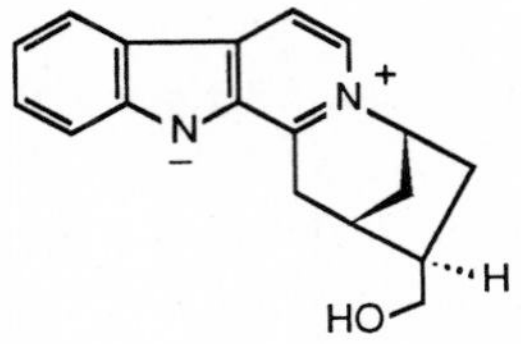

melinonine E (24)

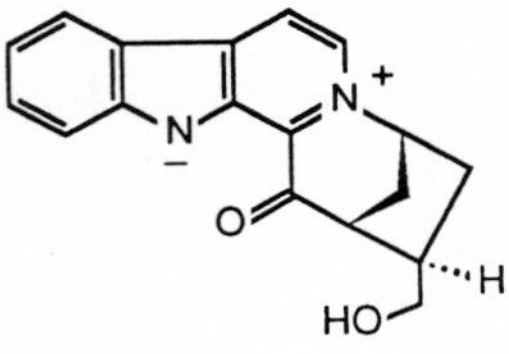

strychnoxanthine (25)

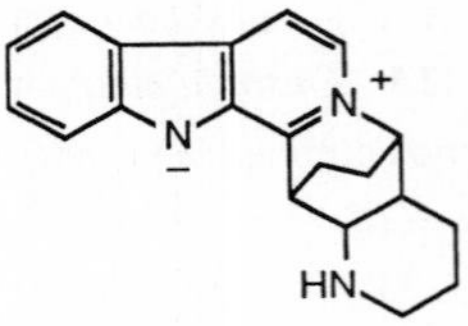

schoberidine (26)

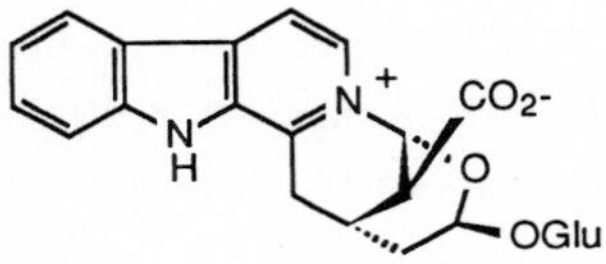

ophiorine A (27a)

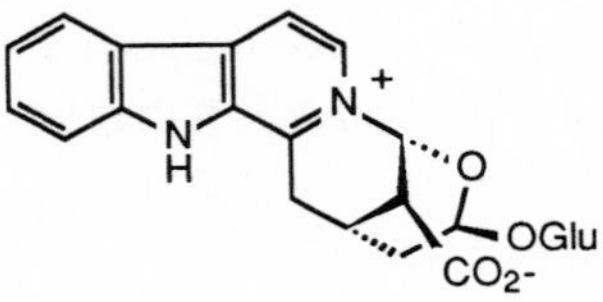

ophiorine B (27b)

In addition to the alkaloids listed in **Schemes 1** and **2**, there are three dimeric indole alkaloids in which one component is of the indolo[2,3-*a*]quinolizine type. These alkaloids are serpentinine (**28**) (*Rauwolfia serpentina* (refs. 15, 34), *Rauwolfia tetraphylla* (ref. 35), *Rauwolfia nitida* (ref. 20)), picrasidine F (**29**) (*Picrasma quassioides* (ref. 36), and afrocurarine (**30**) (*Strychnos usam - barensis* (ref. 37)).

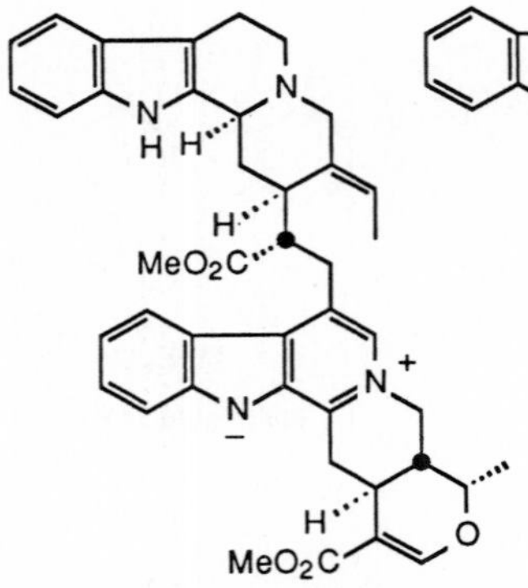

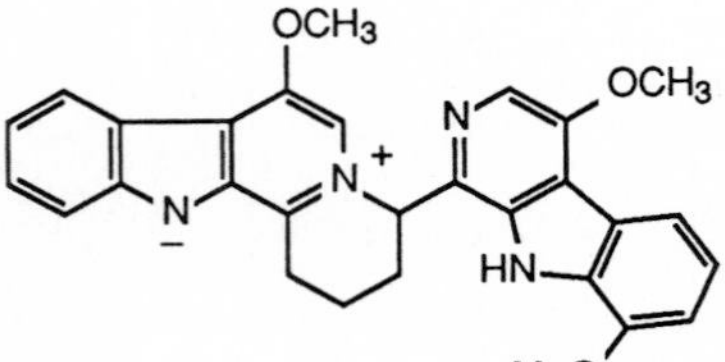

picrasidine F (29)

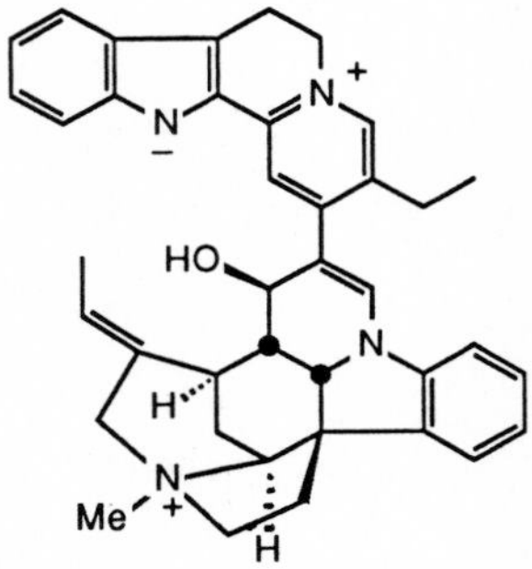

afrocurarine (30)

serpentinine (28)

The closely related zwitterionic indolo[2',3':3,4]pyrido[2,1-*b*]quinazoline alkaloids, such as euxylophorine A (**31**) (ref. 38a), have been reviewed recently (ref. 38b) and will not be covered here. Likewise, isomeric alkaloids such as alkaloid AG-1 (**32**)(ref. 39) and cryptolepine (**33**) (ref. 40), as well as the simple zwitterionic *beta*-carboline alkaloids normelinonine F (**34**) (ref. 41) and melinonine F (**35**) (refs. 5, 41) will be excluded from this review. However, it should be mentioned that melino - nine F (**35**) (ref. 42) as well as several of the indolo[2,3-*a*]quinolizine alkaloids (ref. 43) to be discus - sed have newly discovered antitumor activity (refs. 42, 43), which should inspire renewed synthetic activity in this area.

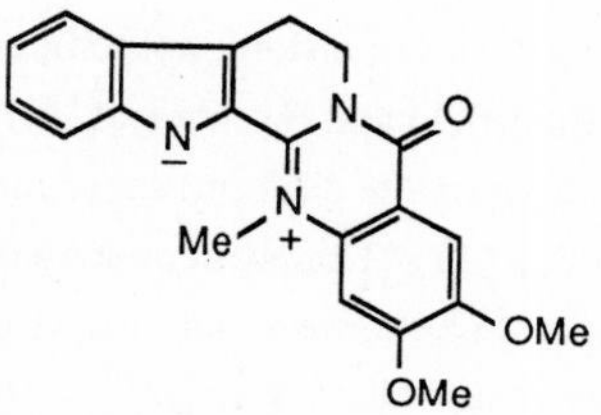

euxylophorine A (**31**)

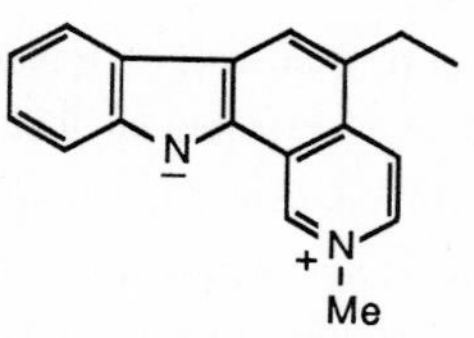

alkaloid AG-1 (**32**)

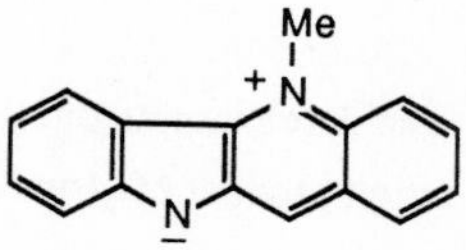

cryptolepine (**33**)

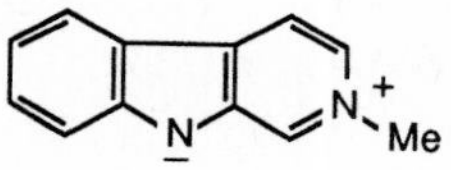

normelinonine F (**34**)

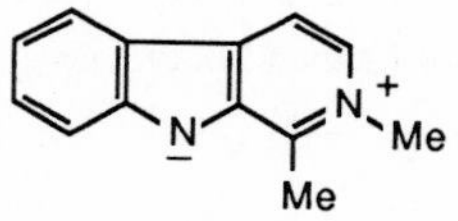

melinonine F (**35**)

2.   CLASSIFICATION OF STRATEGIES

The known synthetic routes to the indolo[2,3-*a*]quinolizine ring system may be categorized into six main strategies, I - VI, for which the key bond formations are indicated. Within each category, the syntheses are arranged chronologically.

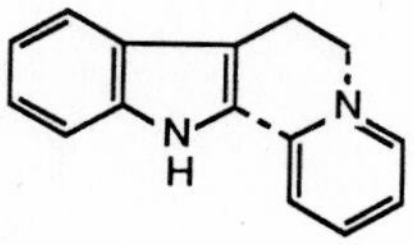

I

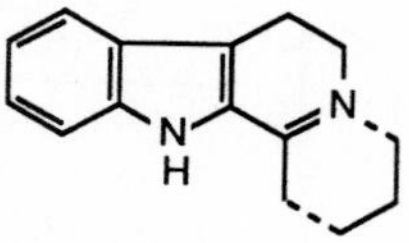

II

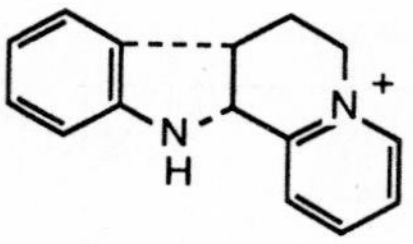

III

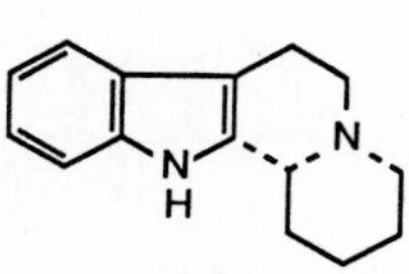

IV

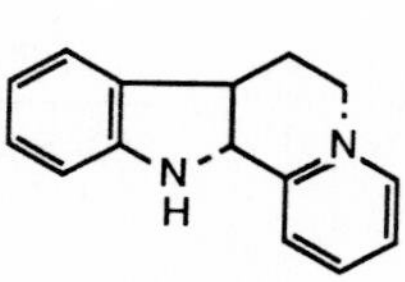

V

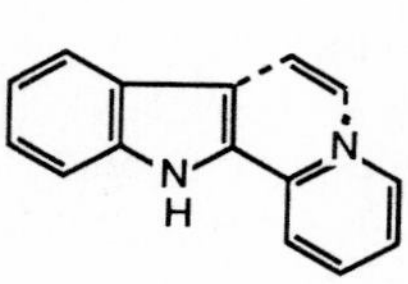

VI

3. TOTAL SYNTHESES

3.1 *Strategy I*

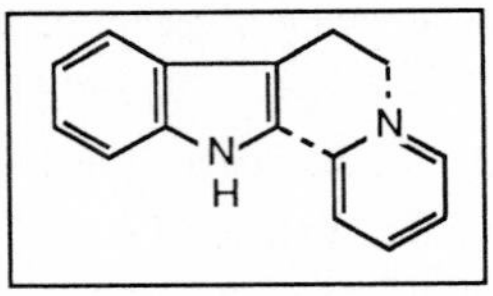

The availability of both 3-(2-substituted-ethyl)indoles (tryptophyl) and the appropriate pyridines and isoquinolines has made this bond-forming strategy the most popular of all.

3.1.1 Elderfield (Alstonilinol, 1958)

The first example of this mode of ring closure, leading to a synthesis of alstonilinol iodide (**43**), a reduction product of alstoniline (**19**), was reported in 1958 by Elderfield and Fischer (ref. 44) (**Scheme 3**). A standard tryptophol synthesis starting with 6-methoxyindole (**36**) and subsequent bromination gave the unstable 2-(6-methoxy-3-indolyl)ethyl bromide (**39**). This was condensed with 5-carbomethoxyisoquinoline (**40**), the synthesis of which is shown in **Scheme 4** (ref. 45), to give isoquinolinium bromide **41**. Application of a very useful and interesting reductive-ring closure with lithium aluminum hydride, first described by Potts and Robinson (ref. 46), gave the pentacyclic tetra-hydroalstonilinol (**42**) in good yield. An efficient oxidation with iodine afforded alstonilinol iodide (**43**), identical with material obtained from alstoniline, via **42** ((a) H$_2$/Pt; (b) LiAlH$_4$). This latter observation confirmed the structure originally proposed for alstoniline (ref. 26).

Scheme 3

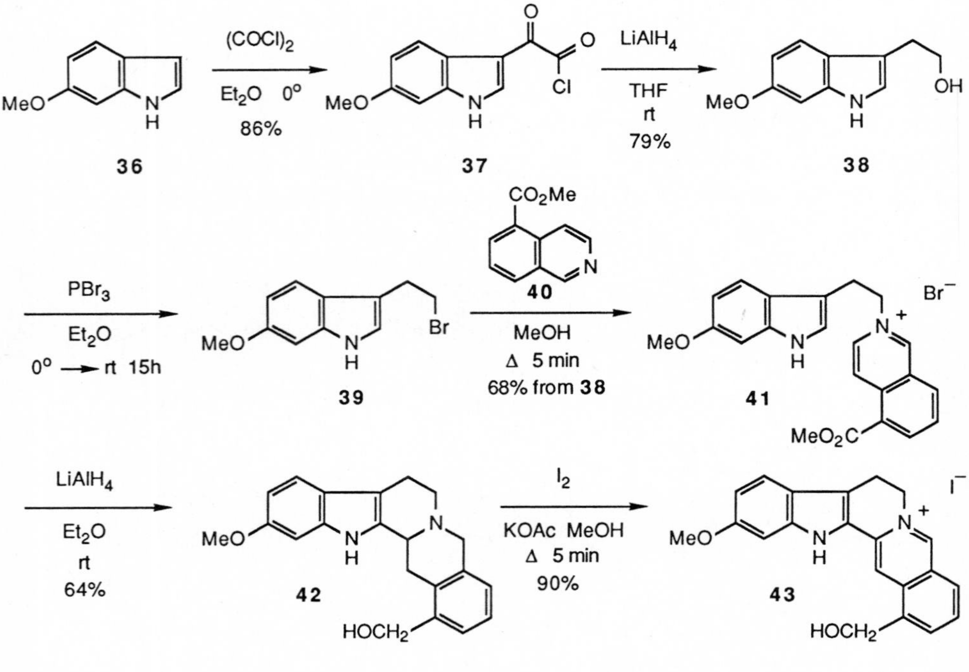

Unfortunately, attempts to utilize this approach in a synthesis of alstoniline itself failed. Thus, oxidation reactions on model compounds similar to **42** and **43** (but lacking the methoxyl group) were unsuccessful. Moreover, attempts to induce cyclization *sans* reduction of the ester group in **41** failed as well.

The crucial 5-carbomethoxyisoquinoline (**40**) was prepared by the method of Tyson (ref. 45) (**Scheme 4**). A Pomeranz-Fritsch isoquinoline synthesis starting with 3-bromobenzaldehyde gave a 1:1 mixture of 5- and 7-bromoisoquinoline. Cyanation and hydrolysis afforded a mixture of carboxy - lic acids, which were separated by fractional crystallization of the sodium salts. The resulting pure acid **47** was converted to 5-carbomethoxyisoquinoline (**40**) in standard fashion.

Scheme 4

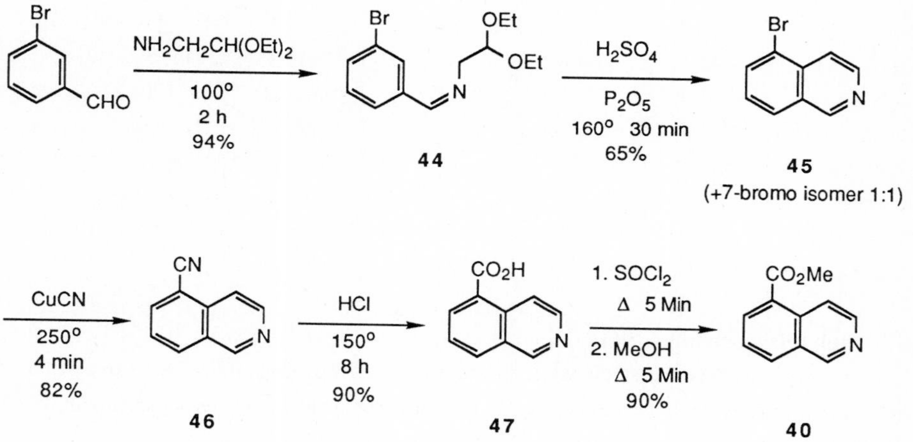

3.1.2 **Thesing** (Flavopereirine, 1959)

The first application of a (modified) Type I strategy to the synthesis of flavopereirine (**3**) was described by Thesing and Festag (ref. 47) (**Scheme 5**). Alkylation of gramine methomethanesulfon - ate (**48**) with N-phenacyl-3-ethylpyridinium bromide (**49**) gave pyridinium salt **50**, which, upon exposure to hydroxide, underwent deacylation to pyridinium salt **51** (ref. 48). Following a procedure of Schöpf (ref. 49), Thesing and Festag hydrogenated **51** to tetrahydropyridine **52**, which, when treated with acid, afforded the tetracyclic indolo[2,3-*a*]quinolizidine derivative **53**. In view of Wenkert's later study (**Scheme 10**), it seems likely that the other isomer was also formed in this reaction. Dehydrogenation with palladium gave flavopereirine (**3**). Unfortunately, yields were not reported and a full account of this work has never appeared.

Scheme 5

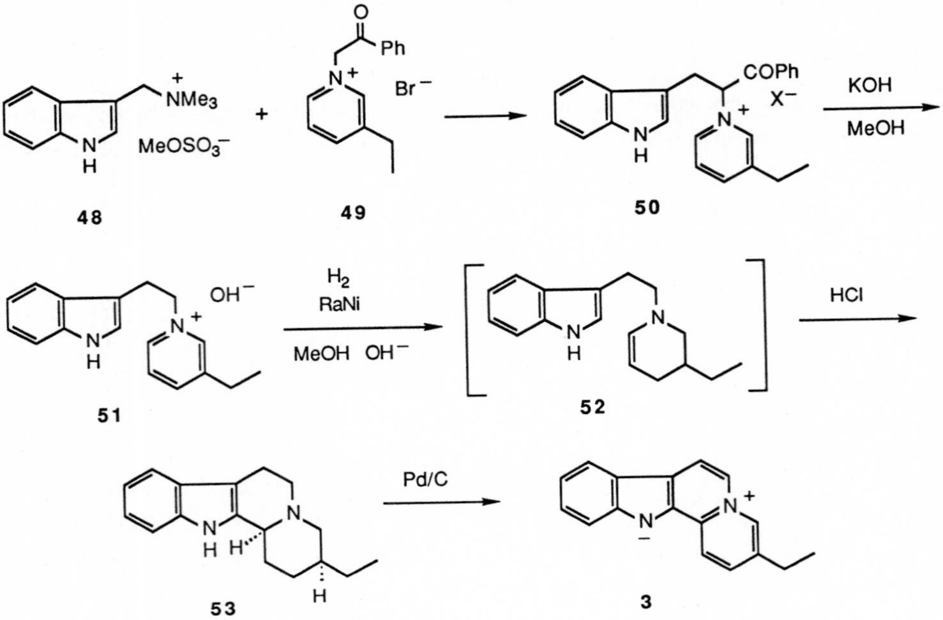

3.1.3 Ban (Flavopereirine, 1961)

An alternative Type I strategy, which has been pioneered by Ban, utilizes a 2-halopyridine in the condensation with tryptophyl bromide, such that subsequent cyclization gives a pyridinium ring rather than a piperidine ring, as was the case earlier (**Schemes 3** and **5**).

In a synthesis of flavopereirine (**3**) (**Scheme 6**), Ban and Seo (ref. 50) converted 5-ethyl-1-methyl-2-pyridone (**54**) (ref. 51) to 2-chloro-5-ethylpyridine (**55**). Reaction of **55** with tryptophyl bromide (**56**), prepared according to an early method using phosphorus tribromide with tryptophol (ref. 52), gave directly the hydrobromide of 6,7-dihydroflavopereirine (**4**). Oxidation of **4** with *ortho*-chloranil (**57**), by the method of Swan (ref. 53), gave flavopereirine (**3**), isolated as the per-chlorate salt.

Scheme 6

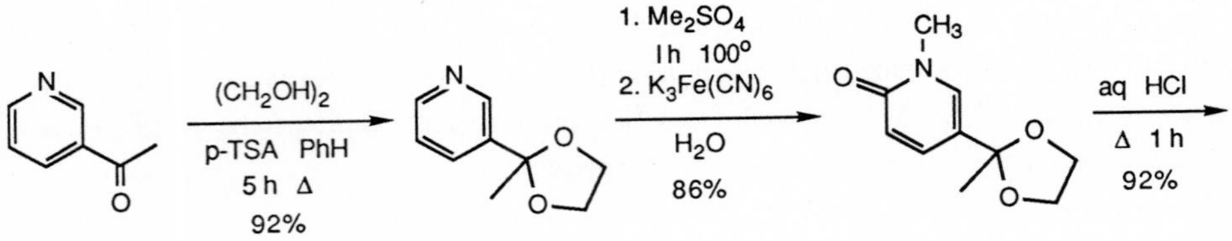

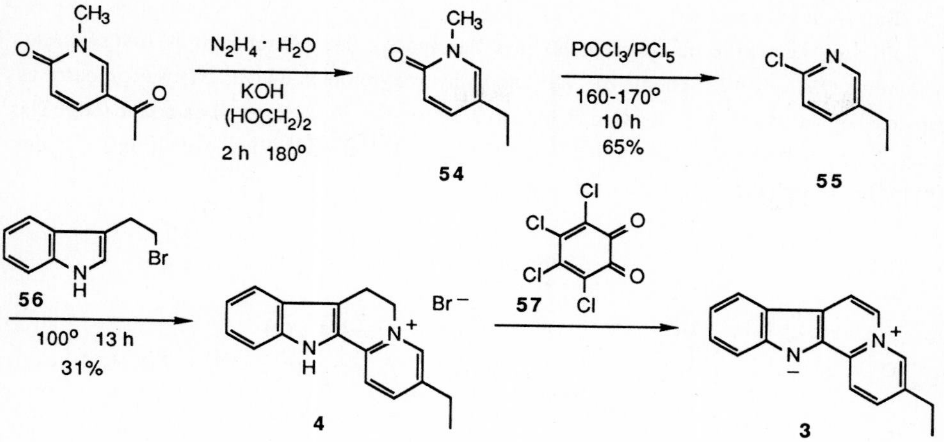

3.1.4 Ban (Sempervirine, 1961)

The same approach (**Scheme 6**) was applied by Ban and Seo (ref. 54) to a synthesis of sem-pervirine (**14**) (**Scheme 7**). Surprisingly, the condensation of tryptophyl bromide (**56**) with 3-chloro-5,6,7,8-tetrahydroisoquinoline (**58**) gave the intermediate chloropyridinium salt **59**, in contrast to what was observed previously by these workers (**Scheme 6**). However, cyclization was achieved to the desired pentacyclic ring system **60** with phosphorus oxychloride in "good yield." In this regard, both aluminum chloride and stannic chloride gave much poorer yields of **60**. Dehydrogenation of **60** with *ortho*-chloranil (ref. 55) gave sempervirine (**14**), isolated as the nitrate salt.

Scheme 7

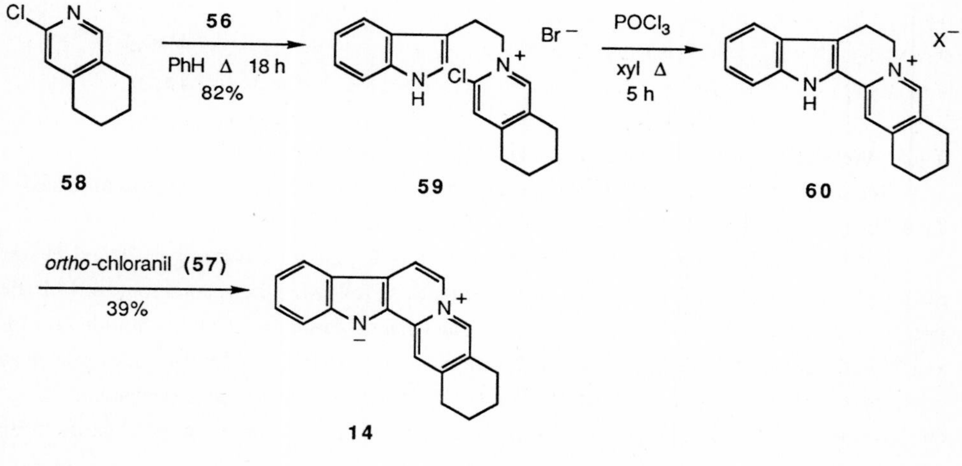

3.1.5 Ban (Alstoniline, 1962)

Following their earlier successes in this area, Ban and Seo (ref. 56) reported the first synthesis of alstoniline (19) in 1962 (Scheme 8). The known hydroxypyridine 64 (ref. 57) was prepared as shown and converted in straightforward fashion to 3-bromo-5-carbomethoxyisoquinoline (68). This was condensed in the usual fashion with 6-methoxytryptophyl bromide (39) to give alstoniline hydro - chloride (19), albeit in minuscule isolated yield.

Scheme 8

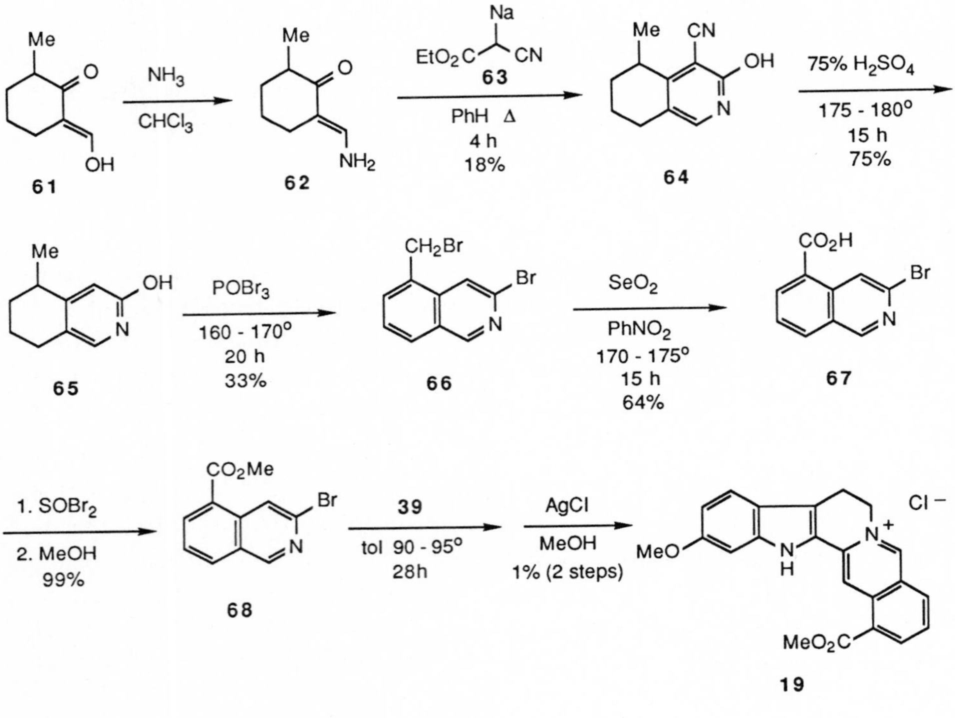

3.1.6 Wenkert (Indolo[2,3-a]quinolizine and Flavopereirine, 1962)

In a series of classic papers, Wenkert and coworkers (refs. 58-60) described several related Type I syntheses of indolo[2,3-a]quinolizine (1) and flavopereirine (3).

In Parts I and II of this series (refs. 58,59), the tricyclic piperidine derivatives 70 and 71 were readily prepared from 3-indoleacetic acid (69) (Scheme 9), following the general route of Elderfield (ref. 61). In Part I of this series (ref. 58), it was found that oxidation of the hydrochloride salts of 70 and 71 with palladium gave directly indolo[2,3-a]quinolizine (1) and flavopereirine (3), isolated as salts, along with the partially oxidized derivatives (72 and 73) and the isomeric compound 74. Unfortunately, the yields of the isolated alkaloids (as picrates, hydrochlorides, or perchlorates) were extremely low.

Scheme 9

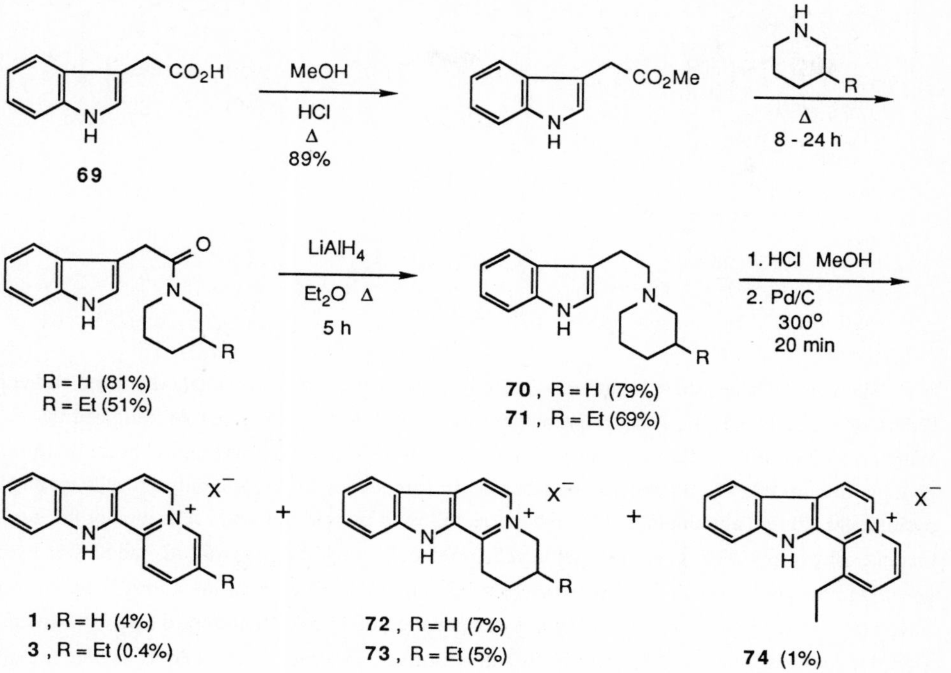

69

R = H (81%)
R = Et (51%)

70, R = H (79%)
71, R = Et (69%)

1, R = H (4%)
3, R = Et (0.4%)

72, R = H (7%)
73, R = Et (5%)

74 (1%)

A better oxidative-cyclization tactic was employed by Wenkert in Part II of his series (ref. 59), in which mercuric acetate was utilized (**Scheme 10**). Thus, piperidine **71** was converted to a mix - ture of octahydroflavopereirines (**53** and **75**) and the isomers **76** and **77** with mercuric acetate in aqueous acetic acid. This oxidation of tertiary amines to immonium ions was developed largely by Leonard (ref. 62), but intensively exploited by Wenkert in the area of indole alkaloid syntheses and transformations. Heating the *trans* isomer **75**·HCl with palladium/charcoal (275°, 10 min) afforded flavopereirine (**3**), isolated as the perchlorate (33% yield).

Scheme 10

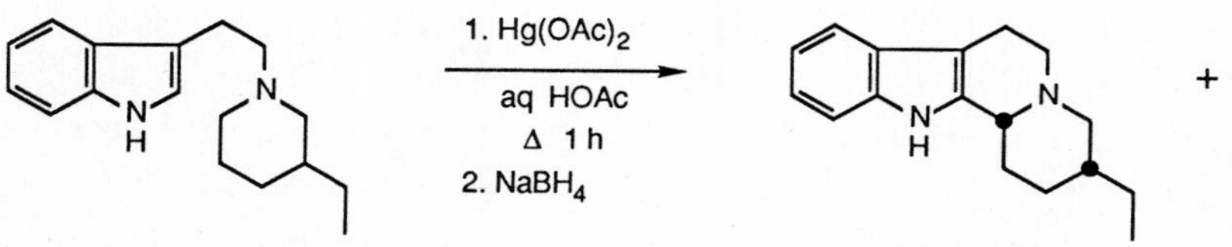

71

53 (21%)

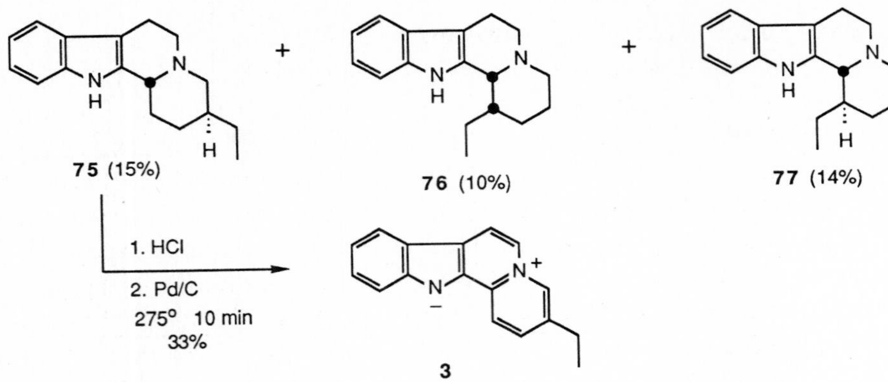

75 (15%)   **76** (10%)   **77** (14%)

1. HCl
2. Pd/C
   275° 10 min
   33%

3

The third Wenkert study in this area (ref. 60) relied on the cyclization method utilized earlier by Elderfield (refs. 44, 61), but, ironically, reported to fail (ref. 61) for the case reinvestigated by Wenkert (**Scheme 11**). Thus, 3-ethylpyridine (**78**) was alkylated with tryptophyl bromide (**56**) to give pyridinium salt **79**. Whereas sodium borohydride reduction of **79** gave only tricyclic tetrahydro-pyridine **80**, lithium aluminum hydride reduction, followed by acid-induced cyclization of the inter-mediate dihydropyridine, gave some tetracyclic product **81**, in addition to **80**. Amine **81** was hydro-genated to the known octahydroflavopereirines (**53** and **75**), thus representing a formal synthesis of flavopereirine (**3**) in view of the earlier dehydrogenation of **75** to **3** (**Schemes 5 and 10**). Inter-estingly, dehydrogenation of **81** afforded 6,7-dihydroflavopereirine (**4**), isolated as its perchlorate.

Scheme 11

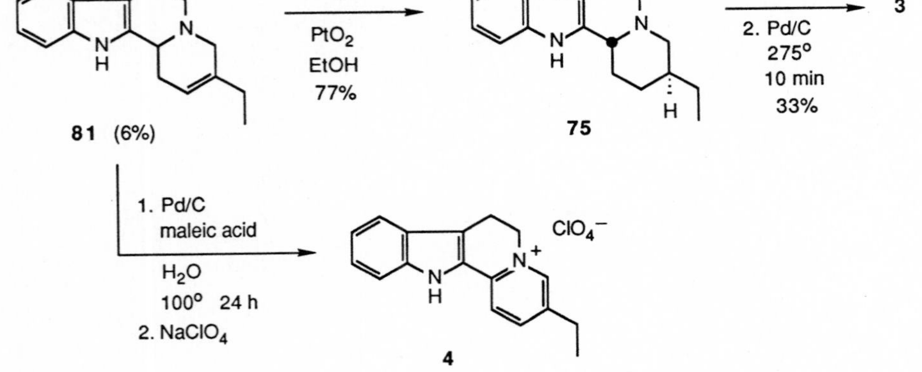

78

56
80° 8 h

79

Br⁻

1. LiAlH₄
   Et₂O
2. 1N HCl

80 (48%)

+

81 (6%)

H₂
PtO₂
EtOH
77%

75

1. HCl
2. Pd/C
   275°
   10 min
   33%

3

1. Pd/C
   maleic acid

   H₂O
   100° 24 h
2. NaClO₄

ClO₄⁻

4

3.1.7 Büchi (Flavocarpine and Flavopereirine, 1962)

In an elegant, complete study, Büchi, Manning, and Hochstein (ref. 8) isolated, characterized, and synthesized flavocarpine (5) (**Scheme 12**). Following a standard method for preparing cyano - pyridines (ref. 63), Büchi and coworkers converted 3-ethylpyridine *N*-oxide into 4-cyano-3-ethyl - pyridine (82). The structure of 82 was established by conversion to known 3,4-diethylpyridine (by a Grignard reaction followed by Wolff-Kishner reduction). Oxidation and chlorination gave a 1:1 mix - ture of chloropyridines 83 and 84. The desired isomer 83, which could be separated from 84 by distillation, was hydrolyzed to amide 85. A Ban condensation (*vide supra*) of 85 with tryptophyl bromide (56) gave the expected tetracycle 86 in low yield. Oxidation of the C-ring with *ortho*-chloranil (57) (ref. 53) and hydrolysis of the amide afforded flavocarpine (hydrochloride) 5. Since 5 can be decarboxylated to flavopereirine (3), this work represents a synthesis of flavopereirine as well.

Scheme 12

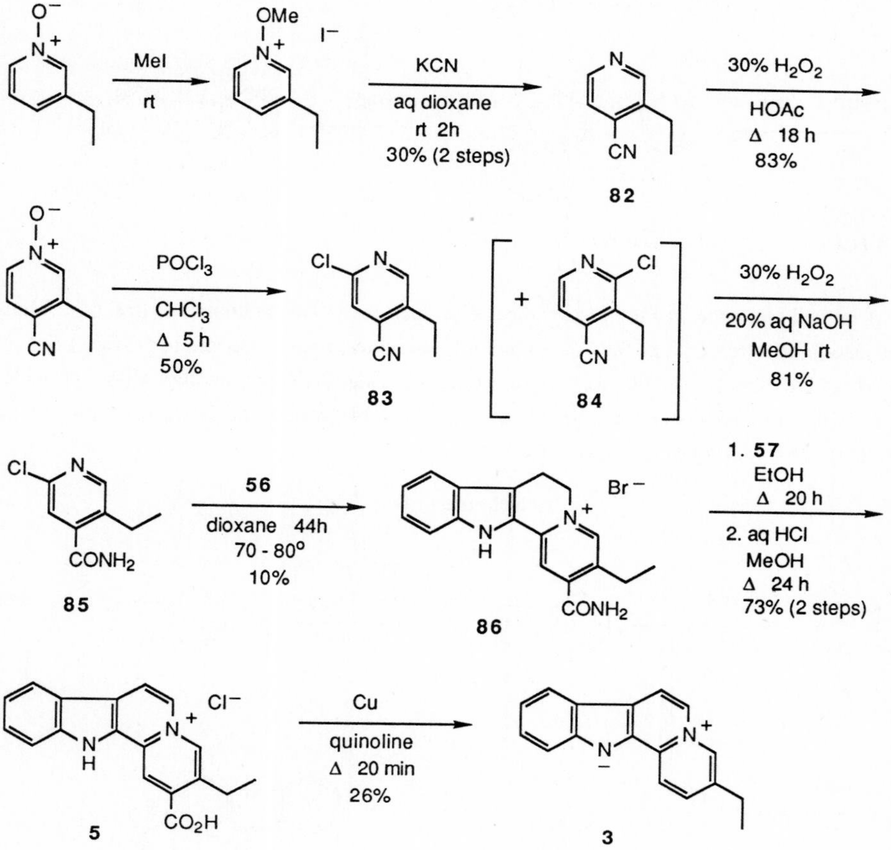

3.1.8 Potts (Flavopereirine and Sempervirine, 1963)

In another reductive cyclization maneuver, Potts and Liljegren (ref. 64) (Scheme 13) synthe - sized flavopereirine (3). Thus, heating a mixture of 3-acetylindole (87), iodine, and 3-ethylpyridine (78) gave keto iodide 88. Reduction of this material with lithium aluminum hydride in tetrahydro - furan (but *not* in ether, which gave no cyclized product) afforded 81. Using the procedure of Le Hir and coworkers (ref. 65), Potts converted 89 into flavopereirine (3).

Scheme 13

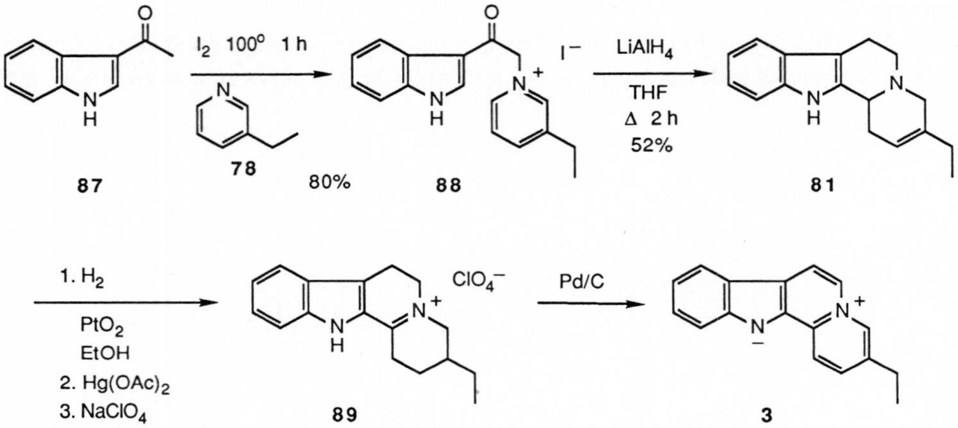

A related sequence was used to synthesize sempervirine (14) (Scheme 14) (ref. 64). Thus, 3-chloroacetylindole (ref. 66) and 5,6,7,8-tetrahydroisoquinoline were condensed to provide a quantita - tive yield of the expected salt 90. Reductive-cyclization under the usual conditions gave 91, which apparently has been previously converted to sempervirine (14) by catalytic dehydrogenation, although the exact method was not reported (ref. 67).

Scheme 14

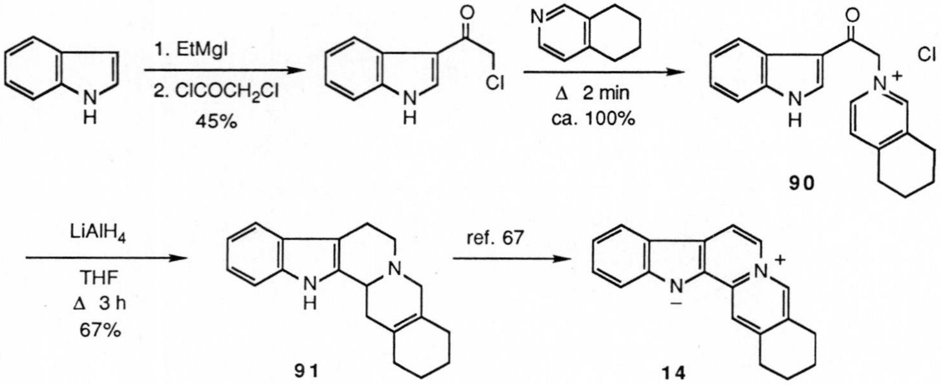

3.1.9 Ban (Desethylflavocarpine, 1968)

The only other published route to a flavocarpine type structure was reported by Ban and Kimura in a synthesis of the desethyl derivative (93) (ref. 68) (Scheme 15). Reaction of tryptophyl bro - mide (56) with methyl 2-chloroisonicotinate gave a fair yield of the anticipated tetracycle 92. Oxida - tion with *ortho*-chloranil (57) and hydrolysis of the ester function led to the hydrochloride salt of desethylflavocarpine. Percolation through an ion exchange resin furnished desethylflavocarpine (93).

Scheme 15

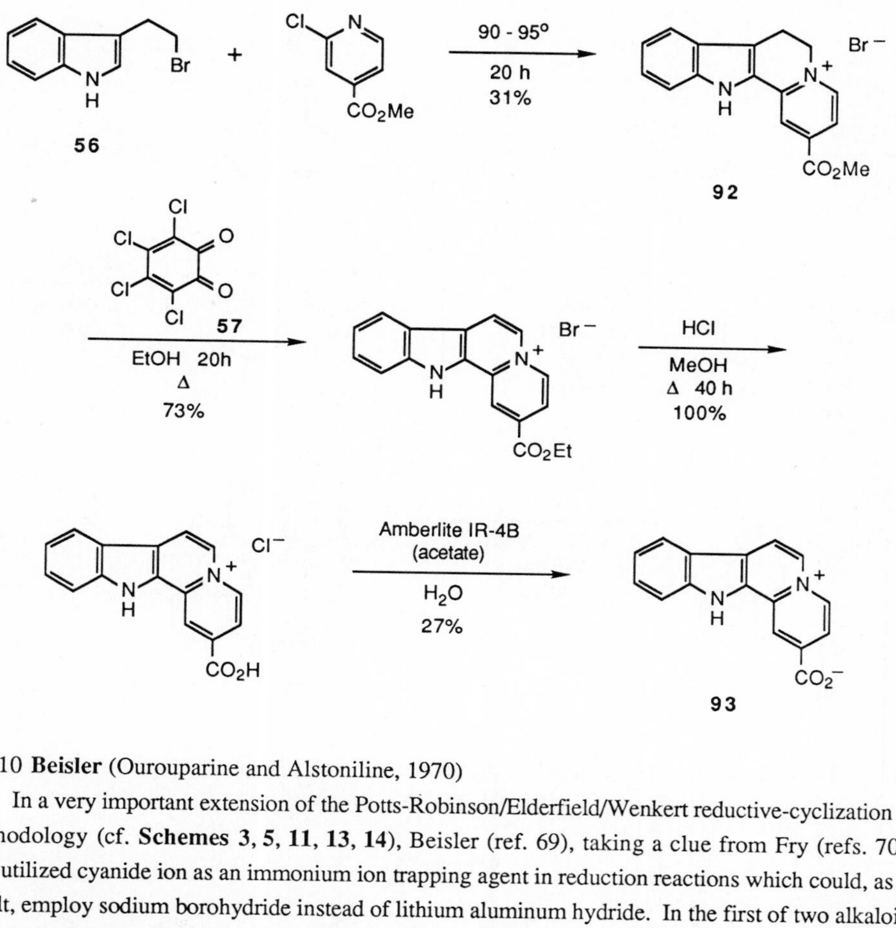

3.1.10 Beisler (Ourouparine and Alstoniline, 1970)

In a very important extension of the Potts-Robinson/Elderfield/Wenkert reductive-cyclization methodology (cf. Schemes 3, 5, 11, 13, 14), Beisler (ref. 69), taking a clue from Fry (refs. 70, 71), utilized cyanide ion as an immonium ion trapping agent in reduction reactions which could, as a result, employ sodium borohydride instead of lithium aluminum hydride. In the first of two alkaloid syntheses, Beisler prepared ourouparine iodide (20) (Scheme 16). Isoquinoline was converted in good yield to 5-carbomethoxyisoquinoline (40), a procedure which represents an improvement over the Tyson synthesis (ref. 45) (Scheme 4). Condensation of 40 with tryptophyl bromide (56) as described (ref. 44) gave salt 94. Upon treatment with sodium borohydride/sodium cyanide and then acid, the salt 94 was smoothly converted into dihydrogambirtannine (96) in very good yield, pre -

sumably via the cyanopiperidine **95**. Oxidation of **96** using a procedure similar to that described by Elderfield and Fischer (ref. 44) afforded ourouparine iodide (**20**). This particular oxidation has also been described by Merlini and Nasini (ref. 72) in a partial synthesis of **20**.

Scheme 16

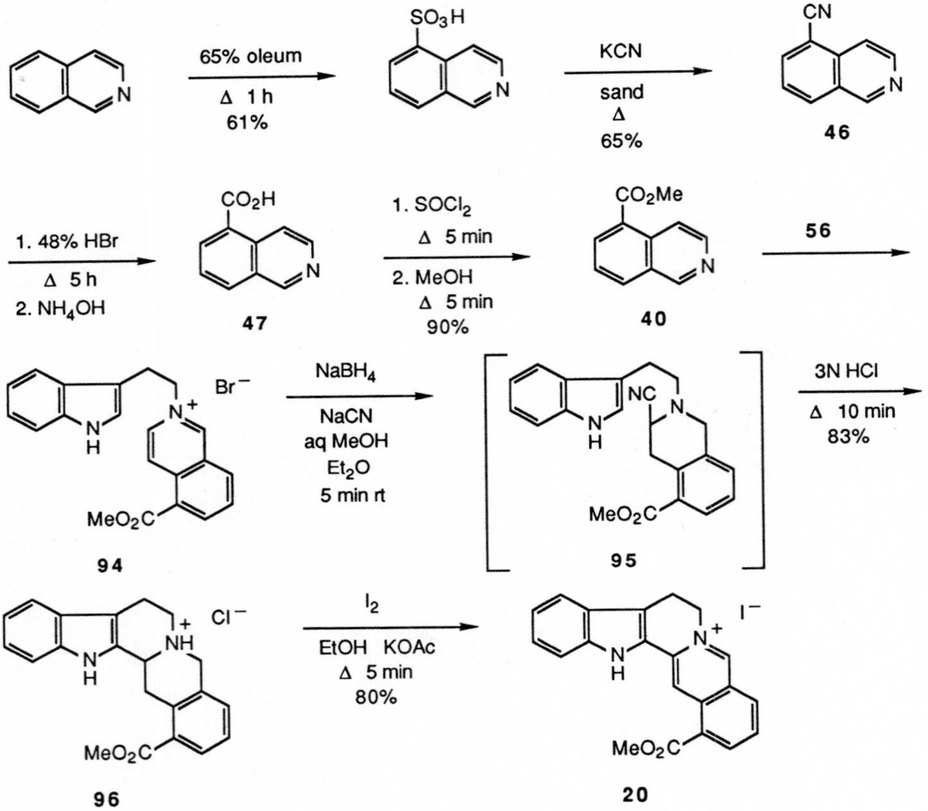

Application of this novel reductive-cyclization to the known isoquinolinium bromide **41** (ref. 44) led to a synthesis of alstoniline (**19**) (**Scheme 17**) (ref. 69b).

Scheme 17

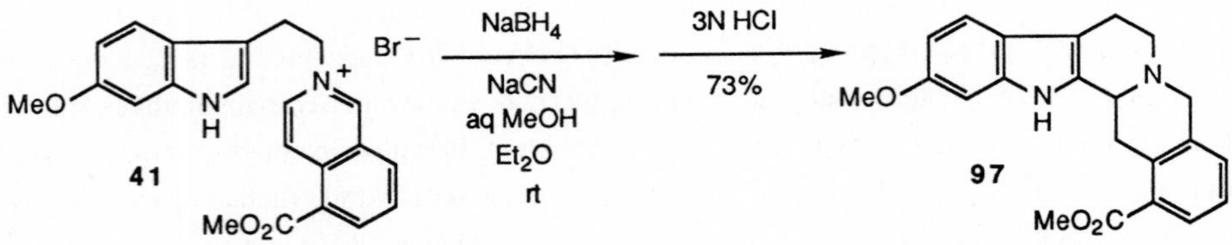

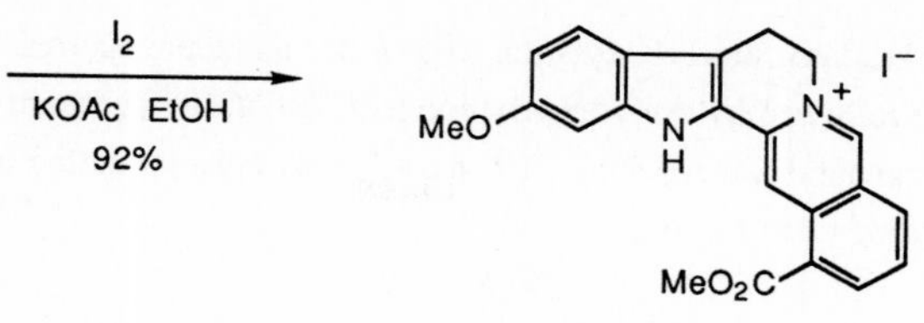

19

3.2 Strategy II

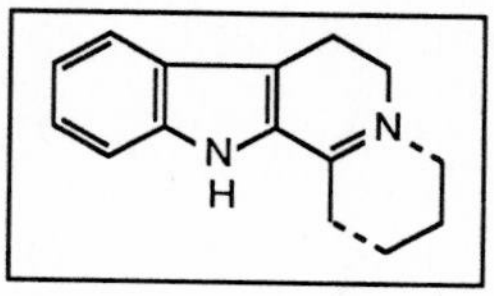

Another logical approach to the indolo[2,3-*a*]quinolizine ring system is a strategy in which a 1-methyl-β-carboline (harman) is used as the foundation for synthesis. Indeed, several research groups have pursued this line of research.

3.2.1 Woodward (*N*-Methylsempervirine, 1949)

Simultaneously with the publication of the structure of sempervirine (**14**) (ref. 14b), Woodward and McLamore announced the synthesis of the *N*-methyl derivative of this alkaloid (ref. 73) (**Scheme 18**). Thus, although the details are lacking, *N*-methyltryptophan (**98**) was converted to *N*-methylhar - man (**99**) by a known procedure (ref. 74) and the latter compound, as the lithium derivative, was con - densed with 2-isopropoxymethylenecyclohexanone (**100**) to give the salt of *N*-methylsempervirine (**101**).

Scheme 18

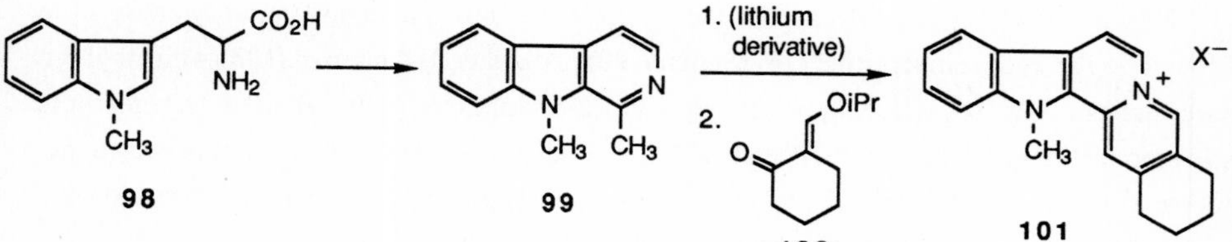

98   **99**   **100**   **101**

Although it has been claimed (ref. 75) that Woodward also synthesized sempervirine (**14**), an account of this has never appeared.

3.2.2 Bradsher (1,2,3,4-Tetradehydrosempervirine, 1963)

Bradsher and Umans reported a synthesis of the fully dehydrogenated pentacycle **107** corres - ponding to sempervirine (ref. 76) (**Scheme 19**). The key β-carboline aldehyde **105** was prepared in five steps from tryptophan, a sequence that featured a Pictet-Spengler reaction with methoxyacetal -

dehyde followed by oxidative decarboxylation to β-carboline ether **103** (ref. 77). Alkylation of **105** with benzyl bromide followed by cyclodehydration (ref. 78) of **106**, brought about by polyphos - phoric acid (PPA), gave the pentacyclic 13*H*-benz[*g*]indolo[2,3-*a*]quinolizinium perchlorate **107** in good overall yield from aldehyde **105**.

Scheme 19

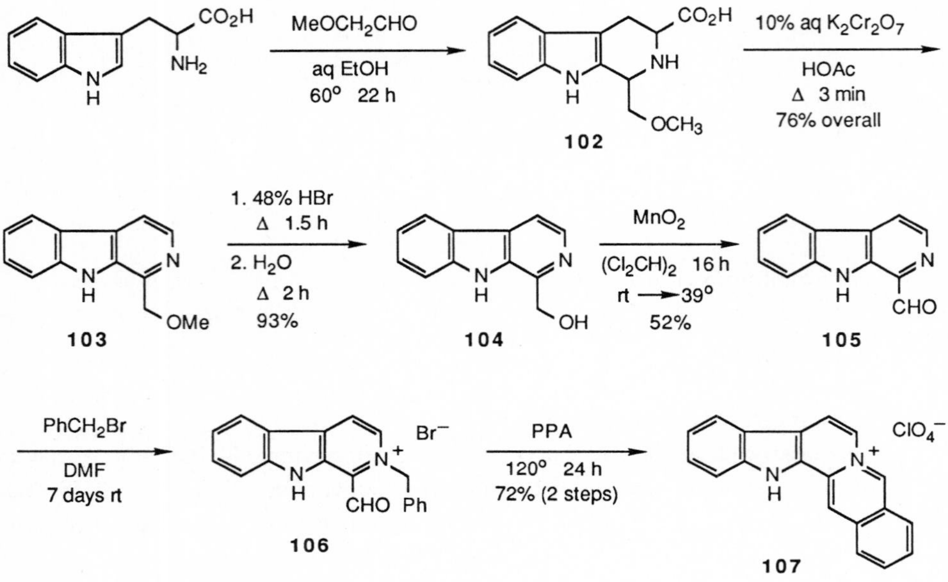

3.2.3 **Potts** (Sempervirine, 1968)

Potts and Mattingly (ref. 79) applied the Westphal (ref. 80) quinolizinium salt synthesis to an efficient synthesis of sempervirine (**14**) (Scheme 20). Alkylation of harman (**108**) with ethyl bro - moacetate gave salt **109**, which, upon heating with 1,2-cyclohexanedione, afforded the pentacyclic ester bromide **110**. Hydrolysis and decarboxylation proceeded smoothly to give sempervirine (hy - drobromide) **14**. Unfortunately, the more direct route to sempervirine involving condensation of harman methiodide with 1,2-cyclohexanedione failed.

Scheme 20

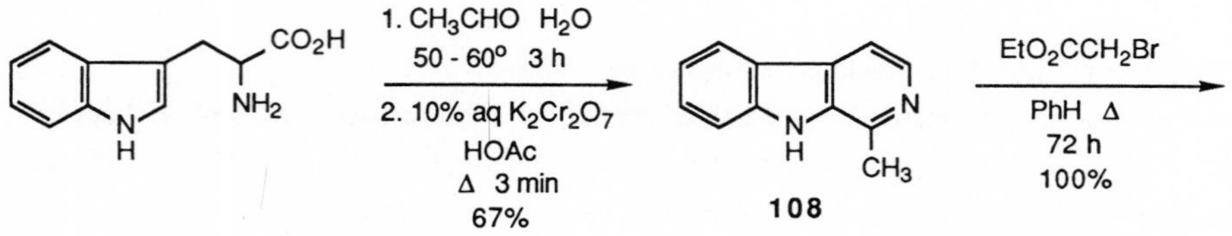

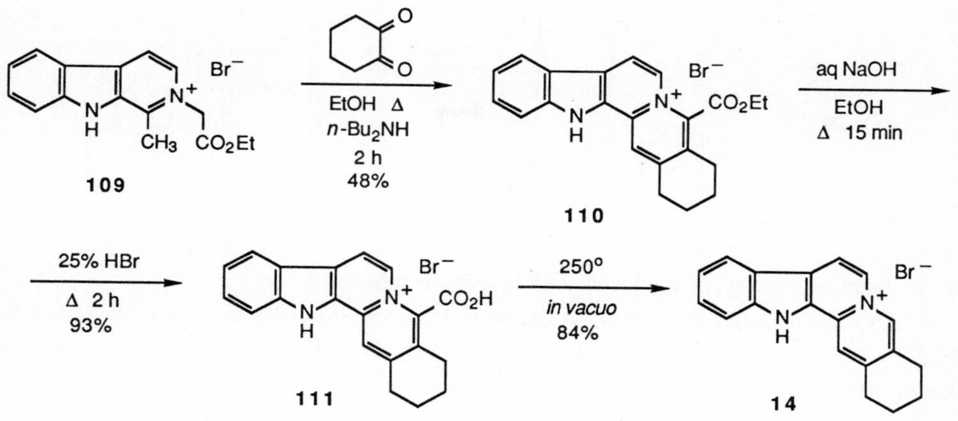

3.2.4 Ninomiya (Flavopereirine, 1978)

In a photochemical Type II strategy, Ninomiya and coworkers (ref. 81) achieved a novel syn -
thesis of flavopereirine (**3**) (**Scheme 21**). Harmalane (**112**) was allowed to react with 2-ethyl-3-
methoxyacryloyl chloride and the resulting unstable enamide **113** was directly photolyzed with a high
pressure mercury lamp to give tetracyclic lactam **115**, after workup with dilute acid. This was con -
verted to flavopereirine (perchlorate) **3** by lithium aluminum hydride reduction and dehydrogenation
with palladium. Interestingly, the photocyclization of the enamide derived from harmalane (**112**) and
2-ethylacryloyl chloride proceeded extremely slowly and afforded only 7.5% cyclized material, thus
indicating that a methoxyl group is necessary for (relatively) efficient photocyclization.

Scheme 21

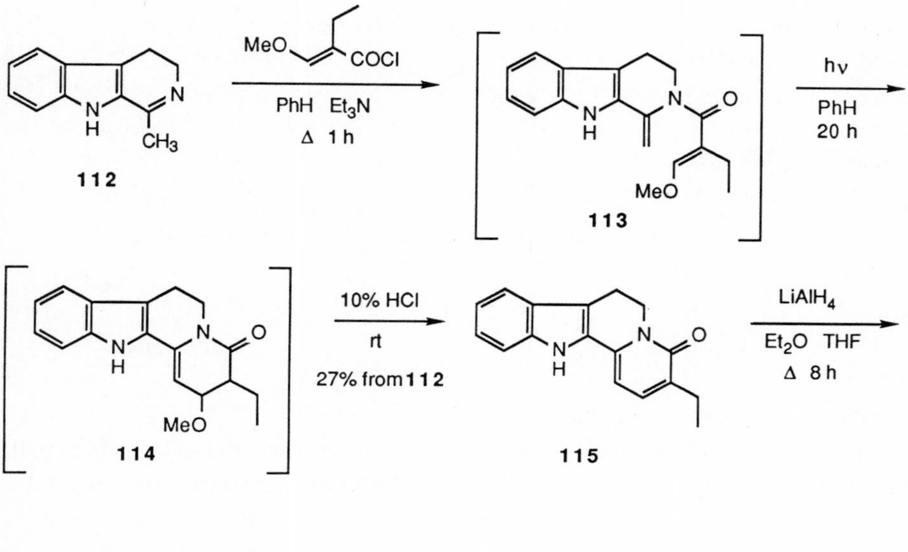

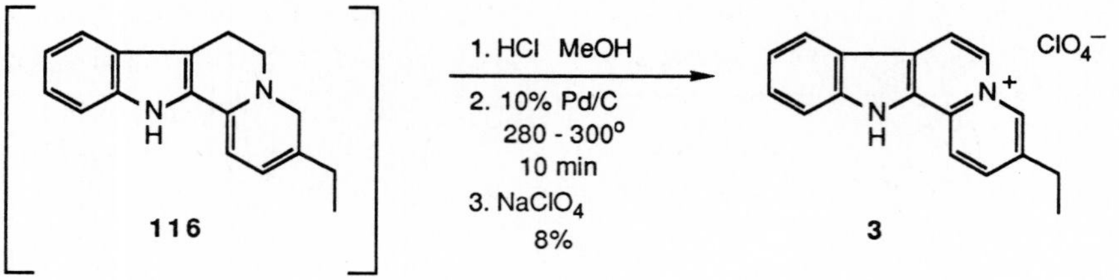

3.2.5 Danieli (Indolo[2,3-a]quinolizine and Flavopereirine, 1980)

Also starting with harmalane (112), Danieli and coworkers (ref. 82) developed very simple and highly efficient syntheses of indolo[2,3-a]quinolizine (1) and flavopereirine (3) (Scheme 22). The key element of this approach relies on the ambident nucleophilic reactivity of harmalane (112), as shown in the tautomeric equilibrium below.

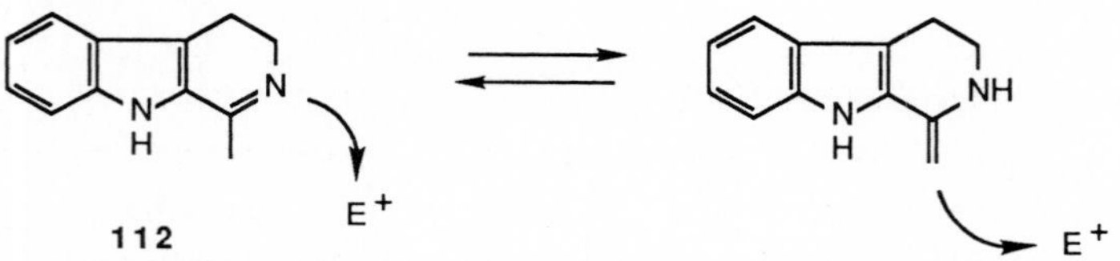

Thus, reaction of harmalane (112) with 1,3-dibromopropane gave in one step the tetracyclic immonium ion salt 117. Stepwise dehydrogenation of 117 afforded indolo[2,3-a]quinolizine (picrate) 1 in excellent overall yield. A similar sequence starting with 1-bromo-2-(bromomethyl) - butane gave flavopereirine (perchlorate) (3) in very good overall yield.

Scheme 22

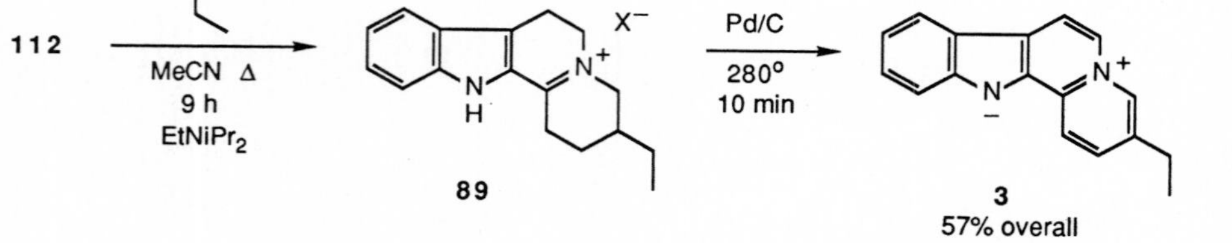

3.2.6 Pakrashi (6,7-Dihydroflavopereirine and Flavopereirine, 1984)

In a recent synthetic effort that incorporates some of each of Ninomiya's and Danieli's Type II strategies (*vide supra*), Pakrashi and coworkers (ref. 83) reported efficient syntheses of both 6,7-di -

hydroflavopereirine (4) and flavopereirine (3) (**Scheme 23**). Thus, taking advantage of the 1,3 - (bis)nucleophilicity of harmalane (**112**), Pakrashi prepared lactam **118** in good yield using ethyl ethoxymethyleneacetoacetate as the 1,3(bis)electrophile. A Wolff-Kishner reduction gave the Ninomiya lactam **115**, which, upon controlled reduction with lithium aluminum hydride, gave 6,7-dihydroflavopereirine **4**. Oxidation with DDQ gave flavopereirine (**3**).

Scheme 23

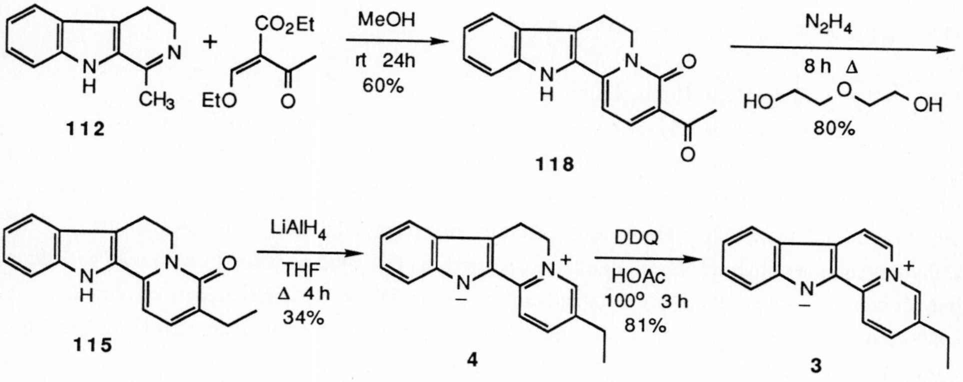

3.2.7 **Szantay** (Flavopereirine, 1963)

In a subvariant of the basic Type II strategy, Szantay and Toke (ref. 84) prepared *trans*-octahy - droflavopereirine (**75**) from β-carboline **119** (**Scheme 24**). Thus, condensation of **119** with enone **120** gave tetracyclic ketone **121**. Wolff-Kishner reduction gave *trans*-octahydroflavopereirine (**75**), which had previously been converted to flavopereirine (**3**) (**Scheme 10**).

Scheme 24

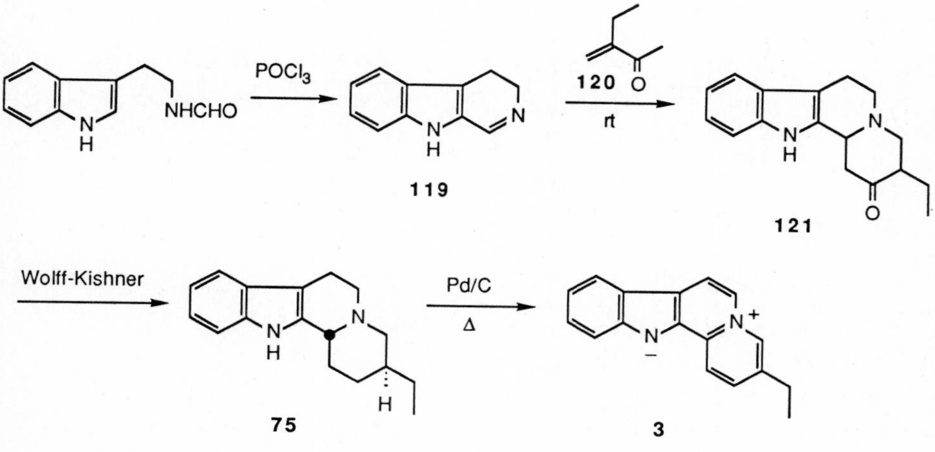

3.3 Strategy III

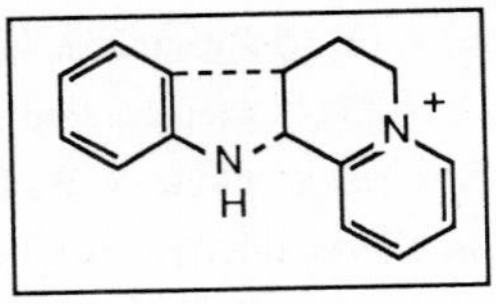

Another popular route to the indolo[2,3-a]quinolizine alkaloids has been via a Fischer-indole cyclization. An inherent advantage of this approach is that ring D is normally fully oxidized, thus avoiding the often difficult dehydrogenation step.

3.3.1 Glover (Indolo[2,3-a]quinolizine, 1958)

Glover and Jones (ref. 85) were the first chemists to describe the synthesis of an indolo[2,3-a] - quinolizine using a Fischer indole synthesis (**Scheme 25**). The known pyridine **122** (refs. 53, 86) was cyclized and converted to phenylhydrazone iodide **124**. This was smoothly converted to 6,7-dihydroindolo[2,3-a]quinolizinium iodide (**125**) by anhydrous ethanolic hydrogen chloride. Dehydrogenation to indolo[2,3-a]quinolizine (hydriodide) (**1**) could be accomplished with palladium, but not with mercuric acetate. By the same sequence of reactions, these researchers converted 3-cyanoisoquinoline into benzindolo[2,3-a]quinolizinium salt **126**, but were unable to effect dehydro - genation to Bradsher's compound **107** (**Scheme 19**) with Pd/C at 330°.

Scheme 25

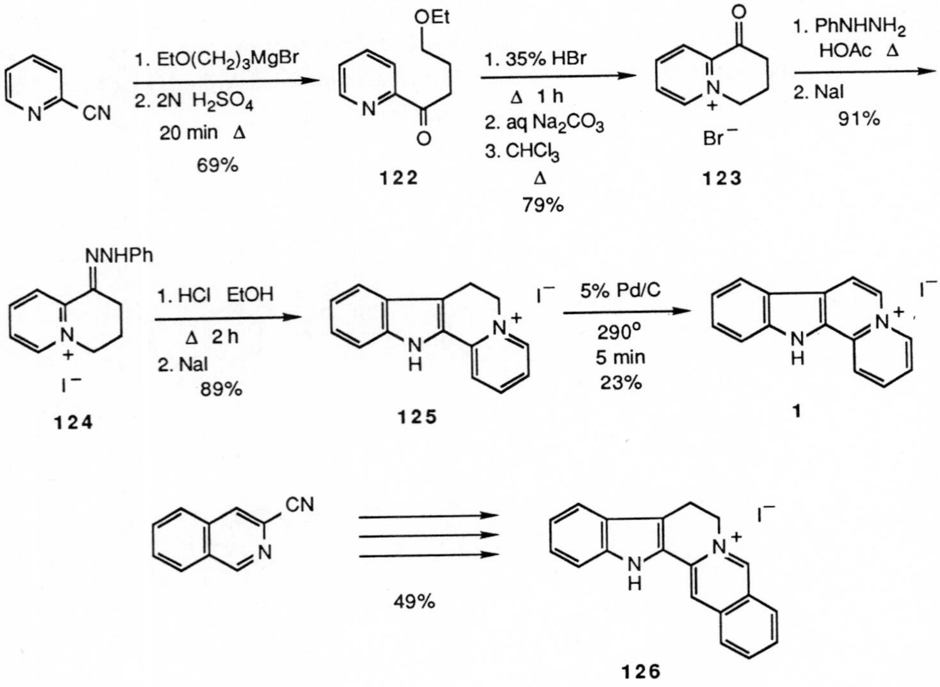

3.3.2 Swan (Indolo[2,3-*a*]quinolizine, Flavopereirine, and Sempervirine, 1958)

Apparently independently, Swan (refs. 53, 55) used exactly the same route as did Glover and Jones to synthesize flavopereirine and sempervirine (**Schemes 26-28**). Thus, in the first paper (ref. 53), Prasad and Swan report the isolation of imine **127**, rather than ketone **122** as described by Glover (ref. 85) and Craig (ref. 86), from the reaction between 2-cyanopyridine and 3-ethoxypropyl magnesium bromide. In any event, cyclization of **127** and a Fischer-indole sequence gave 6,7-dihydroindolo[2,3-*a*]quinolizinium chloride (**125**). Although oxidation of **125** with palladium or *para*-chloranil failed, dehydrogenation of **125** to indolo[2,3-*a*]quinolizine (hydrochloride) (**1**) was achieved with *ortho*-chloranil.

Scheme 26

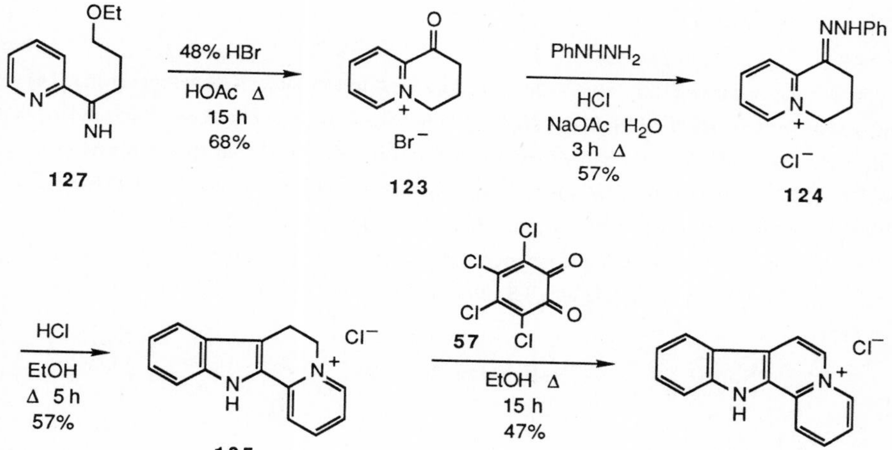

The same procedures were applied to a synthesis of flavopereirine (**3**) (**Scheme 27**), although the preparation of 2-cyano-5-ethylpyridine (**130**) was lengthy. Reaction of **130** with the usual Grignard reagent and the same series of reactions led to dihydroflavopereirine (**4**) and flavopereirine (perchlorate) (**3**).

Scheme 27

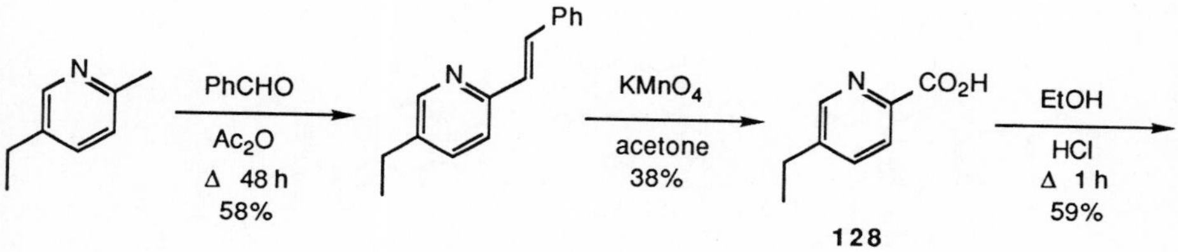

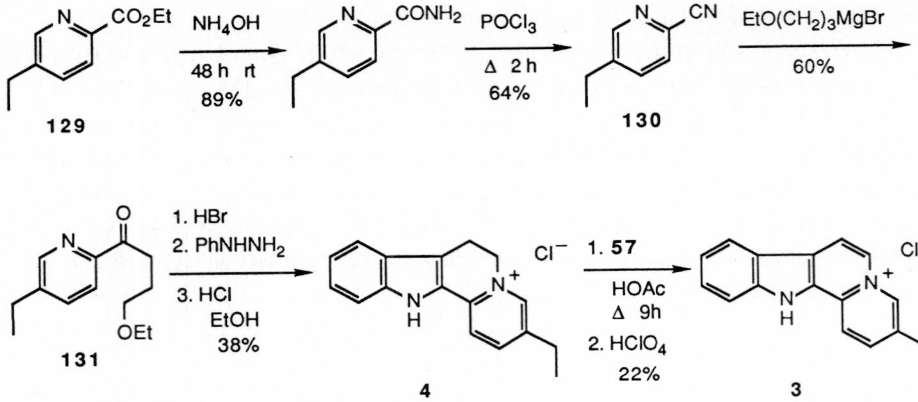

In the second paper (ref. 55), Swan announced the first total synthesis of sempervirine (**14**) using the same Type III strategy (**Scheme 28**). Starting with 5,6,7,8-tetrahydro-3-hydroxyiso - quinoline ester **132**, which was probably prepared in a fashion similar to that shown for **64** in **Scheme 8** (ref. 57), Swan synthesized nitrile **135**, relying on unpublished chemistry of Stevens (ref. 87) (cf., **Scheme 34**) to convert **134** into **135**. Thereafter, the same chemistry was used uneventfully to transform nitrile **135** into sempervirine (nitrate) (**14**). Swan also reported a similar transformation of 3-cyanoisoquinoline into 7,8-dihydro-13H-benz[g]indolo[2,3-a]quinolizinium chloride (**126**).

Scheme 28

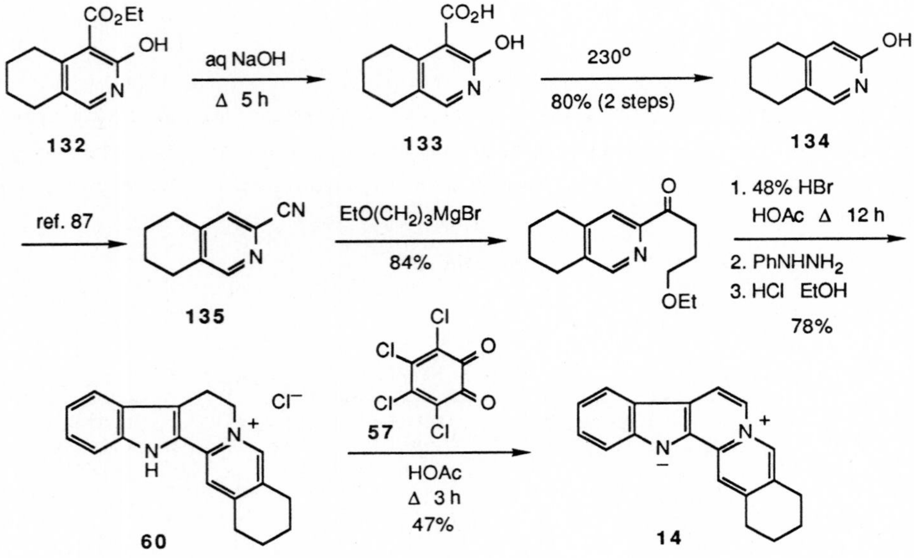

3.3.3 Kaneko (Flavopereirine, 1960)

During studies on the structure of serpentinine (28), Kaneko reported a synthesis of flavopereirine (3) using the same Fischer indole strategy (ref. 88) (Scheme 29) previously done by Swan (ref. 53) (Scheme 27). Thus, picolinic acid 128 was converted into the corresponding acid chloride, which, upon reaction of the requisite organocadmium reagent (made from the Grignard reagent), gave ketone 131. Fischer cyclization produced the unusual result that the tetracycle 136 contained an extra bromine, of unknown location in the molecule. Dehydrogenation and then catalytic hydrogenation (to remove the extra bromine) gave flavopereirine (3). Alternatively, hydrogenation of 136 gave 6,7-dihydroflavopereirine (4).

Scheme 29

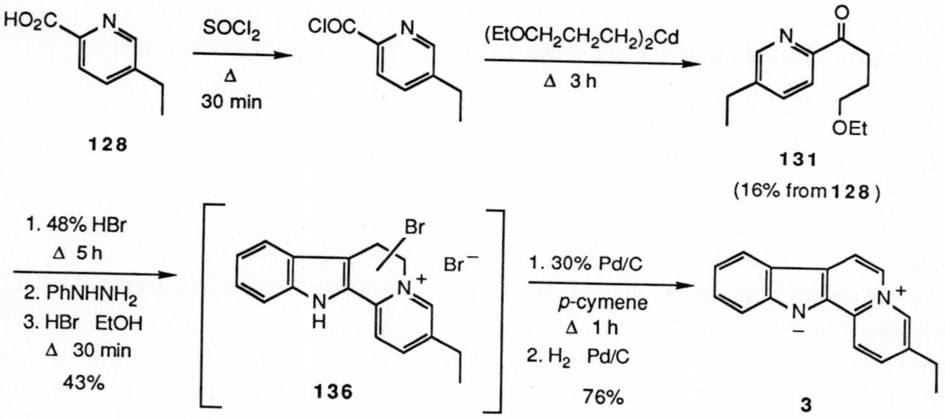

3.3.4 Prasad (1-Methylindolo[2,3-a]quinolizine, 1973)

More recently, Prasad and Shaw (ref. 89) have utilized the same synthetic plan discussed above (Scheme 27) to synthesize the 1-methylindolo[2,3-a]quinolizine analogues 137 and 138, starting from 2-cyano-3-methylpyridine.

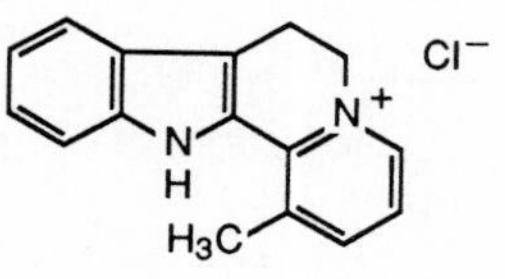

137

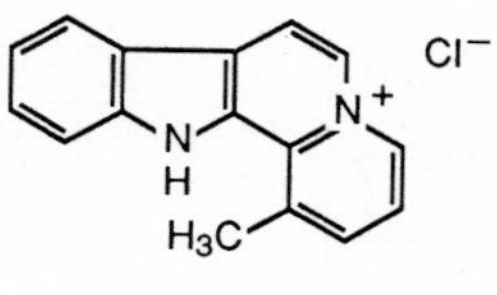

138

3.4 *Strategy IV*

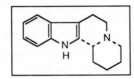

A general strategy that has seen more use in the construction of indolo[2,3-*a*]quinolizidine alka-loids than in the synthesis of indolo[2,3-*a*]quinolizine alkaloids is Strategy IV, in which a Pictet-Spengler or Bischler-Napieralski condensation with tryptamine forms the key step in the synthesis.

3.4.1 Swan (1,2,3,4-Tetrahydroindolo[2,3-*a*]quinolizine, 1952)

In a landmark paper that described one of the first syntheses of the important tetracyclic indolo-[2,3-*a*]quinolizidine ring system (e.g., **139** and **140**), Groves and Swan (ref. 90) reported the syn-thesis of 1,2,3,4-tetrahydroindolo[2,3-*a*]quinolizine (**72**) using a Type IV strategy (**Scheme 30**).

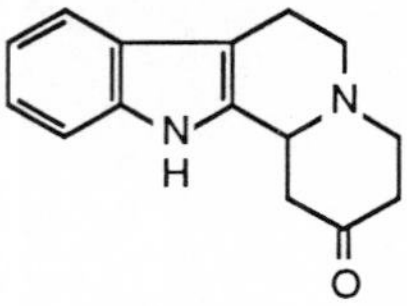

139

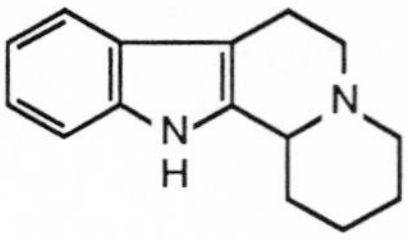

140

A Pictet-Spengler condensation (ref. 91a) between tryptamine hydrochloride (**141**) and 5-hydroxypentanal (**142**) gave hydroxycarboline **143**. Dehydrogenation and cyclization occurred smoothly to give the bromide **146**. Treatment of **146** with hydroxide furnished 1,2,3,4-tetrahydro-indolo[2,3-*a*]quinolizine (**72**). Attempted dehydrogenation of **143** to **144** with lead tetraacetate, chloranil, or Raney nickel failed.

Scheme 30

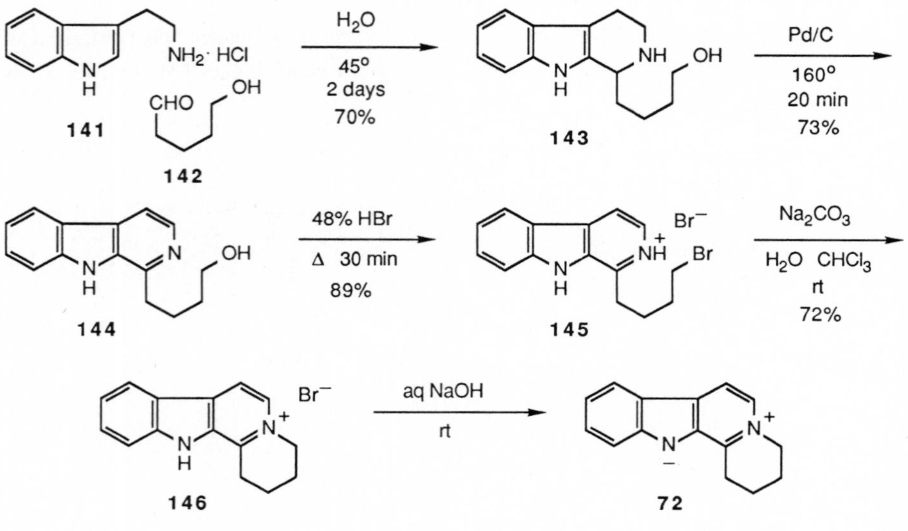

141

142

H_2O
45°
2 days
70%

143

Pd/C
160°
20 min
73%

144

48% HBr
Δ 30 min
89%

145

Na_2CO_3
H_2O $CHCl_3$
rt
72%

146

aq NaOH
rt

72

3.4.2 Le Hir (Flavopereirine, 1958)

One of the first syntheses of flavopereirine (3) utilized a Bischler-Napieralski condensation (ref. 91b) in the key step (**Scheme 31**). Thus, Le Hir and coworkers (ref. 65) prepared bromo ester **148** in a lengthy sequence starting from diethyl ethylmalonate (**147**). Reaction of **148** with tryptamine (**141**) afforded lactam **149**. Cyclization with phosphorus oxychloride gave the tetracyclic immonium salt **89**. Dehydrogenation with palladium afforded flavopereirine (**3**).

Scheme 31

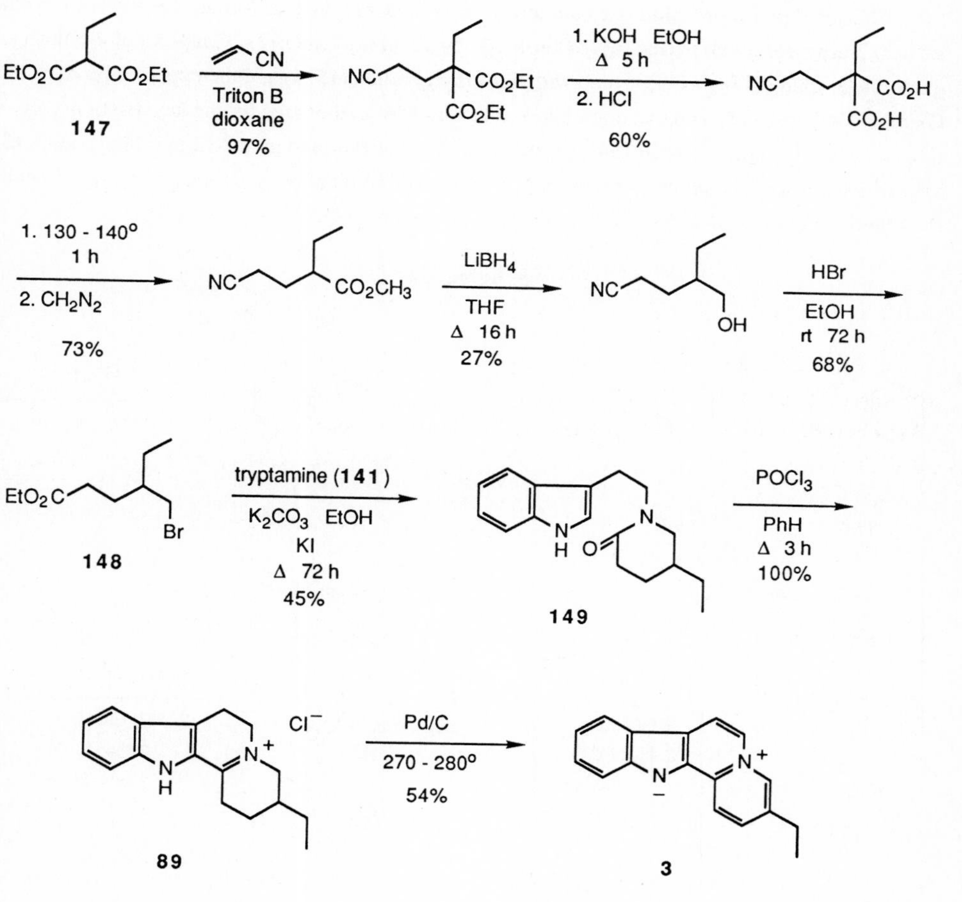

3.5   *Strategy V*

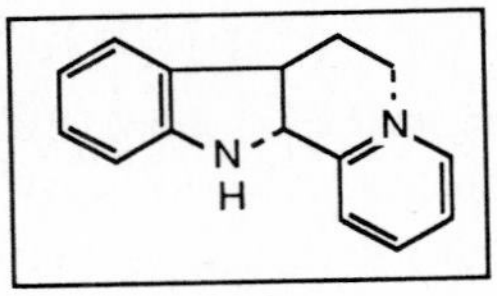

The relatively novel bond disconnections shown in the box above have been utilized by one investigator.

3.5.1   **Takahashi** (6,7-Dihydroflavopereirine, 1971)

The novel ring contraction of a quinoline ring to an indole ring has formed the basis of a synthe - sis of 6,7-dihydroflavopereirine (**4**) (**Scheme 32**) by Takahashi (ref. 92). Coupling of 4-chloro - quinoline N-oxide (**150**) with 2-cyanomethyl-5-ethylpyridine (**151**) gave the expected product **152**. Oxidation to ketone **153** and elaboration to ketolactam **154** set the stage for the key reaction. Hy - drolysis of the lactam moiety in **154** was followed by indolization to give 2-(2-pyridinyl)indole **155** after esterification. Reduction to the primary alcohol and bromination was followed by spontaneous cyclization to give 6,7-dihydroflavopereirine (hydrobromide) (**4**).

Scheme 32

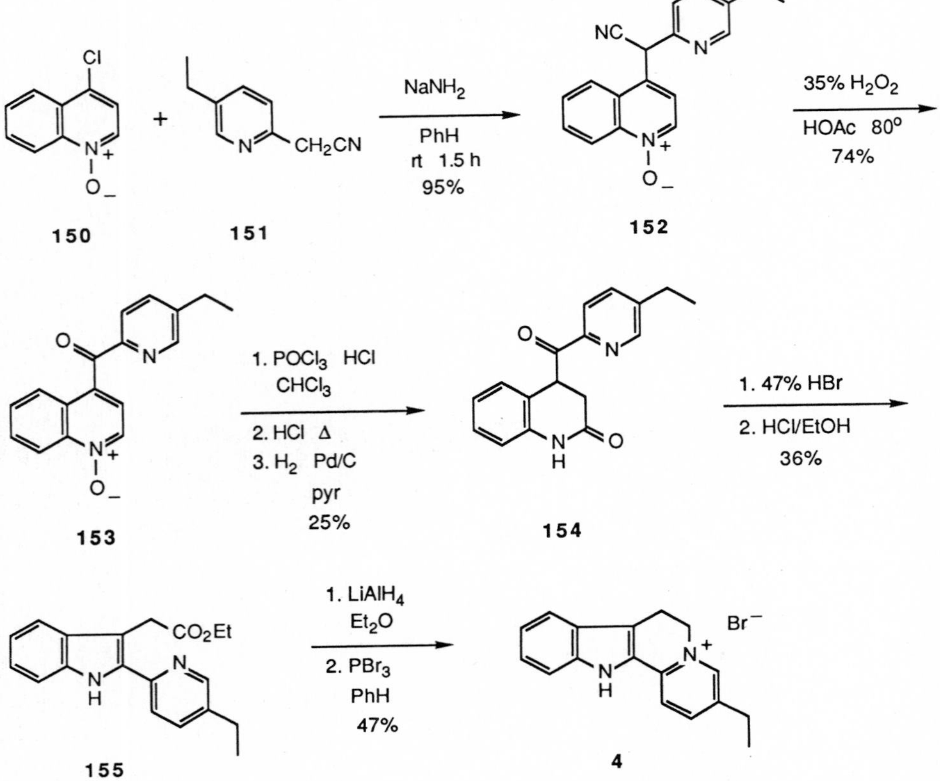

3.6 *Strategy VI*

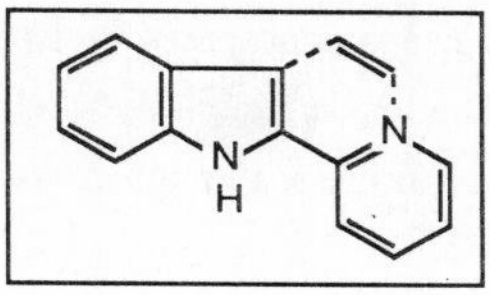

A Type VI strategy may be regarded as the most direct plan for constructing the indolo[2,3-*a*] - quinolizine ring system, since, in principle, it would be possible to avoid ring C or D dehydrogena - tion steps -- reactions that, as we have seen, often proceed in low yield.

3.6.1 Sugasawa (6,7-Dihydroindolo[2,3-*a*]quinolizine, 1956)

Sugasawa and coworkers (ref. 93) utilized a Fischer-indole synthesis to prepare 2-(2-pyridinyl) - indole (**156**) in high yield (**Scheme 33**). A conventional indole-3-acetic acid synthesis via gramine **157** gave acid **158**. Reduction and cyclization, as we saw in **Scheme 32**, gave 6,7-dihydroindolo - [2,3-*a*]quinolizine (hydrobromide) **125**.

Scheme 33

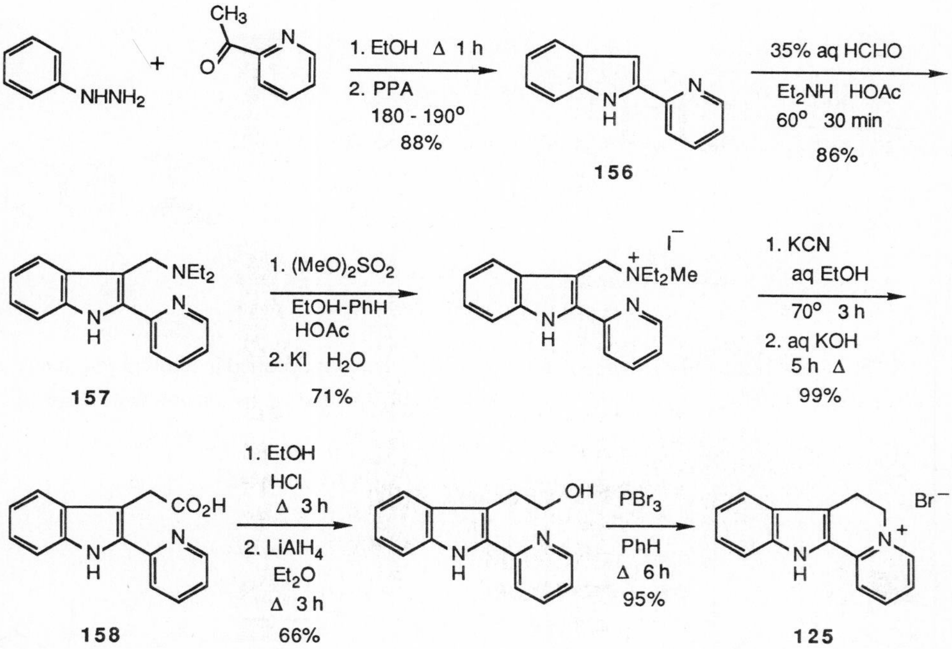

3.6.2 Stevens (Sempervirine Approach, 1970)

A second Type VI strategy directed at sempervirine was unsuccessful (**Scheme 34**). Thus, although Stevens and coworkers (ref. 87) succeeded in synthesizing the key 2-(2-pyridinyl)indole **162**, attempts to close the C-ring with chloroacetaldehyde led only to the isomeric product **163**. The requisite tetrahydroisoquinoline nitrile **135** was prepared by two different methods, with the more efficient one shown in **Scheme 34**. (The less efficient method involved heating **134** with phos - phorus tribromide and then fusing the bromide with cuprous cyanide.) A Grignard reaction with **135** and a Fischer indole synthesis gave **162**.

Scheme 34

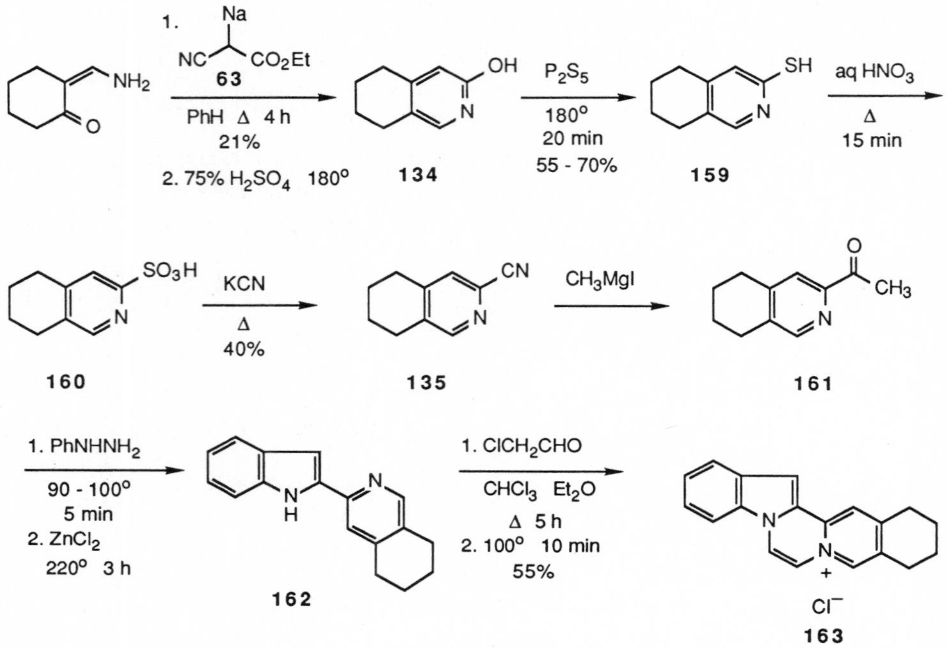

The parent 2-(2-pyridinyl)indole (**156**) also failed to yield the desired indolo[2,3-*a*]quinolizine with chloroacetaldehyde, giving **165** instead (**Scheme 35**). With care, the intermediate carbinol amine **164** could be isolated.

Scheme 35

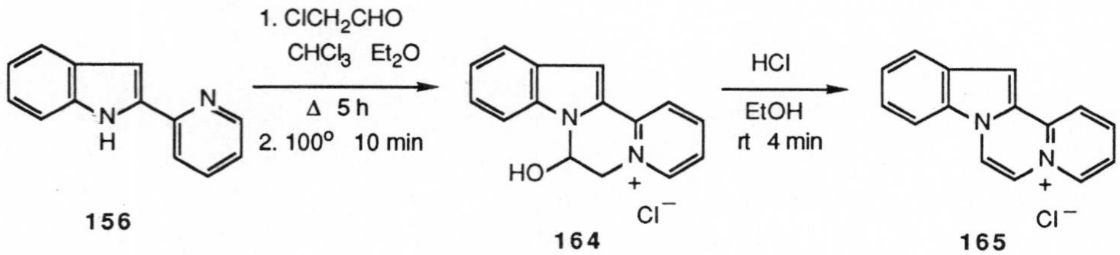

As might be expected, the *N*-benzyl derivative **166** gave the desired tetracyclic product **167**, albeit in minute yield (ref. 87) (**Scheme 36**). Moreover, since attempts to debenzylate a related system were unsuccessful, this strategy was abandoned by Stevens.

Scheme 36

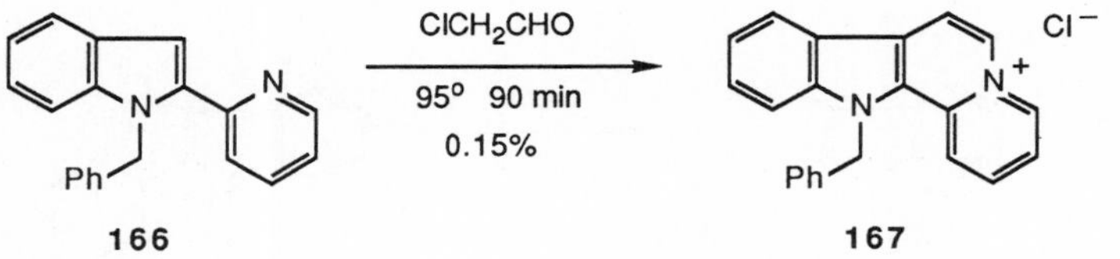

ClCH$_2$CHO

95° 90 min

0.15%

166   **167**

3.6.3 Gribble (Indolo[2,3-*a*]quinolizine, Flavopereirine, Flavocarpine, 6,7-Dihydroflavopereirine, 1987)

The ready availability of 2-(2-pyridinyl)indoles (ref. 93) and the inherent attractiveness of this strategy prompted us to reinvestigate the Type VI strategy first explored by Stevens.

Our strategy exploits the marked *beta*-lithiating ability of the 2-pyridinyl moiety (ref. 94) in that we envisioned that the indolo[2,3-*a*]quinolizine ring system **1** could be crafted from an *N*-protected 3-lithio-2-(2-pyridinyl)indole (**168**) and an appropriate two-carbon bis(1,2)electrophile (**Scheme 37**).

Scheme 37

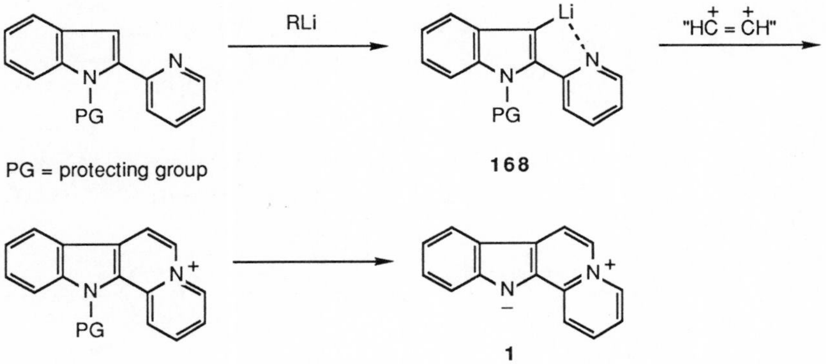

RLi

"HC = CH"

PG = protecting group   **168**

1

Our initial results using 1-phenylsulfonyl-2-(2-pyridinyl)indole (**169**) and simple electrophiles proved to be very encouraging (**Scheme 38**) (ref. 95). Thus, *N*-protection of 2-(2-pyridinyl)indole (**156**) (ref. 93) gave the desired substrate **169**. Treatment of **169** with *n*-butyllithium at -78°C in THF generated the desired 3-lithio species **170**. Quenching with various electrophiles gave the 3-substituted products **171** in fair to good yields.

154

Scheme 38

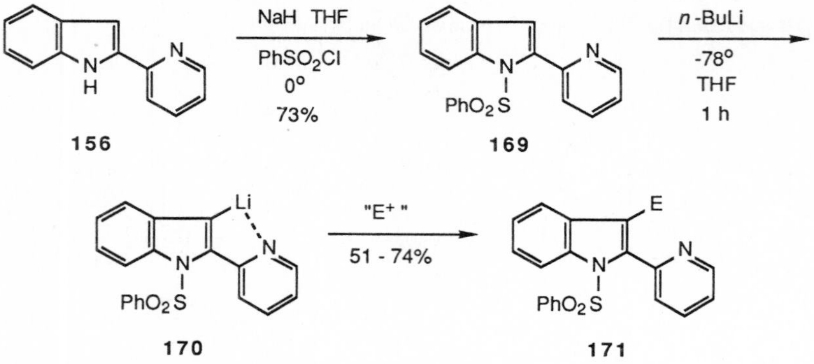

E = CH₃, CO₂H, Me₃Si, CH(Me)OH, CO₂Et, COCH₃

E = CH_3, CO_2H, Me_3Si, $CH(Me)OH$, CO_2Et, $COCH_3$

"E⁺" = CH_3I, CO_2, Me_3SiCl, CH_3CHO, $Cl\,CO_2Et$, $CH_3CON(Me)OMe$

The feared ring-fragmentation (refs. 96, 97) **170** to **172** occurred only at 50°C (**Scheme 39**), a temperature well above that needed for the reactions of **170** with electrophiles. The structure of **172** was established by independent synthesis (ref. 95) (**Scheme 39**).

Scheme 39

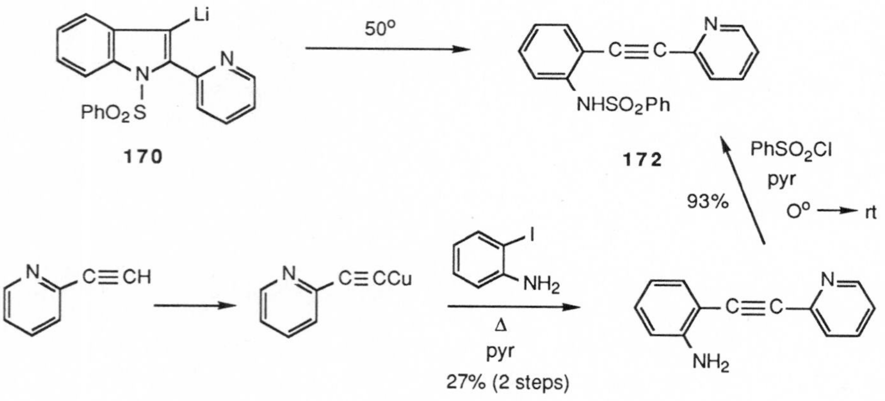

The ideal two-carbon bis(1,2)electrophile for our planned Type VI strategy (**Scheme 37**) was found to be bromoacetaldehyde, which can now be prepared in anhydrous form by the ozonolysis of 1,4-dibromo-*trans*-2-butene (ref. 98). Reaction of 3-lithio-1-phenylsulfonyl-2-(2-pyridinyl)indole (**170**) with bromoacetaldehyde and quenching at low temperature with acetic acid gave bromo alcohol **173** (**Scheme 40**) (ref. 99). Workup of this material afforded the indoloquinolizinium bromide

174, which, upon treatment with base, resulted in cleavage of the *N*-phenylsulfonyl protecting group and dehydration to furnish directly the indolo[2,3-*a*]quinolizine ring system, in 31% overall yield from pyridinylindole **156**.

Scheme 40

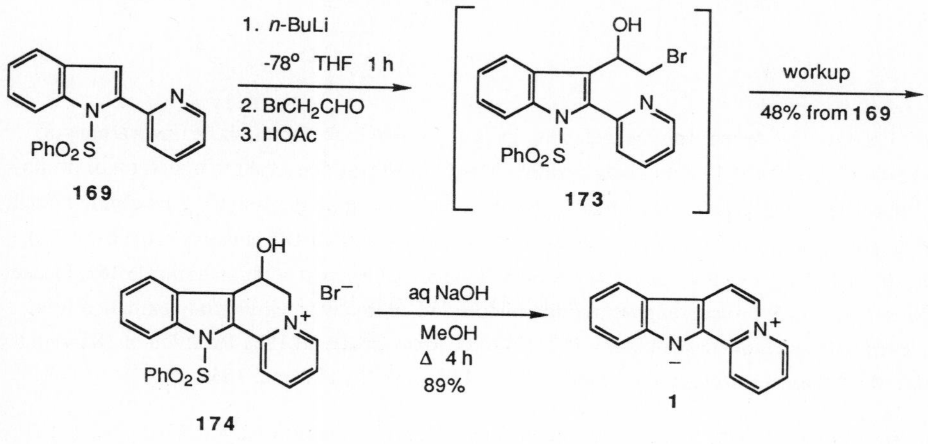

Regarding the synthesis of the requisite 2-(2-pyridinyl)indoles, we have found that the Smith indole synthesis (ref. 100) is nicely applicable to the preparation of these starting materials (**Scheme 41**). Thus, treatment of the dianion of *N*-trimethylsilyl-*o*-toluidine (**175**) with ethyl picolinates **176** and **129** afforded the pyridinylindoles **156** and **177**, respectively, in good yield (ref. 99).

Scheme 41

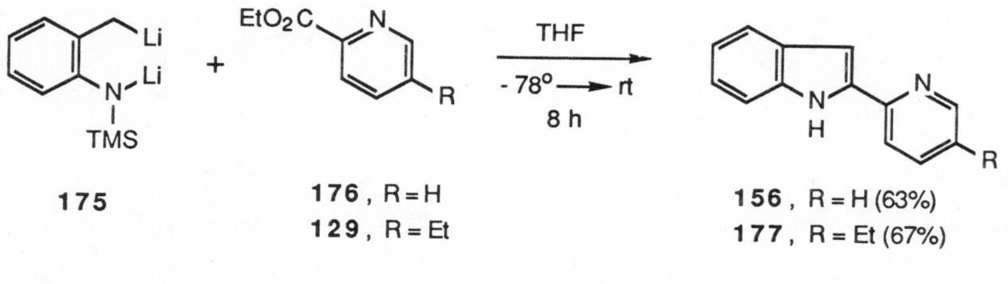

Application of this *beta*-lithiation protocol to the synthesis of flavopereirine (**3**) was readily achieved (**Scheme 42**) (ref. 99). The overall yield of this alkaloid was 33% from **177**.

Scheme 42

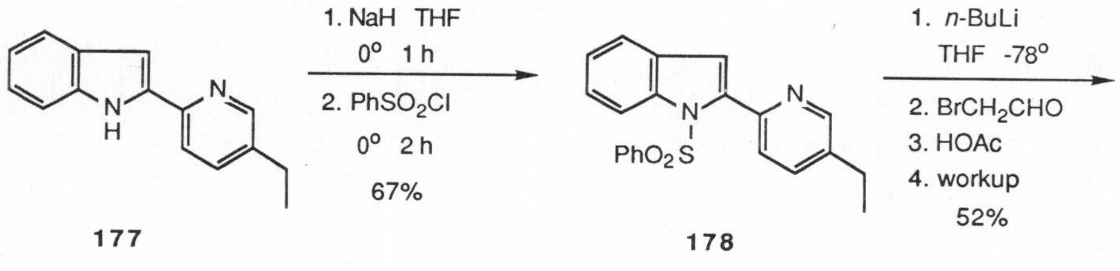

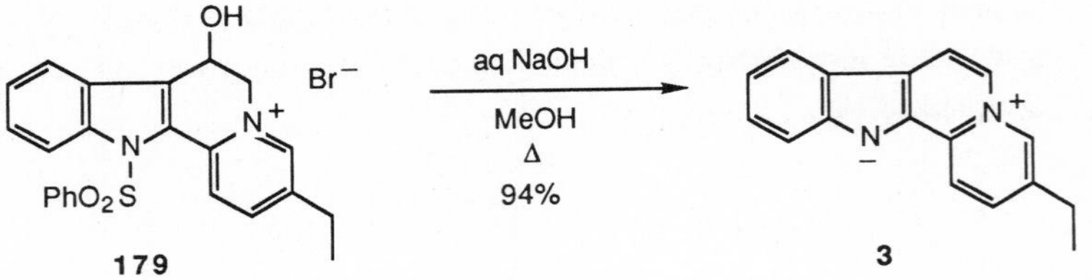

We have also found that this methodology is applicable to the synthesis of flavocarpine (**5**) (**Scheme 43**) (ref. 99). The requisite pyridine *N*-oxide **180** was converted to a mixture of cyano - pyridines (**181**, **182**) using two methods. Whereas the classical route (ref. 101), involving dimethyl sulfate/potassium cyanide, gave a better ratio of the desired isomer **181**, a newer route (ref. 102), using trimethylsilyl cyanide, gave a better overall yield. A Grignard addition to nitrile **181**, Fischer indolization, and *N*-protection gave pyridinylindole **184**. Selective methyl group oxidation was achieved with selenium dioxide (ref. 103) to afford isonicotinic acid **185**. Reaction of **185** with the usual set of reagents afforded flavocarpine (**5**) in 27% overall yield from **185**.

Scheme 43

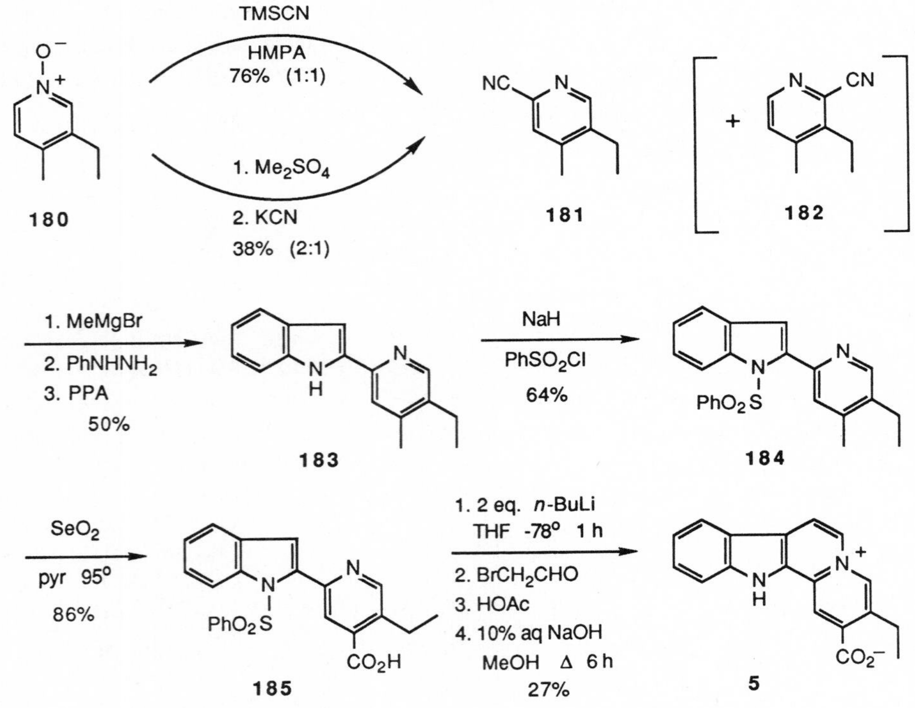

Our general strategy is equally applicable to the synthesis of 6,7-dihydroindolo[2,3-a]quinoli - zine alkaloids. A synthesis of dihydroflavopereirine (4) is shown in **Scheme 44** (ref. 99). Reaction of the 3-lithio species, generated from **178**, with oxirane in the presence of boron trifluoride etherate (ref. 104) -- which is essential for success -- and workup gave alcohol **186**. Hydrolysis and cycliza - tion according to Takahashi (ref. 92) afforded dihydroflavopereirine (hydrobromide) (**4**).

Scheme 44

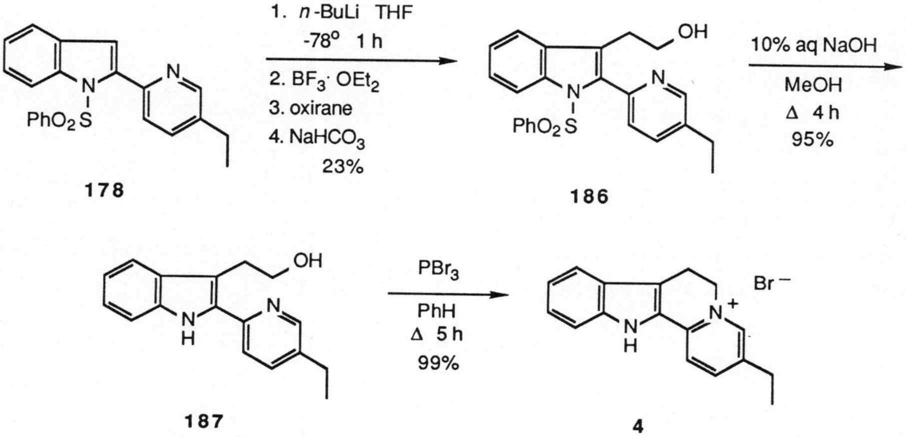

Finally, the availability of pyridinylindole **162** (cf. **Scheme 34**) should allow for the facile construction of sempervirine (**14**) according to our method.

4. PARTIAL SYNTHESES

4.1 *Biomimetic Approaches*

4.1.1 **Husson** (6,7-Dihydroflavopereirine, 1980)

In support of the biogenesis of flavopereirine proposed by Scott (ref. 105), Kan Fan and Husson have succeeded in converting the alkaloid 4,21-dehydrogeissoschizine (**188**) into 6,7-dihy - droflavopereirine (**4**) (ref. 106) (**Scheme 45**). the mechanism presumably involves deprotonation to dienamine **189**, followed by a vinylogous retro-Mannich reaction, and subsequent double bond reor - ganization to give **4**.

Scheme 45

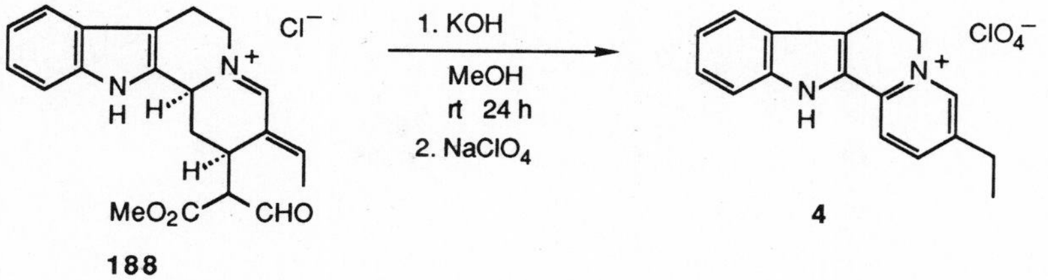

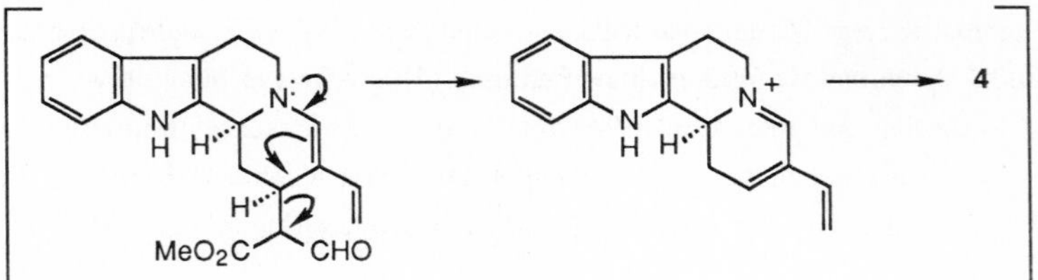

189

4.2 Alkaloid Interconversions

A number of groups have reported the oxidation of alkaloids as a route to the indolo[2,3-a]-quinolizine alkaloids.

We have already cited the Merlini and Nasini (ref. 72) oxidation of (-)-dihydrogambirtannine (**96**) to ourouparine iodide (**20**) with iodine (cf. **Scheme 16**).

It was found that lead tetraacetate smoothly oxidized the C-ring in yohimbine (**190**) and related alkaloids to give the tetradehydro derivative **191** (ref. 107) (**Scheme 46**).

Scheme 46

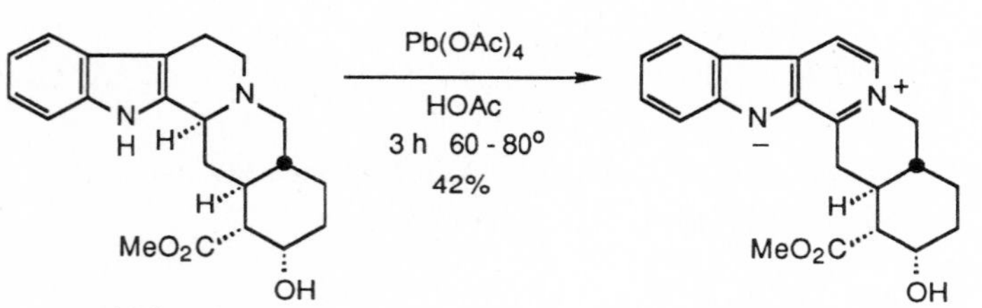

190   **191**

Similarly, bleekerine (**18**) was synthesized from the alkaloid isoreserpiline (**192**) (ref. 25), albeit in very low yield (**Scheme 47**).

Scheme 47

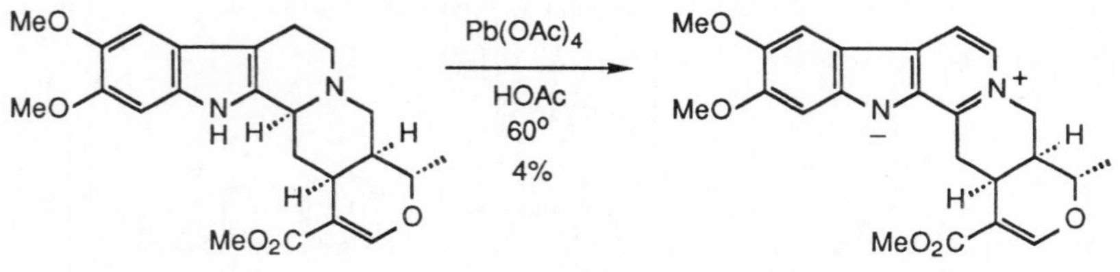

192   **18**

The important alkaloid reserpine (193) was oxidized to 3,4,5,6-tetradehydroreserpine (194) photochemically (ref. 108) or with ceric sulfate (ref. 109) (Scheme 48).

Scheme 48

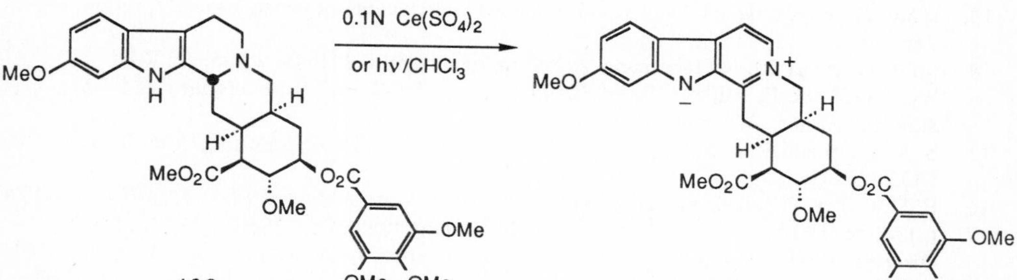

5. CONCLUDING REMARKS

Beginning -- appropriately enough -- with Woodward's synthesis of N-methylsempervirine (Scheme 18), the past forty years have witnessed an array of different synthetic routes to the bio - genetically and theoretically interesting zwitterionic indolo[2,3-a]quinolizine alkaloids. The newly discovered anticancer activity of these compounds will doubtless inspire the synthetic community to improve on existing syntheses of these alkaloids and to develop entirely new methodologies for their construction.

REFERENCES

1. For reviews of indole alkaloid synthesis in general, see (a) J.P. Kutney, in: J. ApSimon (Ed.), The Total Synthesis of Natural Products, Vol. 3, Wiley-Interscience, New York, NY, 1977, Chapter 2; (b) R.T. Brown, in: J.E. Saxton (Ed.), Indoles, Part 4, The Monoterpenoid Indole Alkaloids, John Wiley & Sons, New York, NY, 1983, Chapters 3 and 4; (c) C. Szantay, G. Blasko, K. Honty, and G. Dornyei, in: A. Brossi (Ed.), The Alkaloids, Vol. 27, Academic Press, Orlando, FL, 1986, Chapter 2.
2. For early discussions of these "anhydronium bases," see (a) J.W. Armit and R. Robinson, J. Chem. Soc., 127, 1925, 1604-1608; (b) B. Witkop, J. Am. Chem. Soc., 75, 1953, 3361-3370.
3. R. Kaschnitz and G. Spiteller, Monatsh., 96, 1965, 909-921.
4. (a) M.-M. Janot, R. Goutarel, A. Le Hir, and L.O. Bejar, Ann. Pharm. Fr., 16, 1958, 38-46; Chem. Abs., 52, 1958, 17301b; (b) O. Bejar, R. Goutarel, M.-M. Janot and A. Le Hir, Compt. rend., 244, 1957, 2066-2068.
5. (a) N.A. Hughes and H. Rapoport, J. Am. Chem. Soc., 80, 1958, 1604-1609; (b) E. Bächli, C. Vamvacas, H. Schmid, and P. Karrer, Helv. Chim. Acta, 40, 1957, 1167-1187.
6. G. Massiot, P. Thepenier, M.-J. Jacquier, J. Lounkokobi, C. Mirand, M. Zeches, L. Le Men-Olivier, and C. Delaude, Tetrahedron, 39, 1983, 3645-3656.
7. L. Angenot and A. Denoel, Planta Med., 23, 1973, 226-232; Chem. Abs., 79, 1973, 18924v.
8. G. Büchi, R.E. Manning, and F.A. Hochstein, J. Am. Chem. Soc., 84, 1962, 3393-3397.
9. E. Ali, V.S. Giri, and S.C. Pakrashi, Tetrahedron Lett., 1976, 4887-4890.

10. G.M.T. Robert, A. Ahond, C. Poupat, P. Potier, C. Jolles, A. Jousselin, and H. Jacquemin, *J. Nat. Prod., 46,* **1983**, 694-707.

11. E. Seguin, F. Hotellier, M. Koch, and T. Sevenet, *J. Nat. Prod., 47,* **1984**, 687-691.

12. G.M.T. Robert, A. Ahond, C. Poupat, P. Potier, H. Jacquemin, and S.K. Kan, *J. Nat. Prod., 46,* **1983**, 708-722.

13. More correctly, these alkaloids would be named as derivatives of benz[*g*]indolo[2,3-*a*]quinoli - zine.

14. (a) R. Goutarel, M.-M. Janot, and V. Prelog, *Experientia, 4,* **1948**, 24-25; (b) R.B. Woodward and B. Witkop, *J. Am. Chem. Soc., 71,* **1949**, 379; (c) R. Bentley and T.S. Stevens, *Nature, 164,* **1949**, 141-142.

15. S. Siddiqui and R.H. Siddiqui, *J. Ind. Chem. Soc., 8,* **1931**, 667-680; *9,* **1932**, 539-544; *12,* **1935**, 37-47.

16. E. Haack, A. Popelack, H. Spingler, and F. Kaiser, *Naturwissenschaften, 41,* **1954**, 479-480.

17. (a) M.-M. Janot, R. Goutarel, and A. Le Hir, *Compt. rend., 238,* **1954**, 720-722; (b) C. Djerassi, M. Gorman, A.L. Nussbaum, and J. Reynoso, *J. Am. Chem. Soc., 76,* **1954**, 4463-4465; (c) F.A. Hochstein, K. Murai, and W.H. Boegemann, *J. Am. Chem. Soc., 77,* **1955**, 3551-3554.

18. W.B. Mors, P. Zaltzman, J.J. Beereboom, S.C. Pakrashi, and C. Djerassi, *Chem. & Ind.,* **1956**, 173-174.

19. R.A. Seba, J.S. Campos, and J.G. Kuhlmann, *Bol. inst. vital Brazil, 5,* **1954**, 175-190; *Chem. Abs., 49,* **1955**, 14270.

20. M.A. Amer and W.E. Court, *Phytochemistry, 20,* **1981**, 2569-2573.

21. (a) R.C. Elderfield and A.P. Gray, *J. Org. Chem., 16,* **1951**, 506-523; (b) F.E. Bader, *Helv. Chim. Acta, 36,* **1953**, 215-226.

22. E. Schlittler, H. Schwarz, and F. Bader, *Helv. Chim. Acta, 35,* **1952**, 271-276.

23. B. Uribe Vergara, *J. Am. Chem. Soc., 77,* **1955**, 1864.

24. A. Malik, N. Afza, and S. Siddiqui, *Heterocycles, 16,* **1981**, 1727-1733.

25. M. Sainsbury and B. Webb, *Phytochemistry, 11,* **1972**, 2337-2339.

26. (a) R.C. Elderfield and S.L. Wythe, *J. Org. Chem., 19,* **1954**, 683; (b) R.C. Elderfield and O.L. McCurdy, *J. Org. Chem., 21,* **1956**, 295-296.

27. W.I. Taylor and R. Hamet, *Compt. rend., 262D,* **1966**, 1141-1143.

28. A. Chatterjee and S. Mukhopadhyay, *Ind. J. Chem., 15B,* **1977**, 183-184.

29. S. Sakai, H. Ohtani, H. Ido, and J. Haginiwa, *Yakugaku Zasshi, 93,* **1973**, 483-489; *Chem. Abs., 19,* **1973**, 63538h.

30. R.P. Borris, A. Guggisberg, and M.Hesse, *Helv. Chim. Acta, 67,* **1984**, 455-460.

31. C. Coune, D. Tavernier, M. Caprasse, and L. Angenot, *Planta Med., 50,* **1984**, 93-95; *Chem. Abs., 101,* **1984**, 87496s.

32. A.A. Ibragimov, S. Kh. Maekh, and S. Yu. Yunusov, *Khim. Prir. Soedin, 11,* **1975**, 275-276; *Engl. trans., 11,* **1975**, 293-298.

33. N. Aimi, T. Tsuyuki, H. Murakami, S. Sakai, and J. Haginiwa, *Tetrahedron Lett., 26,* **1985**, 5299-5302.

34. (a) E. Schlittler, H.U. Huber, F.E. Bader, and H. Zahnd, *Helv. Chim. Acta, 37,* **1954**, 1912-1920; (b) H. Irie, K. Ishizuka, S. Kawashima, N. Masaki, K. Osaki, T. Shingu, S. Uyeo, H. Kaneko, and S. Naruto, *J. Chem. Soc., Chem. Commun.,* **1972**, 871.

35. C. Djerassi and J. Fishman, *Chem. & Ind.,* **1955**, 627-629.

36. K. Koike, T. Ohmoto, and K. Ogata, *Chem. Pharm. Bull. Japan, 34,* **1986**, 3228-3236.

37. M. Caprasse, L. Angenot, D. Tavernier, and M.J.O. Anteunis, *Planta Med., 50,* **1984**, 131-133; *Chem. Abs., 102,* **1985**, 218319g.

38. (a) L. Canonica, B. Danieli, P. Manitto, G. Russo, and G. Ferrari, *Tetrahedron Lett.,* **1968**, 4865-4866; (b) J. Bergman, in: A. Brossi (Ed.), The Alkaloids, Vol. 21, Academic Press, New York, NY, 1983, Chapter 2.

39. E.C. Miranda, C.H. Brieskorn, and S. Blechert, *Chem. Ber., 113,* **1980,** 3245-3248.
40. J.E. Saxton, in: R.H.F. Manske (Ed.), The Alkaloids, Vol. VIII, Academic Press, New York, NY, 1965, pp. 19-21.
41. M. Caprasse, C. Coune, and L. Angenot, *J. Pharm. Belg., 38,* **1983,** 135-139; *Chem. Abs., 99,* **1983,** 209852t.
42. R. Bassleer, J.M. Marnette, P. Wiliquet, M.C. DePauw-Gillet, M. Caprasse, and L. Angenot, *Planta Med., 49,* **1983,** 158-161; *Chem. Abs., 100,* **1984,** 98055c.
43. (a) M. Beljanski and M.S. Beljanski, *Exp. Cell. Biol., 50,* **1982,** 79-87; (b) M. Beljanski and M.S. Beljanski, *Oncology, 43,* **1986,** 198-203; (c) R. Bassleer, D. Clermont, J.M. Marnette, M. Caprasse, M. Tits, and L. Angenot, *Ann. Pharm. Fr., 43,* **1985,** 83-88; *Chem. Abs., 103,* **1985,** 171460z.
44. R.C. Elderfield and B.A. Fischer, *J. Org. Chem., 23,* **1958,** 949-953.
45. F.T. Tyson, *J. Am. Chem. Soc., 61,* **1939,** 183-185.
46. K.T. Potts and R. Robinson, *J. Chem. Soc.,* **1955,** 2675-2686.
47. J. Thesing and W. Festag, *Experientia, 15,* **1959,** 127-128.
48. J. Thesing, H. Ramloch, C. Willersinn, and F. Funk, *Angew. Chem., 68,* **1956,** 387-388.
49. C. Schöpf, G. Herbert, R. Rausch, and G. Schröeder, *Angew. Chem., 69,* **1957,** 391-392.
50. Y. Ban and M. Seo, *Tetrahedron, 16,* **1961,** 5-10.
51. S. Sugasawa and M. Kirisawa, *Pharm. Bull. Japan, 3,* **1955,** 190-193.
52. T. Hoshino and K. Shimodaira, *Ann., 520,* **1935,** 19-30.
53. K.B. Prasad and G.A. Swan, *J. Chem. Soc.,* **1958,** 2024-2038.
54. Y. Ban and M. Seo, *Tetrahedron, 16,* **1961,** 11-15.
55. G.A. Swan, *J. Chem. Soc.,* **1958,** 2038-2044.
56. Y. Ban and M. Seo, *J. Org. Chem., 27,* **1962,** 3380-3381; see also *Chem. Abs., 62,* **1965,** 9185g.
57. U. Basu and B. Banerjee, *Ann., 516,* **1935,** 243-248.
58. E. Wenkert and J. Kilzer, *J. Org. Chem., 27,* **1962,** 2283-2284.
59. E. Wenkert and B. Wickberg, *J. Am. Chem. Soc., 84,* **1962,** 4914-4919.
60. E. Wenkert, R.A. Massy-Westropp, and R.G. Lewis, *J. Am. Chem. Soc., 84,* **1962,** 3732-3736.
61. R.C. Elderfield, B. Fischer, and J.M. Lagowski, *J. Org. Chem., 22,* **1957,** 1376-1380.
62. N.J. Leonard and W.K. Musker, *J. Am. Chem. Soc., 82,* **1960,** 5148-5155, and earlier papers in the series.
63. T. Okamoto and H. Tani, *Chem. Pharm. Bull. Japan, 7,* **1959,** 130-131.
64. K.T. Potts and D.R. Liljegren, *J. Org. Chem., 28,* **1963,** 3066-3070.
65. A. Le Hir, M.-M. Janot, and D. van Stolk, *Bull. Soc. Chim. Fr.,* **1958,** 551-556.
66. R. Majima and M. Kotake, *Ber., 55B,* **1922,** 3865-3872.
67. E. Wenkert and R. Roychaudhuri, *J. Am. Chem. Soc., 80,* **1958,** 1613-1619, and footnote 27 therein.
68. Y. Ban and T. Kimura, *Chem. Pharm. Bull. Japan, 16,* **1968,** 549-552.
69. (a) J.A. Beisler, *Tetrahedron, 26,* **1970,** 1961-1965; (b) J.A. Beisler, *Chem. Ber., 103,* **1970,** 3360-3361.
70. E.M. Fry, *J. Org. Chem., 28,* **1963,** 1869-1874.
71. E.M. Fry, *J. Org. Chem., 29,* **1964,** 1647-1650.
72. L. Merlini and G. Nasini, *Gazz. Chim. Ital., 97,* **1967,** 1915-1920.
73. R.B. Woodward and W.M. McLamore, *J. Am. Chem. Soc., 71,* **1949,** 379-380.
74. E. Späth and E. Lederer, *Ber., 63B,* **1930,** 120-125; 2102-2111.
75. J.E. Saxton, *Quart. Rev., 10,* **1958,** 108-147.
76. C.K. Bradsher and A.J.H. Umans, *J. Org. Chem., 28,* **1963,** 3070-3072.
77. H.R. Snyder, S.M. Parmerter, and L. Katz, *J. Am. Chem. Soc., 70,* **1948,** 222-225.
78. C.K. Bradsher, *Chem. Rev., 38,* **1946,** 447-499.

162

79. K.T. Potts and G.S. Mattingly, *J. Org. Chem., 33,* **1968**, 3985-3987.

80. O. Westphal, K. Jann, and W. Heffe, *Arch. Pharm., 294,* **1961**, 37-45.

81. (a) I. Ninomiya, Y. Tada, T. Kiguchi, O. Yamamoto, and T. Naito, *Heterocycles, 9,* **1978**, 1527-1531; (b) I. Ninomiya, Y. Tada, T. Kiguchi, O. Yamamoto, and T. Naito, *J. Chem. Soc. Perkin Trans. 1,* **1984**, 2035-2038.

82. B. Danieli, G. Lesma, and G. Palmisano, *J. Chem. Soc., Chem. Commun.,* **1980**, 860-861.

83. V.S. Giri, B.C. Maiti, and S.C. Pakrashi, *Heterocycles, 22,* **1984**, 233-236.

84. C. Szantay and L. Toke, *Acta Chim. Acad. Sci. Hung., 39,* **1963**, 249-251; *Chem. Abs., 60,* **1964**, 5575.

85. E.E. Glover and G. Jones, *J. Chem. Soc.,* **1958**, 1750-1754.

86. L.C. Craig, *J. Am. Chem. Soc., 56,* **1934**, 1144-1147.

87. R. Bentley, T.S. Stevens, and M. Thompson, *J. Chem. Soc. (C),* **1970**, 791-795.

88. H. Kaneko, *J. Pharm. Soc. Japan, 80,* **1960**, 1374-1378; *Chem. Abs., 55,* **1961**, 6512f.

89. K. B. Prasad and S.C. Shaw, *Ind. J. Chem., 11,* **1973**, 621-623.

90. L.H. Groves and G.A. Swan, *J. Chem. Soc.,* **1952**, 650-661.

91. (a) W.M. Whaley and T.R. Govindachari, *Org. React., 6,* **1951**, 151-190; (b) W.M. Whaley and T.R. Govindachari, *Org. React., 6,* **1951**, 74-150.

92. M. Takahashi, *Itsuu Kenkyusho Nempo,* **1971**, 65-73; *Chem. Abs., 77,* **1972**, 61854t.

93. S. Sugasawa, M. Terashima, and Y. Kanaoka, *Chem. Pharm. Bull. Japan, 4,* **1956**, 16-19; *Chem. Abs., 51,* **1957**, 3593d.

94. H.W. Gschwend and H.R. Rodriquez, *Org. React., 26,* **1979**, 1-360.

95. D.A. Johnson and G.W. Gribble, *Heterocycles, 24,* **1986**, 2127-2131.

96. T.L. Gilchrist, *Adv. Het. Chem., 41,* **1987**, 41-74.

97. G.W. Gribble and M.G. Saulnier, *J. Org. Chem., 48,* **1983**, 607-609.

98. G.A. Kraus and P. Guttschalk, *J. Org. Chem., 48,* **1983**, 2111-2112.

99. G.W. Gribble and D.A. Johnson, *Tetrahedron Lett.,* submitted.

100. (a) A.B. Smith III and M. Visnick, *Tetrahedron Lett., 26,* **1985**, 3757-3760; (b) A.B. Smith III, M. Visnick, J.N. Haseltine, and P.A. Sprengler, *Tetrahedron, 42,* **1986**, 2957-2969.

101. W.E. Feely and E.M. Beavers, *J. Am. Chem. Soc., 81,* **1959**, 4004-4007.

102. H. Vorbrüggen and K. Krolikiewicz, *Synthesis,* **1983**, 316-319.

103. A. Jackson, N.D.V. Wilson, A.J. Gaskell, J.A. Joule, *J. Chem. Soc. (C),* **1969**, 2738-2747.

104. M.J. Eis, J.E. Wrobel, and B. Ganem, *J. Am. Chem. Soc., 106,* **1984**, 3693-3694.

105. A.I. Scott, *Bioorg. Chem., 3,* **1974**, 398-429.

106. C. Kan Fan and H-P. Husson, *Tetrahedron Lett., 21,* **1980**, 4265-4266.

107. M.-M. Janot, R. Goutarel, A. Le Hir, M. Amin, and V. Prelog, *Bull. Chim. Soc. Fr.,* **1952**, 1085-1091.

108. G.E. Wright and T.Y. Tang, *J. Pharm. Sci., 61,* **1972**, 299-300.

109. Th. J. Hakkesteegt, *Pharm. Weekbl., 103,* **1968**, 297-300.

VINYL AZIDES IN NATURAL PRODUCT SYNTHESIS

C. J. MOODY

1. INTRODUCTION

Vinyl azides, readily prepared by a number of methods,[1] undergo a variety of well established reactions.[2] Although vinyl azide (azidoethene) itself was first described in 1910, it is only in the last 20 years that unsaturated azides have been studied in any detail. Much of this work has been concerned with their thermal decomposition reactions in which both vinylnitrenes and 1-azirines play a role.[2,3] Many of these reactions result in good yields of nitrogen containing products, particularly heterocyclic compounds, and can be exploited in synthesis. This review is not intended to be a compilation of all known uses of vinyl azides in synthesis, but concentrates on our own efforts to apply this type of chemistry in the synthesis of heterocyclic natural products. The review will highlight the use of vinyl azides in the synthesis of 7 natural products: the bacterial co-enzyme methoxatin (1), the carbazolequinone alkaloid murrayaquinone-B (2), the isoquinolone alkaloid siamine (3), the isoindolobenzazepine alkaloid lennoxamine (4), the phosphodiesterase inhibitors PDE-I (5) and PDE-II (6), and the closely related potent anti-tumour antibiotic CC-1065 (7).

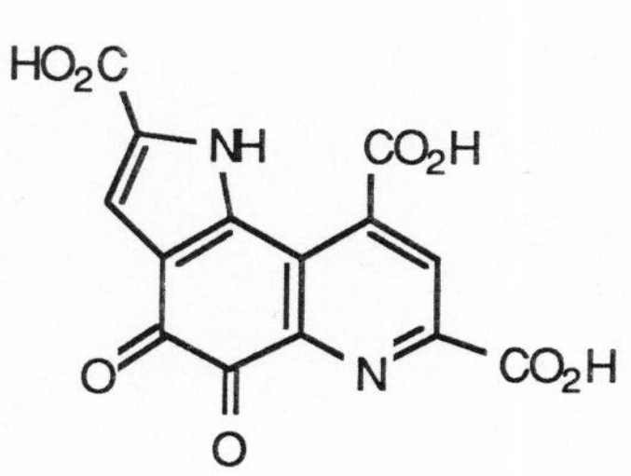

(1)

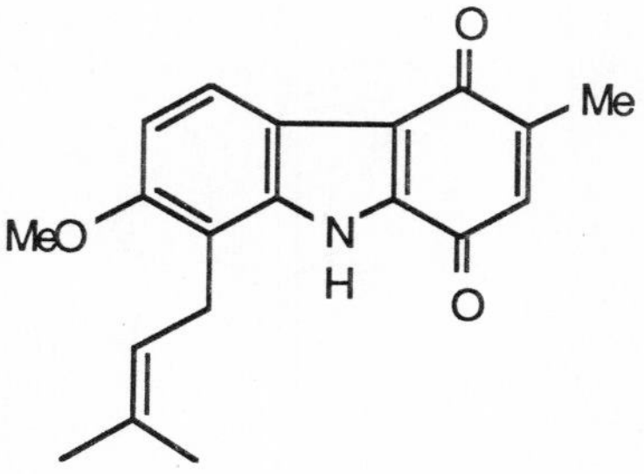

(2)

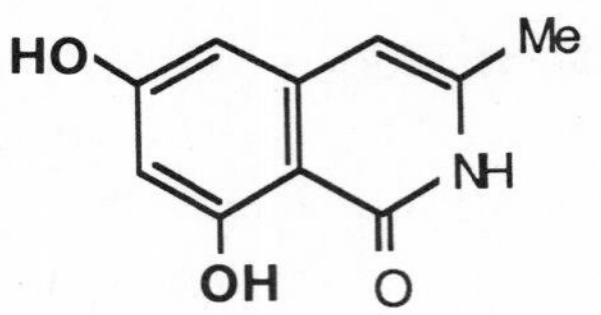
HO
Me
NH
OH
O

(3)

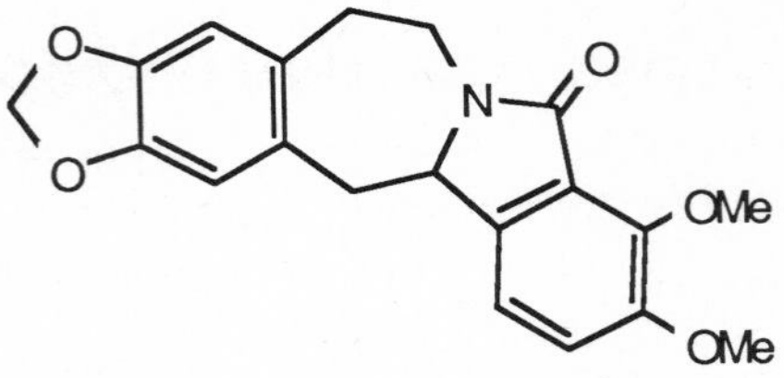
O
O
N
O
OMe
OMe

(4)

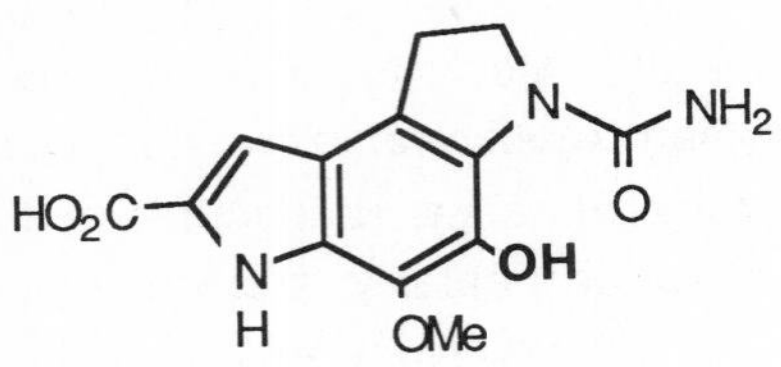
N
NH2
O
HO2C
N
H
OH
OMe

(5)

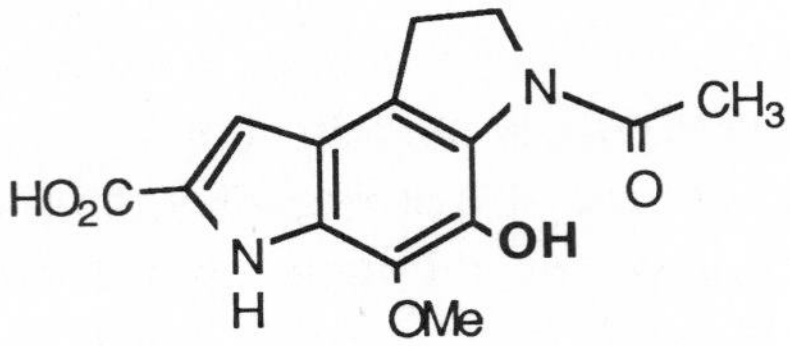
N
CH3
O
HO2C
N
H
OH
OMe

(6)

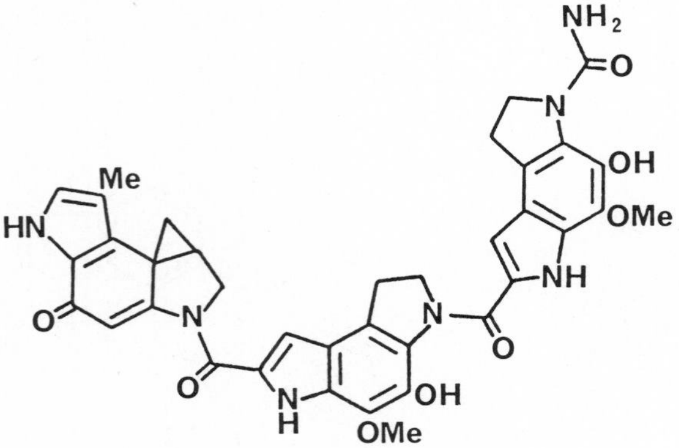
NH2
O
N
OH
OMe
NH
Me
HN
O
N
N
O
O
N
H
OH
OMe

(7)

2. MODEL STUDIES

The vinyl azides that we have used in the synthesis of natural products are mainly of one type: 3-aryl derivatives of 2-azidopropenoate esters (α-azidocinnamates). These vinyl azides, first described by the Austrian group of Hemetsberger and co-workers,[4-6] are easily prepared by condensation of benzaldehydes or heteroaromatic aldehydes with methyl (or ethyl) azidoacetate in the presence of sodium methoxide (or ethoxide) in methanol (or ethanol). In our experience, the most reliable way to carry out this condensation reaction is to add a solution of the aldehyde in the azidoacetate (4 equivalents) to a cold (-15°C) solution of the sodium alkoxide (4 equivalents) in the alcohol. If the aldehyde is insufficiently soluble in the azidoacetate then a *small* amount of alcohol or tetrahydrofuran may be added. On warming to room temperature, the product azidocinnamate often crystallises out of solution, and the excess of azidoacetate is rapidly destroyed under the basic conditions to give red-brown water soluble material that is easily separated from the product.

Although azidocinnamates are often characterisable crystalline solids, they decompose on heating in solution (refluxing toluene or xylene) to give products which are formally derived from the corresponding vinylnitrene. The decomposition pathway involves the intermediacy of an azirine, and

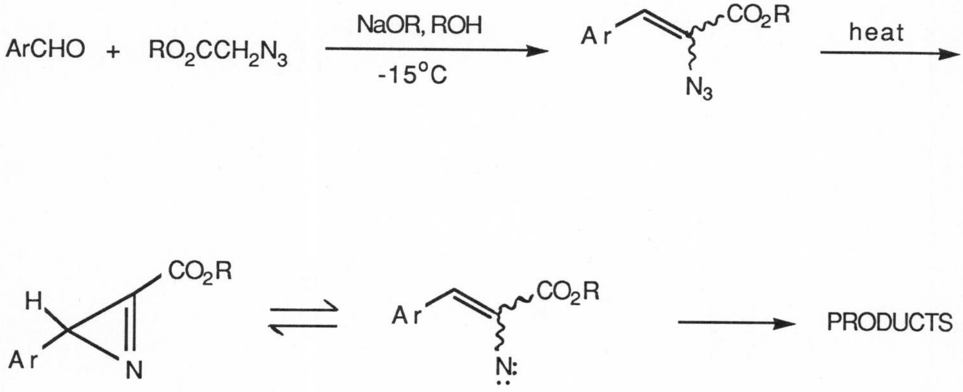

$$\text{ArCHO} + \text{RO}_2\text{CCH}_2\text{N}_3 \xrightarrow[-15°C]{\text{NaOR, ROH}} \overset{\text{CO}_2\text{R}}{\underset{\text{N}_3}{\text{Ar}\diagup\diagdown}} \xrightarrow{\text{heat}}$$

Ar / CO₂R, N₃

Scheme 1

since the azirine is in thermal equilibrium with the vinylnitrene,[7] this provides a mechanism for *cis-trans* isomerisation of the vinyl double bond. Hence the original geometry of the azidocinnamate obtained from the condensation reaction is unimportant (Scheme 1).

The products resulting from the decomposition of azidocinnamates are highly dependent on the nature of the *ortho*-substituent of the aromatic ring. In the absence of substituents that can interact with the vinyl azide/nitrene, the thermolysis of azidocinnamates in boiling toluene or xylene leads to indole-2-carboxylates in good yield. This is an excellent route to indoles, and although it was first reported in 1970,[5] aside from our own work, it has largely been ignored, despite the fact that it involves only 2 steps from simple benzaldehydes. The mechanism presumably involves electrocyclisation of the vinylnitrene, followed by a rapid aromatising [1,5]-hydrogen shift (Scheme 2), rather than a simple insertion into the aromatic C-H bond.

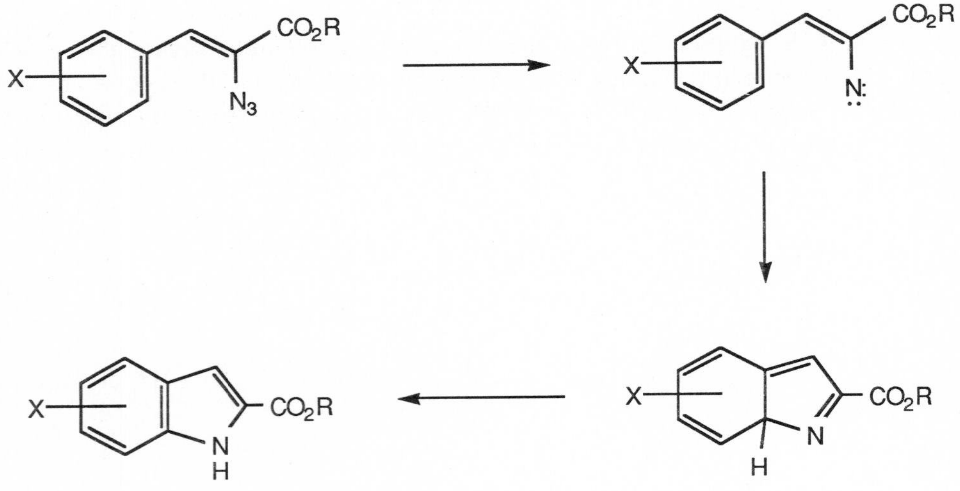

Scheme 2

When a primary or secondary alkyl group is present in the *ortho*-position of the aromatic ring, 6-membered ring formation by "insertion" of the vinylnitrene into a benzylic C-H bond is possible. We first observed this type of reaction in the thermal decomposition of the azidocinnamate (8) derived from mesitaldehyde which gave the isoquinoline (9) (32%).[8] When one *ortho*-position is unsubstituted, indole formation usually supervenes except in the case of the *ortho*-benzyl substituted azidocinnamate (10) which gives some 6-membered ring (11) in addition to the expected ethyl 4-benzylindole-2-carboxylate. The yield of the isoquinoline (11) is markedly increased when the decomposition of the vinyl azide was carried out in the presence of iodine. Similar results were obtained with the azidocinnamate (12) derived from fluorene-1-carboxaldehyde, which gave the interesting azafluoranthene (13) on thermolysis in the presence of iodine. Although, at this stage, the yield of (13) is not synthetically useful (10%), the reaction does represent a novel route to azafluoranthenes, a ring system found in alkaloids such as rufescine and imelutine. The effect of iodine was even more marked in the decomposition of the azidocinnamate derived from 2-methylbenzaldehyde. Whereas thermolysis in refluxing toluene alone gave exclusively ethyl 4-methylindole-2-carboxylate (>95%), thermolysis in the presence of iodine dramatically reduced the yield of the indole, and gave ethyl isoquinoline-3-carboxylate albeit in poor yield. The "insertion" reaction can be extended to the preparation of other [c]-fused pyridines such as thieno- and pyrazolo-pyridines,[8] and we have also used the reaction to prepare ethyl β-carboline-3-carboxylate, a naturally occurring β-carboline of considerable importance.[9]

(8)

(9)

(10)

(11)

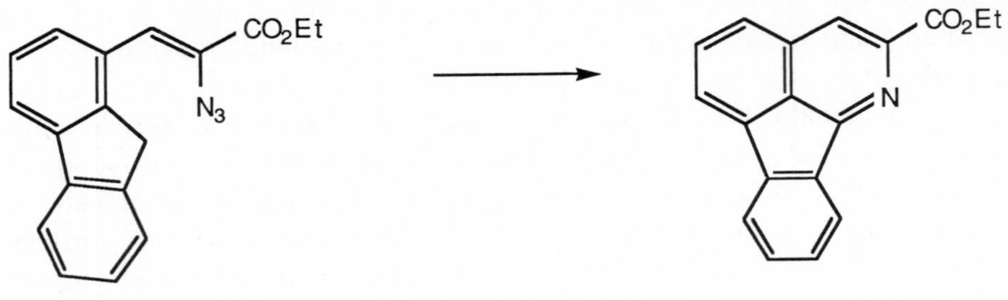

(12) → (13)

Although the thermal decomposition of azidocinnamates bearing *ortho*-alkyl substituents does give isoquinolines, the yields are rather low. However, we have developed a synthetically useful route to isoquinolines based on azidocinnamates, using an intramolecular aza-Wittig reaction.[10] The substrates for the intramolecular aza-Wittig reaction are the azidocinnamates (14, R = Me or Et) bearing an *ortho*-carbonyl substituent. On treatment with triethyl phosphite, the azidocinnamates (14) gave good yields of isoquinolines (15). The reaction is believed to proceed by initial Staudinger reaction to give an iminophosphorane, which was not detected, intramolecular attack on the carbonyl being very rapid. That the intramolecular aza-Wittig reaction in these systems is particularly favourable is demonstrated by the fact that even unreactive ester and acid carbonyl groups both readily participate. Thus the ester (14d) and the acid (14e) gave the 1-ethoxyisoquinoline (15d) and the isoquinolone (15e) respectively on treatment with triethyl phosphite.

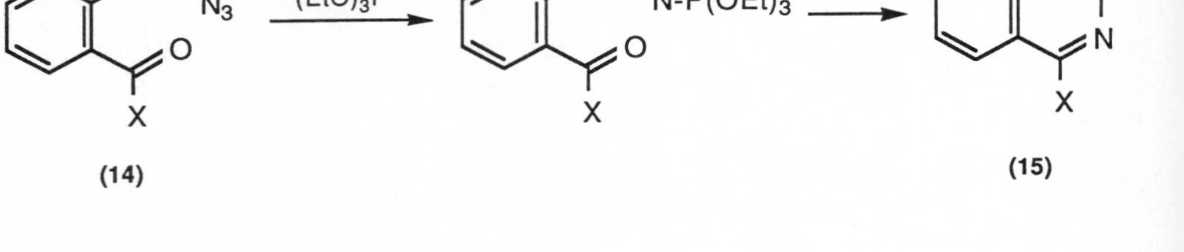

(14)   (15)

a, R = Me, X = H; **b**, R = X = Me; **c**, R = Et, X = Ph; **d**, R = Et, X = OEt; **e**, R = Et, X = OH

The scope of the intramolecular aza-Wittig reaction is not limited to simple bicyclic isoquinolines, and we have been able to use it in an improved synthesis (34%) of the azafluoranthene (13).[10]

Since nitrenes (and azides) react readily with olefinic double bonds, we also investigated the decomposition of azidocinnamates bearing an olefinic *ortho*-substituent in the expectation that the major products might arise from interaction of the nitrene (or azide) with the double bond in preference to the aromatic ring. This proved to be so in several cases,[11-13] and the reaction is a useful route to benzazepines. For example, decomposition of the *ortho*-styryl azidocinnamate (16) gives the 3-benzazepine (17) (37%) and the 1-benzylisoquinoline (18) (36%) as the major products.[11] The exact mechanisms of this and related reactions are discussed in detail in our full papers. When the possibility for such interaction of the intermediate with the double bond is blocked by epoxidation, cyclisation occurs exclusively to the free *ortho*-position of the aromatic ring to give the corresponding 4-substituted indole.

Thus our model studies revealed that the decomposition of azidocinnamates bearing appropriate substituents can lead to the formation of substituted indoles, isoquinolines, and benzazepines, and since many of these reactions proceeded in good yield, the possibility of using them in the synthesis of more complex molecules was investigated.

3. APPLICATION TO NATURAL PRODUCTS

3.1 *Synthesis of Methoxatin* [14]

Methylotrophic bacteria, which can utilise methane or methanol as their sole source of carbon, are of considerable commercial importance since they are used in the production of single cell protein from cheap raw materials. These organisms contain a methanol dehydrogenase which possesses a unique coenzyme, quite different from the usual redox cofactors such as nicotinamide and flavins. The structure of this coenzyme was eventually elucidated as pyrrolo[2,3-f]quinoline-2,7,9-tricarboxlic acid (1), more commonly known by its trivial names methoxatin or PQQ (pyrroloquinoline quinone).

(1)

There has been considerable interest in methoxatin from both the biological and chemical point of view, and this interest has been fuelled by recent discoveries that methoxatin is quite widely distributed in Nature, even occurring in mammalian enzyme systems. This has prompted speculation that methoxatin might be a hitherto unsuspected vitamin. Including our own, there have been 5 total syntheses of methoxatin to date.[14]

Our own synthetic strategy involved, as the key step, the formation of the indole ring from a suitably substituted benzaldehyde using the azidocinnamate method, followed by annelation of the pyridine ring (Scheme 3). The starting benzaldehyde (20) was prepared from commercially available 4-aminosalicylic acid in 67% overall yield by esterification and N-acetylation to give the known methyl 4-acetamido-2-hydroxybenzoate (19), followed by O-benzylation and conversion of the ester into the corresponding aldehyde.

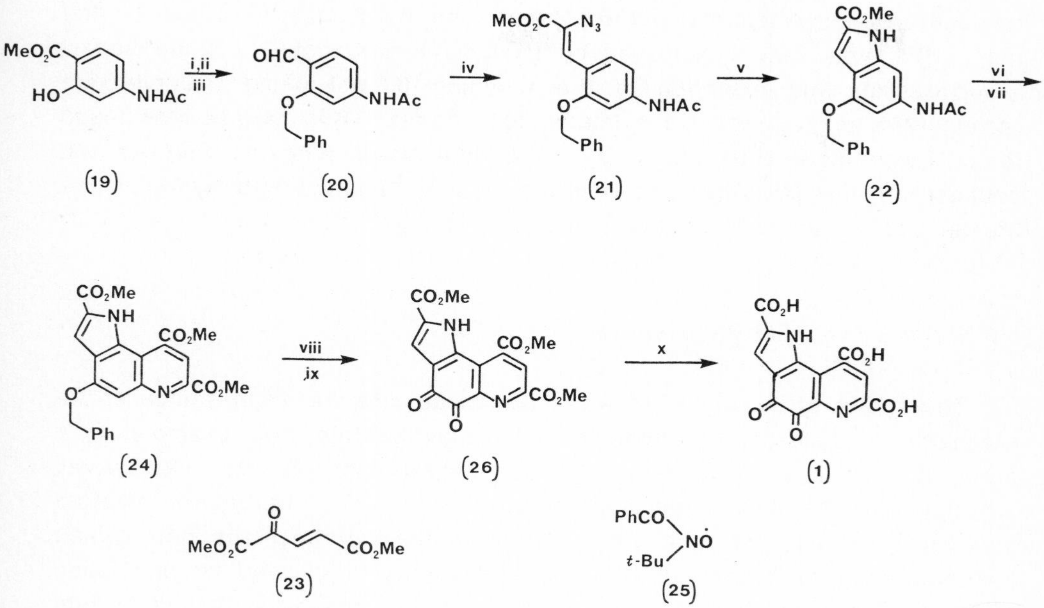

Scheme 3. *Reagents:* i, PhCH₂Br, K₂CO₃, acetone; ii, LiAlH₄, THF; iii, BaMnO₄, CHCl₃; iv, MeO₂CH₂N₃, NaOMe, MeOH; v, xylene, reflux; vi, HCl, MeOH, heat; vii, (23), CH₂Cl₂, then H⁺; viii, H₂, Pd-C, MeOH; ix, (25), CH₂Cl₂-MeOH; x, HC(OMe)₃, MeOH, H⁺, then aq. K₂CO₃, 85°C, then H⁺ to pH 2.5.

The benzaldehyde (20) was condensed with methyl azidoacetate under the usual conditions to give the key azidocinnamate (21) (90%), which on thermolysis in boiling xylene gave the required indole (22) in 82% yield. The indole (22) was elaborated to the tricyclic skeleton of methoxatin by acid cleavage of the acetamide followed by reaction of the resulting 6-aminoindole with commercially available dimethyl 2-oxoglutaconate (23) in a Doebner - von Miller quinoline synthesis. This gave the pyrroloquinoline (24) in 55% overall yield. Although the intermediate 6-aminoindole has 2 possible sites (C-5 and C-7) for cyclisation, there was no trace of the product derived by cyclisation to the indole 5-position. That cyclisation would occur regiospecifically to the indole 7-position was expected on the basis of 'partial bond fixation' in indoles, a phenomenon which also has important consequences in our synthesis of murrayaquinone-B (*q.v.*).

172

Thus using the vinyl azide strategy, the key tricyclic intermediate (24) can be obtained in just 4 steps from a substituted benzaldehyde, and it only remained to introduce the *ortho*-quinone unit by oxidation of the central ring. This was readily achieved by catalytic hydrogenolysis of the benzyl group (89%), and oxidation of the resulting phenol using the nitroxide radical (25) to give methoxatin triester (26) (93%). This reagent was found to be much more satisfactory than the more usual nitroxide Fremy's salt because of its solubility in organic solvents. Finally, hydrolysis of the triester (26) gave methoxatin (1) itself.

3.2 *Synthesis of Murrayaquinone-B* [15]

The shrubs of the genus *Murraya* (Rutaceae) are the major source of the carbazole alkaloids, and have recently provided the first example of a naturally occurring carbazolequinone, murrayaquinone-B (2). Our overall strategy for the first total synthesis of this novel carbazole alkaloid involved the use of the vinyl azide indole forming reaction which incorporated a regioselective Claisen rearrangement, followed by annelation of the third ring employing a new method for the conversion of indoles into carbazoles.

(2)

The starting material was the known 4-(1,1-dimethylallyloxy)-benzaldehyde (27), which was condensed with methyl azidoacetate to give the azidocinnamate (28) (86%). On heating in toluene for 3 h, the vinyl azide decomposed to give the 7-(3,3-dimethylallyl)-6-hydroxyindole (29) (53%) by sequential indole formation and regioselective Claisen rearrangement (Scheme 4). As expected on the basis of earlier results,[16] the Claisen rearrangement of the 6-allyloxyindole derivative was completely regioselective, with no trace of the 5-(3,3-dimethylallyl) isomer being detected. This result is another consequence of 'partial bond fixation' in indoles.

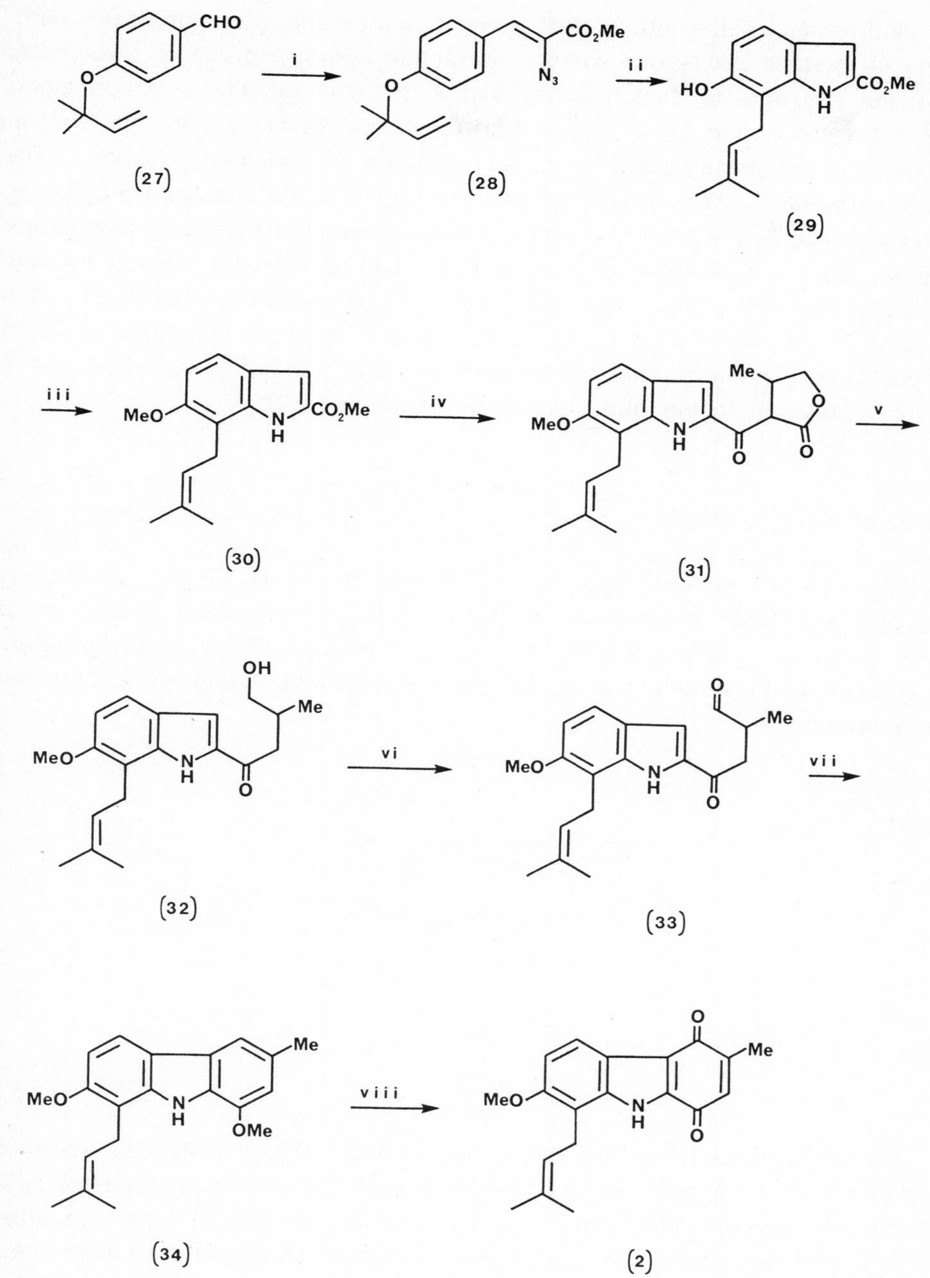

Scheme 4. *Reagents:* i, MeO₂CCH₂N₃, NaOMe, MeOH, -15°C; ii, toluene, reflux; iii, MeI, K₂CO₃, acetone; iv, 4-methylbutyrolactone, NaOMe, dioxan; v, NaOH, aq. dioxan; vi, PCC, CH₂Cl₂; vii, MeOH-BF₃; viii, *h*ν, air, MeOH.

Methylation of the indole (29) using iodomethane and potassium carbonate in acetone proceeded exclusively on the hydroxyl group to give indole (30), the substrate for fusion of the third ring. The remaining carbon atoms required were added by a simple Claisen condensation of the indole ester (30) with 4-methylbutyrolactone in the presence of sodium methoxide. The resulting lactone (31) underwent facile hydrolysis and decarboxylation on heating in aqueous dioxan containing a trace of sodium hydroxide to give the alcohol (32). Oxidation of the alcohol with pyridinium chlorochromate (PCC) in dichloromethane gave the aldehyde (33), which was cyclised to the carbazole (34) on stirring in boron trifluoride - methanol complex at room temperature. The final step of the synthesis was the photo-oxidation of the methoxycarbazole to the quinone, murrayaquinone-B (2).

3.3 *Synthesis of Siamine* [17]

Siamine (3), isolated from the leaves and seeds of *Cassia siamea*, is one of the simpler isoquinolone alkaloids, and we recognised that it could result from an intramolecular aza-Wittig reaction of an appropriately substituted azidocinnamate with the ester substituent acting as a potential methyl group.

(3)

Therefore a route to the protected azidocinnamate (37) was required (Scheme 5). This was prepared from known 3,5-dibenzyloxybenzoic acid (35)*via* the benzaldehyde (36). Thus the benzoic acid (35) was converted into its oxazoline derivative, which was lithiated, and quenched with ethyl chloroformate. The oxazoline group was transformed into the corresponding aldehyde under standard conditions, and finally the newly introduced ester was hydrolysed to the acid (36). This hydrolysis step was necessary in order to prevent problems in the condensation reaction with ethyl azidoacetate, which gave, after re-esterification, the vinyl azide (37), the key substrate for the intramolecular aza-Wittig reaction.

The desired cyclisation reaction proceeded in excellent (94%) yield when the azide (37) was heated with triethyl phosphite in refluxing benzene, and gave the 1-ethoxyisoquinoline-3-ester (38). The synthesis was completed by reduction of the ester and removal of the protecting groups. Although the ester (38) could not be reduced to the 3-methyl-isoquinoline (39) in a single step, this transformation could be achieved in a high yielding indirect sequence (70% overall). Finally, simultaneous cleavage of the ethyl and benzyl ethers with boron tribromide gave siamine (3).

Scheme 5. [Bzl = CH$_2$Ph] *Reagents:* i, SOCl$_2$; HOCH$_2$CMe$_2$NH$_2$; SOCl$_2$; ii, BuLi, DME, -78°C; EtO$_2$CCl; iii, MeI, reflux; NaBH$_4$, MeOH; KOH, aq. dioxan; iv, EtO$_2$CCH$_2$N$_3$, NaOEt, EtOH, -15°C; v, EtI, DBU, MeCN; vi, (EtO)$_3$P, benzene, reflux; vii, NaAl(OCH$_2$CH$_2$OMe)$_2$, THF; viii, SOCl$_2$, benzene; ix, LiAlH$_4$, THF; x, BBr$_3$, CH$_2$Cl$_2$.

3.4 *Synthesis of Lennoxamine* [18]

Members of the plant family Berberidaceae are a rich source of alkaloids, and recent investigations of Chilean barberries have provided the first examples of a new class of alkaloid, the isoindolobenzazepines, exemplified by lennoxamine (**4**).

(**4**)

Our synthesis of this novel alkaloid makes use of vinyl azide chemistry to prepare the key 2-aryl-3-benzazepine intermediate (**44**) (Scheme 6). The precursor to the required vinyl azide (**43**) was the highly substituted stilbene aldehyde (**42**), prepared from 6-bromopiperonal (**40**). The benzaldehyde (**40**) was protected as its ethylene acetal, lithiated, and quenched with dimethylformamide to give the mono-protected dialdehyde (**41**), which underwent Wadsworth - Emmons reaction with dimethyl (3,4-dimethoxy-2-methoxycarbonylbenzyl)phosphonate to give, after acidic hydrolysis of the acetal, the *trans*-stilbene aldehyde (**42**). Condensation of the aldehyde (**42**) with methyl azidoacetate gave the somewhat unstable vinyl azide (**43**). Decomposition of the vinyl azide (**43**) in boiling xylene resulted in selective interaction of the azide/nitrene with the styryl double bond, in preference to the free *ortho*-position of the aromatic ring, to give the 2-aryl-3-benzazepine (**44**) as the major (55%) product.

Treatment of the 3-benzazepine (**44**) with sodium cyanoborohydride in glacial acetic acid resulted in reduction of both double bonds of the 7-membered ring, and cyclisation of the resulting NH onto the ester group of the aromatic substituent to give the required isoindolo[1,2-*b*][3]-benzazepine skeleton (**45**) in 80% yield. The stereochemistry of the tetra-hydroisoindolobenzazepine was assigned as *cis* on the basis of [1]H n.m.r.

spectroscopy. The synthesis was completed by removal of the unwanted ester group at C-6, and this was achieved by reduction to the corresponding aldehyde and rhodium (I) catalysed decarbonylation, hydrolysis and decarboxylation being unsatisfactory. The overall yield of lennoxamine (**4**) from the 2-aryl-3-benzazepine (**44**) was 43%.

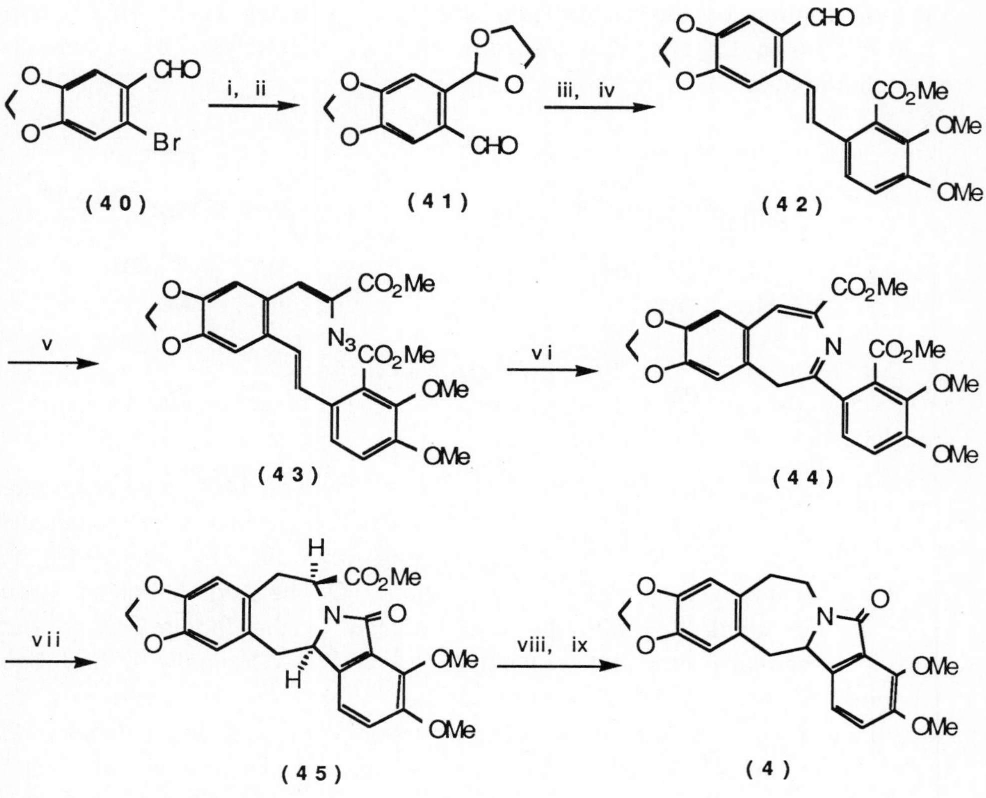

Scheme 6. *Reagents:* i, ethylene glycol, TsOH (cat.), toluene, reflux; ii, n-BuLi, -60°C, ether, then DMF; iii, dimethyl (3,4-dimethoxy-2-methoxycarbonylbenzyl)phosphonate, KOBu^t, THF; iv, dilute HCl; v, MeO$_2$CCH$_2$N$_3$, NaOMe, MeOH - THF; vi, xylene, reflux; vii, NaBH$_3$CN, AcOH; viii, DIBAL, toluene, -70°C; ix, (Ph$_3$P)$_2$PRh(CO)Cl (cat.), 1,3-bis(diphenylphosphino)-propane, xylene, reflux.

3.5 *Synthesis of the Phosphodiesterase Inhibitors PDE-I and PDE-II* [19]

The naturally occurring pyrrolo[3,2-e]indoles, PDE-I (**5**) and PDE-II (**6**), isolated from *Streptomyces* MD769-C6, are inhibitors of cyclic adenosine 3',5'-monophosphate phosphodiesterase.   Pyrroloindoles closely related to PDE-I and PDE-II also make up the central and right-hand sub-units of the potent antitumour antibiotic CC-1065 (**7**) (*q.v.*).   Although both PDE-I and PDE-II have been synthesised by classical routes, they both looked like suitable targets for the vinyl azide approach, whereby *both* pyrrole rings could be fused to a central benzene ring by thermolysis of appropriate azidocinnamates.

(5)   (6)

The starting material was the known benzaldehyde (**46**), easily prepared on a large scale by benzylation and bromination of isovanillin.   Condensation of the aldehyde (**46**) with methyl azidoacetate gave the vinyl azide (**47**), the substrate for the first vinylnitrene cyclisation.   On heating in boiling xylene for 1 h, the azide lost nitrogen and cyclised to the indole-2-carboxylate (**48**) in essentially quantitative yield (Scheme 7).   Originally it had been intended to remove the unwanted 2-ester substituent by hydrolysis and decarboxylation.   However, the decarboxylation step was unsatisfactory due to the instability of the resulting indole under the decarboxylation cond-itions of refluxing quinoline or glycerol.   Therefore an alternative decarb-onylation approach was adopted.   Thus the ester (**48**) was converted into the aldehyde (**49**) by reduction with lithium aluminium hydride, followed by reoxidation with manganese dioxide.   Decarbonylation of the aldehyde (**49**) was achieved by refluxing in mesitylene in the presence of catalytic amounts of bis(triphenylphosphine)(carbonyl)rhodium (I) chloride and 1,3-bis(diphenylphosphino)propane (dppp), and gave the indole (**50**) in 70% yield.

In order to introduce the second vinyl azide side chain, the 4-

bromoindole (50) had to be converted into the corresponding aldehyde. After considerable experimentation, it was discovered that the bromoindole (50) underwent halogen-metal exchange when treated with an excess of t-butyllithium in THF at -78°C *without* protection of the indole nitrogen. Quenching of the aryllithium species with DMF gave the aldehyde (51) in 72% yield, which on condensation with methyl azidoacetate gave the azidocinnamate (52) (71%), the substrate required for the second vinylnitrene cyclisation. Heating the vinyl azide (52) in boiling toluene resulted in an excellent (97%) yield of the key compound, the tricyclic pyrroloindole (53).

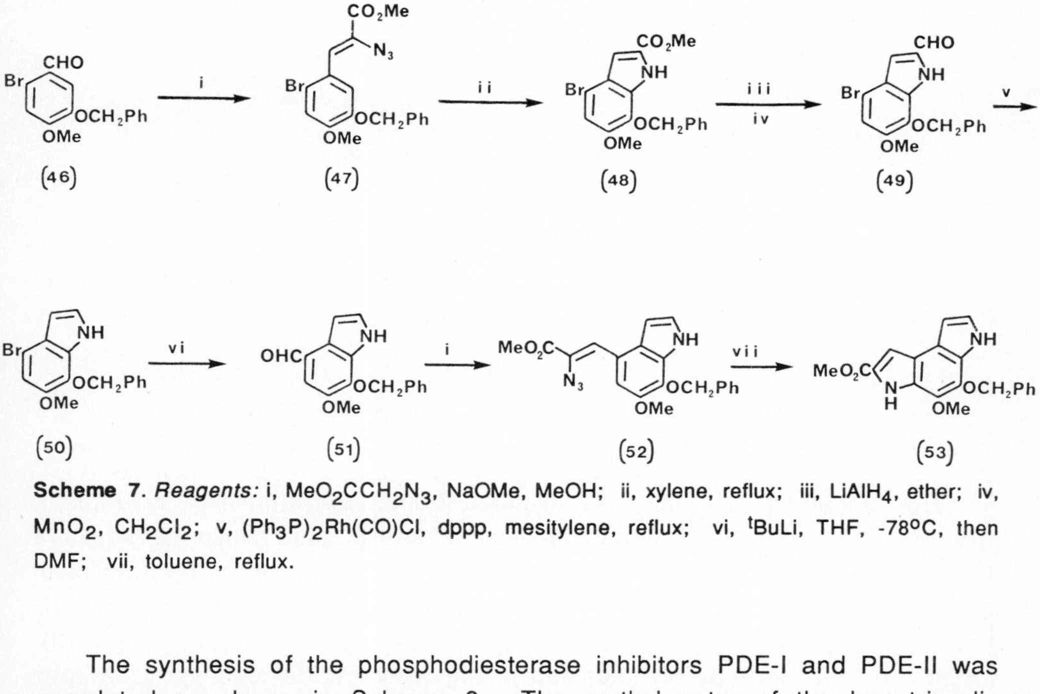

Scheme 7. *Reagents:* i, $MeO_2CCH_2N_3$, NaOMe, MeOH; ii, xylene, reflux; iii, $LiAlH_4$, ether; iv, MnO_2, CH_2Cl_2; v, $(Ph_3P)_2Rh(CO)Cl$, dppp, mesitylene, reflux; vi, ^tBuLi, THF, -78°C, then DMF; vii, toluene, reflux.

The synthesis of the phosphodiesterase inhibitors PDE-I and PDE-II was completed as shown in Scheme 8. The methyl ester of the key tricyclic compound (53) was first transesterified by reaction with benzyl alcohol and sodium benzyloxide in benzene. This not only meant that both the carboxylic acid and phenol protecting groups could be removed by hydrogenolysis in the final step, but also increased the solubility of the ring system. The resulting pyrroloindole benzyl ester was selectively reduced in the more electron rich pyrrole ring by treatment with sodium cyanoborohydride in acetic acid to give the pyrroloindole (54). The carbamoyl group was most conveniently introduced by reaction with trimethylsilyl isocyanate, and subsequent hydrogenolysis of the benzyl groups gave PDE-I (**5**). Similarly, PDE-II (**6**) was obtained by acetylation and hydrogenolysis of the pyrroloindole (**54**).

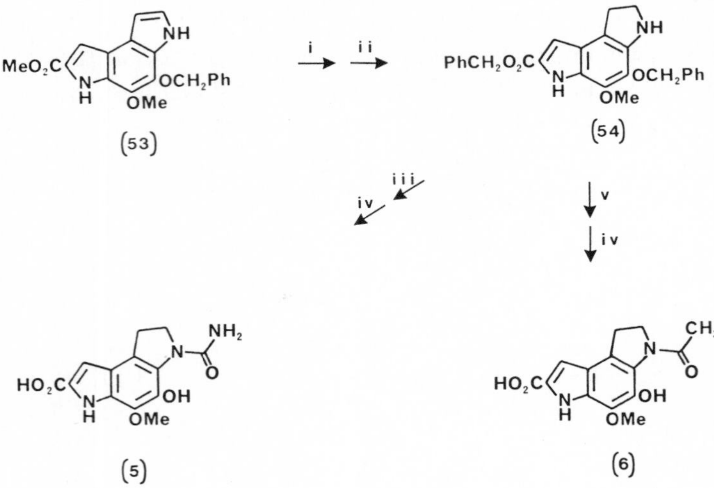

Scheme 8. *Reagents:* i, PhCH$_2$OH, PhCH$_2$ONa, benzene, reflux; ii, NaBH$_3$CN, AcOH; iii, Me$_3$SiNCO, benzene; iv, H$_2$, Pd-C, MeOH; v, Ac$_2$O, pyridine.

3.6 *Formal Synthesis of CC-1065* [20-22]

The antibiotic CC-1065 (**7**), isolated from *Streptomyces zelensis,* is one of the most potent antitumour agents known, and since its structure was determined by the Upjohn Company in 1980, has been the subject of considerable synthetic effort. Since the central and right hand units of CC-1065 are made up of pyrroloindoles very similar to PDE-I (**5**) and PDE-II (**6**), which we had already synthesised using vinyl azide chemistry, a total synthesis of the antibiotic using a similar approach seemed feasible.

As part of the Upjohn Company's work on CC-1065, the alkaline degradation of the antibiotic was studied. This gave two compounds:- the left-hand cyclopropapyrroloindole (**55**) and a compound known as PDE-I dimer (**56**). Subsequently Upjohn were able to show that their own synthetic cyclopropapyrroloindole (**55**) could be coupled to naturally derived dimer (**56**) in a formal synthesis of CC-1065 itself. Hence we concentrated on two aspects: the preparation of the left-hand unit (**55**) using vinyl azide chemistry, and the preparation of PDE-I dimer (**56**) by the coupling of suitable fragments.

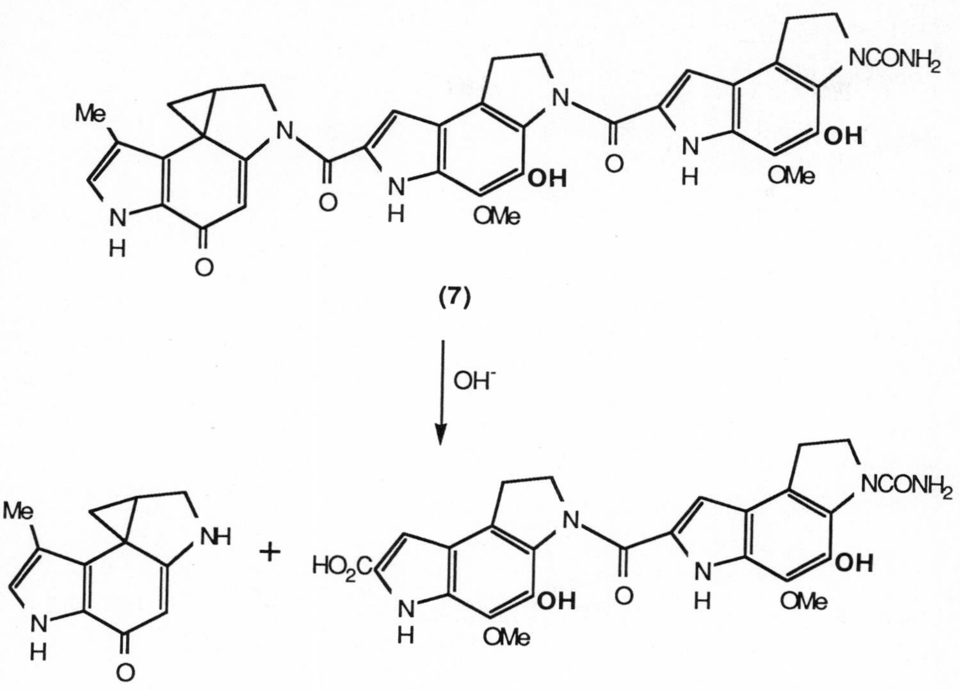

(7)

OH⁻

(55) + (56)

The synthesis of the cyclopropapyrroloindole (55) requires the fusion of 2 β-substituted pyrroles to a central benzene ring, and therefore necessitated a slightly different approach to that used for the PDE compounds. In fact the formation of 3-substituted indoles from vinyl azides of the type ArRC=CHN$_3$ is much less well established than the corresponding route to 2-substituted indoles. Although azides of the required type cannot be prepared in one step from ketones, they can be prepared indirectly. Our starting material was 5-benzyloxy-2-bromo-acetophenone (57), prepared by benzylation and bromination of commercially available 3-hydroxyacetophenone. Reaction of the ketone (57) with dimethyl sulphoxonium methylide in DMSO gave the epoxide (58) in excellent yield. The epoxide was ring opened by reaction with azide ion to give an azidoalcohol, which was dehydrated by treatment with thionyl chloride in pyridine to give the unstable vinyl azide (59). Heating the azide (59) in boiling mesitylene resulted in vinylnitrene cyclisation to give the unstable indole (60), which was immediately converted into its more easily handled benzenesulphonyl derivative (61) (Scheme 9).

Scheme 9. [Bs = SO₂Ph] *Reagents:* i, dimethyl sulphoxonium methylide, DMSO; ii, NaN₃, LiCl, DMF; iii, SOCl₂, pyridine; iv, mesitylene, reflux; v, NaH, THF, PhSO₂Cl; vi, *n*-BuLi, DME, then diethyl oxalate; vii, Ph₃P=CHCl, THF; viii, NaN₃, aq. DMF; ix, H₂, Pd-C; x, ref. 23.

The 4-bromoindole (**61**) was lithiated by reaction with butyllithium in 1,2-dimethoxyethane (DME), and subsequent quenching with diethyl oxalate gave the indole keto-ester (**62**). Reaction of the indole (**62**) with chloro-methylenetriphenylphosphorane, followed by treatment with sodium azide in aqueous DMF gave, *via* the corresponding vinyl chloride, the vinyl azide (**63**), the substrate for the second cyclisation reaction. Heating the azide (**63**) in boiling mesitylene resulted in cyclisation to the desired pyrrolo-indole (**64**) in moderate (43%) yield. Hydrogenolysis of the benzyl group gave the key compound (**65**), a known intermediate in a previous synthesis of the left-hand unit (**55**).[23] The pyrroloindole (**65**) was converted into the cyclopropapyrroloindole (**55**), the left-hand unit of CC-1065, using the conditions described by the Magnus group,[23] involving selective reduction of the ester-bearing double bond, *N*-acetylation, reduction of the ester group, Mitsunobu dehydration to the fused cyclopropane, and deprotection.

The building blocks for the synthesis of PDE-I dimer (56) were the monomeric pyrroloindoles (54) and (66) (Scheme 10). These were prepared from the pyrroloindole (53), the key intermediate in our synthesis of PDE-I and PDE-II. Transesterification of (53), and selective reduction gave the pyrroloindoline benzyl ester (54) as previously described, and hydrolysis of (53) in aqueous methanolic potassium hydroxide gave the corresponding acid (66) in excellent yield. The pyrroloindole acid (66) was coupled to the pyrroloindoline (54) *prior* to the reduction of the unsubstituted pyrrole ring

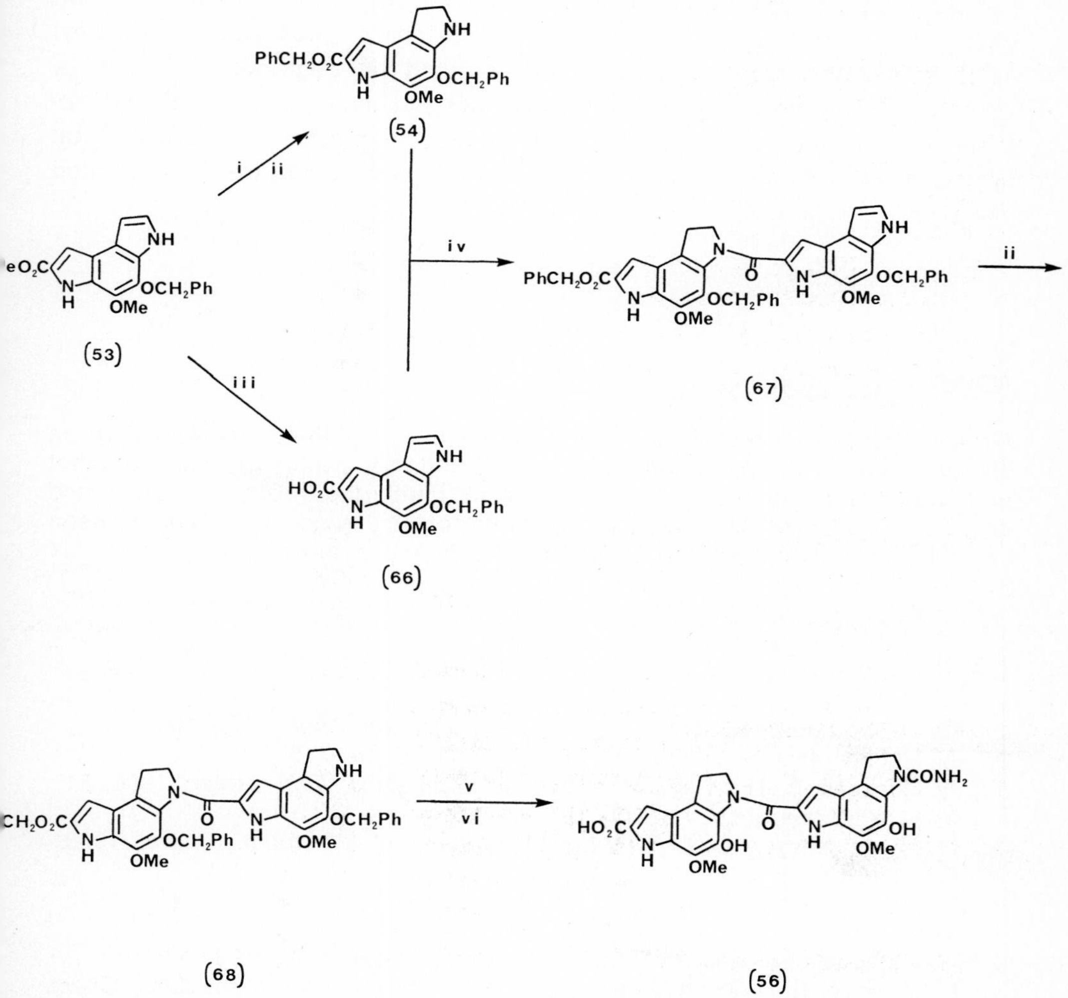

Scheme 10. *Reagents:* i, PhCH$_2$OH, PhCH$_2$ONa, benzene; ii, NaBH$_3$CN, AcOH; iii, KOH, aq. MeOH; iv, CMC, CH$_2$Cl$_2$; v, Me$_3$SiNCO, ClCH$_2$CH$_2$Cl; vi, H$_2$, Pd-C, DMF.

since the non-basic indole nitrogen was not expected to interfere. The coupling reaction was carried out in the presence of the carbodiimide reagent, 1-cyclohexyl-3-(2-morpholinoethyl)carbodiimide metho-4-toluenesulphonate (CMC) in dichloromethane, and gave the dimer (67) in 63% yield. Treatment of (67) with sodium cyanoborohydride in acetic acid resulted in selective reduction of the right hand pyrrole ring to give the protected pyrroloindoline dimer (68) (61%), which was carbamoylated by reaction with trimethylsilyl isocyanate in 1,2-dichloroethane. Finally, hydrogenolysis of the three benzyl groups over palladium-on-carbon in DMF gave PDE-I dimer (56). The synthetic PDE-I dimer produced by the vinyl azide route was identical to material obtained by degradation of natural CC-1065. Therefore this result, taken with Upjohn's coupling of *natural* PDE-I dimer with the left-hand unit (55), completes our formal total synthesis of CC-1065, in which *all 6 ring nitrogen atoms* are incorporated using vinyl azide chemistry.

ACKNOWLEDGEMENTS

I thank Professor Charles Rees, with whom much of this work was done in collaboration, for many stimulating discussions, and for his constant enthusiasm for organic chemistry. I also thank the group of skilful and dedicated co-workers without whom none of this work would have been possible: their names are recorded in the references below.

REFERENCES

1   G. L'abbe and A. Hassner, *Angew. Chem. Int. Edn. Engl.*, 1971, **10**, 98.
2   G. L'abbe, *Angew. Chem. Int. Edn. Engl.*, 1975, **14**, 775.
3   A. Hassner, N. H. Wiegand, and H. E. Gottlieb, *J. Org. Chem.*, 1986, **51**, 3176 and references therein.
4   H. Hemetsberger, D. Knittel, and H. Weidmann, *Monatsh. Chem.*, 1969, **100**, 1599.
5   H. Hemetsberger, D. Knittel, and H. Weidmann, *Monatsh. Chem.*, 1970, **101**, 161.
6   H. Hemetsberger and D. Knittel, *Monatsh. Chem.*, 1972, **103**, 194.
7   K. Isomura, G.-I. Ayabe, S. Hatano, and H. Taniguchi, *J. Chem. Soc., Chem. Commun.*, 1980, 1252.
8   T. L. Gilchrist, C. W. Rees, and J. A. R. Rodrigues, *J. Chem. Soc., Chem. Commun.*, 1979, 627; L. Henn, D. M. B. Hickey, C. J. Moody, and C. W. Rees, *J. Chem. Soc., Perkin Trans. 1*, 1984, 2189.

9   C. J. Moody and J. G. Ward, *J. Chem. Soc. , Perkin Trans. 1,* 1984, 2895.
10 D. M. B. Hickey, A. R. MacKenzie, C. J. Moody, and C. W. Rees, *J. Chem. Soc., Perkin Trans. 1,* 1987, 921.
11 D. M. B. Hickey, C. J. Moody, and C. W. Rees, *J. Chem. Soc. , Perkin Trans. 1*1986, 1113.
12 C. J. Moody and G. J. Warrellow, *J. Chem. Soc. , Perkin Trans. 1,* 1986, 1123.
13 C. J. Moody and G. J. Warrellow, *J. Chem. Soc. , Perkin Trans. 1,* 1987, 913.
14 A. R. MacKenzie, C. J. Moody, and C. W. Rees, *Tetrahedron,* 1986, **42,** 3259 and references therein.
15 T. Martin and C. J. Moody, *J. Chem. Soc., Chem. Commun.,* 1985, 1391.
16 C. J. Moody, *J. Chem. Soc., Perkin Trans. 1,* 1984, 1333.
17 M. Kennedy, C. J. Moody, C. W. Rees, and J. J. Vaquero, *J. Chem. Soc., Perkin Trans. 1,* 1987, 1395.
18 C. J. Moody and G. J. Warrellow, *Tetrahedron Lett.,* submitted for publication.
19 R. E. Bolton, C. J. Moody, C. W. Rees, and G. Tojo, *J. Chem. Soc., Perkin Trans. 1,* 1987, 931.
20 C. J. Moody, M. Pass, C. W. Rees, and G. Tojo, *J. Chem. Soc., Chem. Commun.,* 1986, 1062.
21 R. E. Bolton, C. J. Moody, C. W. Rees, and G. Tojo, *Tetrahedron Lett.,* 1987, **28,** 3163.
22 R. E. Bolton, C. J. Moody, M. Pass, C. W. Rees, and G. Tojo, *J. Chem. Soc., Perkin Trans. 1,* submitted for publication.
23 P. Magnus, T. Gallagher, J. Schultz, Y.-S. Or, and T. P. Ananthanarayan, *J. Am. Chem. Soc.,* 1987, **109,** 2706.

SYNTHESIS OF SOME ISOQUINOLINE ALKALOIDS VIA 8,14-CYCLOBERBINES

Miyoji Hanaoka

1 INTRODUCTION

Protoberberine alkaloids are biogenetic precursors of many related isoquinoline alkaloids (refs. 1, 2) such as benzo[c]-phenanthridine, protopine, phthalideisoquinoline, spirobenzyl-isoquinoline, indenobenzazepine, and rhoeadine alkaloids as shown in Scheme 1. The exact conversion routes from protoberberines to spirobenzylisoquinolines and indenobenzazepines have not yet been clarified in spite of some proposals (ref. 3). One possible candidate for a common intermediate in the above biogenetic conversion might be an 8,14-cycloberbine (ref. 4), the nomen-clature of which is based on the presence of C(8)-C(14) bond in a protoberberine skeleton.

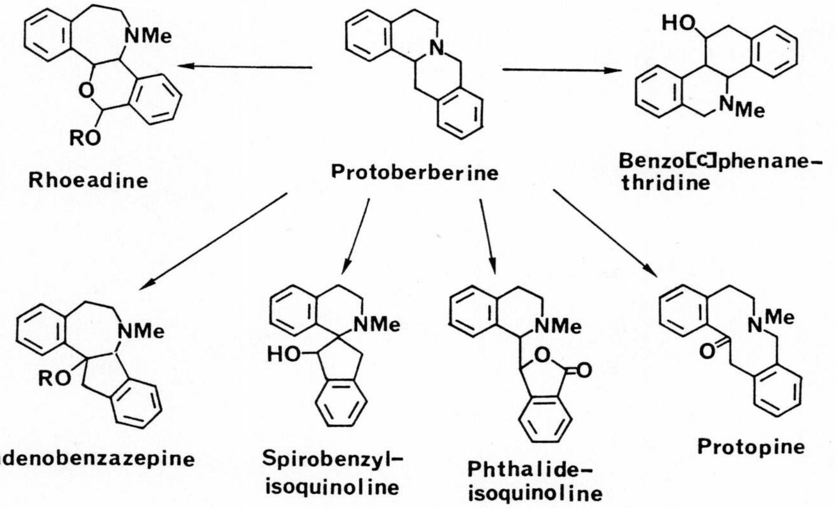

Rhoeadine

Protoberberine

Benzo[c]phenane-thridine

Indenobenzazepine

Spirobenzyl-isoquinoline

Phthalide-isoquinoline

Protopine

Scheme 1

Several chemical conversions involving an 8,14-cycloberbine as a hypothetical intermediate have so far been reported (refs. 5, 6) (Scheme 2), but isolation of such an intermediate has not yet been realized. As this unique 8,14-cycloberbine possesses a reactive aziridine ring in its molecule, it can be easily converted to a spirobenzylisoquinoline, an indenobenzazepine, or a protoberberine skeleton through regioselective C(8)-N, C(14)-N, or C(8)-C(14) bond cleavage, respectively (Scheme 3).

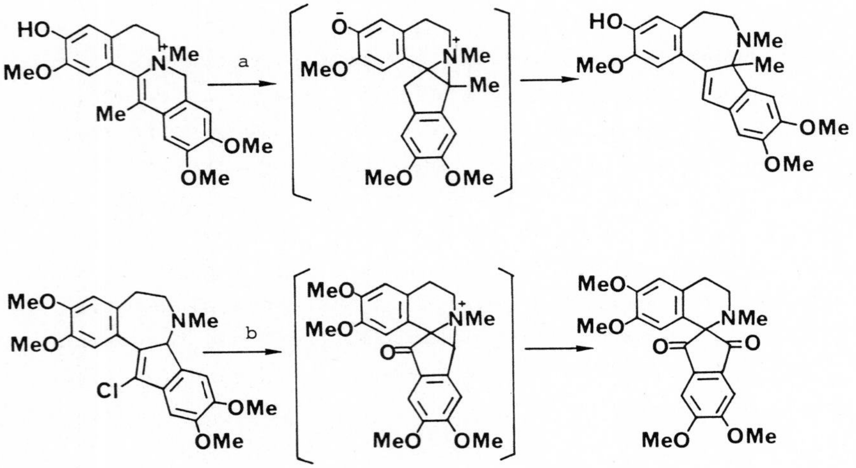

Scheme 2
a, NaOH; b, KMnO$_4$, AcOH, piperidine

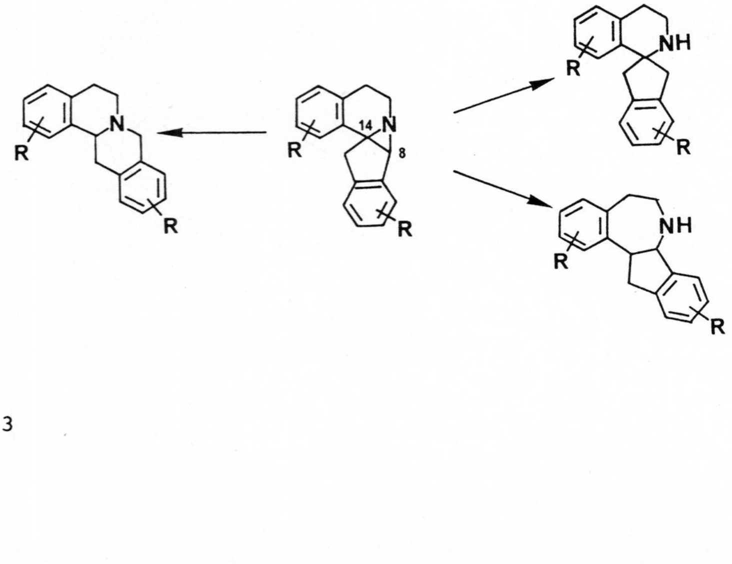

Scheme 3

If an 8,14-cycloberbine could be readily synthesized, a novel synthetic approach to the above alkaloids would be developed. Moreover, a spirobenzylisoquinoline and an indenobenzazepine have been transformed to a phthalideisoquinoline, a rhoeadine, a protoberberine, and a protopine (refs. 1, 2), which enhances the versatility of an 8,14-cycloberbine in the synthesis of various isoquinoline alkaloids.

This paper describes a synthesis of 8,14-cycloberbines and their application to that of protoberberine and related iso-quinoline alkaloids.

2 SYNTHESIS AND REACTION OF 8,14-CYCLOBERBINES

While many methods have been developed for the synthesis of an aziridine ring (ref. 7), however it dose not seem to be so easy to construct a doubly fused aziridine ring such as an 8,14-cyclo-berbine. Photochemical valence isomerization of six-membered heterocyclic betaines to bicyclo[3.1.0]hexane ring systems attracted our attention as a possible approach. Thus isoquinoli-nium betaine can isomerize to indanoaziridine (ref. 8), pyridinium betaine to 6-azabicyclo[3.1.0]hexane (ref. 9), isoquinolinium imide to indanoaziridine (refs. 10, 11), and pyrylium oxide to epoxypentenone(refs. 12, 13) as shown in Scheme 4. Although their isomerization yields are not satisfactory, this type of reaction seems to be suitable to synthesize 8,14-cyclo-berbines by a short route.

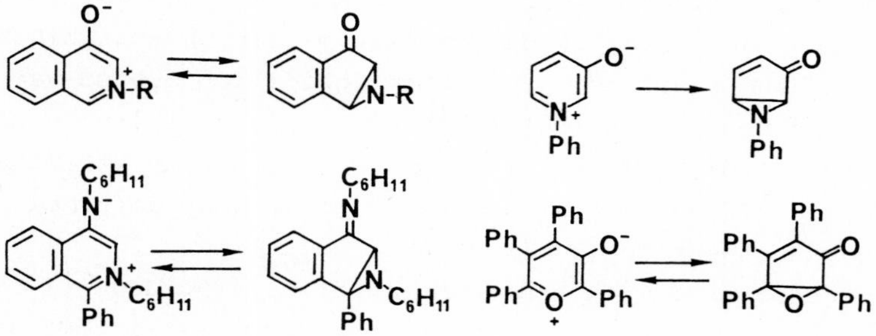

Scheme 4

Thus, our strategy involves the conversion of protoberberines to 13-oxo-8,14-cycloberbines _via_ protoberberinephenolbetaines. The striking advantages of the above method are that the starting protoberberines are easily accessible from nature and conventional synthesis and that the product possesses a carbonyl group convertible to various functionalities.

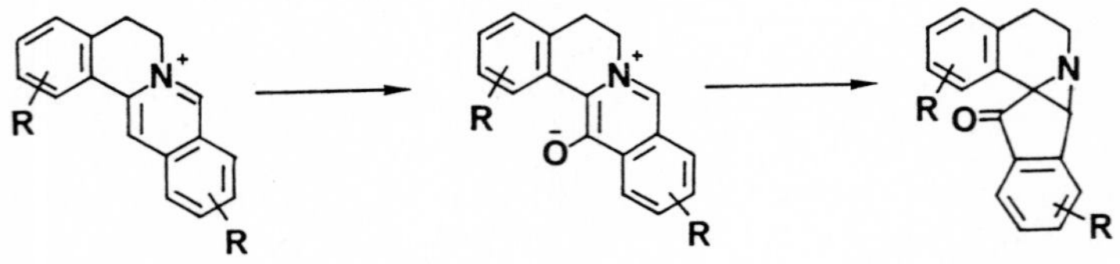

Scheme 5

2.1 Protoberberinephenolbetaines
2.1.1 Synthesis of Protoberberinephenolbetaines

Berberinephenolbetaine (5) was first obtained by Pyman (ref. 14) from acetoneberberine (3) _via_ neoxyberberine acetone (4) through oxidation with potassium permanganate followed by successive treatment with hydrochloric acid and sodium hydroxide (refs. 15, 16). Alternatively 5 can also be synthesized directly from 3 by oxidation with potassium permanganate or osmium tetroxide (ref. 15), and more conveniently and generally from dihydroberberine (6) by oxidation with m-chloroperbenzoic acid in dichloromethane (ref. 17). The latter oxidation method was first developed in the conversion of dihydrocoptisine (7) to coptisine-phenolbetaine (8) (ref. 18). Photooxygenation of 6 in methanol in the presence of rose bengal as a sensitizer also afforded berberinephenolbetaine (5) (refs. 19, 20).

Reduction of 5 with sodium borohydride afforded (+)-ophiocarpine (9) along with its diastereomer, (+)-epiophiocarpine (ref. 20) and similar reduction of the betaine (8) gave (+)-13β-hydroxystylopine (10) (ref. 18).

2.1.2 8-Methoxyprotoberberinephenolbetaines

Oxidation of berberine (1) with potassium ferricyanide

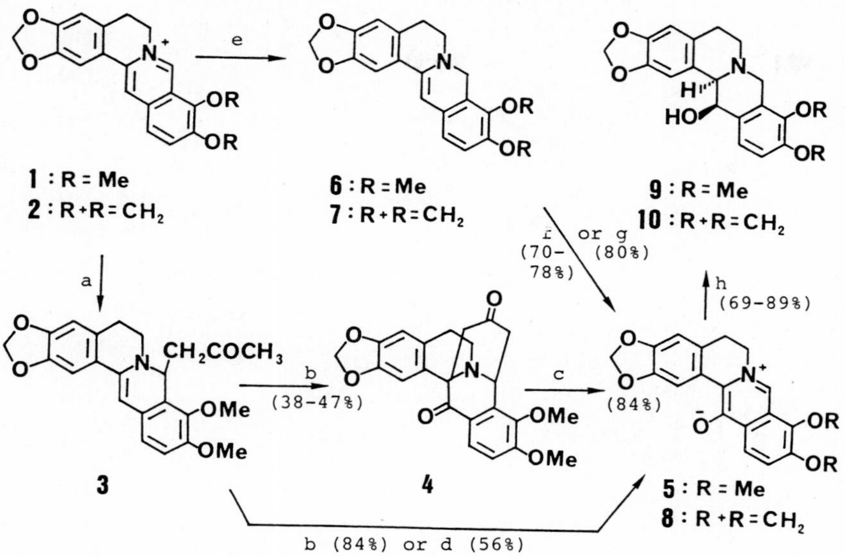

Scheme 6
a, CH_3COCH_3, KOH; b, $KMnO_4$; c, HCl; d, OsO_4; e, $LiAlH_4$; f, mCPBA;
g, hν, O_2, MeOH; h, $NaBH_4$

afforded dimeric oxybisberberine (11). Though its structure is not established, 11 gave 8-methoxyberberinephenolbetaine (12) and 1, on treatment with 10% methanolic hydrogen chloride (refs. 21, 22). On the other hand, irradiation of 1 in methanol in the presence of sodium methoxide and rose bengal in a stream of oxygen gave the labile 13-oxo compound (13), heating of which in methanol afforded 12 (refs. 23, 24). Similarly, on photooxygenation, palmatine and coptisine (2) provided the corresponding 8-methoxybetaines without isolation of the intermediate 13-oxo compounds (refs. 25, 26).

8-Methoxyberberinephenolbetaine (12) has an interesting structural feature, namely a masked carboxylic acid at C-8 and a 1,3-dipole. Hence, it was easily converted to phthalideiso-quinoline alkaloids, (±)-β-hydastine (14) and α-hydrastine (15), on successive treatment with water in ether, sodium borohydride, and methyl iodide (refs. 21, 22). 1,3-Dipolar cycloaddition of various dipolarophiles to 12 underwent smoothly to afford the bridged cycloadducts (16) accompanied with the benzazocines (17)

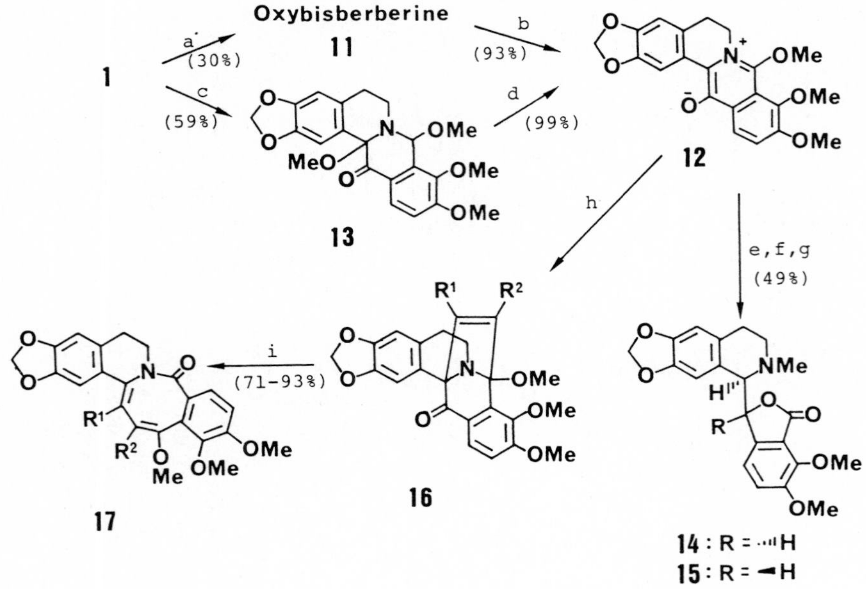

192

Scheme 7
a, $K_3Fe(CN)_6$; b, HCl/MeOH; c, $h\nu$, O_2, NaOMe, rose bengal/MeOH; d,
$\triangle$/MeOH; e, H_2O-Et_2O; f, MeI; g, $NaBH_4$; h, $R^1C{\equiv}CR^2$; i, $\triangle$/EtOH

(ref. 27). The former isomerized readily to the latter on heating
in ethanol. This isomerization cannot be observed in the cycloadduct from berberinephenolbetaine (5) having no methoxyl group at
C-8 (ref. 27).

2.2 8,14-Cycloberbines

2.2.1 Synthesis of 8,14-Cycloberbines

A solution of berberinephenolbetaine (5) in methanol was
irradiated through a Pyrex filter in a stream of nitrogen to
afford the 8,14-cycloberbine (18) in 70% yield (refs. 17, 28). The
reaction can be monitored by disappearance of the reddish orange
color of the starting betaine. The structure of 18 was established
by analysis of its spectral data: a carbonyl band at 1715 /cm in
the IR spectrum, a singlet due to H-8 at 4.00 ppm in the PMR
spectrum, and a doublet and a singlet due to C-8 and C-14 at 40.98
and 49.55 ppm, respectively, in the CMR spectrum.

The product (18) is unexpectedly stable because conrotatory opening of the aziridine ring is thermally disallowed in this ring system, but very labile to light below 300 nm. In fact, on irradiation without the filter 18 reverted back to the starting phenolbetaine (5) in 55% yield through photochemically allowed disrotatory ring opening. Even the colorless spot of 18 on TLC turned reddish orange on irradiation with 245 nm light. The existence of a photo-equilibrium between 18 and 5 was thus confirmed. Such photo-equilibrium has been reported as shown in Scheme 4.

Similarly, the 8-alkylphenolbetaines (19-22) were efficiently converted to the corresponding 8-alkyl-8,14-cycloberbines (23-26): 8-methyl (23, 85%), 8-ethyl (24, 77%), 8-benzyl (25, 37%), 8-allyl (26, 82%) (ref. 17). The present smooth photochemical valence isomerization was remarkable in terms of a high yield probably because of the stability of the products due to the ring system and conjugation with two aromatic rings. Thus, the 8,14-cyclo-berbines, previously reported only as hypothetical reaction intermediates, were actually obtained in hand.

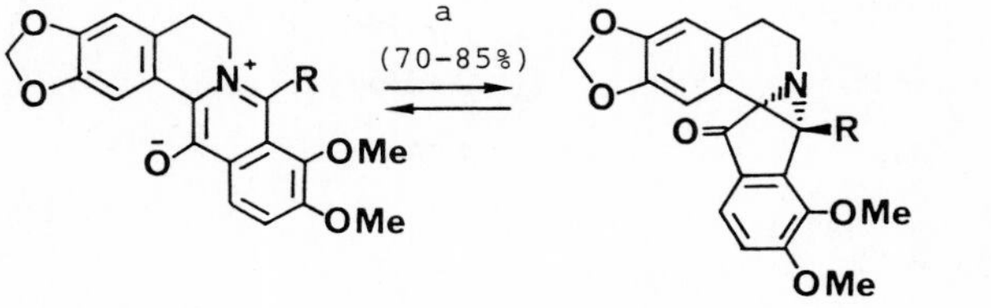

5: R = H
19: R = Me
20: R = Et
21: R = CH₂Ph
22: R = CH₂CH=CH₂

18: R = H
23: R = Me
24: R = Et
25: R = CH₂Ph
26: R = CH₂CH=CH₂

Scheme 8
a, hν/MeOH

Contrary to the above phenolbetaines, irradiation of 8-methoxyberberinephenolbetaine (12) in methanol gave the spiro-benzylisoquinoline (28) in 74% yield instead of the corresponding 8,14-cycloberbine (refs. 17, 29). The direct formation of the spiro compound (28) can be well rationalized in terms of the intermediary 8,14-cycloberbine (27). Due to the substitution of the methoxyl group at C-8 in 27, the C(8)-N bond is so labile that the aziridine ring cleaves immediately to give the spiro compound through the substitution with methanol.

Since 28 possesses a carbonyl and a ketal group on the five-membered ring, it can be transformed into a variety of spirobenzylisoquinolines substituted with various oxygenated functionalities as shown in Scheme 9 (ref. 17). These transformations provide an efficient method for a synthesis of spirobenzylisoquinoline alkaloids and their application is described later.

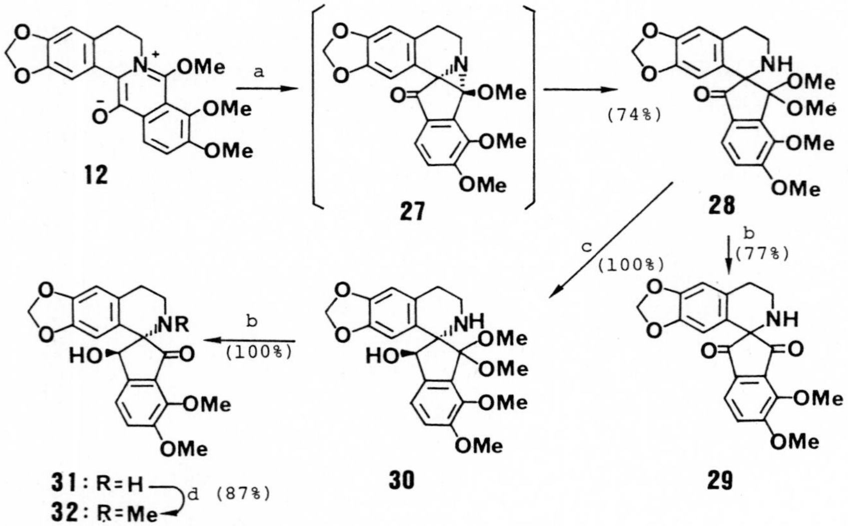

Scheme 9
a, hν/MeOH; b, 10% HCl; c, NaBH$_4$; d, HCHO, HCOOH

2.2.2 Reaction of 8,14-Cycloberbines

Preliminary experiments were carried out in order to find out how to cleave the cycloberbine regioselectively. Treatment of the cycloberbine (18) with ethyl chloroformate afforded the spirobenzylisoquinoline (33) in 70% yield through regioselective C(8)-N bond cleavage. Similar treatment of the 8-methylcycloberbine (23) gave the methylidene-spirobenzylisoquinoline (34) and that of the 8-ethyl derivative (24) afforded the Z- and E-ethylidene-spirobenzylisoquinoline (35 and 36) as well as the oxazolidinone (37) in 3:2:2 ratio (refs. 17, 28).

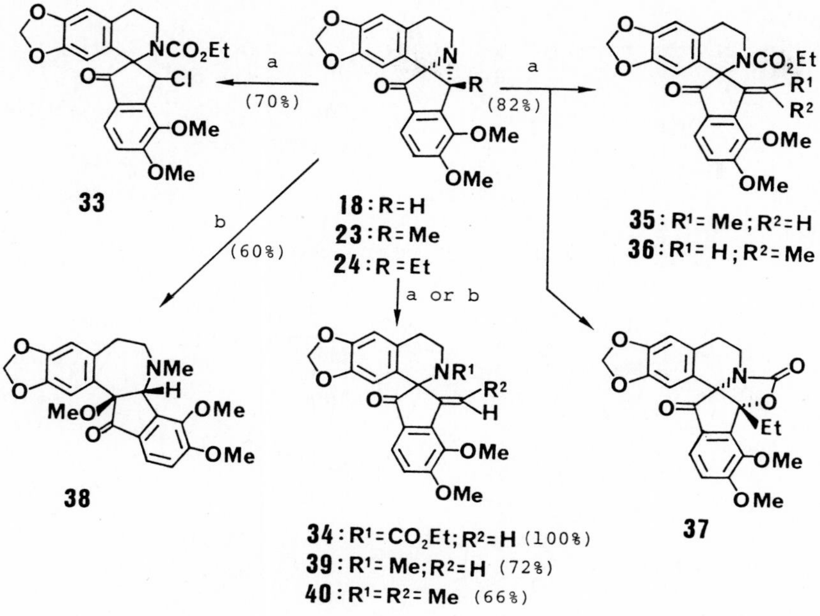

Scheme 10
a, ClCO$_2$Et; b, MeI

Methylation of 18 with methyl iodide in methanol effected regioselective C(14)-N bond cleavage to give the indenobenzazepine (38) in 60% yield. On the other hand, the 8-alkylcycloberbines (23 and 24) underwent N-methylation and subsequent Hofmann elimination to produce the methylidene- and E-ethylidene-spirobenzyliso-quinoline (39 and 40), respectively (refs. 17, 28).

Solvolysis of 18 in water or methanol in the presence of various acids such as hydrochloric acid, sulfuric acid, perchloric acid, trifluoroacetic acid, and p-toluenesulfonic acid afforded both trans- and cis-indenobenzazepines (41; 43 and 42; 44) in 78-94% yield by regioselective C(14)-N bond cleavage. The ratio of both isomers depended on the reaction conditions. The trans- and cis-indenobenzazepines (41; 43 and 42; 44) are kinetically and thermodynamically controlled products, respectively, and there is equilibrium between them. Therefore, the cis-derivative can be easily accessible by isomerization of the trans-derivative (refs. 30-33). Regioselective C(14)-N bond fission occurred probably via

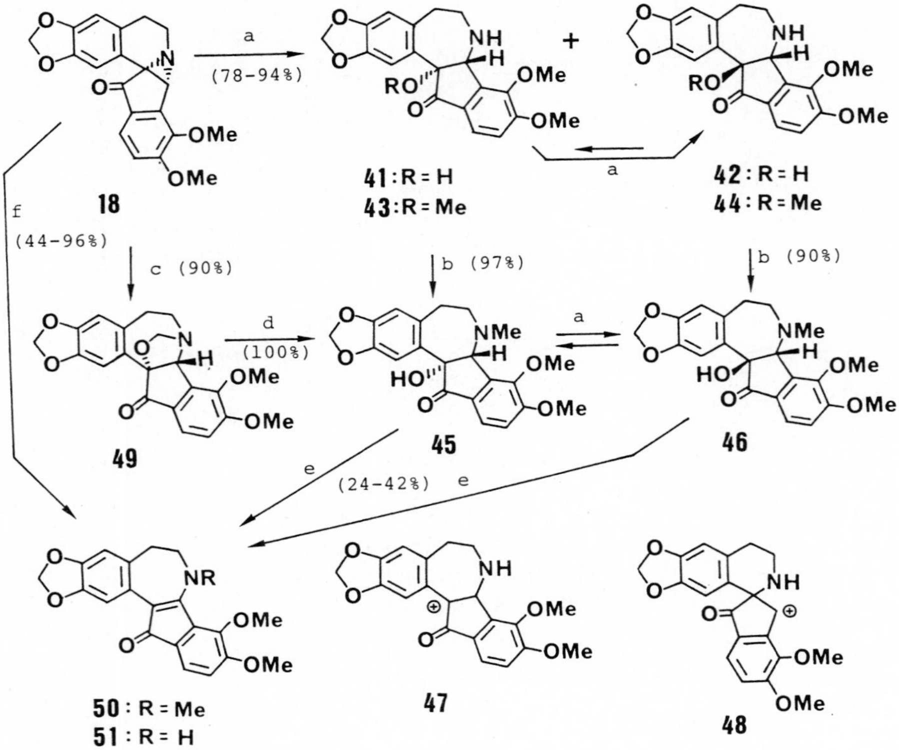

Scheme 11
a, H^+/H_2O or MeOH; b, MeI; c, HCHO; d, $NaBH_3CN$; e, $TiCl_4$ or $BF_3 \cdot Et_2O$; f, I_2/EtOH or AcOH or CF_3CO_2H/benzene or p-TsOH/benzene

the intermediate tertiary carbocation (47), which would be more stable than the alternative secondary carbocation (48) leading to a spirobenzylisoquinoline.

Murugesan et al. developed an elegant method for a synthesis of the trans-derivatives (ref. 34). Treatment of the cycloberbine (18) with formaldehyde gave the oxazolidinone (49), which was reduced with sodium cyanoborohydride to afford stereospecifically the trans-indenobenzazepine (45) in 90% overall yield.

On treatment with Lewis acid such as titanium tetrachloride or boron trifluoride etherate, both trans- and cis-indeno-

benzazepine (45 and 46) afforded the vinylogous amide (50) in 25-
42% yield (refs. 31,33). The corresponding secondary amide (51)
was directly obtained from 18 by treatment with various reagents:
iodine in ethanol (77%), acetic acid (56%), trifluoroacetic acid
in benzene (44%), or p-toluenesulfonic acid in benzene (96%)
(refs. 30, 32, 33).

In contrast to the above results, solvolysis of 8-alkylcyclo-
berbines (23, 24 and 26) in 10% hydrochloric acid or in methanol
in the presence of trifluoroacetic acid afforded exclusively
spirobenzylisoquinolines instead of indenobenzazepines via the
stable tertiary carbocations (52) as shown in Scheme 12 (refs. 35,
36).

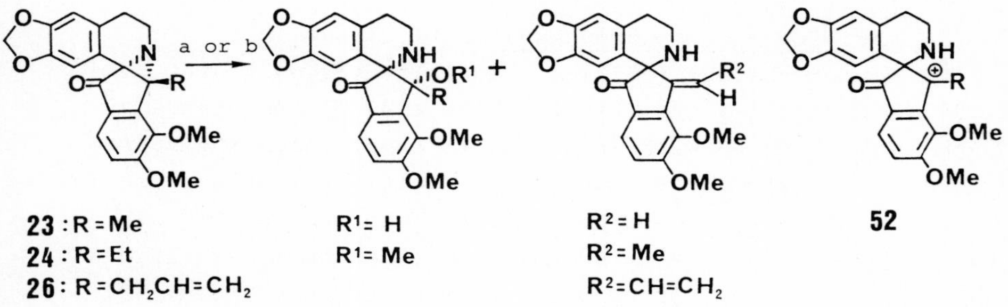

23 : R = Me
24 : R = Et
26 : R = CH₂CH=CH₂

R¹= H
R¹= Me

R²= H
R²= Me
R²= CH=CH₂

52

Scheme 12
a, HCl; b, CF₃CO₂H/MeOH

3   SYNTHESIS OF SPIROBENZYLISOQUINOLINE ALKALOIDS

More than thirty spirobenzylisoquinoline alkaloids (refs.
1, 2, 37-41) have been isolated and are classified into five
groups according to the substituents on the five-membered ring.
Because of an interesting spiro structure of these alkaloids, many
synthetic methods have so far been reported and well reviewed
(refs. 1, 2, 39). A novel method for a synthesis of a spirobenzylisoquinoline skeleton via an 8,14-cycloberbine was developed as
described in Section 2.2 and this method was applied to a
synthesis of almost all types of spirobenzylisoquinoline
alkaloids.

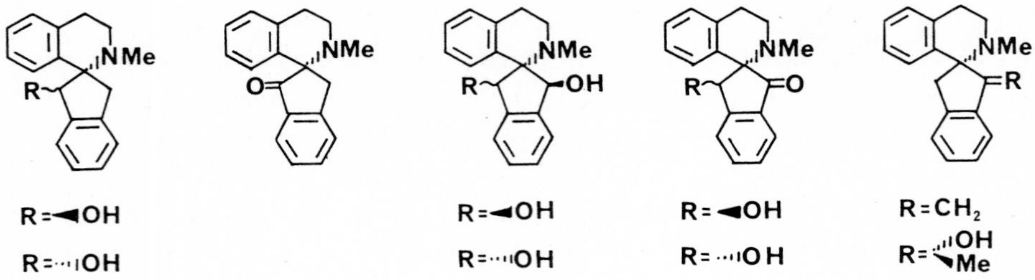

R=◀OH

R=···ıOH

R=◀OH

R=···ıOH

R=◀OH

R=···ıOH

R=CH₂

R=ᵘOH Me

Scheme .13

3.1 Synthesis via 8,14-cycloberbines
3.1.1 Fumaricine and Dihydrofumariline 1

The protoberberine (53) was readily converted to the 8,14-cycloberbine (55) via the phenolbetaine (54) by the standard method. Heating of 55 with ethyl chloroformate gave the spiro-benzylisoquinoline (56) through regioselective cleavage. Dehalo-genation of 56 by hydrogenolysis followed by stereoselective reduction with lithium aluminum hydride provided (±)-fumaricine (58) having a hydroxyl group trans to the nitrogen (refs. 42, 43).

Dihydrofumariline 1 (65) possessing a hydroxyl group cis to the nitrogen cannot be analogously synthesized because hydride reduction of the ketone on five-membered ring in spirobenzyl-isoquinolines afforded usually trans-alcohol due to the hydride attack from the less hindered side as seen in the case of fumaricine. An 8,14-cycloberbine, however, would change the less hindered side due to its rigid structure. Thus, on reduction with sodium borohydride, the cycloberbine (61) derived from the protoberberine (59) afforded stereoselectively the cis-alcohol (62) in 99% yield (ref. 44).

Reductive cleavage of the aziridine ring in the alcohol (62) with sodium cyanoborohydride in the presence of p-toluenesulfonic acid produced the spiro amino-alcohol (63) in 81% yield, N-methyl-ation of which with methyl iodide provided (±)-dihydrofumariline 1 (65) in 56% yield. Alternatively, N-methylation was performed more efficiently via the oxazolidine, namely, treatment of 63 with 37% formaldehyde followed by exposure to sodium cyanoborohydride yielded (±)-dihydrofumariline 1 in 90% yield via the oxazolidine

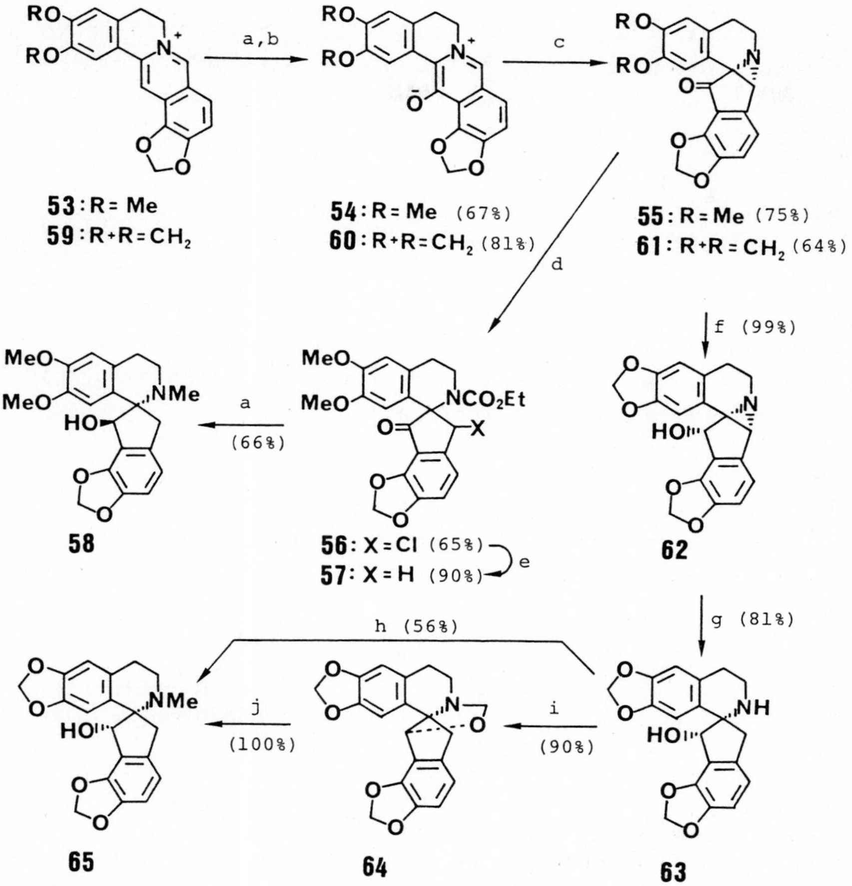

Scheme 14
a, LiAlH$_4$; b, mCPBA; c, hν/MeOH; d, ClCO$_2$Et; e, H$_2$/Pd-C; f, NaBH$_4$; g, NaBH$_3$CN, p-TsOH/THF; h, MeI; i, HCHO; j, NaBH$_3$CN, 10% HCl/MeOH

(64) (ref. 44). Thus we can control the stereochemistry of the hydroxyl group by changing the substrates in hydride reduction.

3.1.2 Ochrobirine and Corydaine

The spirobenzylisoquinoline alkaloids having two substituents in five-membered ring, ($\pm$)-ochrobirine (71) and ($\pm$)-corydaine (74) were stereoselectively synthesized from the protoberberine alkaloid, coptisine (2). Reduction of the cycloberbine (66) derived from coptisinephenolbetaine (8) with lithium tri-$\underline{t}$-butoxyhydride afforded the <u>trans</u>-alcohol (<u>67</u>, 64%) along with the <u>cis</u>-alcohol (<u>68</u>, 20%) (ref. 45). The reverse stereoselectivity with this reagent in comparison with the previous reduction with sodium borohydride is well interpreted by the initial formation of the complex from the nitrogen of the cycloberbine with this reagent followed by intramolecular hydride attack of the complex to the carbonyl group.

On treatment with ethyl chloroformate, the <u>trans</u>-alcohol (<u>67</u>) underwent regioselective C(8)-N bond cleavage and simultaneous oxy-functionalization at C-8 with correct stereochemistry to produce the oxazolidinone (<u>69</u>) in 75% yield. Basic hydrolysis of <u>69</u> afforded the diol (<u>70</u>), which was methylated with methyl iodide to give (<u>+</u>)-ochrobirine (<u>71</u>) (ref. 45).

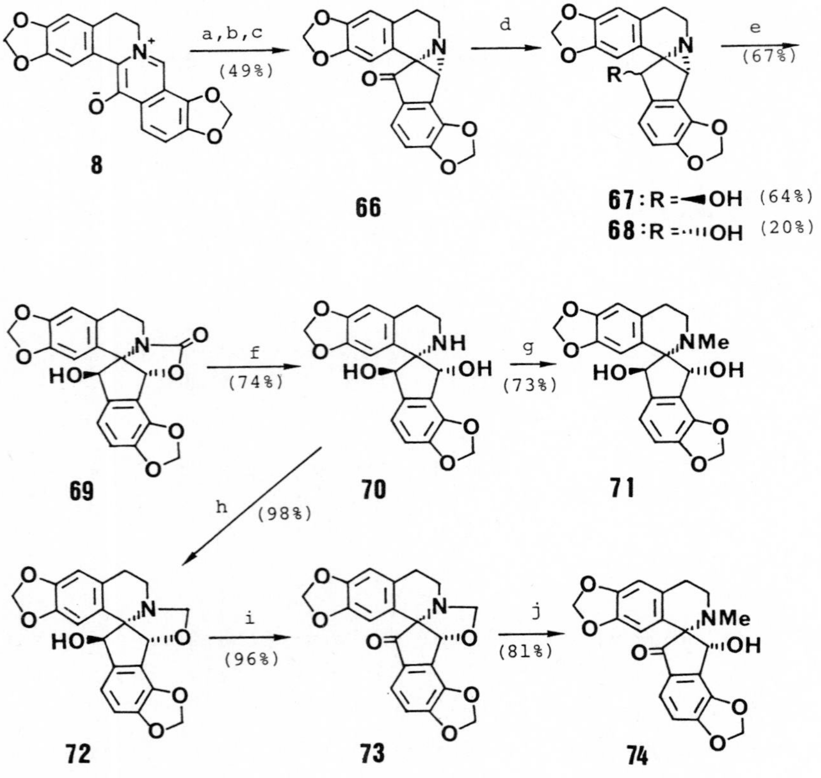

Scheme 15

a, LiAlH$_4$; b, mCPBA; c, hν/MeOH; d, LiAlH(OBu^t)$_3$; e, ClCO$_2$Et; f, KOH; g, MeI; h, HCHO; i, Ag$_2$CO$_3$; j, NaBH$_3$CN

Concomitant protection of the nitrogen and the cis-hydroxyl group in 70 was achieved by reaction with formaldehyde to give the oxazolidine (72) in 98% yield. Oxidation of the unprotected trans-hydroxyl group in 72 with silver carbonate on Celite provided the keto-oxazolidine (73) in 96% yield. Treatment of 73 with sodium cyanoborohydride effected both N-methylation and deprotection of the cis-hydroxyl group to produce (±)-corydaine (74) in 81% yield (ref. 46).

3.1.3 Raddeanamine and Ochotensimine

Ochotensimine (82) is the first spirobenzylisoquinoline alkaloid to be synthesized through the Wittig reaction of the corresponding ketone (83) (refs. 47, 48). Raddeanamine (81) having a tertiary cis-hydroxyl group and a methyl group in five-membered ring appears to be easily accessible to synthesis through the methylation of the same ketone (83). The methylation reagent, however, would approach to the carbonyl group from the same side of the nitrogen as in the case of reduction with hydride resulting in a product with a trans-hydroxyl group. This was verified by the fact that methylation of the analogous ketone (84) with methyl-lithium gave the trans-alcohol (85) (ref. 49). Therefore, another method should be developed for a synthesis of raddeanamine. As described in Section 2.2.2, nucleophilic substitution of 8-alkyl-cycloberbines effected regioselective C(8)-N bond cleavage to give exclusively spirobenzylisoquinolines. Therefore, an intramolecular nucleophilic opening of the aziridine ring in 8-methylcycloberbine would be expected to afford the cis-alcohol by stereoselective control.

Methylation of the protoberberine (53) with methylmagnesium iodide followed by oxidation with m-chloroperbenzoic acid gave the phenolbetaine (76), irradiation of which afforded the cycloberbine (77). Sodium borohydride reduction of 77 did not occur with complete stereoselectivity due to the presence of the extra methyl group on the aziridine ring, affording the cis-alcohol along with the trans-alcohol. Subsequent treatment of this mixture (78) with formaldehyde afforded a mixture of two diastereomeric oxazolidines (79) in 79% yield in a ratio of 7.8:1. Deoxygenation of 79 was accomplished by chlorination with thionyl chloride followed by hydrogenolysis with tri-n-butyltin hydride in the presence of azabisisobutyronitrile to yield the product (80) in 52% yield. The product (80) was reduced with sodium cyanoborohydride to produce

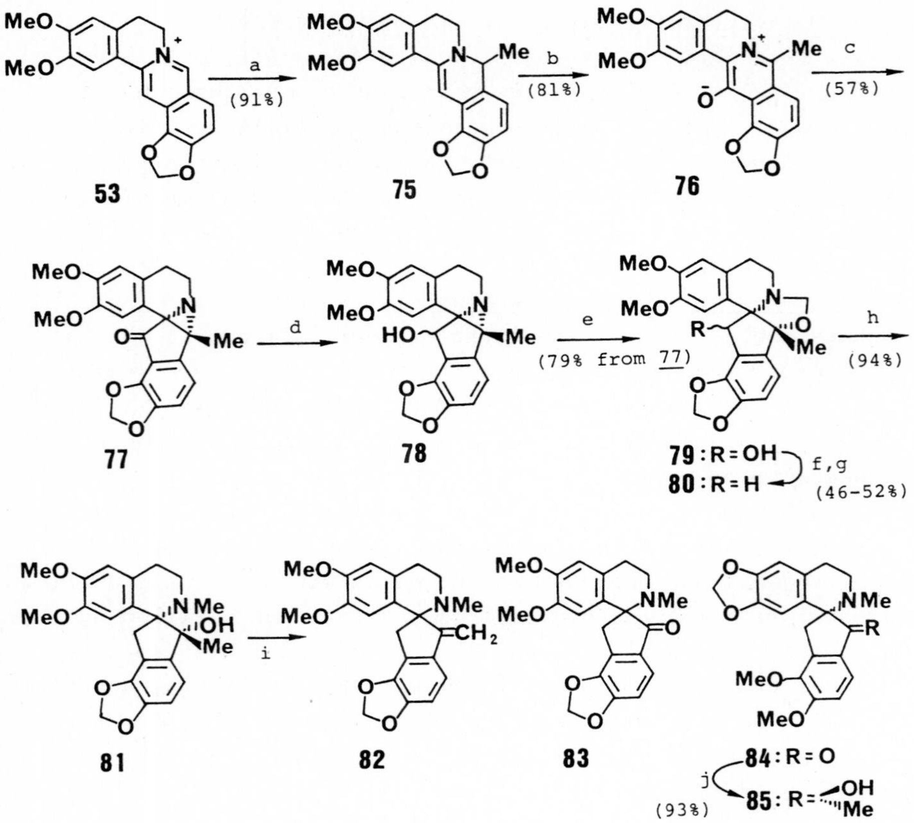

Scheme 16
a, MeMgI; b, mCPBA; c, hν/MeOH; d, NaBH₄; e, HCHO; f, SOCl₂;
g, n-Bn₃SnH; h, NaBH₃CN; i, Δ, KHSO₄; j, MeLi

(±)-raddeanamine (81) in 94% yield (ref. 49). Raddeanamine has
already been transformed to ochotensimine (82) by dehydration
(ref. 50).

3.2 Synthesis from 8-Methoxyprotoberberinephenolbetaines

Direct formation of the spirobenzylisoquinoline by irradiation of 8-methoxyberberinephenolbetaine was described in Section
2.2.2. This method was effectively applied to a synthesis of
various spirobenzylisoquinoline alkaloids having two oxygenated
substituents and found to be complementary to the previous method
via 8,14-cycloberbines described in Section 3.1.

3.2.1 Sibiricine, Sewercinine, Raddeanone, Raddeanine, and Raddeanidine

Photooxygenation of the protoberberine (59) in methanol in the presence of sodium methoxide afforded the 8-methoxyphenol-betaine (86) in 66% yield, which was further irradiated to give the spirobenzylisoquinoline (87) in 64% yield. Reduction of 87 with sodium borohydride gave stereoselectively the trans-alcohol (88) quantitatively. Eschweiler-Clarke reaction of 88 effected N-methylation and concomitant deketalization to provide (±)-sibiricine (89) in 91% yield (ref. 51). Thus, (±)-sibiricine was stereoselectivly synthesized from the corresponding protoberberine in four steps. (±)-Sibiricine (89) was reduced with sodium boro-hydride to afford (±)-sewercinine (90) (ref. 51).

Similarly, the protoberberine (53) was converted efficiently and stereoselectively to the spirobenzylisoquinoline alkaloids, (±)-raddeanone (94) and (±)-raddeanine (96) as shown in Scheme 17.

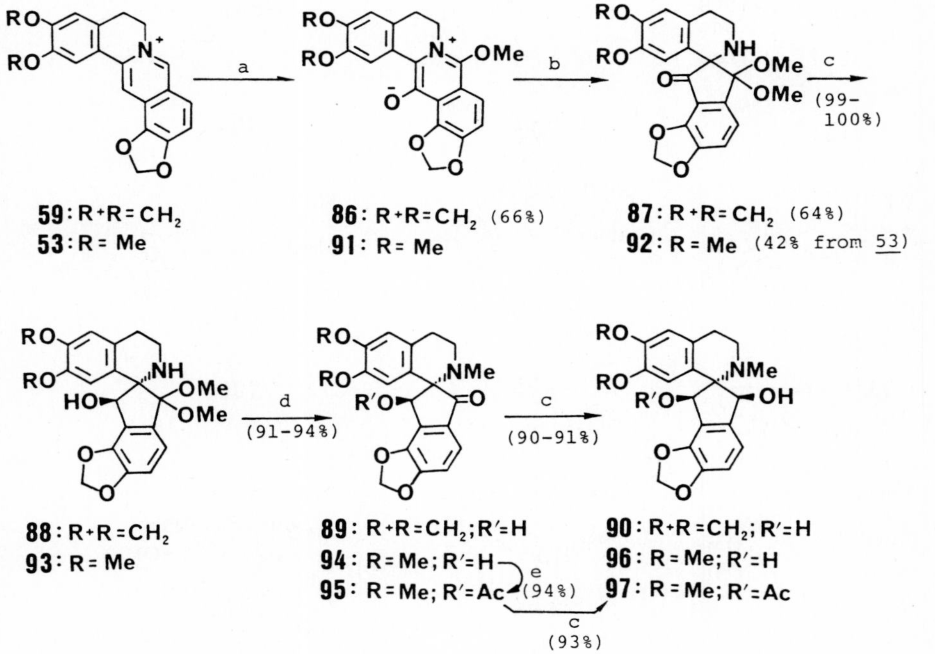

59: R+R=CH₂
53: R= Me

86: R+R=CH₂ (66%)
91: R= Me

87: R+R=CH₂ (64%)
92: R= Me (42% from 53)

88: R+R=CH₂
93: R= Me

89: R+R=CH₂;R′=H
94: R=Me;R′=H
95: R=Me;R′=Ac (94%)

90: R+R=CH₂;R′=H
96: R=Me;R′=H
97: R=Me;R′=Ac

(93%)

Scheme 17
a, hν, O₂, NaOMe, rose bengal/MeOH; b, hν/MeOH; c, NaBH₄;
d, HCHO, HCOOH; e, Ac₂O, pyr.

Acetylation of <u>94</u> followed by reduction with sodium borohydride gave (±)-raddeanidine (<u>97</u>) (ref. 52).

3.2.2 <u>Yenhusomidine and Yenhusomine</u>

Yenhusomidine (<u>108</u>) possessing a <u>cis</u>-hydroxy group is a diastereomer of raddeanone (<u>94</u>) and was synthesized through the

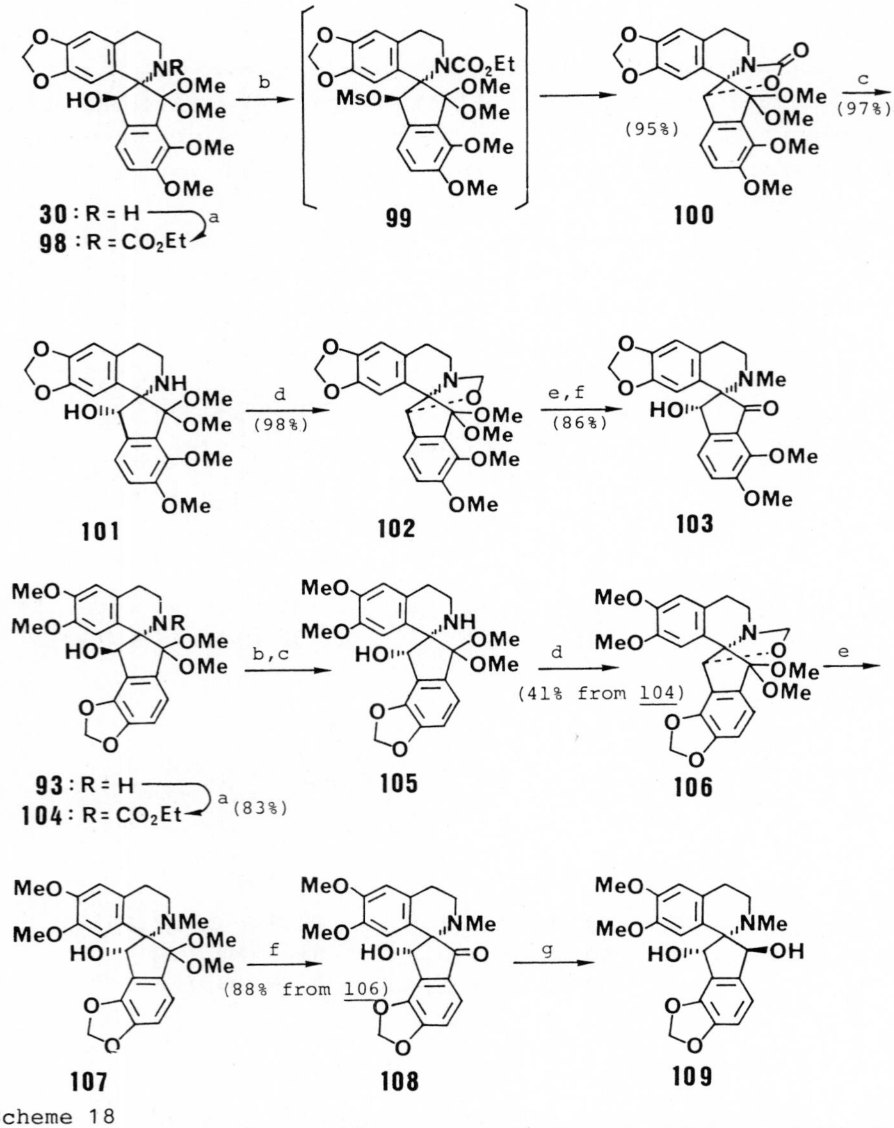

Scheme 18

a, ClCO$_2$Et; b, MsCl, Et$_3$N/THF; c, KOH/MeOH; d, HCHO; e, NaBH$_3$CN; f, HCl; g, NaBH$_4$

inversion of a hydroxyl group by neighboring group participation of the urethane group.

In a preliminary experiment, the inversion of the hydroxyl group was tried. The spirobenzylisoquinoline (30) derived from berberine was transformed to the urethane (98), the hydroxyl group of which could not be inverted by a usual method such as the Mitsunobu reaction. The urethane (98) was treated with methane-sulfonyl chloride in the presence of triethylamine to give the oxazolidinone (100) in 95% yield via the methanesulfonate (99) through intramolecular substitution with the ethoxycarbonyl group. Hydrolysis of 100 with potassium hydroxide gave successfully the inverted diastereomer (101) of the starting spiro compound (30) in 97% yield. The product (101) was converted to the yenhusomidine analogue (103) (ref. 53).

The above method was applied to a synthesis of yenhusomidine. The alcohol (93) was treated with ethyl chloroformate to give the urethane (104) in 83% yield. Sequential treatment of 104 with methanesulfonyl chloride, potassium hydroxide, and formaldehyde caused formation of the oxazolidinone ring, hydrolysis, and formation of oxazolidine ring, respectively, to produce the oxazolidine (106) in 41% overall yield via the inverted cis amino-alcohol (105). Reduction of 106 with sodium cyanoborohydride gave the ketal (107), which was hydrolyzed to afford (+)-yenhusomidine (108) in 88% yield (ref. 53). This alkaloid has already been converted to (+)-yenhusomine (109) (ref. 54).

3.3   Synthesis from Indenobenzazepines

An interesting rearrangement of an indenobenzazepine into spirobenzylisoquinoline via an 8,14-cycloberbine was reported, though the intermediate was not isolated (ref 55).

Treatment of O-methylfumarofine (110) with trifluoroacetic anhydride in pyridine, followed by ammonium hydroxide work-up afforded the spirobenzylisoquinoline (112) in 86% yield through the 8,14-cycloberbinium cation (111). The product was reduced with sodium borohydride to give (+)-raddeanine (96).

Similarly both diastereomers (45 and 46) of indenobenz-azepines afforded the same spirobenzylisoquinoline (115) in excellent yield. The newly produced hydroxyl group is always trans to the nitrogen and stereochemistry at C-14 of the starting indenobenzazepine is indifferent to this rearrangement. Thus, the

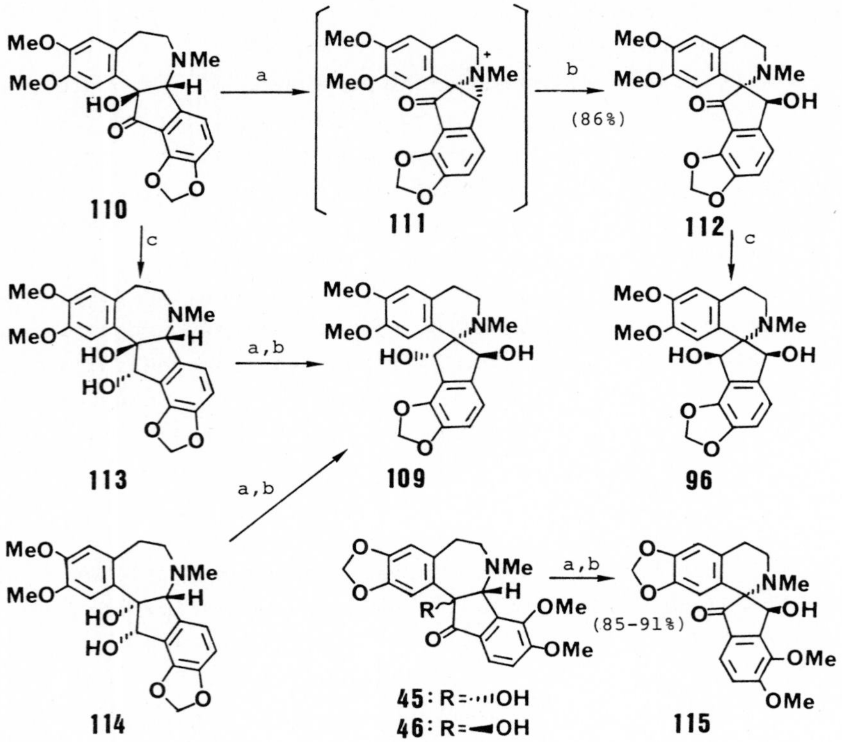

Scheme 19
a, $(CF_3CO)_2O$, pyr.; b, NH_4OH; c, $NaBH_4$

diol (113) derived from O-methylfumarofine (110) afforded (±)-yenhusomine (109), which was also obtained from the diastereomer (114).

4 SYNTHESIS OF INDENOBENZAZEPINE ALKALOIDS

Indenobenzazepine alkaloids do not possess an isoquinoline skeleton in their molecules, but they belong to the isoquinoline alkaloids since they are probably derived biogenetically from protoberberine alkaloids. Indenobenzazepine alkaloids were recently isolated (ref. 56) and some alkaloids assigned earlier as spirobenzylisoquinoline structure were revised to indenobenz-azepine alkaloids (refs. 57, 58). Some of these alkaloids were synthesized via 8,14-cycloberbines.

4.1 Synthesis from Spirobenzylisoquinolines

4.1.1 Lahorine and Lahoramine

Rearrangement of a spirobenzylisoquinoline to an indenobenz-azepine skeleton was first developed by Irie et al. in their synthesis of rhoeadine alkaloids (Section 5) (refs. 59, 60). This rearrangement was utilized to a synthesis of indenobenzazepine alkaloids.

Dihydroparfumidine (58) derived from the spirobenzyliso-quinoline alkaloid, parfumidine (116) was treated with methane-sulfonyl chloride and triethylamine to afford the indenobenz-azepine (118) in 75% yield probably via the 8,14-cycloberbine (117). Dehydrogenation of 118 with iodine in ethanol provided lahoramine (119) in 20% yield. Similarly dihydrofumariline 2 (121) was converted to lahorine (124) (ref. 56).

116 : R = Me
120 : R + R = CH$_2$

58 : R = Me
121 : R + R = CH$_2$

117 : R = Me
122 : R + R = CH$_2$

118 : R = Me (75%)
123 : R + R = CH$_2$

119 : R = Me (20%)
124 : R + R = CH$_2$

Sheme 20
a, NaBH$_4$; b, MsCl, Et$_3$N/THF; c, I$_2$/EtOH

4.1.2 O-Methylfumarofine

Fumarofine and O-methylfumarofine had been assigned as spirobenzylisoquinoline alkaloids (131 and 129), respectively (ref. 61), however, recently their structures were revised to the indenobenzazepine structures (130 and 110), respectively, on the basis of spectroscopic reinvestigation and total synthesis of O-methylfumarofine (ref. 57).

The spirobenzylisoquinoline (83) was reduced with sodium borohydride to give the trans-alcohol (125, 52%) along with the cis-alcohol (126, 14%). Treatment of the former with methanesulfonyl chloride and triethylamine as in the previous section gave the indenobenzazepine (127) in 68% yield. The product was oxidized with osmium tetroxide leading stereoselectively to the cis-fused cis-diol (128) in 71% yield as a result of the reagent attack from the less hindered side. The diol (128) was further oxidized with pyridinium chlorochromate to provide (±)-O-methylfumarofine (110).

4.1.3 Fumaritrine and Fumaritridine

The structures of fumaritrine and fumaritridine were also revised from the spirobenzylisoquinoline systems (136 and 137) (refs. 62, 63) to the indenobenzazepine systems (132 and 135), respectively, and synthesized from spirobenzylisoquinoline alkaloids through skeletal rearrangement (ref. 58).

Dihydroparfumidine (58) and dihydroparfumine (133) having the

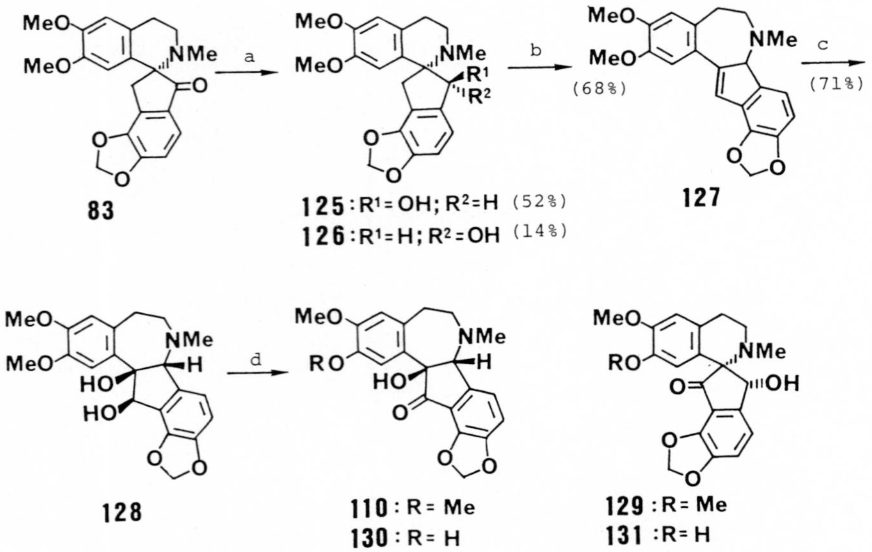

Scheme 21
a, NaBH$_4$; b, MsCl, Et$_3$N/THF; c, OsO$_4$; d, PCC

trans-alcohol were treated with trifluoroacetic anhydride and then quenched with methanol to afford fumaritrine (132) and fumaritridine (135), respectively, in 40-45% yield. The rearrangements proceeded probably through 8,14-cycloberbines (117 and 134), these reactions being completely reverse to those described in Section 3.3 in the synthesis of the spirobenzylisoquinoline alkaloids from the indenobenzazepine alkaloids.

Similarly, the fumaritrine analogue (138) was prepared from the corresponding spirobenzylisoquinoline trans-alcohol (125) and this synthesis confirmed the position of the methylenedioxy group in ring D of fumaritrine (ref. 58).

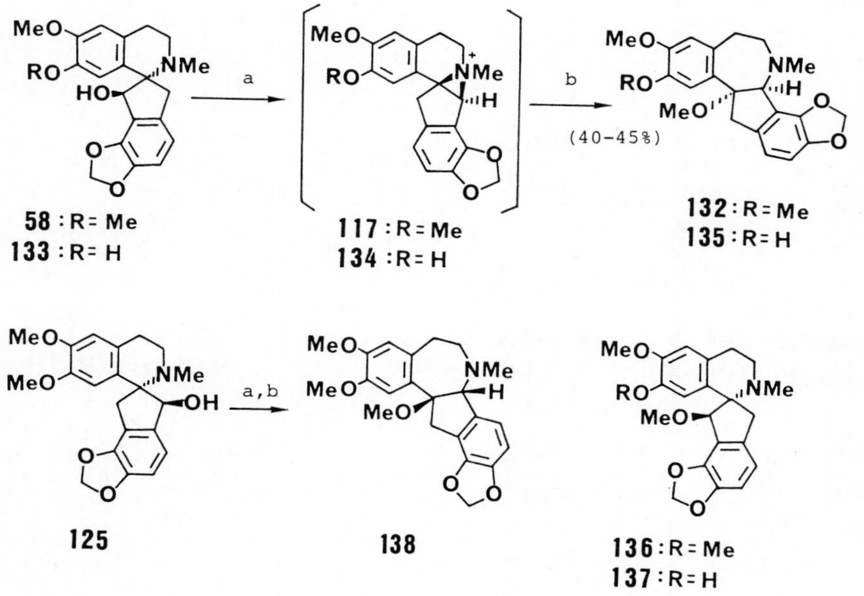

58 : R= Me
133 : R= H

117 : R= Me
134 : R= H

(40-45%)

132 : R= Me
135 : R= H

125

138

136 : R= Me
137 : R= H

Scheme 22
a, (CF$_3$CO)$_2$O; b, MeOH

4.2   Synthesis via 8,14-cycloberbines
4.2.1   Fumarofine and O-Methylfumarofine

Oxidation of dihydroprotoberberine (139) with m-chloroperbenzoic acid followed by irradiation in methanol afforded the 8,14-cycloberbine (141) in 52% yield via the phenolbetaine (140). On treatment with methanesulfonic acid in aqueous tetrahydrofuran,

141 was converted to the cis- and trans-indenobenzazepines (142 and 143) in 92% yield in a ratio of 2:1 through regioselective C(14)-N bond cleavage. N-Methylation of the mixture with methyl iodide gave only the cis N-methyl derivative (144) while the trans-indenobenzazepine (143) was recovered unchanged. The latter was very difficult to be methylated and gave the N-methyl derivative (145) in poor yield even under drastic conditions with dimethyl sulfate.

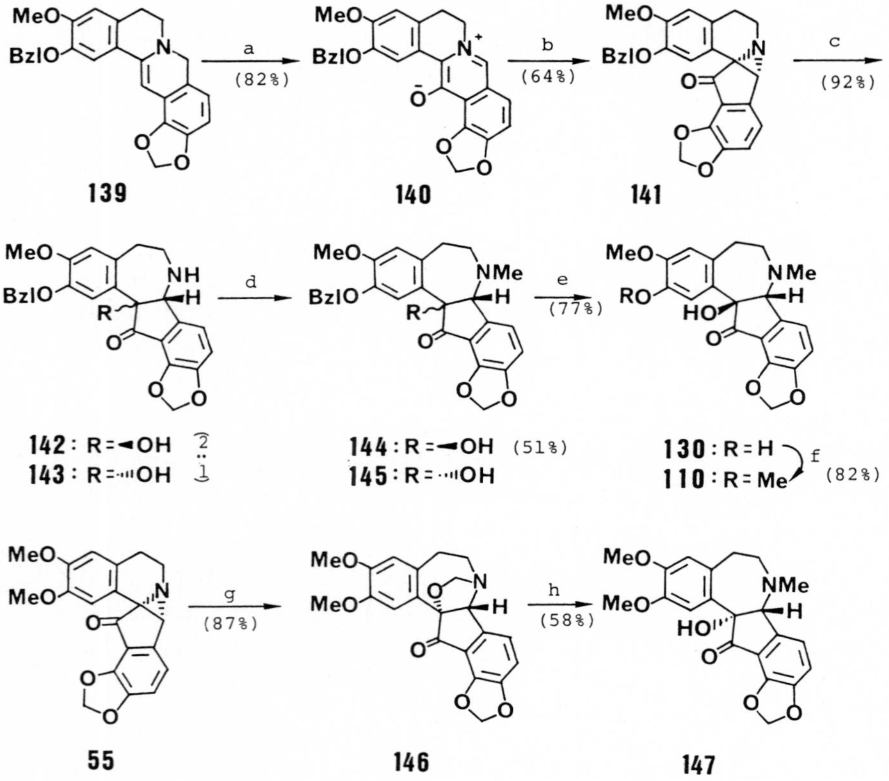

Scheme 23
a, mCPBA; b, hν/MeOH; c, MsOH/aq. THF; d, MeI; e, H₂, Pd-C/MeOH;
f, CH₂N₂; g, HCHO; h, NaBH₃CN

In contrast to the previous observation that generally a trans-indenobenzazepine isomerizes easily to the corresponding cis-isomer, the present trans-isomers (143 and 145) resisted isomerization.

Debenzylation of the _cis_-derivative (144) afforded fumarofine (130) in 77% yield, O-methylation of which with diazomethane provided O-methylfumarofine (110) (ref. 64).

The 8,14-cycloberbine (55) was treated with formaldehyde to afford the oxazolidine (146) in 87% yield. The product was reduced with sodium cyanoborohydride to provide the O-methylfumarofine analogue (147) with _trans_-fused structure. This synthesis supported unequivocally the _cis_-fused structure of fumarofine and O-methylfumarofine (ref. 64).

4.2.2 Fumaritrine

Reduction of the 8,14-cycloberbine (18) derived from berberine with sodium borohydride afforded stereoselectively the _cis_-alcohol (148). Treatment of 148 with _p_-toluenesulfonic acid in methanol followed by methylation with methyl iodide afforded the

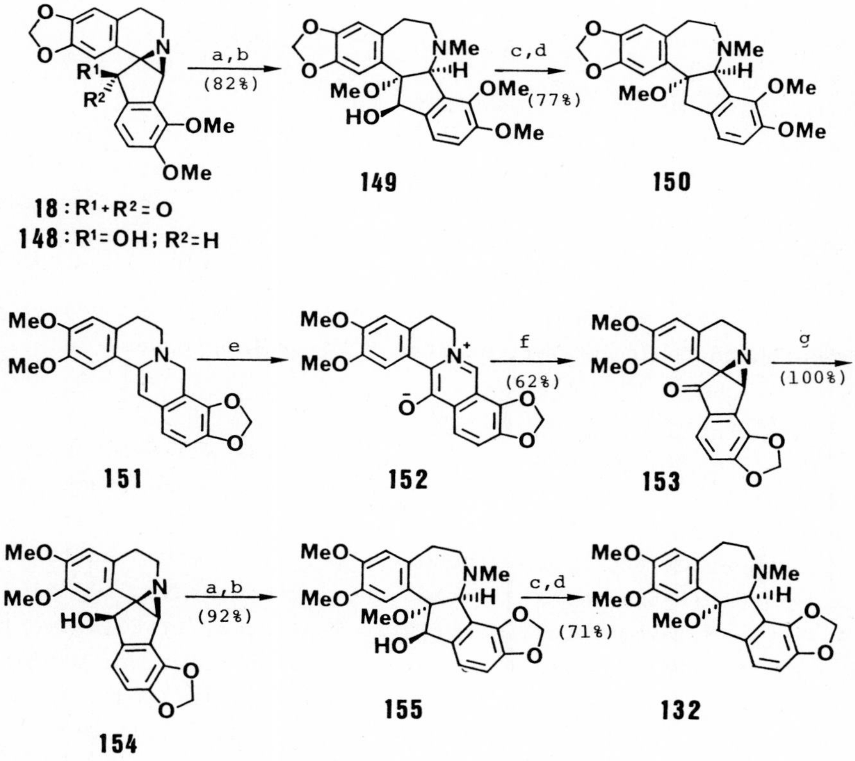

18 : R¹+R² = O
148 : R¹ = OH; R² = H

Scheme 24
a, p-TsOH/MeOH; b, MeI; c, MsCl; d, NaBH$_4$/(MeOCH$_2$)$_2$; e, mCPBA;
f, hν/MeOH; g, NaBH$_4$/MeOH

cis-fused indenobenzazepine (149) in 82% yield as the sole product. In order to remove the hydroxy group, the product (149) was methanesulfonated with methanesulfonyl chloride and the resulting methanesulfonate was treated with sodium borohydride in refluxing dimethoxymethane to furnish the fumaritrine analogue (150) having the correct stereochemistry in 77% yield (ref. 65).

On the basis of the above preliminary results, an efficient synthesis of fumaritrine was accomplished. Epiberberinephenol-betaine (152) derived from dihydroepiberberine (151) was irradiated in methanol to produce the 8,14-cycloberbine (153) in 62% yield. Reduction of 153 with sodium borohydride afforded quantitatively and stereoselectively the cis-alcohol (154) as in the case described earlier. Solvolysis of 154 in methanol in the presence of p-toluenesulfonic acid followed by N-methylation afforded stereoselectively the cis-indenobenzazepine (155) in 92% yield. Deoxygenation of 155 according to the above procedure provided (+)-fumaritrine (132) in 71% yield (ref. 65).

5   SYNTHESIS OF RHOEADINE ALKALOIDS

Rhoeadine alkaloids have an interesting stucture possessing a benzazepine fused with a six-membered cyclic acetal or hemiacetal and are shown to be biosynthesized from protoberberine alkaloids. The rhoeadine alkaloids including their synthesis have been reviewed (refs. 1, 2, 37-40, 66) and their list was published (ref. 67). Almost alkaloids were synthesized from indenobenz-azepines which could be derived from 8,14-cycloberbines.

5.1   From Spirobenzylisoquinolines

The first synthesis of rhoeadine alkaloids was accomplished from the spirobenzylisoquinolines (158) prepared from condensation of the phenethylamine (156) and the indanediones (157) (refs. 47, 60). Treatment of the trans-alcohols (160) with methanesulfonyl chloride and triethylamine afforded the indenobenzazepines (162 and 163) via 8,14-cycloberbines (161). The former (162) isomerized to the latter (163) on standing in an alkaline solution. Oxidation of the former with osmium tetroxide gave the diols (164). Treatment of the diol (164a) with sodium periodate followed by sodium borohydride afforded (+)-rhoeagenine diol (166a) (ref. 60). The product has already been converted to (+)-rhoeagenine (167) (ref. 68).

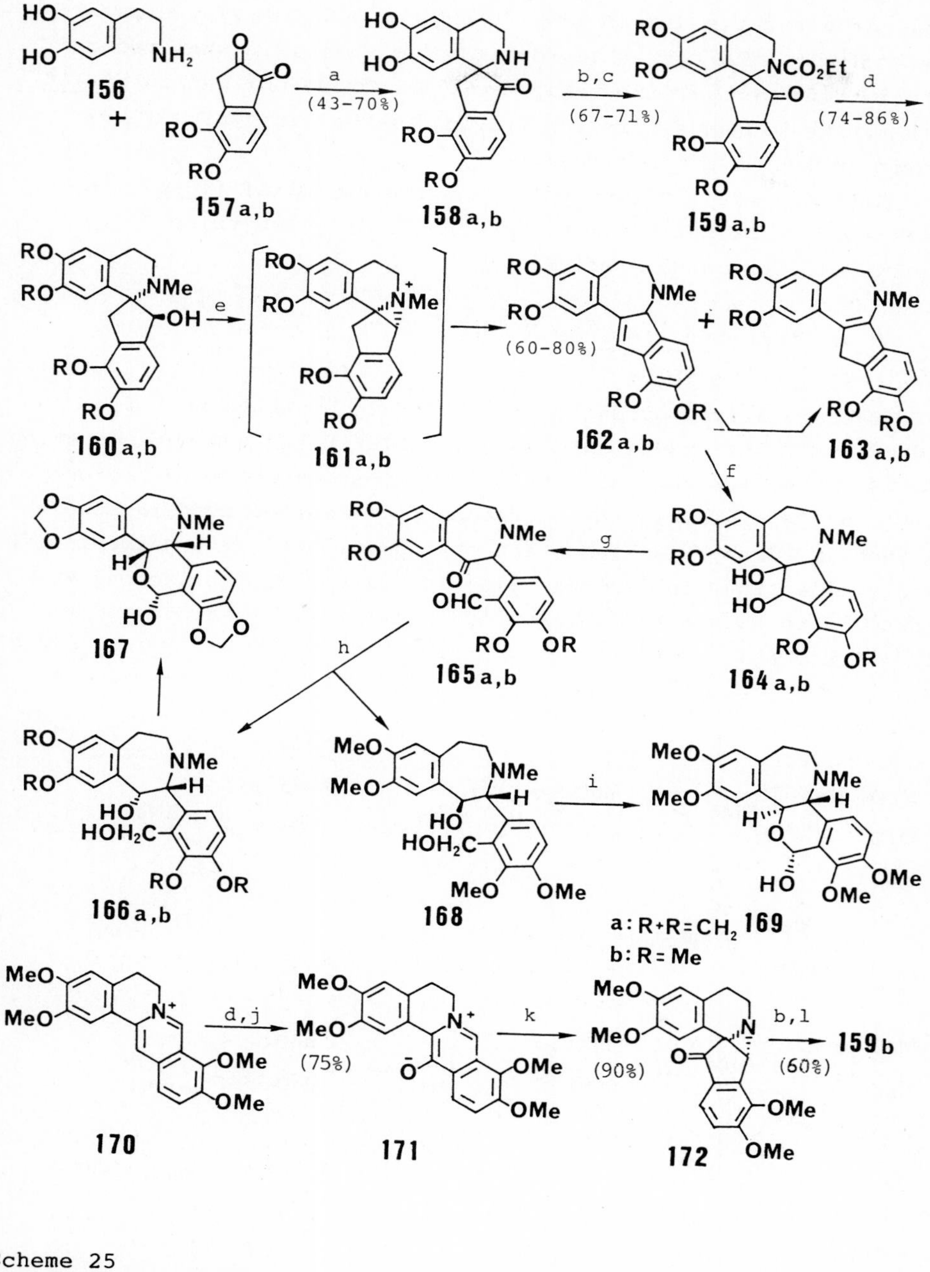

Scheme 25
a, Δ /EtOH; b, ClCO₂Et; c, CH₂I₂, K₂CO₃/DMSO or MeI,
K₂CO₃/acetone; d, LiAlH₄; e, MsCl, Et₃N; f, OsO₄; g, HIO₄;
h, NaBH₄; i, MnO₂; j, mCPBA; k, hν/MeOH; l, H₂, Pd-C

Similarly the diol (164b) was transformed to (±)-alpinigenine diol (168) and its diastereomer (166b) in 10 and 50% yields, respectively. The former was obtained in 40% yield when reduced with lithium perhydro-9b-boraphenalenylhydride. Oxidation of 168 with active manganese dioxide provided (±)-alpinigenine (169) albeit in 10% yield (ref. 60).

Recently the starting spirobenzylisoquinoline (159b) was prepared from palmatine (170) via the 8,14-cycloberbine (172) as depicted in Scheme 25 (ref. 43).

5.2 From Indenobenzazepines
5.2.1 cis-Alpinigenine

Base-catalyzed oxidation of the indenobenzazepine (163b) afforded the vinylogous amide (173) in 79% yield. Photooxygenation of 173 in the presence of rose bengal afforded the keto-lactone (175) in 37% yield via the dioxetane (174) as depicted in Scheme 26. Sodium borohydride reduction of 175 followed by treatment with acid yielded the cis-lactone (176) in 90% yield. The lactone was reduced with diisobutylaluminum hydride to (±)-cis-alpinigenine (177) (ref. 69).

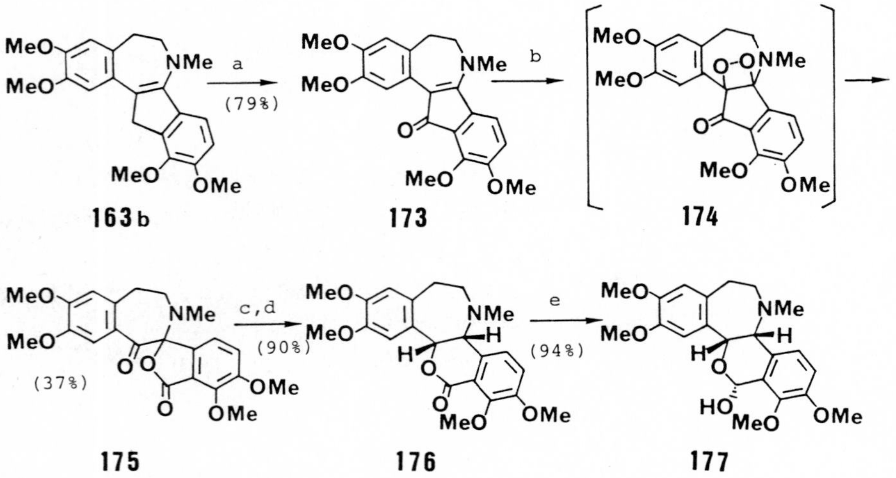

Scheme 26
a, O$_2$, Triton B, pyr.; b, hν, O$_2$, rose bengal; c, NaBH$_4$; d, HCl;
e, DIBAL

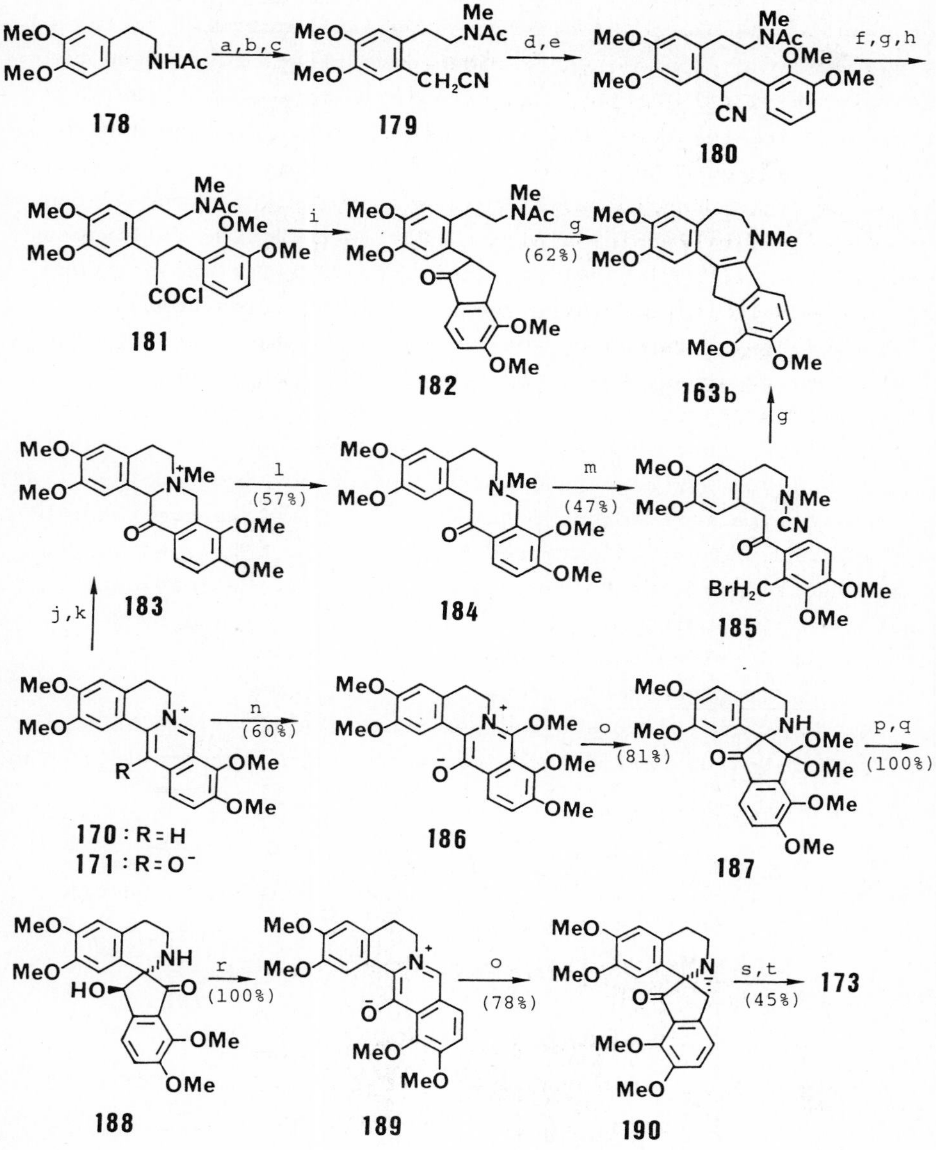

Scheme 27
a, MeI, NaH; b, HCHO, HCl; c, NaCN/DMSO; d, 2,4-
dimethoxybenzaldehyde; e, Na-Hg/EtOH; f, HCl/EtOH; g, KOH;
h, SOCl₂; i, AlCl₃; j, LiAlH₄; k, Me₂SO₄; l, Zn/AcOH; m, BrCN;
n, hν, O₂, NaOMe, rose bengal/MeOH; o, hν/MeOH; p, NaBH₄; q, HCl;
r, 10% NaOH; s, p-TsOH/benzene; t, Me₂SO₄, NaH

The starting indenobenzazepine (163b) is identical with that obtained in the preceding section and was alternatively prepared by two different routes. One is the conventional route from the phenethylamine (178) through chloromethylation, alkylation, Friedel-Crafts reaction, and cyclization (ref. 69). The second one is the route starting from the protoberberine alkaloid, palmatine (170). Palmatinephenolbetaine (171) was converted to the 13-oxo-compound (183) by lithium aluminum hydride reduction followed by N-methylation with dimethyl sulfate (ref. 70). The oxo-compound was reduced with zinc to give the ten-membered ring compound (184), which was cleaved by von Braun reaction and then cyclized by alkaline treatment to afford the product (163b) (ref. 71). The third one is a synthesis of the vinylogous amide (173) starting from palmatine through D-ring inversion (see Section 6). Thus palmatine (170) was converted to the spirobenzylisoquinoline (188) via 8-methoxypalmatinephenolbetaine (186). Alkaline treatment of 188 afforded the D-ring inverted phenolbetaine (189), which was irradiated to give the 8,14-cycloberbine (190). Regioselective cleavage of 190 using p-toluenesulfonic acid followed by N-methylation provided the indenobenzazepine (173) (ref. 25).

5.2.2   cis-Alpinigenine Analogue

Oxidation of the trans-indenobenzazepine (45) (see Section 2.2.2) with sodium periodate afforded the keto-lactone (191). The

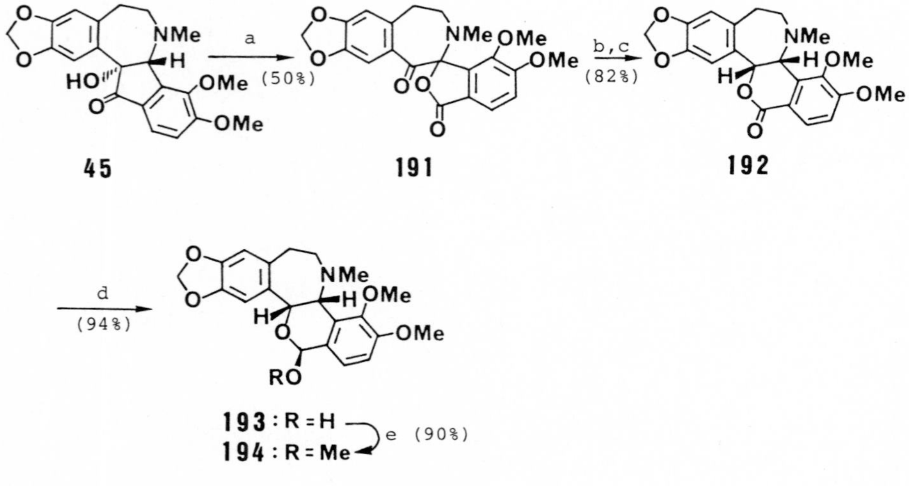

Scheme 28
a, NaIO$_4$/HCl-MeOH; b, NaBH$_4$; c, HCl; d, DIBAL; e, HC(OMe)$_3$

product was converted to the cis-alpinigenine analogue (193) and its methyl derivative (194) (ref. 34) through the reaction similar to that described in Section 5.2.1. The stereochemistry of the final products depicted in Scheme 28 was not completely clarified.

6. SYNTHESIS OF PROTOBERBERINE ALKALOIDS

Protoberberine alkaloids occupy a central position in the biosynthesis of various related isoquinoline alkaloids as shown in Scheme 1 and many synthetic approaches to these alkaloids have been developed (ref. 72). In this section only the methods involving 8,14-cycloberbines are described. Although 8,14-cyclo-berbines have been derived from the corresponding protoberberines, they provide a clue to modification of the starting proto-berberines.

6.1 From Spirobenzylisoquinolines

Spirobenzylisoquinolines (195a and 195b) derived from the phenethylamines and the indanedione were treated with potassium t-butoxide in dimethyl sulfoxide to afford dihydroprotoberberin-8-ones (197a and 197b) in 20 and 41% yields, respectively. The reaction probably proceeded through 8,14-cycloberbines (196a,b) followed by their fragmentation (ref. 73).

In connection with the above transformation, another conversion of a spirobenzylisoquinoline to a protoberberine by photolysis is also described (ref. 74). Irradiation of the spiro-benzylisoquinoline (195b) in dry tetrahydrofuran afforded the

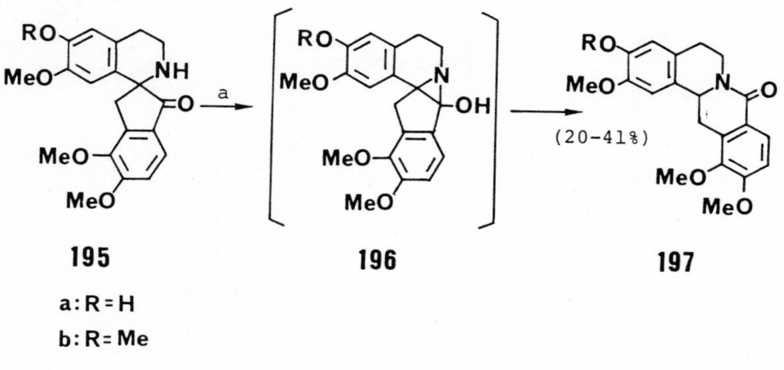

195

a: R = H
b: R = Me

196

197

Scheme 29

a, KOBu^t/DMSO

protoberberine (198) and oxyprotoberberine (199) in 80 and 10%
yields, respectively. Similarly the spirobenzylisoquinoline (200)
afforded the protoberberine (201) and the lactam (202) in 85 and
8% yields, respectively. Reduction of the former with sodium
borohydride furnished (±)-xylopinine.

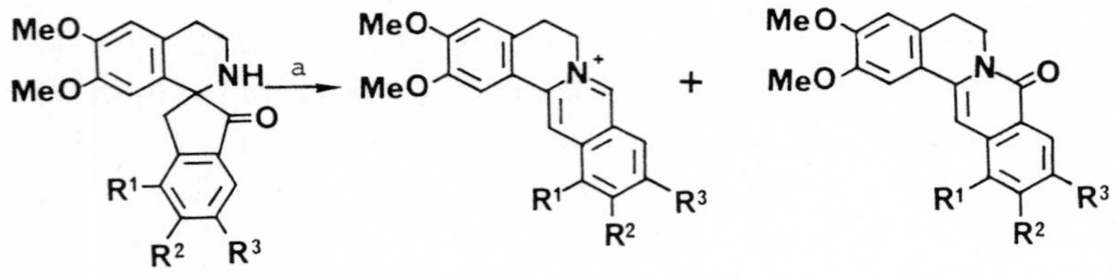

195b: R^1=R^2=OMe; R^3=H   198 (80%)   199 (10%)
200 : R^1=H; R^2=R^3=OMe   201 (85%)   202 (8%)

Scheme 30
a, hν/THF

6.2 Ring D Inversion of Protoberberines via Spirobenzyl-
isoquinolines

Ring D inversion seems to be one of key steps in the
biogenetic transformation from the most abundant 9,10-oxygenated

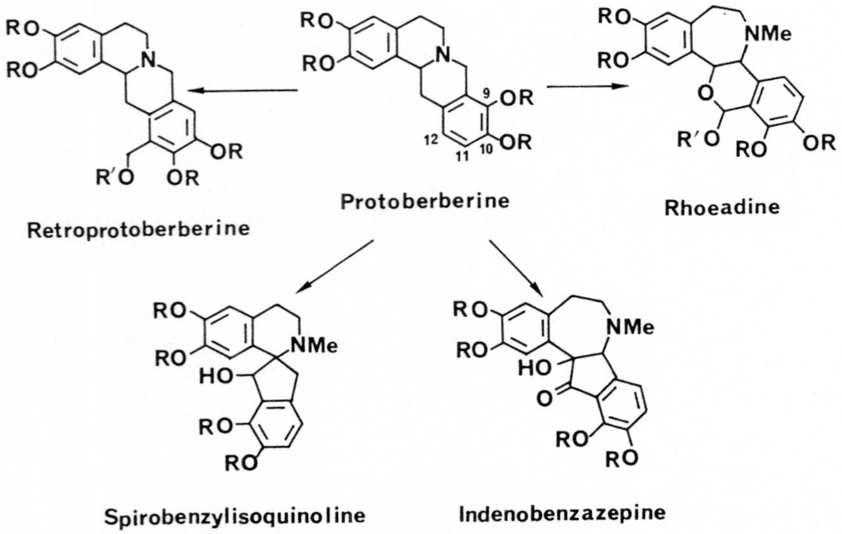

Retroprotoberberine   Protoberberine   Rhoeadine

Spirobenzylisoquinoline   Indenobenzazepine

Scheme 31

protoberberine alkaloids to related alkaloids such as rhoeadine, retroprotoberberine, spirobenzylisoquinoline, and indenobenz-azepine alkaloids as shown in Scheme 31. In fact, a spirobenzyl-isoquinoline alkaloid, (+)-fumaricine (58) (refs. 42, 43) and a rhoeadine alkaloid, (+)-cis-alpinigenine (177) (ref. 75) have been prepared from non-natural 11,12-oxygenated protoberberines, which can be regarded as the ring D inverted products of 9,10-oxygenated protoberberines.

Three methods were developed for the transformation of naturally occurring 9,10-oxygenated protoberberines into non-naturally occurring 11,12-oxygenated protoberberines through ring D inversion via spirobenzylisoquinolines (refs. 76, 77).

The cis-alcohol cycloberbine (148) derived stereoselectively from berberine (1) was treated with ethyl chloroformate to give the oxazolidinone (203), the chloride of which was converted to the ketone via the alcohol by successive treatment with silver nitrate in aqueous tetrahydrofuran and pyridinium dichromate in dimethylformamide. Heating of the keto-oxazolidinone (205) with 10% aqueous sodium hydroxide in ethanol effected hydrolysis of the oxazolidinone, retro-aldol reaction, cyclization, and subsequent dehydration to provide the 11,12-oxygenated phenolbetaine (208) in quantitative yield via the hydroxy-ketone (206) and the keto-aldehyde (207).

The second and more convenient method was accomplished starting from the spirobenzylisoquinoline (31), which was obtained in 73% overall yield from 8-methoxyberberinephenolbetaine (12) derived from berberine. Treatment of 31 with 10% aqueous sodium hydroxide in ethanol provided the expected phenolbetaine (208) in 95% yield.

The third method involved the photochemical transformation of a spirobenzylisoquinoline to a protoberberine described in Section 6.1. Hydrogenolysis of the spirobenzylisoquinoline (33) derived from berberine over 5% palladium-charcoal afforded the dehalo-genated ketone (209) which was hydrolyzed to the amino-ketone (210) in 73% yield. Its irradiation in tetrahydrofuran provided the 11,12-oxygenated protoberberine (211) in 43% yield.

6.3 Synthesis of 13-Methylprotoberberine Alkaloids

Introduction of an alkyl group at the C-13 position of protoberberine has been achieved by the reaction of 8-acetonyl-

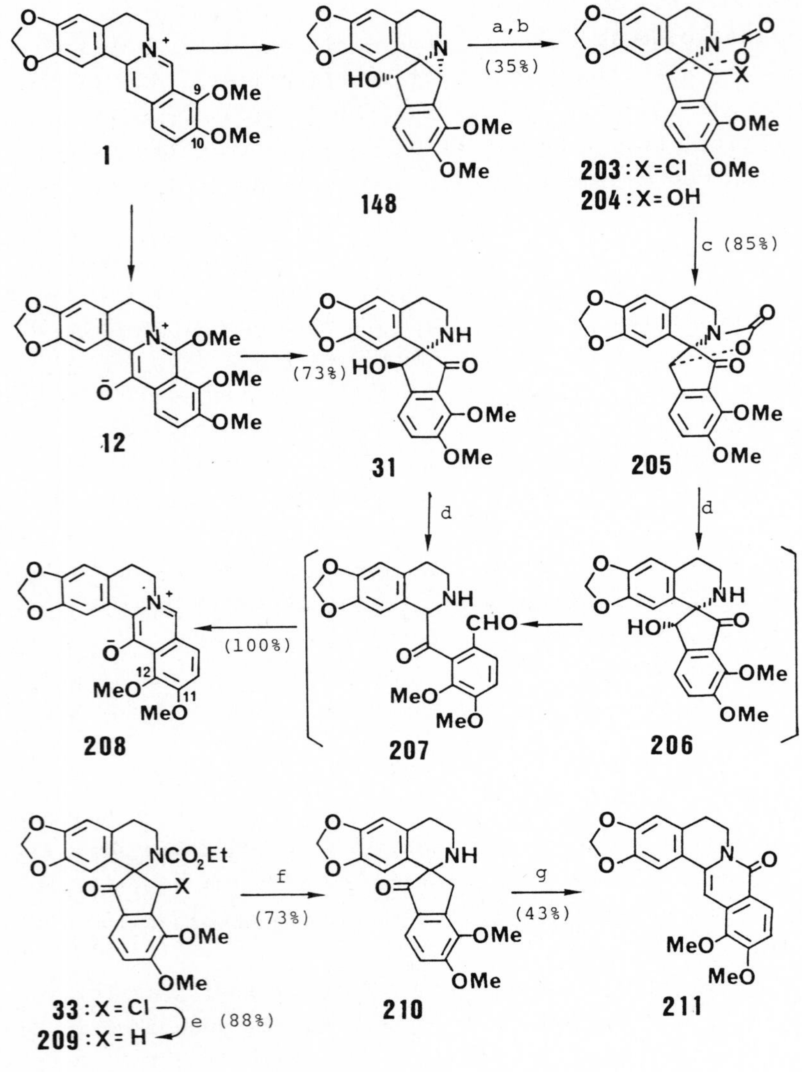

Scheme 32
a, ClCO$_2$Et; b, AgNO$_3$/aq. THF; c, PDC; d, 10% NaOH, e, H$_2$, Pd-C;
f, 20% KOH; g, hν/THF

7,8-dihydro derivative with alkyl halides (ref. 78) or reaction of
7,8-dihydro derivative with formaldehyde (ref. 79), though yields
are not always satisfactory and the 13-alkylation products (212)
are always contaminated with the starting quaternary proto-

berberines recovered. The second method cannot be applied to ethylation with acetaldehyde.

A novel procedure for 13-alkylation *via* 8,14-cycloberbines has been developed (ref. 80). The Wittig reaction of the cyclo-berbines (18, 66, and 172) derived from protoberberines (1, 2, and 170) with methylidene-triphenylphosphorane produced the 13-methylene cycloberbines (213a,b,c) in high yield. Irradiation of 213 resulted in electrocyclic reaction to yield 13-methylproto-berberines (212a,b,c) in high yield without contamination of the starting protoberberines. Reduction of 212a,b,c with sodium borohydride gave (±)-thalictricavine (214a), (±)-tetrahydro-corysamine (214b), and (±)-corydaline (214c) in excellent yield.

13-Ethyl and 13-propyl derivatives were also obtained by a similar procedure.

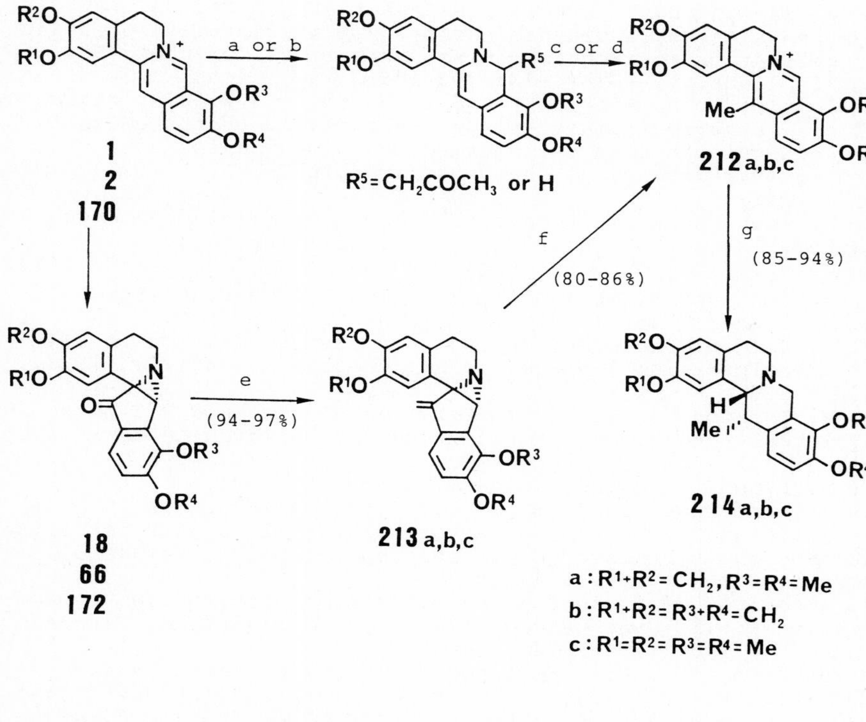
a or b
c or d
R^5=CH_2COCH_3 or H
1
2
170
212 a,b,c
f
(80-86%)
g
(85-94%)
e
(94-97%)
18
66
172
213 a,b,c
214 a,b,c
a : R^1+R^2=CH_2, R^3=R^4=Me
b : R^1+R^2=R^3+R^4=CH_2
c : R^1=R^2=R^3=R^4=Me

Scheme 33
a, CH_3COCH_3, KOH; b, $LiAlH_4$; c, MeI; d, HCHO/AcOH; e, $Ph_3P{=}CH_2$; f, hν/MeOH; g, $NaBH_4$

REFERENCES

1   M. Shamma, The Isoquinoline Alkaloids Chemistry and
    Pharmacology, Academic Press, New York, 1972.
2   M. Shamma and J. L. Moniot, Isoquinoline Alkaloids Research
    1972-1977, Plenum Press, New York, 1978.
3   M. Shamma and C. D. Jones, A model for the biogenesis of the
    spirobenzylisoquinoline alkaloids, J. Am. Chem. Soc., 91(14)
    (1969) 4009-4010.
4   Dr. H. G. Kiryakov, Higher Medical Institute, Bulgaria, also
    proposed an 8,14-cycloberbine as a biogenetic intermediate;
    private communication (1980).
5   M. Shamma and J. F. Nugent, The protoberberine⟶
    spirobenzylisoquinoline⟶ dibenzocyclopent[b]azepine
    rearrangement, Tetrahedron, 29(10) (1973) 1265-1272.
6   T. Kametani, T. Ohsawa, S. Hirata, M. S. Premila, M. Ihara and
    K. Fukumoto, Studies on the syntheses of heterocyclic
    compounds. DCXCII. A novel synthetic route to phthalide-
    isoquinoline and spirobenzylisoquinoline type alkaloids, Chem.
    Pharm. Bull., 25(2) (1977) 321-326.
7   P. Padwa and A. D. Woolhouse, Aziridines and fused-ring
    derivatives, in: A. R. Katritzky and C. W. Rees (Eds.),
    Comprehensive Heterocyclic Chemistry The Structure, Reactions,
    Synthesis and Uses of Heterocyclic Compounds, Vol.7, W.
    Lwowski (Ed.), Pergamon Press, Oxford, 1984, pp. 47-93.
8   P. E. Hansen and K. Undheim, N-Qnaternary compounds. Part XL.
    Syntheses of 1a,6a-dihydroindeno[1,2-b]azirin-6(1H)-ones and
    isomerisation to isoquinolinium derivatives, J. Chem. Soc.
    Perkin Trans. 1, (3) (1975) 305-308.
9   N. Dennis, A. R. Katritzky and H. Wilde, 1,3-Dipolar character
    of six-membered aromatic rings. Part XXVII. Photochemically
    induced valence bond tautomerism and dimerisation of 3-oxido-
    1-phenylpyridinium, J. Chem. Soc. Perkin Trans. 1, (21) (1976)
    2338-2343.
10 J. W. Lown and K. Matsumoto, Thermally disallowed valence
    tautomerization of an indano[1,2-b]aziridine to an
    isoquinolinium imine, J. Org. Chem., 36(10) (1971) 1405-1413.
11 J. W. Lown and K. Matsumoto, Reactions of carbonyl ylides,
    Can. J. Chem., 49(21) (1971) 3443-3455.
12 E. F. Ullman, Photochemical valence tautomerization of 2,4,6-
    triphenylpyrylium 3-oxide, J. Am. Chem. Soc., 85(21) (1963)
    3529-3530.
13 J. M. Dunaton and P. Yates, Pyrylium 3-oxide and 4,5-epoxy-2-
    cyclopentenone valence tautomerism, Tetrahedron Letters, (10)
    (1964) 505-507.
14 F. L. Pyman, Isoquinoline derivatives. Part VI.
    Neooxyberberine, J. Chem. Soc., 99 (1911) 1690-1699.
15 J. Iwasa and S. Naruto, Structure of so-called neoxyberberine
    acetone and synthesis of its related compounds, Yakugaku
    Zasshi, 86(6) (1966) 534-538.
16 T. Takemoto and Y. Kondo, Reactions of protoberberine-type
    alkaloids. II. Synthesis of 13-hydroxy derivatives, Yakugaku
    Zasshi, 82(10) (1962) 1413-1417.
17 M. Hanaoka, C. Mukai, K. Nagami, K. Okajima and S. Yasuda,
    Chemical transformation of protoberberines. V. Photochemical
    valence isomerization of berberinephenolbetaines to 8,14-
    cycloberbines, versatile aziridine derivatives for a novel and
    efficient entry to spirobenzylisoquinolines and benzindeno-
    azepines, Chem. Pharm. Bull., 32(6) (1984) 2230-2240.

18 P. W. Jeffs and J. D. Scharver, 13β-Hydroxystylopine. Structure and synthesis, J. Org. Chem., 40(5) (1975) 644-647.

19 Y. Kondo, H. Inoue and J. Imai, Concentration dependent photoxidation of 13-hydroxyberberinium phenolbetaine and formation of stable epidioxy compound, Heterocycles, 6(7) (1977) 953-957.

20 Y. Kondo, J. Imai and H. Inoue, Reaction of protoberberine-type alkaloids. Part 12. A facile method for regiospecific oxygenation and excited oxidative ring-cleavage of berberine alkaloids, J. Chem. Soc. Perkin Trans. 1, (4) (1980) 911-918.

21 J. L. Moniot and M. Shamma, The conversion of berberine into (+)-α- and (+)-β-hydrastine, J. Am. Chem. Soc., 98(21) (1976) 6714-6715.

22 J. L. Moniot and M. Shamma, Conversion of berberine into phthalideisoquinolines, J. Org. Chem., 44(24) (1979) 4337-4342.

23 M. Hanaoka, C. Mukai and Y. Arata, Photo-oxygenation of berberine. A novel conversion of berberine to (+)-ophiocarpine, (+)-α- and (+)-β-hydrastine, Heterocycles, 6(7) (1977) 895-898.

24 M. Hanaoka, C. Mukai and Y. Arata, Chemical transformation of protoberberines. III. Convenient synthesis of 8-methoxy-berberinephenolbetaine by photooxygenation of berberine. A novel conversion óf berberine into (+)-ophiocarpine, (+)-epiophiocarpine, (+)-α-hydrastine, and (+)-β-hydrastine, Chem. Pharm. Bull., 31(3) (1983) 947-952.

25 M. Hanaoka, M. Inoue, N. Kobayashi and S. Yasuda, Chemical transformation of protoberberines. XII. A novel synthesis of rhoeadine alkaloids. An alternative synthesis of a key intermediate, benzindenoazepine, for a synthesis of (+)-cis-alpinigenine and (+)-cis-alpinine from palmatine, Chem. Pharm. Bull., 35(3) (1987) 980-985.

26 M. Hanaoka, unpublished data.

27 M. Hanaoka, A. Wada, S. Yasuda, C. Mukai and T. Imanishi, Synthesis of the 8H-isoquino[2,1-b][2]benzazocine system via 1,3-dipolar cycloaddition of 8-methoxyberberinephenolbetaine, Heterocycles, 12(4) (1979) 511-514.

28 M. Hanaoka, S. Yasuda, K. Nagami, K. Okajima and T. Imanishi, Photochemical valence tautomerization of berberinephenol-betaines to 8,14-cycloberbines, versatile aziridine derivatives for spirobenzylisoquinolines, Tetrahedron Letters, (39) (1979) 3749-3752.

29 M. Hanaoka and C. Mukai, Photo-oxygenation of 8-methoxy-berberinephenolbetaine. A novel synthesis of spirobenzyl-isoquinoline system from berberine, Heterocycles, 6(12) (1977) 1981-1984.

30 G. Blaskó, V. Elango, N. Murugesan and M. Shamma, Conversion of a protopine into an indenobenzazepine. Stereoselective formation of cis- and trans-B/C fused indenobenzazepines, J. Chem. Soc. Chem. Commun., (24) (1981) 1246-1248.

31 M. Hanaoka, M. Inoue, K. Nagami, Y. Shimada and S. Yasuda, Conversion of berberine into benzindanoazepines via 8,14-cycloberbines, Heterocycles, 19(2) (1982) 313-318.

32 M. Hanaoka, M. Inoue, S. Sakurai, Y. Shimada and S. Yasuda, Efficient conversion of protoberberines into benzindeno-azepines. A formal synthesis of (+)-cis-alpinigenine and (+)-cis-alpinine, Chem. Pharm. Bull., 30(3) (1982) 1110-1112.

33 M. Hanaoka, S. K. Kim, M. Inoue, K. Nagami, Y. Shimada and S. Yasuda, Chemical transformation of protoberberines. VII.

224

Efficient conversion of protoberberines into benzindeno-
azepines via 8,14-cycloberbines, Chem. Pharm. Bull., 33(4)
(1985) 1434-1443.

34 N. Murugesan, G. Blaskó, R. D. Minard and M. Shamma, An
efficient conversion of berberine into a rhoeadine via an
indenobenzazepine, Tetrahedron Letters, 22(33) (1981) 3131-
3134.

35 M. Hanaoka, S. Sakurai, Y. Sato and C. Mukai, Simple
conversion of 8-alkyl-8,14-cycloberbines to spirobenzylisoquinolines by regioselective C-N bond cleavage, Heterocycles,
19(12) (1982) 2263-2268.

36 M. Hanaoka, S. K. Kim, S. Sakurai, Y. Sato and C. Mukai,
Chemical transformation of protoberberines. XIV. Acid-
catalyzed cleavage of 8-alkyl-8,14-cycloberbines. A simple
method for the preparation of N-unsubstituted spirobenzylisoquinolines, Chem. Pharm. Bull., 35(8) (1987) 3155-3165.

37 R. H. F. Manske, Papaveraceae alkaloids, in: R. H. F. Manske
(Ed.), The Alkaloids Chemistry and Physiology, Vol. X,
Academic Press, New York, 1968, pp. 467-483.

38 F. Šantavý, Papaveraceae alkaloids, in: R. H. F. Manske (Ed.),
The Alkaloids Chemistry and Physiology, Vol. XII, Academic
Press, New York, 1970, pp. 333-454.

39 M. Shamma, The spirobenzylisoquinoline alkaloids, in: R. H. F.
Manske (Ed.), The Alkaloids Chemistry and Physiology, Vol.
XIII, Academic Press, New York, 1971, pp. 165-188.

40 F. Šantavý, Papaveraceae alkaloids. II, in: R. H. F. Manske
and R. G. A. Rodrigo (Eds.), The Alkaloids Chemistry and
Physiology, Vol. XVII, Academic Press, New York, 1979, pp.
385-544.

41 R. M. Preisner and M. Shamma, The spirobenzylisoquinoline
alkaloids, J. Nat. Products, 43(3) (1980) 305-318.

42 M. Hanaoka, S. Yasuda, Y. Hirai, K. Nagami and T. Imanishi, A
novel synthesis of (±)-fumaricine, Heterocycles, 14(10) (1980)
1455-1456.

43 M. Hanaoka, K. Nagami, Y. Hirai, S. Sakurai and S. Yasuda,
Chemical transformation of protoberberines. VIII. A novel
synthesis of (±)-fumaricine and a formal synthesis of (±)-
alpinigenine, Chem. Pharm. Bull., 33(6) (1985) 2273-2280.

44 M. Hanaoka, M. Iwasaki and C. Mukai, A highly stereoselective
synthesis of (±)-dihydrofumariline-1, Tetrahedron Letters,
26(7) (1985) 917-920.

45 M. Hanaoka, S. Sakurai, T. Ohshima, S.Yasuda and C. Mukai,
Transformation of protoberberines into spirobenzylisoquinolines. Stereoselective conversion of coptisine into
(±)-ochrobirine, Chem. Pharm. Bull., 30(9) (1982) 3446-3449.

46 M. Hanaoka, A. Ashimori and S. Yasuda, A total synthesis of
(±)-corydaine from coptisine, Heterocycles, 22(10) (1984)
2263-2264.

47 H. Irie, T. Kishimoto and S. Uyeo, The total synthesis of
(±)-ochotensine and related compounds, J. Chem. Soc. (C),
(24) (1968) 3051-3057.

48 S. McLean, M.-S. Lin and J. Whelan, The total synthesis of
(±)-ochotensimine, Tetrahedron Letters, (20) (1968) 2425-2428;
S. McLean, M.-S. Lin and J. Whelan, Total synthesis of
spirobenzylisoquinoline alkaloids: (±)-ochotensimine, Can. J.
Chem., 48(6) (1970) 948-954.

49 M. Hanaoka, M. Kohzu and S. Yasuda, A first and stereoselective synthesis of (±)-raddeanamine, Chem. Pharm. Bull.,
33(9) (1985) 4113-4115.

50 T. Kametani, M. Takemura, M. Ihara and K. Fukumoto, Studies on the syntheses of heterocyclic compounds. Part 682. Six new isoquinoline alkaloids from Corydalis ochotensis var. raddeana, J. Chem. Soc. Perkin Trans. 1, (4) (1977) 390-393.

51 M. Hanaoka, M. Kohzu and S. Yasuda, A novel and highly stereoselective synthesis of (+)-sibiricine, Chem. Pharm. Bull., 33(6) (1985) 2621-2623.

52 M. Hanaoka, unpublished data.

53 M. Hanaoka, unpublished data.

54 B. C. Nalliah, D. B. MacLean, H. L. Holland and R. Rodrigo, A versatile synthesis of spirobenzylisoquinoline and phthalideisoquinoline alkaloids. Conversion of a phthalide-isoquinoline to spirobenzylisoquinolines, Can. J. Chem., 57(13) (1979) 1545-1549.

55 G. Blaskó, N. Murugesan, A. J. Freyer, D. J. Gula, B. Şener and M. Shamma, The indenobenzazepine-spirobenzylisoquinoline rearrangement; stereocontrolled syntheses of (+)-raddeanine and (+)-yenhusomine, Tetrahedron Letters, 22(33) (1981) 3139-3142.

56 G. Blaskó, S. F. Hussain, A. J. Freyer and M. Shamma, A new class of isoquinoline alkaloids: The indenobenzazepines, Tetrahedron Letters, 22(33) (1981) 3127-3130.

57 G. Blaskó, N. Murugesan, S. F. Hussain, R. D. Minard and M. Shamma, Revised structure for fumarofine, an indenobenzazepine type alkaloid, Tetrahedron Letters, 22(33) (1981) 3135-3138.

58 G. Blaskó, N. Murugesan, A. J. Freyer, R. D. Minard and M. Shamma, Revised structures for fumaritridine and fumaritrine: Two indenobenzazepine type alkaloids, Tetrahedron Letters, 22(33) (1981) 3143-3146.

59 H. Irie, S. Tani and H. Yamane, Total synthesis of the alkaloid rhoeadine, J. Chem. Soc. Chem. Commun., (24) (1970) 1713.

60 H. Irie, S. Tani and H. Yamane, Total synthesis of the alkaloids rhoeadine and alpinigenine, J. Chem. Soc. Perkin Trans. 1, (23) (1972) 2986-2990.

61 C. K. Yu, J. K. Saunders, D. B. MacLean and R. H. F. Manske, The structure of fumarofine, Can. J. Chem., 49(18) (1971) 3020-3024.

62 N. M. Mollov, H. G. Kirjakov and G. I. Yakimov, Two new spiroisoquinoline alkaloids from Fumaria, Phytochemistry, 11(7) (1972) 2331-2332.

63 Kh. G. Kiryakov and P. P. Panov, Alkaloids of Fumaria vaillantii, Dokl. Bolg. Akad. Nauk, 24(9) (1971) 1191-1194.

64 M. Hanaoka, A. Ashimori, H. Yamagishi and S. Yasuda, Transformation of protoberberines into benzindenoazepines. A total synthesis of fumarofine and O-methylfumarofine, Chem. Pharm. Bull., 31(6) (1983) 2172-2175.

65 M. Hanaoka, M. Iwasaki, S. Sakurai and C. Mukai, Transformation of protoberberines into benzindenoazepines. Stereo-selective synthesis of (+)-fumaritrine, Tetrahedron Letters, 24(36) (1983) 3845-3848.

66 H. Rönsch, Rhoeadine alkaloids, in: A. Brossi (Ed.), The Alkaloids Chemistry and Pharmacology, Vol. 28, Academic Press, New York, 1986, pp. 1-93.

67 C. T. Montgomery, B. K. Cassels and M. Shamma, The rhoeadine alkaloids, J. Nat. Products, 46(4) (1983) 441-453.

68 F. Šantavý, J. L. Hruban, L. Dolejš, V, Hanuš, K. Bláha and A. D. Cross, Constitution of rhoeadine and isorhoeadine, Coll. Czech. Chem. Commun., 30(10) (1965) 3479-3500.

226

69 K. Orito, R. H. Manske and R. Rodrigo, Photosensitized
    oxidation of an enaminoketone. The total synthesis of a
    rhoeadine alkaloid, J. Am. Chem. Soc., 96(6) (1974) 1944-1945.
70 Cf. B. Nalliah, R. H. F. Manske, R Rodrigo and D. B. MacLean,
    A photolytic protoberberine⟶ Spirobenzylisoquinoline
    rearrangement, Tetrahedron Letters, (29) (1973) 2795-2798.
71 B. Nalliah, R. H. Manske and R. Rodrigo, Transformation of 13-
    oxoprotoberberinium metho salts III: Biogenetically patterned
    conversions to rhoeadines, Tetrahedron Letters, (33) (1974)
    2853-2856.
72 T. Kametani, The total syntheses of isoquinoline alkaloids,
    in: J. ApSimon (Ed.), The Total Synthesis of Natural Products,
    Vol. 3, John Wiley & Sons, Inc., New York, 1977, pp. 1-272.
73 D. Greenslade and R. Ramage, Rearrangement of spirobenzyl-
    isoquinoline to protoberberine systems. A comparison of base
    induced and photochemical processes, Tetrahedron, 33(9) (1977)
    927-930.
74 H. Irie, K. Akagi, S. Tani, K. Yabusaki and H. Yamane,
    Synthesis of the alkaloid, xylopinine and the related
    compound by photo-induced rearrangement of spiroisoquinoline,
    Chem. Pharm. Bull., 21(4) (1973) 855-857.
75 S. Prabhakar, A. M. Lobo, M. R. Tavares and I. M. C. Oliveira,
    Total synthesis of the alkaloids (+)-alpinigenine and (+)-cis-
    alpinigenine, J. Chem. Soc. Perkin Trans. 1, (4) (1981) 1273-
    1277.
76 M. Hanaoka, M. Inoue, M. Takahashi and S. Yasuda, Ring D
    inversion of protoberberine alkaloids. Conversion of berberine
    into non-naturally occurring 11,12-oxygenated protoberberines,
    Heterocycles, 19(1) (1982) 31-36.
77 M. Hanaoka, M. Inoue, M. Takahashi and S. Yasuda, Chemical
    Transformation of Protoberberines. VI. Ring D inversion of
    protoberberine alkaloids. Conversion of berberine into non-
    naturally occurring 11,12-oxygenated protoberberines via
    spirobenzylisoquinolines, Chem. Pharm. Bull., 32(11) (1984)
    4431-4436.
78 S. Naruto and H. Kaneko, Constituents of Corydalis sps. VIII.
    Synthesis of dehydrocorydaline derivatives, Yakugaku Zasshi,
    92(8) (1972) 1017-1023 and references cited therein.
79 Z. Kiparissides, R. H. Fichtner, J. Poplawski, B. C. Nalliah
    and D. B. MacLean, A regiospecific synthesis of protoberberine
    alkaloids, Can. J. Chem., 58(24) (1980) 2770-2779 and
    references cited therein.
80 M. Hanaoka, S. Yoshida and C. Mukai, A novel and efficient
    synthesis of 13-methylprotoberberine alkaloids, J. Chem. Soc.
    Chem. Commun., (18) (1985) 1257-1258.

THE SYNTHESIS OF 1-AZABICYCLIC ALKALOIDS

Y. NISHIMURA

1. INTRODUCTION

The synthesis of 1-azabicyclic alkaloids described in this chapter covers the synthesis of naturally occurring pyrrolizidine and indolizidine alkaloids. This arrangement should be advantageous, because many synthetic methods are applicable to the construction of nuclei of both types of alkaloids. These two classes of alkaloids constitute a very large family of natural products which are widely isolated from plants, insects, animals, oceanic lives and secondary metabolites of microbes. The pyrrolizidine alkaloids possess the 1-azabicyclo[3.3.0]octane skeleton, while the indolizidine alkaloids have the 1-azabicyclo[3.4.0]nonane skeleton.

Pyrrolizidine

Indolizidine

They are often functionalized at a variety of structural sites and demonstrate a wide range of potent biological and pharmacological effects. Due to the various pharmacological and biological activities and the intriguing chemical structures, these two classes of alkaloids have attracted considerable synthetic interest.

A review published in 1979 by D. J. Robins (1) covered the literature (up to 1977) concerning synthesis of necine bases, spectroscopic studies and biogenesis of pyrrolizidines. Since then two reviews (2,3) have been published on the chemistry, biosynthesis and pharmacology of alkaloids, necine bases and necic acids in 1982 by D. J. Robins (covering the literature up to early 1981) and in 1985 by J. T. Wróbel (covering the literature between 1981 and 1984). A review published in 1978 by F. J. Swinbourne (4) covered developments in indolizine and indolizidine chemistry up to early 1977. A series of Annual Reports including pyrrolizidine and indolizidine alkaloids has been introduced (5,6). The intense interest in the chemistry, synthesis and pharmacology of pyrrolizidine and indolizidine alkaloids has grown during the last decade. The greatest advances during the last 10 years have been in the stereoselective and chiral syntheses based on the new methodologies and strategies. The present review discusses mainly the synthesis of natural alkaloids, focussing on the progress in the stereoselective, enantioselective and chiral synthesis. This review coveres the literature that appeared within 1976-the middle of 1987.

2. SYNTHESES OF PYRROLIZIDINE ALKALOIDS

2.1 Syntheses of Necine Bases

a) Intramolecular alkylation and acylation

This group of methods is general and useful for preparation of pyrrolizidines. Several routes to pyrrolizidines are available from various types of
starting materials. The typical approach by N-alkylation to necine base was
introduced by Kochetkov et al. (7) in their synthesis of (±)-trachelanthamidine
(1) in 1960.

(a) NaOEt; H$_2$/Raney Ni (b) HBr; SOCl$_2$/MeOH (c) NH$_3$/MeOH (d) LiAlH$_4$

Kametani and co-workers (8) have reported a new synthetic way to the
stereoselective syntheses of (±)-retronecine (2) and (±)-turnefrocidine (3) by
the regioselective [3,3]sigmatropic rearrangement and a sulpheno-cycloamination
with an addition of benzenesulphenyl chloride to olefinic amine followed by
base induced ring closure. Treatment of 3-pyrrolidinone **4** with cis-butendiol
produced **6** with the formation of two chiral centers in a regioselective manner
via the chair-like transition state 5. The amino-olefine 7 was treated with
benzenesulphenyl chloride and base to give a common intermediate 8 to 2 and 3
via an episulfonium intermediate. Compound 8 was transformed to 3 and 2 by
reduction and by a syn-elimination of sulfoxide followed by reduction, respectively.

(a) TsOH (cat.), Na$_2$SO$_4$, xylene, Δ (b) NaBH$_4$, MeOH; NaH, BzlBr, THF, Δ;
KOH, HO(CH$_2$)$_2$OH, Δ (c) HCl/Et$_2$O-MeOH; PhSCl, CH$_2$Cl$_2$; K$_2$CO$_3$, NaI, MeCN, Δ
(d) Raney Ni, EtOH, Δ; H$_2$/PdCl$_2$, MeOH-CHCl$_3$ (e) HCl/Et$_2$O-MeOH; mCPBA,
CH$_2$Cl$_2$; xylene, Δ ; Li/liq. NH$_3$, THF

Macdonald et al. (9) have synthesized (±)-supinidine (12) by a regioselective N-1—C-2 vicinal annulation of a 1,3-dihaloalkane onto a 3-(hydroxymethyl)-3-pyrroline 9 system. Regioselective C-alkylation of the derived
allylic organometallic 10 via a five-membered internal chelate at the C-2
position followed by intramolecular N-alkylation resulted in 12 .

(a) NaBH₃CN, MeOH; BzCl, py, DMAP (cat.); DBU, PhH, Δ; LiAlH₄, Et₂O
(b) LiTMP, THF, then Br(CH₂)₃Cl (c) MeLi, THF

The optically active (−)-rosmarinecine (16) and (−)-isoretronecanol (20)
were prepared from D-glucosamine by Tatsuta et al. (10). Grignard reaction of
3-C-formyl-α-D-xyrofuranoside 13 derived from D-glucosamine with allylmagnesium bromide afforded the threo amido alcohol 14 by a chelation-controlled
allylation. Compound 14 was transformed to 16 by the intramolecular N-acylation and N-alkylation. (−)-Isoretronecanol (20) was synthesized from the
ditrityl derivative 17 in a similar way.

(a) CH₂=CHCH₂MgBr, Et₂O, Δ (b) NaIO₄, KMnO₄, H₂O/t-BuOH, then K₂CO₃;
CH₂N₂, Et₂O; MOMCl, i-Pr₂EtN, THF; MsCl, py (c) BH₃·Me₂S; 0.5N HCl
(d) Ph₃P=CHCO₂Me, PhCH₃; H₂/Pd-C, THF, AcOH; Amberlyst 15, MeOH; MsCl, py
(e) BH₃·Me₂S, THF; KOAc, DMSO (f) SOCl₂, Δ; H₂/Raney Ni, EtOH; NH₃/MeOH

Tufariello has introduced a new synthetic approach to necine base using
the 1,3-dipolar cycloaddition of five-membered nitrone to alkenes for the
stereo-control as a key step. The initial work (11) of the synthesis of (±)-
supinidine (12) from 1-pyrroline 1-oxide has been extended by Tufariello and
Lee (12) to the synthesis of (±)-retronecine (2) by the use of substituted
nitrones. The protected α-keto-nitrone 22 was generated regiospecifically by

the mercuric oxide mediated oxidation of 21. This selectivity may be related to a diminution of eclipsing interactions (i.e., to a more favorable dihedran angle relationship). The regiospecific addition of nitrone 22 to methyl γ-hydroxycrotonate provided isoxazolidine 23. Hydrogenolysis of the nitrogen-oxygen bond of the mesylate of 23 generated the pyrrolizidine 24. The base mediated elimination of the mesylate of 24, followed by hydrolysis of the ketal and reduction, afforded α,β-unsaturated ester 25, which was converted into 2 by reduction with lithium alanate. The intermediate 24 was also effectively used for the synthesis of (±)-croalbinecine (28) (13).

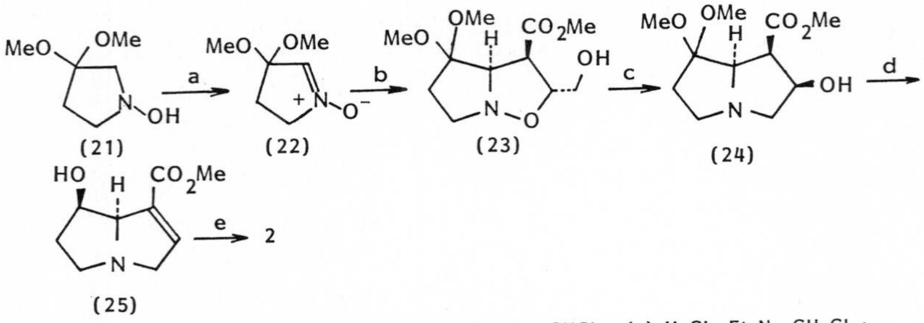

(21) (22) (23) (24)

(25)

(a) HgO, NH$_4$OH, CH$_2$Cl$_2$ (b) HOCH$_2$=CHCO$_2$Me, CHCl$_3$ (c) MsCl, Et$_3$N, CH$_2$Cl$_2$; H$_2$/Pd-C, MeOH (d) MsCl, Et$_3$N, HCl, DME; NaBH$_4$, MeOH (e) LiAlH$_4$, THF

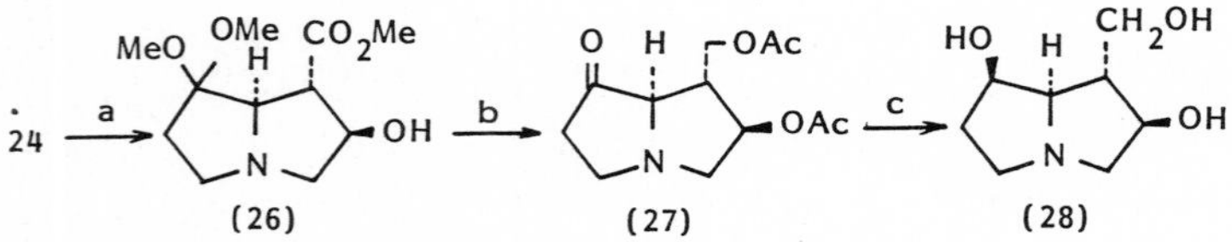

(26) (27) (28)

(a) NaOMe, MeOH (b) LiAlH$_4$; Ac$_2$O; CF$_3$CO$_2$H, aq. NaHCO$_3$ (c) Adams cat., AcOH

An analogous route to (±)-tussilagin (32) and (±)-isotussilagin (33) has been reported by Röder et al. (14). The regiospecific 1,3-dipolar cycloaddition of 1-pyrroline 1-oxide (29) to a mixture of cis and trans methyl 3-methyl-4-hydroxy-2-butenoate afforded an endo-adduct, isoxazolidine 30. Compound 30 was converted into 32 and 33 by an analogous procedure described by Tuferiello (12,13).

(29)   b ⎰(30) R=H
         ⎱(31) R=Ms

(a) CHCl₃   (b) MsCl, py   (c) H₂/Pd-C, MeOH

Kakisawa et al. (15) have been described a related approach to (±)-iso-
retronecanol (37). The regiospecific 1,3-dipolar cycloaddition of l-pyrroline
l-oxide (29) to dihydrofuran resulted stereoselectively in an exo-adduct 34.
Isoxazolidine 34 was transformed to 37 by hydrogenolysis of nitrogen-oxygen
bond followed by regioselective N-alkylation.

(29)   (34)   c ⎰(35) R=OTMS   (37)
                                            ⎱(36) R=I

(a) PhH   (b) LiAlH₄, THF, Δ; E₃N, TMSCl   (c) TMSI, CHCl₃   (d) BzlMe₃NF

Keck and Nickell (16) have presented an another methodology to (±)-helio-
tridine (46), together with (±)-retronecine (2). This synthesis depends upon
the intramolecular [4+2]cycloaddition by intramolecular transfer of the acylnitroso dienophile. Addition of the dienal (39) derived from acetylenic ester
38 to the lithium enolate of 40 afforded an alcohol 41. The intramolecular
transfer of the acylnitroso dienophile effected by thermolysis of the protected
alcohol to give the 1,2-oxazine derivative 42. No asymmetric induction at C-7
position (pyrrolizidine numbering) by assistance of the bulky ter-butyl-
dimethylsilyloxy group has occurred in this process. Reductive cleavage of
the nitrogen-oxygen bond of 42 gave a hydroxylactam 43, and the mesylate of 43
was cyclized by base-induced intramolecular alkylation to yield the easily
separable diastereomeric pyrrolizinones (44 and 45). Removal of the protecting groups in 40 followed by reduction of the lactam-carbonyl group resulted
in 2. A similar sequence with the diastereomer 45 furnished 46.

(a) $(CH_2=CH)_2CuLi$, THF; DIBAL, Et_2O; MnO_2, PhH (b) LDA, THF/HMPA
(c) TBDMSCl, imidazole, DMF; PhH, 80°C (d) Na/Hg, Na_2HPO_4, Et_2O (e) MsCl,
Et_3N, CH_2Cl_2; LDA, THF (f) py·p-TsOH, MeOH; Bu_4NF, THF; $LiAlH_4$, THF, Δ

Nishimura et al. (17) have introduced the first enantiodivergent strategy
in their syntheses of (+)-retronecine (55) and its enantiomer, (-)-retronecine
(56), from D-glucose. This approach to (+)-retronecine and its enantiomer involves, as the key step, the introduction of the allylic alcohol part at either
C-1 or C-7 of the same intermediate, pyrrolizidine meso diol 49, by differentiation of each position. The pyrrolidinefuranose 47 prepared from D-
glucose was transformed into pyrrolidine triol 48 by Wittig reaction of 47 followed by (methoxyethoxy)methylation and hydroboration. The mesylate of 48 was
cyclized by catalytic reduction to yield a pivotal intermediate, pyrrolizidine
meso diol 49. Selective removal of protecting groups in 49 by either hydrogenolysis or acid hydrolysis provided 1-hydroxy and 7-hydroxy derivatives (50
and 51) by differentiation of C-1 and C-7 position. The transformation of 50
and 51 to (+)-retronecine (55) and its enantiomer (56) involve hydroxymethyl-
ation at carbon bearing an α-sulfur substituent followed by sulfoxide-base
elimination. Treatment of the mesylate of 50 with sodium thiophenolate afforded sulfide (52), which was converted into sulfoxide upon oxidation.
Benzyloxymethylation of the sulfoxide, followed by syn elimination of sulfoxide
furnished the olefin 54. Removal of protecting groups in 54 resulted in (+)-
retronecine (55). Analogous treatment of 51 afforded unnatural (-)-retronecine
(56).

New synthetic routes by intramolecular alkylation to necine base have been
reported. These strategies based on two carbon homologation at C-8 (pyrrolizidine numbering) of pyrrolidine nucleus and a subsequent carbon-carbon bond
formation between C-1 and C-2 by alkylation. Kraus and Neuenschwander (18)
used successfully amidoalkylation and base-mediated intramolecular alkylation
in their synthesis of (±)-trachelanthamidine (1), together with (±)-isoretro-
necanol (37). Amidoalkylation of N-2-bromoethylsuccinimide derivative 57 with

ethylacetoacetate gave amide ester 58, which was cyclized to yield a 1:4
mixture of pyrrolizidine lactams (59 and 60). Lactam 59 was converted to 37
by reduction with lithium alanate. Similar treatment of 60 gave 1.

R=Bzl,   R'=MEM

(a) H$_2$/Raney Ni, EtOAc/EtOH; benzyl S-4,6-dimethylpyrimid-2-yl thio-
carbonate, Et$_3$N, MeOH; LiCl, DMF; Bu$_3$SnH, PhCH$_3$; HCl/MeOH; NaH,
BzlCl, DMF; 3 M HCl/AcOH (b) CH$_3$PPh$_3$Br, n-BuLi, THF; MEMCl,
i-Pr$_2$NH, CH$_2$Cl$_2$; 9-BBN, THF, then 2 M NaOH, H$_2$O$_2$ (c) MsCl, py;
H$_2$/Raney Ni, EtOAc/MeOH (d) H$_2$/Raney Ni, EtOH, Δ (e) 3 M HCl
(f) MsCl, py; NaSPh, DMF (g) HCl/MeOH, Et$_2$O; mCPBA, CH$_2$Cl$_2$;
LDA, THF/HMPA, then PhCH$_2$OCH$_2$Cl (h) xylene, Δ (i) 3 M HCl;
Li/liq. NH$_3$, THF (j) Li/liq. NH$_3$, THF

(59) R=β-CO$_2$Et
(60) R=α-CO$_2$Et

(a) NaBH$_4$, EtOH, HCl (b) MeCOCH$_2$CO$_2$Et, AlCl$_3$, CH$_2$Cl$_2$ (c) NaOEt, EtOH
(d) LiAlH$_4$, Et$_2$O

Kametani et al. (19) have described an efficient and novel synthesis of
(±)-trachelanthamidine (1) and (±)-supinidine (12) based on the ring opening
reaction of an aziridinium salt, and a subsequent intramolecular alkylation.
Treatment of the α,β-unsaturated ester 61 with an excess of aziridine afforded
the pyrrolidine derivative 63 via the aziridinium salt (62). Intramolecular
alkylation of 63 with lithium diisopropylamide (LDA) resulted in thermodynamically more stable 1-carboethoxypyrrolizidine (64), which was converted
to 1 by a procedure described by Borch (20). On the other hand, when compound

63 was treated with LDA and subsequently treated with phenylselenyl chloride, two diastereomeric selenides (65 and 66) were formed. Oxidative elimination of each diastereomer yielded the same α,β-unsaturated 1-carboethoxypyrrolizidine ester 67, which was transformed into 12 by a method previously used by Robins (21).

(a) neat, 0°C   (b) LDA, THF   (c) LDA, THF, then PhSeCl   (d) mCPBA, CH$_2$Cl$_2$

A synthesis of optically active (+)-retronecine (55) from (R)-(+)-maleic acid (68) has been reported by Niwa and Yamada et al. (22) by utilizing novel intramolecular Wittig reaction and intramolecular alkylation. The cyclic imide bromoacetate 69 prepared from 68 was converted into lactone lactam 70 by intramolecular Wittig reaction in a one-pot procedure, and a subsequent catalytic reduction. Selective reduction of the lactam carbonyl group in compound 70 was successfully carried out by using a modified method of Raucher (23) to yield the pyrrolidine 71 through the thiolactam derivative. Treatment of lithium enolate of 71 with phenylselenenyl chloride, and followed by acid hydrolysis afforded the selenide alcohol 72, which was cyclized by the three-step sequence in a one-pot procedure to yield the key intermediate, tricyclic lactone 73. Compound 73 was transformed into 55 by reduction, and subsequent sulfoxide-base elimination.

A new approach to (±)-trachelanthamidine (1) and (±)-isoretronecanol (37) by intramolecular N-acylation has been described by Pinnick and Chang (24). Thiopyrrolidone prepared from pyrrolidone was converted into the enamine ester 74 by the two-step procedure developed by Eschenmoser (25). Treatment of compound 74 with LDA followed by ethyl bromoacetate gave the diester, which was cyclized with potassium hydride to yield the unsaturated lactam 76. Catalytic reduction of 76 afforded lactam ester 77. Conversion of compound 77 into the key intermediate 78 was successfully carried out by selective reduction using phosphoryl chloride/sodium borohydride. Epimerization of compound 78 upon heating with base gave the more stable diastereomer 79 (25). Reduction of 78 and 79 with LiAlH$_4$ resulted in 37 and 1, respectively.

(a) AcCl, Δ, then H$_2$NCH$_2$CH$_2$OH, and then AcCl, Δ; dry HCl/EtOH; (CH$_2$)$_3$COCl,
py, Et$_2$O; BrCH$_2$COBr, py, Et$_2$O (b) Ph$_3$P, MeCN, then Et$_3$N; 5% Rh/Alumina,
EtOAc (c) p-methoxyphenylthiophosphine sulfide, PhCH$_3$, Δ; Et$_3$OBF$_4$, CH$_2$Cl$_2$,
then NaBH$_3$CN, MeOH (d) LDA, THF, then PhSeCl; 6N HCl (e) n-BuLi, THF,
then TsCl, and then LDA, HMPA (f) LiAlH$_4$, THF; H$_2$O$_2$, AcOH

(a) P$_4$O$_{10}$, xylene; BrCH$_2$CO$_2$Et, CH$_2$Cl$_2$; Ph$_3$P/xylene, t-BuOK/t-BuOH (b) LDA,
THF, then BrCH$_2$CO$_2$Et (c) KH, THF (d) H$_2$/Pd-C, EtOH (e) POCl$_3$, NaBH$_4$,
DME/EtOH (f) LiAlH$_4$, THF (g) NaOMe, MeOH

Shono et al. (26) have demonstrated an another synthetic routes to (±)-
trachelanthamidine (1) and (±)-isoretronecanol (37) by 1,2-annulation on a
pyrrolidine ring by utilizing anodically prepared 1-(benzyloxycarbonyl)-2-
methoxypyrrolidine, followed by intramolecular N-acylation. α-Methoxylated
carbamate 80 was prepared by the anodic oxidation of N-(benzyloxycarbonyl)-
pyrrolidine. Compound 80 was converted into triester derivative 81 by the
acid-catalized generation of a carbomethoxy iminium ion, and a subsequent
carbon-carbon bond-forming reaction with trimethyl ethanetricarboxylate.
Successive treatment of 81 by base hydrolysis, decarboxylation, and esterification resulted in diester 82. Hydrogenolysis of compound 82 afforded two
diastereomeric mixture of 83 and 84, which were separable by chromatography.

Compound 83 was transformed into 1 by reduction with LiAlH$_4$. Similar treatment of 84 gave 37.

(a) -2e, MeOH (b) MeCO$_2$CH$_2$CH(CO$_2$Me)$_2$, TiCl$_4$, CH$_2$Cl$_2$ (c) NaOH, EtOH/H$_2$O; AcOH, H$_2$O, Δ; conc. H$_2$SO$_4$/EtOH, Δ (d) H$_2$/Raney Ni, EtOH (e) LiAlH$_4$, THF

b) Condensation

A typical method of this group is Geissman's route (27) to (±)-retronecine (2) using Dieckmann condensation of diester generated from bicyclic lactone 85 as a key step in 1962. Due to its practical and useful method for producing pyrrolizidines, the synthetic route to necine base by Dieckmann condensation has continued to receive considerable synthetic attention.

(a) NaBH$_4$, EtOH; Ba(OH)$_2$, H$_2$O; 1 N HCl; BrCH$_2$CO$_2$Et, Na$_2$CO$_3$, EtOH
(b) KOEt, PhH (c) H$_2$/Pt, H$_2$O; Ba(OH)$_2$, H$_2$O; HCl/EtOH (e) LiAlH$_4$, THF

(a) EtONa, EtOH, then NaBH$_4$; Ac$_2$O, py, cat. DMAP (b) t-BuOK (c) DIBAL

A modified Geissman's route to (±)-retronecine (2) has been reported by Narasaka et al. (28). The Dieckmann condensation of bicyclic lactone 85, and subsequent sodium borohydride reduction and acetylation afforded β-acetoxy ester **87**. The ester **87** was converted into an unsaturated ester **88** on base treatment, and the successive diisobutylaluminum hydride reduction resulted in **2**.

Rüeger and Benn (29) have achieved the first enantioselective synthesis of all of (-)-isoretronecanol (**20**), (-)-trachelanthamidine (**94**), (-)-petasinecine (**97**), and (-)-supinidine (**100**) from L-proline using Dieckmann condensation as a key step. The homologation of N-Cbz-L-proline without loss of chirality was carried out by two conventional procedures: (a) Arndt-Eistert homologation, and (b) a sequence of reduction, tosylation, cyanide displacement, and ethanolysis to give the ester **89**. N-Alkylation of **89** afforded the diester **90**, which was submitted to Dieckmann condensation to yield the pyrrolizidine enol ester **91**. Reduction of **91** by the two-step sequence (Rh/Al$_2$O$_3$, and then LiAlH$_4$) gave **20**. The endo-ester **92**, upon base treatment, yielded the thermodynamically more stable 1-carboethoxypyrrolizidine **93**, and the successive reduction with LiAlH$_4$ resulted in **94**. Reduction of the double-bond of **91** with NaBH$_3$CN, and the subsequent dehydration and reduction of the ester group with DIBAH afforded **100**, while catalytic reduction of **91** gave a separable diastereomeric mixture of **95** and **96**. Each of **95** and **96**, upon reduction with LiAlH$_4$, yielded **97** and its C-1 epimer **98**, respectively. All of synthetic (-)-isoretronecanol (**20**), (-)-trachelanthamidine (**94**), (-)-petasinecine (**97**), and (-)-supinidine (**100**) had enantiomeric excesses greater than 80%.

(a) BrCH$_2$CO$_2$Et, Na$_2$CO$_3$ (b) NaOEt; AcOH/H$_2$O (c) H$_2$/Rh-Al$_2$O$_3$, AcOH/H$_2$O
(d) NaOEt, EtOH (e) LiAlH$_4$ (f) NaBH$_3$CN, H$_2$O (g) MsCl, Et$_3$N (h) DIBAL
(i) H$_2$/Pt, AcOH/H$_2$O

238

An analogous enantioselective synthetic route to (+)-retronecine (55), (+)-croalbinecine (107), (+)-platynecine (109) has been announced by Rüeger and Benn (30). The strategy is based on the synthesis of (±)-retronecine by Geissman (27), which was improved by Narasaka et al. (28). The Geissman's bicyclic lactone 101 in optically active form was prepared from 4-hydroxy-L-proline by Arndt-Eistert homologation, followed by dehydration and iodolactonization. The bicyclic lactone 101 was converted into 55, 107, 109, and C-1 epimer of (+)-croalbinecine (105) by the similar sequence used in their previous enantioselective synthesis of necine bases (29).

(a) KOEt, PhCH$_3$ (b) H$_2$/PtO$_2$, AcOH/H$_2$O (c) LiAlH$_4$ (d) cat. NaOEt, EtOH
(e) Ac$_2$O, py DMAP; NaH, THF; HCl/EtOH (f) H$_2$/Rh-Al$_2$O$_3$, AcOH/H$_2$O

(a) NH$_2$OH·HCl, py; MsCl, py; activ. Zn, BrCH$_2$CO$_2$Me, THF, Δ (b) DBU, CH$_2$Cl$_2$; NaBH$_3$CN, MeOH, HCl; BzlCl, Et$_3$N, CH$_2$Cl$_2$ (c) CF$_3$CO$_2$H; 1,1'-thiocarbonyl-diimidazole, py, THF, Δ; Bu$_3$SnH, PhH, Δ (d) H$_2$/Pd-C, EtOH, HCl

Carbohydrate precursor has been converted into (1R,5R)-6-aza-2-oxabicyclo-[3.3.0]octan-3-one hydrochloride (114), a precursor of necine base in optically active form by Buchanan et al. (31). The several sequences of the readily-available 2,3-O-isopropylidene-D-erythrose (110) resutled in 114, which could be converted efficiently into (+)-retronecine and other pyrrolizidines.

Miyano et al. (32) have reported a convenient short way to a mixture of (±)-heliotridane (117) and (±)-pseudoheliotridane (118). The synthesis involves, as a key step, the transformation of γ-(N-2-pyrrolidinonyl)-α-methyl-butyric acid (115) to the unsaturated pyrrolizidine 116 by cyclodehydration accompanied with decarboxylation. Catalytic reduction of enamine 116 afforded 117 with 90% isomeric purity, while reduction of 116 resulted in 118 with 66% isomeric purity.

(a) Na   (b) soda lime, Δ   (c) H$_2$/PtO$_2$, Et$_2$O   (d) HCO$_2$H

(a) (EtO)$_2$P(O)CH$_2$CO$_2$Et, NaH; DIBAL   (b) Ti(OPr^i)$_4$, (+)-diethyl L-tartrate
(c) [MeO(CH$_2$)$_2$O]$_2$AlH   (d) PhCOCl, Et$_3$N; potassium phthalimide, Ph$_3$P,
EtO$_2$CN=NCO$_2$Et; K$_2$CO$_3$, MeOH; H$_2$NNH$_2$·H$_2$O   (e) HClO$_4$ (f) NaBH$_3$CN   (h) DMSO,
(COCl)$_2$, Et$_3$N; Ph$_3$P=CH(CH$_2$)$_3$Me; H$_2$/Pt$_2$O

An enantioselective synthetic route to an ant venom alkaloid (127) have been described by Takano et al. (33). The strategy involves, as the key stage, an enantioselective Sharpless epoxidation of the alkene 119, and a subsequent regioselective opening of the epoxide 120. The diol 121 was converted into the amino alcohol 122, with inversion of configuration, by Mitsunobu reaction. The pyrrolinium perchlorate 123 was then formed by treatment of 122 with perchloric acid. Reduction of 123 with NaBH$_3$CN afforded a mixture of (ca 1:1) of pyrrolizidine 125 and 126. The stereochemical outcome is attributed to the participation of the hydroxy group, which directs delivery of hydride ion or cyanide ion from the si-face of the imino group via boronate complex 124. Swern oxidation of 125, and followed by Wittig reaction and catalytic hydrogenation resulted in 127.

Ishibashi et al. (34) have synthesized (−)-trachelanthamidine (94) from L-proline, using olefin cyclization initiated by an α-thiocarbocation as the key step. The olefin 128 prepared from L-prolinol was effectively converted into the key sulphoxide 130 by deprotection, followed by (methylthio)acetylation and oxidation. Cyclization of 130 was carried out by treatment with trifluoroacetic anhydride to afford a diastereomeric mixture of pyrrolizidine 133. The regiospecificity of cyclization of the cation 131 is assumed that the carbonyl group in the newly formed ring plays an important role in the stereoelectronic preference of the reaction. The stereospecificity is attributed to an attack of the olefin bond on the cation 131 via a transition state which avoids steric repulsion between the substituent at C-1 and the C-1—C-8 bond in 132. Oxidation of 133 followed by reduction gave the lactam 134, which was transformed into 94 by oxidative cleavage of the vinyl group followed by reduction.

(a) NaOH, HOCH$_2$CH$_2$OH/H$_2$O; MeSCH$_2$COCl, Et$_3$N, Et$_2$O (b) NaIO$_4$, MeOH/H$_2$O
(c) (CF$_3$CO)$_2$O, CH$_2$Cl$_2$ (d) NaIO$_4$, MeOH/H$_2$O; aluminium amalgam, THF/H$_2$O
(e) OsO$_4$, NaIO$_4$, THF/H$_2$O; LiAlH$_4$, THF, Δ

c) Transannular reaction

This route to pyrrolizidines is based on transannular interaction between tertiary amino nitrogen and ketone carbonyl of 1-azacyclooctane derivative with oxygen function at the C-5 position. 1-Benzyl-1-azacyclooctan-5-one 135 exhibited infrared absorption at 1693 cm^{-1}, indicative of transannular inter-

action. Perchlorate 136 are in the transannular form, as evidenced by the lack
of infrared absorption in the carbonyl region (1620–1800 cm^{-1}) and by the
presence of absorption in the hydroxyl region ($\sim$3290 cm^{-1}).

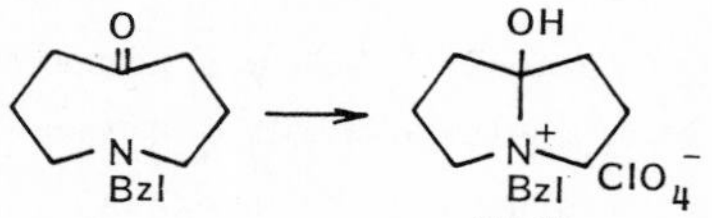

(135) (136)

A transannular reaction was utilized effectively for the stereospecific
synthesis of (±)-isoretronecanol (37) by Leonard and his coworker (35) in 1969.
Tertiary amino nitrogen and ketone carbonyl in an eight-membered amino-ketone
138 are close together by transannular interaction. The diester 137 was subjected to Dieckmann cyclization to yield amino-ketone 138, which was converted
to perchlorate 139. Hydrogenation of the perchlorate 139 by two-steps sequence
gave 37. Leonard's route by transannular reaction is widely applied for the
synthesis of various pyrrolizidine skeletones.

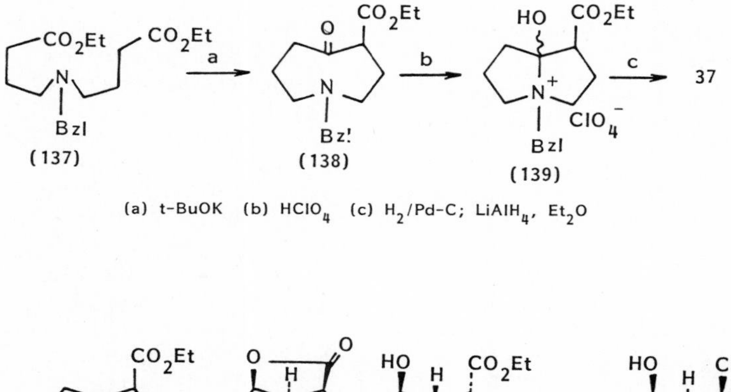

(137) (138) (139)

(a) t-BuOK (b) HClO$_4$ (c) H$_2$/Pd-C; LiAlH$_4$, Et$_2$O

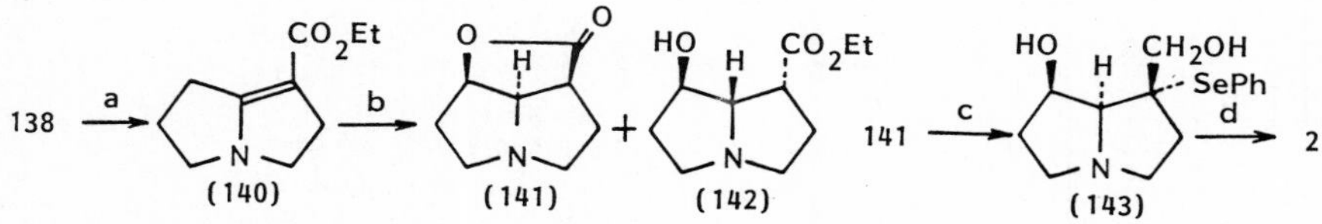

(140) (141) (142) (143)

(a) H$_2$/Pd-C, EtOH (b) LDA, THF, then MoO$_5$·py·HMPA (c) LDA, THF, then
(PhSe)$_2$; LiAlH$_4$, THF (d) H$_2$O$_2$, AcOH, heat

Yamada et al. (36) have been reported a convenient synthesis of (±)-
retronecine (2) via Leonard's intermediate 138. γ-Hydroxylation of the
lithium enolate of the known unsaturated ester 140, and followed by catalytic
hydrogenation gave the tricyclic lactone 141, together with the hydroxy ester

142. Phenylselenylation of 141, and a subsequent reduction and selenium-based elimination reported by Robins (21) resulted in 2.

The first synthesis of (±)-otonecine (149) has been achieved by Yamada et al. (37). Michael addition of thiophenolate anion to the α,β-unsaturated ester 140 afforded the tricyclic lactone 144. The lactone, upon treatment with merculic diacetate, gave the hydroxy lactone, which was converted into the methiodide 146. Treatment of 146 with sodium hydride and phenylselenyl chloride afforded the lactone 147. The crucial step in this synthesis, selective reductin of the lactone 147 was successfully carried out by treatment with the hydride reducing agent in the presence of Lewis acid. The product was acetylated to give the diacetate 148. Oxidative elimination of phenylseleno group and removal of the acetyl groups yielded 149.

(a) LiSPh (b) Hg(OAc)₂, AcOH/H₂O (c) MeI, CH₃COCH₃ (d) NaH, PhSeCl
(e) DIBAL, Et₂AlCl, Ac₂O, py (f) H₂O₂; NaOMe, MeOH

(a) K₂CO₃, aq. MeOH (b) H₂/Pt (c) LiAlH₄, THF, Δ (d) CH(OMe)₃, MeOH, cat.
H₂SO₄; LiAlH₄, THF, Δ; BzlCl (e) Br₂/MeOH; Et₄NOAc, CH₃COCH₃, Δ; NH₃/MeOH
(f) dihydropyran, CSA, CH₂Cl₂; LiAlH₄, THF, Δ; PCC, NaOAc, CH₂Cl₂; HCl/MeOH

Ban et al. (38) have reported a criss-cross annulation to the synthesis of (±)-heliotridane (117) and (±)-dihydrodesoxyotonecine 158. This sequential process involves a formation of carbinolamine 152, followed by a ring-opening of retroaldol type and a transannular cyclization of 153 (R=H). When the protected secondary amine (151, R=Bzl) is used as a substrate, controlled criss-cross annulation is occurred to give the eight-membered ketolactam 154. Compound 150, upon base treatment, gave the pyrrolizidine 155, and a successive reduction yielded 117. Compound 151 afforded the ketolactam 154 on base treatment, and a successive sequence of a conversion of protecting group, hydroxylation, removal of protecting group and methylation resulted in 158.

An another transannulation approach to (±)-heliotridane (117) has been achieved by Garst et al. (39) by the hydroboration-carbon monoxide insertion of bis-olefinic amine. Hydroboration of 159 followed by cyanidation gave the azacyclooctanone 160, and a successive sequence of methylation afforded 117.

(a) thexylborane, THF, then KCN, and then H_2O_2, 1M KOH (b) lithium hexa-methyldisilazide, THF, then MeI; H_2/Pd-C, $HClO_4$

d) Iminium ion initiated cyclization

This methodology consists of the cyclization by intramolecular nucleophilic attack to five-membered iminium or acyliminium ion generated intermediaryly. In the last decade, due to many extensive efforts for development of iminium ion cyclization to pyrrolizidine, this group of method now occupy a prominent position in the pyrrolizidine synthesis.

Leonard et al. (40) have achieved the synthesis of (±)-trachelanthamidine (1) along the view of plant alkaloid biogenesis in 1960. This route seems to proceed via iminium cyclization.

(a) phosphate buffer (pH 7) (b) $NaBH_4$

Takano et al. (41) have reported a short analogous route to trachelanthamidine in both racemic and optically active form. Mannich reaction was carried out by treatment of 162 with acid to give the crude Mannich base 163, which was converted into (±)-trachelanthamidine (1) by reduction. Asymmetric cyclization into (+)-trachelanthamidine 164 have been achieved in the presence of pyridinium (d)-camphor-10-sulfonate with an optically purity of 33%.

(a) pyridinium (d)-comphor-10-sulfonate   (b) NaBH₄

A new route to (±)-trachelanthamidine (1) and (±)-isoretronecanol (37) by acyliminium cyclization has been described by Speckamp et al. (42). The acyliminium ion 166 generated from the ethoxy-lactam 165 by acid treatment cyclized to afford, after hydrolysis, a 4:1 epimeric mixture (167 and 168). Terminal phenylthio group is considered to stabilize an exocyclic vinyl cationic intermediate to lead the preferential formation of the five-membered ring. Each of epimeric mixture (167 and 168), upon reduction, yielded 1 and 37.

(a) HCO₂H; 2M HCl   (b) LiAlH, THF

Hart et al. (43) have achieved a new synthetic route to (±)-trachelanthamidine (1) and (±)-supinidine (12) via aza-Cope rearrangement followed by cyclization of acyliminium ion. Treatment of the amide 169 with formic acid gave the pyrrolizidinones (170 and 171) along with the indolizidinone 172. The reaction proceeds stereoselectively in every step. After completion of saponification of 170 to 171, and conversion of protecting group, the C-2 sidechain was degrated to give the iodide 173. The iodide 173, upon reduction, yielded 1. Alternatively, the iodide 173 was transformed into 12 by dehydrohalogenation and reduction.

(a) [structure]—MgBr (b) HCO$_2$H (c) NaOH, MeOH/H$_2$O (d) H$_2$/Pd-C, EtOH; Ac$_2$O,
py; HgO, I$_2$, CCl$_4$ (e) n-Bu$_3$SnH; LiAlH$_4$ (f) DBU, PhH (g) LiAlH$_4$

A related strategy (44) has been applied to the synthesis of (-)-hasta-
necine 176 and its enantiomer 177 using maleic acid as a chiral educt. The
optically active amides ((+)-175 and (-)-175) were prepared by coupling the
amine 174 with (R)-(+)- and (S)-(-)-maleic acid, respectively. Each of them
was converted into 176 and 177 by analogous procedures mentioned above.

(a) (R)-(+)-maleic acid, or (S)-(-)-maleic acid, CH$_2$Cl$_2$, Δ (b) NaBH$_4$, MeOH;
HCO$_2$H; NaOH, MeOH/H$_2$O (c) H$_2$/Pd-C; Ac$_2$O, py; HgO, I$_2$, CCl$_4$; n-Bu$_3$SnH;
LiAlH$_4$

The elegant enantioselective routes to seven pyrrolizidines from a single
precursor have been achieved by Chamberlin et al. (45). The key steps in this
synthesis involves an acyliminium ion initiated cyclization of ketene dithioacetal. The amide 178 derived from (S)-maleic acid was transformed into the
optically active key precursor 179, in 68% yield, via cyclization of acyliminium ion generated under basic condition along with the diastereomer in 3%
yield. The efficient relative asymmetric induction during the cyclization step
arise from effective block of the α-face of acyliminium ion by the acetoxy
group. After removal of acetyl group, a controlled migratin of the double bond
to the endocyclic position 180 carried out via a deprotonation-protonation reaction. Hydrolysis of the dithiane 180 followed directly by reduction of both

the aldehyde and the lactam carbonyl groups afforded (+)-heliotridine (181).
The catalytic reduction of 181 with Raney Ni was proceeded stereoselectively
to yield (-)-dihydroxyheliotridane 182. Hydrolysis of 179 gave lactame diester
183, which was converted into (+)-hastanecine (177) by reduction. Aluminium
hydride reduction of both the ester and lactam carbonyl groups in 179, followed
by hydrolysis, yielded the ester alcohol 187, which was effectively transformed
into the C-7 epimer 188 by two-step sequence of swern oxidation and catalytic
reduction. Further reduction of 188 afforded (+)-turneforcidine (189). Efficient preparation of the C-1 diastereomer, (-)-platynecine (191), was carried
out by base treatment of 188 and reduction of the resulting tricyclic lactone
190. Conversion of tricyclic lactone 190 into (+)-retronecine (55) was accomplished by the procedure reported by Robins (21) and Yamada (36) for the
racemic form.

(a) NaBH$_4$, MeOH (b) MsCl, Et$_3$N (c) MeONa; LDA, then MeOH (d) HgCl$_2$,
CH$_3$CN/H$_2$O, CaCO$_3$; LiAlH$_4$ (e) Raney Ni (f) HgCl$_2$, HCl/MeOH (g) LiAlH$_4$
(h) AlH$_3$ (i) LDA, then MeOH (j) HgCl$_2$, CH$_3$CN/H$_2$O, CaCO$_3$; NaBH$_4$
(k) (COCl)$_2$, DMSO, Et$_3$N; H$_2$/Pt (l) EtONa, EtOH, Δ (m) LDA, then PhSe$_2$Ph;
LiAlH$_4$; H$_2$O$_2$, AcOH

This work based on their previous synthesis of (±)-supinidine (12) and (±)-
trachelanthamidine (1) using the ketene dithioacetal as a terminator for
cationic cyclization (46).

(a) succinimide, Ph$_3$P, EtO$_2$CNNCO$_2$Et; NaBH$_4$, MeOH (b) MsCl, py, Et$_3$N (c) LDA, HMPA, then MeOH (d) BF$_3$·Et$_2$O, HgO, THF/H$_2$O; LiAlH$_4$, THF (e) HClO$_4$, H$_2$O, THF, MeCN

The key step in the synthesis of (±)-trachelanthamidine (1) and (±)-iso-retronecanol (37) by Blum et al. (47) involved an anodic functionalization of the tertiary amide followed by an intramolecular amidoalkylation. An anodic methoxylation of the ester 192 gave the methoxylated amide 193, which was converted into a diastereomeric mixture of pyrrolizidine ester (196 and 197) via cyclization of acyliminium ion generated with Lewis acid. Reduction of each of 196 and 197 afforded 37 and 1, respectively. Alternatively, 1 and 37 were synthesized starting from the diester 194 by anodic methoxylation, intra-molecular amidoalkylation, dealkoxycarbonylation and reduction.

(192) R=H
(194) R=CO$_2$Me

(193) R=H
(195) R=CO$_2$Me

(196) R^1=H, R^2=CO$_2$Me
(197) R^1=CO$_2$Me, R^2=H

196 $\xrightarrow{\text{c}}$ 1
197 $\xrightarrow{\text{c}}$ 37

(a) −2e, MeOH (b) TiCl$_4$ (R=H), or AlCl$_3$ (R=CO$_2$Me) (c) LiAlH$_4$ (R=H), or LiCl, DMSO, LiAlH$_4$ (R=CO$_2$Me)

New routes to (±)-isoretronecanol by both acyliminium ion and free radical cyclization using allylstannanes as terminators have been reported by Keck et al. (66, see Section j)).

e) Cycloaddition

This strategy involves intermolecular [3+2] and intramolecular [4+1]cyclo-addition based on the 1,3-dipolar cycloaddition. This is one of the shortest and most efficient routes to pyrrolizidines. Based on the fundamental works, a two-step stereospecific route to (±)-isoretronecanol (37) have been achieved

the pyrrolizidine ethyl ester 218 as a mixture of endo- and exo-isomers. Isomerization of 218 with LDA afforded a single exo-isomer 79. Reduction of 79
gave (±)-trachelanthamidine (1). The 1,2-unsaturated pyrrolizidine ester 67
was prepared from 218 by procedure reported by Robins (49). Compound 67 represents the formal synthesis of (±)-supinidine (12) and (±)-isoretronecanol
(37).

The very similar approach to pyrrolizidine has been carried out using new
strategy of intramolecular [4+1]pyrroline annulation by Hudlicky et al. (52)
and Pearson (53), independently. The key intermediate, azidodiene 219 in a
mixture of E and Z, was prepared by each of groups by different way. In the
route of Hudlicky et al., flash pyrolysis of 219 was carried out at ∿500°C to
give a mixture of imine (222) and tetrahydropyrrolizidines (223 and 224) (ap-
prox. 50:40:10) in 86% yield. Intramolecular 1,3-dipolar cycloaddition of 219
proceeds through triazoline 220 and vinyl aziridine 221, giving the 1,5-homo-
dienyl shift product 222 and the pyrrolizidines 223 and 224. Compound 224
represents a formal synthesis of (±)-supinidine (12). Pearson has also reported the same result by pyrolysis of 219.

(a) Zn/Cu, Et$_2$O, Δ, BrCH$_2$CHCHCO$_2$Et (b) Ac$_2$O, py, DMAP; DME, DBU
(c) lithium enolate of CH$_3$CH=CHCO$_2$Et; MsCl, Et$_3$N; AcOH, THF/H$_2$O (d) MsCl,
Et$_3$N; NaN$_3$, DMSO (e) flash pyrolysis ∿500°C

f) Ring opening of cyclopropane

The key step in a new fasinating strategy to (±)-isoretronecanol (37) by
Stevens et al. (54) involved an acid-catalyzed rearrangement of cyclopropyl
imine. Cyclopropanation of phenylthio acetonitrile (225) with ethylene di-
chloride, followed by selective reduction of nitrile group afforded cyclopropyl aldehyde (227). Schiff base formation of 227 with acetal amine 228 gave
the key cyclopropyl imine 239. Rearrangement of 229 was best accomplished by
refluxing in xylene containing suspended ammonium chloride yielded the stabilized five-membered endocyclic enamine 230. Enamine 230 gave the pyrroli-

zidine 231 on acid treatment, and the subsequent reduction and desulfurization resulted in 37.

(a) LDA, HMPA (b) DIBAL (d) NH$_4$Cl, xylene, Δ (e) HCl/MeOH, H$_2$O (f) LiAlH$_4$, then Raney Ni

An analogous rearrangement of cyclopropylimine has been used by Pinnick et al. (55) in their synthesis of (±)-trachelanthamidine (1) and (±)-isoretronecanol (37). A successful synthesis of the crucial cyclopropyl imine 232 was developed from ethyl acetoacetate by several steps. Rearrangement of cyclopropyl imine 232 was carried out by procedure reported by Stevens (54) to give the unsaturated pyrrolizidine 140. Compound 140 was converted into 1 and 37 by the known procedure.

(a) NaH, BuLi, then CH$_2$=CHCH$_2$Br (b) BrCH$_2$CH$_2$Br, K$_2$CO$_3$, DMF; HOCH$_2$CH$_2$OH, H$^+$ (c) O$_3$; NaBH$_4$; MsCl, Et$_3$N (d) potassium phthalimide; NH$_2$NH$_2$; H$^+$ (e) NH$_4$Cl, xylene, Δ (f) H$_2$/Pd-C (g) LiAlH$_4$ (h) EtONa; LiAlH$_4$

Danishefsky et al. (56) have developed new kinetically controlled stereospecific routes to mono- and dihydroxypyrrolizidines by the intramolecular nucleophilic opening of diactivated cyclopropanes. The principles in their synthesis are those of (i) cis insertion of carbenoid from diazomalonate to double bond in cyclopropanation and (ii) inversion of configuration in the ring opening of cyclopropane. Internal cyclopropanation of the E diazomalonate 233 was achieved by treatment with copper bronze. Treatment of cyclopropane

a:b:c:d=2:14:2.2:1 (preparative GLC). Compound **243a** was identified by comparison with a natural specimen.

An analogous nonstereoselective approach to (5E,8Z)-3-(1-non-8-enyl)-5-((E)-1-pro-1-enyl)pyrrolizidine (**244d**) has been reported (60).

a: (5Z,8Z) b: (5Z,8E)
c: (5E,8E) d: (5E,8Z)

(244)

(a) H$_2$O$_2$, NaOH (b) NH$_4$OAc, NaBH$_3$CN (c) KSeCN, MeOH, Δ

(245) (246) (247) (248) (249) (250) (251)

(a) MeC(OEt)$_3$, pivalic acid; LiAlH$_4$, Et$_2$O; p-MeC$_6$H$_4$SO$_2$Cl, py, then KCN, Me$_2$SO; DIBAL, Et$_2$O (b) CHCl$_3$ (c) AgBF$_4$, CH$_2$Cl$_2$ (d) MeC(O)CH=CH$_2$, THF (e) H$_2$/PdCl$_2$, EtOH (f) CrO$_3$, py; HSCH$_2$CH$_2$SH, BF$_3$·Et$_2$O, CH$_2$Cl$_2$; Raney Ni, EtOH

New stereoselective route to 3,5-disubstituted pyrrolizidine from thief ant has been achieved using, as key steps, 1,3-dipolar cycloaddition of nitrone and intramolecular reductive amination by Lathbury and Gallagher (61). Cyclization of E allenic oxime **245** with silver tetrafluoroborate gave the nitrone **246**, which was cyclized with methyl vinyl ketone to afford a diastereomeric mixture of isoxazolidines (**247** and **248**). Hydrogenation of each of **247** and **248** yielded pyrrolizidine alcohols (**249** and **250**), respectively. Reduction of double bond, N–O bond cleavage, and reductive cyclization were occurred simultaneously in one pot procedure. Oxidation of each of these

alcohols followed by formation of 1,3-dithiolane and desulphurisation gave
(±)-3α,5α,8β-hexahydro-3-heptyl-5-methyl-1H-pyrrolizine (251).

h) Wittig reaction

The reactions between Wittig reagents and pyrrole-carbonyl compounds provide pyrrolizines, the precursors to pyrrolizidines. Two types of Wittig reactions involving phosphonium salts and using cumulated ylides are available for this synthesis.

The key step in new approach to (±)-isoretronecanol (37) and (±)-trachelanthamidine (1) by Muchowski et al. (62) involved the nucleophilic opening of the activated cyclopropane ring, and the successive intramolecular Wittig reaction. Reaction of sodium succinimide with cyclopropylphosphonium salt gave the unsaturated pyrrolizidinone ester 252. Catalytic reduction gave the endo-ester 253 with high stereoselectivity. Reduction of 253 with LiAlH$_4$ yielded 37. Compound 253 could be converted into 1 by epimerization and reduction.

(a) xylene, Δ   (b) H$_2$/PtO$_2$, AcOH   (c) LiAlH$_4$

The same route mentioned above to (±)-isoretronecanol (37) and (±)-trachelanthamidine (1) has been described by Flitsch et al. (63).

Bohlmann et al. (64) have reported the synthesis of 5,7a-didehydrohelio-tridin-3-on (256) isolated from Senecio species. Bestmann reaction of the disubstituted pyrrole derivative 254 with triphenylphosphoranylideneketen gave the pyrrolizinone 255. Catalytic hydrogenation of one double bond, followed by removal of protecting group in compound 255 yielded 256.

(a) xylene, 135°C   (b) H$_2$/Pd-Ba$_2$SO$_4$, THF   (c) p-TsOH, THF/H$_2$O

The methylthio-substituted α-carbamoyl radical as an initiator for radical olefin cyclization has been included in formal synthesis of (±)-pseudohelio-tridane (118) by Ishibashi et al. (68). Treatment of the chloride 273 with n-Bu₃SnH and AIBN gave pyrrolizidine lactam 274 in 60% yield together with the reduction product 272 (24%). Desulfurization afforded the 1α-methyl-lactam 275 in a yield of 80% and a trace amount of 1β-methyl isomer 276. Reduction of each isomers leading to 118 and 117 have been described by Mori et al. (64).

j) Miscellaneous methods

Robins et al. (69) have described a biogenetically patterned synthesis of (±)-trachlanthamidine (1) using enzymes and physiological conditions. Homo-spermidine (277) was incubated in phosphate buffer (pH 7) with pea seedling diamine oxidase and catalase. After six days, reduction of the presumed inter-mediate 278 with NaBH₄ gave 1 in 40% yield.

(a) diamine oxidase and catalase of pea seeding, phosphate buffer (pH 7)

(b) NaBH₄, MeOH

(a) CbzCl; ClCO₂Et; NaBH₄; PCC; PPh₃CH₃Br, n-BuLi; HBr/AcOH, then ClCH₂COCl, and then KI (b) Pd(PPh₃)₄, CH₃CN (c) AgOAc; LiAlH₄ (d) B₂H₆, THF; H₂O₂, aq. NaOH; HCl, MeOH (e) H₂/PtO₂; LiAlH₄, THF

Palladium catalyzed cyclization of α-haloamide with internal double bond was involved as key step in syntheses of pyrrolizidines by Mori et al. (64). The key N-iodoacetyl-2-vinylpyrrolidine (279) was effectively prepared from L-proline. Treatment of 279 with Pd(PPh₃)₄ in CH₃CN afforded pyrrolizidine derivatives, 280, 281 and 282 in a yield of 29, 14 and 7%, respectively. Cat-alytic hydrogenation of 282 followed by reduction with LiAlH₄ gave (±)-Helio-tridane (117) and (±)-Pseudoheliotridane (118). Compound 280 was converted into (±)-trachelanthamidine (1) by treatment with AgOAc, and subsequent reduction. Hydroboration of 281 followed by acid treatment yielded (±)-iso-retronecanol (37).

A new entry using intramolecular carbenoid displacement reaction to pyr-rolizidines has been reported by Kametani et al. (71). The requisite diazo-compound 283 was prepared from succinimide in a good overall yield. Intra-molecular carbenoid displacement reaction was carried out by heating of 283 in benzene in the presence of a catalytic amount of rhodium acetate to give di-astereomeric mixture of sulphide 284. Reductive desulphurisation of 284 with Raney Ni gave the lactam esters (285 and 286) in a ratio of 1:1. Reduction of 284 with LiAlH$_4$ afforded the sulphide alcohol 287, which was converted into the allyl acetate 288. Compounds 285, 286 and 288 represent the formal synthe-ses of (±)-trachelanthamidine (1), (±)-isoretronecanol (37) and (±)-supinidine (12).

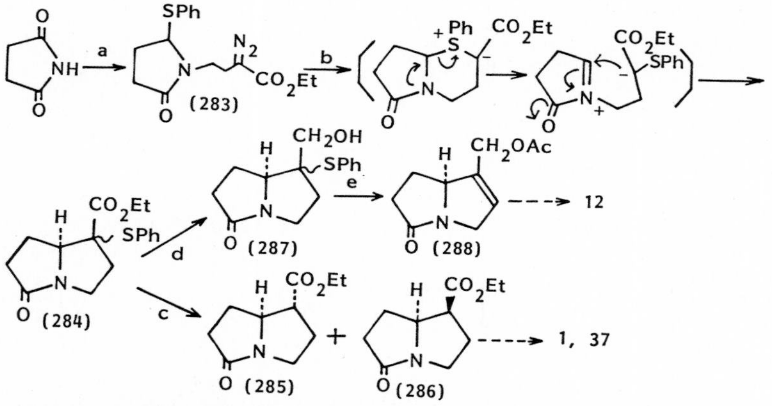

(a) NaH, Br(CH$_2$)$_3$CO$_2$Et, DMF; NaBH$_4$, EtOH; HCl, EtOH; PhSH, p-TsOH; NaH, HCO$_2$Et, PhH; p-TsN$_3$, Et$_3$N, CH$_2$Cl$_2$ (b) Rh$_2$(OAc)$_4$, PhH, Δ (c) Raney Ni, EtOH
(d) LiAlH$_4$, Et$_2$O (e) mCPBA, CH$_2$Cl$_2$, then PhCH$_3$, Δ; Ac$_2$O, Et$_3$N, DMAP, CH$_2$Cl$_2$

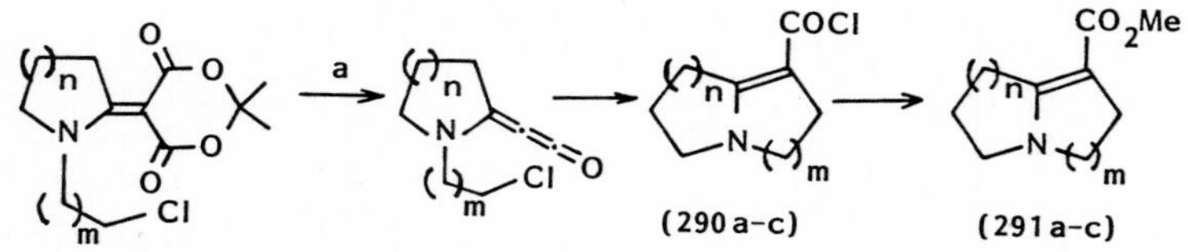

a: n=m=1
b: n=1, m=2
c: n=m=2

(a) flash vaccuum pyrolysis at 570°C/10^{-4} torr, then Et$_3$N/MeOH

(a) $(CH_3)_2CHMgBr$, Et_2O $-78°C \rightarrow RT$ (b) BzlCl, NaH, DMF; AcOH/3M HCl; Ac_2O, py (c) $NaBH_4$, EtOH, Δ; $NaIO_4$, $KMnO_4$, Na_2CO_3, t-BuOH/H_2O
(d) H_2/Pd-C, MeOH

c) C_8-Acids

All eight stereoisomers of monocrotalic acid have been synthesized by Matsumoto et al. (77). Cis-hydroxylation of dimethyl (±)-cis-2,3,4-trimethyl-2-pentenedioate afforded epimeric γ-lactone esters (313 and 314) in a ratio of 5:1. The major epimer 313 was hydrolyzed to give (±)-monocrotalic acid (315). Optical resolution of 315 with brucine afforded natural (−)-monocrotalic acid (2R,3R,4R) and its enantiomer (2S,3S,4S). The (2R,3R,4S)- and (2S,3S,4R)-acids of 316 were prepared analogously from 314. Resolution of (±)-trans-2,3,4-trimethyl-2-pentenedioic acid with cinchonidine gave two epimers. The configuration of each epimer was assigned by correlation with (R)-(−)-2-phenylpropanoic acid. Methylation of (4S)-acid 317 followed by cis-hydroxylation and hydrolysis afforded (2R,3S,4S)- and (2S,3R,4S)-isomers (318a and 318b) of monocrotalic acid. Similarly, (4R)-acid was transformed into (2S,3R,4R)- and (2R,3S,4R)-isomers of monocrotalic acid. The stereochemistry of all eight stereoisomers was also confirmed by transformation into the known methyl (R)-(+)-anhydromonocrotalate (319a) and its (S)-(−)-enantiomer 319b.

(a) $KMnO_4$, $MgSO_4$, MeOH/H_2O (b) HCl (c) CH_2N_2; $KMnO_4$, $MgSO_4$, MeOH/H_2O; HCl (d) $POCl_3$, py

Vedejs and Larsen (78) have described a practical synthesis of (±)-fulvinic acid (325). The thermodynamically unstable $\underline{E}$ enol silane 321 ($\underline{E}:\underline{Z}$=11:1) was prepared from dibromide 320 by bromocarbenoid method. The selective [2+2]cycloaddition of methyl ketene with 321 afforded meso-cyclobutanone 322. Compound 322 was converted into 325 by bromocarbenoid ring expansion followed by oxidative cleavage of resulting silyl enol ether 324.

d) C_{10}-Acids

Non-stereocontrolled syntheses of mixtures of diastereomeric racemates of senecivernic acid (326) and nemorensic acid (327) have been reported (79). These acids were made no isolations and no comparisons with natural specimens.

(a) Me$_2$CuLi, Et$_2$O; CO$_2$; CH$_2$N$_2$ (b) mCPBA, Li$_2$CO$_3$, CH$_2$Cl$_2$ (c) LDA, THF, then CH$_3$CHO (d) 2-fluoro-1-methylpyridinium p-toluenesulfonate, Et$_3$N, CH$_2$Cl$_2$ (e) LiOH, THF/H$_2$O

A stereoselective synthesis of (±)-integerrinecic acid (334) have been achieved by Narasaka et al. (80). Treatment of 2-methyl-2-cyclopentenone (328) with (dimethylcopper)lithium, followed by carboxylation of resulting enolate and esterification gave the 2,3-cis-dimethyl β-ketoester 329. Baeyer-Villiger

benzoate **360**. Methanolysis of **360**, and successive acid treatment and ruthenium tetraoxide oxidation gave the lactone alcohol, which was esterified to give (+)-lactone ester **361**. Conversion of (+)-**361** into (+)-integerrinecic acid lactone ester **362** was carried out by a modified procedure reported by Narasaka (80). Transformation of (+)-**362** into **363** is known in the literature (81).

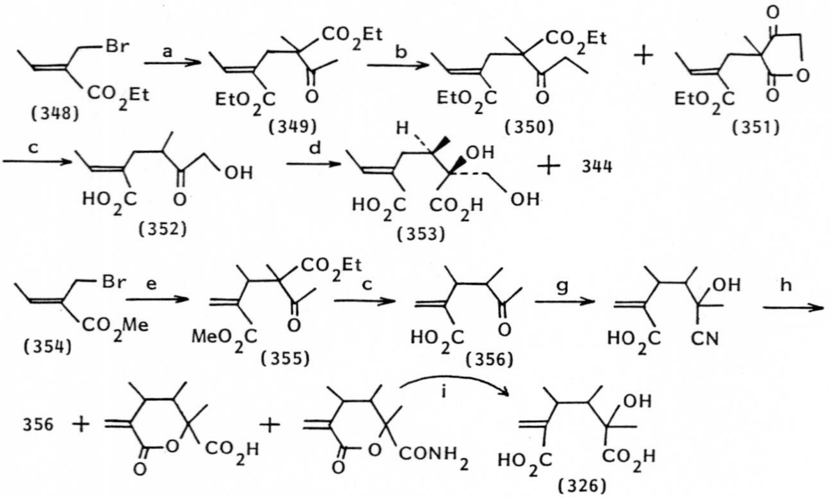

(348) (349) (350) (351)

(352) (353) + 344

(354) (355) (356)

356 + + (326)

(a) $CH_3COCH(Me)CO_2Et$, NaOEt, EtOH (b) Br_2, $AlCl_3$, Et_2O; AgOAc, AcOH
(c) 1 M KOH, then H^+ (d) NaCN, H_2O; 3.5 M H_2SO_4; conc. HCl; 2 M KOH; Zeocarb H^+ (e) MeCOCH(Me)CO_2Et, NaH, THF (g) NaCN, H_2O; H^+ (h) conc. HCl (i) 0.6 M Ba(OH)$_2$; Zeocarb H^+

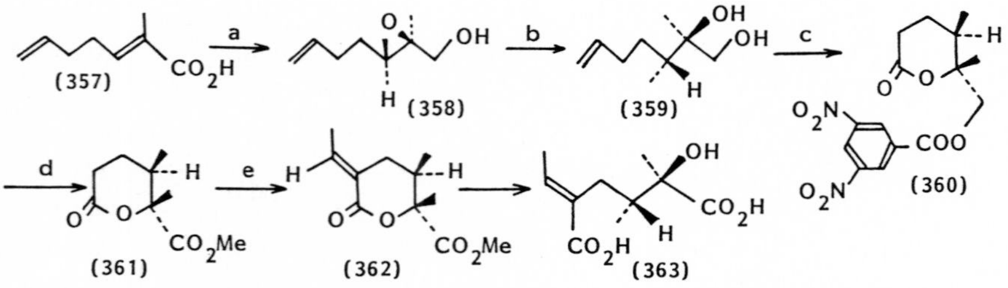

(357) (358) (359) (360)

(361) (362) (363)

(a) $LiAlH_4$, THF; t-BuOOH, (+)-DET, (i-PrO)$_4$Ti, CH_2Cl_2 (b) $AlMe_3$, hexane
(c) 3,5-dinitrobenzoyl chloride, py; $RuCl_3/NaIO_4$; cat. p-TsOH, PhH, Δ (d) MeONa, MeOH; cat. p-TsOH; $RuCl_3/NaIO_4$; CH_2N_2 (e) LDA, THF, then CH_3CHO; MsCl, py; DBU, PhH, Δ

White _et al._ (86) have achieved a stereocontrolled synthesis of natural enantiomer of integerrinecic acid lactone (372) using (R)-(-)-3-hydroxy-2-methylpropionate as a chiral source. A chelation-controlled Grignard reaction of the ketone **364**, derived from methyl (R)-(-)-3-hydroxy-2-methylpropionate,

with vinylmagnesium bromide gave a 4:1 mixture of the desired vinyl alcohol
365 and its diastereomer 366. These were separated as their carbonates, and
the major carbonate 367 was transformed into iodo acetate 368. The relative
configuration of 368 was confirmed by transformation into the lactone methyl
ester 370 known in racemic form (80). Condensation of the lithium enolate
369 with acetaldehyde, and successive acetylation afforded the lactone acetate
371. Oxidative cleavage of vinyl group in 371, followed by base-catalyzed
elimination, yielded a single E isomer 372.

(a) SEMCl, i-Pr$_2$NEt, CH$_2$Cl$_2$; LiAlH$_4$, Et$_2$O; (COCl)$_2$, DMSO, CH$_2$Cl$_2$, then Et$_3$N;
MeMgBr, Et$_2$O; (COCl)$_2$, DMSO, CH$_2$Cl$_2$, then Et$_3$N (b) CH$_2$=CHMgBr, THF
(c) Bu$_4$NF, HMPA; CO(Im)$_2$, PhCH$_3$ (d) MeONa, MeOH; TsCl, py; NaI, 2-BuOH;
Ac$_2$O, DMAP, CH$_2$Cl$_2$ (e) LDA, THF (f) aq. NH$_4$Cl; cat. RuCl$_3$·3H$_2$O, NaIO$_4$,
MeCN/CCl$_4$/H$_2$O; CH$_2$N$_2$ (g) CH$_3$CHO; Ac$_2$O, Et$_3$N, cat. DMAP, CH$_2$Cl$_2$ (h) cat.
RuCl$_3$·3H$_2$O, NaIO$_4$, MeCN/CCl$_4$/H$_2$O, DBU, CH$_2$Cl$_2$

2.3 Syntheses of Pyrrolozidine Alkaloids

a) Monoester and diester pyrrolizidines

The first total syntheses of the simplest group of monoester pyrrolizidines
were achieved in 1969 (87). Trachelanthamine (373), viridiflorine (374) and
lindelofine (375) were prepared by transesterification of methyl ester of necic
acid with monohydroxy-necine base.

(373) R^1= α-H, R^2=OH, R^3=H
(374) R^1= α-H, R^2=H, R^3=OH
(375) R^1= β-H, R^2=OH, R^3=H

Selective esterification of pyrrolizidine diol has proved more difficult.
Culvenor et al. (88) synthesized monoester pyrrolizidine diol, heliotrine (376),
by coupling 1-(chloromethyl)-heliotridine with heliotric acid sodium salt.
Intermedine (377) was also prepared analogously.

270

A total synthesis of (±)-indicine (394) by coupling of racemic monopro-
tected retronecine 392 and acetonide of racemic trachelanthic acid 391 has been
described by Vedejs et al. (50b). Esterification of 392 with 391 was carried
out by procedure reported by Piper (90) to yield 3:1 mixture of diastereomers
in favor of the natural diastereomer, (±)-product 393, which was confirmed by
removal of o-nitrobenzyl group to form 394 as the preferential diastereomer.

(a) DCC, DMAP, PhCH₃; HCl   (b) hv, THF

Nishimura et al. (17,76) have achieved the first completely enantioselective syntheses of indicine N-oxide (381), intermedine N-oxide (388) and their
enantiomers (395 and 396) via regiospecific esterification of synthetic retronecine (55) and its enantiomer (56) with synthetic (−)-trachelanthic acid
acetonide 379 and its enantiomer 387 by analogous method described by Piper
(90), followed by oxidation.

(a) DCC, DMAP, PhCH₃; HCl; mCPBA, Me₂CO

b) Macrocyclic pyrrolizidine dilactones

Due to its structural complicacy, conformational flexibility and instability with basic nature, macrocyclic pyrrolizidine dilactone is the challenging synthetic task in recent years.

The pioneer work in this area has been expressed by Robins et al. (93) in
1980. The unnatural 11-membered pyrrolizidine dilactone 398 was reconstructed

from (+)-retronecine (55) and 3,3-dimethylglutaric anhydride. The mixture of 7- and 9-monoesters 397 was cyclized intramolecularly <u>via</u> the corresponding 2-pyridinethiol ester prepared by Corey-Mukaiyama method.

(a) $CHCl_3$ (b) 2,2'-dithiobipyridyl, Ph_3P; DMF, Δ

Based on this method, the authors (94) have synthesized a natural 11-membered dilactone, dicrotaline (401), and its C-13 epimer (402). The 1:1 mixture of 7- and 9-monoesters 400 was prepared from retronecine and tri-methylsilyl ether of 3-hydroxy-3-methylglutaric acid anhydride 399. The intramolecular cyclization of 400 yielded 401 and its diastereomer 402. The absolute configuration of synthetic 401 was established as 13<u>S</u> by its degradation to optically active (<u>R</u>)-mevalonolactone (403).

(a) $CHCl_3$, then H^+ (b) 2,2'-dithiobipyridyl, Ph_3P; DMF, Δ (c) H_2/PtO_2, AcOH; Na/liq. NH_3

A synthesis of crobarbatine acetate and its diastereomer (407 and 408) having an unsymmetrical necic acid has been described by Meinwald <u>et al</u>. (95). (±)-<u>Trans</u>-β-methyl-γ-carboxy-γ-valerolactone (404) was converted to the <u>tert</u>-butyl thioester 405 by treatment with dimethylaluminum <u>tert</u>-butyl sulfide followed by acetylation of hydroxy group. Condensation of 405 and retronecine (55) was carried out by procedure reported by Crout (89). Treatment of the imidazolite of 405 with 55 in the presence of catalytic amount of NaH gave the 9-monoester 406. Intramolecular cyclization with copper(I) trifluoromethane-sulfonate-benzene complex gave two diastereomeric products (407 and 408). Since

A stereocontrolled synthesis of the natural enantiomer of integerrimine (431) has been accomplished by White et al. (86). The protected hydroxy acid 429 was prepared from the synthetic homochiral integerrinecic acid lactone (372) by an analogous procedure in the literature (28). Coupling 429 with mono-silylated (+)-retronecine 412, and a successive intramolecular lactonization were carried out by the similar method described by Vedejs (78) to give the monosilylated dilactone 430. Removal of protecting group of 430 resulted in 431.

(a) Me₃Si(CH₂)₂OH, 2-chloro-1-methylpyridium iodide,Et₃N; LiOH, H₂O₂, THF/H₂O; TBDMSOTf, 2,6-lutidine, CH₂Cl₂; AcOH/THF/H₂O (b) (EtO)₂POCl, Et₃N, THF, then 412, BuLi, THF, cat. DMAP; NH₄F, MeOH; MsCl, Et₃N, CH₂Cl₂; Bu₄NF, MeCN
(c) aq. HF, MeCN

(a) DCC, CH₂Cl₂ (b) Bu₂SnO, PhH (c) PhH, 0°C → RT (d) 2,4,6-trichlorobenzoyl chloride, Et₃N, THF, then PhCH₃, Δ

The enantioselective synthesis of optically active (−)-integerrimine (431) has been described by Yamada et al. (96). The synthetic (+)-integerrinecic acid methylthiomethyl ether 432 was converted into the cyclic anhydride 433. The cyclic stannoxane 434 was prepared by treatment of (+)-retronecine (55) with dibutyltin oxide. Regioselective monoesterification was effected by ad-dition of 433 to a solution of 434 in benzene to give the 9-monoester 435.

Lactonization was carried out by addition of the mixed anhydride of 435 to a refluxing toluene containing DMAP. Deprotection of the resulting integerrimine methylthiomethyl ether afforded 431.

2.4 Pharmacological and Biological Studies

The pyrrolizidine alkaloids constitute an exceptionally large class of natural occurring materials which demonstrate a broad range of pharmacological activities. Many pyrrolizidine alkaloids have been associated with severe pneumovascular and hepatotoxicities and with carcinogenic, mutagenic and teratogenic activities. The pyrrolic metabolites from pyrrolizidine alkaloids produced in the liver are highly reactive alkylating agents, and alkylation of a range of nucleophiles is believed to be responsible for the toxicity. Form- ation of toxic pyrrole metabolites is related to $\Delta^{1,2}$-unsaturation in the pyr- rolizidine nucleus and esterification at C-9 of necine. Toxicity increases with both of the substitution of bulky group at α-carbon of necic acid and esterification at C-7 of necine.

In 1976, Kugelman et al. (97) found that indicine N-oxide (381) exhibited the antitumor activity against W-256 carcinosarcoma, leukemia L-1210, leukemia P-388, leukemia P-1534 and melanoma B-16 in vivo without the hepatotoxicity normally associated with other pyrrolizidine alkaloids. Indicine N-oxide was the first pyrrolizidine alkaloid to be selected for clinical trials against neoplastic diseases. It was found to be effective against advanced gastero- intestinal cancer, and in cases of leukemia and melanoma (98-102).

Zalkow et al. (103-105) have compared several natural and semisynthetic monoester pyrrolizidine N-oxide with indicine N-oxide (381) in a screen with P-388 lymphocytic leukemia in mice. Echinatine N-oxide (436), europine N- oxide (437) and three diastereomers of necic acid part of indicine N-oxide (438, 388 and 439) were less active than 381, whereas, semisynthetic compounds 440 and 441 showed more potent activity.

(436) $R^1=\alpha$-OH, $R^2=R^5$=OH, R^3=CH(CH$_3$)$_2$, R^4=H

(437) $R^1=\beta$-OH, R^2=OH, R^3=C(OH)(CH$_3$)$_2$, R^4=OCH$_3$, R^5=H

(438) $R^1=\beta$-OH, $R^2=R^5$=OH, R^3=CH(CH$_3$)$_2$, R^4=H

(439) $R^1=\beta$-OH, R^2=CH(CH$_3$)$_2$, $R^3=R^4$=OH, R^5=H

(440) R=(2R and 2S)-COC(OH)(C$_6$H$_5$)Et

(441) R=(2R,3R and 2S,3S)-COC(OH)(CH$_3$)C(OH)CH$_3$

major product. Oxidation of 457 yielded the diketone 458. The conversion of
458 to 459 is known in the literature (111).

(a) CHCl₃, Δ; HCl, THF (b) MsCl (c) Zn, AcOH

The only known nuphar indolizidine having the gross structure corresponding to 464 (stereochemistry unknown) is present in castoreum. LaLonde et
al. (112) have synthesized compound 464 via epoxypiperidone. The key epoxypiperidone 461 was prepared from cyclopentanone 460 through the Beckman rearrangement of the corresponding oxime followed by epoxidation of alkenyl side
chain. Formation of the indolizidone 462 from 461 was promoted by treatment
with NaH in refluxing benzene. Transformation of 462 to the nuphar indolizidine
464 was carried out by the sequences of the removal of hydroxymethyl group,
the introduction of 3-furyl group, and reduction.

(a) H₂NOH·HCl, py; PCl₅, Et₂O, then NaOH; mCPBA, CH₂Cl₂ (b) NaH, phH, Δ
(c) CrO₃, H₂SO₄, Me₂CO; CH₂N₂, Et₂O; Pb(OAc)₄-Cu(OAc)₂·H₂O, py, Δ; H₂/Pd,
MeOH; NaCNBH₃, MeOH (pH 3) (d) 3-Furyl Li, Et₂O; NaCNBH₃, MeOH (pH 3)

Asymmetric epoxidations have been effectively used in enantioselective
synthesis of swainsonine (474) by Sharpless et al. (113). Asymmetric epoxidation of the allylic alcohol 465 employing (-)-diisopropyl tartrate (Dipt)
yielded the crystalline epoxy alcohol 466 in 95% ee. Conversion of 466 into
the allylic alcohol 468 was accomplished in eight steps via reduction of
acetoxy sulfide 467 with LAH followed by Swern oxidation without epimerization
at the α-position of 468. The second asymmetric epoxidation of 469 employing

(-)-Dipt resulted in the epoxy alcohol 470 in the homogeneity of ≥321:1. After two carbon homologation of 470, the removal of tosyl protecting group of 471 gave the pyrrolidine hydroxy ester 472, which was transformed into a mixture of cis- and trans-fused bicyclic quaternary ammonium salts 473. Removal of benzyl groups in 473 and a successive desilylation afforded 474.

(a) NaH, (E)-ClCH₂CH=CHCH₂Cl; NaOAc; K₂CO₃ (b) (-)-Dipt, Ti(Oi-Pr)₄, TBHP
(c) PhSH, t-BuOH, NaOH; BzlBr, NaH, n-Bu₄NI, THF; mCPBA; Ac₂O, (CF₃CO)₂O,
2,6-lutidine (d) LiAlH₄; (COCl)₂, DMSO, DBU; (EtO)₂P(O)CH₂CO₂Et, NaH; DIBAL
(e) DCC, DMSO, C₅H₅NHOTf, then Ph₃PCHCO₂Et; KO₂CN=NCO₂K, py, AcOH
(f) C₁₀H₈/Na, DME (g) t-Bu(Me)₂SiOTf, Et₃N; DIBAL; MsCl, Et₃N (h) H₂/Pd;
Dowex 50W-X8, MeOH

(a) PhCH₂NH₂; LiAlH₄; trifluoroacetylation; TBDMSCl, imdiazole; MsCl, py; Bu₄NF,
THF; MeONa, MeOH (b) NaBH₄, EtOH (c) (COCl)₂, DMSO, Et₃N; lithio t-butylacetate
(d) hydrogenolysis, then TFA/H₂O (c) DIBAL

An enantiospecific synthesis of (+)-castanospermine (480) by Ganem et al.
(114) has established its absolute configuration. Reduction of the epoxide
475 with NaBH₄ afforded a 1:1 mixture of piperidine 476 and azepane 477 via
intramolecular nucleophilic attack of amine generated to both carbon atoms of
epoxide ring. Oxidation of 476 followed by treatment with lithio t-butylacetate gave a 1:1 mixture of separable diastereomers of 478. The less polar

diastereomer was converted into lactam 479 upon hydrogenolysis and acid treatment. Reduction of 479 with DIBAL resulted in 480. 1-Epi-castanospermine was prepared similarly from the more polar isomer.

Suami et al. (115) have achieved a stereospecific synthesis of swainsonine (474) by a convergent method employing carbohydrate as a chiral source. The key acyclic carbohydrate intermediate 482 was derived from methyl 3-acetamido-4,6-O-benzylidene-3-deoxy-α-D-glucopyranoside (481). Cyclization of 482 by treatment with aqueous NaOH gave the pyrrolidine 483, which was converted into the α,β-unsaturated ester 484 via Honer-Emmons olefination of the corresponding aldehyde. Hydrogenation of 484 followed by intramolecular N-acylation with ethanolic KOH yielded the lactam 485. Reduction of 485 with LiAlH$_4$ and a successive removal of protecting groups resulted in 474.

The analogous syntheses of (-)-8-epi-swainsonine (486), (-)-1,8-di-epi-swainsonine (487) and (-)-8a-epi-swainsonine (488) have also been described by the authors (116).

(a) MsCl, py; HCl/MeOH; AcONa/aq. MeOCH$_2$CH$_2$OH; acetylation, HCl; acetylation; NaOMe/MeOH; HSCH$_2$CH$_2$SH, HCl; Ph$_3$CCl, py; BzlCl, NaH, DMF; de-tritylation; TsCl, py (b) aq. NaOH, Δ (c) HgCl$_2$, CaCO$_3$, then EtO$_2$CCH$_2$P(O)(OEt)$_2$, NaH (d) H$_2$/Raney Ni; KOH/Et$_2$OH-H$_2$O (e) LiAlH$_4$; H$_2$/Pd(OH)-C

A similar route to swainsonine (474) by Richardson et al. (117) has also involved conversion of an amino-sugar derivative 489 into an indolizidinone 492 via formation of a pyrrolidine 490.

A double-cyclization process has been involved in the enantiospecific synthesis of swainsonine (474) from D-mannose by Hashimoto et al. (118). Transformation of the epoxyamine conjugated ester 493 to 474 was effectively carried out via both stepwise and one-step fashion using a new application of sodium borohydride reduction of conjugated ester and lactam in trifluoroethanol (TFE).

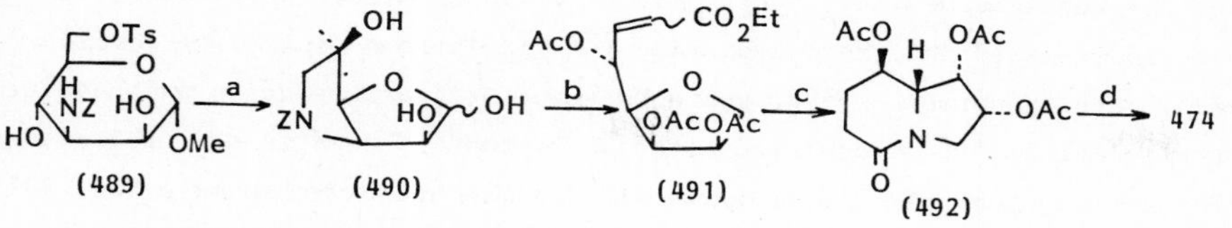

(489)   (490)   (491)   (492)   474

(a) H$_2$/Pd-C; NaOAc, EtOH, Δ; CbzCl; HCl   (b) HSCH$_2$CH$_2$SH, HCl; acetylation;
HgCl$_2$, CdCO$_3$; Ph$_3$P=CHCO$_2$Et   (c) H$_2$/Pd-C   (d) BH$_3$/DMSO; NaOMe

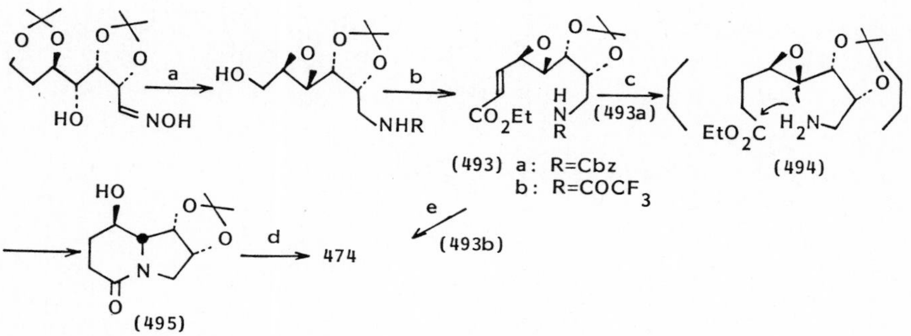

(493) a: R=Cbz
b: R=COCF$_3$ (494)

(493a)   (493b)

(495)   474

(a) LiAlH$_4$, THF; CbzCl, THF/H$_2$O, or (CF$_3$CO)$_2$O, CH$_2$Cl$_2$; MsCl, py; TsOH, MeOH/H$_2$O,
then Amberlite IRA-400 (OH$^-$)   (b) Collins oxidation, then Ph$_3$P=CHCO$_2$Et, THF   (c) NaBH$_4$,
EtOH/TFE; H$_2$/Pd-C, EtOH, then refluxing   (d) NaBH$_4$, EtOH/TFE; 6 N HCl, THF
(e) NaBH$_4$, EtOH/TFE; 6 N HCl, THF

The similar double-cyclization route to castanospermine (480) from D-mannose has been employed by the authors (119).

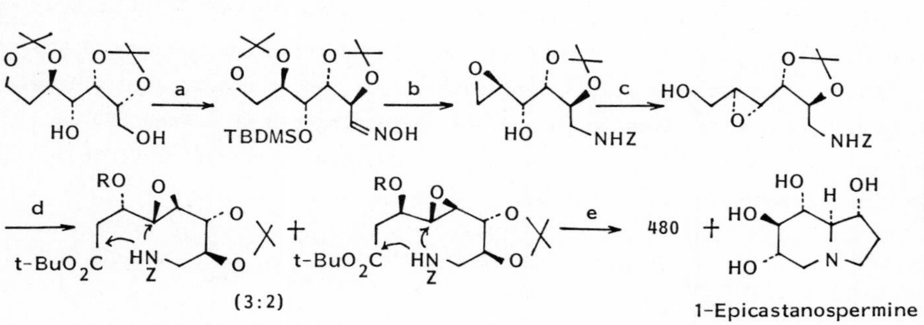

(3:2)   480   1-Epicastanospermine

(a) BzCl, py; TBDMSCl, imidazole, DMF; 1 N NaOH, MeOH; DMSO, DCC, TFA, py,
PhH; K$_2$CO$_3$, MeOH; H$_2$NOH·HCl, NaHCO$_3$, EtOH/H$_2$O   (b) LiAlH$_4$, THF; CbzCl,
THF/H$_2$O; TsOH, MeOH/H$_2$O; n-Bu$_4$NF, THF; MsCl, py; MeONa, MeOH   (c) MeONa,
MeOH   (d) CrO$_3$·2py, CH$_2$Cl$_2$, then t-butyl lithioacetate; TBDMSCl, imidazole, DMF
(e) H$_2$/Pd-C, EtOH, then refluxing; borane·THF, Δ; 6 N HCl, THF, Δ

Kibayashi et al. (120) have described the stereospecific synthesis of
(±)-gephyrotoxin 223AB (503) via an acyl nitro Diels-Alder reaction. Treat-

ment of the key intermediate, hydroxamic acid 496 with tetrapropylammonium
(meta)periodate afforded the 1,2-oxazine 498 <u>via</u> intramolecular [4+2]cyclo-
addition. Hydrogenation of 498, and subsequent Grignard reaction and reduction
gave the isoxazole 499. Reductive cleavage of the N-O bond in 499, followed
by treatment with benzyl chloroformate yielded the hydroxy carbamate 500, 501
and 502 in a ratio of 1.5:1:1. Compound 500 was transformed into 503 upon
mesylation and hydrogenation.

(496)   (497)   (498)

(499)   (500) R¹=Cbz, R²=H   (503)
         (501) R¹=H, R²=Cbz
         (502) R¹=R²=Cbz

(a) H$_2$NOH, KOH, MeOH   (b) n-PrN(IO$_4$), CHCl$_3$ (c) H$_2$/Pd-C, MeOH; n-PrMgBr,
Et$_2$O; NaBH$_3$CN, MeOH   (d) Zn, AcOH/H$_2$O; CbzCl, aq. Na$_2$CO$_3$, CHCl$_3$ (e) MsCl;
H$_2$/Pd-C, MeOH

b) Condensation

The synthesis of (±)-elaeokanine A (506) and (±)-elaeokanine C (459) by
Tufariello et al. (121) has been described by use of 1,3-dipolar cycloaddition
of nitrone and aldol condensation. The key β-aminoketone 505 was prepared <u>via</u>
regio- and stereoselective cycloaddition of 29 and 1-pentene. Compound 505
was converted into a separable 4:1 mixture of 506 and an unsaturated aldehyde
507 by treatment with acrolein in benzene containing t-BuOK. When 505 was
exposed to acrolein followed by concentrated hydrochloric acid to afford a
separable 3:1 mixture of 459 and 507.

(504)   (505)   (507)   459

(506)   (507)

(a) 110°C, sealed tube   (b) H$_2$/Pd-C; Jones oxidation   (c) acrolein, PhH, t-BuOK
(d) acrolein, CH$_2$Cl$_2$, then conc. HCl

283

The authors (122) have reported an analogous approach to (±)-isoelaeo-carpicine **508** using a nitrone cycloaddition to generate a β-aminoketone and its annulation with acrolein.

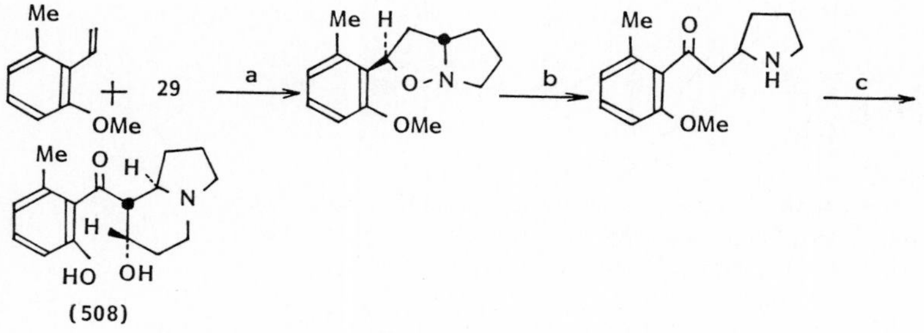

(508)

(a) PhCH$_3$, 95°C (b) H$_2$/PtO$_2$, EtOH; Jones oxidation (c) acrolein, then conc. HCl

The acylative ring closure of exocyclic vinylogous urethane has been in-volved as a key step in the synthesis of ipalbidine by Howard et al. (123). The key vinylogous urethane **510** was prepared by extrusion of sulfur of the S-alkylated iminium salt of the N-substituted thiolactam **509**. Hydrolysis of **510** followed by treatment with methyl chloroformate gave a mixed anhydride, which was irreversibly cyclized to the enone ester **511**. Removal of the methoxy-carbonyl group and a subsequent reduction with LiAlH$_4$ yielded the amino ketone **512**. Transformation of **512** into ipalbidine (**513**) is known in the literatures (124).

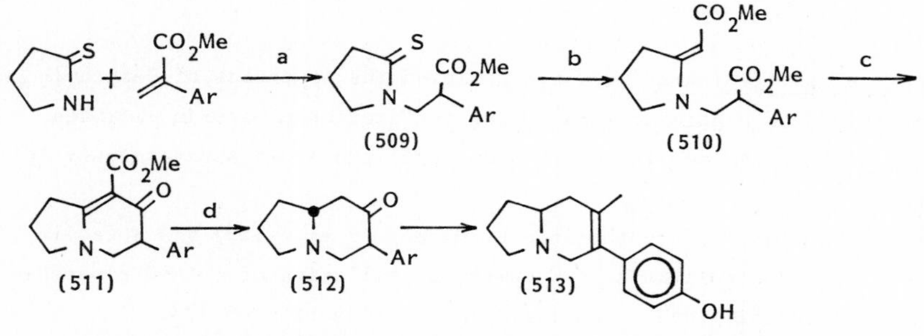

(509)   (510)

(511)   (512)   (513)

(a) NaOH, THF (b) BrCH$_2$CO$_2$Me, THF; PPh$_3$, Et$_3$N, MeCN (c) NaOH, H$_2$O; cat. Bu$_4$NF, ClCO$_2$Me, THF (d) aq. KOH, N$_2$; HCl,H$_2$O; LiAlH$_4$, THF

A general approach to Elaeocarpus alkaloids via an acylative ring closure of vinylogous urethane has been presented by the authors (125). The vinylogous amide **514** was elaborated to Elaeocarpus alkaloid precursors **516** via specific formation of the lithium enolate of enol acetate **515**.

(a) BrCH$_2$CO$_2$Et, MeCN; PPh$_3$, Et$_3$N, MeCN (b) aq. NaOH, Δ; Ac$_2$O, MeCN
(c) aq. KOH, Δ, then HCl, Δ; AcCl, AgClO$_4$, MeCN, N$_2$ (d) NaBH$_4$, MeCN;
AcOH, Δ (e) MeLi, THF, then n-C$_3$H$_7$COCN

The vinylogous amide **514** has also been used as an intermediate to (±)-elaeokanines A, B and C (**506**, **517** and **459**) by Watanabe **et al.** (126).

(a) LiAlH$_4$, THF/Et$_2$O (b) (CH$_2$OH)$_2$, p-TsOH, PhH; LiAlH$_4$, THF/Et$_2$O; NCS, DMS,
PhCH$_3$/CH$_2$Cl$_2$ (c) n-PrMgBr, THF/Et$_2$O, Δ; Jones oxidation; aq. HBr (d) (CH$_2$SH)$_2$,
BF$_3$·Et$_2$O, AcOH (e) aq. HBr; LiAlH$_4$, THF/Et$_2$O (f) MeI, THF, NeCN, HCl, Δ
(g) n-PrMgBr, THF/Et$_2$O, Δ (h) Jones oxidation

Herbert **et al.** (127 and 128) have achieved the syntheses of O-methyl-ipalbidine (**519**) and septicine (**454**) using cyclization of enamine-ketones. Cyclization of the enamine-ketones (**518** and **520**) proceeded spontaneously in dry methanol.

δ-Coniseine (**524**) has synthesized by Miyano **et al.** (129) **via** a facile synthesis of dehydroindolizidine **522** by dry distillation of γ-(N-2-piperidi-nonyl)butyric acid (**521**) over soda-lime and the reduction of **523**.

The synthesis of (±)-septicine (**454**) and (±)-ipalbidine (**513**) has been described by Kibayashi **et al.** (130 and 131) using 1,3-dipolar cycloaddition and intramolecular aldol condensation as key steps. The keto-amide **526**, the intermediate for cyclization, was prepared from a diastereomeric mixture of pyrroloisoxazoles **525** by successive sequences of hydrogenolysis and oxidation. Conversion of **526** into **454** was accomplished by the similar procedure described by Bhakuni (132) via aldol condensation with alcoholic KOH. 513 was also synthesized analogously.

(a) MeOH, 1 M phosphate buffer (pH 7.25)   (b) 4-methoxyphenylacetaldehyde, PhH
(c) dry MeOH, then NaBH$_4$

(a) Na, [cyclobutanone]O, H$^+$ (b) soda lime, Δ   (c) HClO$_4$ (d) LiAlH$_4$

(a) PhCH$_3$, Δ   (b) H$_2$/Pd-C   (c) 3,4-(MeO)$_2$C$_6$H$_3$CH$_2$COCl, K$_2$CO$_3$, CHCl$_3$; partial hydrolysis; Collins oxidation   (d) alcoholic KOH; LiAlH$_4$-AlCl$_3$, THF/H$_2$O

Shono et al. (26) have demonstrated the synthesis of (±)-elaeokanine A (506) and (±)-elaeokanine C (459) by the similar approach used for the synthesis of pyrrolizidine alkaloid (See 2.1.a)).

(a) TiCl$_4$ (b) (HOCH$_2$)$_2$, p-TsOH; NH$_2$NH$_2$, KOH; BrCH$_2$CH$_2$⟨O⟩ , NaOH (c) HCl
(d) NaOH, Δ

c) Iminium ion initiated cyclization

An α-acyliminium cyclization has been used as a key step in the synthesis of elaeokanine B (517) by Speckamp et al. (133). The key intermediate, α-oxy olefine 528, was prepared via oxidation reduction coupling of the enone alcohol 527 with succinimide. The acid-catalyzed cyclization of 528 was best achieved upon HCl/MeOH treatment to give the chloride 529. Dehydrohalogenation of 529 with DABCN followed by reduction with NaBH$_4$ afforded the known lactam alcohol, which was converted into elaeokanine B (517) by reduction with DIBAL.

(a) LiAlH$_4$, THF; CrO$_3$, py, CH$_2$Cl$_2$, HCl/MeOH (b) oxidation reduction coupling with succinimide; ethylenedioxy(bis)-trimethylsilane, TMSOTf; NaBH$_4$, H$^+$, then base work-up (c) HCl/MeOH (d) DABCN, PhCH$_3$, Δ; NaBH$_4$; DIBAL, Et$_2$O

(a) MeMgI; SOCl$_2$, py, THF; mCPBA, CH$_2$Cl$_2$ (b) Et$_2$O, Δ (c) KOH, MeOH/H$_2$O
(d) paraformaldehyde, EtOH, then (+)-compher-10-sulfonic acid, EtOH, Δ

A new approach to chiral dendrobatid toxin 251D using stereospecific iminium ion-vinylsilane cyclization has been achieved by Overman et al. (134).
Cyclization of the key amino alcohol 530, derived from L-proline, to (+)-251D
(531) was carried out by treatment with paraformaldehyde, and subsequent refluxing the resultant oxazolidine with (+)-comphor-10-sulfonic acid.

The analogous route to chiral pumiliotoxin B (534) has also described via
Wittig reaction of the aldehyde 532 and threo selective reduction of resulting
ketone 533. The method is a potentially general one for forming unsaturated
azacyclic rings leading to the pumiliotoxin A alkaloids.

(a) Ph₃P=C(Me)COC(Me)OSiPh₂t-Bu, CH₂Cl₂, Δ   (b) LiAlH₄, THF

(a) HCO₂H (b) NaOH, MeOH/H₂O; NaH, CS₂, MeI; n-BuSnH; P₂S₅ (c) BrCH₂COEt;
PPh₃, Et₃N (d) NaBH₃CN, MeOH; H₂Cr₂O₇ (e) HS(CH₂)₃SH, HCl (f) Li/EtNH₂

The stereoisomers of gephyrotoxin-223AB (543 and 544) have been synthesized by formic acid induced cyclization of carbinolamide by Hart et al. (135).
The known imide 535 was reduced to carbinolamide 536, which was cyclized with
formic acid to give the formate 537. Treatment of the thiolactam 538 with l-

bromo-2-butanone, Et_3N and PPh_3 afforded the vinylogous amide 539, which was converted to a isomeric mixtures of amino ketone 540 by reduction and an oxidative workup. Thioketals 541 and 542, after separation, were transformed into 543 and 544 upon reduction.

Chamberlin et al. (46b) have demonstrated the synthesis of elaeokanine A (506) using acyliminium ion cyclization, used in the synthesis of pyrrolizidine alkaloids (See 2.1.d)).

(a) MsCl, Et_3N (b) $LiAlH_4$; LDA, $Me(CH_2)_2I$ (c) $HgCl_2$, H_2O

d) Diels-Alder reaction

A new synthetic strategy to indolizidine alkaloid via the intramolecular imino Diels-Alder reaction has extensively been developed by Weinreb et al. (136). An acyl imine dienophile was generated thermally in situ from a methylol acetate precursor in the syntheses of δ-coniceine (524), elaeokanine A (506), elaeokanine B (517) and slaframine (553). The authors have chosen δ-coniceine (524) as a model target to test the basic principles of the approach. The carboxamide 546 was prepared from the readily available diene ester 545 by treatment with dimethylaluminum amide. The amide 546 was converted into the methylol 547 on treatment with aqueous formaldehyde and sodium hydroxide, and 547 was acetylated without purification to give the crystalline acetate 548. The transformation of 548 into the bicyclic lactam 550 via the unstable acyl imine 549 was best effected by passing a toluene solution of 548 through a column of glass helices heated at 370-390°C. Catalytic hydrogenation followed by reduction with BH_3 afforded 524.

a { (545) X=OEt   (547) R=H
    (546) X=NH$_2$ (548) R=Ac

(549)   (550)   524

(a) Me_2AlNH_2, PhH (b) aq. HCHO, NaOH, glyme; Ac_2O, py (c) passing $PhCH_3$ soln. through glass column at 370-390°C (d) H_2/Pd-C; BH_3THF

The methodology has been applied to synthesis of elaeokanine A (506) <u>via</u> elaeokanine B (517). In this case, due to the possible instability of the diene portion of 551, the "masked" form was employed as this part of the molecule.

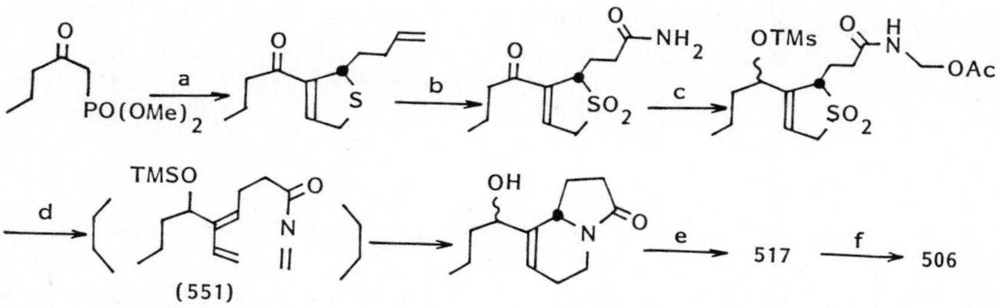

(551)

(a) $CH_2=CH(CH_2)_2CHO$, piperidine, Ac_2O, PhH, Δ; mercaptoaldehyde, Et_3N, CH_2Cl_2, Δ
(b) mCPBA; CrO_3/H_5IO_6, aq. Me_2CO; $ClCO_2Et$, Et_3N, then anhydrous NH_3
(c) $ClCH_2SCH_3$, THF; $NaBH_4/CeCl_3$, MeOH; Ac_2Hg, AcOH; TSMCl, py, $(Me_3Si)_2NH$
(d) thermal elimination; $HCl/MeOH-H_2O$ (e) DIBAL, THF (f) DMSO, TFAA, CH_2Cl_2

The fungal toxin, slaframine (553), has also been synthesized from the diene aldehyde 552.

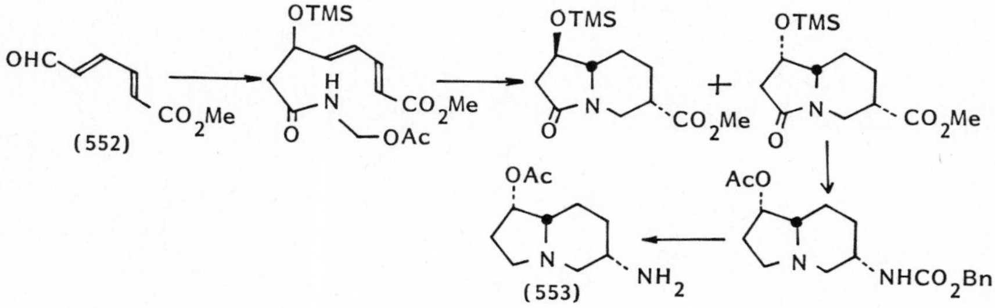

(552)

(553)

e) Reductive amination

Fleet <u>et al</u>. (137) has described an enantiospecific synthesis of swainsonine (474) by intramolecular reductive amination. The protected 4-azido-mannose 554 was prepared from D-mannose with overall retention of configuration. The two-carbon homologation at C-6 of mannose, followed by reduction afforded the equivalent of an amino-dialdehyde, which was converted into 474 by two intramolecular reductive aminations.

(a) Ph$_2$t-BuSiCl, imidazole, then Me$_2$CO/Me$_2$C(OMe)$_2$, CSA; PCC, CH$_2$Cl$_2$, then NaBH$_4$, EtOH; TFAA, py, CH$_2$Cl$_2$; NaN$_3$, DMF; Bu$_4$NF, THF (b) PCC, CH$_2$Cl$_2$, then Ph$_3$P=CHCHO (c) H$_2$/Pd-C, MeOH (d) H$_2$/Pd-black, AcOH; TFA; D$_2$O

(a) LDA, THF, then 531 (b) AgBF$_4$, THF, then Zn(BH$_4$)$_2$ (c) MeMgI, Et$_2$O
(d) H$_2$/Pd-C, MeOH, 1 M HCl

(a) MeMgBr, Et$_2$O; H$_2$/Pd-C, MeOH; Zn, AcOH/H$_2$O (b) CbzCl, aq. Na$_2$CO$_3$, CH$_2$Cl$_2$; CrO$_3$·2py, CH$_2$Cl$_2$ (c) H$_2$/Pd-C, MeOH (d) CbzCl, aq. Na$_2$CO$_3$, CH$_2$Cl$_2$; TMSI, CH$_2$Cl$_2$ (e) MeOH, rt

The first asymmetric synthesis of (–)-monomorine I (562) has been achieved via the chiral 2-cyano-6-oxazolopiperidine synthon 555 by Husson et al. (138). This synthesis of 562 having the 3S,5R,9R absolute configuration establishes the absolute configuration of natural (+)-monomorine I as 3R,5S,9S. The synthon 555 was previously prepared from (–)-phenylglycinol, glutaraldehyde and

KCN by Robinson-Schopf type condensation. Alkylation of **555** with the iodo ketal **556**, and a subsequent selective cleavage of the cyano group afforded the oxa-zolopiperidine **559** having the 2S configuration. The stereospecificity implied an elimination-addition mechanism wherein hydride ion attacked to a preferred iminium conformer **558** from the axial direction under complete stereoelectronic control. Compound **559** was alkylated with MeMgI to give a separable 4:1 mixture of the cis alcohol **560** and its trans isomer **561**. Hydrogenation of **560** under acidic condition afforded **562**.

The strategy (120), developed in the synthesis of (±)-gephyrotoxin 223AB (**503**), has been applied to a stereoselective synthesis of (±)-monomorine I (**563**) by Kibayashi et al. (139, See 3.a)).

f) Transannular reaction

δ-Coniceine (**524**) has been synthesized via transannular reductive closure of azacyclononanone **564** used in the synthesis of pyrrolizidine by Garst et al. (140).

(564)

(a) Me$_2$CHCMe$_2$BH$_2$, THF, then KCN, and then TFAA,H$_2$O$_2$ (b) H$_2$/Pd, MeOH

(565) R=Me or n-Bu

(566)

(567)

(568)

(a) Me$_2$CuLi or n-Bu$_2$CuLi; Jones oxidation; H$_2$NOH·HCl   (b) CH$_2$Cl$_2$, py, then p-TsCl, and then aq. K$_2$CO$_3$/THF   (c) Hg(OAc)$_2$, H$_2$O/THF, then NaBH$_4$ (d) LiAlH$_4$

A stereospecific transannular cyclization, induced by mercuric acetate, has been used for the synthesis of alkyl indolizinones **568** by Wilson et al. (141). Beckmann rearrangement of alkylcyclooctenones **565** gave the 3-alkyl-2H-azonin-2-ones (**566**). Cyclization of **566** with Hg(OAc)$_2$, followed by reduction gave the alkyl indolizinone **567**. The stereochemical control was caused by steric interactions between the alkyl side chain and the transannular hydrogens. The strategy could be used for the synthesis of natural indolizidines.

The controlled crisscross annulation has been applied to the synthesis of the only one known nuphar indolizidine (569) (stereochemistry unknown) by Ban et al. (38b, See 2.1.c)).

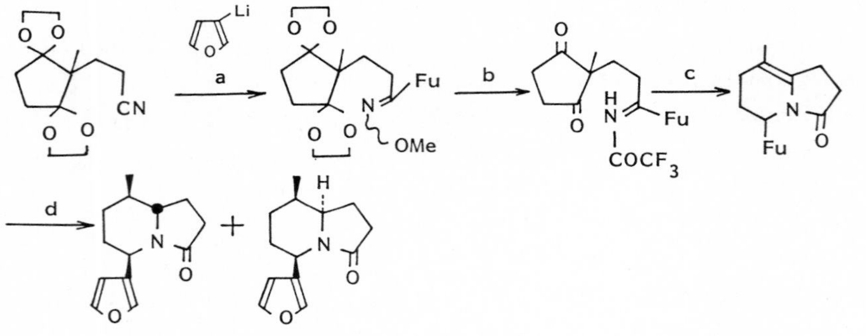

(569a) (4:1) (569b)

(a) H_2O; NH_2-OMe (b) $LiAlH_4$, THF; TFAA, py; aq. TFA (c) LiOH, aq. MeOH; aq. HCl/MeOH (d) $NaBH_3CN$, MeOH; $LiAlH_4$, THF

g) Radical cyclization

An α-acylamino radical cyclization has been used for the synthesis of (±)-coniceine (524) by Hart et al. (65, See 2.1.i)).

The stereoselective homolytic cyclization has been involved in the synthesis of (±)-gephyrotoxin-223AB (503) by Broka et al. (142). An aminyl radical heterocyclization of chloramines (570 and 571) was best effected by using the copper(I) chloride-copper(II) chloride redox system to give a diastereomeric chloroindolizidines 572. Reduction of 572 with Bu_3SnH afforded 503 and its epimer 573.

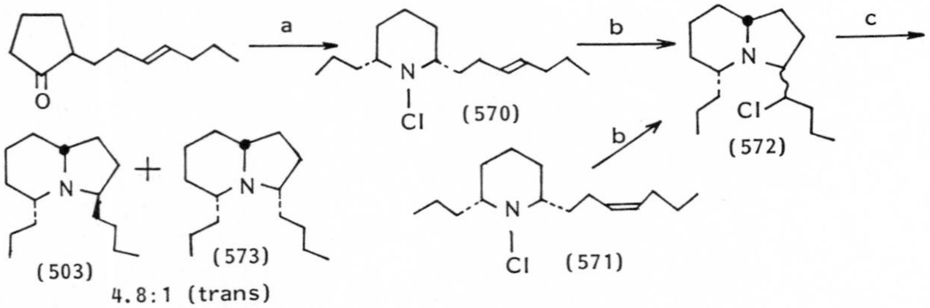

(503)   (573)
      4.8:1 (trans)
      3.3:1 (cis)

(a) $H_2NOH \cdot HCl$; MsCl, Et_3N, CH_2Cl_2; Al(n-Pr)$_3$, CH_2Cl_2/PhCH$_3$; DIBAL, CH_2Cl_2;
NCS (b) CuCl/CuCl$_2$, THF/AcOH/H_2O (c) Bu_3SnH, AIBN, PhH, Δ

h) Miscellaneous methods

The application of the acid-catalyzed rearrangement of cyclopropylimine to the synthesis of (±)-ipalbidine (513) and (±)-septicine (454) has been described by Stevens et al. (143, See 2.1.f)).

(a) LDA, THF, Δ; (HOCH$_2$)$_2$, TsOH, PhH; H$_2$/Raney Ni, EtOH, NH$_3$ (b) PhH, Δ, -H$_2$O (c) NH$_4$Cl, xylene, Δ (d) HCl, MeOH, (MeO)$_3$H; H$_2$/Raney Ni, EtOH; 1 N HCl, CH$_2$Cl$_2$ (e) MeLi (f) ArLi

A stereoelectronically controlled nucleophilic capture of a tetrahydropyridinum salt has been involved as a key step in the synthesis of (±)-monomorine I (563) by Stevens et al. (144). Reductive cyclization of the nitroketone 574 afforded the pyrrolidine 575, which was isolated as its oxalate salt. Acid hydrolysis of 575, followed by basic work-up gave the unstable endocyclic enamine 576, which was directly converted into 563 upon reduction with NaCNBH$_3$.

(a) CH$_2$=CHCHO, Mg, THF; MnO$_2$ (b) CH$_3$(CH$_2$)$_4$NO$_2$, tetramethylguanidine (c) H$_2$/Pd-C, HO$_2$CCO$_2$H (d) H$_3$O$^+$, then OH$^-$ (e) NaBH$_3$CN (pH 3.8-5.4)

The analogous route to gephyrotoxin 223 (**577**) has been described by the authors (145).

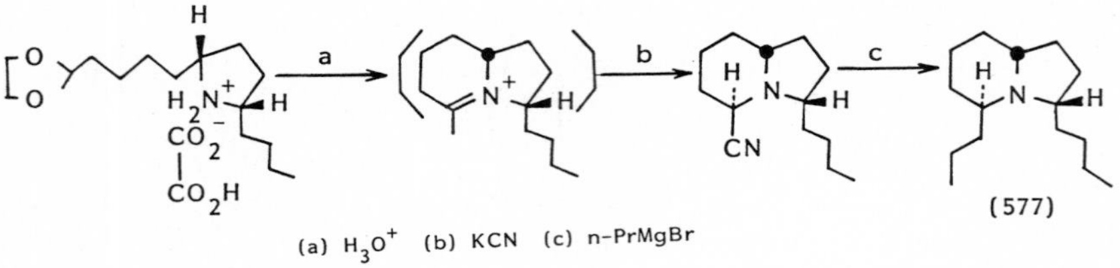

a
b
c
(577)
(a) H_3O^+ (b) KCN (c) n-PrMgBr

The synthesis of (_E_)-alkylidene analogues of pumiliotoxin A has been reported _via_ Lewis acid catalyzed intramolecular ene cyclization by Overman _et al_. (146). Cyclization of unsaturated ketones **578** was best effected by using freshly sublimed $AlCl_3$. Cyclization of **578a** gave alkylidene indolizidine **579** and chloride **580**, while cyclization of **578b** and **578c** afforded the corresponding alkylidene isomers. The stereoselectivity depends on the steric bulk of the alkylidene side chain.

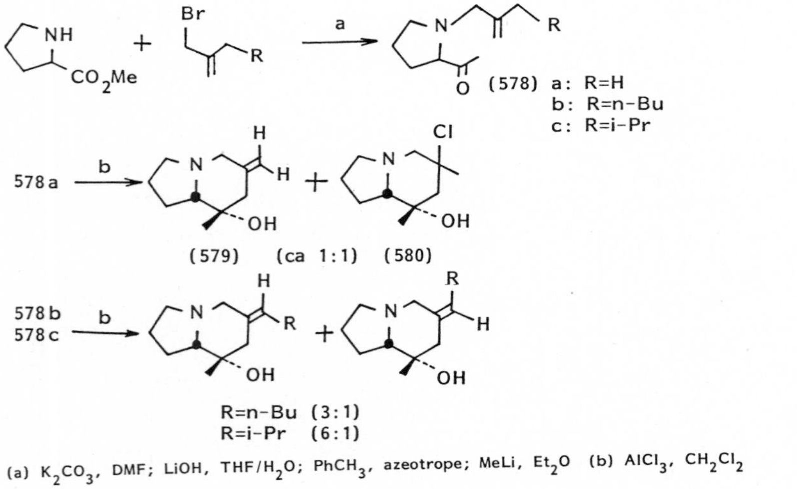

a
(578) a: R=H
b: R=n-Bu
c: R=i-Pr
578a
b
(579) (ca 1:1) (580)
578b
578c
b
R=n-Bu (3:1)
R=i-Pr (6:1)
(a) K_2CO_3, DMF; LiOH, THF/H_2O; $PhCH_3$, azeotrope; MeLi, Et_2O (b) $AlCl_3$, CH_2Cl_2

The Robinson-Schopf type condensation has effectively been used in enantiospecific synthesis of the natural (-)-gephyrotoxin-223AB (**583**) by Husson _et al_. (147, See 3.c)). Treatment of the chiral cis amine **581** with HCl and KCN afforded the aminonitrile **582**, which was transformed into **583** and _R_ epimer **584** on reaction with BuMgBr.

(a) LDA, THF, BrCH$_2$CH$_2$-CH$\langle$O$\rangle$; AgBF$_4$, THF, then Zn(BH$_4$)$_2$ (b) PrMgBr, Et$_2$O; H$_2$/Pd-C, MeOH (c) 1 N HCl, KCN, CH$_2$Cl$_2$/H$_2$O (d) n-BuMgBr, Et$_2$O

REFERENCES

1 D.J. Robins, Advances in pyrrolizidine chemistry, in: Adv. Heterocycl. Chem., Vol. 24, Academic Press, New York, 1979, pp. 247-291.
2 D.J. Robins, The pyrrolizidine alkaloids, Prog. Chem. Org. Nat. Prod., (1982) 115-203.
3 J.T. Wróbel, Pyrrolizidine alkaloids, in: The Alkaloids, Vol. 26, Academic Press, New York, 1985, pp. 327-385.
4 F.J. Swinbourne, Advances in indolizine chemistry, in: Adv. Heterocycl. Chem., Vol. 23, Academic Press, New York, 1978, pp. 103-170.
5 D.H.G. Grout, Pyrrolizidine alkaloids, in: The Alkaloids, Specialist Periodical Reports, Vol. 6 and 7, Chem. Soc., London, 1976-1977; D.J. Robins, Pyrrolizidine alkaloids, in: The Alkaloids, Specialist Periodical Reports, Vol. 8-13, Chem. Soc., London, 1978-1983; D.J. Robins, Pyrrolizidine alkaloids, Nat. Prod. Rep., (1984) 235-243; D.J. Robins, Pyrrolizidine alkaloids, ibid, (1985) 213-220.
6 J.A. Lamberton, Indolizidine alkaloids, in: The Alkaloids, Specialist Periodical Reports, Vol. 11-13, Chem. Soc., London, 1981-1983; J.A. Lamberton, Indolizidine alkaloids, Nat. Prod. Rep., (1984) 245-246; M.F. Grundon, Indolizidine alkaloids, Nat. Prod. Rep., (1985) 235-243.
7 N.K. Kochetkov, A.M. Likhosherstov, and E.I. Budovskii, Pyrrolizidine alkaloids. I. Synthesis of 1-hydroxymethylpyrrolizidine (dl-trachelantha-midine), Zh. Obshch. Khim., 30 (1960) 2077-2082; Chem. Abstr., 55 (1961) 7386i.
8 T. Ohsawa, M. Ihara, and K. Fukumoto, Stereoselective total synthesis of necine bases, (±)-retronecine and (±)-turneforcidine, Heterocycles, 19(11) (1982) 2075-2077; Novel synthesis of the pyrrolizidine skeleton by sulfeno-cycloamination. Total synthesis of (±)-retronecine and (±)-turneforcidine, J. Org. Chem., 48 (1983) 3644-3648.
9 T.L. Macdonald and B.A. Narayanan, Pyrrolizidine alkaloid synthesis. (±)-Supinidine, J. Org. Chem., 48 (1983) 1129-1131.
10 K. Tatsuta, H. Takahashi, Y. Amemiya, and M. Kinoshita, Stereoselective total synthesis of pyrrolizidine alkaloid bases: (-)-rosmarinecine and (-)-isoretronecanol, J. Am. Chem. Soc., 105 (1983) 4096-4097.
11 J.J. Tufariello and J.P. Tette, Synthesis of the alkaloidal base (±)-supinidine, J. Chem. Soc., Chem. Commun., (1971) 469-470.
12 J.J. Tufariello and G.E. Lee, Functionalized nitrones. A highly stereo-selective and regioselective synthesis of dl-retronecine, J. Am. Chem. Soc., 102 (1980) 373-374.

13   J.J. Tufariello and K. Winzenberg, A nitrone-based synthesis of the
     pyrrolizidine alkaloid croalbinecine, Tetrahedron Lett., 27 (1986) 1645-
     1648.
14   E. Röder, H. Wiedenfeld, and E-J. Jost, Synthese der Pyrrolizidine (±)-
     Tussilagin und (±)-Isotussilagin, Arch. Pharm., 317 (1984) 403-407.
15   T. Iwashita, T. Kusumi, and H. Kakisawa, A synthesis of dl-isoretronecanol,
     Chem. Lett., (1979) 1337-1340; Syntheses of isoretronecanol and lupinine,
     J. Org. Chem., 47 (1982) 230-233.
16   G.E. Keck and D.G. Nickell, Synthetic studies on pyrrolizidine alkaloids.
     1. (±)-Heliotridine and (±)-retronecine via intramolecular dienophile
     transfer, J. Am. Chem. Soc., 102 (1980) 3632-3634.
17   Y. Nishimura, S. Kondo, and H. Umezawa, Synthetic studies on pyrrolizidine
     alkaloid antitumor agents. Enantioselective synthesis of retronecine and
     its enantiomer from D-glucose, J. Org. Chem., 50 (1985) 5210-5214.
18   G.A. Kraus and K. Neuenschwander, Amidoalkylation in organic synthesis.
     1. Total synthesis of isoretronecanol and trachelanthamidine, Tetrahedron
     Lett., 21 (1980) 3841-3844.
19   T. Kametani, K. Higashiyama, H. Otomasu, and T. Honda, A short and novel
     synthesis of the pyrrolizidine alkaloids, (±)-supinidine and (±)-trache-
     lanthamidine, Heterocycles, 22 (1984) 729-731.
20   R.F. Borch and B.C. Ho, A new synthesis of the pyrrolizidine alkaloids.
     (±)-Isoretronecanol and (±)-trachelanthamidine, J. Org. Chem., 42 (1977)
     1225-1227.
21   D.J. Robins and S. Sakdarat, Synthesis of the pyrrolizidine base, (±)-
     supinidine, J. Chem. Soc., Perkin I, (1979) 1734-1735.
22   H. Niwa, Y. Miyachi, O. Okamoto, Y. Uosaki, and K. Yamada, Total synthesis
     of optically active integerrimine, a twelve-membered dilactonic pyrroli-
     zidine alkaloid of retronecine type. II. Enantioselective synthesis of
     (+)-retronecine, Tetrahedron Lett., 27 (1986) 4605-4608.
23   S. Raucher and P. Klein, A convenient method for the selective reduction
     of amides to amines, Tetrahedron Lett., 21 (1980) 4061-4064.
24   H.W. Pinnick and Y-H. Chang, New approaches to the pyrrolizidine ring
     system: Total synthesis of (±)-isoretronecanol and (±)-trachelanthamidine,
     J. Org. Chem., 43 (1978) 4662-4663.
25   S. Brandange and C. Lundin, Synthesis of endo- and exo-1-ethoxycarbonyl-
     pyrrolizidine, Acta Chem. Scand., 25 (1971) 2447-2450.
26   T. Shono, Y. Matsumura, K. Uchida, K. Tsubata, and A. Makino, Efficient
     synthesis of pyrrolizidine and indolizidine alkaloids utilizing anodically
     prepared α-methoxy carbamates as key intermediates, J. Org. Chem., 49
     (1984) 300-304.
27   T.A. Geissman and A.C. Waiss, Jr., The total synthesis of (±)-retronecine,
     J. Org. Chem., 27 (1962) 139-142.
28   (a) K. Narasaka, T. Sakakura, T. Uchimaru, K. Morimoto, and T. Mukaiyama,
     A total synthesis of (±)-integerrimine, Chem. Lett., (1982) 455-458.   (b)
     K. Narasaka, T. Sakakura, T. Uchimaru, and D. Guedin-Vuong, Total synthe-
     sis of a macrocyclic pyrrolizidine alkaloid, (±)-integerrimine, utilizing
     an activable protecting group, J. Am. Chem. Soc., 106 (1984) 2954-2961.
29   H. Rüeger and M. Benn, The synthesis of (-)-isoretronecanol, (-)-trache-
     lanthamidine, and (-)-supinidine from (S)-proline, Heterocycles, 19 (1982)
     1677-1680; The enantioselective synthesis of (+)-retronecine, (-)-
     platynecine, and (+)-croalbinecine and its C-1 epimer, ibid., 20 (1983)
     235-237.
30   H. Rüeger and M. Benn, A synthesis of (+)-2-oxa-6-azabicyclo[3.3.0]octan-
     3-one (the Geissman-Waiss lactone): A synthon for some necines, Hetero-
     cycles, 19 (1982) 23-25; The synthesis of (1R,2S,8S)- and (1S,2S,8S)-1-
     hydroxymethyl-2-hydroxypyrrolizidine: petasinecine and its C-1 epimer,
     ibid., 20 (1983) 1331-1334.
31   J.G. Buchanan, G. Singh, and R.H. Wightman, An enantiospecific synthesis
     of (+)-retronecine and related alkaloids, J. Chem. Soc., Chem. Commun.,
     (1984) 1299-1300.

32 S. Miyano, S. Fujii, O. Yamashita, N. Toraishi, K. Sumoto, F. Satoh, and
    T. Masuda, A convenient synthesis of 1-methylpyrrolizidines [(±)-helio-
    tridane and (±)-pseudoheliotridane], J. Org. Chem., 46 (1981) 1737-1738.
33 S. Takano, S. Otaki, and K. Ogasawara, Enantioselective synthesis of an
    ant venom alkaloid (-)-[3S-(3β,5β,8α)]-3-heptyl-5-methylpyrrolizidine, J.
    Chem. Soc., Chem. Commun., (1983) 1172-1174.
34 H. Ishibashi, H. Ozeki, and M. Ikeda, Synthesis of optically active (-)-
    trachelanthamidine from L-prolinol, J. Chem. Soc., Chem. Commun., (1986)
    654-655.
35 N.J. Leonard and T. Sato, A transannular route for stereospecific synthe-
    sis. (±)-Isoretronecanol, J. Org. Chem., 34 (1969) 1066-1070.
36 H. Niwa, A. Kuroda, and K. Yamada, A convenient synthesis of (±)-retro-
    necine, Chem. Lett., (1983) 125-126.
37 H. Niwa, Y. Uosaki, and K. Yamada, Total synthesis of (±)-otonecine, a
    necine base of pyrrolizidine alkaloids, Tetrahedron Lett., 24 (1983)
    5731-5732.
38 (a) T. Ohnuma, M. Tabe, K. Shiiya, and Y. Ban, A novel annulation method
    for the synthesis of some nitrogen-containing heterocycles: The synthesis
    of (±)-heliotridane and (±)-nuphar indolizidine, Tetrahedron Lett., 24
    (1983) 4249-4252.   (b) T. Ohnuma, M. Nagasaki, M. Tabe, and Y. Ban, An
    efficient synthesis of medium-sized ketolactams through controlled criss-
    cross annulation: A synthesis of (±)-dihydrodesoxyotonecine, ibid., 24
    (1983) 4253-4256.
39 M.E. Garst, J.N. Bonfiglio, and J. Marks, Hydroboration-carbon monoxide
    insertion of bis-olefinic amine derivatives. Synthesis of δ-coniceine,
    pyrrolizidine, (±)-heliotridane, and (±)-pseudoheliotridane, J. Org. Chem.,
    47 (1982) 1494-1500.
40 N.J. Leonard and S.W. Blum, Laboratory realization of the Robinson-Schopf
    scheme of alkaloid synthesis. The pyrrolizidine alkaloids, J. Am. Chem.
    Soc., 82 (1960) 503-504.
41 S. Takano, N. Ogawa, and K. Ogasawara, A simple route to (±)- and (+)-
    trachelanthamidine, Heterocycles, 16 (1981) 915-919.
42 (a) P.M.M. Nossin and W.N. Speckamp, Directing and stabilizing effects of
    substituted acetylenes in heterocyclisation. Efficient synthesis of pyrro-
    lizidine alkaloids, Tetrahedron Lett., (1979) 4411-4414.   (b) W.N.
    Speckamp, Recent developments in alkaloid synthesis, Recueil Review, 100
    (1981) 345-354.
43 D.J. Hart and T-K. Yang, Rearrangements of N-acyl-2-aza-1,5-hexadienes:
    Application to syntheses of trachelanthamidine and supinidine, Tetra-
    hedron Lett., 27 (1982) 2761-2764.
44 (a) D.J. Hart and T-K. Yang, Enantioselective syntheses of (-)-hastanecine
    and its enantiomer using malic acid as a chiral educt, J. Chem. Soc., Chem.
    Commun., (1983) 135-136.   (b) idem., N-Acyliminium ion rearrangements:
    Generalities and application to the synthesis of pyrrolizidine alkaloids,
    J. Org. Chem., 50 (1985) 235-242.
45 (a) A.R. Chamberlin and J.Y.L. Chung, Synthesis of optically active pyrro-
    lizidine diols: (+)-heliotridine, J. Am. Chem. Soc., 105 (1983) 3653-3656.
    (b) idem., Enantioselective synthesis of seven pyrrolizidine diols from a
    single precursor, J. Org. Chem., 50 (1985) 4425-4431.
46 (a) A.R. Chamberlin and J.Y.L. Chung, The ketene thioacetal group as a
    cationic cyclization terminator. A synthesis of the pyrrolizidine ring
    system, Tetrahedron Lett., 23 (1982) 2619-2622.   (b) A.R. Chamberlin, H.D.
    Nguyen, and J.Y.L. Chung, Cationic cyclizations of ketene dithioacetals.
    A general synthesis of pyrrolizidine, indolizidine, and quinolizidine
    alkaloid ring systems, J. Org. Chem., 49 (1984) 1682-1688.
47 Z. Blum, M. Ekström, and L-G. Wistrand, Total synthesis of DL-isoretro-
    necanol and DL-trachelanthamidine using anodic methoxylation in a key
    functionalizing step, Acta Chemi. Scand., B38 (1984) 297-302.
48 M.T. Pizzorno and S.M. Albonico, Stereospecific synthesis of 1-substi-
    tuted pyrrolizidines, J. Org. Chem., 39 (1974) 731.

298

49 (a) D.J. Robins and S. Sakdarat, Synthesis of the 8β-pyrrolizidine bases (+)-isoretronecanol, (+)-luburnine, and (+)-supinidine, J. Chem. Soc., Chem. Commun., (1979) 1181–1182. (b) idem., Synthesis of optically active pyrrolizidine bases, J. Chem. Soc., Perkin I, (1981) 909–913.

50 (a) E. Vedejs and G.R. Martinez, Stereospecific synthesis of retronecine by imidate methylide cycloaddition, J. Am. Chem. Soc., 102 (1980) 7993–7994. (b) E. Vedejs, S. Larsen, and F.G. West, Nonstabilized imidate ylides by the desilylation method: A route to the pyrrolizidine alkaloids retronecine and indicine, J. Org. Chem., 50 (1985) 2170–2174.

51 Y. Terao, N. Imai, K. Achiwa, and M. Sekiya, A 1,3-dipolar cycloaddition of alicyclic methylene iminium ylide to form pyrrolizidine nuclei and its application to synthesis of pyrrolizidine alkaloids, Chem. Pharm. Bull., 30 (1982) 3167–3171.

52 T. Hudlicky, J.O. Frazier, and L.D. Kwart, Intramolecular [4+1]pyrroline annulation approach to pyrrolizidine alkaloids. Formal total synthesis of (±)-supinidine, Tetrahedron Lett., 26 (1985) 3523–3526.

53 W.H. Pearson, Synthesis of fused pyrrolines by the intramolecular cyclo-addition of azides. Application to the pyrrolizidine alkaloids, Tetra-hedron Lett., 26 (1985) 3527–3530.

54 R.V. Stevens, Y. Luh, and J-T. Sheu, General methods of alkaloid synthe-sis. XII. The total synthesis of (±)-isoretronecanol and (±)-δ-coniceine, Tetrahedron Lett., 42 (1976) 3799–3802.

55 H.W. Pinnick and Y-H. Chang, Cyclopropyl imine rearrangement: Total syn-thesis of (±)-isoretronecanol and (±)-trachelanthamidine, Tetrahedron Lett., (1979) 837–840.

56 (a) S. Danishefsky, R. McKee, and R.K. Singh, Kinetically controlled total syntheses of dl-trachelanthamidine and dl-isoretronecanol, J. Am. Chem. Soc., 99 (1977) 4783–4788. (b) S. Danishefsky, Electrophilic cyclo-propanes in organic synthesis, Acc. Chem. Res., 12 (1979) 66–72.

57 S. Danishefsky, R. McKee, and R.K. Singh, Stereospecific total synthesis of dl-hastanecine and dl-dihydroxyheliotridane, J. Am. Chem. Soc., 99 (1977) 7711–7713.

58 R.F. Borch and B.C. Ho, A new synthesis of the pyrrolizidine alkaloids (±)-isoretronecanol and (±)-trachelanthamidine, J. Org. Chem., 42 (1977) 1225–1227.

59 T.H. Jones, M.S. Blum, H.M. Fales, and C.R. Thompson, (5Z,8E)-3-Heptyl-5-methylpyrrolizidine from a thief ant, J. Org. Chem., 45 (1980) 4778–4780.

60 T.H. Jones, R.J. Highet, A.W. Don, and M.S. Blum, Alkaloids of the ant Chelaner antarcticus, J. Org. Chem., 51 (1986) 2712–2716.

61 D. Lathbury and T. Gallagher, Stereoselective synthesis of pyrrolizidine alkaloids via substituted nitrones, J. Chem. Soc., Chem. Commun., (1986) 1017–1018.

62 J.M. Muchowski and P.H. Nelson, The reaction of carboalkoxycyclopropyltri-phenylphosphonium salts with imide anions: A three-step synthesis of (±)-isoretronecanol, Tetrahedron Lett., 21 (1980) 4585–4588.

63 W. Flitsch and P. Wernsmann, A diastereoselective route to (±)-isoretro-necanol, Tetrahedron Lett., 22 (1981) 719–722.

64 W. Klose, K. Nickisch, and F. Bohlmann, Synthese von 5,7a-Didehydrohelio-tridin-3-on, dem Grundkörper einer neuen Gruppe von Pyrrolizidin-Alka-loiden, Chem. Ber., 113 (1980) 2694–2698.

65 D.J. Hart and Y-M. Tsai, New methods for alkaloid synthesis: α-Acylamino radical cyclizations, J. Am. Chem. Soc., 104 (1982) 1430–1432.

66 G.E. Keck and E.J. Enholm, Synthetic studies on pyrrolizidine alkaloids. II. Intramolecular additions of radicals and electrophiles to allyl-stannanes as methods for ring closure, Tetrahedron Lett., 26 (1985) 3311–3314.

67 J-C. Gramain, R. Remuson, and D. Vallee, Intramolecular photoreduction of α-keto esters. Total synthesis of (±)-isoretronecanol, J. Org. Chem., 50 (1985) 710–712.

68 H. Ishibashi, T. Satoh, M. Irie, S. Harada, and M. Ikeda, New entry to γ-butyrolactams by free radical cyclization of N-allyl-α-chloro-α-

(methylthio)acetamides. Formal total synthesis of (±)-pseudoheliotridane, Chem. Lett., (1987) 795-798.

69 D.J. Robins, A biogenetically patterned synthesis of the pyrrolizidine alkaloid trachelanthamidine, J. Chem. Soc., Chem. Commun., (1982) 1289-1290.

70 M. Mori, N. Kanda, I. Oda, and Y. Ban, New synthesis of heterocycles by use of palladium catalyzed cyclization of α-haloamide with internal double bond, Tetrahedron, 41 (1985) 5465-5474.

71 T. Kametani, H. Yukawa, and T. Honda, Synthesis of pyrrolizidine alkaloids, (±)-trachelanthamidine, (±)-isoretronecanol, and (±)-supinidine, by means of an intramolecular carbenoid displacement (ICD) reaction, J. Chem. Soc., Chem. Commun., (1986) 651-652.

72 H. Dhimane, J-C. Pommelet, J. Chuche, G. Lhommet, M. Haddad, Cyclization of aminomethyleneketenes by intramolecular displacement of chlorine: Synthesis of bicyclic enaminoesters, Tetrahedron Lett., 28 (1987) 885-886.

73 D.H.G. Crout and D. Whitehouse, Absolute configuration of 2,3-dihydroxy-3-methylpentanoic acid. An intermediate in the biosynthesis of isoleucine, and its identity with the esterifying acid of the pyrrolizidine alkaloid strigosine, J. Chem. Soc., Perkin I, (1977) 544-549.

74 W. Ladner, Stereoselektive Synthese von α-Alkyl-α,β-dihydroxycarbonsäure-ester-Derivaten: Synthese der (+)-Viridiflorsäure, Chem. Ber., 116 (1983) 3413-3426.

75 R.S. Glass and M. Shanklin, Stereoselective reduction of α-alkoxy-β-keto esters. Synthesis of (±)-trachelanthic acid and (±)-viridifloric acid, Synth. Commun., 13(7) (1983) 545-552.

76 Y. Nishimura, S. Kondo, and H. Umezawa, The total syntheses of (-)-indicine N-oxide and intermedine N-oxide, Tetrahedron Lett., 27 (1986) 4323-4326; Total syntheses of indicine N-oxide, intermedine N-oxide, and their enantiomers using carbohydrate as a chiral educt, Bull. Chem. Soc. Jpn., 60 (1987) 4107-4113.

77 T. Matsumoto, M. Takahashi, and Y. Kashihara, Pyrrolizidine alkaloids. The synthesis and absolute configuration of all stereoisomers of monocrotalic acid, Bull. Chem. Soc. Jpn., 52 (1979) 3329-3336.

78 E. Vedejs and S.D. Larsen, Total synthesis of d,l-fulvine and d,l-crispatine, J. Am. Chem.Soc., 106 (1984) 3030-3032.

79 (a) U. Pastewka, H. Wiedenfeld, and E. Röder, Synthese von racem. 5-Hydroxy-3,4-dimethyl-1-hexen-2,5-dicarbonsäure (Seneciverninsäure), Arch. Pharm. (Weinheim), 313 (1980) 785-790. (b) E. Röder, H. Wiedenfeld, and M. Frisse, Zur Synthese der 5-Hydroxy-2,4-dimethyl-1-hexen-1,5-dicarbonsäure (Nemorensinsäure), ibid., 313 (1980) 803-806.

80 K. Narasaka and T. Uchimaru, Stereoselective synthesis of (±)-integerrinecic acid, Chem. Lett., (1982) 57-58.

81 C.C.J. Culvenor and T.A. Geissman, The total synthesis of senecic and integerrinecic acid, J. Am. Chem. Soc., 83 (1961) 1647-1652.

82 S.E. Drewes and N.D. Emslie, Necic acid synthons. Part 1. Total synthesis of integerrinecic acid, J. Chem. Soc., Perkin Trans. I, (1982) 2079-2083.

83 F. Ameer, S.E. Drewes, R. Hoole, P.T. Kaye, and A.T. Pitchford, Necic acid synthons. Part 5. Total synthesis of (±)-retronecic acid and related compounds via zinc-mediated coupling of halogeno-esters, J. Chem. Soc., Perkin Trans. I, (1985) 2713-2717.

84 F. Ameer, S.E. Drewes, M.W. Houston-McMillan, and P.T. Kaye, Necic acid synthons. Part 6. Regiocontrolled S reactions of 2-bromomethyl-2-alkenoate esters in the total synthesis of necic acid systems, S. Afr. J. Chem., 39 (1985) 57-63.

85 H. Niwa, Y. Miyachi, Y. Uosaki, and K. Yamada, Total synthesis of optically active integerrimine, a twelve-membered dilactonic pyrrolizidine alkaloid of retronecine type. I. Enantioselective synthesis of the protected (+)-integerrinecic acid, Tetrahedron Lett., 27 (1986) 4601-4604.

86 J.D. White and S. Ohira, Synthesis of the macrolactone pyrrolizidine alkaloid integerrimine, J. Org. Chem., 51 (1986) 5492-5494.

87 N.K. Kochetkov, A.M. Likhosherstov, and V.N. Kulakov, The total synthesis of some pyrrolizidine alkaloids and their absolute configuration, Tetrahedron, 25 (1969) 2313-2323.

88 (a) C.C.J. Culvenor, A.T. Dann, and L.W. Smith, Recombination of amino alcohols and acids derived from pyrrolizidine alkaloids, Chem. & Ind., (1959) 20-21. (b) C.C.J. Culvenor and L.W. Smith, The alkaloids of Amsinckia species: a. intermedia Fisch. & Mey., a. hispida (Ruiz. & Pav.) Johnst. and a. lycopsoides Lehm, Aust. J. Chem., 19 (1966) 1955-1964.

89 W.M. Hoskins and D.H.G. Crout, Pyrrolizidine alkaloid analogues. Preparation of semisynthetic esters of retronecine, J. Chem. Soc., Perkin I, (1977) 538-543.

90 J.R. Piper, P. Kari, and Y.F. Shealy, Synthesis of ^{3}H-labelled indicine N-oxide, J. Labelled Compd. Radiopharm., 18 (1981) 1579-1591.

91 J.A. Glinski and L.H. Zalkow, The synthesis of heliotridine and related alkaloids, Tetrahedron Lett., 26 (1985) 2857-2860.

92 (a) L.T. Gelbaum, M.M. Gordon, M. Miles, and L.H. Zalkow, Semisynthetic pyrrolizidine alkaloid antitumor agents, J. Org. Chem., 47 (1982) 2501-2504. (b) L.H. Zalkow, J.A. Glinski, L.T. Gelbaum, T.J. Fleischmann, L.S. McGowan, and M.M. Gordon, Synthesis of pyrrolizidine alkaloids indicine, intermedine, lycopsamine, and analogues and their N-oxide. Potential antitumor agents, J. Med. Chem., 28 (1985) 687-698.

93 D.J. Robins and S. Sakdarat, Pyrrolizidine alkaloids. Synthesis of 13,13-dimethyl-1,2-didehydrocrotalanine, J. Chem. Soc., Chem. Commun., (1980) 282-283.

94 J.A. Devlin and D.J. Robins, Synthesis and stereochemistry of dicrotaline, a macrocyclic pyrrolizidine alkaloid, J. Chem. Soc., Chem. Commun., (1981) 1272-1274.

95 J. Huang and J. Meinwald, Synthesis of crobarbatine acetate. A macrocyclic pyrrolizidine alkaloid ester, J. Am. Chem. Soc., 103 (1981) 861-867.

96 H. Niwa, Y. Miyachi, Y. Uosaki, and K. Yamada, Total synthesis of optically active integerrimine, a twelve-membered dilactonic pyrrolizidine alkaloid of retronecine type. III. Regioselective elaboration of the unsymmetrical twelve-membered dilactone and total synthesis of (-)-integerrimine, Tetrahedron Lett., 38 (1986) 4609-4610.

97 M. Kugelman, W-C. Liu, M. Axelrod, T.J. McBride, and K.V. Rao, Indicine N-oxide: The antitumor principle of Heliotropium indicum, Lloydia, 39 (1976) 125-128.

98 J.S. Kovach, M.M. Ames, G. Powis, C.G. Moertel, R.G. Hahn, and E.T. Creagan, Toxicity and pharmacokinetics of a pyrrolizidine alkaloid, indicine N-oxide, in humans, Cancer Res., 39 (1979) 4540-4544.

99 L. Letendre, W.A. Smithson, G.S. Gilchrist, E.O. Burgert, C.H. Hoagland, M.M. Ames, G. Powis, and J.S. Kovach, Activity of indicine N-oxide in refractory acute leukemia, Cancer, 47 (1981) 437-441.

100 D.S. Poster, S. Bruno, J. Penta, and J.S. Macdonald, Indicine N-oxide: A new antitumor agent, Cancer Treat Rep., 65 (1981) 53-56.

101 W.C. Nichols, C.G. Moertel, J. Rubin, A.J. Schutt, and J.C. Britell, Phase II trial of indicine N-oxide (INDI) in patients with advanced colorectal carcinoma, Cancer Treat. Rep., 65 (1981) 337-339.

102 M.M. Ames, J.S. Miser, W.S. Smithson, P.F. Coccia, C.S. Hughes, and D.M. Davis, Pharmacokinetic study of indicine N-oxide in pediatric cancer patients, Cancer Chem. Pharm., 10 (1982) 43-46.

103 L.H. Zalkow, S.Bonetti, L. Gelbaum, M.M. Gordon, B.B. Patil, A. Shani, and D. Van Derver, Pyrrolizidine alkaloids from middle eastern plants, J. Nat. Prod., 42 (1979) 603-614.

104 L.T. Gelbaum, M.M. Gordon, M. Miles, and L.H. Zalkow, Semisynthetic pyrrolizidine alkaloid antitumor agents, J. Org. Chem., 47 (1982) 2501-2504.

105 L.H. Zalkow, J.A. Glinski, L.T. Gelbaum, T.J. Fleischmann, L.S. McGowan, and M.M. Gordon, Synthesis of pyrrolizidine alkaloids indicine, inter-

medine, lycopsamine, and analogues and their N-oxide. Potential antitumor agents, J. Med. Chem., 28 (1985) 687-694.

106 W.K. Anderson, Activity of bis-carbamoyloxymethyl derivatives of pyrroles and pyrrolizines against human tumor xenografts in nude mice, Cancer Res., 42 (1982) 2168-2170.

107 Y. Nishimura, Syntheses of pyrrolizidine alkaloid and lignan lactone glycoside antitumor agents and their biological activities, J. Synth. Org. Chem. Jpn. (Yuki Gosei Kagaku Kyokai Shi), 45 (1987) 873-887.

108 T.L. Macdonald, Indolizidine alkaloid synthesis. Preparation of the pharaoh ant trail pheromone and gephyrotoxin 223 stereoisomers, J. Org. Chem., 45 (1980) 193-194.

109 T. Iwashita, M. Suzuki, T. Kusumi, and H. Kakisawa, Synthesis of dl-septicine, Chem. Lett., (1980) 383-386.

110 H. Otomasu, N. Takatsu, T. Honda, and T. Kametani, Alternative total synthesis of elaeocarpus alkaloids, Heterocycles, 19 (1982) 511-514.

111 N.K. Hart, S.R. Johns, J.A. Lamberton, Elaeocarpus alkaloids. V. Alkaloids of Elaeocarpus kaniensis, Aust. J. Chem., 25 (1972) 817-835.

112 R.T. LaLonde, N. Muhammad, C.F. Wong, and E.R. Sturiale, Extension of a nuphar piperidine synthesis to quinolizidines and an indolizidine, J. Org. Chem., 45 (1980) 3664-3671.

113 K.B. Sharpless, Recent applications of asymmetric epoxidation in syntheses of biological active molecules, the invited lecture at the 103rd Annual Meeting of Pharmaceutical Society of Japan (Tokyo), Abst. (1983), 67; C.E. Adams, F.J. Walker, K.B. Sharpless, Enantioselective synthesis of swainsonine, a trihydroxylated indolizidine alkaloid, J. Org. Chem., 50 (1985) 422-424.

114 R.C. Bernotas and B. Ganem, Total syntheses of (+)-castanospermine and (+)-deoxynojirimycin, Tetrahedron Lett., 25 (1984) 165-168.

115 T. Suami, K. Tadano, and Y. Iimura, Total synthesis of (-)-swainsonine, an α-mannosidase inhibitor isolated from Swainsona canescens, Chem. Lett., (1984) 513-516.

116 K. Tadano, Y. Iimura, Y. Hotta, C. Fukabori, and T. Suami, Syntheses of (-)-8-epi-swainsonine and (-)-1,8-di-epi-swainsonine, stereoisomers of physiologically interesting indolizidine alkaloid, swainsonine, Bull. Chem. Soc. Jpn., 59 (1986) 3885-3892; K. Tadano, Y. Hotta, M. Morita, T. Suami, B. Winchester, and I.C. di Bello, Synthesis of (-)-8a-epi-swainso-nine, (1S,2R,8R,8aS)-octahydro-1,2,8-indolizinetriol, Chem. Lett., (1986) 2105-2108.

117 M.H. Ali, L. Hough, and A.C. Richardson, A chiral synthesis of swainsonine from D-glucose, J. Chem. Soc., Chem. Commun., (1984) 447-448; Synthesis of the indolizidine alkaloid swainsonine from D-glucose, Carbohyd. Res., 136 (1985) 225-240.

118 H. Setoi, H. Takeno, and M. Hashimoto, Enantiospecific total synthesis of (-)-swainsonine: New applications of sodium borohydride reduction, J. Org. Chem., 50 (1985) 3948-3950.

119 H. Setoi, H. Takeno, and M. Hashimoto, Total synthesis of (+)-castanosper-mine from D-mannose, Tetrahedron lett., 26 (1985) 4617-4620.

120 H. Iida, Y. Watanabe, and C. Kibayashi, Stereospecific total synthesis of (±)-gephyrotoxin 223AB, J. Am. Chem. Soc., 107 (1985) 5534-5535.

121 J.J. Tufariello and SK. Asrof Ali, Elaeocarpus alkaloids. The synthesis of dl-elaeokanine-A and dl-elaeokanine-C, Tetrahedron Lett., No. 46 (1979) 4445-4448.

122 J.J. Tufariello and SK. Asrof Ali, Elaeocarpus alkaloids. Synthesis using nitrones, J. Am. Chem. Soc., 101 (1979) 7114-7116.

123 A.S. Howard, G.C. Gerrans, and J.P. Michael, Use of vinylogous urethane in alkaloid synthesis: Formal synthesis of ipalbidine, J. Org. Chem., 45 (1980) 1713-1715.

124 (a) T.R. Govindachai, A.R. Sidhaye, and N. Viswanathan, Synthesis of ipalbidine, Tetrahedron, 26 (1970) 3829-3831. (b) R.V. Stevens and Y. Luh, General methods of alkaloid synthesis. XIII. The total synthesis of (±)-ipalbidine and (±)-septicine, Tetrahedron Lett., (1977) 979-982.

125 A.S. Howard, G.C. Gerrans, and C.A. Meerholz, Vinylogous urethanes in alkaloid synthesis: Formal syntheses of elaeocarpus alkaloids, Tetrahedron Lett., 21 (1980) 1373–1374.

126 T. Watanabe, Y. Nakashita, S. Katayama, and M. Yamaguchi, Elaeocarpus alkaloids. The synthesis of (±)–elaeokanine A, (±)–elaeokanine B, and (±)–elaeokanine C, Heterocycles, 14 (1980) 1433–1436.

127 J.E. Cragg, S.H. Hedges, and R.B. Herbert, Synthesis of the alkaloids julandine and ipalbidine–use of silicon (IV) chloride, Tetrahedron Lett., 22 (1981) 2127–2130.

128 J.E. Cragg, R.B. Herbert, F.B. Jackson, C.J. Moody, and I.T. Nicolson, Phenanthroindolizidine and related alkaloids: Synthesis of tylophorine, septicine, and deoxytylophorinine, J. Chem. Soc., Perkin Trans. I, (1982) 2477–2485.

129 S. Miyano, S. Fujii, O. Yamashita, N. Toraishi, and K. Sumoto, Synthesis of $\Delta^{1[9]}$– and/or $\Delta^{8[9]}$–dehydroindolizidines and related compounds (1), J. Heterocyclic Chem., 19 (1982) 1465–1468.

130 H. Iida, M. Tanaka, and C. Kibayashi, Synthesis of (±)–septicine and (±)–tylophorine by regioselective [3+2]cycloaddition, J. Chem. Soc., Chem. Commun. (1983) 271–272.

131 H. Iida, Y. Watanabe, and C. Kibayashi, Formal synthesis of (±)–ipalbidine, Chem. Lett., (1983) 1195–1196.

132 V.K. Mangla and D.S. Bhakuni, A new synthesis of septicine, a secophenanthroindolizidine alkaloid, Indian J. Chem., 19B (1980) 748–749.

133 B.P. Wijnberg and W.N. Speckamp, α–Acyliminium ion synthesis of elaeokanine B, Tetrahedron Lett., 22 (1981) 5079–5082.

134 (a) L.E. Overman and K.L. Bell, Enantiospecific total synthesis of dendrobatid toxin 251D. A short chiral entry to the cardiac-active pumiliotoxin A alkaloids via stereospecific iminium ion-vinysilane cyclizations, J. Am. Chem. Soc., 103 (1981) 1851–1853. (b) L.E. Overman, K.L. Bell, and F. Ito, Enantioselective total syntheses of pumiliotoxin B and pumiliotoxin 251D. A general entry to the pumiliotoxin A alkaloids via stereospecific iminium ion-vinylsilane cyclization, J. Am. Chem. Soc., 106 (1984) 4192–4201.

135 D.J. Hart and Y-M. Tsai, Stereoselective indolizidine synthesis: Preparation of stereoisomers of gephyrotoxin-223AB, J. Org. Chem., 47 (1982) 4403–4409.

136 (a) S.M. Weinreb, N.A. Khatri, and J. Shringarpure, Alkaloid synthesis by the intramolecular imino Diels-Alder reaction. δ-Coniceine and tylophorine, J. Am. Chem. Soc., 101 (1979) 5073–5074. (b) H.F. Schmitthenner and S.M. Weinreb, Total synthesis of elaeokanine A, J. Org. Chem., 45 (1980) 3372–3373. (c) N.A. Khatri, H.F. Schmitthenner, Synthesis of indolizidine alkaloids via the intramolecular imino Diels-Alder reaction, J. Shringarpure, and S.M. Weinreb, J. Am. Chem. Soc., 103 (1981) 6387–6393. (d) R.A. Gobao, M.L. Bremmer, and S.M. Weinreb, A new total synthesis of the fungal toxin slaframine, J. Am. Chem. Soc., 104 (1982) 7065–7068.

137 G.W.J. Fleet, M.J. Gough, and P.W. Smith, Enantiospecific synthesis of swainsonine, (1S,2R,8R,8aR)-1,2,8-trihydroxyoctahydroindolizine, from D-mannose, Tetrahedron Lett., 25 (1984) 1853–1856.

138 J. Royer and H-P. Husson, Asymmetric synthesis. 2. Practical method for the asymmetric synthesis of indolizidine alkaloids: Total synthesis of (−)-monomorine I, J. Org. Chem., 50 (1985) 670–673.

139 H. Iida, Y. Watanabe, and C. Kibayashi, A stereoselective synthesis of the ant trail pheromone (±)-monomorine I, Tetrahedron Lett., 27 (1986) 5513–5514.

140 M.E. Garst and J.N. Bonfiglio, Hydroboration-carbonylation of bisolefinic amines: A facile synthesis of δ-coniceine, Tetrahedron Lett., 22 (1981) 2075–2076.

141 S.R. Wilson and R.A. Sawicki, Indolizidine synthesis by transannular cyclization, J. Heterocyclic Chem., 19 (1982) 81–83.

142 C.A. Broka and K.K. Eng, A short total synthesis of (±)-gephyrotoxin-223AB, J. Org. Chem., 51 (1986) 5043–5045.

143 (a) R.V. Stevens and Y. Luh, General methods of alkaloid synthesis. XIII. The total synthesis of (±)-ipalbidine and (±)-septicine, Tetrahedron Lett., (1977) 979–982. (b) R.V. Stevens, General methods of alkaloid synthesis, Acc. Chem. Res., 10 (1977) 193–198.

144 R.V. Stevens and A.W.M. Lee, Studies on the stereochemistry of nucleophilic additions to tetrahydropyridinium salts. A stereospecific total synthesis of (±)-monomorine I, J. Chem. Soc., Chem. Commun., (1982) 102–103.

145 R.V. Stevens and A.W.M. Lee, Studies on the stereochemistry of nucleophilic additions to tetrahydropyridinium salts. A stereospecific total synthesis of one of the stereoisomers of gephyrotoxin 223, J. Chem. Soc., Chem. Commun., (1982) 103–104.

146 L.E. Overman and D. Lesuisse, The synthesis of indolizidines by intramolecular ene cyclizations. Preparation of (E)-alkylidene analogs of pumiliotoxin A, Tetrahedron Lett., 26 (1985) 4167–4170.

147 (a) J. Royer and H-P. Husson, Asymmetric synthesis III. Enantiospecific synthesis of the natural 3R,5R,9R(-)gephyrotoxin-223AB, Tetrahedron Lett., 26 (1985) 1515–1518. (b) H-P. Husson, A new approach to the asymmetric synthesis of alkaloids, J. Nat. Prod., 48 (1985) 894–906.

SYNTHESIS OF POTENT ANTITUMOR ALKALOIDS ISOLATED FROM *SESBANIA DRUMMONDII* SEEDS

F. MATSUDA AND S. TERASHIMA

1. Introduction

A number of leguminous plants belonging to the genus *Sesbania* native to the Gulf Costal Plains of the U.S.A. from Florida to Texas are notorious for toxicity of their seeds to livestock and fowl. Efforts to identify the toxic principle of these species have led to isolation of a variety of saponins and sapogenins, none of which was uncovered to be toxic.[1] Powell *et al.* had reported that alcoholic extracts of the seeds of *S. drummondii*, *S. vesicaria*, and *S. punicea* were markedly cytotoxic against KB cells *in vitro* and showed significant inhibitory activity against P388 murine leukemia *in vivo*.[2] Further their investigation directed towards identification of the antileukemic principles of *S. drummondii* seed resulted in the isolation and structure elucidation of potent antitumor alkaloids, sesbanimide A (**1**) and sesbanimide B (**2**).[3] Sesbanimide A (**1**), the major and most active component, exhibited IC_{50}

Fig 1

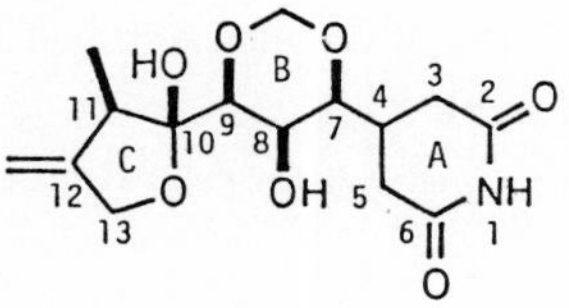

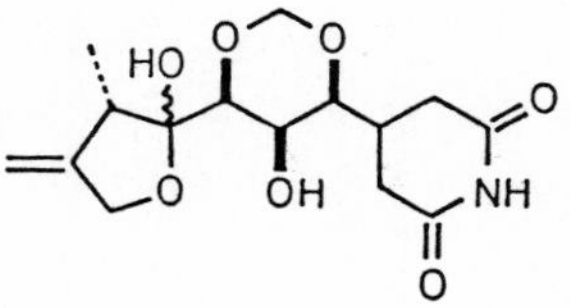

value of 7.7×10^{-3} µg/ml against KB cells *in vitro* and T/C values of 140–181% in 8–32 µg/kg dose level against P388 murine leukemia *in vivo*. Sesbanimide B (**2**), the C–11 epimer of **1**, also showed considerable antitumor activity, although which was inferior to that of **1**.[3c] On the other hand, **1** was also obtained from aqueous ethanol extracts of *S. punicea* seeds native to South Africa as the toxic principle by Gorst–Allman *et al.*, monitoring the fractionation by bioassay for acute toxicity in 1 day-old chickens.[4]

Sesbanimides have unique tricyclic structures in which the three characteristic rings, glutarimide (A-ring), 1,3-dioxane (B-ring), and tetrahydrofuran (C-ring), are linked by two single bonds. Although the structure of 1 including its relative stereochemistry had been established by X-ray crystallographic analysis,[3a] its absolute configuration had not been determined. Thus, their remarkable antitumor activity and novel structures in addition to lack of determination of their absolute stereochemistry distinguish these molecules as unusually interesting targets for total synthesis and a number of synthetic studies on sesbanimides have been reported.[5-8] The authors started the program directed toward the total synthesis of these interesting alkaloids with the aim to determine their absolute configurations and to elucidate the structure-activity relationships. Our efforts culminated in the first total synthesis of natural (+)-sesbanimide A (1) and (-)-sesbanimide B (2) starting from readily available D-xylose, concluding the absolute configurations of 1 and 2 as shown in Fig 1.[6] Other two total syntheses of the antipodes of sesbanimides were also reported by Pandit et al.[7] and Schlessinger et al.[8] In both of those syntheses, D-sorbitol was employed as the starting material and the same conclusion with regard to the absolute configurations of sesbanimides was obtained. This review deals with these three independently accomplished total syntheses and other synthetic studies performed on these novel alkaloids.

2. Synthetic Strategy

From retrosynthetic perspective on sesbanimides, the most logical strategy to construct the five chiral centers involved in these novel alkaloids in an optically active form was anticipated to be introduction of the three significant asymmetric centers at the C-7, C-8, and C-9 positions from an appropriate sugar. This is because both of the C-11 epimers had been isolated and the configurations at the C-10 position were expected to be governed by an equilibrium between the two possible anomers. The authors selected D-xylose as

Fig 2

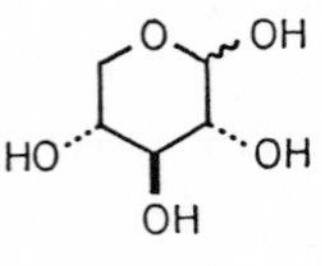

D-xylose 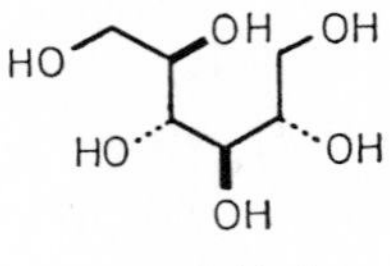   D-sorbitol 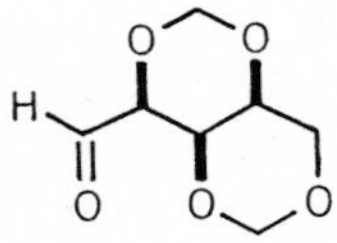   3 the starting material since the three asymmetric centers of D-xylose just correspond to the three contiguous asymmetric centers of sesbanimides and relatively easily available L-xylose can be used if the synthesis of the antipodes is required.

In the syntheses by Pandit *et al.* and Schlessinger *et al.*, both of their synthetic schemes started from the known aldehyde (**3**)[9] being readily available from D-sorbitol. The aldehyde (**3**) already carries the 1,3-dioxane ring correspnding to the B-ring system of sesbanimides and the selective cleavage of the unnecessary 1,3-dioxane ring formed by acetalization of the primary and secondary hydroxyl groups could be predicted.[9a,c] However, such synthetic strategy seems to be disadvantageous since synthesis of the antipodes starting from not readily available L-sorbitol is virtually impossible and **3** is obtainable only in a low overall yield from D-sorbitol. Eventually, the natural sesbanimides have not been synthesized from sorbitol.

Information obtained from the structure determination of sesbanimides revealed that the C-ring system is very unstable to base and the B-ring system is most stable among the three rings.[3b] Therefore, it appeared reasonable to carry out the synthesis in the sequence of (1) construction of the B-ring system from an appropriate sugar in an optically active form, (2) formation of the A-ring system, (3) introduction of the C_5-unit into the AB-ring system, and (4) formation of the C-ring system. Each of the three total sysntheses including ours has been accomplished according to this synthetic scheme. Then, in what follows, the synthetic strategies and tactics so far reported will be discussed concerning formation of the AB-ring system in an optically active form, introduction of the C_5-unit into the AB-ring system, and construction of the C-ring system.

3. Construction of the AB-Ring System

The authors newly developed an efficient and straightforward synthetic method to construct the B-ring system from D-xylose. Since it was well known that direct acetalization of D-xylose did not give the required 2,4-O-methylenexyloside but 1,2;3,5-di-O-methylenexylofuranoside,[10] selective protection of the C-3 and C-5 hydroxyl groups (corresponding to the C-8 and C-10 positions of sesbanimides) was required at the first stage of the synthesis. Preliminary experiments carried out with some D-xylose derivatives disclosed that the known diol (**4**),[11] obtained from D-xylose in 2 steps and in a good overall yield, was quite suitable for this purpose (**Fig 3**). Considering the subsequent synthtic scheme, the two hydroxyl groups of **4** (the C-8 and C-9 positions of sesbanimides) were protected in forms of benzyl ethers to afford the dibenzyl ether (**5**). After acidic removal of the acetonide group , treatment

Fig 3

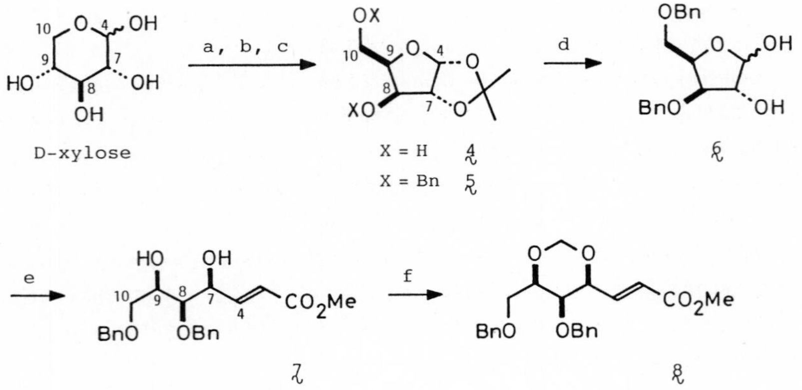

a) 36N-H$_2$SO$_4$, CuSO$_4$, Me$_2$CO, rt, 25 h b) 0.06N-HCl, rt, 1 h, 71%
(2 steps) c) 1) NaH, THF, reflux, 15 min 2) BnCl, Bu$_4$N•Br,
reflux, 5 min, 92% (2 steps) d) 12N-HCl, AcOH, rt, 5 min, 73%
e) Ph$_3$P=CHCO$_2$Me, PhMe, reflux, 30 sec, 92% f) TMSOTf, 2,6-Lu,
(MeO)$_2$CH$_2$, 0 °C, 15 min, 79%.

Fig 4

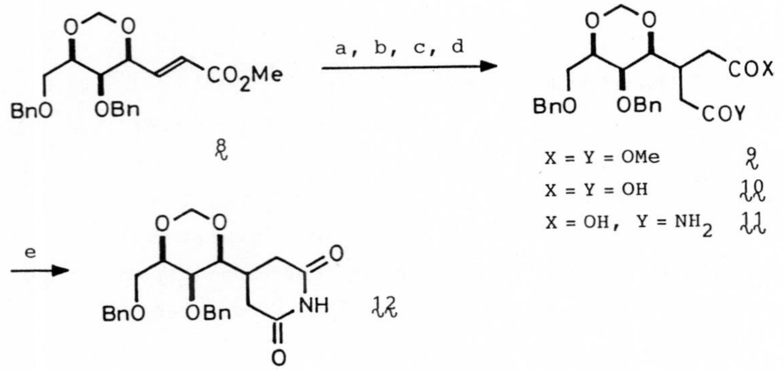

a) NaCH(CO$_2$Me)$_2$, Bu$_4$N•Br, THF, 40 °C, 12 h b) NaCl, H$_2$O, DMSO,
160 °C, 1 h, 89% (2 steps) c) 1N-KOH, MeOH, rt, 48 h d) 1)
MeOCOCl, Et$_3$N, THF, -20 °C, 3 h 2) NH$_3$ (gas), 0 °C, 30 min,
e) NaOAc, Ac$_2$O, 100 °C, 20 min, 51% (4 steps).

of the produced lactol (**6**) with the stabilized ylide resulted in simultaneous opening of the furanoside ring and carbon chain elongation, producing the enester (**7**). Exposure of **7** to trimethylsilyl trifluoromethanesulfonate in dimethoxymethane in the presence of 2,6-lutidine effected construction of the 1,3-dioxane ring to yield **8**.

Micheal addition of the sodium salt of dimethyl malonate for introducing the C_2-unit cleanly took place in the presence of a catalytic amount of tetrabutylammonium bromide in tetrahydrofuran (**Fig 4**). Subsequent demethoxycarbonylation of the resulting Micheal adduct in brine-dimethylsulfoxide gave the diester (**9**). The diester (**9**) was converted into the glutarimide (**12**) corresponding to the AB-ring system of sesbanimides, by the sequence (1) hydrolysis of the two methoxycarbonyl groups, (2) activation of the diacid (**10**) with methyl chloroformate in a form of the glutaric anhydride and subsequent ammonolysis, and (3) glutarimide formation by treatment of the resulting amide-ester (**11**) with acetic anhydride in the presence of sodium acetate as a buffer.

On the other hand, Pandit *et al.* and Schlessinger *et al.* commenced their synthetic schemes from the known aldehyde (**3**) prepared from D-sorbitol. However, **3** could be obtained only in a low overall yield from D-sorbitol as

Fig 5

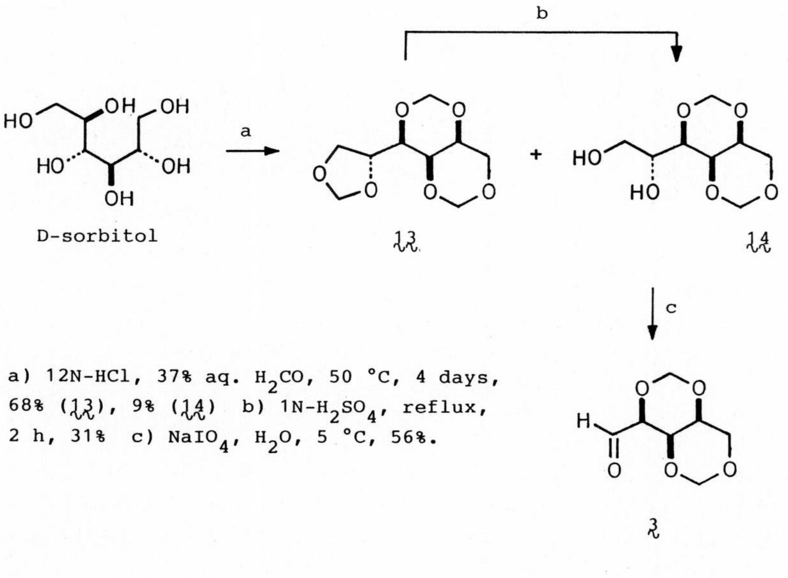

a) 12N-HCl, 37% aq. H_2CO, 50 °C, 4 days, 68% (**13**), 9% (**14**) b) 1N-H_2SO_4, reflux, 2 h, 31% c) $NaIO_4$, H_2O, 5 °C, 56%.

described below (**Fig 5**). Thus, direct acetalization of D-sorbitol gave the trimethylenesorbitol (**13**) as the major product. After removal of **13** by recrystallization, the desired crystalline dimethylenesorbitol (**14**) was isolated from the mother liquor.[9a] Acidic hydrolysis of **13** gave a further amount of **14** in a low yield.[9b] Periodate oxidation of **14** obtained as the minor product gave **3**. In both syntheses, the glutarimide ring was constructed at the next stage of the synthesis according to almost the same methodology as ours, since **3** already involves the 1,3-dioxane ring corresponding to the B-ring system of sesbanimides.

In the synthesis by Pandit *et al.*, after elongation of carbon chain by treating **3** with the stabilized ylide, Micheal addition of *t*-butyl carbamoylacetate in the presence of potasium *t*-butoxide was found to effect simultaneous formation of the glutarimide ring to afford the ester (**15**) (**Fig 6**).

Fig 6

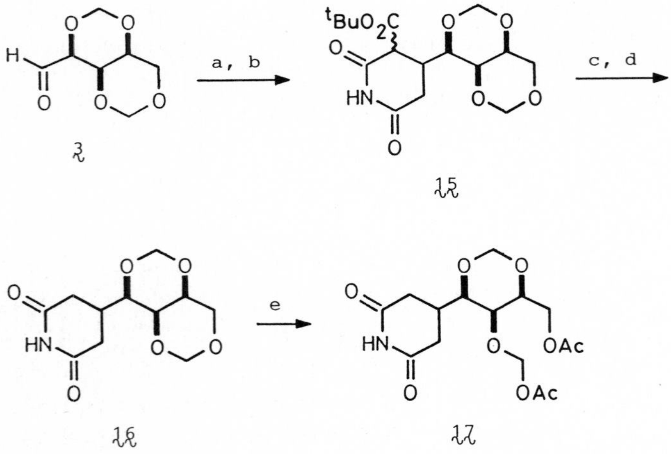

a) $Ph_3P=CHCO_2Me$, PhMe   b) $H_2NCOCH_2CO_2{}^tBu$, KO^tBu
c) CF_3CO_2H, rt   d) DMF, reflux, 45% (4 steps)
e) $36N-H_2SO_4$, Ac_2O, AcOH, 89%.

Decarboxylation followed by selective cleavage of the unnecessary 1,3-dioxane ring of the dimethyleneacetal (**16**) in acetic acid and acetic anhydride gave the glutarimide (**17**).[9a]

Fig 7

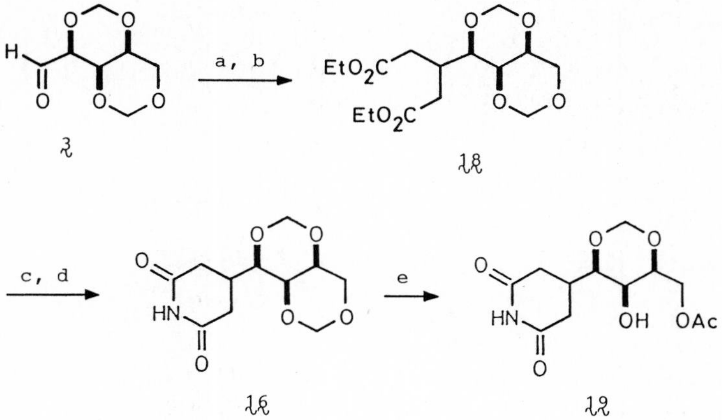

a) Na(^{i}PrO)$_2$POCHCO$_2$Et, 87% b) MgO$_2$CCHCO$_2$Et, 77%
c) NH$_4$OH, 50 °C, 24 h d) 155→210 °C, 40 min, 68%
(2 steps) e) (CF$_3$CO)$_2$O, AcOH, 22 °C, 6 h, 87%.

Schlessinger *et al.* introduced the carbon chain by olefimation of **3** with the sodium salt of the phosphonate (**Fig 7**). Micheal addition of magnesium monoethyl malonate to the resulting enester accompanyed concurrent decarboxylation, giving the diester (**18**). The glutarimide ring was constructed by the reaction of **18** with ammonium hydroxide followed by pyrolysis. Selective hydrolysis of the useless 1,3-dioxane ring was carried out by treatment with a mixture of acetic acid and trifluoroacetic anhydride to yield the glutarimide (**19**).[9c]

Other approaches to the AB-ring system of sesbanimides have been also reported (**Fig 8**). Fleet *et al.* employed the known acetonide (**20**) derived form D-glucose as the starting material.[5b] After opening of the furanoside ring, formation of the 1,3-dioxane ring produced **21** as the common intermediate. An enantiomeric pair of the glutarimides (**22** and **23**) was synthesized from **21** by elaborating the glutarimide ring from an aldehyde created from either the C-5 or C-1 carbon of D-glucose. In the synthetic studies of Shibuya, the glutarimide (**25**) was formed at the early stage of the synthesis from the known aldehyde (**24**) prepared from D-glucose.[5c] Opening of the furanoside ring followed by acetalization effected construction of the 1,3-dioxane ring to give **26**. Methods for constructing the AB-ring system employed in these two synthetic studies are

312

Fig 8

Fleet et al.

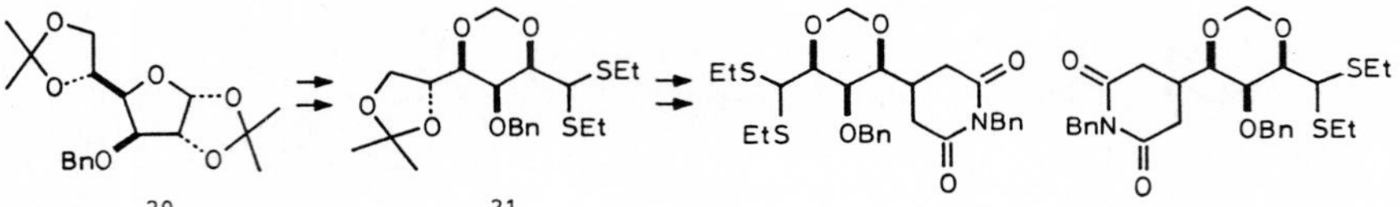

22   21   22   23

Shibuya

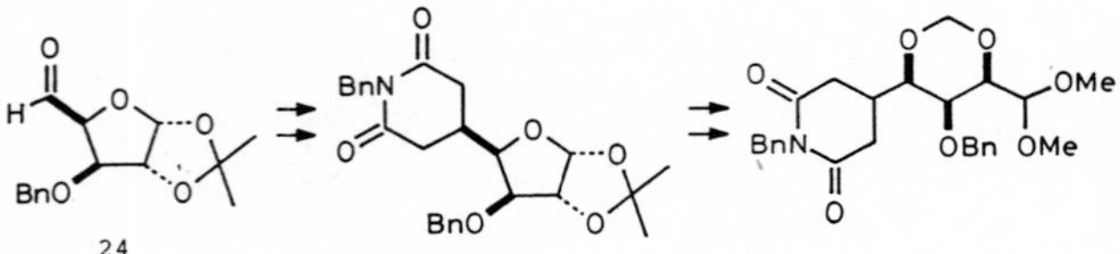

24   25   26

Roush et al.

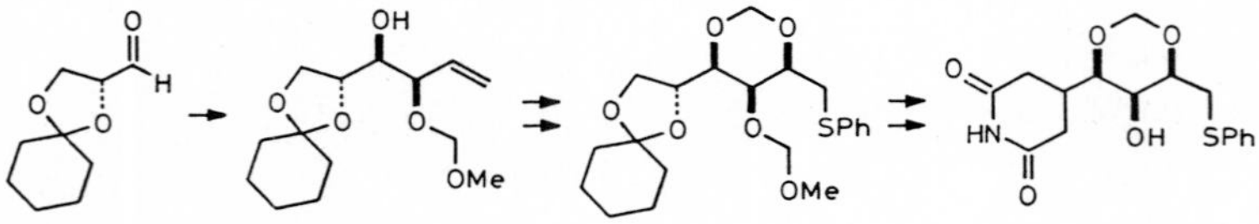

27   28   29   30

Fig 9

Tomioka, Koga

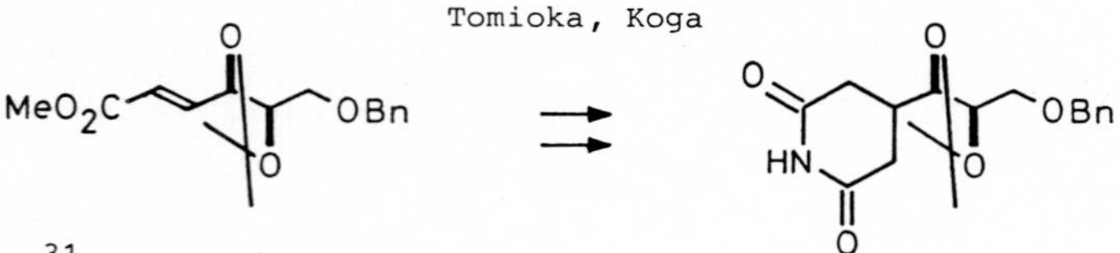

31   32

Just et al.

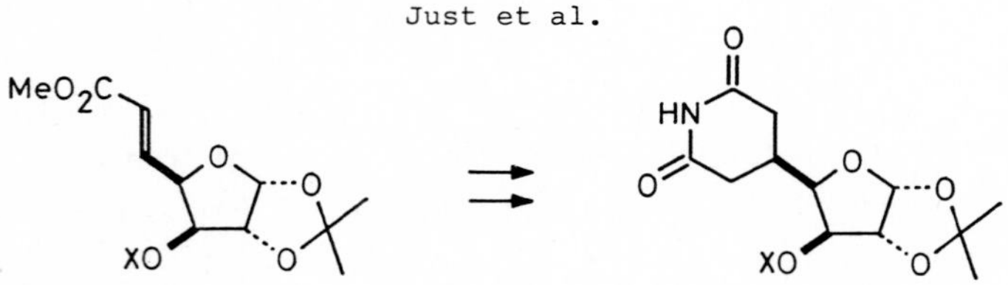

33   34

$X = Si^{t}BuPh_2$

very similar to ours. Roush *et al.* controlled the stereochemistry at the C-7, C-8, and C-9 positions of sesbanimides in an acyclic system.[5g] Thus, the three contiguous asymmetric centers of sesbanimides were introduced by stereoselective addition of the allylboronate to the glyceraldehyde (**27**) followed by selective epoxidation of the resulting homoallylic alcohol (**28**). Formation of the 1,3-dioxane ring afforded **29**, which was converted into the glutarimide (**30**). Kinoshita *et al.* also reported the synthesis of the AB-ring system only briefly.[5d]

Two groups devised other methods of constructing the carbon skeleton of the A-ring system which proceed without decarboxylation (**Fig 9**). Tomioka and Koga converted the enester (**31**) derived from D-tartaric acid into the glutarimide (**32**) employing novel Micheal addition of lithiated trimethylsilylacetonitirile as a key step.[5a] Just *et al.* developed Micheal addition reaction of the organic radical generated from iodoacetamide, which converted successfully the enester (**33**) prepared from D-glucose into the glutarimide (**34**).[5e]

4. Introduction of the C_5-Unit into the AB-Ring System

In order to construct the C-ring system, introduction of the C_5-unit into the C-10 position is required. Thus, the authors first converted the glutarimide (**12**) into the aldehyde (**39**) for this purpose (**Fig 10**). After reductive cleavage

Fig 10

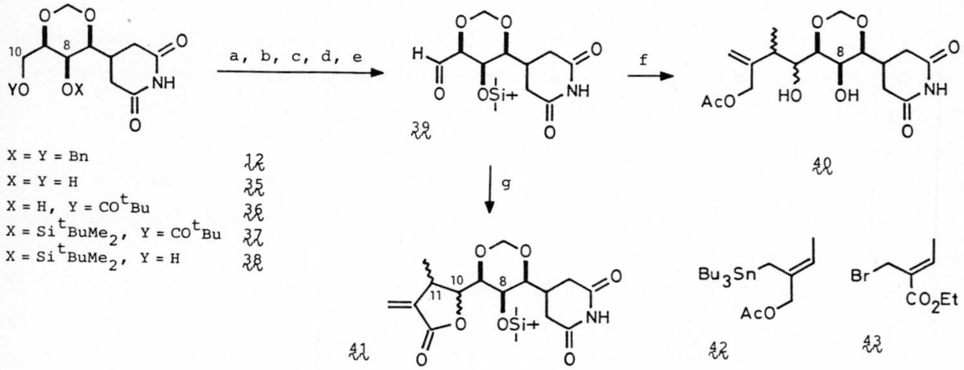

X = Y = Bn   **12**
X = Y = H   **35**
X = H, Y = CO^{t}Bu   **36**
X = Si^{t}BuMe$_2$, Y = CO^{t}Bu   **37**
X = Si^{t}BuMe$_2$, Y = H   **38**

a) H$_2$ (5 atm), Pd-C, AcOH, MeOH, rt, 2 h, 95%   b) ^{t}BuCOCl, Py, 0 °C, 2.5 h, 91%
c) ^{t}BuMe$_2$SiOTf, 2,6-Lu, CH$_2$Cl$_2$, rt, 10 min, 86%   d) ^{i}Bu$_2$AlH, CH$_2$Cl$_2$, -78 °C, 1 h, 87%
e) CrO$_3$·2Py, CH$_2$Cl$_2$, rt, 10 min, 84%   f) **42** (10 eq.), BF$_3$·OEt$_2$, CH$_2$Cl$_2$, -78 °C, 2 h, 65%
g) **43** (1.5 eq.), Zn, THF, reflux, 6 min, 73%.

of the two benzyl ethers, the primary and secondary hydroxyl groups present in
the resulting diol (35) were sequentially protected in forms of pivalate ester
and t-butyldimethylsilyl ether, respectively, to yield the silylpivalate (37).
Other protective groups for the secondary hydroxyl group were also examined.
However, methoxymethylation or 2-(trimethylsilyl)ethoxymethylation of the
pivalate (36) afforded low yields of the products since the starting material
(36) was unstable under the conditions for protection. Reductive cleavage of
the pivalate ester cleanly occurred without disruption of the glutarimide ring
by treating with diisobutylaluminumhydride, affording the alcohol (38). Collins
oxidation of 38 readily produced 39.

Interestingly, the glutarimide carbonyl groups exhibited absorption bands at
1670 cm^{-1} in the IR spectra of 37, 38, and 39. In contrast, in those of 12, 35,
and 36, absorptions due to the glutarimide carbonyl groups appeared at the
ordinary wave numbers around 1705 cm^{-1}. Detailed comparisons of the coupling
patterns recorded in the 400 MHz ^{1}H NMR spectra of 12, 36, 37, and 38 revealed
that in cases of 37 and 38 the glutarimide rings take the same distorted
conformations and, in contrast, the glutarimide rings of 12 and 36 are in the
stable chair conformations. Furthermore, comparing the IR and 400 MHz ^{1}H NMR
spectra of related compounds (40, 41, 53, 1, 2, 54, 55, and 58), it became
evident that the glutarimide rings of the silylated compounds where the C–8
hydroxyl groups are protected in forms of t-butyldimethylsilyl ethers always
take the common distorted conformations.

It is noteworthy that the bulky silyl ether distorted the remote glutarimide
ring rather than the proximate 1,3–dioxane ring. Such distortion may account for
the unusual lability of the t-butyldimethylsily ether at the C–8 position under
acidic conditions. Due to this reason, Lewis acid catalyzed addition reaction
of the allylstannane (42) to the aldehyde (39), the most promising method
developed during our model study, accompanied desilylation to only produce the
diol (40). After several experimentations to introduce the C$_5$–unit into 39, it
was finally found that Reformatsky reaction employing the bromoester (43)[12]
proceeded smoothly without any loss of the labile t-butlydimethylsilyl group to
give the lactone (41).

In the syntheses by Pandit et al. and Schlessinger et al., introduction of
the C$_5$–unit into the AB–ring system was achieved by employing the compounds
which bear other protective groups at the C–8 position. Pandit et al. found
that reductive cleavage of the benzylidene ring of 44 obtained from the
glutarimide (17) by successive hydrolysis and benzylidation, by triethylsilane
proceeded in a regioselective manner to yield the alcohol (45) in which the C–8
hydroxyl group was protected in the form of benzyl ether (Fig 11). After
oxidation, the aldehyde (46) thus obtained was allowed to react with the

Fig 11

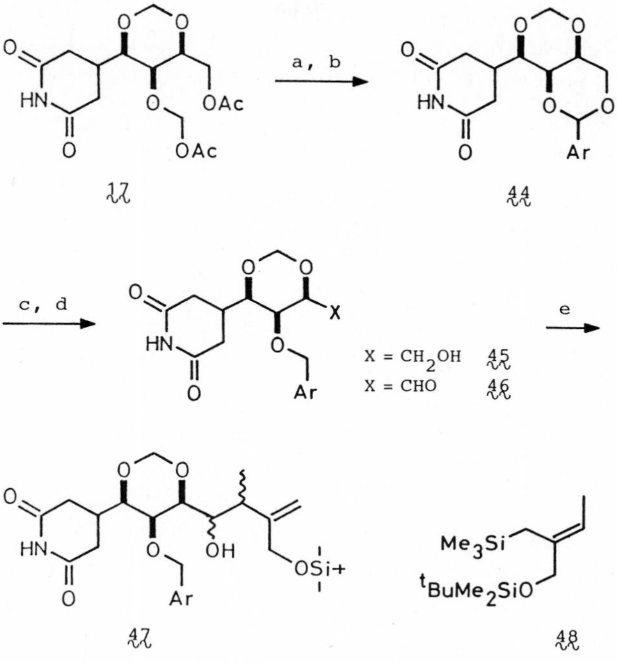

$Ar = C_6H_3(OMe)_2$

a) NaOMe, MeOH, -15 °C, 86% b) $(MeO)_2C_6H_3CHO$,
96% c) Et_3SiH, Et_2AlCl, CH_2Cl_2, -55 °C, 61%
d) CrO_3, Ac_2O, Py, rt, 65% e) **48**, $BF_3 \cdot OEt_2$
CH_2Cl_2, -90 °C, 50%.

allylsilane (**48**) in the presence of borontrifluoride etherate as Lewis acid to afford the alcohol (**47**).

Schlessinger *et al.* protected the C-8 hydroxyl group as a *t*-butyldiphenylsilyl ether using recently developed *t*-butyldiphenylsilyl trifluoromethanesulfonate[13] (**Fig 12**). After reductive removal of the acetate ester with diisobutylalumiumhydride, Swern oxidation of the resulting alcohol (**49**) gave the aldehyde (**50**). Addition reaction of the allylstannane (**52**) to **50** catalyzed by borontrifluoride etherate proceeded without a loss of the *t*-butyldiphenylsilyl group protecting the C-8 hydroxyl group, to afford the

Fig 12

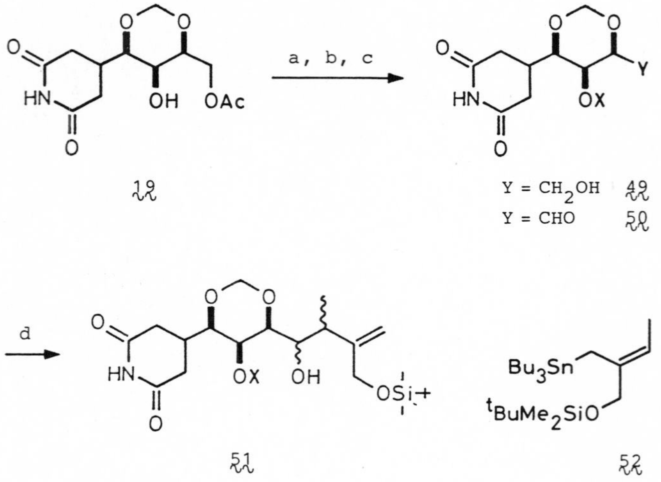

Y = CH$_2$OH **49**
Y = CHO **50**

X = Si^{t}BuPh$_2$

a) ^{t}BuPh$_2$SiOTf, 2,6-Lu, CH$_2$Cl$_2$, 100% b) ^{i}Bu$_2$AlH,
THF, -78 → 0 °C, 97% c) Swern Ox., 97% d) **52**
(2.0 eq.), BF$_3$·OEt$_2$, CH$_2$Cl$_2$, -78 °C, 5.5 h, 83%.

alcohol (**51**).

5. Construction of the C—Ring System: Total Synthesis of Sesbanimides

In our total synthesis, reduction of the *exo*-methylene-γ-lactone moiety was necessary at the next stage. Since direct reduction of the lactone (**41**) to the diol (**53**) with a hydride reagent gave unsatisfactory results, transformation of **41** to **53** was attempted in a stepwise manner (**Fig 13**). Thus, treatment of **41** with diisobutylaluminumhydride yielded the corresponding lactol, which without isolation was further reduced with sodium borohydride in the presence of cerium (III) chloride,[14] affording **53**. The diol (**53**) was derived to an almost 1:1 mixture of **1** and **2** by the sequence of (1) selective protection of the primary hydroxyl group of **53**, (2) Collins oxidation of the remaining secondary hydroxyl group, and (3) removal of the two silyl groups.

The mixture of **1** and **2** could be readily separated by silica gel TLC. The less polar epimer and its diacetate were identical with natural sesbanimide A

(1) and the authentic diacetate (54) derived from natural 1 by our hands, respectively, in all respects (mp, mmp, optical rotation, 400 MHz ^{1}H NMR, IR, MS, and TLC mobilities with several different solvent systems). Furthermore, synthetic and natural 1 exhibited almost the same magnitude of activity in P388

Fig 13

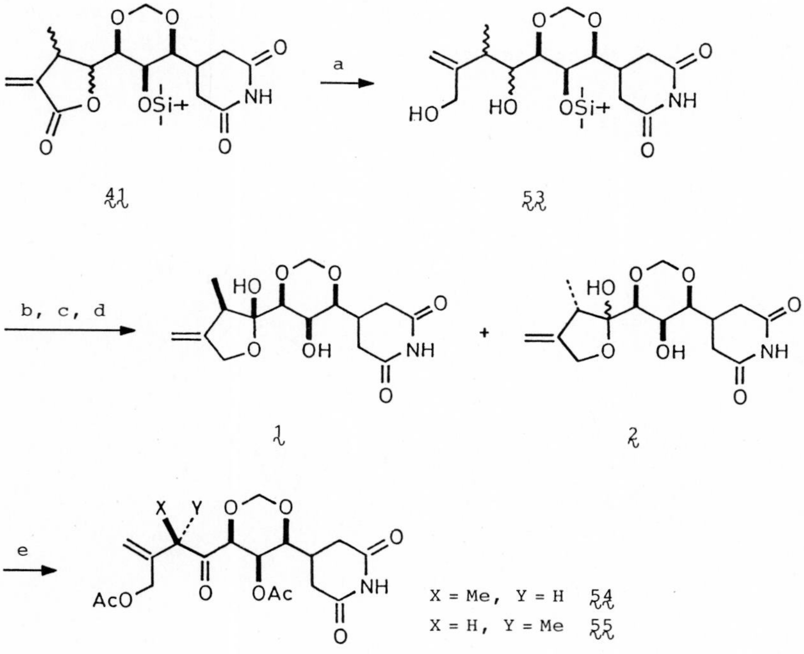

a) 1) ^{i}Bu$_2$AlH, CH$_2$Cl$_2$, -78 °C, 1 h 2) NaBH$_4$, CeCl$_3\cdot$7H$_2$O, MeOH, 0 °C, 10 min, 73% (2 steps) b) ^{t}BuPh$_2$SiCl, Imidazole, DMF, rt, 40 min c) CrO$_3\cdot$2Py, CH$_2$Cl$_2$, rt, 30 min d) Bu$_4$N$\cdot$F, THF, rt, 10 min, 16% (1, 3 steps), 19% (2, 3 steps) e) Ac$_2$O, Py, rt, 12 h, 53% (54), 61% (55).

murine leukemia *in vitro* cytotoxicity assay. Accordingly, our synthesis of natural sesbanimide A (1) obviously confirmed its absolute configuration. On the other hand, the more polar epimer and its diacetate were found to show the 400 MHz ^{1}H NMR spectra identical with those of natural sesbanimide B (2) and its diacetate (55), respectively. Other spectral data (IR and MS) of the synthetic compounds further supported their structures. Since 2 has been isolated from the same plants as that giving 1, its absolute configuration can be tentatively assinged as shown. Weak cytotoxicity was also observed for synthetic 2 in P388 murine leukemia *in vitro* assay. Therefore, sesbanimide B (2) synthesized by us was anticipated to be identical with natural 2 even if comparison of the optical rotation could not be carried out due to lack of the reported rotation value.

In the total synthesis of Pandit *et al.*, the major epimer of the alcohol (47) was oxidized after separation of stereoisomers resulting from the C-11 methyl group. The ketone thus produced was converted to unnatural (−)-sesbanimide A (56) upon removal of the protecting groups (**Fig 14**).

Fig 14

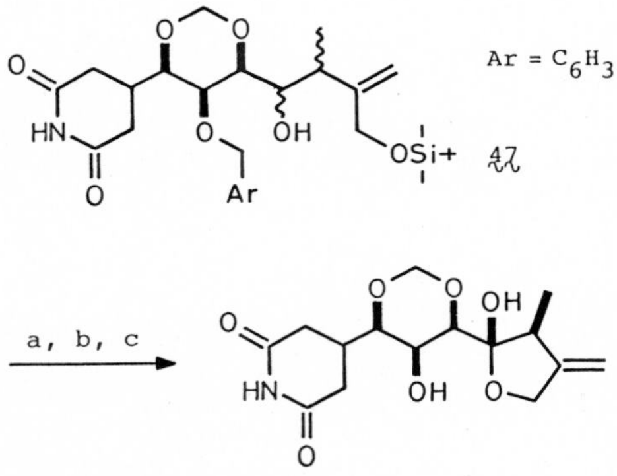

$Ar = C_6H_3(OMe)_2$

a) CrO_3, Ac_2O, Py, rt, 60% b) DDQ, CH_2Cl_2, rt c) AcOH, THF, H_2O, rt, 50% (2 steps).

Similarly, Schlessinger *et al.* also separated the stereoisomers of the alcohol (51) concerning the C-11 methyl group. Swern oxidation and deprotection converted each of the stereoisomers into unnatural (−)-sesbanimide A (56) and (+)-sesbanimide B (57) (**Fig 15**). In this case, the acidic conditions employed to cleave the *t*-butyldiphenylsilyl ether were weaker than those usually required to remove a *t*-butyldiphenylsilyl group.[15] This may suggest that the *t*-butyldiphenylsilyl ether present at the C-8 position distorts the glutarimide ring in a similar manner to those suggested for 37, 38, 39, 41, and 53.

Fig 15

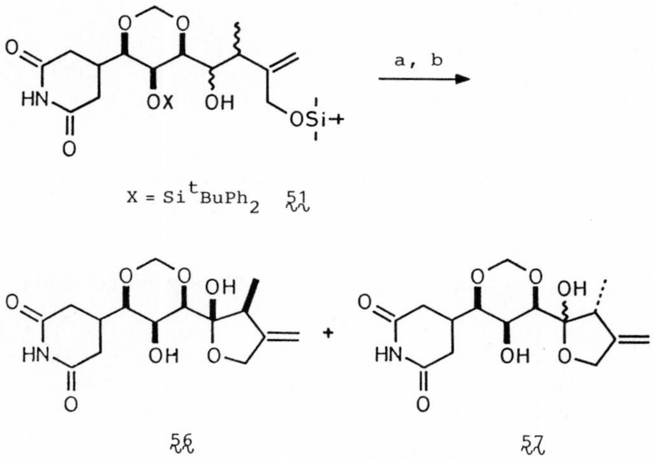

X = Si^{t}BuPh$_2$ **51**

56 + **57**

a) Swern Ox. b) AcOH, THF, H$_2$O, 22 °C,
4.3 h, 89% (**56**, 2 steps), 89% (**57**, 2 steps).

In connection with introduction of the C$_5$-unit into the AB-ring system and formation of the C-ring system, Rama Rao *et al.* further reported the model study employing Zaitzev reaction in which the allylbromide was reacted with the aldehyde in the presence of activated zinc dust.

6. Antitumor Activity of Sesbanimides and Its Related Compounds

As described above, the authors have achieved the first total synthesis of natural (+)-sesbanimide A (**1**) and (−)-sesbanimide B (**2**). Next, the structure-activity relationships were studied by employing various structural types of the sesbanimide congeners which could be synthesized by application of the method developed in the course of our total synthesis. It should be underscored that studies on the contribution of each of the three rings of sesbanimides to their antitumor activity are only feasible by using the artificial compounds obtained by organic synthesis.

At first, antitumor activity of the lactone (**58**) and the diol (**40**) which involve all the carbon framwork of sesbanimides, was investigated (**Table 1**). Although considerable cytotoxicity was observed for **58** in P388 murine leukemia

Table 1 Antitumor Activity Against P388 Murine Leukemia

IC$_{50}$ 3.3 × 10^{-5} µg/ml
T/C 181% (32 µg/kg)[1]

IC$_{50}$ 3.1 × 10^{-2} µg/ml
T/C 161% (3.8 µg/kg)[1]

IC$_{50}$ 9.0 × 10^{-3} µg/ml
T/C 95% (10 mg/kg)

IC$_{50}$ 11 µg/ml

Table 2 *In Vitro* Antitumor Activity Against P388 Murine Leukemia

Compound	IC$_{50}$ (µg/ml)	Compound	IC$_{50}$ (µg/ml)
59	1.6	65	0.39
X = H 60	19	X = H 66	2.6
X = OMe 61	22	X = OMe 67	1.1
X = F 62	5.8	X = F 68	0.49
63	>25	69	0.95
64	>25	70	>30

in vitro assay, **58** shows no significant antitumor activity against P388 murine leukemia *in vivo*. The diol (**40**) exhibited no marked cytotoxicity in *in vitro* assay. Accordingly, it was anticipated that the C-ring system plays a key role for sesbanimides to exhibit antitumor activity. Thus, the ketones (**59–62**) which equilibrate with the hemiacetals corresponding to the C—ring system, were prepared by applying the explored synthetic method to the various aldehydes. Similarly, the tetrahydronfurans (63 and 64) consisting of the BC—ring system were synthesized from the trimethylenesorbitol (13) which was obtained as the major product by direct acetalization of D-sorbitol as described previously. All of these compounds (59-62) showed no significant antitumor activity when subjected to P388 murine leukemia *in vitro* assay, though slight cytotoxicity was observed for the corresponding lactones (65-69) (**Table 2**). Finally, antitumor activity of the AB—ring system was examined. Comparing antitumor activity of sesbanimide A (**1**) and sesbanimide B (**2**) with those of **58** and **40**, it was anticipated that the C-10 ketonic function is indespensable to antitumor activity. Accordingly, the ketone (**70**) was synthesized from the aldehyde (**39**), the key intermediate of our sesbanimides synthesis, by the use of allylmagnesium chloride. Contrary to our expectation, no marked cytotoxicity was observed for **70** in P388 murine leukemia *in vitro* assay.

Summing up the results accumulated by the studies on structure—activity relationships, it appears evident that sesbanimides show marked antitumor activity owing to cooperation of their A— and C—ring and lack of the either rings causes complete loss of activity.

7. Summary

As described above, comparison of the three independent total syntheses of sesbanimides reveals the similarity of their synthetic strategies and tactics. Interestingly, the most characteristic feature of these syntheses seems to be introduction of the C_5—unit into the AB—ring system, the most difficult step in the total synthesis. Sesbanimides are reported to show serious toxicity along with potent antitumor activity. This toxicity seems to be main disadvantage for further development of sesbanimides as antitumor agents. The authors hope that these synthetic studies will be helpful to overcome this disadvantage.

Acknowledgement

We are grateful to Dr. P.G. Powell, U.S. Department of Agriculture, for providing us with an authentic sample and spectral data of **1, 54,** and **55**. We also thank Dr. K. Sakai, Misses K. Yamada and N. Hida, Sagami Chemical Research Center, Drs. S. Tsukagoshi and T. Tashiro, Cancer Chemotherapy Center, Japanese Foundation for Cancer Research, for evaluation of antitumor activity. Thanks are given to Dr. M. Kawasaki and Miss M. Ohsaki, Sagami Chemical Research Center, for their technical assistance.

References

1) J.M. Kingsbury, *"Poisonous Plants of the United States and Canada,"* Prentice-Hall, Englewood Cliffs, New Jersey, p353-357 (1964).

2) R.G. Powell, C.R. Smith, Jr., R.V. Madrigal, *Planta Med.,* **30,** 1-8 (1976); R.G. Powell, C.R. Smith, Jr., *J. Nat. Prod.,* **44,** 86-90 (1981).

3) a) R.G. Powell, C.R. Smith, Jr., D. Weisleder, G.K. Matsumto, J. Clardy, J. Kozlowski, *J. Am. Chem. Soc.,* **105,** 3739-3741 (1983). b) R.G. Powell, C.R. Smith, Jr., D. Weisleder, *Phytochemistry,* **23,** 2789-2796 (1984). c) P.G. Powell, C.R. Smith, Jr., *USP,* 4,534,327 (1985).

4) C.P. Gorst-Allman, P.S. Steyn, R. Vleggaar, *J. Chem. Soc., Perkin Trans. I,* 1311-1314 (1984).

5) a) K. Tomioka, K. Koga, *Tetrahedron Lett.,* **25,** 1599-1600 (1984). b) G.W.J. Fleet, T.K.M. Sing, *J. Chem. Soc., Chem. Commun.,* 835-837 (1984). c) M. Shibuya, *Heterocycles,* **23,** 61-63 (1985). d) T. Kinoshita, K. Okamoto, J. Clardy, *Synthesis,* 402-403 (1985). e) G. Sacripan, C. Tan, G. Just, *Tetrahedron Lett.,* **26,** 5643-5646 (1985). f) A.V. Rama Rao, J.S. Yadav, A.M. Naik, A.G. Chaudhary, *Tetrahedron Lett.,* **27,** 993-994 (1986). g) W.R. Roush, M.R. Michaelides, *Tetrahedron Lett.,* **27,** 3353-3356 (1986).

6) F. Matsuda, M. Kawasaki, S. Terashima, *Tetrahedron Lett.,* **26,** 4639-4642 (1985); F. Matsuda, S. Terashima, *Tetrahedron Lett.,* **27,** 3407-3410 (1986).

7) M.J. Wanner, G.-J. Koomen, U.K. Pandit, *Heterocycles,* **22,** 1483-1487 (1984); M.J. Wanner, N.P. Willard, G.-J. Koomen, U.K. Pandit, *J. Chem. Soc., Chem. Communm.,* 396-397 (1986); M.J. Wanner, N.P. Willard, G.-J. Koomen, U.K. Pandit, *Tetrahedron,* **43,** 2549-2556 (1987).

8) R.H. Schlessinger, J.L. Wood, *J. Org. Chem.,* **51,** 2621-2623 (1986).

9) a) A.T. Ness, R.M. Hann, C.S. Hudson, *J. Am. Chem. Soc.,* **66,** 665-670 (1944). b) E.J. Bourne, L.F. Wiggins, *J. Chem. Soc.,* 517-521 (1944). c) E.J. Bourne, J. Burdon, J.C. Tatlow, *J. Chem. Soc.,* 1864-1870 (1959).

10) C.A. Lobry de Bruyn, W.A. van Ekenstein, *Recueil Trav. Chim. Pays-Bas,* **22,** 159-165 (1903).

11) P.A. Levene, A.L. Raymond, *J. Bio. Chem.,* **102,** 317-330 (1933).

12) H.M.R. Hoffmann, J. Rabe, *Angew. Chem. Int. Ed. Eng.,* **22,** 795-796 (1983).

13) A.R. Basindale, T. Stout, *J. Organmet. Chem.,* **271,** C1-C3 (1984).

14) J.-L. Luche, *J. Am. Chem. Soc.,* **100,** 2226-2227 (1978); A.L. Gemal, J.-L. Luche, *J. Am. Chem. Soc.,* **103,** 5454-5459 (1981).

15) T.W. Greene, *"Protective Groups in Organic Synthesis,"* John Wiley & Sons, New York, p47-48 (1981).

THE SYNTHESIS OF PYRROLIDINE-CONTAINING NATURAL PRODUCTS VIA [3+2] CYCLOADDITIONS

WILLIAM H. PEARSON

1. INTRODUCTION

A substantial number of natural products have a pyrrolidine ring as part of their constitution. Synthetic methods which allow the rapid construction of the pyrrolidine ring are particularly useful. More specifically, methods which form more than one ring bond in a single operation are most efficient. Conceptually, there are six possible two-bond constructions which may be grouped into two general categories; namely, the [4+1] and [3+2] approaches (Figure 1). Although the [4+1] approach has been utilized successfully by our group (ref. 1) as well as others (ref. 2), this review will concentrate on the [3+2] construction. The disconnection highlighted by the box appears to be the most generally useful, although examples of alternative [3+2] methods exist (ref. 3).

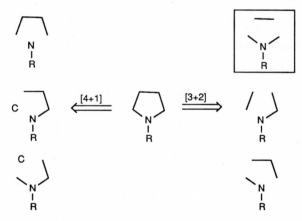

Figure 1. The six possible two-bond disconnections of the pyrrolidine ring.

Reactions in which two ring bonds are formed in one operation most readily fall into the cycloaddition category. Whereas the Diels-Alder reaction has been of major importance in the synthesis of both carbocyclic and heterocyclic six-membered rings, the isoelectronic [π4s+π2s] cycloadditions to form five-membered rings has only recently begun to achieve a useful status for the synthesis of natural products. For the [3+2] approach considered in this review, we require an olefin as the 2π-electron component and a 2-azaallylic anion system as the 4π-electron fragment (Figure 2). When X is a lone pair, the cycloaddition is that of a 2-azaallyl anion, leading to a pyrrolidine anion as the product (reaction a). When X is a substituent H or R, the reaction is that of a neutral 1,3-dipole (most commonly an azomethine ylide), leading to a neutral pyrrolidine (reaction b). In both cases, a 2-azaallylic anion orbital is undergoing reaction. This review will cover both of these scenarios for pyrrolidine ring construction, focusing on their use in natural products synthesis.

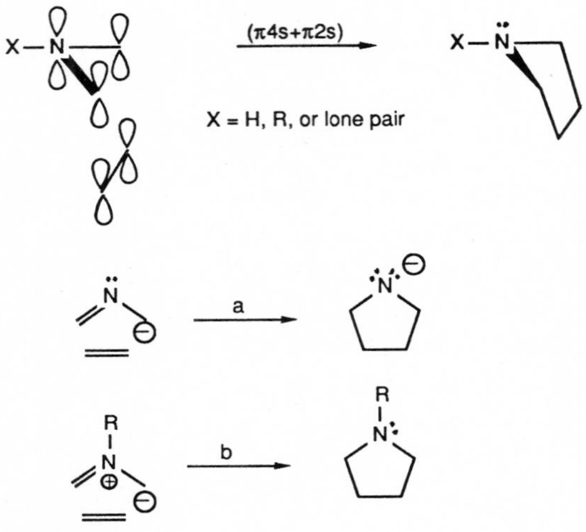

Figure 2. Concerted [3+2] cycloadditions of 2-azaallyl fragments.

2. THE 1,3-DIPOLAR CYCLOADDITION APPROACH

Azomethine ylides have proven to be exceptionally useful for the synthesis of pyrrolidines (ref. 4). Their utility in natural products synthesis is reviewed below and is organized by method of dipole generation. Related 1,3-dipoles (e.g., munchnones and oxidopyridiniums) will also be covered.

2.1 Generation of Azomethine Ylides

The major methods for the generation of azomethine ylides are shown in Figure 3. They are:

(1) Desilylation of N-(trimethylsilyl)methyliminium ions (eq. 1).
(2) Ring opening of aziridines (eq. 2).
(3) Azomethine ylides derived from α-amino acids and their esters (eqs. 3,4).
(4) Treatment of amine N-oxides with strong base (eq. 5).

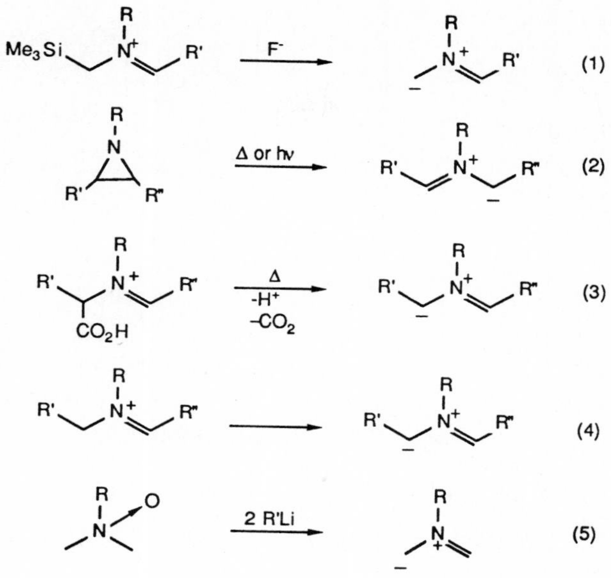

Figure 3. Major routes to azomethine ylides.

2.2 Desilylation of N-(Trimethylsilyl)methyl Iminium Ions

By far the most useful method for the generation of azomethine ylides for natural products synthesis is the fluoride ion triggered desilylation of N-(trimethylsilyl)methyl iminium ions (eq. 1, Figure 3). Vedejs pioneered this area and has recently published an excellent review on the subject (ref. 5). The key features of this approach are the generation of azomethine ylides under mild conditions (i.e., room temperature) and without the necessity of activating/stabilizing groups. The following applications to alkaloid synthesis attest to the utility of this method.

Rapid access to the pyrrolizidine alkaloids has been demonstrated by Vedejs (refs. 5,6) in his syntheses of retronecine **4** and indicine **5** (Figure 4). O-Methylation of lactam **1** gave an intermediate iminium ion, which was desilylated with cesium fluoride to generate azomethine ylide **2**. In situ trapping of **2** with methyl acrylate followed by methanol loss gave **3**, which could be converted to the alkaloids **4** and **5**.

Later, Sekiya (ref. 7) reported a similar approach to the pyrrolizidine alkaloids, beginning with 1-pyrroline trimer **6** (Figure 5). Trachelanthamidine, supinidine, and isoretronecanol could be prepared from intermediate **7**.

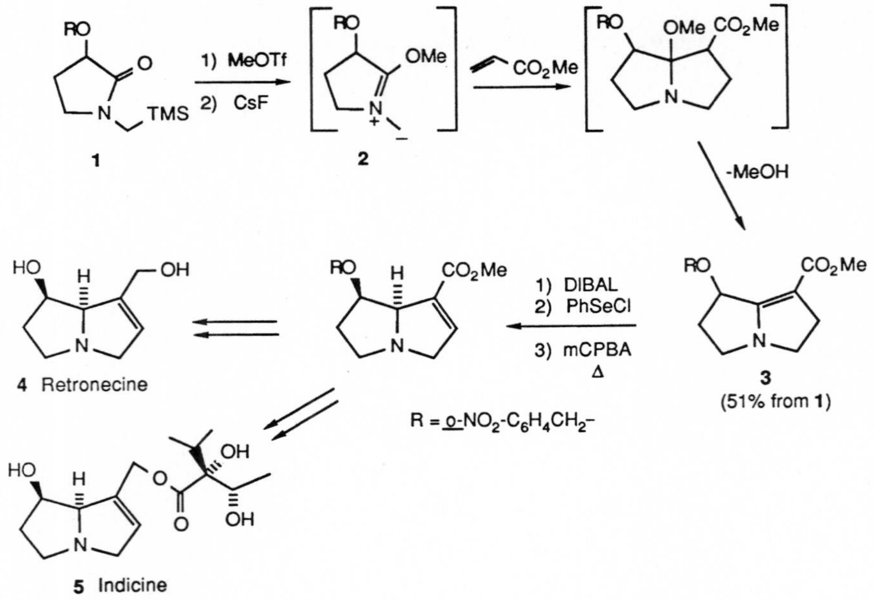

Figure 4. Vedejs' pyrrolizidine alkaloid synthesis.[5,6]

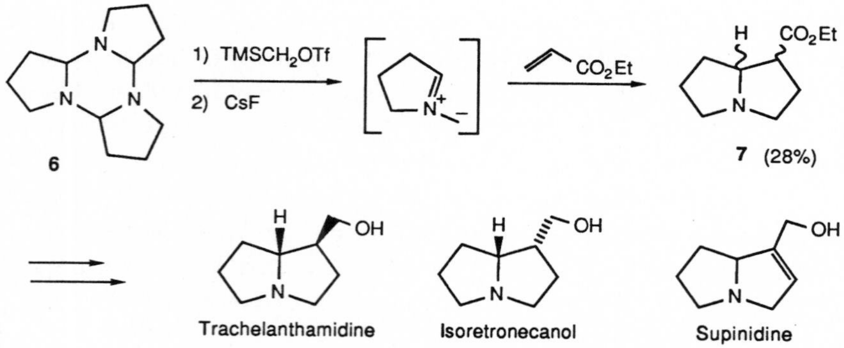

Figure 5. Sekiya's pyrrolizidine alkaloid synthesis.[7]

Related syntheses of non-naturally occurring materials with the pyrrolizidine (refs. 6c, 8) or indolizidine (refs. 6c, 7) skeleta using the iminium ion desilylation route have been published.

Livinghouse (ref. 9) has prepared the neurotoxic physostigmine alkaloid eserethole 9 by an intramolecular cycloaddition of the imidate methylide 8 (Figure 6).

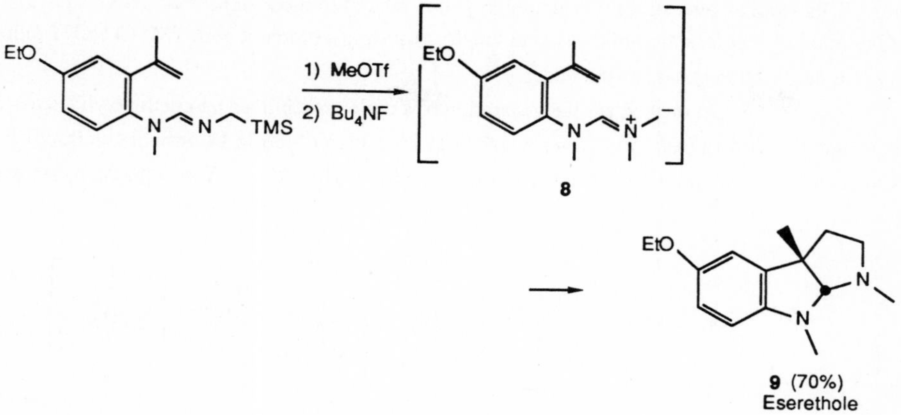

Figure 6. Livinghouse's eserethole synthesis.[9]

Livinghouse (refs. 9c, 10) has also been successful in the assembly of the erythrinane skeleton, represented by **11** and **12** (Figure 7). The carbonyl group present in the intermediate azomethine ylide **10** was necessary to prevent 1,5-hydrogen shift, which was the major pathway with ylide **13** (ref. 9c).

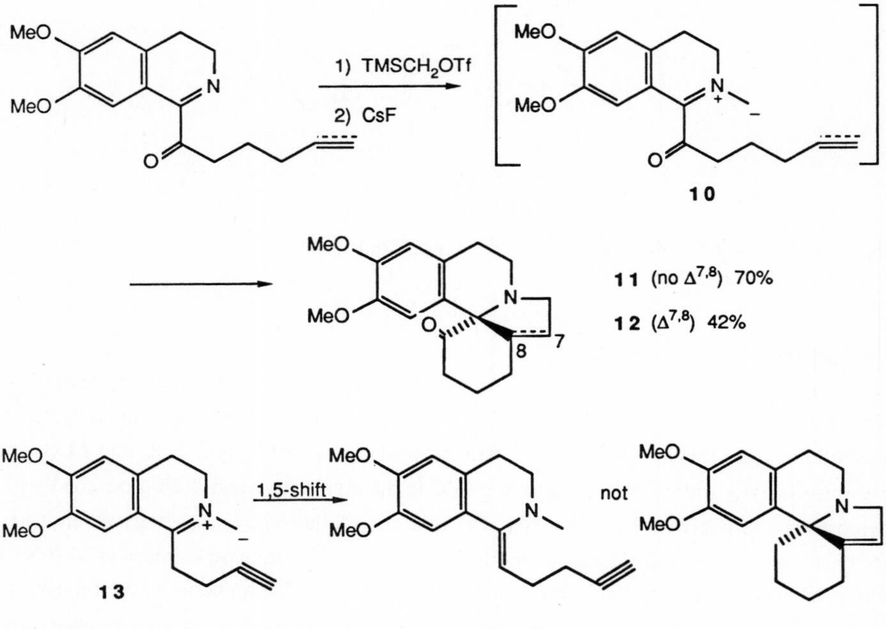

Figure 7. Livinghouse's synthesis of the erythrinane skeleton.[9c,10]

Padwa has also studied azomethine ylides related to those described above (ref. 11). N-Alkylation of imidates, thioimidates, and vinylogous versions thereof with TMSCH2OTf followed by CsF desilylation generates the ylides.

A very useful method for the generation of simple unstabilized azomethine ylides has been developed by Padwa (ref. 12). Silver ion assisted ionization of nitrile **14** provides an iminium ion, which suffers desilylation by fluoride ion, generating ylide **15**. This approach has seen

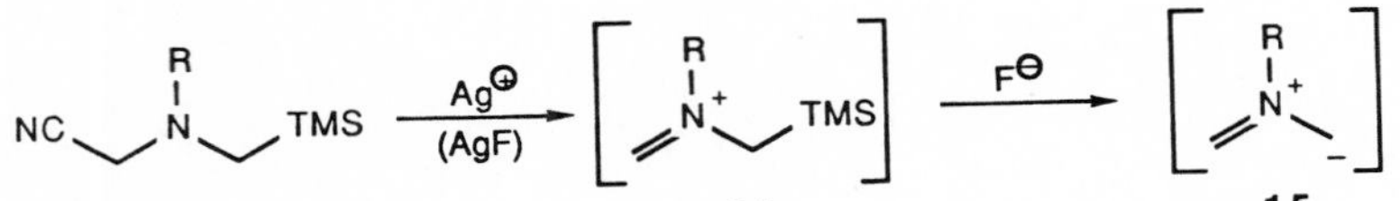

application to natural products synthesis by Parker and Padwa (ref. 13) in their preparation of the Reniera isoindole **16** (Figure 8).

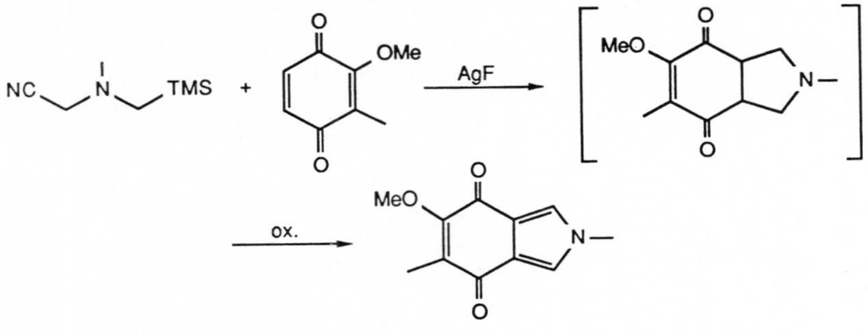

Figure 8. Parker and Padwa's synthesis of the Reniera isoindole.[13]

2.3 Ring Opening of Aziridines

The thermal and photochemical ring opening of aziridines has been reviewed (ref. 4). More recently, DeShong (ref. 14) has published the first example of the use of an aziridine-derived azomethine ylide for the synthesis of a natural product, namely, allo-kainic acid **21** (Figure 9). Thermolysis of aziridine **17** in toluene at 175°C in the presence of enone **18** proceeded with high stereoselectivity to afford the cycloadduct **19**. Five steps followed by base-catalyzed epimerization of **20** gave allo-kainic acid **21** and its epimer **22**. An attempt to prepare kainic acid **24** from the Z-enone **23** was unsuccessful due to Z → E isomerization of the enone at the relatively high temperature required for aziridine opening. Only pyrrolidines with a 3,4-trans relationship were observed.

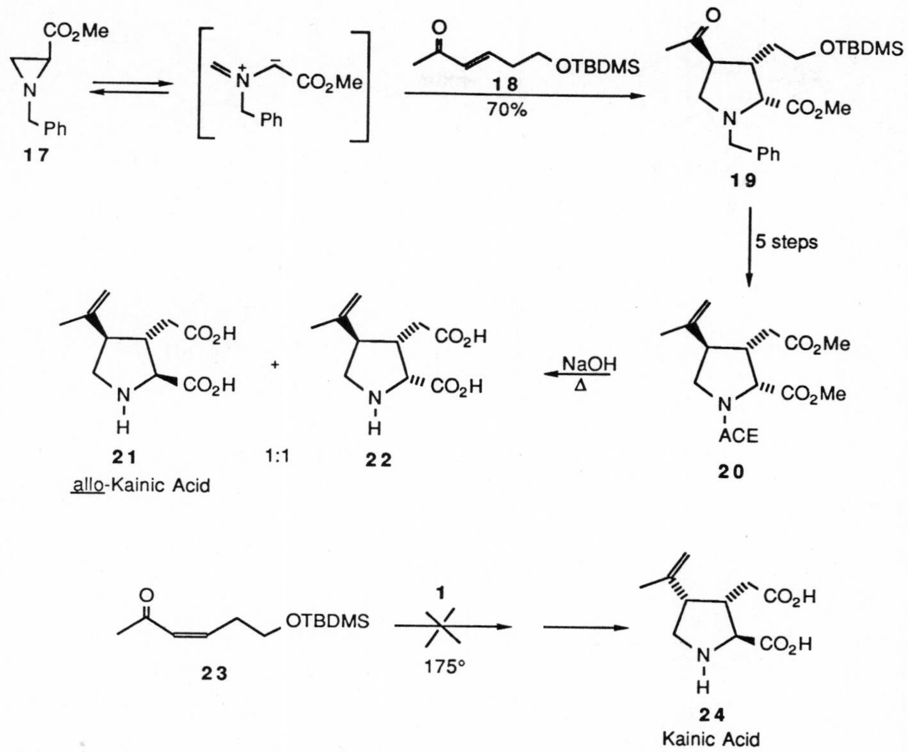

Figure 9. DeShong's <u>allo</u>-kainic acid synthesis.[14]

Takano (ref. 15) has prepared a related kainoid amino acid, acromelic acid A (**27**), by a similar route (Figure 10). Intramolecular dipolar cyclization of the azomethine ylide derived from **25** led to a single isomer **26**. Several steps, including an epimerization of an amino ester as in DeShong's work, gave **27**.

Other intramolecular azomethine ylide cycloadditions using the aziridine route have been reported by Padwa (ref. 16), DeShong (ref. 14b), Wenkert (ref. 17) and Eberbach (ref. 18). These examples illustrate the potential of aziridines for the synthesis of natural products or other pharmaceutically interesting bicyclic pyrrolidines (Figure 11). Note that high temperatures are not required for ring opening if the aziridine is sufficiently activated. Also, the favorable entropic nature of intramolecular cyclizations allows cycloaddition to occur with unactivated olefins, which has often been a stumbling block for intermolecular azomethine ylide cycloadditions.

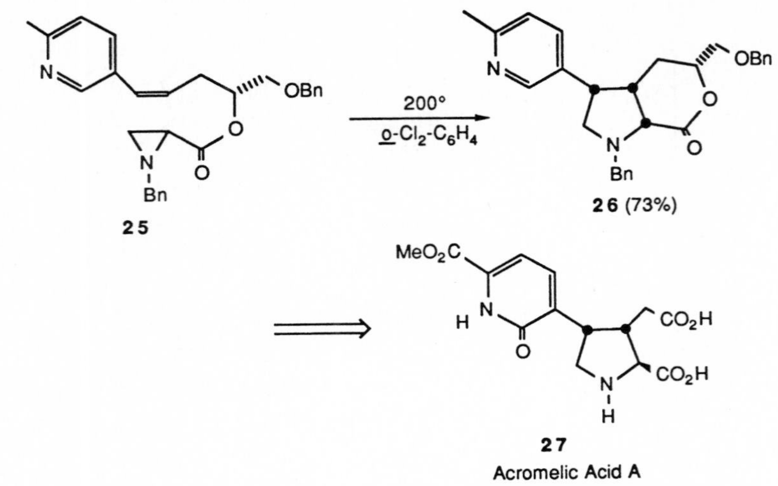

Figure 10. Takano's acromelic acid A synthesis.[15]

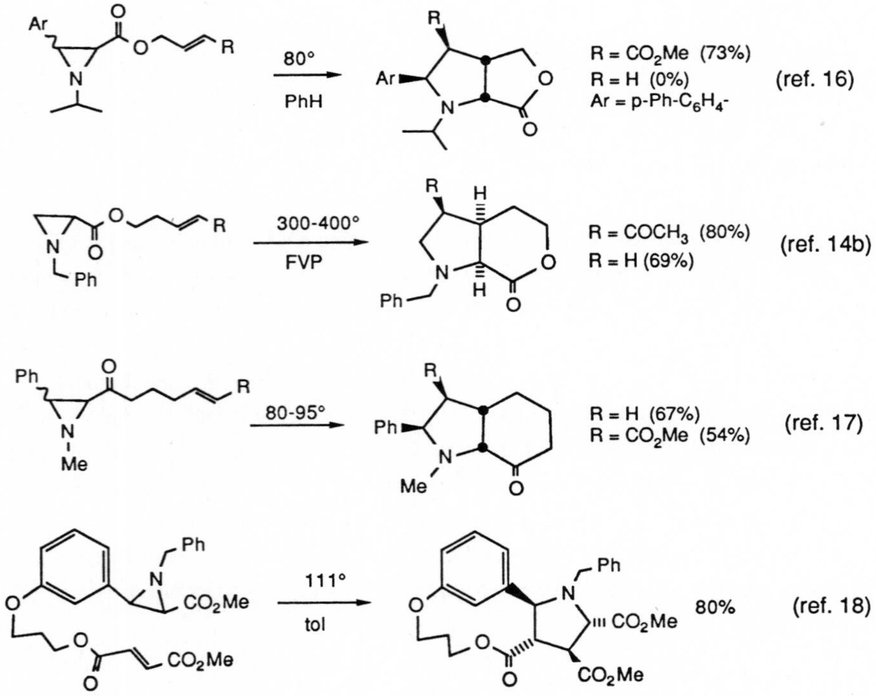

Figure 11. Intramolecular azomethine ylide cycloadditions via aziridine thermolysis.

Finally, DeShong has prepared an analog (**28**) of swainsonine (**29**) by the aziridine strategy (refs. 14b, 19) (Figure 12). This compound shows α-mannosidase inhibition activity similar to that of swainsonine.

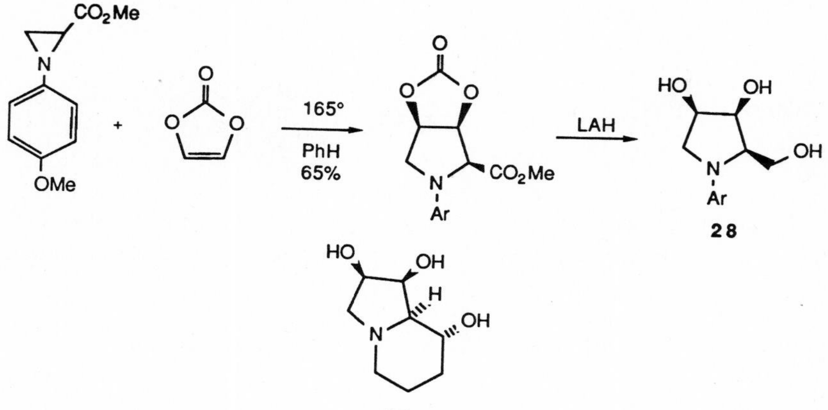

Figure 12. DeShong's synthesis of an α-mannosidase inhibitor.[14b,19]

To summarize, the aziridine route to azomethine ylides has been around for many years, but has only recently begun to be important for natural products synthesis. The disadvantage of the relatively high temperatures required for ring opening may be lessened by activating substituents and/or intramolecular cycloadditions.

2.4 Azomethine Ylides Derived from α-Amino Acids and Their Esters

α-Amino acids and their esters may be converted to azomethine ylides by condensation with aldehyes under various conditions. These methods may be grouped loosely into three categories. First, imines derived from α-amino acid esters may undergo a formal thermal 1,2-H shift to provide an ylide (eq. 6). Second, and closely related, is the condensation of N-alkyl α-amino acid esters with aldehydes (eq. 7). Finally, free N-alkyl α-amino acids or their trimethylsilyl esters undergo decarboxylation to azomethine ylides (eq. 8). These methods will each be discussed below, with particular attention to their application to natural products synthesis.

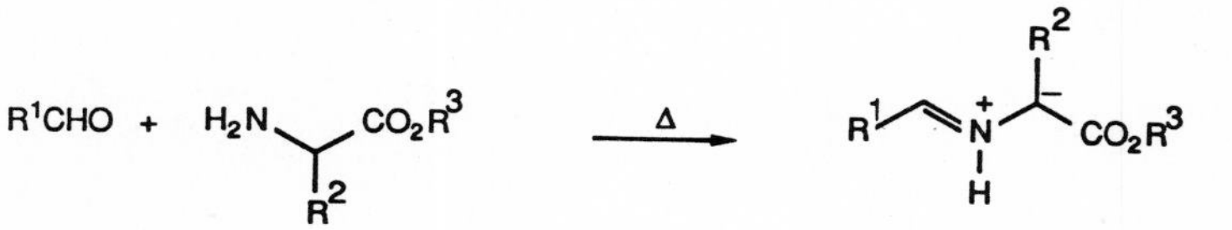

$$R^1CHO \ + \ H_2N \underset{R^2}{\overset{}{\diagup}} CO_2R^3 \quad \xrightarrow{\Delta} \quad R^1 \diagdown \overset{+}{\underset{H}{N}} \overset{R^2}{\underset{}{\diagup}} CO_2R^3 \qquad (eq. \ 6)$$

$$R^1CHO + R^2NH-\underset{R^3}{\overset{}{C}}H-CO_2R^4 \xrightarrow{\Delta} R^1-CH=\overset{+}{\underset{R^2}{N}}-\underset{R^3}{\overset{}{C}}H-CO_2R^4 \quad \text{(eq. 7)}$$

$$R^1CHO + R^2NH-\underset{R^3}{\overset{}{C}}H-CO_2H \xrightarrow[-H_2O]{\Delta \atop -CO_2} R^1-CH=\overset{+}{\underset{R^2}{N}}-\underset{R^3}{\overset{R}{C}}H \quad \text{(eq. 8)}$$

The prototropic generation of 1,3-dipoles (eq. 9) was first reported by Grigg (ref. 20) and Joucla (ref. 21a). The ease of generation of the dipole depends on the basicity of the central atom

$$X=\overset{..}{Y}-ZH \rightleftharpoons X=\overset{\overset{H}{|}}{\underset{}{Y}}\overset{\oplus}{}-\overset{\ominus}{Z} \quad \text{(eq. 9)}$$

Y and the pK_a of the ZH proton. For the generation of azomethine ylides (refs. 20-27, eq. 10), the formal 1,2-H shift proceeds at reasonable temperatures when imines derived from α-amino acid esters are used.

$$R-CH=N-\underset{R'}{\overset{}{C}}H-CO_2R'' \xrightarrow{\Delta} R-CH=\overset{\oplus}{\underset{\underset{H}{|}}{N}}-\overset{\ominus}{\underset{R'}{C}}-CO_2R'' \xrightarrow{A=B} R-CH\overset{A-B}{\underset{\underset{H}{N}}{}}\underset{R'}{\overset{}{C}}-CO_2R'' \quad \text{(eq. 10)}$$

Successful cycloaddition normally requires a dipolarophile bearing an electron-withdrawing group, although intramolecular versions may circumvent this requirement (refs. 22-25), for example, eq. 11.

$$\xrightarrow[\text{reflux}]{\text{xylene}} \quad \text{(eq. 11)}^{22}$$

A special feature of this method for azomethine ylide generation is the direct production of N-unsubstituted pyrrolidines, in contrast to most other methods (ref. 28). Limitations include the requirement for an electron-withdrawing substituent (for pK_a reasons) and the temperatures required for the 1,2-shift, although Grigg (ref. 22b) has observed cycloadditions at room temperature in the presence of an acid catalyst.

Although strictly speaking, Grigg's method has not been used for natural products synthesis to date, the very closely related version using (N-alkyl)-α-amino acid esters (eq. 7) has seen such activity. Confalone (refs. 24-26) has provided the only examples of the application of this methodology to alkaloid synthesis, although Joucla (ref. 21) has assembled the basic pyrrolizidine and indolizidine skeleta in a similar fashion.

A short synthesis of sceletium alkaloid A4 (32) has been reported by Confalone (ref. 25, Figure 13). Heating the aldehyde 30 with sarcosine ethyl ester in xylene at 180° provided the intramolecular cycloadduct 31. Rapoport's procedure (ref. 29) was required to remove the ester function after it had served its purpose, providing the target alkaloid 32.

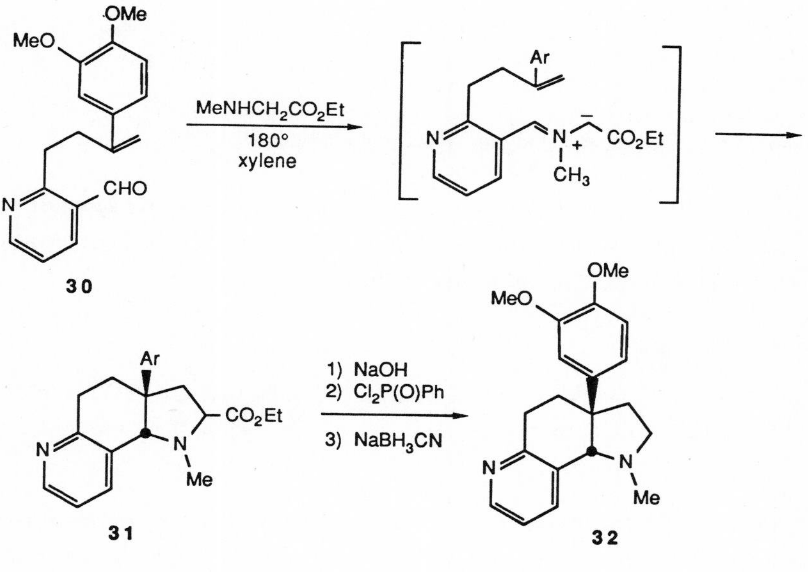

Figure 13. Confalone's sceletium alkaloid A4 synthesis.[25]

A model study for the synthesis of the antihypertensive alkaloid lycorenine 36 has also been reported by Confalone's group (Figure 14, ref. 26). Warming the aldehyde 33 with sarcosine ethyl ester in refluxing xylene containing a catalytic amount of camphorsulfonic acid gave the two regioisomeric cycloadducts 34 and 35. The dithiane serves a double purpose. For the synthesis of lycorenine, carbonyl functionality at that position would be useful. The dithiane also provides a nonenolizable aldehyde, which prevents competing condensation reactions.

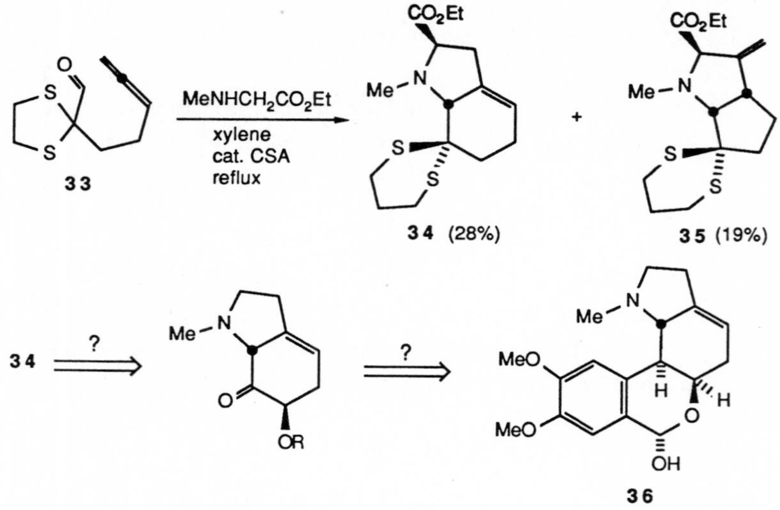

Figure 14. Confalone's model study for lycorenine synthesis.[26]

Other methods using α-amino acid esters and their derivatives have appeared (refs. 5, 30, 31) which utilize basic conditions to generate the 1,3-dipole (eq. 12). A major limitation of this approach is the necessity that R^1 and R^2 have no enolizable protons.

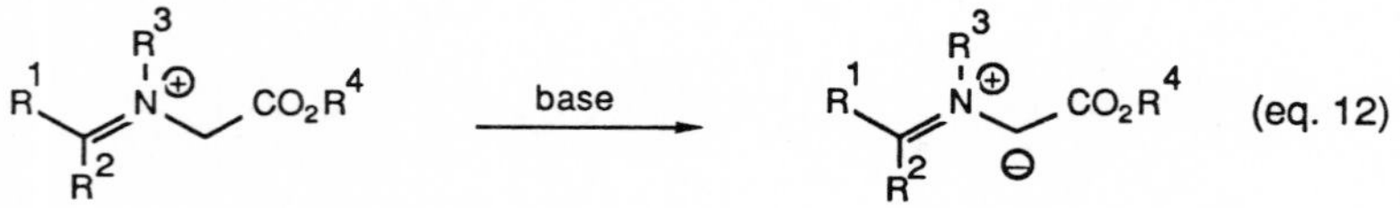

(eq. 12)

Kraus (ref. 30) has reported the synthesis of α-allo-kainic acid 21 by such a deprotonative approach (Figure 15). In this case, the enolization problem is solved by using an aromatic thiazole ring 37 as R^1/R^3. Combination of 37 with enone 39 in the presence of triethylamine gave the adduct 40, presumably via dipole 38. Several synthetic transformations afforded the natural product. Recall that DeShong (ref. 14) also prepared 21 by an azomethine cycloaddition (Figure 9).

Tsuge (ref. 31) has studied the related pyridinium methylides, which are derived from pyridinium salts by deprotonation with a mild base. Dipolar cycloaddition provides the tetrahydroindolizin-7-one skeleton (eq. 13).

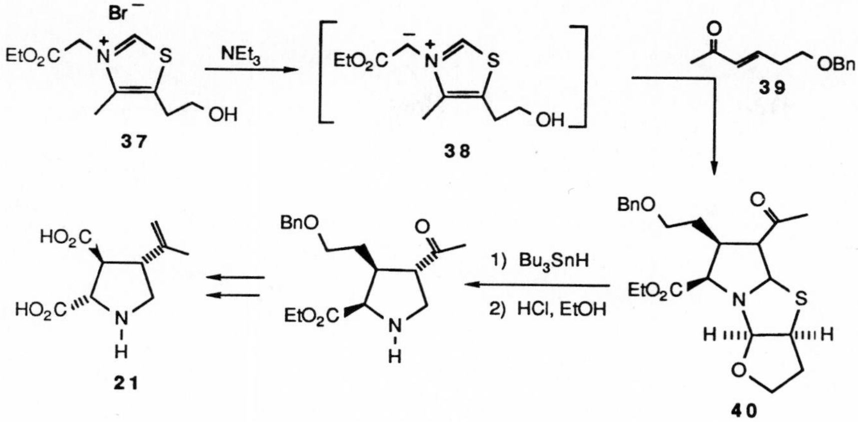

Figure 15. Kraus' α-allo-kainic acid synthesis.[30]

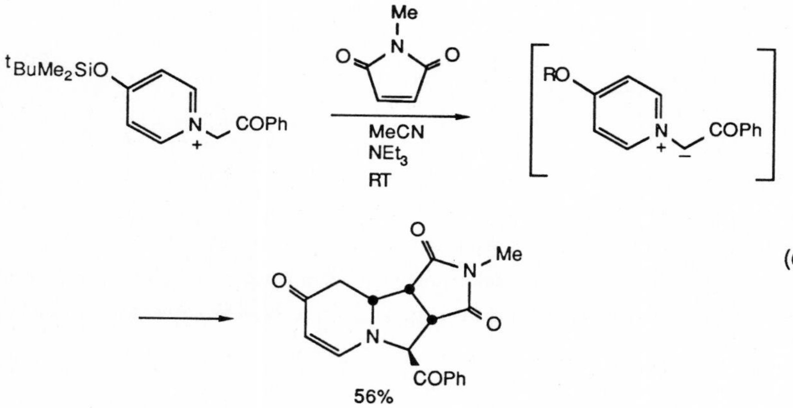

(eq. 13)

Achiwa (ref. 32) has assembled the pyrrolizidine and indolizidine skeleta by deprotonation of N-(phenylthio)methyl-α–amino acid esters (eq. 14).

Unstabilized azomethine ylides may be derived from imines of free α-amino acids or their trimethylsilyl esters. In 1970, Rizzi (ref. 33) proposed the intermediacy of an azomethine ylide in the carbonyl-assisted decarboxylation of sarcosine (eq. 15). Since that time, several groups have exploited this method for 1,3-dipolar cycloadditions (refs. 34-39). A key feature of this technique is the ability to generate unstabilized azomethine ylides by a simple procedure which does not require exotic reagents. A drawback has been its capricious behavior in cases where an enolizable imine is initially generated. Confalone, in particular, has noted a number of examples (particularly those using unactivated dipolarophiles) which fail or proceed in low yield, whereas the α–amino acid ester method works well in those cases (vide supra) (refs. 24-26, 35). However, as

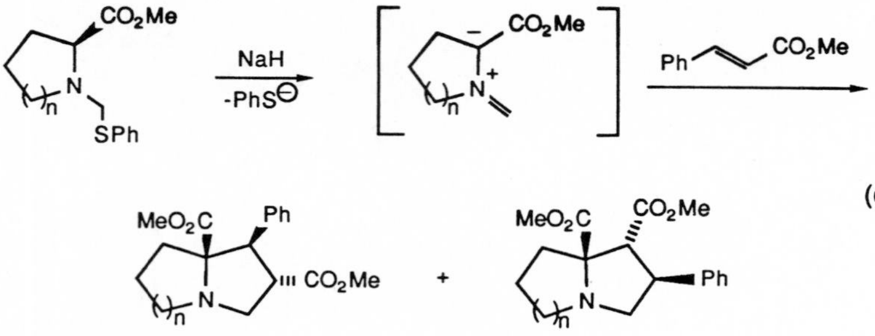

(eq. 14)

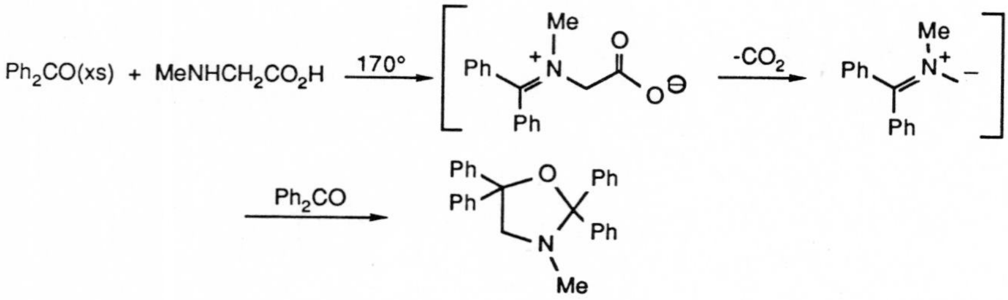

(eq. 15)

mentioned earlier, the α-amino acid ester method generates pyrrolidines bearing a carboalkoxy group, which must be removed by a three-step procedure if it is not desired. Tsuge (ref. 38) has shown that the decarboxylative method works quite well when an <u>activated</u> dipolarophile is used, even when an enolizable carbonyl compound is employed (e.g., eq. 16). Another particularly

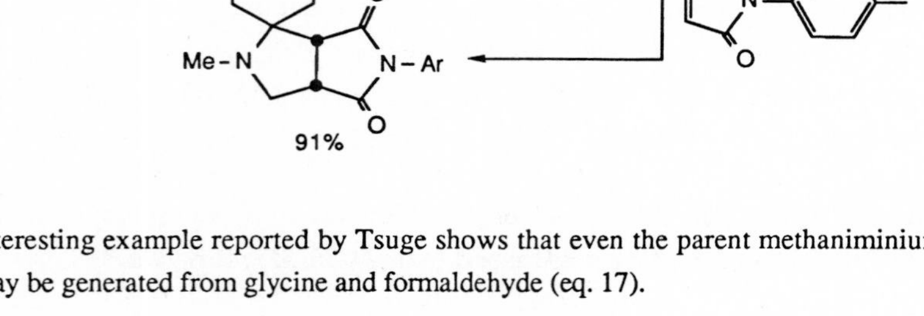

(eq. 16)

interesting example reported by Tsuge shows that even the parent methaniminium methylide **41** may be generated from glycine and formaldehyde (eq. 17).

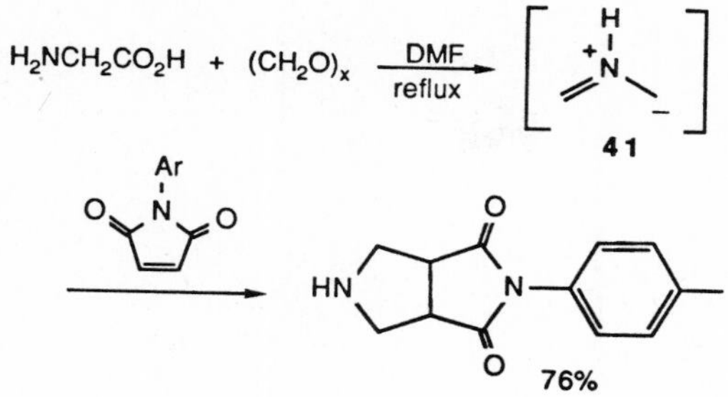

(eq. 17)

Application of the decarboxylative method to natural products synthesis has been reported by Confalone's group (refs. 25, 35) where α–lycorane (45) has been synthesized by an intramolecular cycloaddition (Figure 16). Heating a mixture of aldehyde 42 with N-benzylglycine and hexamethyldisilazane gave 44 as a single isomer, presumably via the unstabilized ylide 43. This was readily transformed to α-lycorane 45. Pelizzoni (ref. 36) has published an approach to the pyrrolizidine skeleton using similar chemistry.

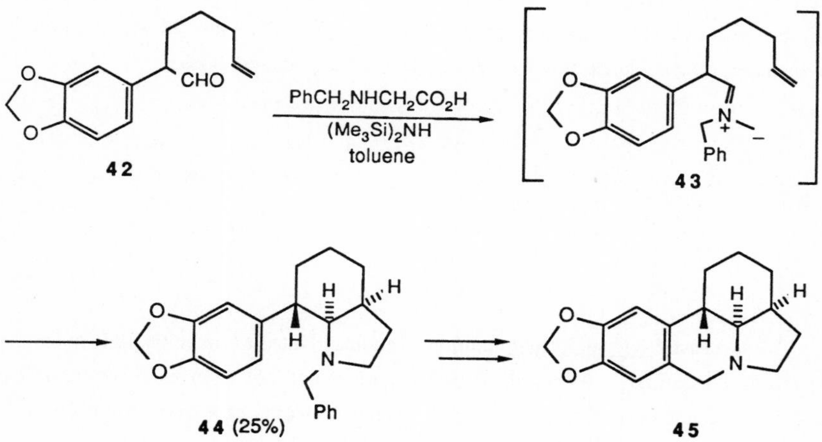

Figure 16. Confalone's α-lycorane synthesis.[35]

Joucla (ref. 40) has reported a route to azomethine ylides which is closely related to the decarboxylative method, namely the cycloreversion of oxazolidines (eq. 18). Generally, flash vacuum thermolysis is required. Eq. 19 shows the assembly of the basic skeleton of the pyrrolizidine alkaloids.

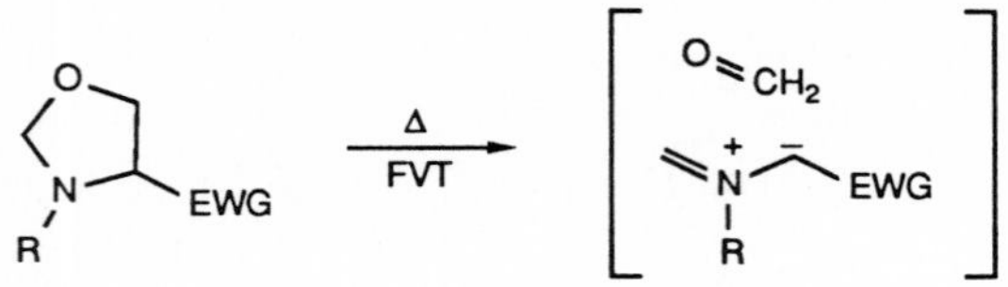

338

(eq. 18)

98%

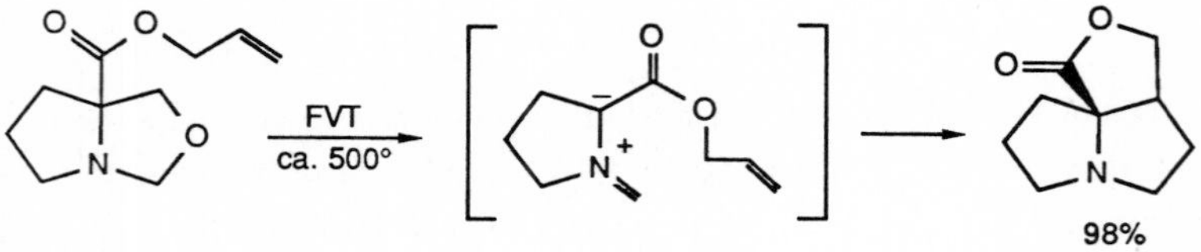

(eq. 19)

2.5 Treatment of Amine N-Oxides with Strong Base

In 1983, Roussi and coworkers reported a fascinating method for the generation of unstabilized azomethine ylides (refs. 41, 42, eq. 20), where, for example, trimethylamine N-oxide was simply treated with lithium diisopropyl amide (LDA). The ylide **46** was found to be highly reactive, undergoing cycloaddition with unactivated alkenes in good yield.

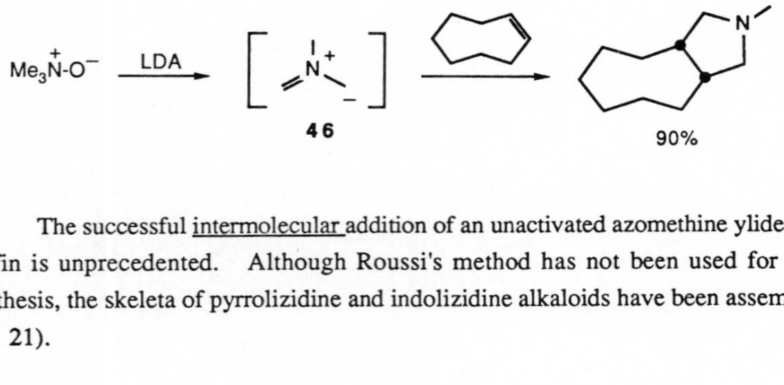

$Me_3\overset{+}{N}\text{-}O^-$ LDA

46

90%

(eq. 20)

The successful intermolecular addition of an unactivated azomethine ylide to an unactivated olefin is unprecedented. Although Roussi's method has not been used for natural products synthesis, the skeleta of pyrrolizidine and indolizidine alkaloids have been assembled (refs. 42b,c) (eq. 21).

LDA

n = 1, 79%
n = 2, 92%

(eq. 21)

2.6 Related Dipolar Cycloadditions

A number of natural products syntheses have been accomplished using 1,3-dipoles related to azomethine ylides. These will be briefly surveyed below.

Huisgen (ref. 43) prepared the pyrrolizidine skeleton in 1970 using his munchnones, which are essentially azomethine ylides. Pizzorno and Albonico (ref. 44) were the first to prepare a natural pyrrolizidine with these dipoles. Their short and efficient synthesis of the pyrrolizidine alkaloid isoretronecanol **49** is shown in Figure 17. Refluxing N-formyl-L-proline in acetic anhydride with ethyl propiolate gave the pyrrole **48** as a single isomer, presumably via munchnone **47**. Two reduction steps afforded racemic isoretronecanol **49**.

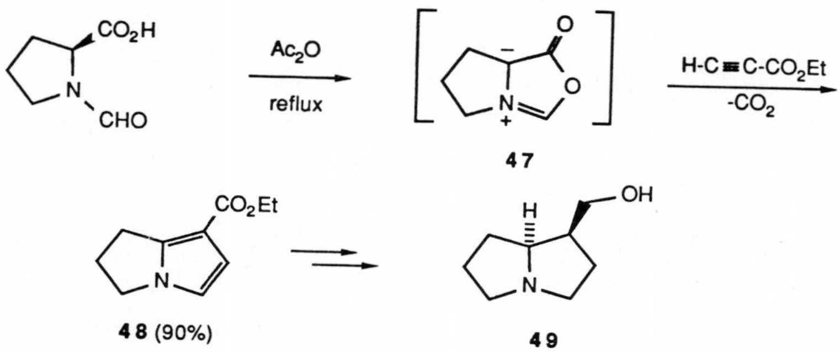

Figure 17. Pizzorno and Albonico's synthesis of isoretronecanol.[44]

Robins later used a similar route to prepare all six 1-hydroxymethyl pyrrolizidine alkaloids in optically pure form (ref. 45, Figure 18). The N,O-diformyl derivative **50** of natural (-)-4-hydroxy-L-proline was heated with ethyl propiolate in refluxing acetic anhydride to afford the pyrrole **51** in 80% yield. This key intermediate was then transformed into the alkaloids (+)-isoretronecanol **52**, (-)-trachelanthamidine **53**, (+)-laburnine **54**, (-)-isoretronecanol **55**, (+)-supidine **56**, and (-)-supinidine **57**.

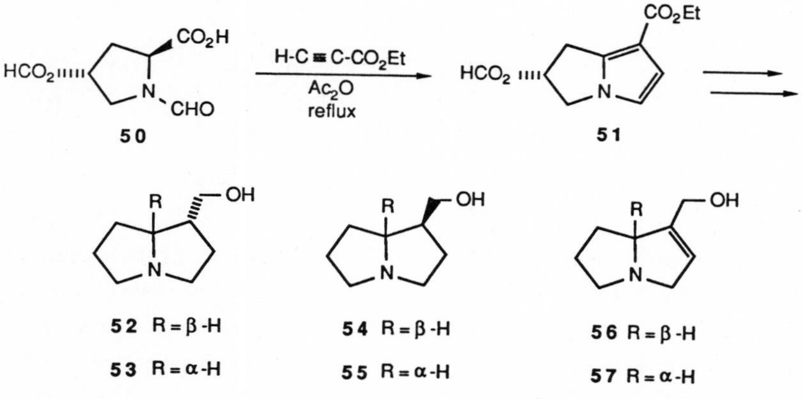

Figure 18. Robins' synthesis of optically pure pyrrolizidine alkaloids.[45]

Rebek and Hershenson (refs. 46-48) have attacked the mitomycin problem using munchnones. Rebek (ref. 47) has prepared mitosene **61** by the route outlined in Figure 19. Munchnone **59** was generated from **58** with acetic anhydride. In situ cycloaddition with DMAD gave the adduct **60**. Dieckman cyclization followed by a number of steps afforded mitosene **61**.

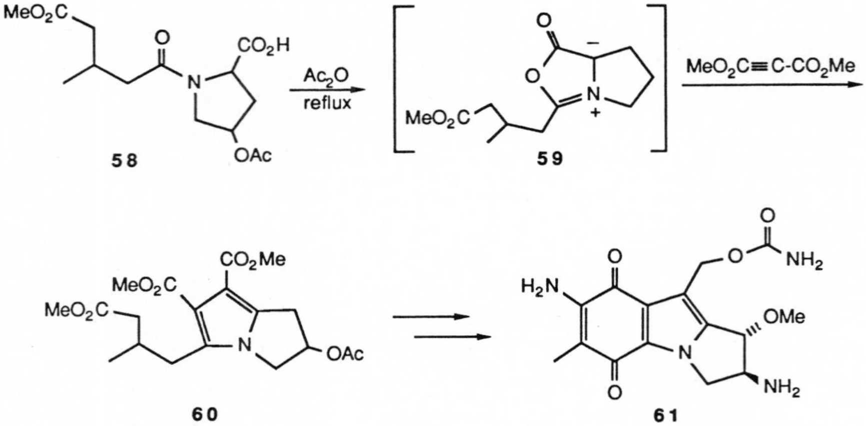

Figure 19. Rebek's synthesis of a mitosene.[47]

Oxidopyridiniums are related to azomethine ylides (refs. 49, 50) and undergo 1,3-dipolar cycloaddition reactions (eq. 22). Katritzky (ref. 49) has shown that the cycloadducts may be

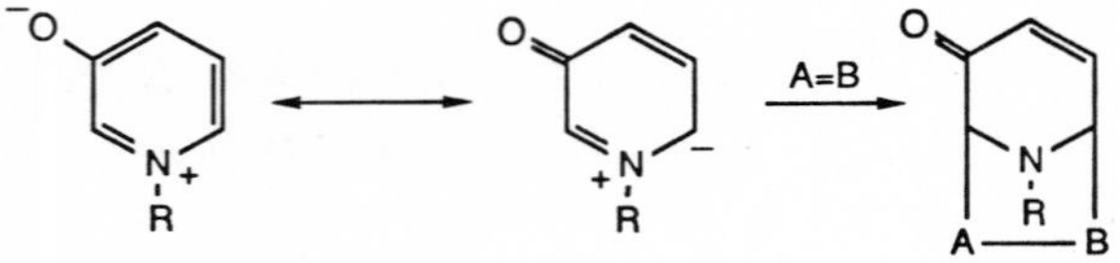

(eq. 22)

converted into tropones. Tamura (ref. 50) has used this method to prepare the naturally occurring tropolones stipitatic acid **65** and hinokitiol **69** (Figure 20). Dihydropyridones **62** and **66** were oxidized to oxidopyridiniums **63** and **67**, respectively. These underwent smooth dipolar cycloaddition with ethyl propiolate or methyl acrylate, respectively, providing adducts **64** and **68**. Katritzky's method for unravelling the adducts to tropolones was used, affording the natural materials **65** and **69**, after a few additional steps.

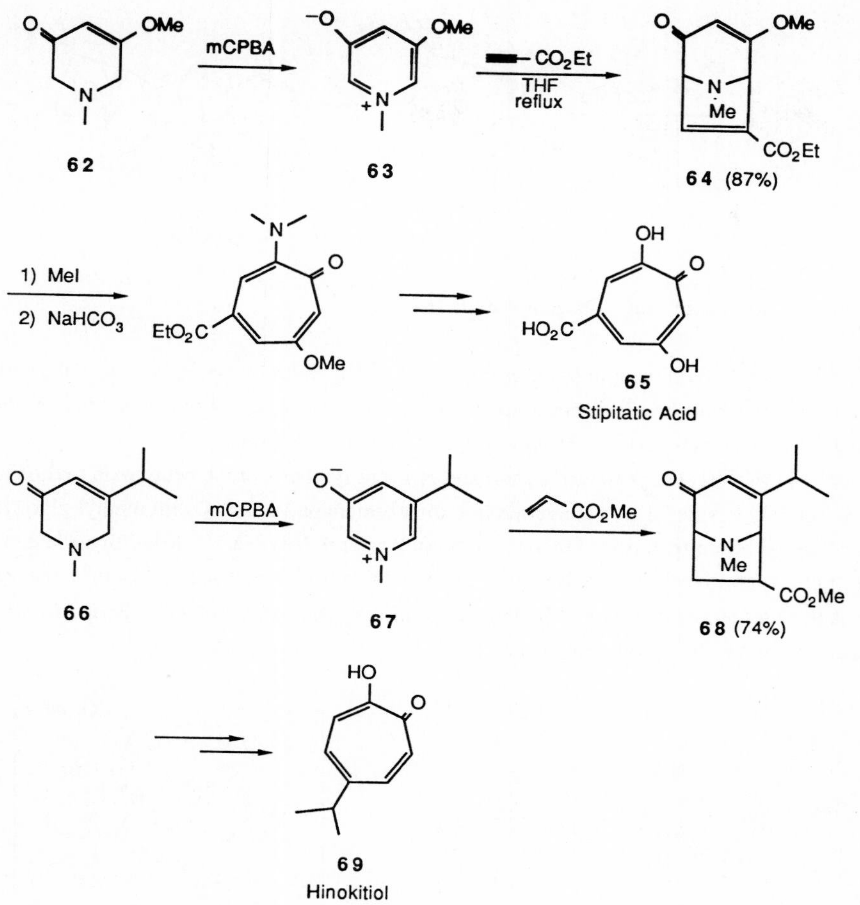

Figure 20. Tamura's synthesis of naturally occurring tropolones stipitatic acid and hinokitiol.[50]

Closely related to the oxidopyridiniums are the oxidopyraziniums (ref. 51) which still have an azomethine ylide substructure (eq. 23). Joule (ref. 51) has studied 1,5-dimethyl-3-oxido-pyrazinium **70** as a model study for the synthesis of antitumor metabolites such as quinocarcin **72** (Figure 21). Cycloaddition with methyl acrylate gives **71** as the correct regio- and stereoisomer, with functionality in place which should be suitable for elaboration to quinocarcin.

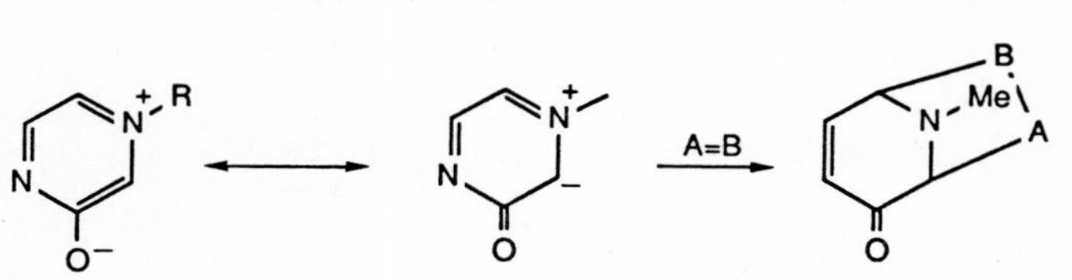

(eq. 23)

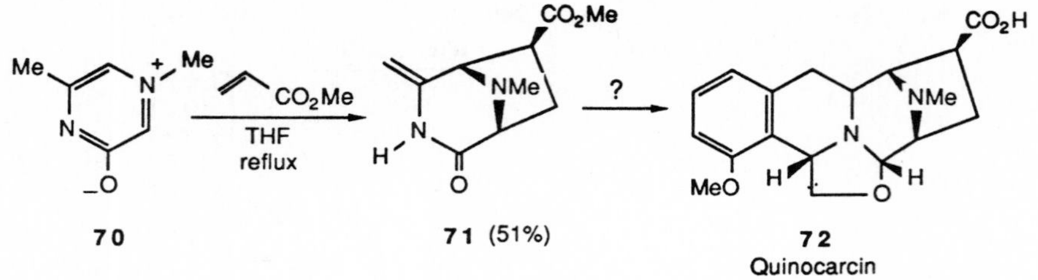

Figure 21. Joule's model study for quinocarcin.[51]

Although azomethine imines differ from azomethine ylides by the replacement of a carbon with nitrogen, we will briefly present applications of their cycloadditions to natural products synthesis, since there has been recent activity in this closely related area.

Jacobi (ref. 52) has presented a short and efficient synthesis of the neurotoxin saxitoxin **76** (Figure 22). Hydrazide **73** was converted into the azomethine imine **74** with methyl glyoxylate hemiacetal and boron trifluoride etherate. Intramolecular 1,3-dipolar cycloaddition yielded the pyrazolidine derivative **75** as one isomer. Saxitoxin **76** was obtained after a few efficient steps. Once again, the intramolecular 1,3-dipolar cycloaddition has demonstrated its power in rapidly constructing complex polycyclic molecules.

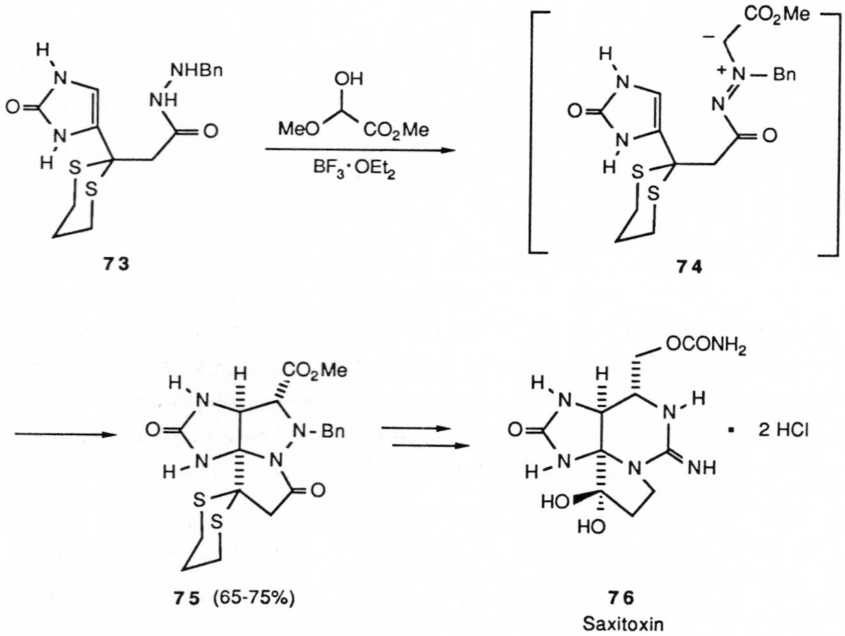

Figure 22. Jacobi's saxitoxin synthesis.[52]

Ranganathan (ref. 53) has reported a short synthesis of withasomnine using a sydnone (azomethine imine) (eq. 24).

| Withasomnine 18% | iso-Withasomnine 51% |

(eq. 24)

Jungheim (ref. 54) has prepared a very exciting class of γ-lactam analogues of β-lactam antibiotics using azomethine imines. For example, cycloaddition of pyrazolidinium ylide **77** with diallyl acetylene dicarboxylate provided the bicyclic pyrazolidinone **78** (eq. 25).

(eq. 25)

77   **78** (67%)

3. THE 2-AZAALLYL ANION APPROACH

The simplest 2-azaallylic system which may behave as a 4π-electron partner in a $[\pi 4s + \pi 2s]$ cycloaddition is that of the 2-azaallyl anion. Whereas allyl anions are generally reluctant to undergo concerted cycloadditions with olefins (refs. 55, 56), replacement of the central carbon with the more electronegative nitrogen has a favorable effect, since the reaction proceeds by a shift of electron density from the terminal carbons of the allylic system to the central nitrogen in the product.

In 1929, Ingold (ref. 57) proposed the intermediacy of 2-azaallyl anions in the alkoxide-promoted isomerization of N-(arylidene)benzyl amines (eq. 26). In 1949, Hauser (ref. 58) described the first irreversible generation of a 2-azaallyl anion, namely, 1,1,3-triphenyl-2-azaallylpotassium (eq. 27). However, it was Kauffmann in 1970 who first demonstrated the ability of these anions to undergo cycloaddition with olefins (ref. 59), for example, eq. 28. Kauffmann published a review on this area in 1974 (ref. 55).

(eq. 26)

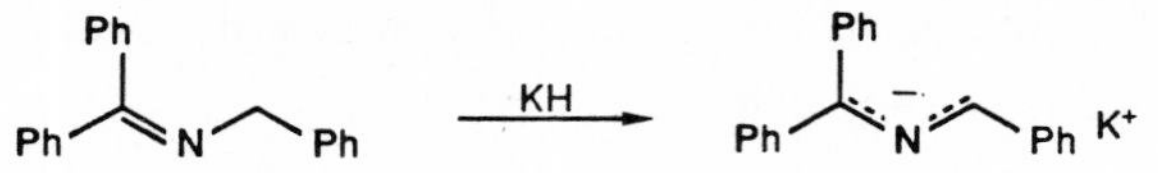

$$(eq.\ 27)$$

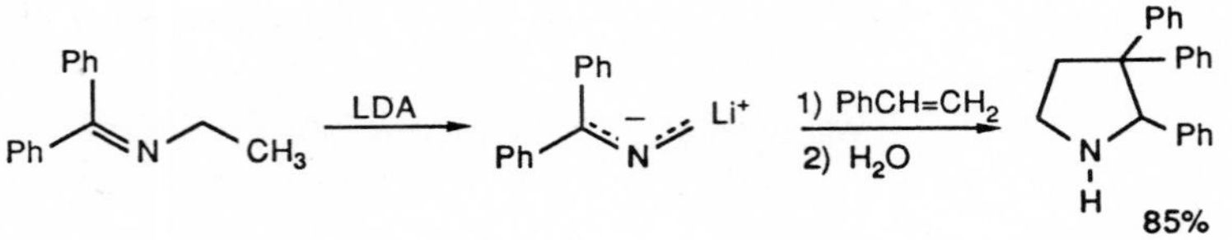

$$(eq.\ 28)$$

The major methods for the generation of 2-azaallyl anions are summarized in Figure 23. They are:

(1) Deprotonation of imines (eq. 29).

(2) Ring opening of N-metalloaziridines (eq. 30).

(3) Anionic cycloreversion of N-lithioimidazolidines (eq. 31).

(4) Cleavage of carbon-silicon or carbon-tin bonds (eq. 32).

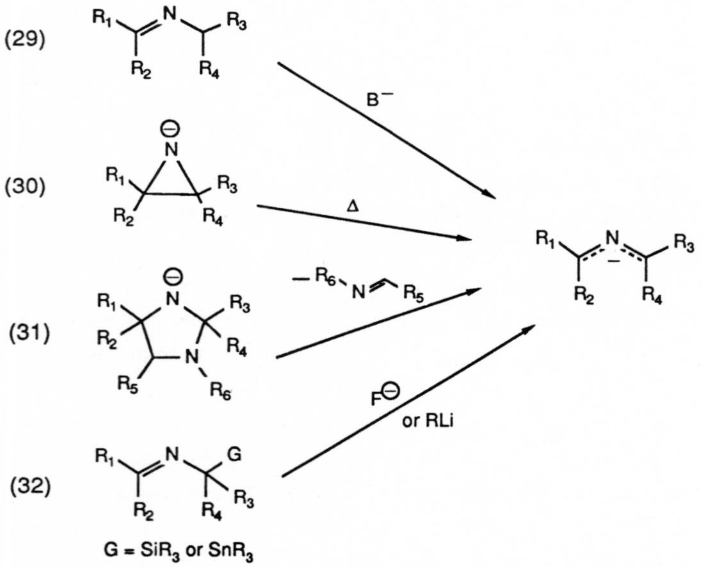

Figure 23. Methods for the generation of 2-azaallyl anions.

Another limited source of such anions is the thermal fragmentation of 1,3,5-triazolidine anions (ref. 55).

Prior to our own work in the area, anion formation/cycloaddition was generally limited to those examples bearing at least two aryl groups (ref. 60). Methods (1) and (2) were used to prepare such stabilized anions. We have shown (ref. 61) that all four methods may be used to prepare monoaryl-substituted anions, and more importantly, that method (4) may be used to generate <u>unstabilized</u> anions (ref. 62). These results will be discussed in later sections.

We will focus on the cycloaddition of 2-azaallyl anions with olefins, thereby providing pyrrolidines. However, other unsaturated groups are successful "anionophiles" (eq. 33), for example, imines (refs. 55, 63-65), azo compounds (ref. 55), nitriles (refs. 55, 66), heterocumulenes (refs. 55, 65, 67), allenes (refs. 63b, 68), dienes (refs. 55, 64a, 69), trienes (ref. 70) and acetylenes (refs. 55, 63b, 68a, 71). In some cases, non-cyclic products are isolated (e.g., with carbonyl compounds and some imines).

$$\text{A} \overline{\overline{}} \text{B} \; = \; R_2C=CR_2, RC \equiv CR, RCH=NR, RN=NR,$$
$$RC \equiv N, RX=C=Y, \text{dienes, cycloheptatriene,}$$
$$\text{alkenes, benzyne.}$$

Although the cycloaddition of 2-azaallyl anions onto olefins is potentially a very powerful method for the synthesis of pyrrolidinic natural products, it has, until recently, been very limited in scope due to the methods available for anion generation. Obviously, a requirement for two or more aryl groups on the anion limits the utility of the method. Furthermore, unactivated olefins have only recently been shown to be useful in these cyclizations (refs. 61, 72). Prior to that time, only activated olefins such as styrene and stilbene had been used. What follows is a brief account of the methods for anion generation, focusing on recent work. Anion generation is a major frontier in 2-azaallyl anion chemistry and will be important in opening new pathways for application to synthesis. To date, 2-azaallyl anion cycloadditions have not been used to prepare a naturally occurring compound (ref. 99).

3.1 Deprotonation of Imines

The majority of literature reports concerning 2-azaallyl anions utilize imine deprotonation for their generation. However, this method is limited in scope by the types of imines required; namely, those which may not form the preferred 1-azaallyl anions (metalloenamines), as shown in eq. 34.

Two strategies are available to prevent 1-azaallyl anion formation. First, there are those imines which have no "enolizable" hydrogens. For example, N-(benzylidene)benzyl amine may be deprotonated with lithium diisopropyl amide (LDA) to gain access to 1,3-diphenylazaallyllithium, the most common anion in cycloaddition studies (eq. 35). Other diaryl- or triarylazaallyllithium

$$Ph\diagdown N \diagup Ph \xrightarrow{\text{LDA}} Ph \diagdown \overset{..}{N} \diagup Ph \ Li^+ \qquad \text{(eq. 35)}$$

anions have been generated this way. The second variety of imines are those bearing a strongly anion stabilizing group [ASG, e. g., CO_2R, $P(O)(OR)_2$], which acidifies the desired hydrogens (eq. 36)

$$R \diagdown N \diagup ASG \xrightarrow{\text{Base}} R \diagdown N \diagup ASG \qquad \text{(eq. 36)}$$

R may be enolizable
ASG = CO_2R', $P(O)(OR')_2$, etc.

Di- or triarylazaallyl anions have seen the most activity in cycloaddition studies (ref. 55). In fact, all of the cycloadditions mentioned above (see eqs. 28 and 33) were carried out with this type of anion. In order to boost 2-azaallyl anions to the status of generally synthetically useful, this limitation must be overcome. For example, pyrrolidine-containing natural products rarely (if ever) have two or more aromatic substituents in these positions on the ring. Furthermore, the apparent requirement for a stabilizing group on the anionophile has not been conductive to maximum generality.

We have extended the imine deprotonation route to include monoaryl-substituted 2-azaallyl anions (ref. 61). In particular, a number of intramolecular examples have been reported (for example, eq. 37) which demonstrates the potential of 2-azaallyl anion methodology for the preparation of bicyclic pyrrolidines. These compounds are common structural subfeatures in many alkaloids. Of additional interest is the successful use of unactivated olefins in these cycloadditions. A two-step procedure has been used to remove the phenyl group from 79 (ref. 73), but yields are low. Therefore, other methods for anion generation must be developed.

$$\xrightarrow[\text{RT}]{\text{LDA} \atop \text{THF}} \xrightarrow{\text{H}_2\text{O}} \qquad \text{(eq. 37)}$$

79 (63%)
a: β-Ph
b: α-Ph   a:b = 5.1:1

2-Azaallyl anions bearing a strong anion stabilizing group have a rich chemistry of their own, but not particularly in the area of cycloadditions. For example, they participate in alkylations, aldol condensations and Michael additions (refs. 74-86). "Cycloaddition" with activated olefins (refs. 84-86) have been reported, but most often, pyrrolidines are not directly formed. Acyclic adducts or intermediates are formed which may then cyclize to pyrrolidines. Some examples show stereoselectivities which may indicate a cycloaddition-like component. An example of a cyclization of an activated olefin with a stabilized 2-azaallyl anion is illustrated by Dehnel's work (eq. 38, ref. 86).

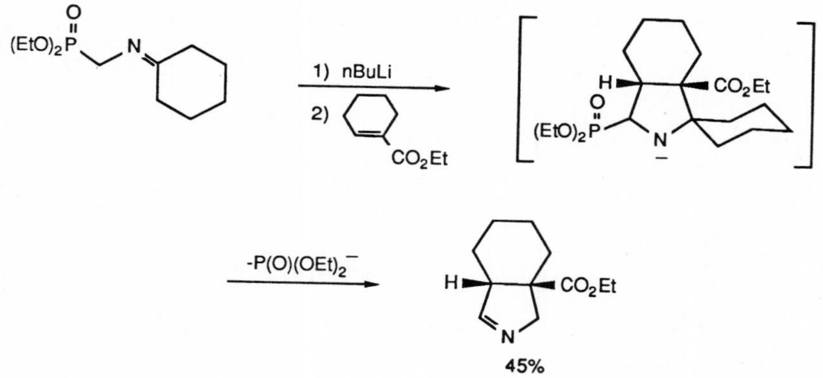

(eq. 38)

Cycloadditions with less activated anionophiles do not succeed, even when carried out intramolecularly. For example, we have prepared anions **80** and **81**, which do not cyclize (ref. 87). Proof that the anions were indeed present was verified by quenching with benzaldehyde, which produced 2-azadienes in good yield (eq. 39). These dienes were found to undergo intramolecular Diels-Alder reaction in very low yield (ref. 88). Apparently, these 2-azaallyl anions are too stabilized to undergo cycloadditions with all but the most activated anionophiles. This would be a substantial limitation to natural products synthesis.

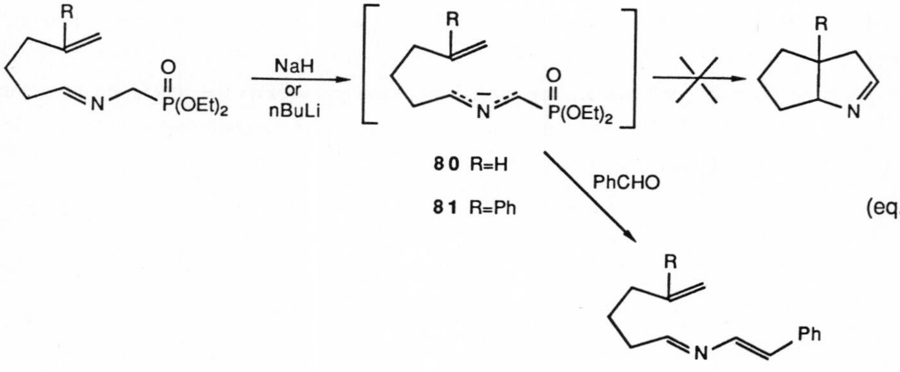

80 R=H

81 R=Ph

PhCHO

(eq. 39)

3.2 Ring Opening of N-Metalloaziridines

A particularly clean way to generate 1,3-diphenylazaallyllithium is by the thermal con-rotatory ring opening of cis- or trans-N-lithio-1,3-diphenylaziridine (ref. 55) (eq. 40). The advantage over the imine deprotonation method is that no diisopropylamine is present in solution. We (refs. 61, 89) and others (ref. 90) have found that certain nonstabilized 2-azaallyl anions may be isomerized to the more stable 1-azaallyl anions with proton sources as weak as dialkylamines. Most cycloadditions by this method use cis-1,3-diphenylaziridine **82** as the starting material, which presumably generates the sickle-form of the anion upon ring opening. However, the temperature required (30-40°C) does not allow its observation, since the barrier to isomerization to the more stable W-form is ca. 17-19 kcal/mol (ref. 91).

(eq. 40)

With regard to less substituted aziridines, Kauffmann (ref. 55) reports that the ring opening of N-lithio-2-phenylaziridine proceeds "slowly even at 65°" in THF but does not provide any details as to how this was determined or how the anion was trapped. This example is of interest since it would be one of the first cases of a monoaryl-substituted 2-azaallyl anion (refs. 91, 92). We have investigated the ring opening and cycloaddition of N-lithio-2-arylaziridines, for example, eq. 41 (ref. 89). In agreement with Kauffmann's work, the lithium salt of aziridine **83** does not open at all after 24 h at 66°C in THF. Conditions for ring opening were found to be 110°C in benzene in a sealed tube, providing an 81% yield of the bicyclic 1-pyrroline **84**. Cycloaddition is accompanied by deposition of lithium hydride on the sides of the vessel. Unfortunately, metal salts of aziridines lacking aryl groups (e.g., **85**) fail to undergo ring opening at useful temperatures.

Thus, ring opening of N-metalloaziridines is to date ony useful for the generation of anions stabilized by one or more aryl substituents. We are currently examining alternative conditions to promote ring opening of less activated aziridines.

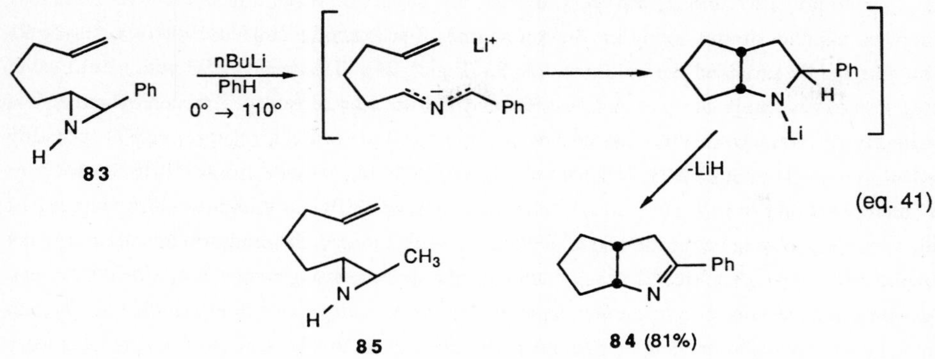

(eq. 41)

3.3 Anionic Cycloreversion of N-Lithioimidazolidines

During the course of our studies on imine deprotonations (Section 3.1, ref. 61), we observed the reversible formation· of substantial amounts of dimers such as **88** and **89** (eq. 42). For example, deprotonation of imine **86** with LDA was slow enough at room temperature to allow the initially formed anion **87** to undergo cycloaddition onto the C=N bond of starting material. The dimers **88** and **89** were shown to be regioisomeric N-lithioimidazolidines. Although they are initially present in solution in considerable amounts, they may cyclorevert to **86** and **87**, leading eventually and irreversibly to the cycloadduct **90**.

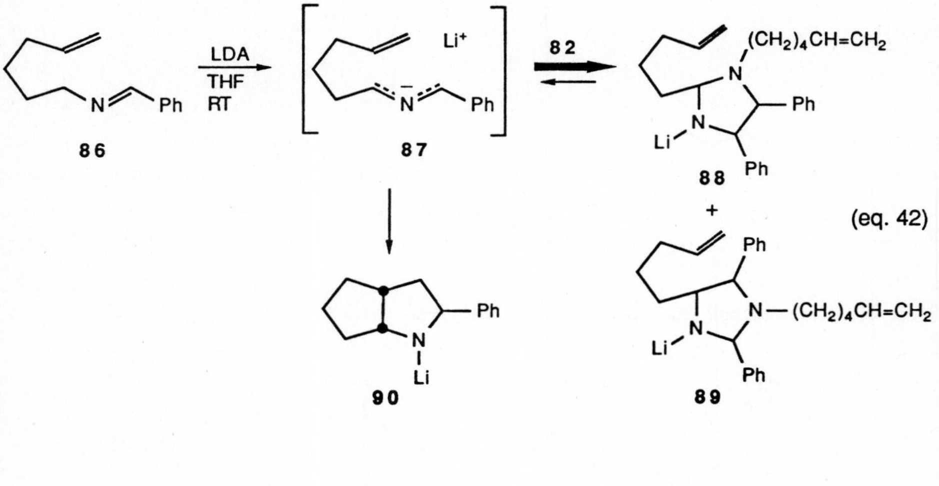

(eq. 42)

This novel cycloreversion (refs. 61, 93, 94) suggested to us a new route to 2-azaallyl anions which does not rely on imine deprotonations. For example, condensation of aldehyde **91** with diamine **92** provided the imidazolidine **93** (Figure 24). Treatment of **93** with nBuLi gave, after aqueous workup, the pyrrolidines **79a** and **79b** in a 1:>20 ratio. The stereochemistry is strikingly different from that obtained by imine deprotonation (<u>vide supra</u>, eq.37), possibly indicating a different geometry of the anion formed by the two methods. The barrier to sickle → W interconversion in <u>mono</u>aryl 2-azaallyl anions may be sufficiently high to allow trapping of either form, depending on the method of generation. Alternatively, the imidazolidine route may not be proceeding through a "free" 2-azaallyl anion, and may be exerting some type of memory effect. To test for the presence of a free anion, we prepared the optically active imidazolidine (+)-**94**, and found racemic pyrrolidine **95** upon treatment with nBuLi, tentatively assigned the stereochemistry shown (eq. 43). This is most consistent with the free 2-azaallyl anion mechanism.

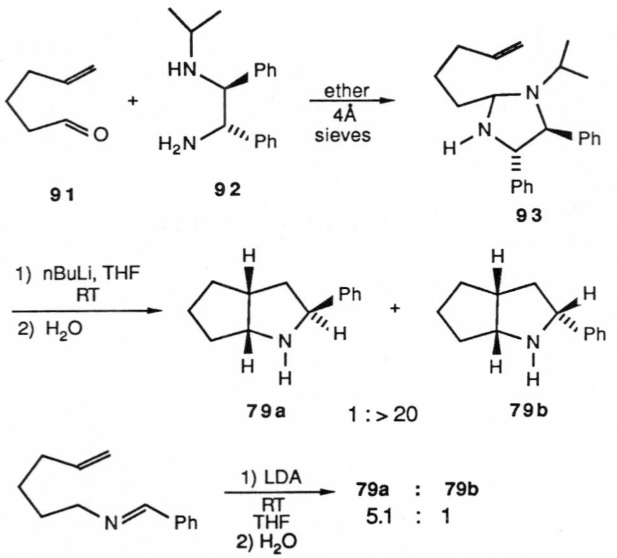

Figure 24. The imidazolidine route to 2-azaallyl anions.

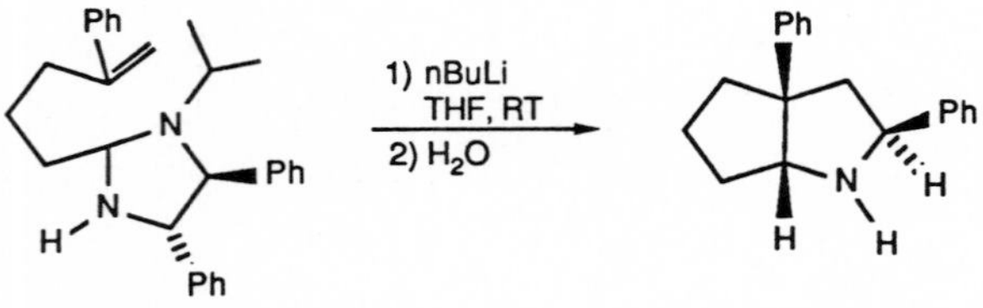

The successful fragmentation of imidazolidines to 2-azaallyl anions has been found to be very dependent on the ring substitution. Examples **93, 94** and **96** are successful. Replacement of the isopropyl substituent with phenyl (i.e., **97**) completely shuts down the cycloreversion, and only starting material is recovered, even upon refluxing the THF solution. Examples which would generate a 2-azaallyl anion bearing no phenyl group also fail to cyclorevert (e.g., **98**). We are currently studying solvent, counter-ion and stereochemical effects in an attempt to generalize this method for 2-azaallyl anion formation.

3.4 Cleavage of Carbon-Silicon or Carbon-Tin Bonds

Carbon-silicon and carbon-tin bonds may often serve as latent carbanions. Treatment with fluoride ion or an alkyllithium reagent, respectively, serves to liberate an anion. We hoped to use this methodology to generate 2-azaallyl anions, as outlined in eqs. 44 and 45.

Our early work centered on attempts to cyclize the N-(trimethylsilyl)methyl imine **99** (ref. 95, eq. 46). All attempts with a variety of fluoride sources led only to low yields of protodesilylated material in a very slow reaction, with no trace of cycloadduct.

Tsuge (ref. 96) has recently had more success with examples bearing a phenyl group (eq. 47).

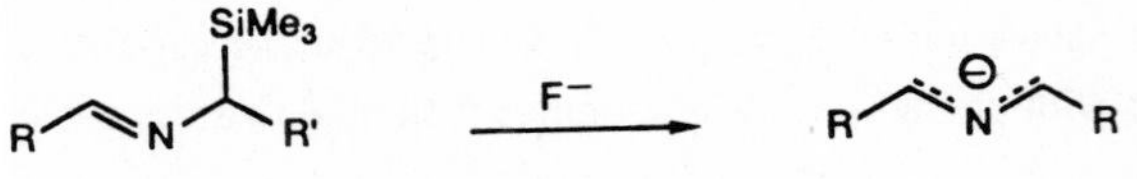
SiMe3
R
N
R'
F⁻
R
N
R

(eq. 44)

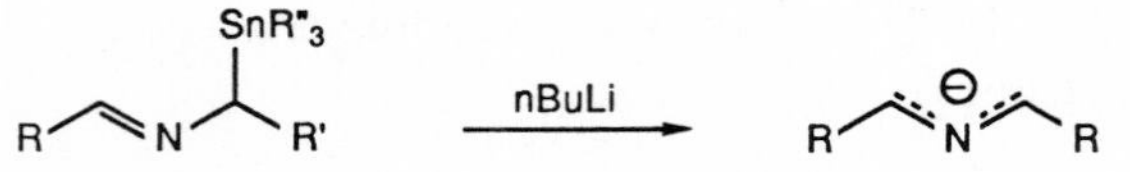
SnR"3
R
N
R'
nBuLi
R
N
R

(eq. 45)

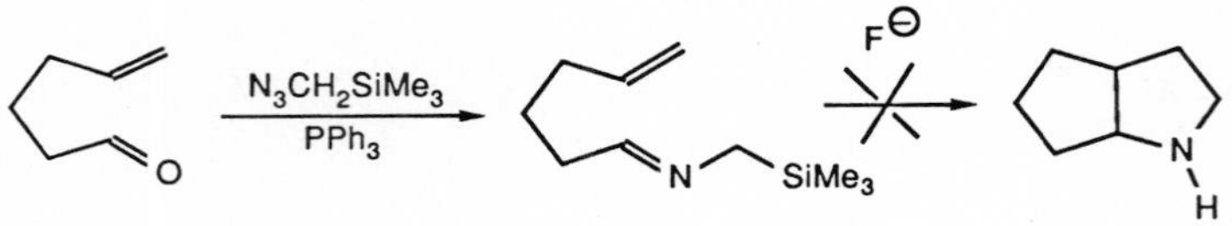
O
N3CH2SiMe3
PPh3
N
SiMe3
F
N
H

99

(eq. 46)

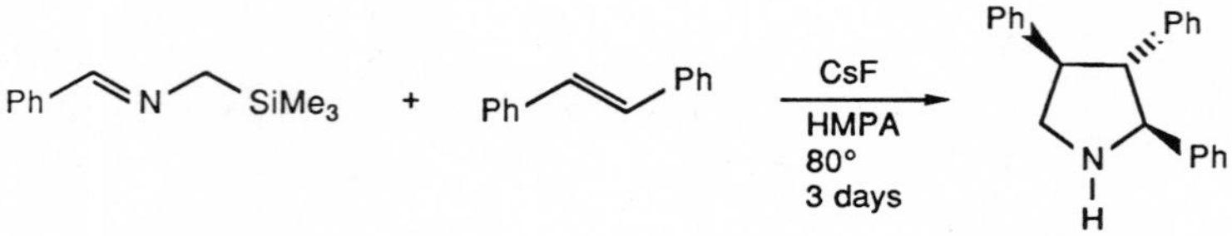
Ph
N
SiMe3
+
Ph
Ph
CsF
HMPA
80°
3 days
Ph
Ph
N
H
Ph

80%

(eq. 47)

We next turned to the carbon-tin bond, which is known to be easily converted to carbon-lithium bonds at low temperature (refs. 62, 97). For example, imine **100** was prepared by a Standinger reaction (ref. 98) between an aldehyde and (azidomethyl)-tri-n-butylstannane. Treatment of **100** with n-BuLi at -78°C afforded the 2-azaallyl anion which cyclized in high yield to the bicyclic pyrrolidine **101** (eq. 48, ref. 100). Intermolecular examples have also been observed. Particularly noteworthy is the generation, for the first time, of an unstabilized 2-azaallyl anion. We have also observed cyclizations of this type which are complete in 5-10 minutes at -78°C! In the future, we hope to show the generality of this method through natural products synthesis.

$$\text{(eq. 48)}$$

100

101 (83-95%)

4. SUMMARY

The cycloaddition of 2-azaallyl anions could conceivably become a useful method for the assembly of pyrrolidinic natural products. No examples of natural products synthesis with this method have appeared to date (ref. 99), presumably due to the long-standing requirement for aryl substitution on the anion. However, that requirement has recently been lifted through organo-stannane chemistry (eq. 48), and we hope to see renewed interest in the synthetic utility of these intriguing anions.

Azomethine ylides and related 1,3-dipoles have proven to be exceedingly useful for the synthesis of naturally occurring pyrrolidines. The dipolar cycloaddition approach has become the method of choice for rapid construction of five-membered ring heterocycles. However, 2-azaallyl anions promise to be a useful complement to azomethine ylides. Whereas azomethine ylides generally require dipolarophiles substituted with electron-withdrawing substituents, 2-azaallyl anions do not. The anions are much more reactive 4π-partners in such cycloadditions, as evidenced by a number of examples which work with unactivated olefins (ref. 61 and work described above). They also provide N-unsubstituted pyrrolidines directly after aqueous workup, in contrast to most azomethine ylide cycloadditions. Of course, if N-substituted pyrrolidines are desired, reaction of the N-lithiopyrrolidines with a suitable electrophile may be carried out.

Frontiers for 2-azaallyl anion chemistry include examination of the scope and stereo-chemistry of their cycloadditions, understanding their structure, and exploring their use in natural products synthesis.

354

REFERENCES

1. (a) Pearson, W. H. *Tetrahedron Lett.* **1985**, *26*, 3527-3530. (b) Pearson, W. H.; Celebuski, J. E.; Poon, Y.-F.; Dixon, B. R.; Glans, J. H. *Ibid.* **1986**, *27*, 6301-6304.

2. (a) Schultz, A. G.; Dittami, J. P.; Myong, S. O.; Sha, C.-K. *J. Am. Chem. Soc.* **1983**, *105*, 3273 and earlier work cited therein. (b) Schultz, A. G.; Staib, R. R.; Eng, K. K. *J. Org. Chem.* **1987**, *52*, 2968-2972. (c) Naruta, Y.; Arita, Y.; Nagai, N.; Uno, H.; Maruyama, K. *Chem. Lett.* **1982**, 1859. (d) Hudlicky, T.; Frazier, J. O.; Seoane, G.; Tiedje, M.; Seoane, A.; Kwart, L. D.; Beal, C. *J. Am. Chem. Soc.* **1986**, *108*, 3755. (e) Hudlicky, T.; Frazier, J. O.; Kwart, L. D. *Tetrahedron Lett.* **1985**, *26*, 3523. (f) Boulaajaj, S.; LeGall, T.; Vaultier, M.; Gree, R.; Toupet, L.; Carrie, R. *Ibid.* **1987**, *28*, 1761.

3. (a)Trost, B. M.; Bonk, P. J. *J. Am. Chem. Soc.* **1985**, *107*, 1778-1781. (b) Knochel, P. A.; Normant, J. F. *Tetrahedron Lett.* **1985**, *26*, 4455.

4. Lown, W. J. in "1,3-Dipolar Cycloaddition Chemistry", Padwa, A. (Ed.), Vol. 1, Wiley, New York, 1984, Ch. 6, pp. 663-732.

5. Vedejs, E.; West, F. G. *Chem. Rev.* **1986**, *86*, 941-955.

6. (a) Vedejs, E.; Martinez, G. R. *J. Am. Chem. Soc.* **1980**, *102*, 7993-7994. (b) Vedejs, E.; Larsen, S.; West, F. G. *J. Org. Chem.* **1985**, *50*, 2170-2714. See also: (c) Vedejs, E.; West, F. G. *Ibid.* **1983**, *48*, 4773-4774.

7. Terao, Y.; Imai, N.; Achiwa, K.; Sekiya, M. *Chem. Pharm. Bull.* **1982**, *30*, 3167-3171.

8. Belloir, P. F.; Laurent, A.; Mison, P.; Lesniak, S.; Bartnik, R. *Synthesis* **1986**, 683-686.

9. (a) Livinghouse, T.; Smith, R. *J. Chem. Soc., Chem. Commun.* **1983**, 210-211. (b) Livinghouse, T.; Smith, R. *J. Org. Chem.* **1983**, *48*, 1554-1555. (c) Smith, R.; Livinghouse, T. *Tetrahedron* **1985**, *41*, 3559-3568.

10. Westling, M.; Smith, R.; Livinghouse, T. *J. Org. Chem.* **1986**, *51*, 1159-1165.

11. (a) Padwa, A.; Chen, Y.-Y.; Koehler, K. F.; Tomas, M. *Bull. Soc. Chim. Belg.* **1983**, *92*, 811-817. (b) Padwa, A.; Haffmanns, G.; Tomas, M. *Tetrahedron Lett.* **1983**, *24*, 4303-4306. (c) Padwa, A.; Haffmanns, G.; Tomas, M. *J. Org. Chem.* **1984**, *49*, 3314-3322.

12. (a) Padwa, A.; Chen, Y.-Y. *Tetrahedron Lett.* **1983**, *24*, 3447-3450. (b) Padwa, A.; Chen, Y.-Y.; Chiacchio, U.; Dent, W. *Tetrahedron* **1985**, *41*, 3529-3535.

13. (a) Parker, K. A.; Cohen, I. D.; Padwa, A.; Dent, W. *Tetrahedron Lett.* **1984**, *25*, 4917-4920. (b) Padwa, A.; Chen, Y.-Y.; Dent, W.; Nimmesgern, H. *J. Org. Chem.* **1985**, *50*, 4006-4014.

14. (a) DeShong, P.; Kell, D. A. *Tetrahedron Lett.* **1986**, *27*, 3979-3982. (b) DeShong, P.; Kell, D. A.; Sidler, D. R. *J. Org. Chem.* **1985**, *50*, 2309-2315. For an earlier and closely related approach to alpha-allo-kainic acid, see Kraus in Section 2.4, Figure 15, ref. 30.

15. Takano, S.; Iwabuchi, Y.; Ogasawara, K. *J.Am. Chem. Soc.* **1987**, *109*, 5523.

16. Padwa, A.; Ku, H. *J. Org. Chem.* **1979**, *44*, 255-261.

17. Wenkert, D.; Ferguson, S. B.; Porter, B.; Qvarnstrom, A.; McPhail, A. T. *J. Org. Chem.* **1985**, *50*, 4114-4119.

18. Eberbach, W.; Fritz, H.; Heinze, I.; von Laer, P.; Link, P. *Tetrahedron Lett.* **1986**, *27*, 4003-4006.

19. DeShong, P.; Sidler, D. R.; Kell, D. A.; Aronson, N. N., Jr. *Tetrahedron Lett.* **1985**, *26*, 3747-3748.

20. (a) Grigg, R.; Kemp, J.; Sheldrick, G.; Trotter, J. *J. Chem. Soc., Chem. Commun.* **1978**, 109-111. For an excellent overview of this area, see: (b) Grigg, R.; Gunaratne, H. Q. N.; Kemp, J. *J. Chem. Soc., Perkin Trans. 1* **1984**, 41 and references therein.

21. (a) Joucla, M.; Hamelin, J. *Tetrahedron Lett.* **1978**, 2885-2888. (b) Joucla, M.; Mortier, J.; Hamelin, J. *Ibid.* **1985**, *26*, 2775-2778 and earlier references cited therein.

22. (a) Grigg, R.; Jordan, M.; Malone, J. F. *Tetrahedron Lett.* **1979**, 3877. (b) Grigg, R.; Gunaratne, H. Q. N. *Ibid.* **1983**, *24*, 4457.

23. Armstrong, P.; Grigg, R.; Jordan, M. W.; Malone, J. F. *Tetrahedron* **1985**, *41*, 3547-3558.

24. Confalone, P. N.; Huie, E. M. *J. Org. Chem.* **1983**, *48*, 2994-2997.

25. Confalone, P. N.; Huie, E. M. *J. Am. Chem. Soc.* **1984**, *106*, 7175-7178.

26. Confalone, P. N.; Earl, R. A. *Tetrahedron Lett.* **1986**, *27*, 2695-2698.

27. Yamada, S.; Hongo, C.; Yoshioka, R.; Chibata, I. *J. Org. Chem.* **1983**, *48*, 843.

28. For good leading references on the direct or indirect production of N-unsubstituted pyrrolidines from azomethine ylide cycloadditions, see: Kraus, ref. 30.

29. Dean, R. T.; Padgett, H. C.; Rapoport, H. *J. Am. Chem. Soc.* **1976**, *98*, 7448.

30. (a) Kraus, G. A.; Nagy, J. O. *Tetrahedron Lett.* **1981**, *22*, 2727. (b) Kraus, G. A.; Nagy, J. O. *Ibid.* **1983**, *24*, 3427-3430. (c) Kraus, G. A.; Nagy, J. O. *Tetrahedron* **1985**, *41*, 3537-3545. See also: Potts, K. T.; Choudhury, D. R.; Westby, T. R. *J. Org. Chem.* **1976**, *41*, 187.

31. Tsuge, O.; Kanemasa, S.; Takenaka, S. *J. Org. Chem.* **1986**, *51*, 1853-1855 and references cited therein.

32. Imai, N.; Nemot, M.; Terao, Y.; Achiwa, K.; Sekiya, M. *Chem. Pharm. Bull.* **1986**, *34*, 1080-1088.

33. Rizzi, G. P. *J. Org. Chem.* **1970**, *35*, 2069-2072.

34. (a) Grigg, R.; Aly, M. F.; Sridharan, V.; Thianpatanagul, S. *J. Chem. Soc., Chem. Commun.* **1984**, 182-183. (b) Grigg, R.; Thianpatanagul, S. *Ibid.* **1984**, 180-181 and references cited therein.

35. Wang, C.-L.; Ripka, W. C.; Confalone, P. N. *Tetrahedron Lett.* **1984**, *25*, 4613-4616 and references cited therein.

36. Forte, M.; Orsini, F.; Pelizzoni, F. *Gazz. Chim. Ital.* **1985**, *115*, 569-571.

37. (a) Joucla, M.; Mortier, J. *J. Chem. Soc., Chem. Commun.* **1985**, 1566. For a related reaction, see: (b) Joucla, M.; Mortier, J. *Tetrahedron Lett.* **1987**, *28*, 2973.

38. Tsuge, O.; Kanemasa, S.; Ohe, M.; Takenaka, S. *Chem. Lett.* **1986**, 973-976.

39. Eschenmoser, A. *Chem. Soc. Rev.* **1976**, *5*, 377.

40. (a) Joucla, M.; Mortier, J. *Tetrahedron Lett.* **1987**, *28*, 2973. (b) Joucla, M.; Mortier, J.; Bureau, R. *Ibid.* **1987**, *28*, 2975.

41. Beugelmans, R.; Negron, G.; Roussi, G. *J. Chem. Soc., Chem. Commun.* **1983**, 31.

42. (a) Beugelmans, R.; Benadjila-Iguertsira, L.; Chastanet, J.; Negron, G.; Roussi, G. *Can. J. Chem.* **1985**, *63*, 725-734. (b) Chastanet, J.; Roussi, G. *Heterocycles* **1985**, *23*, 653-659. (c) Chastanet, J.; Roussi, G. *J. Org. Chem.* **1985**, *50*, 2910-2914.

43. (a) Bayer, H. O.; Gotthardt, H.; Huisgen, R. *Chem. Ber.* **1970**, *103*, 2356. (b) Huisgen, R.; Gotthardt, H.; Bayer, H. O.; Schafer, F. C. *Ibid.* **1970**, *103*, 2611.

44. (a) Pizzorno, M. T.; Albonico, S. M. *J. Org. Chem.* **1974**, *39*, 731. For a closely related synthesis of a butterfly pheromone using the same dipole, see: (b) Pizzorno, M. T.; Albonico, S. M. *Chem. Ind. (London)* **1978**, 349.

45. Robins, D. J.; Sakdarat,S. *J. Chem. Soc., Chem. Commun.* **1979**, 1181.

46. (a) Hershenson, F. M. *J. Org. Chem.* **1975**, *40*, 1260-1264. (b) Hershenson, F. M. *J. Heterocycl. Chem.* **1979**, *16*, 1093-1095.

47. (a) Rebek, J., Jr.; Gehret, J.-C. *Tetrahedron Lett..* **1977**, 3027-3030. (b) Rebek, J., Jr.; Shaber, S. H.; Shue, Y.-K.;Gehret, J.-C.; Zimmerman, S. *J. Org. Chem..* **1984**, *49*, 5164-5174.

48. For a related system, see: Laudreé, D.; Lancelot, J.-C.; Robba, M. *Tetrahedron Lett.* **1985**, *26*, 1295-1296.

49. Katritzky, A. R.; Takeuchi, Y. *J. Am. Chem. Soc.* **1970**, *92*, 4134-4136. See also ref. 51 for leading references.

50. Tamura, Y.; Saito, T.; Kiyokawa, H.; Chen, L.-C. *Tetrahedron Lett.* **1977**, 4075-4078.

51. Kiss, M.; Russell-Maynard, J.; Joule, J. A. *Tetrahedron Lett.* **1987**, *28*, 2187-2190 and references therein.

52. Jacobi, P. A.; Martinelli, M. J.; Polanc, S. *J. Am. Chem. Soc.* **1984**, *106*, 5594-5598.

53. (a) Ranganathan, D.; Bamezai, S. *Tetrahedron Lett.* **1985**, *26*, 5739-5742. (b) Ranganathan, D.; Bamezai, S. *Synth. Commun.* **1985**, *15*, 259-265.

54. Jungheim, L. N.; Sigmund, S. K. *J. Org. Chem.* **1987**, *52*, 4007-4013 and earlier references cited therein.

55. Kauffmann, T. *Angew. Chem., Int. Ed. Engl.* **1974**, *13*, 627-639.

56. (a) Kauffmann, T. *Top. Curr. Chem.* **1980**, *92*, 109. For recent investigations and excellent leading references, see: (b) Beak, P.; Wilson, K. D. *J. Org. Chem.* **1986**, *51*, 4627-4639. See also references 6, 9 in: (c) Boger, D. L.; Brotherton, C. E. *J. Am. Chem. Soc.* **1986**, *108*, 6695-6713.

57. Ingold, C. K.; Shoppee, C. W. *J. Chem. Soc.* **1929**, 1199.

58. Hauser, C. R.; Flur, I. C.; Kantor, S. W. *J. Am. Chem. Soc.* **1949**, *71*, 294.

59. Kauffmann, T.; Berg, H.; Köppelmann, E. *Angew. Chem., Int. Ed. Engl.* **1970**, *9*, 380.

60. 2-Azaallyl anions bearing strongly electron-withdrawing groups are well known (for example, see ref. 74), but generally do not undergo [3+2] cycloadditions.

61. Pearson, W. H.; Walters, M. A.; Oswell, K. D. *J. Am. Chem. Soc.* **1986**, *108*, 2769-2771.

62. Pearson, W. H.; Szura, D. P.; Harter, W. G., submitted for publication.

63. (a) Vo-Quang, L.; Vo-Quang, Y. *Tetrahedron Lett.* **1977**, 3963. (b) Vo-Quang, L.; Vo-Quang, Y. *J. Het. Chem.* **1982**, *19*, 145.

64. (a) Dryanska, V.; Popandova-Yambolieva, K.; Ivanov, C. *Ibid.* **1979**, 443. (b) Dryanska, V.; Ivarov, C.; Krusteva, R. *Synthesis* **1984**, 1038.

65. Gracheva, R. A.; Potapov, V. M.; Sivov, N. A.; Sivova, L. I. *J. Org. Chem. USSR* **1982**, *17*, 1963.

66. (a) Kauffmann, T.; Busch, A.; Habersaat, K.; Köppelmann, E. *Angew. Chem., Int. Ed. Engl.* **1973**, *12*, 569. (b) Kauffmann, T.; Busch, A.; Habersaat, K.; Köppelmann, E. *Chem. Ber.* **1983**, *116*, 492.

67. (a) Kauffmann, T.; Eidenschink, R. *Ibid.* **1977**, *110*, 651. (b) Kauffmann, T.; Eidenschink, R. *Angew. Chem., Int. Ed. Engl.* **1973**, *12*, 568.

68. (a) Vo-Quang, L.; Vo-Quang, Y. *Tetrahedron Lett.* **1978**, 4679. (b) *Ibid.* **1980**, 939. (c) Vo-Quang, L.; Vo-Quang, Y.; Povet, M. J.; Simonnin, M. P. *Ibid.* **1981**, *37*, 4343.

69. Kauffmann, T.; Eidenschink, R. *Chem. Ber.* **1977**, *110*, 645.

70. Bower, D. J.; Howden, M. E. H. *J. Chem. Soc., Perkin 1* **1980**, 672.

71. Vo-Quang, L.; Gaessler, H.; Vo-Quang, Y. *Angew. Chem., Int. Ed. Engl.* **1981**, *20*, 880.

72. Kamata, K.; Terashima, M. *Heterocycles* **1980**, *14*, 205.

73. Oxidation of benzylic amines with ruthenium tetraoxide under acidic conditions provides amino acids: (a) Hill, R. K.; Prakash, S. R.; Zydowsky, T. M. *J. Org. Chem.* **1984**, *49*, 1666. (b) Ayres, D. C. *J. Chem. Soc., Perkin Trans. 1* **1978**, 585. The amino acid may then be decarboxylated by Rapoport's method (ref. 29).

74. Stork, G.; Leong, A. Y. W.; Touzin, A. M. *J. Org. Chem.* **1976**, *41*, 3491.

75. (a) Yamada, S.; Oguri, T.; Shioiri, T. *J. Chem. Soc., Chem. Commun.* **1976**, 136. (b) Oguri, T.; Shioiri, T.; Yamada, S. *Chem. Pharm. Bull.* **1977**, *25*, 2287.

76. Duhamel, P.; Eddine, J. J.; Valnot, J.-Y. *Tetrahedron Lett.* **1984**, *25*, 2355 and references cited therein.

77. Minowa, N.; Hirayama, M.; Fukatsu, S. *Ibid.* **1984**, *25*, 1147.

78. (a) Hoppe, I.; Schöllkopf, U.; Tölle, R. *Synthesis* **1983**, 789. (b) Schöllkopf, U.; Schutze, R. *Liebigs Ann. Chem.* **1987**, 45-49 and references cited therein.

79. McIntosh, J. M.; Leavitt, R. K. *Tetrahedron Lett.* **1986**, *27*, 3839-3842.

80. Beak, P.; Zajdel, W. J.; Reitz, D. B. *Chem. Rev.* **1984**, 471-523.

81. (a) Böhme, E. H. W.; Applegate, H. E.; Ewing, J. B.; Funke, P. T.; Puar, M. S.; Dolfini, J. E. *J. Org. Chem.* **1973**, *38*, 230. (b) Firestone, R. A.; Christensen, B. G. *J. Chem. Soc., Chem. Commun.* **1976**, 288.

82. Bey, P.; Vevert, J. P. *Tetrahedron Let.* **1977**, 1455.

358

83. O'Donnell, M. J.; Boniece, J. M.; Earp, S. E. *Ibid.* **1978**, 2641.

84. Grigg, R.; Kemp, J.; Malone, J.; Tangthongkum, A. *J. Chem. Soc., Chem. Commun.* **1980**, 648-650.

85. Fouchet, B.; Joucla, M.; Hamelin, J. *Tetrahedron Lett.* **1981**, *22*, 3397-3400.

86. (a) Dehnel, A.; Lavielle, G. *Tetrahedron Lett.* **1980**, *21*, 1315. (b) Rabilla, C.; Dehnel, A.; Lavielle, G. *Can. J. Chem.* **1982**, *60*, 926.

87. Pearson, W. H.; Szura, D. P., unpublished results.

88. See also: Cheng, Y.-S.; Ho, E.; Mariano, P. S.; Ammon, H. L. *J. Org. Chem.* **1985**, *50*, 5678-5686 and references cited therein.

89. Pearson, W. H.; Rosen, M. K., unpublished results.

90. Smith, J. K.; Bergbreiter, D. E.; Newcomb, M. *J. Org. Chem.* **1985**, *50*, 4549.

91. Young, R. N.; Ahmad, M. A. *J. Chem. Soc., Perkin Trans. 2* **1982**, 35.

92. Popowski, E. *Z. Chem.* **1975**, *14*, 275.

93. Pearson, W. H.; Harter, W. G.; Walters, M. A., unpublished results.

94. For reviews on anionic cycloreversions, see: (a) Bianchi, G.; DeMicheli, C.; Gandolfi, R. *Angew. Chem., Int. Ed. Engl.* **1979**, *18*, 721. (b) Bianchi, G.; Gandolfi, R. in "1,3-Dipolar Cycloaddition Chemistry," Padwa, A. (Ed.), Vol. 2, Wiley, New York, 1984, Ch. 14, pp. 451-542.

95. Pearson, W. H.; Walters, M. A.; Robertson, S., unpublished results.

96. (a) Tsuge, O.; Kanemasa, S.; Hatada, A.; Matsudo, K. *Bull. Chem. Soc. Jpn.* **1986**, *59*, 2537. (b) Tsuge, O.; Kanemasa, S.; Yamada, T.; Matsuda, K. *J. Org. Chem.* **1987**, *52*, 2523.

97. Pereyre, M.; Quintard, J.-P.; Rahm, A., "Tin in Organic Synthesis," Butterworths, London, 1987.

98. Staudinger, H.; Meyer, J. *Helv. Chim. Acta* **1919**, *2*, 635.

99. We have very recently achieved the synthesis of mesembrane by an intramolecular 2-azaallyl anion cycloaddition: Pearson, W. H.; Szura, D. P.; Degan, S., unpublished results.

100. Pearson, W. H.; Szura, D. P.; Harter, W. G., submitted.

TOTAL SYNTHESIS OF NITROGEN-CONTAINING NATURAL PRODUCTS VIA CYCLOADDITION REACTIONS

CHIHIRO KIBAYASHI

1 INTRODUCTION

Thermally induced heterocycloaddition such as 1,3-dipolar cycloaddition (ref. 1) and hetero Diels-Alder reaction (refs. 2, 3), both 6π electron pericyclic reactions in which the two new σ-bonds are formed simultaneously, provide a versatile route to a wide variety of heterocyclic compounds. Of these pericyclic reactions, only the [3 + 2] dipolar cycloaddition of nitrones (ref. 4) and the Diels-Alder reactions involving nitroso compounds as heterodienophiles (ref. 5) can afford cyclic hydroxylamines as shown in eqns. (1) and (2), in which both the nitrogen and oxygen atoms are ring members. A synthetically important feature common to these two types of cycloadditions is introduction of nitrogen and oxygen functionality at the same time with a high degree of regiochemical and stereochemical control. Subsequent reduction of these [3 + 2] and [4 + 2] cycloadducts viz. tetrahydro-1,2-oxazoles (isoxazolidines) and tetrahydro-1,2-oxazines, respectively, can easily be achieved by using a wide variety of reducing agents, such as zinc in acetic acid, sodium or aluminum amalgam in ethanol, or hydrogen over catalysts, and normally result in cleavage of the N–O bond leading to the formation of corresponding acyclic γ- and δ-amino alcohols as shown in the following eqns. (1) and (2).

These amino alcohols thus generated can be utilized as key building blocks for the construction of a variety of nitrogen-containing natural products including alkaloids. In the early 1980s, when we began a program to develop general utility of the cycloaddition–N-O bond cleavage approach to natural product synthesis, this approach had only received isolated attention although both nitrone and nitroso cycloadditions themselves were well known reactions. However, the popularity of the nitrone usage in the synthesis of alkaloids and related natural products has been sharply increased in recent years due to development of mechanistic and synthetic aspects of 1,3-dipolar cycloaddition.

The purpose of this chapter is to review our recent work on the development of both nitrone and nitroso methodologies for preparing nitrogen-containing natural products including alkaloids, antibiotics, and a pheromone.

2 INDOLIZIDINE ALKALOIDS

Our initial efforts in the area of the indolizidine alkaloid synthesis were centered around using cyclic nitrones in the 1,3-dipolar cycloaddition since cyclic nitrones have an advantage in being fixed in the E configuration over acyclic nitrones which exist in E/Z equilibrium.

2.1 Tylophorine and Septicine

We first chose tylophorine (10), the major alkaloid of Tylophora species from Sri Lanka (ref. 6), as a synthetic target which possesses as a parent system the phenanthroindolizidine nucleus. Tylophorine has been synthesized in racemic form through numerous approaches (ref. 7), and recently two optically active preparations have been reported (ref. 8). All of these syntheses have been accomplished by utilizing synthetic routes starting from phenanthrene derivatives except for two routes (refs. 7d, 7f) involving oxidative coupling.

Our approach (ref. 9) to tylophorine and its biogenetic congener, septicine (9), based on the dipolar [3 + 2] cycloaddition, involved reaction between 1-pyrroline 1-oxide (2) and 3,4-dimethoxystyrene (1) to give a diastereomeric mixture of 3 and 4 in a ratio of 5:2 (by GC), favoring the trans isomer 3 in nearly quantitative yield (Fig. 1). Catalytic hydrogenation of this mixture resulted in N-O bond cleavage to afford the amino alcohol 5 (86%) as a mixture of two isomers (α- and ß-OH in formula 5). Acylation of 5 afforded the amide alcohol 6 along with a minor amount of the amide ester 7. The crude mixture of the products was hydrolysed under alkaline conditions to produce 6 in 70% yield based on 5. Oxidation with Collins reagent followed by intramolecular aldol condensation converted 6 into the lactam 8 in 65%

Fig. 1 (a) toluene, reflux; (b) H$_2$, Pd/C, MeOH; (c) 3,4-(MeO)$_2$-C$_6$H$_3$CH$_2$COCl, K$_2$CO$_3$, CH$_2$Cl$_2$, r.t., then K$_2$CO$_3$, MeOH, H$_2$O, reflux; (d) CrO$_3$·2Py, CH$_2$Cl$_2$, r.t.; (e) KOH, EtOH, reflux; (f) AlH$_3$, THF-Et$_2$O, r.t.; (g) hν, I$_2$, CH$_2$Cl$_2$.

yield. Reduction of **8** with alane afforded (±)-septicine (**9**) (ref. 10) in 88% yield, which was subjected to UV irradiation to provide (±)-tylophorine (**10**) in 43% yield.

2.2 Ipalbidine

The Ipomoea alkaloids, isolated from the seeds of <u>Ipomoea</u> <u>alba</u> and <u>Ipomoea</u> <u>muricata</u>, comprise two glycoside alkaloids ipalbine (**11**) (ref. 11) and ipomine (**12**) (refs. 12-14) and their aglycone ipalbidine (**13**) (ref. 10). A unique structural feature of this class of alkaloids is that the C-methyl group is found on the hexahydroindolizine nucleus. They are the only examples of naturally occurring indolizidine alkaloids (ref. 15). We envisaged a synthetic approach to ipalbidine (**13**) via the bicyclic ketone (**15**) which can be directly transformed into O-methylipalbidine (**14**), a synthetic precursor of ipalbidine (ref. 16).

11 R = H

12 R = p-coumaroyl

13 R = H

14 R = Me

15

As a part of preliminary experiments (ref. 17) aimed at constructing the bicyclic ketone system by the nitrone based approach, cycloaddition of the nitrone **2** with the 2-phenyl-3-butenoate **16** was carried out to give three adducts **17**, **18**, and **19** in 17, 31, and 32% yields, respectively (Fig. 2). Reductive treatment of the minor product **17** (6:5 diastereomeric mixture) with zinc in aqueous acetic acid resulted in the formation of **20** which possesses the equivalent functionality on the indolizidine ring system to that possessed by the indolizidinone **15**.

Because of low selectivity in the formation of the cycloadduct **17**, we abandoned the above approach to the bicyclic ketone intermediate **15**. Thus an alternative nitrone based approach to **15** was next elaborated as shown in Fig. 3 (ref. 18).

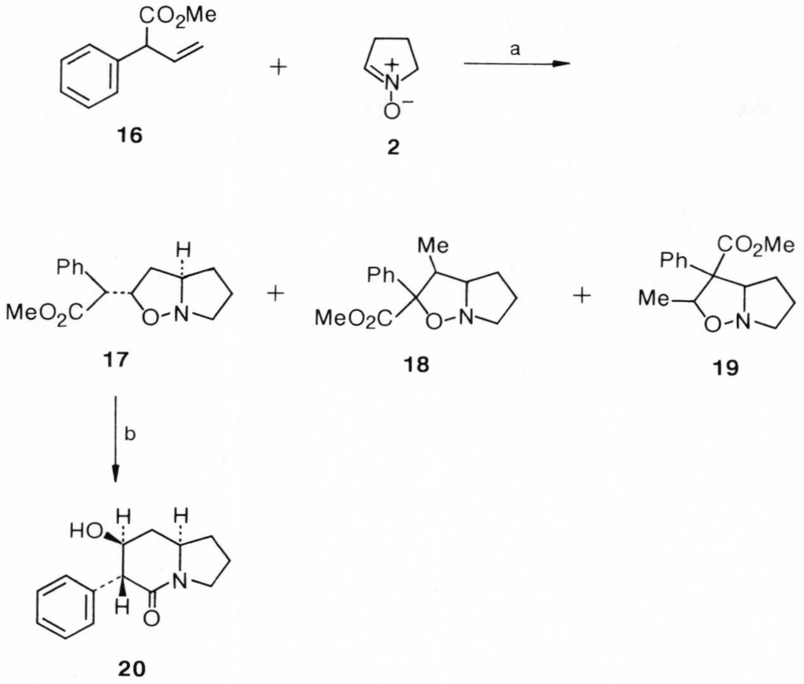

Fig. 2 (a) toluene, reflux; (b) Zn, aq. AcOH, 120 °C.

The reaction of the nitrone 2 with p-methoxy(allyl)benzene (21) in refluxed toluene produced the cycloadduct 22 in 70% yield as a single diastereoisomer. The high regioselectivity and stereoselectivity observed in this reaction are rationalized on the basis of the preference for dipole LUMO control (ref. 19) and exo-oriented transition state (ref. 20), respectively, when non-conjugated, electron-rich olefins such as 21 are used as dipolarophiles. Subsequent N-O bond cleavage of 21 by hydrogenation afforded the γ-amino alcohol 23 (82%), which was heated with formic acid in toluene to give N,O-diformylated compound 24. For selective hydrolysis of the O-formate group, 24 was exposed to ammonia in methanol to give the N-formylcarbinol 25 in 69% yield from 23. Collins oxidation of 25 yielded the formylketone 26 (72%).

For the construction of the indolizidine skeleton, we then considered application of internal aldol condensation to 26. Compared with the case in the aldol cyclization previously employed in the tylophorine (and also julandine, see below) synthesis [eqn. (3)], the type of aldol reaction required in the present case [eqn. (4)] is unusual as the amide C=O group is normally unreactive as the electrophilic partner (ref. 21). After several unsuccessful

Fig. 3 (a) toluene, reflux; (b) H$_2$, Pd/C, MeOH; (c) HCO$_2$H, toluene, reflux; (d) NH$_3$, MeOH, r.t.; (e) CrO$_3$·2Py, CH$_2$Cl$_2$, r.t.; (f) Al($\underline{t}$-BuO)$_3$, xylene, reflux; (g) Li, NH$_3$, THF, -78 °C; (h) refs. 10d and 16a.

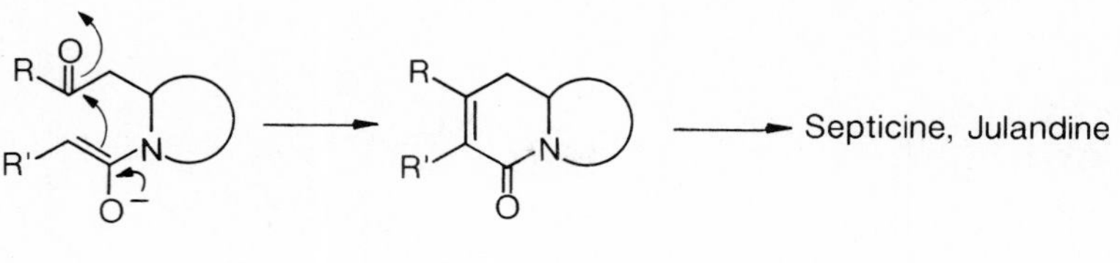

————► Septicine, Julandine   (3)

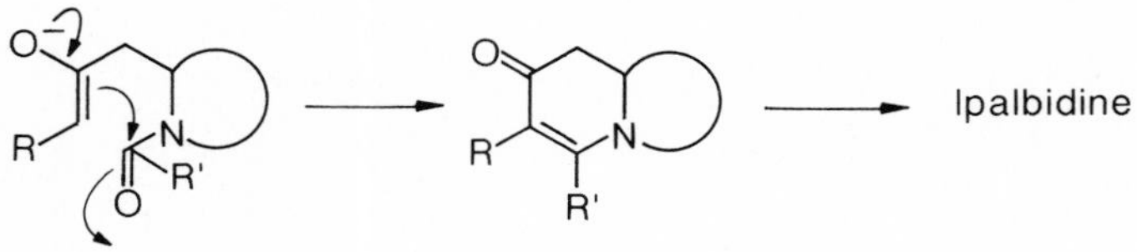

————► Ipalbidine   (4)

attempts, however, the intramolecular aldol cyclization was attained when heated with aluminum tert-butoxide in xylene, providing the bicyclic enaminone 27 albeit in low yield (36%). Eventually selective reduction of the olefinic double bond of 27 was achieved by the Birch procedure to afford the objective bicyclic ketone 15 in 54% yield. Since compound 15 has already been converted into (±)-ipalbidine (13) by two steps (refs. 10d, 16a), our preparation of the ketone intermediate 15 constitutes a formal total synthesis of racemic 13.

3 QUINOLIZIDINE ALKALOIDS

3.1 Cryptopleurine and Julandine

The nitrone methodology described for the indolizidine alkaloids was next planned to apply to the synthesis of a quinolizidine alkaloid, cryptopleurine (28) (ref. 22), first isolated from an Australian laurel Cryptocarya pleurosperma (ref. 23). Cryptopleurine forms together with cryptopleuridine (29) a rare group of alkaloids with the trans-fused phenanthroquinolizidine ring system and has been shown to possess unique and interesting biological properties including cytotoxic (ref. 24), vesicant (ref. 23), and antitumor (ref. 25) activities, and, thus, is responsible for considerable recent synthetic attention (ref. 26).

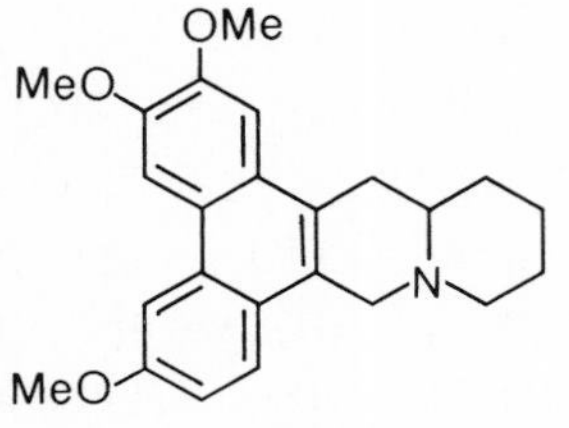

28

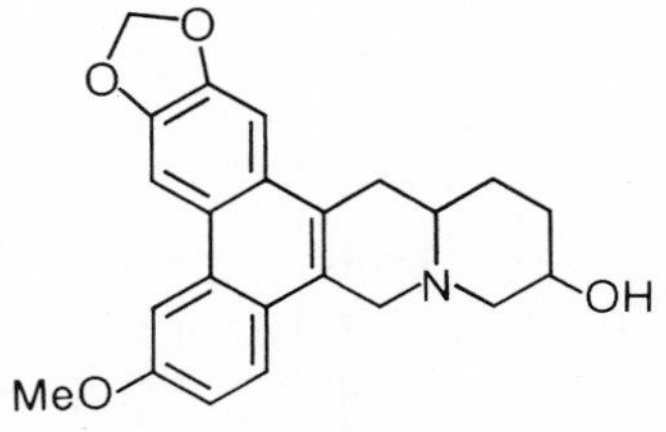

29

Fig. 4 (a) toluene, reflux; (b) Zn, aq. AcOH, 50 °C; (c) p-MeO-C$_6$H$_4$CH$_2$COCl, K$_2$CO$_3$, MeCN, r.t. then K$_2$CO$_3$, H$_2$O, MeOH, reflux; (d) CrO$_3$·2Py, CH$_2$Cl$_2$, r.t.; (e) NaOEt, EtOH, reflux; (f) LiAlH$_4$, THF, reflux; (g) hν, I$_2$, dioxane.

Our synthesis of cryptopleurine (**28**) and its seco base julandine (**36**) began with the 1,3-dipolar cycloaddition reaction of the cyclic nitrone **30** with 3,4-dimethoxystyrene (**1**) (Fig. 4) (ref. 27). The cycloadducts, actually a 20:1 mixture of _trans_ and _cis_ isomers **31** and **32**, so obtained (93%) was sub-jected to reductive N-O bond cleavage with zinc in aqueous acetic acid to give the cyclic amino alcohol **33** in 90% yield. Acylation of **33** afforded **34** along with the N,O-diacylated product, however, in practice, the crude mixture of these products was, without separation, subjected to brief treatment under alkaline condition to produce **34** as a single product in 64% yield from **33**. Oxidation of **34** with Collins reagent followed by intramolecular aldol condensa-tion yielded the bicyclic lactam **35** (55% yield from **34**) which on LiAlH$_4$ reduc-tion provided (±)-julandine (**36**) (ref. 28). Irradiation of the bicyclic lactam **35** in the presence of iodine yielded two fluorescent photocyclisation products **37** and **38** in 48 and 14% yield, respectively. The major product **37** was converted to (±)-cryptopleurine (**28**) by LiAlH$_4$ reduction in 95% yield.

3.2. Lasubine I and Subcosine I

Other target molecules which we decided to synthesize are the new quinolizidine alkaloids lasubine I (**39**) and subcosine I (**40**) with the _cis_-fused

39 Lasubine I

40 Subcosine I

4I Lasubine II

42 Subcosine II

quinolizidine ring system, which have been isolated (ref.29) from the leaves of <u>Lagerstroemia</u> <u>subcostata</u> Koehne along with lasubine II (**41**) (ref. 30) and subcosine II (**42**) with the <u>trans</u>-fused quinolizidine ring system. Our first synthesis of **39** and **40** based on nitrone cycloaddition (ref. 30) has been achieved as follows (ref. 31).

1,3-Cycloaddition between 1-(3,4-dimethoxyphenyl)butadiene (**43**), a 9:5 mixture of the <u>E</u> and <u>Z</u> isomers, and the nitrone **30** gave the corresponding <u>E</u> and <u>Z</u> cycloadducts **44** (<u>trans</u>/<u>cis</u> = 10:3) and **45** (<u>trans</u>/<u>cis</u> = 5:1) in 49 and 22% yields, respectively (Fig. 5). With the <u>E</u> isomer **44** in hand, we attempted to transform it directly to lasubine I (**39**). Thus hydrogen chloride was added to the olefin moiety of **44** to generate the chloride as a diastereomeric mixture of **46** and **47**, which, without separation, underwent hydrogen-

Fig. 5 (a) toluene, reflux; (b) HCl, CHCl$_3$, 0–5 °C; (c) H$_2$, Pd/C, Py-MeOH.

ation to afford (±)-lasubine I (**39**) in 44% yield from **44** along with (±)-2-epilasubine II (**48**) as a minor product (14% yield).

The formation of both alkaloids **39** and **48** must arise via both chlorides **46** and **47**, respectively (Fig. 6). Subsequent reductive N-O bond cleavage furnishing the amino alcohols **49** and **50** followed by in situ ring closure via S_N2 displacement may afford lasubine I (**39**) and 2-epilasubine II (**48**), respectively. In the case of the reaction leading to lasubine I, the initially formed conformer **39'**, which has a <u>trans</u>-fused quinolizidine ring and an axial aryl group should be thermodynamically disfavored because of strong non-bonded interactions between the aromatic hydrogens at C-2' and/or C-6' and the two axial quinolizidine hydrogens at C-2 and/or C-6. Thus **39'** epimerizes to the thermodynamically more stable <u>cis</u>-fused quinolizidine conformation **39** by nitrogen inversion.

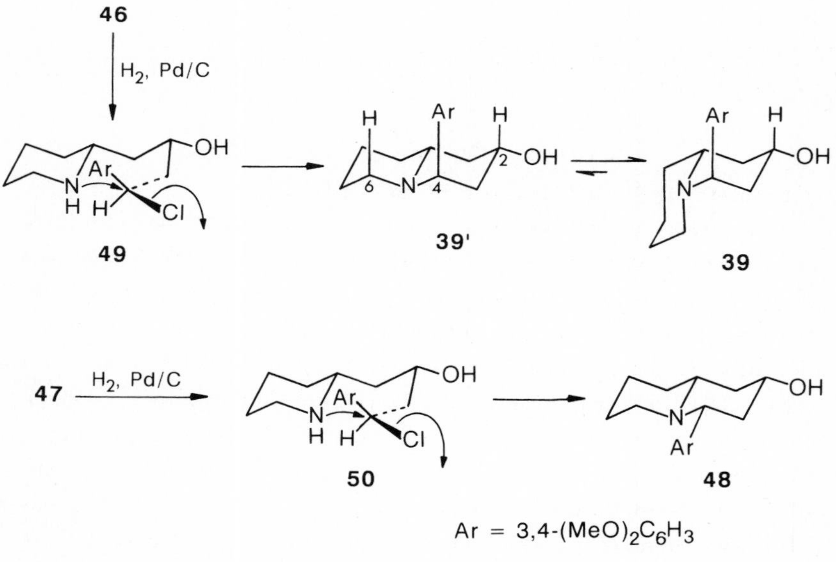

Fig. 6 Formation of lasubine I (**39**) and 2-epilasubine II (**48**).

Treatment of the lithium salt of synthetic lasubine I (**39**) with 3,4-di-methoxycinnamic anhydride gave (±)-subcosine I (**40**) in 48% yield. Thus the first total synthesis of (±)-subcosine I was achieved by utilizing 1,3-dipolar cycloaddition of a nitrone as a key step.

39	**40**

(a) n̲-BuLi, THF, -78 °C; (b) 3,4-dimethoxycinnamic anhydride, THF, r.t.

4 (+)-NEGAMYCIN AND (-)-3-EPINEGAMYCIN

In the area of cycloaddition reactions, inducing the chirality under control of chiral auxiliaries has represented a continuing challenge during the last decade (ref. 32). Our next goal in the program of nitrone methodology was the development of efficient asymmetric induction with chiral nitrones and its application to the synthesis for (+)-negamycin in natural form. Negamycin (**51**) (ref. 33), a structurally unique peptide-like natural product which exhibits striking activity against Gram-negative bacteria, including Pseudo-monas aerginosa (ref. 34), has attracted considerable synthetic interest (refs. 35, 36) since its structure elucidation in 1971 (ref. 36). Our strategy for the chiral synthesis of (+)-**51** is outlined retrosynthetically in Fig. 7. The key step envisioned would involve an enantioselective 1,3-dipolar cycloaddition of a nitrone with an appropriate chiral auxiliary to the allylamine. This cycloaddition would simultaneously create two new asymmetric centers adapt-able to the 3R̲,5R̲ stereochemistry of (+)-**51**.

Of reported chiral inductor groups in asymmetric nitrone cycloaddition (ref. 38), carbohydrate derivatives initially developed by Vasella (ref. 39) seemed to be most applicable. Accordingly, our first objective was to develop the efficient chiral N-glycosyl nitrones and to demonstrate acceptable dia-stereoselection during the cycloaddition.

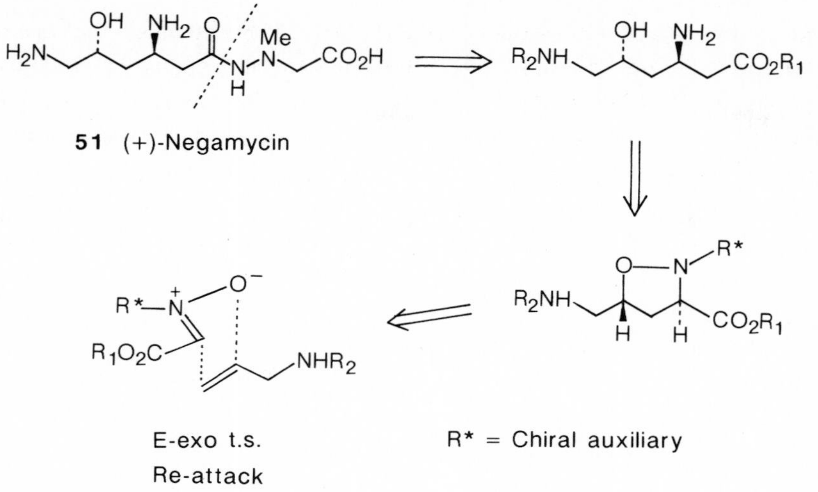

Fig. 7   Retrosynthetic analysis of (+)-negamycin (51).

4.1 Asymmetric 1,3-Dipolar Cycloaddition

Treatment of known 1-O-benzyl-D-<u>ribo</u>-furanose (**52**) with 1,1-dimethoxy-cyclohexane was followed by methylation and hydrogenation, affording **53** (70% overall yield), which was then quantitatively converted to the oxime **54**.   The nitrone **55**, generated in situ by the reaction of **54** with methyl glyoxylate, was allowed to react with the allylamine derivative **56** (toluene, reflux) to furnish an inseparable mixture of the (3<u>R</u>,5<u>R</u>)-<u>trans</u> (**57a**) and (3<u>S</u>,5<u>R</u>)-<u>cis</u> (**57b**) adducts in total yield of 91%.   The mixture was subjected to LiAlH$_4$ reduction, hydrolysis to remove the chiral auxiliary (diluted HCl, MeOH), and N-benzylation to give the <u>trans</u> (3<u>R</u>,5<u>R</u>)-**58a** and <u>cis</u> (3<u>S</u>,5<u>R</u>)-**58b** alcohols in a ratio of 4:9 (total overall yield: 46% from **57a** + **57b**), after chromatographic separation.   The ee of the products was determined as the Mosher's (+)-MTPA ester (ref. 40), proving to be 73.6% (Fig. 8) (ref. 41).

This result indicates that the asymmetric induction observed with D-ribosyl nitrone **55** is less than satisfactory.   A much better result was obtained by employing the D-glosyl template shown in Fig. 9 (ref. 42).   D-Gulono-γ-lactone (D-**59**) was converted to 2,3:5,6-O-cyclohexylidene-D-<u>gulo</u>-furanose (D-**60**) by treatment with 1,1-dimethoxycyclohexane followed by DIBAL reduction in 88% yield from D-**59**.   Compound D-**60** was converted quantitatively to the oxime D-**61**.   The nitrone D-**62**, generated in situ by the reaction of D-**61** with methyl glyoxylate as a mixture of <u>E</u> and <u>Z</u> isomers, was allowed to react with the allylamine derivative **56** to produce a mixture of the 3<u>R</u>,5<u>R</u>-<u>trans</u> (D-**63a**) and 3<u>S</u>,5<u>R</u>-<u>cis</u> (D-**63b**) adducts in 84% yield.   After

removal of the D-gulosyl auxiliary group by acid hydrolysis, the diastereomeric mixture of the products was subjected to N-benzylation followed by
LiAlH$_4$ reduction to provide the chromatographically separable <u>trans</u> alcohol
(3<u>R</u>,5<u>R</u>)-**58a** and <u>cis</u> alcohol (3<u>S</u>,5<u>R</u>)-**58b** in a ratio of 2:3 (55% overall yield
from D-**63a** + D-**63b**). Thus utilization of the D-gulosyl chiral template in this
process resulted in a highly stereobiased synthesis of **58a** and **58b** in 93.7% ee
and 94.2% ee [determined as the (+)-MTPA ester (ref. 40)], respectively.

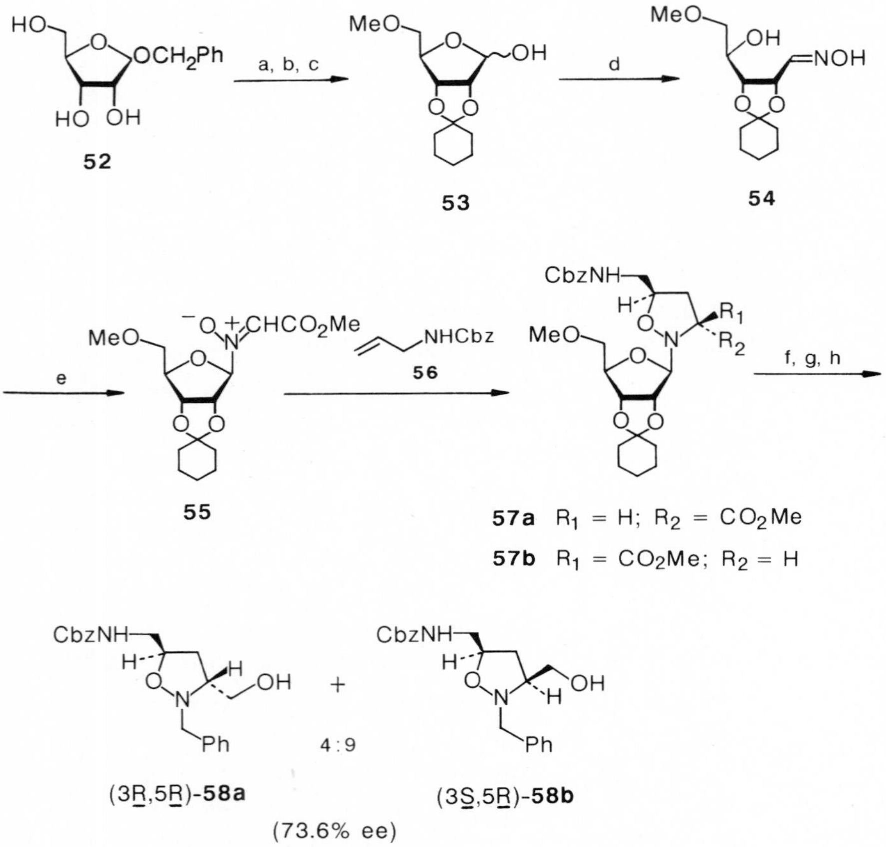

Fig. 8 (a) 1,1-dimethoxycyclohexane, TsOH, benzene, reflux; (b) MeI,
NaH, glyme, 40 °C; (c) H$_2$, Pd/C, MeOH; (d) NH$_2$OH•HCl, Py, r.t.;
(e) methyl glyoxylate, toluene, MS 4A, reflux; (f) HCl-MeOH, 40 °C;
(g) LiAlH$_4$, Et$_2$O, r.t.; (h) PhCH$_2$Br, K$_2$CO$_3$, DMF, 50 °C.

Fig. 9 (a) 1,1-dimethoxycyclohexane, TsOH, benzene, reflux; (b) DIBAL, toluene-THF, -78 °C; (c) NH$_2$OH•HCl, Py, r.t.; (d) methyl glyoxylate, toluene, reflux; (e) HCl-MeOH, 90 °C; (f) PhCH$_2$Br, K$_2$CO$_3$, DMF, 50 °C; (g) LiAlH$_4$, Et$_2$O, r.t.

Both trans (D-63a) and cis (D-63b) products obtained in this cyclo-addition using the nonconjugated olefin 56 as the dipolarophile must arise (applying Diels-Alder terminology) from the exo transition state since endo transition state would be greatly restricted by suffering from unfavorable steric interactions between the CH_2NHCbz group in the incoming dipolarophile **56** and the furan ring oxygen atom of the nitrone D-**62**; the <u>E</u> isomer of the nitrone D-**62** thus yields the <u>trans</u> adduct D-**63a**, while the <u>Z</u> isomer yields the <u>cis</u> adduct D-**63b**. The facial selectivity observed in this cycloaddition with both <u>E</u> and <u>Z</u> nitrones may be interpreted in terms of "O-endo" transition state model (A) (ref. 39b) as shown in Fig. 10, wherein, by analogy to recent reports (ref. 43), the electron-donating group (secondary alkyl) rather than the polar group (alkoxy) is perpendicular to the plane of the nitrogen-carbon double bond to permit the maximum orbital overlap of the participating centers, leading to the favored <u>re</u> face approach at the prochiral olefin. An alternative "O-exo" transition state model (B) should be disfavored due to serious nonbonded interactions between the furan ring oxygen and the $CHCO_2Me$ group (ref. 44). A similar approach to a prochiral diene has been observed in pericyclic cycloaddition reaction of chiral sugars (ref. 45).

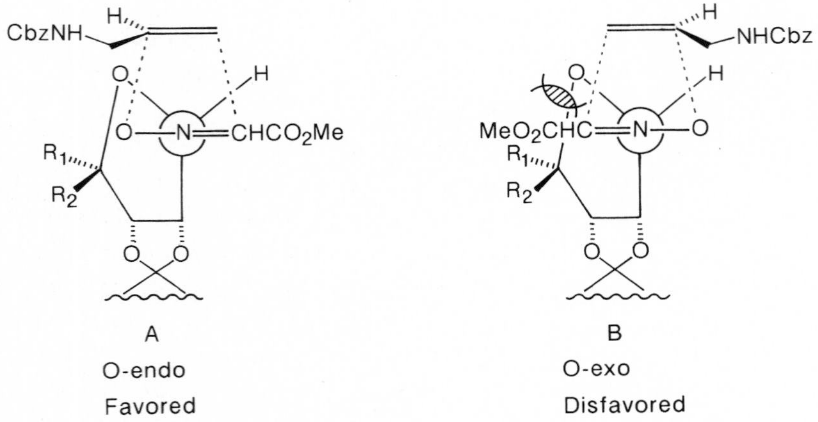

A
O-endo
Favored

B
O-exo
Disfavored

Fig. 10

Due to the above demonstrated capacity of the D-gulosyl auxiliary to create the chiral centers with high selectivity, opposite enantiomeric chiral induction by the asymmetric nitrone cycloaddition was then investigated. In this regard, as an easily available and inexpensive enantiomeric chiral template the L-gulosyl nitrone L-**62** was prepared from D-glucuronolactone (**64**) according to Fig. 11. The cycloaddition of L-**62** with **56** (affording a mixture of L-**63a** and L-**63b**) followed by separation by silica gel chromatography gave the cis adduct L-**63b**, which was converted to the alcohol (3R,5S)-**58b** (ref. 46). The product (3R,5S)-**58b** obtained was identical in all respects except the sign of optical rotation with the product (3S,5R)-**58b** previously obtained from the D-gulosyl nitrone D-**62** (Fig. 9).

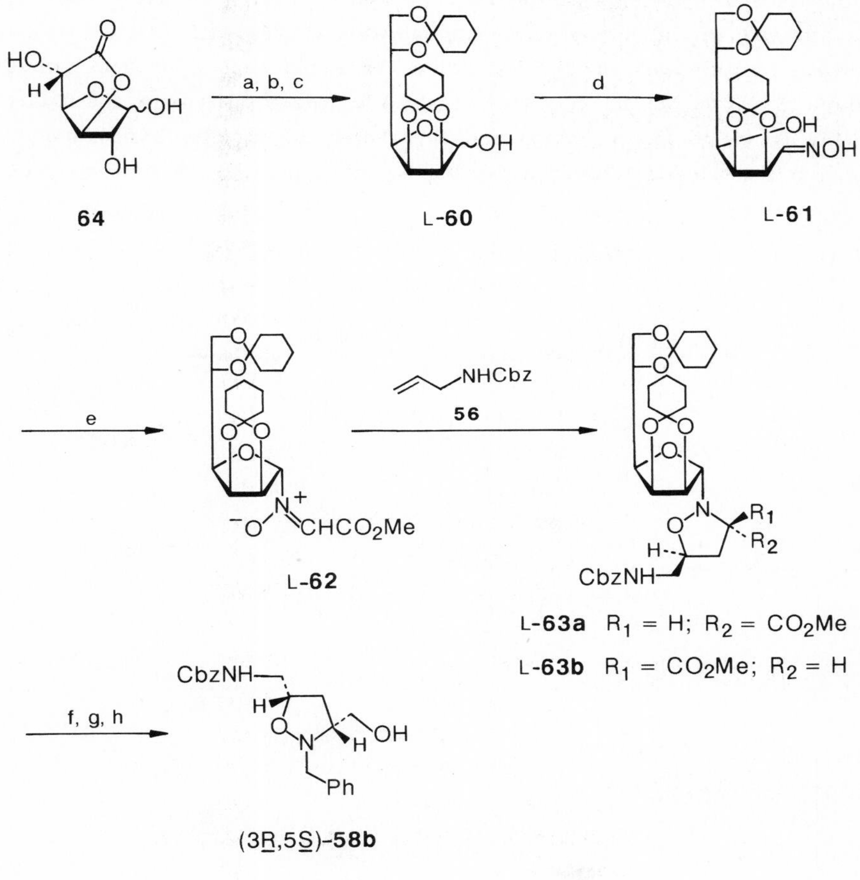

Fig. 11   (a) H_2, PtO_2, MeOH; (b)–(h) the same as (a)–(g) in Fig. 9.

4.2 Synthesis of (+)-Negamycin

The required C_3, C_5-asymmetric centers for (+)-negamycin were thus established via high level of asymmetric induction. The six step synthesis of (+)-negamycin utilizing the <u>trans</u> alcohol (3R,5R)-58a as a chiral precursor was executed as follows (Fig. 12). Tosylation of (3R,5R)-58a followed by substitution (NaCN, Me$_2$SO) gave the nitrile 65a in 72% overall yield. Compound 65a was converted to the carboxylic acid 66a in 79% yield via ethanolysis followed by saponification. Condensation of 66a with benzyl (1-methylhydrazino)acetate was carried out using the mixed carboxylic acid anhydride method (EtCO$_2$Cl, Et$_3$N) (ref. 47) to lead to the formation of the hydrazide 67a in 67% yield. Deprotection and N–O bond cleavage by hydrogenolysis at the same time, followed by purification by silica gel chromatography provided

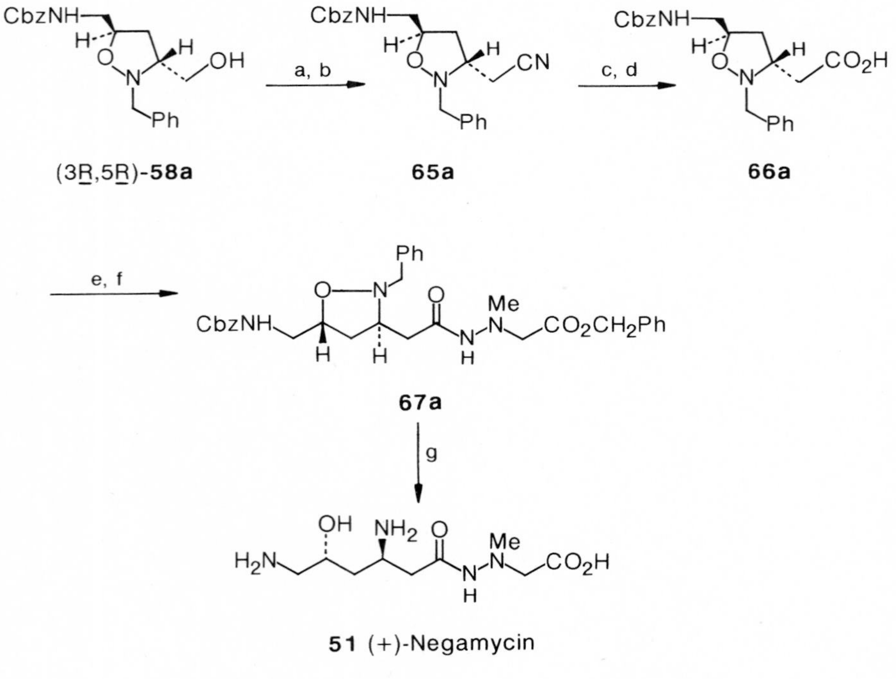

(3R,5R)-58a   65a   66a

67a

51 (+)-Negamycin

Fig. 12   (a) TsCl, (i-Pr)$_2$NEt, Et$_3$N, CH$_2$Cl$_2$, r.t.; (b) NaCN, Me$_2$SO, 50 °C; (c) HCl-EtOH, r.t.; (d) aq. NaOH, MeOH, r.t.; (e) ClCO$_2$Et, Et$_3$N, toluene, 0 °C; (f) benzyl (1-methylhydrazino)acetate, r.t.; (g) H$_2$ (3 atm), Pd/C, aq. AcOH-MeOH.

(+)-negamycin (**51**) in 75% yield. This material was found to be identical to natural negamycin by TLC, ^{1}H NMR, and $[\alpha]_D$ [+2.3° for synthetic **51**; +2.5° for natural **51** (ref. 37)]. The antibacterial activity of synthetic (+)-negamycin against a variety of bacteria was proved to be identical to those reported for natural negamycin and also to be effective to Pseudomonas aerginosa harboring multiple drug resistant plasmid kr102 (ref. 48).

4.3 Synthesis of (-)-3-Epinegamycin

Next we planned to apply this methodology to the synthesis of optically active 3-epinegamycin (**68**) by transformation of the (3S,5R)-cis isomer **58b** (Fig. 13). Compound (3S,5R)-**58b** was converted in four steps to the carboxylic acid **66b**, which was then worked up in a manner similar to that described for **65a**, giving rise to the hydrazide **67b** in 34.4% overall yield from (3S,5R)-**58b**. Hydrogenolysis of **67b** followed by silica gel chromatography afforded (-)-3-epinegamycin (**68**) in 65% yield. In antibacterial tests (-)-3-epinegamycin was virtually inactive against a variety of Gram-positive and Gram-negative bacteria (ref. 49).

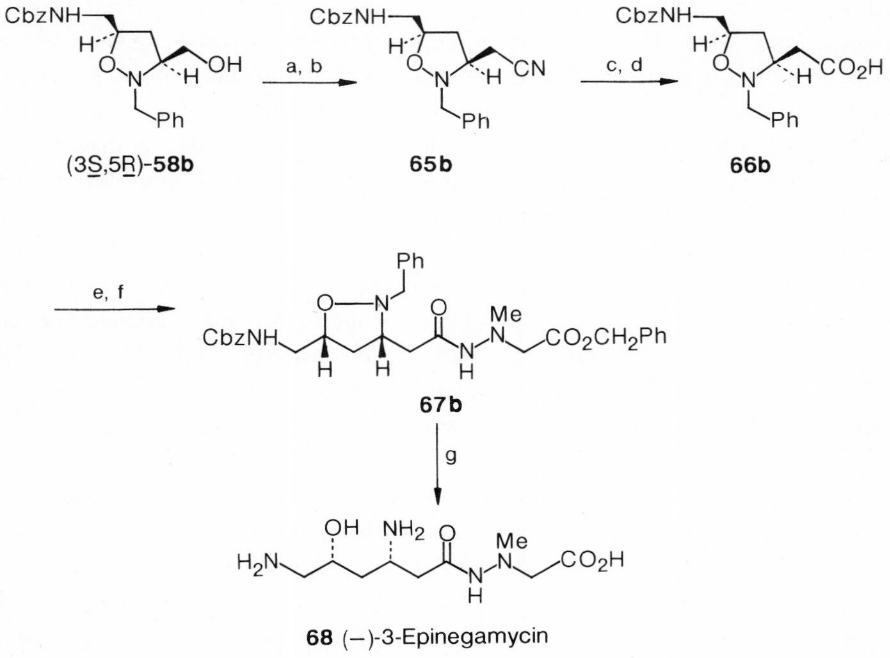

Fig. 13   (a)-(g) the same as in Fig. 12.

5 TROPANE ALKALOIDS

The tropane alkaloids occur as esters of relatively simple organic carboxylic acids with amino alcohols (alkamines) which are all hydroxylated derivatives of tropane (69), i.e., tropine (70), pseudotropine (71), and nortropine (72) as monohydroxylated alkamines. Because of their pharmaceu-

69 R_1 = Me; R_2 = R_3 = H
70 R_1 = Me; R_2 = H; R_3 = OH
71 R_1 = Me; R_2 = OH; R_3 = H
72 R_1 = R_2 = H; R_3 = OH
73 R_1 = Me; R_2, R_3 = O

tical significance and the presence of an unusual ring system, this class of alkaloids have been the subject of intensive stereochemical, biogenetical, and synthetic activities. In particular, a great deal of synthetic work on natural and nonnatural tropane bases has been carried out with the aim of investigating their pharmacological activity. The earliest synthetic approach to a tropane base was described by Willstätter (ref. 50). This approach to tropinone (73) in a multistage synthesis was followed by a more lucid and practical Robinson synthesis (ref. 51). Since these classical syntheses of tropinone (73), a number of general synthetic methods for the preparation of some tropanes have been reported. However, except for two instances of new approaches to tropane alkaloids via [3 + 2] cycloaddition by Tufariello (ref. 52) and [3 + 4] cycloaddition by Noyori (ref. 53), eqns. (5) and (6), respectively, efficient methods for the preparation of naturally occurring tropane alkaloids are limited.

We anticipated that a hetero Diels-Alder cycloaddition using nitroso compounds as dienophiles could be a successful new approach in accessing heterocyclic natural products. Our initial investigations focused on this area, which had been little explored at the time our work began (ref. 54). Thus attention was turned to developing a facile new route for the elaboration of the tropane ring system by utilizing a [4 + 2] nitroso cycloaddition and its application to the synthesis of the naturally occurring tropane alkaloids as depicted by eqn. (7) (ref. 55).

Pseudotropine   (5)

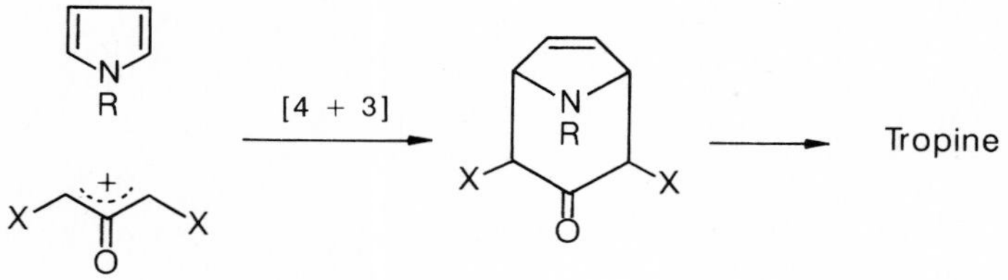

Tropine   (6)

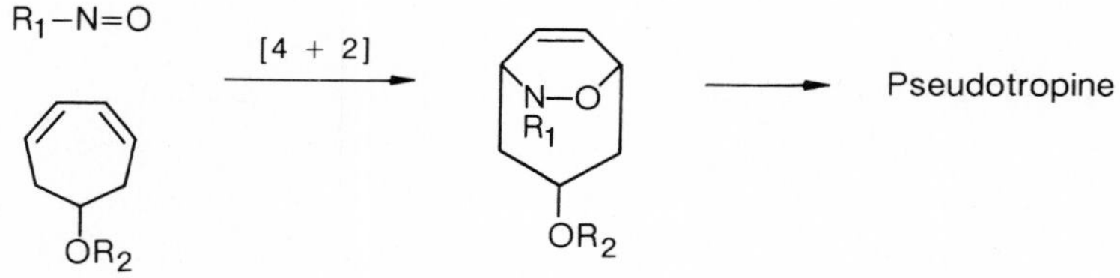

Pseudotropine   (7)

5.1 N-Benzoylnortropane

As our first model we chose N-benzoylnortropane (**79**) to investigate construction of the tropane ring system based on [4 + 2] nitroso cycloaddition with a 1,3-cycloheptadiene. A search of the literature indicated that only one example of a Diels-Alder cycloaddition of a nitroso compound with a seven-membered ring diene has been reported (ref. 56). In view of this, the present study of tropane synthesis was initiated by the examination of the nitroso Diels-Alder reaction of 1,3-cycloheptadiene (**74**) (Fig. 14). Thus reaction of **74** with the acylnitroso compound **75** generated in situ from benzo-hydroxamic acid by oxidation with tetrapropylammonium metaperiodate (ref. 57) was carried out at room temperature, affording the [4 + 2] cycloadduct **76** in 85% yield. Reductive N–O bond cleavage of **76** with sodium amalgam in ethanol gave the amide alcohol **77** (77%), which was converted to the <u>trans</u> chloride **78** by hydrogenation followed by treatment with thionyl chloride and triethylamine (64%). The desired N-benzoylnortropane (**79**) was obtained in 87% yield when the chloride **78** was treated with potassium tert-butoxide in HMPA–benzene (1:1).

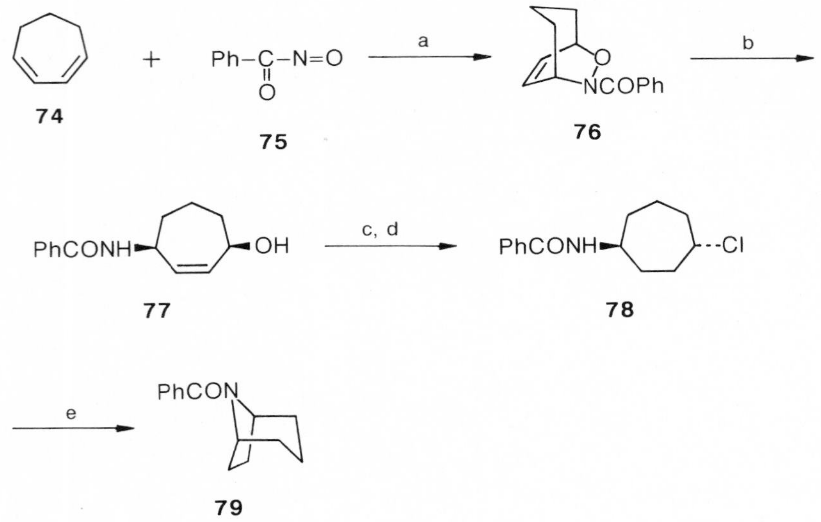

Fig. 14 (a) DMF-CHCl$_3$, r.t.; (b) Na-Hg, EtOH, Na$_2$HPO$_4$, 0°C → r.t.; (c) H$_2$, Pd/C, MeOH; (d) SOCl$_2$, Et$_3$N, CHCl$_3$, r.t.; (e) t-BuOK, HMPA-benzene, r.t.

5.2 Tropane

An alternative approach for the constructing the tropane ring system according to [4 + 2] nitroso cycloaddition utilizing 1-chloro-1-nitrosocyclo-hexane (**80**) as a dienophile is as follows (Fig. 15). The cyclic diene **74** was reacted with **80** in an ethanol–carbon tetrachloride solution at -20 °C for 10 h then at -10 °C for 48 h to yield the cycloadduct as its hydrochloride **81** in 68% yield. Catalytic hydrogenation of **81** afforded the amino alcohol **82** (isolated as its hydrochloride) which was then subjected to selective acylation with benzyl or ethyl chloroformates to provide the desired carbamates **83** or **84** in 80% and 84% yields, respectively.

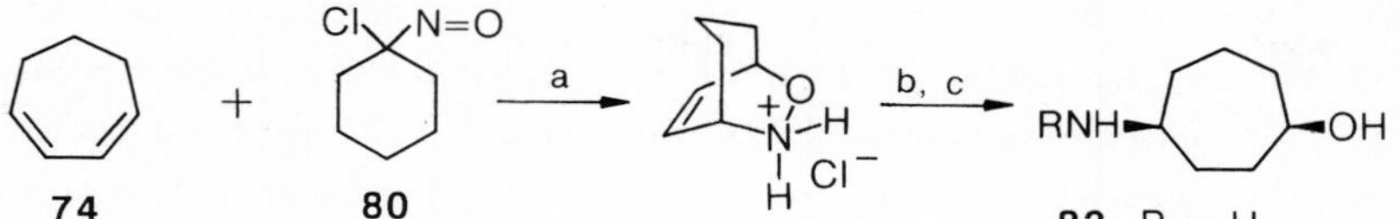

74 **80** **81**

82 R = H
83 R = Cbz
84 R = CO_2Et

Fig. 15 (a) EtOH-CCl_4, -10 °C; (b) H_2, Pd/C, MeOH; (c) benzyl (for **83**) [or ethyl (for **84**)] chloroformate, aq. K_2CO_3, $CHCl_3$, 0 °C → r.t

When the benzyl carbamate **83** was treated with thionyl chloride, the results varied widely depending upon the reaction conditions employed. Thus treatment with thionyl chloride in chloroform in the absence of a base gave a 1:1 mixture of the dehydration products **86a** and **87a** in 61% yield along with the chloride **85a** in 12% yield. When the reaction was carried out in the presence of 1.3 equiv of pyridine at room temperature the chloride **85a** was obtained in 32% yield along with the 1:1 olefin mixture of **86a** and **87a** (6%) and the dicycloheptylsulfite **88a** (18%). On heating in the presence of excess pyridine, **85a** was obtained in improved yield (55%), together with the olefin mixture of **86a** and **87a** (4%) and **88a** (18%). Otherwise, the reaction of ethyl carbamate **84** with thionyl chloride and pyridine (1.9 equiv) in chloroform at reflux generated the chloride **85b** (65%) accompanied, by the 1:1 mixture of olefins **86b** and **87b** (11%) and the dicycloheptyl sulfite **88b** (5%).

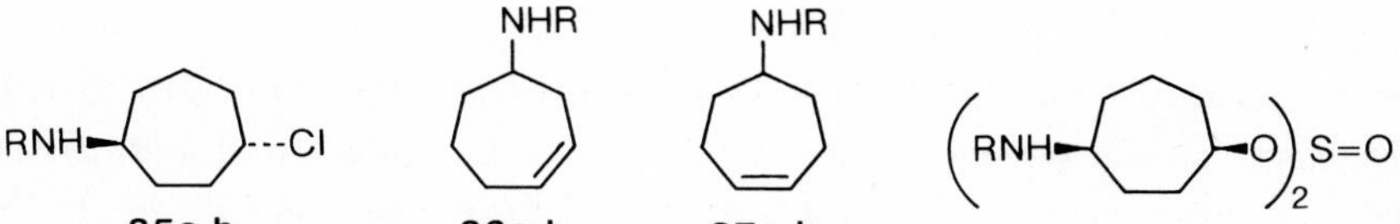

85a,b **86a,b** **87a,b** **88a,b**

a R = Cbz; **b** R = CO_2Et

Recent studies have demonstrated (refs. 58-60) the potential synthetic utility of C–N bond formation via heteromercuration of carbamate derivatives of unsaturated amines. This prompted us to utilize the dehydration products **86a** and **87a** in the tropane synthesis via intramolecular heteromercuration. Thus the 1:1 mixture of **86a** and **87a** was treated with mercuric acetate followed by reduction with NaBH$_4$ to furnish N-carbobenzoxynortropane (**93**) in 45% yield (based on reacted starting material). The product **93** must result from reductive demercuration of both organomercurials **91** and **92** initially formed from the asymmetrical and symmetrical olefins **86a** and **87a**, respectively, via cationic intermediates **89** and **90** as depicted in Fig. 16. Thus separation of the olefin mixture is not necessary since both the olefins can equally be utilized as the reaction substrates in this cyclization.

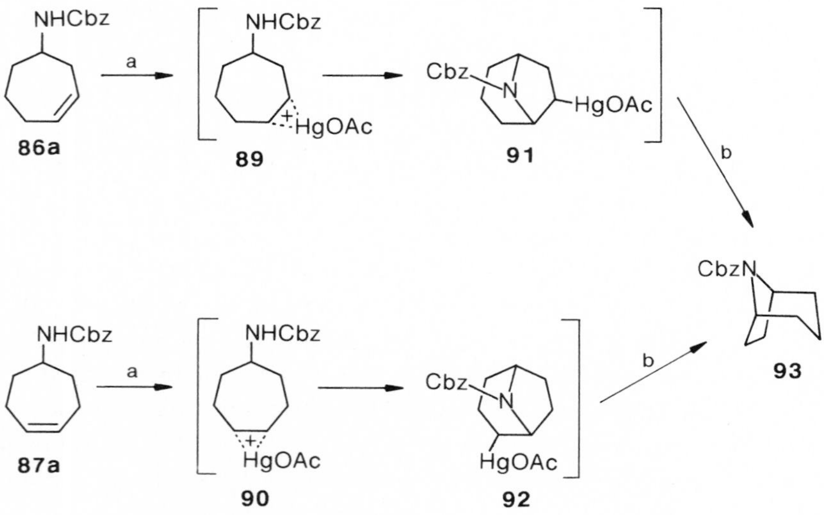

Fig. 16 (a) Hg(OAc)$_2$, aq. THF, r.t.; (b) NaBH$_4$, aq. NaOH, r.t.

Alternatively, the chloride **85a** was directly converted into N-carbobenzoxynortropane (**93**) in satisfactory yield (78%) when subjected to base-induced intramolecular "amidocyclization" as previously described for the preparation of N-benzoylnortropane (**79**) (Fig 14). In a similar reaction involving the chloride **85b**, N-(ethoxycarbonyl)nortropane (**94**) was formed in 78% yield. The synthesis of tropane (**69**) was achieved by LiAlH$_4$ reduction of the benzyl carbamate **93** or ethyl carbamate **94** in 72 or 71% yield, respectively (Fig. 17).

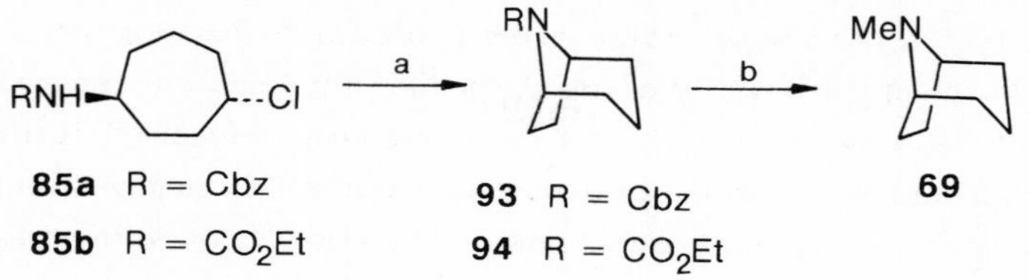

RNH$\blacktriangleright$---Cl $\xrightarrow{a}$ (RN) $\xrightarrow{b}$ (MeN)

85a R = Cbz

85b R = CO$_2$Et

93 R = Cbz

94 R = CO$_2$Et

69

Fig. 17   (a) t-BuOK, HMPA-benzene, r.t.; (b) LiAlH$_4$, THF, reflux.

5.3 Pseudotropine and Tropacocaine

Having developed the method for the preparation of the tropanes based on a nitroso Diels-Alder approach, we sought to apply this to the synthesis

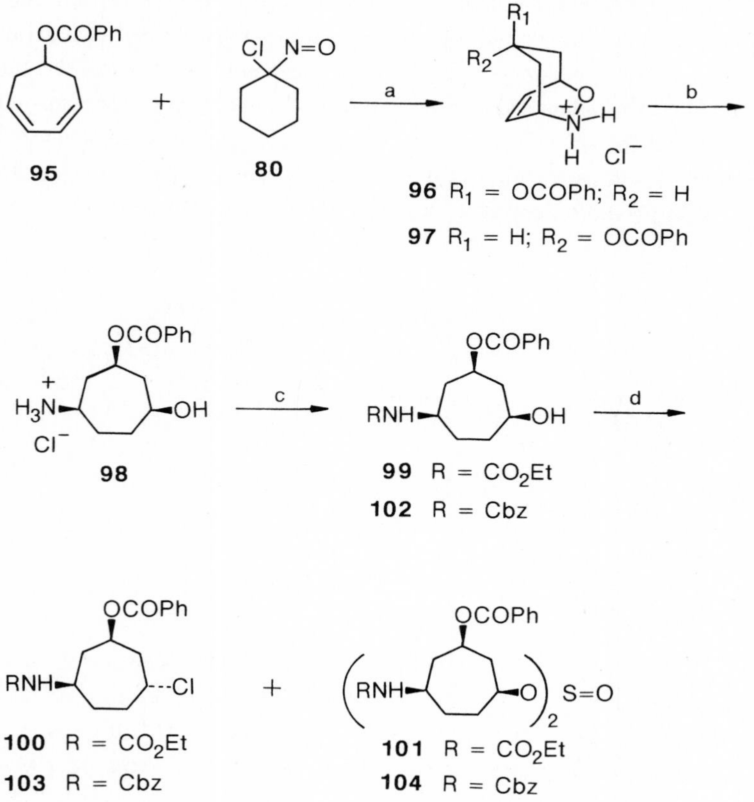

OCOPh + Cl N=O $\xrightarrow{a}$ R$_2$ R$_1$ ... O $\overset{+}{N}$—H ... H Cl$^-$ $\xrightarrow{b}$

95

80

96 R$_1$ = OCOPh; R$_2$ = H

97 R$_1$ = H; R$_2$ = OCOPh

OCOPh H$_3$$\overset{+}{N}$$\blacktriangleright$ ---OH Cl$^-$ $\xrightarrow{c}$ OCOPh RNH$\blacktriangleright$ ---OH $\xrightarrow{d}$

98

99 R = CO$_2$Et

102 R = Cbz

OCOPh RNH$\blacktriangleright$ ---Cl + (RNH$\blacktriangleright$ OCOPh ---O)$_2$ S=O

100 R = CO$_2$Et

103 R = Cbz

101 R = CO$_2$Et

104 R = Cbz

Fig. 18   (a) EtOH-CCl$_4$, -10 °C; (b) H$_2$, Pd/C, MeOH; (c) ethyl (for 99) [or benzyl (for 102)] chloroformate, aq. K$_2$CO$_3$, CHCl$_3$, 0 °C $\rightarrow$ r.t.; (d) SOCl$_2$, Py, CHCl$_3$, reflux.

of the naturally occurring tropane alkaloids pseudotropine (**71**) and tropa-cocaine (**109**) (ref. 61), which exhibit marked local anesthetic action.

From the above results, the most reliable approach to these alkaloids seemed to be the one involving cycloaddition with 1-chloro-1-nitrosocyclo-hexane (**80**) followed by base-induced amidocyclization as outlined in Fig. 18. Thus the reaction of 3,5-cycloheptadienyl benzoate (**95**) with the nitroso compound **80** generated a 4:1 mixture of the oxazabicyclononene hydrochlo-rides favoring exo form **96** over endo form **97**. The stereochemistry of these cycloadducts was verified with their free bases on the basis of their ^{1}H NMR spectra. The signal attributed to the 3-endo proton adjacent to the benzoyl-oxy group in the spectrum of the major adduct (free base of **96**) occurs at significantly higher field (δ 4.95) than usual, indicating that it must lie within the shielding cone of the C-6–C-7 double bond, while the 3-exo proton signal (δ 5.57) suffers much less from a shielding effect of the double bond.

The major cycloadduct **96**, readily separable from **97** by recrystallization, underwent catalytic hydrogenation to give the amino alcohol hydrochloride **98** which was in turn selectively N-acylated with ethyl chloroformate affording the carbamate **99** in 84% overall yield. Chlorination of **99** with thionyl chloride and excess of pyridine in chloroform at reflux yielded the chloride **100** (55%) along with a byproduct assigned the dicycloheptyl sulfite **101** (28%). Similar-

EtOCON⟩—[ ]—OCOPh
105

MeN⟩—[ ]—OH
71

OCOPh
RNH—[ ]—CI

100 R = CO₂Et
103 R = Cbz

a

b

a

CbzN⟩—[ ]—OCOPh
106

c, d

RN⟩—[ ]—OCOPh
108 R = H
109 R = Me

Fig. 19 (a) t-BuOK, HMPA-benzene, r.t.; (b) LiAlH₄, THF, reflux; (c) H₂, Pd/C, MeOH; (d) HCO₂H, HCHO, reflux.

ly, treatment of **102**, prepared from **98**, with thionyl chloride and pyridine gave **103** (36%) and **104** (38%).

The consequent tropane ring elaboration was pursued via base-induced amidocyclization of the chloride **100** under the previously described conditions (<u>t</u>-BuOK, HMPA-benzene, room temperature), resulting in the formation of **105** (46%) (Fig. 18). Reduction with LiAlH$_4$ converted **105** to pseudotropine (**71**) (67%). Furthermore, the chloride **103** was subjected to amidocyclization as well to give **106**. Deprotection by catalytic hydrogenation converted **106** into N-nortropacocaine (**108**), which subsequently underwent the Eschweiler-Clarke reaction with formic acid and formaldehyde to provide desired tropacocaine (**109**) in 71% yield from **106** (Fig. 19).

6 INTRAMOLECULAR NITROSO DIELS-ALDER REACTION

The methodology, based on nitroso Diels-alder reaction, that proved useful in the synthesis of tropane alkaloids also seemed to open an attractive route to some other alkaloids. Compared to the intramolecular imino Diels-Alder reaction (ref. 3c), the intramolecular variant of the nitroso Diels-Alder reaction has received far less attention (ref. 54b, 62), despite the enormous potential it holds for alkaloid synthesis. With this in mind, we proceeded to examine the application of the intramolecular nitroso Diels-Alder cycloaddition in the synthesis of alkaloids possessing saturated nitrogen heterocyclic ring systems.

6.1 Gephyrotoxin 223AB

Gephyrotoxin (GTX) 223AB (**110**), one of the neurotoxin alkaloids

110

isolated in minute quantity from skin extracts of neotropical poison-dart frogs (family Dendrobatidae) (refs. 63, 64), has attracted a vast amount of interest because of its unusual biological characteristics. Numerous groups have thus been involved in the development of methodology for the total synthesis of this molecule. These efforts have recently resulted in the synthesis of **110** (ref. 65) and its stereoisomers (ref. 66). Our approach to **110** involving an acylnitroso Diels-Alder reaction, which took place in an intramolecular fashion (Fig. 20) (ref. 67), is distinguished from these previous efforts (ref. 68) by the fact that the entire sequence provided single stereoisomers in the desired sense and thus met with no separation problems.

The hydroxamic acid **111**, a precursor to the acylnitroso compound **112**, was formed when the ester, prepared from 4-pentanal via twofold Wittig reactions, was exposed to hydroxylamine in methanolic KOH at room temperature in 81% yield. Oxidation of **111** by the periodate in chloroform (0–5 °C)

Fig. 20 (a) n̲-Pr$_4$N(IO$_4$), CHCl$_3$, 0–5 °C; (b) H$_2$, Pd/C, MeOH; (c) n̲-PrMgBr, Et$_2$O, 0 °C→r.t.; (d) NaCNBH$_4$, HCl-MeOH (pH 3.8–5.4), 0 °C→r.t.

generated **112** in situ. Under the reaction conditions, **112** underwent intramolecular [4 + 2] cycloaddition to give as the single product the 1,2-oxazine derivative **113** in 82% yield. Hydrogenation of **113** provided **114** in 90% yield. Then **114** was treated with the Grignard reagent at ambient temperature to give the enamine **115** (89%), which was subsequently subjected to reduction with NaCNBH$_3$ in methanol at pH 3.8–5.4 resulting in the exclusive formation of **116** (70%).

The ^{13}C NMR spectrum of **116** at 24 °C showed pairs of resonances for each of the carbons in the molecule. The relative intensity of each doublet was nearly unit independent of temperature and the doublet collapsed into a single line at high temperature depending on its peak separation. These observations strongly indicate that **116** exists in conformational equilibrium between **116A** and **116B** (eqn. 8) due to nitrogen inversion with nearly equal population energy barrier ($\Delta G^{\ddagger}$), estimated to be about 8.0 kcal mol^{-1}.

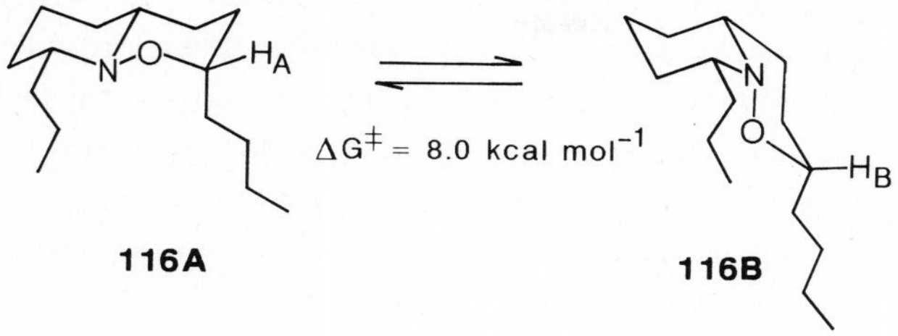

$$\Delta G^{\ddagger} = 8.0 \text{ kcal mol}^{-1}$$

(8)

116A   **116B**

The ^{1}H NMR spectra in pyridine-d$_5$ of **116** at 27.5 °C showed two sets of multiplets at δ 3.87 ($W_{1/4}$ = 23.9 Hz) and 3.72 ($W_{1/4}$ = 31.6 Hz) with an integration ratio of 10:9 for the C-2 proton, indicative of the equatorial

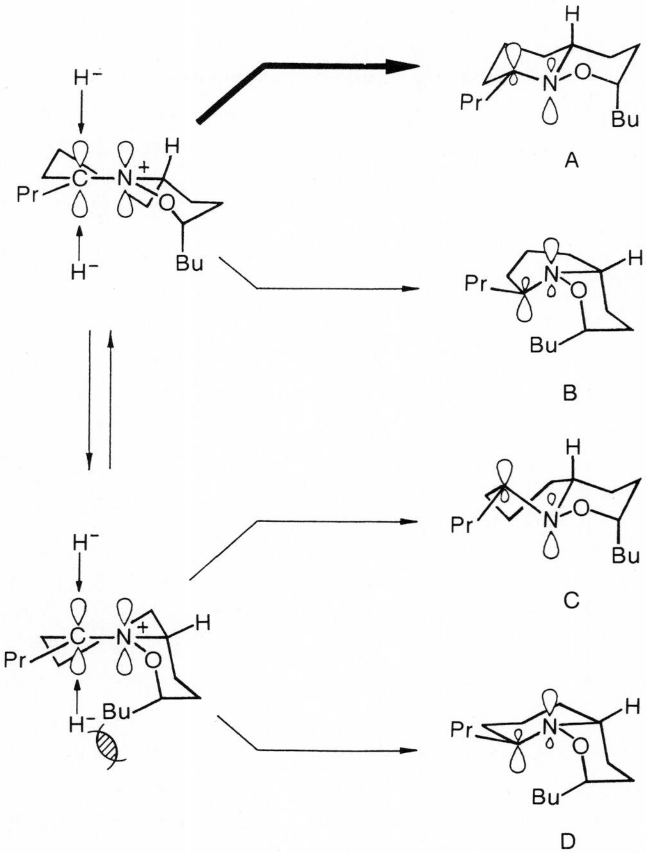

Fig. 21

hydrogen in **116A** and the axial hydrogen in **116B**, respectively. These C-2 proton signals converged to a single resonance centered at δ 3.88 at 100 °C.

The exclusive formation of **116** can be rationalized to be a result of stereoelectronically controlled reaction of the transient iminium salt generated from **116** under acidic conditions. Due to inversion at nitrogen, there are thus four possible transition states A–D (Fig. 21) which maintain maximum orbital overlap with respect to the approaching hydride ion and the developing nitrogen lone pair. Two of these transition states (A and D) are of more stable chair shape, with the latter being disfavored due to a strong <u>peri</u> interaction between the butyl group and the incoming hydride ion. On the other hand, A can accommodate the entering hydride ion without steric interference of the 2-alkyl side chain thereby leading to **116**.

With the required stereochemistry thus established, the only requirement in order to complete the synthesis was constructing the pyrrolidine moiety of the target molecule. Reductive cleavage of the N–O bond in **116** gave the amino alcohol **117** (85%). Exposure of this material to benzyl chloroformate in an alkaline solution furnished the hydroxy carbamate **118** (36%), along with **119** (24%) and **120** (25%), the latter two products of which could easily be hydrogenated back to the amino alcohol **117**. Thus actual yield of **118** based

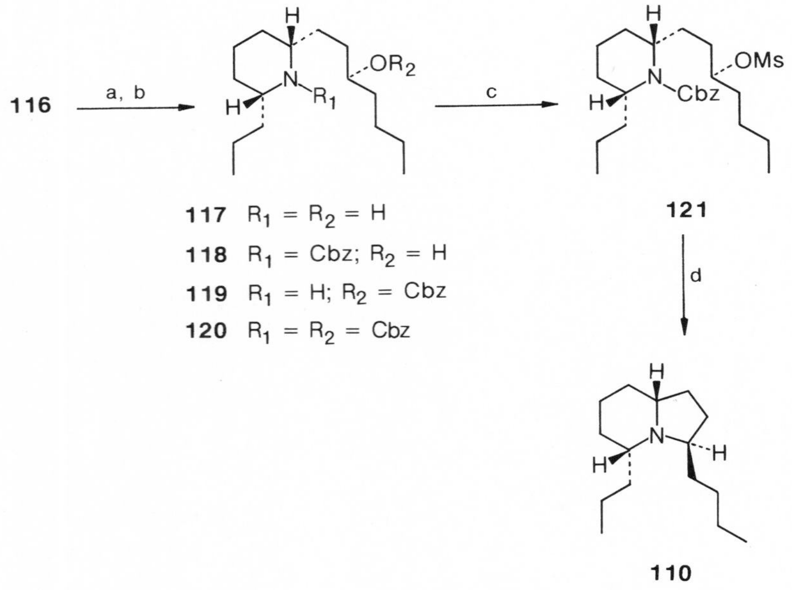

117	$R_1 = R_2 = H$
118	$R_1 = Cbz;\ R_2 = H$
119	$R_1 = H;\ R_2 = Cbz$
120	$R_1 = R_2 = Cbz$

121

110

Fig. 22 (a) Zn, aq. AcOH, 60 °C; (b) benzyl chloroformate, aq. Na$_2$CO$_3$, CHCl$_3$, r.t.; (c) MsCl, Et$_3$N, CH$_2$Cl$_2$, -10 °C; (d) H$_2$, Pd/C, MeOH.

on recovered **117** was 71%. Finally, the hydroxy carbamate **118** was converted to the mesylate **121** (83%) which upon hydrogenation provided GTX 223AB (**110**) in 81% yield as the sole product (Fig. 22).

6.2 Monomorine I

From the above survey of results the synthetic potential of the oxazino-lactam intermediate **114** in hand, we envisaged to synthesize monomorine I (**130**) which constitutes of an extension of the methodology based on intra-molecular nitroso Diels-Alder cycloaddition. The relative stereochemistry of this substance, isolated as one of the trail pheromones from Pharaoh ants (<u>Monomorium pharaonis</u> L.) (ref. 69), has been determined by nonstereoselective synthesis (ref.70). More recently, a stereospecific synthesis of racemic **130** (ref. 71) and a chiral synthesis of the (-)-enantiomer of natural **130** (ref. 72) were reported.

Our synthesis was initiated with the Grignard reaction of the bicyclic 1,2-oxazine **114** using methylmagnesium bromide to generate the unstable

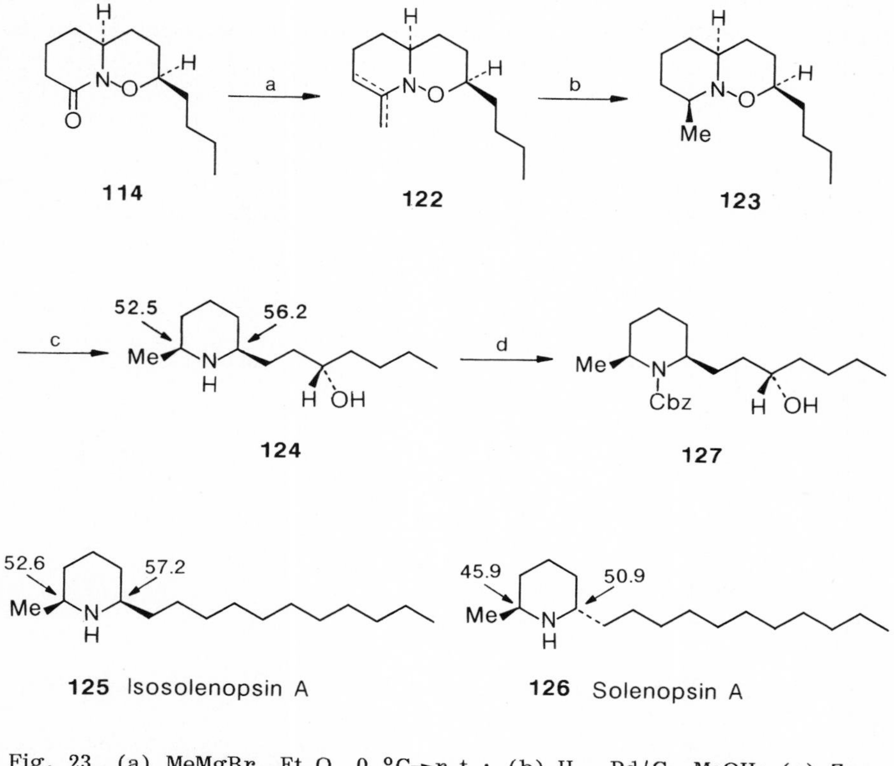

Fig. 23 (a) $MeMgBr$, Et_2O, 0 °C→r.t.; (b) H_2, Pd/C, MeOH; (c) Zn, aq. AcOH, 60 °C; (d) benzyl chloroformate, aq. Na_2CO_3, CH_2Cl_2, r.t.

enamine **122**, which was then immediately hydrogenated leading to a single isomer **123** in 70% overall yield. Compound **123** was subjected to N–O bond cleavage by treatment with zinc in aqueous acetic acid to give the amino alcohol **124** in 68% .yield. The <u>cis</u> relationship of C-2 and C-6 alkyl substituents in **124** was confirmed by the ^{13}C NMR spectrum showing 2-C and 6-C signals at δ 56.2 and 52.5, respectively, in good agreement with those reported for the fire-ant vernom isosolenopsin A (**125**) rather than those reported for solenopsin A (**126**) (ref. 73). Treatment of **124** with benzyl chloroformate (1.5 equiv) afforded the benzyl carbamate **127** (53%), accompanied by the formation of the mixture of the O-acyl (6%) and N,O-diacyl (32%) derivatives of **124** which could readily be reverted to starting **124** by hydrogenolysis in quantitative yield. Thus the actual yield of **127** based on recovered **124** was 85% (Fig. 23).

When exposed to iodotrimethylsilane at room temperature, **127** underwent C–O bond cleavage along with iodination to produce the silyl ester **128**. In situ cyclization was carried out by treating **128** in methanol at room temperature to provide (±)-monomorine I (**130**) and its C-3 epimer (**131**) in 42 and 40% yields from **127**, respectively. An appreciable improvement of the dia-

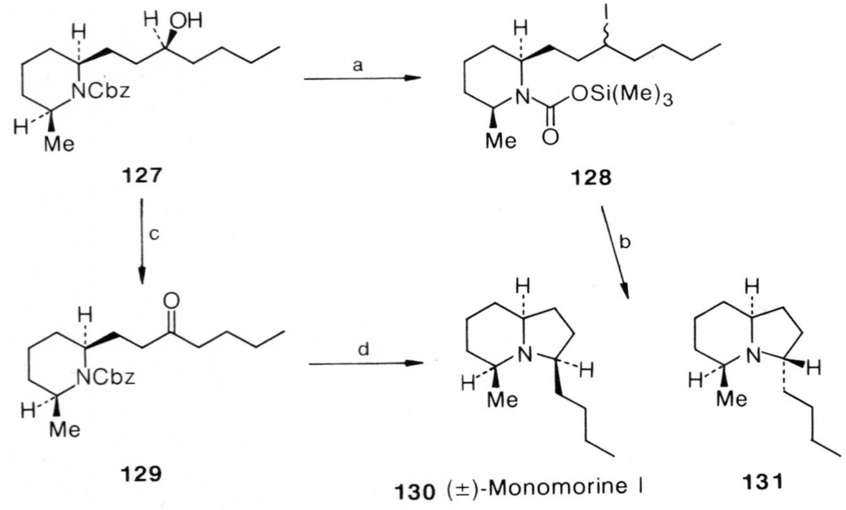

Fig. 24 (a) Me$_3$SiI, CH$_2$Cl$_2$, r.t.; (b) MeOH, r.t.; (c) CrO$_3$·2Py, CH$_2$Cl$_2$, r.t.; (d) H$_2$, Pd/C, MeOH.

stereoselectivity in the cyclization to monomorine I was obtained by the following sequence. With compound **127** in hand, the ketone **129** was prepared by Collins oxidation in 94% yield. On catalytic hydrogenation of **129**, (±)-monomorine I (**130**) was obtained in 71% yield, along with (±)-3-epimonomorine I (**131**) as a minor isomer (15%) (Fig. 24).

6.3 Dihydropinidine

In order to demonstrate further the utility of the above described sequence for the synthesis of gephyrotoxin 223AB (**110**) and monomorine I (**130**) involving intramolecular hetero Diels-Alder reaction followed by highly stereoselective introduction of the alkyl group to the amide we then explored an extension of these studies to the synthesis of dihydropinidine (**141**) (ref. 74).

Fig. 25   (a) $\underline{n}$-Pr$_4$N(IO$_4$), CHCl$_3$, -10 °C; (b) H$_2$, Pd/C, MeOH; (c) MeMgBr, Et$_2$O, 0 °C→r.t.; (d) H$_2$, Pd/C, MeOH; (e) HCO$_2$H, toluene, reflux; (f) PBr$_3$, CH$_2$Cl$_2$, r.t.; (g) H$_2$, Pd/C, Et$_3$N, MeOH; (h) 10% HCl, reflux.

The synthetic route utilizing this sequence for the elaboration of the cis-2,6-dialkylated piperidine is illustrated in Fig. 25. Intramolecular cyclo-addition of the acylnitroso compound 133, produced in situ from the corresponding hydroxamic acid 132 by periodate oxidation, yielded the oxazino-lactam 134. Compound 134 was converted to the enamine 136 by catalytic hydrogenation followed by treatment with methylmagnesium bromide. Catalytic hydrogenation of 136 resulted in both olefin hydrogenation and N-O bond cleavage providing the amino alcohol 137 as a single diastereomer. Confirmation that 137 bears the 2 and 6 alkyl substituents in a cis relationship was made by the ^{13}C NMR spectrum in the same manner as described above for the amino alcohol 124 (Fig. 23), a synthetic precursor to monomorine I. The signals for the 2-C and 6-C ring carbons appearing at δ 52.2 and 56.9 matched well with those reported for isosolenopsin A (125) (ref. 73).

The amino alcohol 137 was selectively N-formylated to give the formamide 138, deoxygenation of which was achieved by bromination with PBr$_3$ followed by hydrogenation to yield 140. Removal of the protective group by acid treatment afforded (±)-dihydropinidine (141) (Fig. 25).

REFERENCES

1. For reviews of 1,3-dipolar cycloaddition, see: (a) R. Huisgen, Angew. Chem., Int. Ed. Engl., 2, 565, 633 (1963). (b) A. Padwa, Angew. Chem. Int. Ed., Engl., 15, 123 (1976). (c) W. Oppolzer, Ibid., 16, 10, (1977). (d) A. Padwa, Ed., "1,3-Dipolar Cycloaddition Chemistry," Wiley Interscience, New York, 1984, Vols. 1 and 2.
2. For reviews of heterodiene cycloadditions, see: (a) J. H. Hamer and M. Ahmad in "1,4-Cycloaddition Reactions," J. H. Hamer, Ed., Acadimic Press; New York, 1967. (b) G. Desimoni and G. Tacconi, Chem. Rev., 75, 651 (1975). (c) D. L. Boger, Tetrahedron, 39, 2869 (1983).
3. For reviews of heterodinophile cycloadditions see: (a) S. M. Weinreb and J. I. Levin, Heterocycles, 12, 949 (1979). (b) S. M. Weinreb and R. S. Staib, Tetrahedron, 38 3127 (1982). (c) S. M. Weinreb, Acc. Chem. Res., 18, 16 (1985). (d) ref. 2c.
4. For reviews of 1,3-dipolar cycloaddition of nitrones see: (a) D. S. C. Black, R. F. Crozier, and V. C. Davis, Synthesis, 205 (1975). (b) J. J. Tufariello, Acc. Chem. Res., 11, 396 (1979) and also ref. 1.
5. For reviews of the Diels-Alder reaction of nitroso compounds see: G. W. Kirby, Chem. Soc. Rev., 6, 1 (1977) and also ref. 3.
6. Y. D. Phillipson, I. Tezcan, and P. J. Hylands, Planta Med., 25, 301 (1974).
7. (a) T. R. Govindachari, M. V. Lakshmikantham, and S. Rajadurai, Tetrahedron, 14, 284 (1961). (b) R. B. Herbert and C. J. Moody, J. Chem. Soc. D., 121 (1970). J. E. Cragg, R. B. Herbert, F. B. Jackson, C. J. Moody, and I. T. Nicolson, J. Chem. Soc., Perkin Trans. 1, 2477 (1982). (c) B. Chauncey and E. Gellert, Aust. J. Chem., 23, 2503 (1970). (d) A. J. Liepa and R. E. Summons, J. Chem. Soc., Chem. Commun., 826 (1977). (e) S. M. Weinreb, N. A. Khatri, and J. Shringarpure, J. Am. Chem. Soc., 101, 5073 (1979). N. A. Khatri, H. F. Schmitthenner, J. Shringarpure, and S. M. Weinreb, Ibid., 103, 6387 (1981). (f) V. K. Mangla and D. S. Bhakuni, Tetrahedron, 36, 2489 (1980). (g) M. Ihara,

393

M. Tsuruta, K. Fukumoto, and T. Kametani, J. Chem. Soc., Chem. Commun., 1159 (1985). (h) J. E. Nordlander and F. G. Njoroe, J. Org. Chem., **52**, 1617 (1987).

8. (a) H. Rapoport and T. F. Buckley, J. Org. Chem., **48**, 4222 (1983). (b) J. E. Nordlander and F. G. Njoroge, Ibid., **52**, 1627 (1987).

9. (a) H. Iida, M. Tanaka, and C. Kibayashi, J. Chem. Soc., Chem. Commun., 271 (1983). (b) H. Iida, Y. Watanabe, M. Tanaka, and C. Kibayashi, J. Org. Chem., **49**, 2412 (1984).

10. For syntheses of septicine, see: (a) J. H. Russel and H. Hunziker, Tetrahedron Lett., 4035 (1969). (b) T. R. Govindachari and N. Viswanathan, Tetrahedron, **26**, 715 (1970). (c) R. B. Herbert, F. B. Jackson, and I. T. Nicolson, J. Chem. Soc., Chem. Commun., 450 (1976) and the latter paper in ref. 7b. (d) R. V. Stevens and Y. Luh, Tetrahedron Lett., 979 (1977). (e) T. Iwashita, M. Suzuki, T. Kusumi, and H. Kakisawa, Chem. Lett., 383 (1980).

11. J. M. Gouley, R. A. Heacock, A. G. McInnes, B. Nikolin, and D. G. Smith, Chem. Commun., 709 (1969).

12. A. M. Dawidar, F. Winternitz, and S. R. Johns, Tetrahedron, **33**, 1733 (1977).

13. V. M. Chari, M. Jordan, and H. Wagner, Planta Med., **34**, 93 (1978).

14. The position of the p-coumaroyl group in ipomine previously published (4-position) (ref. 12) has been revised and is that represented in the present formula 12 (6-position) (ref. 13). In both the papers (refs. 11, 12) cited, the ipalbidinyl portion in the formula for ipomine was depicted incorrectly.

15. (a) T. R. Govindachari and N. Viswanathan, Heterocycles, **11**, 587 (1978). (b) I. R. C. Bick and W. Sidhaye in "The Alkaloids," R. H. F. Manske, Ed., Academic Press, New York, 1981, Vol. XIX, Chapter 3.

16. For syntheses of ipalbidine, see: (a) T. R. Govindachari, A. R. Sidhaye, and N. Viswanathan, Tetrahedron, **26**, 3829 (1970). (b) A. E. Wick, P. A. Bartlett, and D. Dolphin, Helv. Chim. Acta, **54**, 513 (1971). (c) ref. 9d. (d) S. H. Hedges and R. B. Herbert, J. Chem. Res., (S), 1 (1979); (M), 413. (e) A. S. Howard, G. C. Gerrans, and J. P. Micael, J. Org. Chem., **45**, 1713 (1980). (f) S. J. Danishefsky and C. Vogel, Ibid., **51**, 3915 (1986). (g) C. W. Jofford, T. Kubota, and A. Zaslana, Helv. Chim. Acta, **69**, 2048 (1986).

17. H. Iida, Y. Watanabe, and C. Kibayashi, Chem. Pharm. Bull., **33**, 351 (1985).

18. H. Iida, Y. Watanabe, and C. Kibayashi, Chem. Lett., 1195 (1983); J. Chem. Soc., Perkin Trans. 1, 261 (1985).

19. (a) K. N. Houk, J. Sims, R. E. Duke, Jr., R. W. Storozier, and J. K. George, J. Am. Chem. Soc., **95**, 7287 (1973). (b) K. N. Houk, J. Smith, C. R. Watts, and L. J. Lukus, Ibid., **95**, 7301 (1973).

20. (a) R. Grée, F. Tonnard, and R. Carrié, Tetrahedron Lett., 453 (1973). (b) J. J. Tufariello and Sk. Asrof Ali, Ibid., 4647 (1978).

21. A. T. Nielsen and W. J.Houlihan in "Organic Reactions," Vol. 16, John Wiley, New York, 1968, pp. 1-438.

22. E. Gellert and N. V. Riggs, Aust. J. Chem., **7**, 113 (1954); E. Gellert, Ibid., **9**, 489 (1956); H. Hoffman, Aust. J. Expt. Biol. Med. Sci., **30**, 541 (1952).

23. I. S. de la Lande, Aust. J. Exp. Biol. Med. Sci., **26**, 181 (1948).

24. N. R. Farnsworth, N. K. Hart, S. R. Johns, J. A. Lamberton, W. Messmer, Aust. J. Chem., **22**, 1805 (1969).

25. E. Gellert and R. Rudzats, J. Med. Chem., **7**, 361 (1964); G. R. Donaldson, M. R. Atkinson, and A. W. Murray, Biochem. Biophys. Res. Commun., **31**, 104 (1968); J. L. Hatwell and B. J. Abbott, Adv. Pharmacol. Chemother., **7**, 117 (1969).

26. (a) C. K. Bradsher and H. Berger, J. Am. Chem. Soc., **79**, 3287 (1957); **80**, 930 (1958). (b) P. Marchini and B. Belleau, Can. J. Chem., **36**, 581 (1958). (c) J. M. Paton, P. L. Pauson, and T. S. Stevens, J. Chem.

394

Soc. C, 1309 (1969). (d) E. Kotani, M. Kitazawa, and S. Tobinaga, Tetrahedron, 3027 (1974). (e) R. B. Herbert, J. Chem. Soc., Chem. Commun., 794 (1978). (f) ref. 7a.

27. H. Iida, C. Kibayashi, Tetrtahedron Lett., 22, 1913 and ref. 9a.

28. For previous syntheses of julandine, see: refs. 26c and 26e.

29. K. Fuji, T. Yamada, E. Fujita, and H. Murata, Chem. Pharm. Bull., 26, 2515 (1978).

30. For syntheses of of the related alkaloids based on nitrone methodology, see: (a) S. Takano and K. Shishido, J. Chem. Soc., Chem. Commun., 941 (1982). (b) S. Takano and K. Shishido, Heterocycles, 19, 1439 (1982). (c) K. Shishido, K. Tanaka, K. Fukumoto, and T. Kametani, Tetrahedron Lett., 24, 2783 (1983). (d) R. W. Hoffmann and A. Endesfelder, Liebigs Ann. Chem., 1823 (1986), 1177 (1985).

31. (a) H. Iida, M. Tanaka, and C. Kibayashi, J. Chem. Soc., Chem. Commun., 1143 (1983). (b) J. Org. Chem., 49, 1909 (1984).

32. For recent reviews of asymmetric cycloaddition reactions, see: (a) L. A. Paquette in "Asymetric Synthesis," J. D. Morrison, Ed., Academic Press, Orlando, Fl., 1984, Vol. 3, Part B, p. 455. (b) W. Oppolzer, Angew. Chem. Int. Ed. Engl., 23, 876 (1984).

33. M. Hamada, T. Takeuchi, S. Kondo, Y. Ikeda, H. Naganawa, K. Maeda, and H. Umezawa, J. Antibiot., 23, 170 (1970).

34. T. Korzybski, Z. Kowszyk-Gindifer, and W. Kurytowicz, "Antibiotics," American Society of Microbiology, Washington, D. C., 1978, Vol. 1, pp 343-346.

35. For syntheses of racemic negamycin, see: (a) W. Streicher, H. Reinshagen, and F. Turnowsky, J. Antibiot., 31, 725 (1978). (b) A. Pierdet, L. Nédélec, V. Delaroff, and A. Allais, Tetrahedron, 36, 1763 (1980). (c) G. Pasquet, D. Boucherot, and W. R. Pilgrim, Tetrahedron Lett., 21, 931 (1980).

36. For syntheses of (+)-negamycin in natural form, see: (a) S. Shibahara, S. Kondo, K. Maeda, H. Umezawa, and M. Ohno, J. Am. Chem. Soc., 94, 4353 (1972). (b) Y.-F. Wang, T. Izawa, S. Kobayashi, and M. Ohno, Ibid., 104, 6465 (1982).

37. S. Kondo, S. Shibahara, S. Takahashi, K. Maeda, H. Umezawa, and M. Ohno, J. Am. Chem. Soc., 93, 6305 (1971).

38. For cycloaddition with chiral nitrones, see: (a) W. Oppolzer and M. Petrzilka, Helv. Chim. Acta, 61, 2755 (1978). (b) C. Belzecki and I. Panfil, J. Org. Chem., 44, 1212 (1979). (c) P. M. Wovkulich and M. R. Uskoković, J. Am. Chem. Soc., 103, 3956 (1981). (d) C. M. Tice and B. Gannem, J. Org. Chem., 48, 5048 (1983). (e) T. Kametani, T. Nagashima, and T. Honda, Ibid., 50, 2327 (1985) and ref. 39.

39. (a) A. Vasella, Helv. Chim. Acta, 60, 426 (1981). (b) A. Vasella, Ibid., 60, 1273 (1977). (c) A. Vasella and R. Voeffray, J. Chem. Soc., Chem. Commun., 97 (1981). (d) A. Vasella and R. Voeffray, Helv. Chim. Acta, 65, 1134 (1981). (e) A. Vasella, R. Voeffray, J. Pless, and R. Huguenin, Ibid., 66, 1241 (1983).

40. J. A. Dale and D. L. Dull, and H. S. Mosher, J. Org. Chem., 34, 2543 (1969).

41. Unpublished results from H. Iida, K. Kasahara, and C. Kibayashi.

42. H. Iida, K. Kasahara, and C. Kibayashi, J. Am. Chem. Soc., 108, 4647 (1986).

43. (a) P. DeShong and J. M. Leginus, J. Am. Chem. Soc., 105, 1686 (1983). (b) V. Jäger, R. Schohe, and E. F. Paulus, Tetrahedron Lett., 24, 5501 (1983). (c) K. N. Houk, R. M. Susan, Y.-D. Wu, N. G. Rondan, V. Jäger, R. Schohe, and F. R. Fronczek, J. Am. Chem. Soc., 106, 3880 (1984). (d) K. N. Houk, H.-Y. Duh, Y.-D. Wu, and S. R. Moses, Ibid., 108, 2754 (1986).

44. Chirality induction in this manner has been interpreted in terms of kinetic anomeric effect (ref. 39a).

45. S. J. Danishefsky, C. J. Marring, M. R. Barbachy, and B. E. Segmuller,

J. Org. Chem., **49**, 4565 (1984).

46. Unpublished results from H. Iida, K. Kasahara, and C. Kibayashi.
47. J. R. Vaughan, Jr. and R. L. Osato, J. Am. Chem. Soc., **74**, 676 (1952).
48. M. Kono, K. O'hara, H. Ohmiya, H. Iida, C. Kibayashi, and K. Kasahara, Jpn. J. Antibiot., **39**, 247 (1986).
49. Unpublished results from H. Iida, K. Kasahara, and C. Kibayashi.
50. H. L. Holmes in "The Alkaloids," R. H. F. Manske, Ed., Academic Press, New York, 1950, Vol. I, Chapter 6.
51. R. Robinson, J. Chem. Soc., **111**, 762 (1917).
52. J. J. Tufariello and E. J. Trybulski, J. Chem. Soc., Chem. Commun., 720 (1973). J. J. Tufariello, G. B. Mullen, J. J. Tegeler, E. J. Trybulski, S. C. Wang, and Sk. Asrof Ali, J. Am. Chem. Soc., **101**, 2435 (1979).
53. R. Noyori, Y. Baba, and Y. Hayakawa, J. Am. Chem. Soc., **96**, 3336 (1974). Y. Hayakawa, Y. Baba, Y. Makino, and R. Noyori, Ibid., **100**, 1786 (1978).
54. For recent entries into natural products utilizing nitroso Diels-Alder reactions, see: (a) N. J. Leonard and A. J. Playtis, J. Chem. Soc., Chem. Commun., 133 (1972). (b) G. E. Keck and D. G. Nickell, J. Am. Chem. Soc., **102**, 3632 (1980). (c) G. E. Keck and R. R. Webb, II, J. Org. Chem., **47**, 1302 (1982). (d) J. E. Boldwin, P. D. Bailey, G. Gallacher, K. A. Singleton, and P. M. Wallace, J. Chem. Soc., Chem. Commun., 1049 (1983); J. E. Baldwin, P. D. Bailey, G. Gallacher, M. Otsuka, K. A. Singleton , and P. M. Wallace, Tetrahedron, **40**, 3695 (1948). (e) A. Defoin, H. Fritz, G. Geffroy, and J. Streith, Tetrahedron Lett., **27**, 4727 (1986).
55. H. Iida, Y. Watanabe, and C. Kibayashi, Tetrahedron Lett., **25**, 5091 (1984); J. Org. Chem., **50**, 1818 (1985).
56. H. Hart, S. K. Ramaswami, and R. Willer, J. Org. Chem., **44**, 1 (1979).
57. G. E. Keck and S. A. Fleming, Tetrahedron Lett., 4736 (1978).
58. D. L. J. Clive, V. Farina, A. Singh, C. K. Wong, W. A. Kiel, and S. M. Menchen, J. Org. Chem., **45**, 2120 (1980).
59. (a) K. E. Harding and S. R. Burks, J. Org. Chem., **46**, 3920 (1981). (b) K. E. Harding and S. R. Burks, Ibid., **49**, 40 (1984).
60. S. Danishefsky, E. Taniyama, and R. R. Webb, II, Tetrahedron Lett., **24**, 11 (1983).
61. S. R. Johns, J. A. Lamberton, A. A. Sioumis, Aus. J. Chem., **24**, 2399 (1971).
62. G. E. Keck, Tetrahedron Lett., 4767 (1978).
63. J. W. Daly, G. B. Brown, M. Mensah-Dwumah, and C. W. Myers, Toxicon, **16**, 163 (1978).
64. Th. F. Spande, J. W. Daly, D. J. Hart, Y.-M. Tsai, T. L. Macdonald, Experientia, **37**, 163 (1981).
65. For syntheses of gephyrotoxin 223AB, see: (a) T. L. Macdonald, J. Org. Chem., **45**, 193 (1980). (b) J. Royer and H.-P. Husson, Tetrahedron Lett., **26**, 1515 (1985). (c) C. A. Broka and K.K. Eng, J. Org Chem., **51**, 5045 (1986).
66. (a) R. V. Stevens and A. W. M. Lee, J. Chem. Soc., Chem. Commun., 133 (1982). (b) D. J. Hart and Y.-M. Tsai, J. Org. Chem., **47**, 4403 (1982).
67. H. Iida, Y. Watanabe, and C.Kibayashi, J. Am. Chem. Soc., **107**, 5535 (1985).
68. Refs. 66 and 67. For unpublished syntheses of **110** and a complehensive review in this area, see: J. W. Daly and Th. F. Spande in "The Alkaloids II: Chemical and Biological Perspectives," S. W. Pelletier, Ed., Wiley-Interscience, New York, 1986, Vol. 4, Chapter 1.
69. F. J. Ritter, T. E. M. Rotgens, E. Talman, P. E. J. Verwiel, and F. Stein, Experientia, **29**, 530 (1973); F. J. Ritter and C. J. Persoons, Neth. J. Zool., **25**, 261 (1975).
70. (a) J. E. Oliver and P. E. Sonnet, J. Org. Chem., **39**, 2662 (1974). (b) P. E. Sonnet and J. E. Oliver, J. Heterocycl. Chem., **12**, 289 (1975).

(c) P. E. Sonnet, D. A. Natzel, and R. Mendoza, Ibid., **16**, 1040 (1980).

71. (a) R. V. Stevens and A. W. M. Lee, J. Chem. Soc., Chem. Commun., 102 (1982). (b) R. Yamaguchi, E. Hata, T. Matsuki, and M. Kawanishi, J. Org. Chem., **52**, 2049 (1987).

72. J. Royer and H.-P. Husson, J. Org. Chem., **50**, 670 (1985).

73. Y. Moriyama, D.-D. Huynh, C. Monneret, and Q.-K. Huu, Tetrahedron Lett., 825 (1977).

74. Unpublished results from Y. Watanabe and C. Kibayashi. For syntheses of dihydropinidine, see (a) R. K. Hill and T. Yuri, Tetrahedron, **33**, 1569 (1977). (b) M. Bonin, J. R. Romero, D. S. Grierson, and H-P. Husson, Tetrahedron Lett., **23**, 3369 (1982). (c) L. Guerrier, J. Royer, D. S. Grierson, and H-P. Husson, J. Am. Chem. Soc., **105**, 7754 (1983).

SYNTHETIC APPROACHES TO COMPLEX NUCLEOSIDE ANTIBIOTICS

Philip Paul Garner

1 INTRODUCTION

The chemical study of substances collectively termed as nucleosides has proven to be most fruitful over the years. Progress made in this field has helped to lay the foundation of modern molecular biology - especially regarding our understanding of molecular genetics. In light of the pivotal role which nucleosides play in a wide variety of life processes, it is perhaps not so surprising that this class of compounds also includes a number of agents which exhibit antibiotic properties (refs. 1-3). As new and ever more complex nucleoside antibiotics continue to be discovered, the need to efficiently prepare and/or modify these substances provides the impetus for development of novel synthetic strategies and methodologies (refs. 2 & 4).

Structures for eight representative examples of what we would term as complex nucleoside antibiotics are shown in Fig. 1. It will be seen that each molecule consists of a purine or pyrimidine base attached to a carbohydrate moiety of varying complexity and may also include an unusual aminoacid appendage as well. Obviously, the magnitude of the synthetic challenge associated with each target will be a function of both structure at the monosaccharide level as well as the size of the final array which may necessarily incorporate sensitive functionality such as glycosidic linkages. This article will detail recent progress (including both synthetic approaches as well as total syntheses) which has been made towards the asymmetric synthesis of targets **1-8** with a particular emphasis on the problem of controlling of developing stereochemistry. Syntheses of nucleoside antibiotics which have relied primarily on the more traditional modification of carbohydrates will not be dealt with here. Likewise, the synthesis of important targets which incorporate atypical "bases" or come under the heading of C-nucleosides are not to be covered. It is my intention to provide an overview of advances that are rapidly being made in this field as the power of modern synthetic organic chemistry is being applied to these challenging and biologically important natural products.

Traditionally, the synthesis of nucleoside antibiotics has involved modification of readily accessible carbohydrate precursors so as to utilize the carbon skeleton and resident chirality already present in the starting

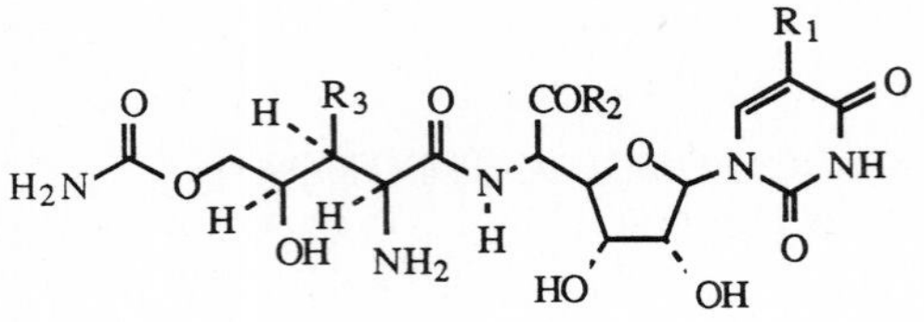

1 POLYOXINS

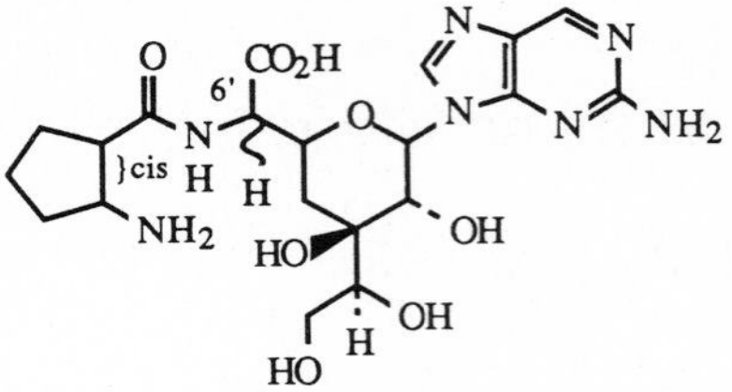

2 AMIPURIMYCIN

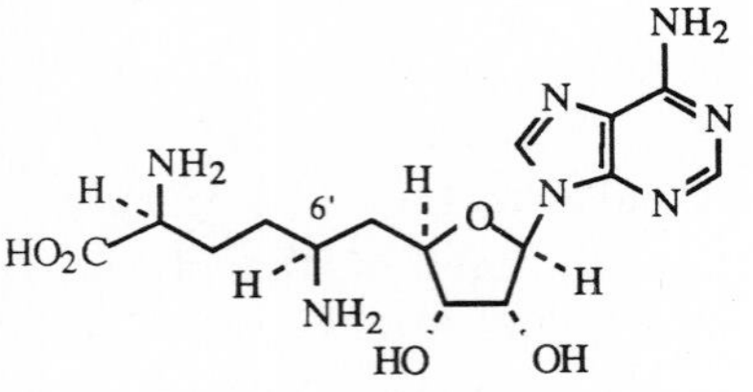

4 SINEFUNGIN

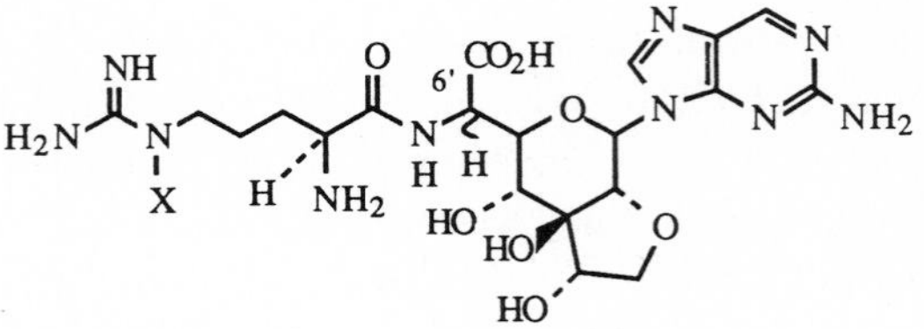

3 MIHARAMYCINS

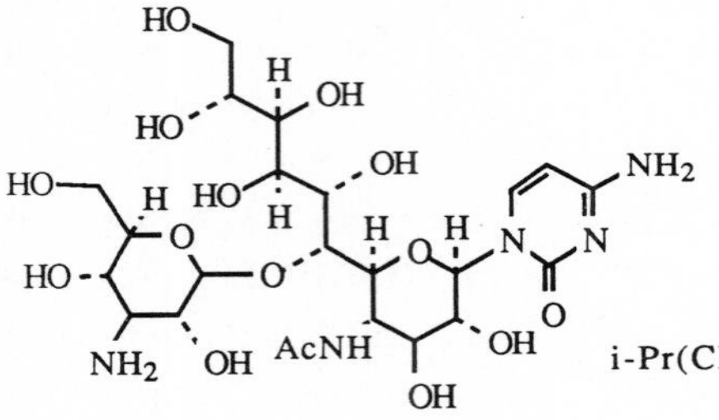

5 HIKIZIMYCIN

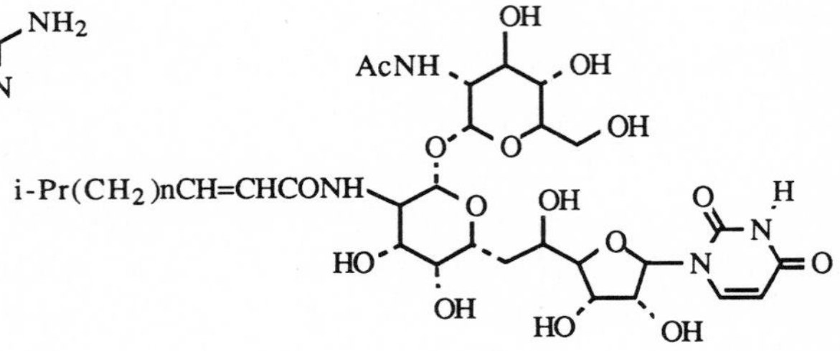

6 TUNICAMYCINS

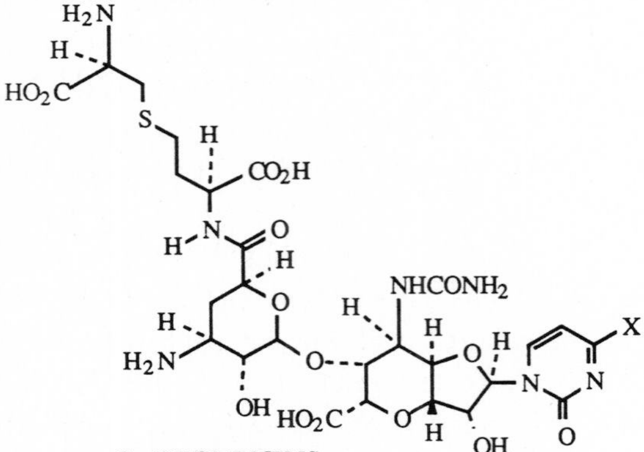

8 EZOMYCINS

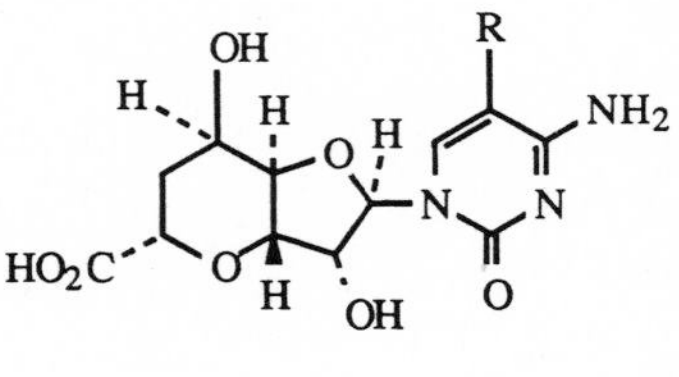

7 OCTOSYL ACIDS

Fig. 1. Representative nucleoside antibiotics.

sugar. This approach has considerable merit when it can be applied in an efficient manner but is often precluded when the target molecules incorporate very unusual residues into their structures. Such instances provide a convincing rationale for the ongoing development of totally synthetic routes to carbohydrates (ref. 5). Since we are dealing with enantiomerically pure targets that may contain numerous stereocenters, the proposed routes should necessarily involve at least some aspect of asymmetric synthesis *per se*. Only when the preparation of constituent monosaccharides and aminoacids is secure can attention be turned to the task of linking them together in a controlled manner - a problem which is not always to be taken lightly.

2 THE POLYOXINS

The polyoxins **1** comprise a family of agricultural antibiotics isolated from *Streptomyces cacaoi* var. *asoensis* that exhibit marked and selective activity against phytopathogenic fungi (ref. 6). They function by inhibiting chitin synthetase, presumably by mimicking the natural substrate, uridine diphosphate N-acetylglucosamine (UDP-GlcNAc). The first (and only) total synthesis of a dipeptidyl polyoxin was achieved by Kuzuhara et al using the traditional carbohydrate-based strategy mentioned above (refs. 7-9) but the length of this synthesis (ca. 40 steps) illustrates the usual drawback of such an approach. A nonselective synthesis of the basic nucleoside aminoacid constituent "uracil polyoxin C" from uridine has also been reported (ref. 10).

More recent efforts have focused on the use of a single tartrate-derived aldehyde **9** for the synthesis both the α-aminoaldonic acid and α-aminouronic residues present in **1**. This unified approach requires that the additional chiral centers be introduced with some degree of stereocontrol. Thus, in his approach to thymine polyoxin C (Ref. 11), Mukaiyama utilized a solvent dependent, (Z)-selective Wittig reaction to assemble the butenolide **11** (Fig. 2). *cis*-Hydroxylation of this compound using $KMnO_4$ proceeded exclusively from the less hindered α-face as required leading to **12** after acetonide formation and desilylation. The nitrogen functionality was then introduced at C-5 via azide displacement of a 2-pyridinium alkoxide (generated in situ) and yielded compound **13**. After lactone reduction and formation of the anomeric acetates **14**, nucleoside formation was effected using the silyl Hilbert-Johnson methodology developed by Vorbrueggen (ref. 12). Deacetylation followed by acetonide formation led to the intermediate **16** which had previously been converted to thymine polyoxin C **17** by Kuzuhara and thus constituted a formal synthesis of this compound.

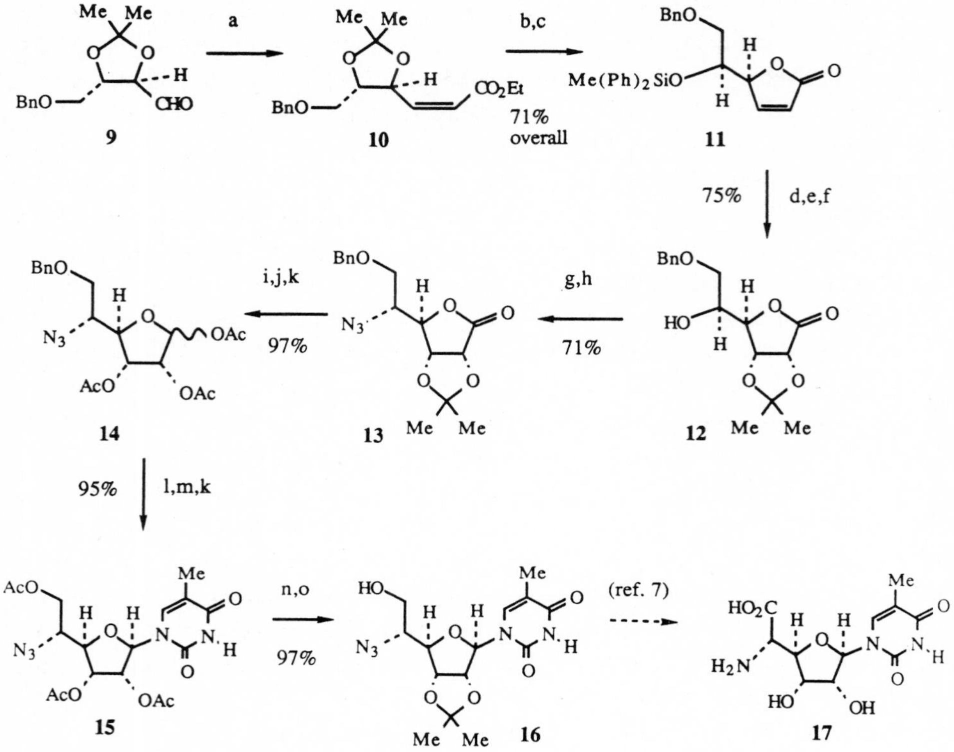

(a) Ph3PCHCO2Et, MeOH, R.T. (b) 3N HCl (c) Ph2(Me)SiCl, 2,6-lutidine (d) KMnO4, dicyclohexano-18-crown-6 (e) Me2C(OMe)2, cat. p-toluene sulfonic acid (f) KF, cat. $^nBu_4N^+HSO_4^-$, -42°C (g) 1-methyl-2-fluoro-pyridinium tosylate, Et3N (h) LiN3, hexamethylphosphoric triamide (i) diisobutylaluminum hydride, -78°C (j) 70% acetic acid, 80°C (k) acetic anhydride, C5H5N (l) 2,4-bis(trimethylsiloxy)-5-methylpyrimidine, trimethylsilyl triflate, reflux (m) BBr3, -42°C (n) NH3/MeOH (o) Me2C(OMe)2, cat. p-toluenesulfonic acid, Me2CO.

Fig. 2. Mukaiyama's formal synthesis of "thymine polyoxin C".

The same aldehyde **9** also served as starting material for Mukaiyama's approach to 5-O-carbamoylpolyoxamic acid as well (Fig. 3). Addition of a lithium acetylide derivative to **9** proceeded with excellent *erythro*-selectivity to give propargylic alcohol **18** when $TiCl_2(O^iPr)_2$ was used as a catalyst. This stereochemical result is consistent with addition via a Felkin-Anh transition state (ref. 13) in the presence of a non-chelating catalyst. The nitrogen function was again introduced via an azide displacement, but this time of a propargylic tosylate to yield compound **19**. Reduction of the azide group with lithium aluminum hydride followed by N-protection with a *tert*-butoxycarbonyl (BOC) moiety resulted in formation of the amine derivative **20**. Dissolving metal reduction released the primary alcohol and allowed introduction of the carbamate moiety via a standard two step procedure to give the primary urethane **21**. Oxidative degradation of the vinyl group using $KMnO_4$ resulted in production of the α-aminoacid derivative **22** but its conversion to and comparison with 5-O-carbamoylpolyoxamic acid itself was not formally completed.

An actual synthesis of 5-O-carbamoylpolyoxamic acid via the aldehyde **9** has recently been reported by workers at Schering-Plough (Ref. 14). Their route began with a Wadsworth-Horner-Emmons reaction which gave the (E)-α,β-usaturated ester **23** (Fig. 4). This compound was reduced to an allylic alcohol and then converted to the trichloromethyl imidate derivative **24** which underwent a facile [3,3] sigmatropic rearrangement to give a (1:1) mixture of the diastereomeric trichloroacetamides **25**. These were separately transformed into 5-O-carbamoyl polyoxamic acid **27** and its (2R) epimer "iso"-**27** via the protected aminoacids **26** and "iso"-**26** though neither of these compounds matched Mukaiyama's intermediate **22**. Product **27** was, however, shown to possess the correct stereochemistry by an X-ray analysis of a crystalline γ-lactone derivative.

Such a convergent approach to the polyoxins might be improved by utilizing a chiral synthon that already incorporates the nitrogen functionality present in targets **17** and **27**. Thus in our own synthesis of 5-O-carbamoyl-polyoxamic acid **27** (ref. 15), we have employed a serine-derived oxazolidine aldehyde **28** (ref. 16) as a penaldic acid equivalent (ref.17) which incorporates the eventual α-aminoacid moiety into a substituted oxazolidine ring (Fig. 5). First, the carbon backbone was assembled by addition of vinyl magnesium bromide to the aldehyde **28**. Though inconsequential at this stage, the major alcohol product **29** was shown to possess the *erythro* stereochemistry, again as a consequence of a preferred Felkin-Anh transition state. This compound was converted to

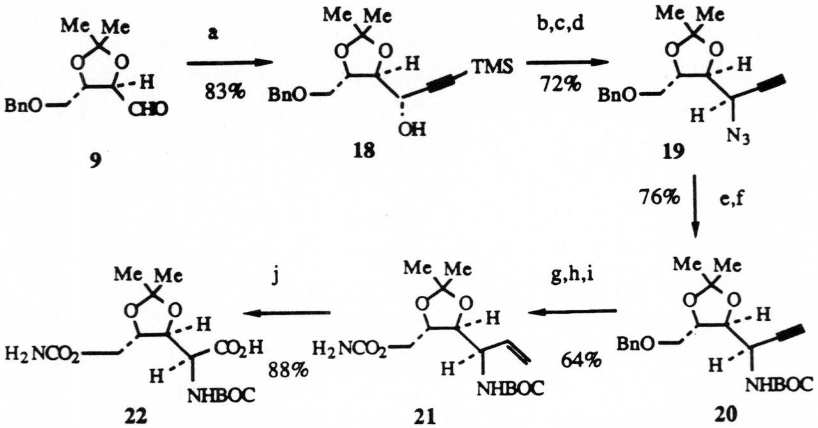

(a) LiCCSiMe$_3$, TiCl$_2$(O^{i}Pr)$_2$ (b) p-toluenesulfonyl chloride, C$_5$H$_5$N (c) LiN$_3$, hexamethylphosphoric triamide (d) NH$_4$F, cat. ^{n}Bu$_4$N$^+$HSO$_4^-$ (e) lithium aluminum hydride (f) *tert*-butyl S-(4,6-dimethylpyrimidin-2-yl) thiolcarbonate, Et$_3$N (g) Na/NH$_3$ (h) 4-nitrophenyl chloroformate, C$_5$H$_5$N (i) NH$_3$/MeOH (j) KMnO$_4$, H$_2$O-Me$_2$CO.

Fig. 3.   Mukaiyaima's approach to 5-O-carbamoylpolyoxamic acid.

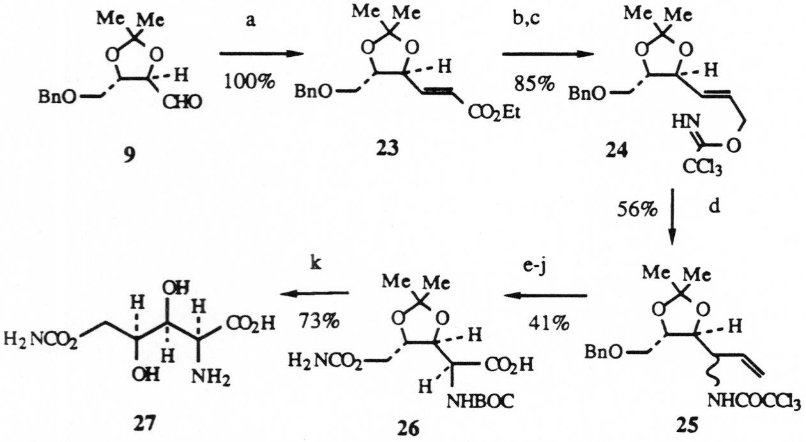

(a) (EtO)$_2$P(O)CH$_2$CO$_2$Et, NaH (b) diisobutylaluminum hydride (c) Cl$_3$CCN, cat. NaH (d) xylene reflux (e) NaOH (f) di-*tert*-butyl dicarbonate (g) Na/NH$_3$ (h) 4-nitrophenyl chloroformate, Et$_3$N (i) NH$_3$/MeOH (j) NaIO$_4$, cat. RuCl$_3$, MeCN-CCl$_4$-H$_2$O (k) MeOH-trifluoroacetic acid.

Fig. 4.   The Schering-Plough synthesis of 5-O-carbamoylpolyoxamic acid.

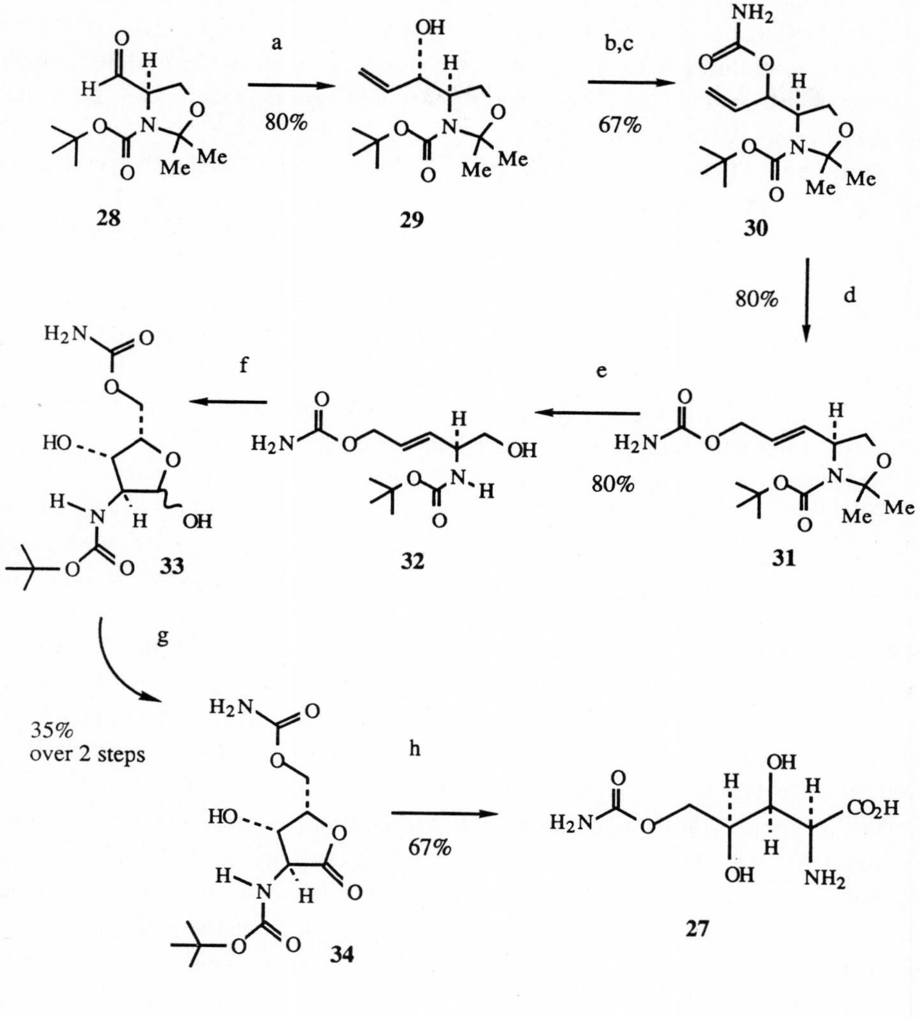

(a) CH$_2$=CHMgBr, tetrahydrofuran, -78°C (b) 4-nitrophenyl chloroformate, C$_5$H$_5$N (c) NH$_3$/MeOH (d) PdCl$_2$(CH$_3$CN)$_2$ (e) p-toluenesulfonic acid, MeOH (f) KMnO$_4$, H$_2$O-Me$_2$CO, CO$_2$ (g) H$_2$NCONHBr (h) 1N HCl

Fig. 5. Our synthesis of 5-0-carbamoylpolyoxamic acid.

the secondary urethane **30** which underwent a facile Pd(II)-catalyzed allylic rearrangement resulting in the production of the primary urethane **31**. The oxazolidine ring could be selectively cleaved under mildly acidic conditions which left the N-*tert*-butoxycarbonyl group intact and gave the primary homoallylic alcohol **32**.

Treatment of this compound with $KMnO_4$ under buffered conditions resulted in the quantitative formation of a lactol mixture **33** which resisted further oxidation by permanganate but could be converted to a (2:1) mixture of γ-lactone **34** and its (3S,4R) diastereomer, **"iso"-34**, by treatment with N-bromourea. Hydrolysis of **34** with aqueous HCl yielded an equilibrium mixture of 5-O-carbamoyl polyoxamic acid **27** and a minor amount of its corresponding γ-lactone as their HCl salts. The high field ^1H NMR spectrum of our synthetic material was identical to that obtained with an authentic sample of 5-O-carbamoyl-polyoxamic acid.

Aldehyde **28** may also be used to construct the α-aminouronic acid residue **17** as indicated by our preliminary work shown in Figure 6. In this case addition of a lithium acetylide to **28** also occurs with high *erythro* (ie. Felkin-Anh) selectivity yielding the propargylic alcohol **35** which then was semi-hydrogenated to give the (Z)-allylic alcohol **36**. *cis*-Hydroxylation with OsO_4 occured almost exclusively from the *si*-face as would be expected if the reactive conformation was directed by the allylic C-O bond according to Kishi's empirical rules (ref. 18). These results can be rationalized on the basis of a preferred starting material conformation that is reflected in the (early) transition state. Treatment of the γ-hydroxy acetal **37** with mild acid promoted cyclization to the anomeric methyl furanosides **38**. Once again the oxazolidine system could be selectively hydrolyzed to give **39** and this material was converted to the protected α-aminoacids **40** by RuO_4 oxidation followed by diazomethane esterification. It remains only to introduce the pyrimidine moiety using some variation Vorbrueggen's method (vide supra) to give "thymine polyoxin C" **17** after deprotection. Suitably protected derivatives of both **17** and **27** could then be used to assemble the target polyoxin J **1** by way of the usual peptide coupling procedures.

3 AMIPURIMYCIN AND THE MIHARAMYCINS

Amipurimycin and the miharamycins are structurally similar agricultural antibiotics, isolated from *Streptomyces novoguineensis* and *Streptomyces miharaensis* respectively (refs. 19 & 20). Both of these compounds display good activity against *Pyricularia oryzae*, the causative agent for rice blast disease. The proposed structures for these two compounds were based on a combination of chemical and spectroscopic

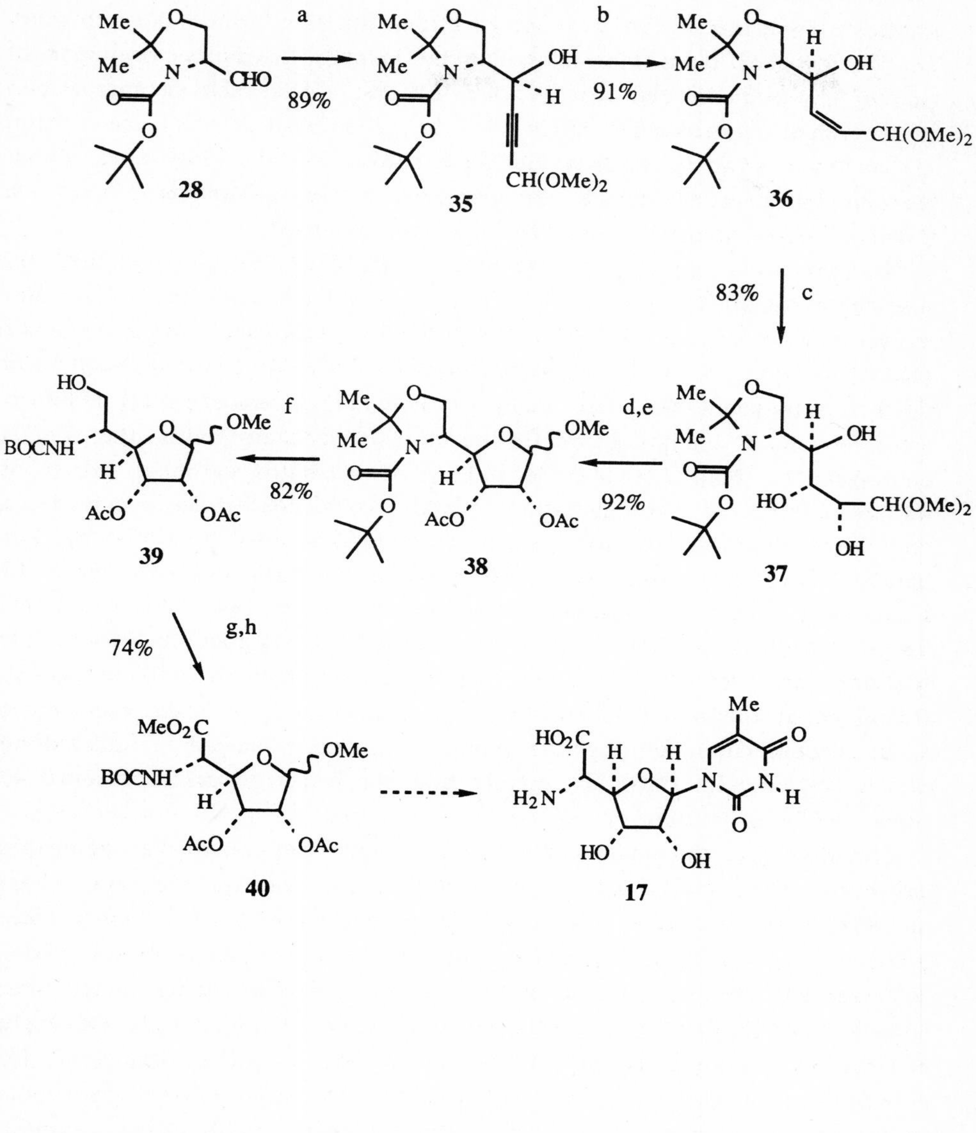

(a) LiCCCH(OMe)$_2$, tetrahydrofuran, -78°C (b) Pd on BaSO$_4$, quinoline (c) cat. OsO$_4$, 4-methylmorpholine N-oxide (d) cat. pyridinium p-toluenesulfonate (e) acetic anhydride, C$_5$H$_5$N, cat. 4-dimethylamino-pyridine (f) 2% HCl in MeOH (g) cat. RuO$_2$·H$_2$O, NaIO$_4$ (h) CH$_2$N$_2$.

Fig. 6. Our approach to "thymine polyoxin C".

(primarily NMR) evidence. Among the points left unresolved by both studies was the relative configuration at C-6' and the absolute configuration of these molecules thereby making flexibility an important part of any contemplated strategy. We felt that a just such an approach to these complex nucleosides might also be accomplished using the general serine-based strategy already outlined above for the polyoxins. Other than our own efforts which are described below, no synthetic work on either of these unusual targets has yet been reported.

We began by applying Danishefsky's versatile (vide infra) method for the stereocontrolled synthesis of pyrans via cyclocondensation to our own oxazolidine aldehyde **41** (= **ent-28**) (Fig. 7). Thus electron-rich diene **42** reacted smoothly with **41** in the presence of ZnCl$_2$ to give a good yield of the *threo* pyranone **43** after acidic workup. This stereochemical result is ascribed to a chelation controlled transition state with the Zn(II) coordinated to both the aldehyde and the N-*tert*-butoxycarbonyl group of **41**. As expected, non-chelating (Felkin-Anh) conditions result in a complimentary *erythro* selective process adding some flexibilty to this approach (Ref. 21). Base-catalyzed conjugate addition of MeOH to **43** resulted in an epimeric mixture of α-hydroxyketones **44** with the desired 2α–OH isomer predominating. This mixture undergoes a very facile hydroxyl-directed Wittig reaction with Ph$_3$PCHCO$_2$Et to produce the (E)-trisubstituted olefin **45** selectively (Ref. 22). Reduction of the ester group with diisobutylaluminum hydride produced an exocyclic allylic alcohol **46** which reacted with OsO$_4$ exclusively from the β-face to give compound **47** after per-benzoylation.

Unmasking of the latent α-aminoacid moiety was readily accomplished using the methodology already described. Thus, selective cleavage of the 2,2-dimethyloxazolidine ring was accomplished with HCl impregnated silica gel and gave the primary alcohol **48**. In keeping with our earlier observations, the acid sensitive BOC group remained intact under these mild conditions. Oxidation of **48** with RuO$_4$ cleanly produced a carboxylic acid which was esterified with diazomethane to give compound **49**. Removal of the N-BOC group with trifluoroacetic acid followed by peptide coupling with racemic acid **50** gave a good yield of two peptide diastereomers **51** and **52**. It should be noted that since we don't know the absolute stereochemistry of this subunit in the natural product, both diastereomers may be needed for comparison.

In preparation for glycosylation, the methyl pyranosides **51/52** were transformed into the more reactive anomeric acetates **53/54** using a reagent made up of wet acetic anhydride + H$_2$SO$_4$. We are now

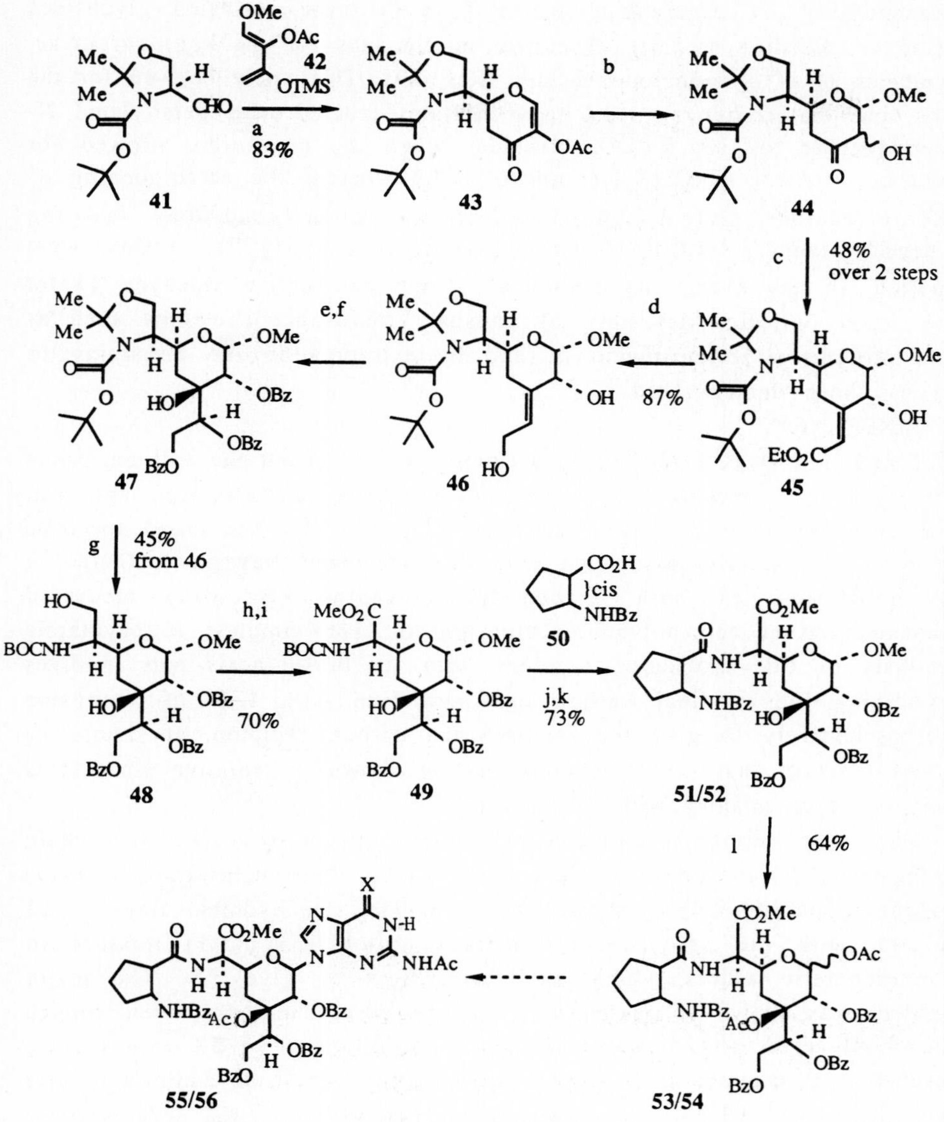

(a) **42**, ZnCl$_2$, CH$_2$Cl$_2$ (b) K$_2$CO$_3$/MeOH, 0°C (c) Ph$_3$PCHCO$_2$Et, MeCN, reflux (d) diiisobutylaluminum hydride (e) Cat. OsO$_4$, 4-methylmorpholine N-oxide (f) benzoyl chloride, C$_5$H$_5$N, cat. 4-dimethylaminopyridine (g) HCl/silica gel (h) cat. RuO$_2$·H$_2$O, NaIO$_4$ (i) CH$_2$N$_2$ (j) trifluoroacetic acid (k) **50**, 1-(3-dimethylaminopropyl)-3-ethylcarbodiimide·HCl, hydroxybenzotriazole (l) wet acetic anhydride+H$_2$SO$_4$, -20°C.

Fig. 7. Our approach to amipurimycin.

concentrating on attaching a purine base to these activated glycosides using a procedure recently developed in our labs for the *regioselective* synthesis of N^9-guanine nucleosides (Ref. 23). This study showed for the first time that *O-benzoylated* sugar acetates reacted with persilylated N-acetylguanine to give β-N^7-nucleosides when the conditions allowed for "kinetic" control (SnCl$_4$, acetonitrile, R.T.) whereas the corresponding β-N^9-nucleosides were obtained selectively when conditions favoring "thermodynamic" control (trimethylsilyl triflate, (CH$_2$Cl)$_2$, reflux) were applied. In any event, the usefulness of the oxazolidine aldehyde **41** for the stereocontrolled assembly of unusual aminosugar skeletons such as that present in amipurimycin **2** (and by extrapolation the miharamycins **3**) has been demonstrated.

4 SINEFUNGIN

Sinefungin **4** is a nucleoside antibiotic isolated from the soil bacterium *Streptomyces griseolus* that exhibits antifungal, anthelmintic, antiviral, and even antitumor properties (see: ref. 2, pp 19-23). The broad spectrum of biological activity associated with this substance may be attributed to its inhibition of both S-adenosyl-methionine (AdoMet) mediated transmethylation and polyamine biosynthesis. The simplest retrosynthetic analysis of this molecule coincides with its biosynthesis and involves combining the terminal carbon of L-ornithine with C-6' of adenosine stereoselectively to give the required ten carbon skeleton. It should be noted however that the (6'S) configuration shown is tentative since it is only based on analogy with AdoMet itself.

The first reported synthesis of sinefungin **4** involved the chain extension of an adenine derivative using two nitroaldol (Henry) condensations (ref. 24) as shown in Fig. 8. Thus the hydrated aldehyde **57** reacted with the anion of nitromethane to give a (3:1) mixture of diastereomeric aldols which were acetylated to give **58**. The major product was shown to possess the *erythro* stereochemistry even though the chiral center at C-5' was to be removed. Treatment of **58** with NaBH$_4$ resulted in β-Elimination of HOAc followed by 1,4-hydride addition to give the nitroalkane **59**. A second nitroaldol reaction with the L-homoserine derived aldehyde **60** (Z = CO$_2$Bn) afforded a mixture of aldol products **61** which was reduced as described above to give the C-6' epimers **62**. This mixture was subjected to a sequence that involved acetonide hydrolysis, protecting group hydrogenolysis, reduction of the nitro group, and finally N-debenzoylation to give a (3:1) mixture of sinefungin **4** and its C-6' isomer as indicated by aminoacid analysis and comparison with authentic material.

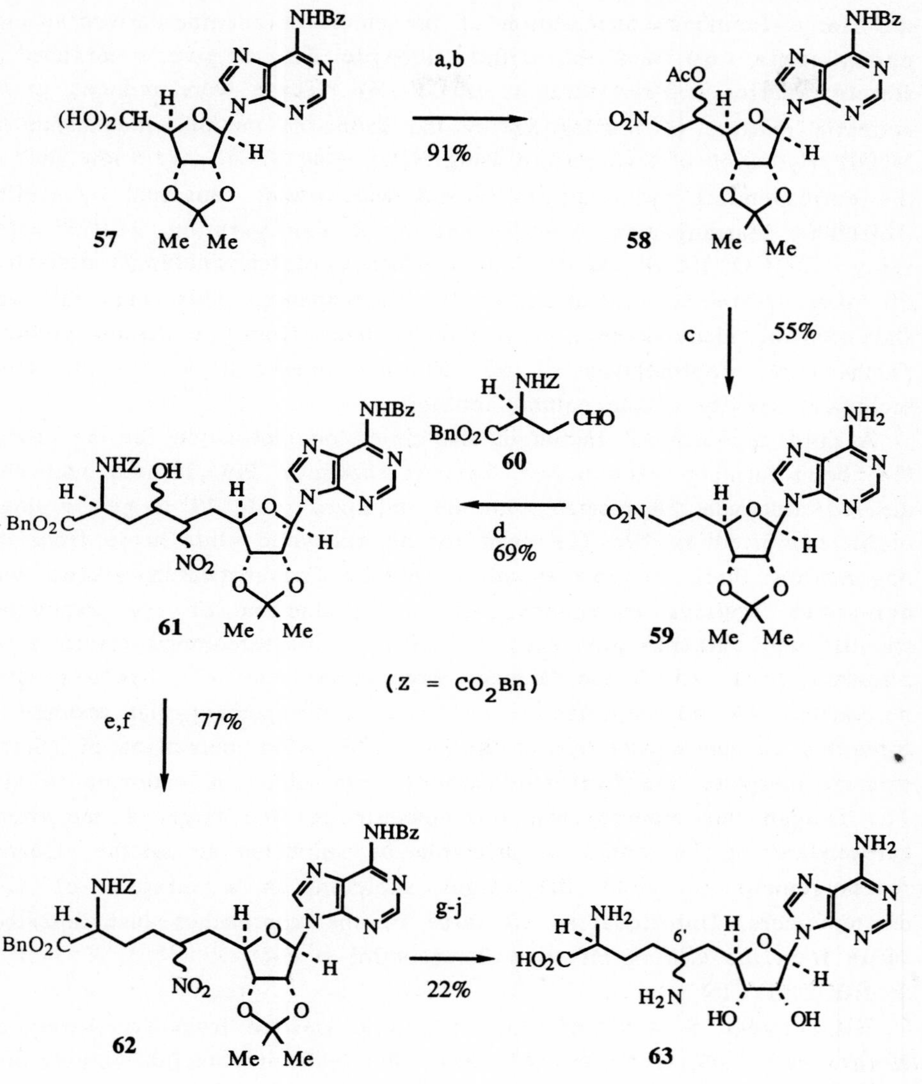

(a) MeNO$_2$, KO^{t}Bu   (b) acetic anhydride, HClO$_4$ (c) NaBH$_4$ (d) **60**, KO^{t}Bu
(e) acetic anhydride, C$_5$H$_5$N, 4-dimethylaminopyridine   (f) NaBH$_4$ (g) aq.
trifluoroacetic acid   (h) 10% Pd on carbon/H$_2$ (-OBn→-OH) (i) PtO$_2$/H$_2$
(-NO$_2$→-NH$_2$) (j) NH$_4$OH.

Fig. 8.   Moffatt's synthesis of sinefungin.

An alternative synthesis (Fig. 9) of sinefungin **4** (ref. 25) began with the Horner-Emmons condensation of the anion of adenine-derived cyanophosphonate **64** and the chiral aldehyde **65** to give a mixture of stereoisomeric α,β-unsaturated nitriles **66**. These were reduced to an epimeric mixture of nitriles **67** by the action of metallic magnesium in MeOH. Hydration of this mixture using H_2O_2 + aq NaOH gave a low yield of the corresponding epimeric amides **68** which were separated by HPLC. Hoffmann rearrangement with retention of configuration was effected using $(CF_3CO_2)IC_6H_5$ and led to production of the sinefungin derivative **69** after treatment with di-*tert*-butyl dicarbonate. This material was shown to be identical to a derivative prepared from the natural product. Furthermore, deprotection of **69** afforded sinefungin **4** with the same biological activity as the natural antibiotic.

A third approach to sinefungin also relied on olefination for the crucial C-C bond forming reaction (ref. 26). As shown in Fig. 10, the adenosine derived aldehyde **70** reacted with the stabilized ylide **71** to give a single olefin identified as **72**. The need for an exocyclic ylide arose from the observation that the corresponding acyclic, trisubstituted alkene was unreactive towards hydrogenation. Thus reduction of the exocyclic, trisubstituted alkene proceeded to give a diastereomeric mixture of butyrolactones **73**. These lactones were opened to the γ-hydroxy acid derivatives **74** and subjected to a Curtius rearrangement that resulted in formation of the diastereomeric urethanes **75**. After conversion of **75** to a primary mesylate, the final two carbons were added as malonate to give **76**. Though this material was not converted to the target **4**, the *pro-S* carboxylate of **76** could in principle be subjected to another Curtius rearrangement to yield the target skeleton as a mixture of C-6' diastereomers. Unfortunately all three of the approaches just described suffer from the lack of effective stereocontrol at C-6'.

5 HIKIZIMYCIN

Hikizimycin **5** is a nucleoside antibiotic isolated from *Streptomyces longissimus* and *Streptomyces A-5* that exhibits useful anthelmintic activity against a number of parasitic vectors (ref. 27 - see also: ref. 2, pp 77-79). An interesting property that this substance shares with certain other pyrimidine nucleoside antibiotics is its inhibition of protein synthesis at the stage of peptide bond formation. The characteristic structural feature of this molecule is an unusual eleven carbon aminosugar residue which is incorporated into a disaccharidyl nucleoside array. Synthetic work on hikizimycin **5** has focused on construction of this

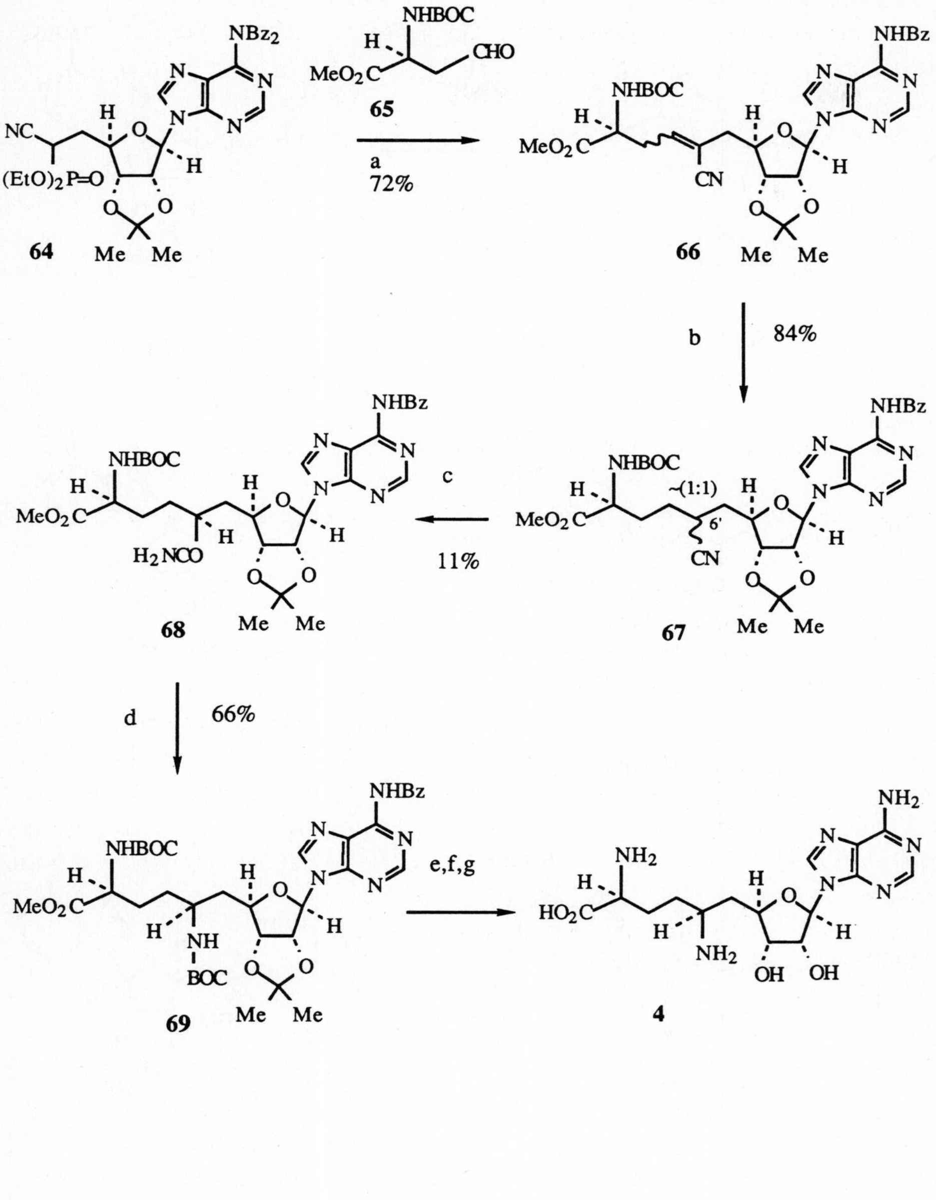

(a) **65**, Mg(OMe)$_2$ (b) Mg metal/MeOH   (c) H$_2$O$_2$/NaOH   (d) (CF$_3$CO$_2$)$_2$IC$_6$H$_5$ (e) K$_2$CO$_3$ (f) trifluoroacetic acid   (g) HCO$_2$H.

Fig. 9. Fourrey's synthesis of sinefungin.

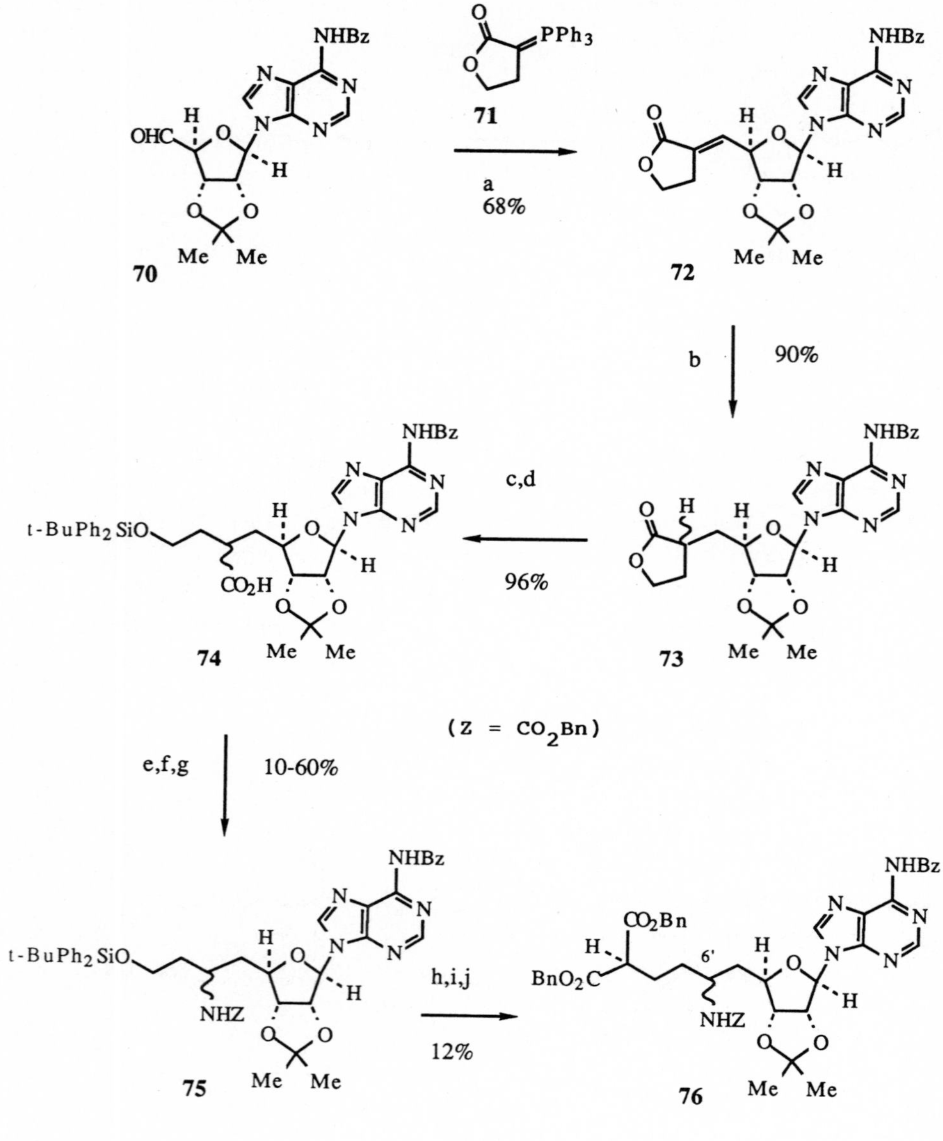

NHBz

71

a
68%

NHBz

70   Me Me

72   Me Me

b   90%

NHBz

c,d

t-BuPh₂SiO

CO₂H

96%

NHBz

74   Me Me

73   Me Me

(Z = CO₂Bn)

e,f,g   10-60%

NHBz

t-BuPh₂SiO

NHZ

h,i,j

12%

NHBz

75   Me Me

CO₂Bn

BnO₂C   6'

NHZ

76   Me Me

(a) **71**, C₆H₆-CH₂Cl₂, reflux (b) PtO₂/H₂/60psi (c) NaOH (d) ᵗBuPh₂SiCl, imidazole (e) KOᵗBu/MeOH (f) (COCl)₂ (g) NaN₃, BnOH (h) ⁿBu₄N⁺F⁻ (i) methanesulfonyl chloride, Et₃N (j) dibenzyl malonate, NaH.

Fig. 10. Secrist's approach to sinefungin.

undecose residue hikosamine, which has itself been isolated as its peracetylated methyl glycoside after degradation of the intact antibiotic.

One obvious approach for the synthesis of this substance would entail linking suitably functionalized five and six carbon sugar derivatives together in a stereocontrolled manner. In this sense the proposed synthesis would involve more or less classical modification of the common carbohydrate precursors at the outset. Thus in his synthesis of methyl peracetyl α-hikosaminide (Ref. 28), Secrist again planned on a strategy wherein a Wittig reaction between two carbohydrate-derived partners would form the crucial C-C linkage.

He first prepared the requisite phosphonium ylide **77** from D-arabinose using standard carbohydrate transformations. The aldehyde component **78**, was prepared from D-galactose using a sequence that included introduction of nitrogen at C-4 with inversion via azide displacement of an axial mesylate. These two units were coupled together (Fig. 11) by an (unstabilized) Wittig reaction to give only the (Z)-disubstituted alkene **79** (in spite of the presence of LiI in the reaction mixture). However, all attempts to achieve a net *trans*-hydroxylation across the double bond of **79** failed. This problem was overcome by first reducing the azide function to an amine with LAH followed by photochemical olefin isomerization (Z:E=3:2), chromatographic purification of the desired (E)-isomer, and N-acetylation to afford **80**. Now *cis*-Hydroxylation of **80** using OsO4 resulted in the facile and exclusive formation of diol **81**. This stereochemical result is consistent with Kishi's empirical rules for the osmylation of allylic ethers (vide supra) and is probably being reinforced by both allylic C-O bonds present in **80**. Finally, compound **81** was hydrolyzed and hydrogenolyzed in one pot then peracetylated to give methyl peracetyl α-hikosaminide **82** which was shown to be identical with naturally derived material.

A totally synthetic route to the racemic modification of **82** was reported by Danishefsky (ref. 29) and relied on his iterative cyclocondensation strategy for higher monosaccharide synthesis. The sequence began (Fig. 12) with a lanthanide catalyzed cycloaddition of diene **84** to furfural **83** to give the *cis*-disubstituted pyranone **85** after acidic workup, a result consistent with an *endo* mode of cycloaddition. Ketone reduction using NaBH4-CeCl3 (axial hydride attack) followed by stereoselective *cis*-hydroxylation from the less hindered α-face gave compound **86** after acetonide formation. The furan moiety was then transformed into the aldehyde **87** by ozonolysis (with reductive workup) followed by controlled oxidation of the primary alcohol.

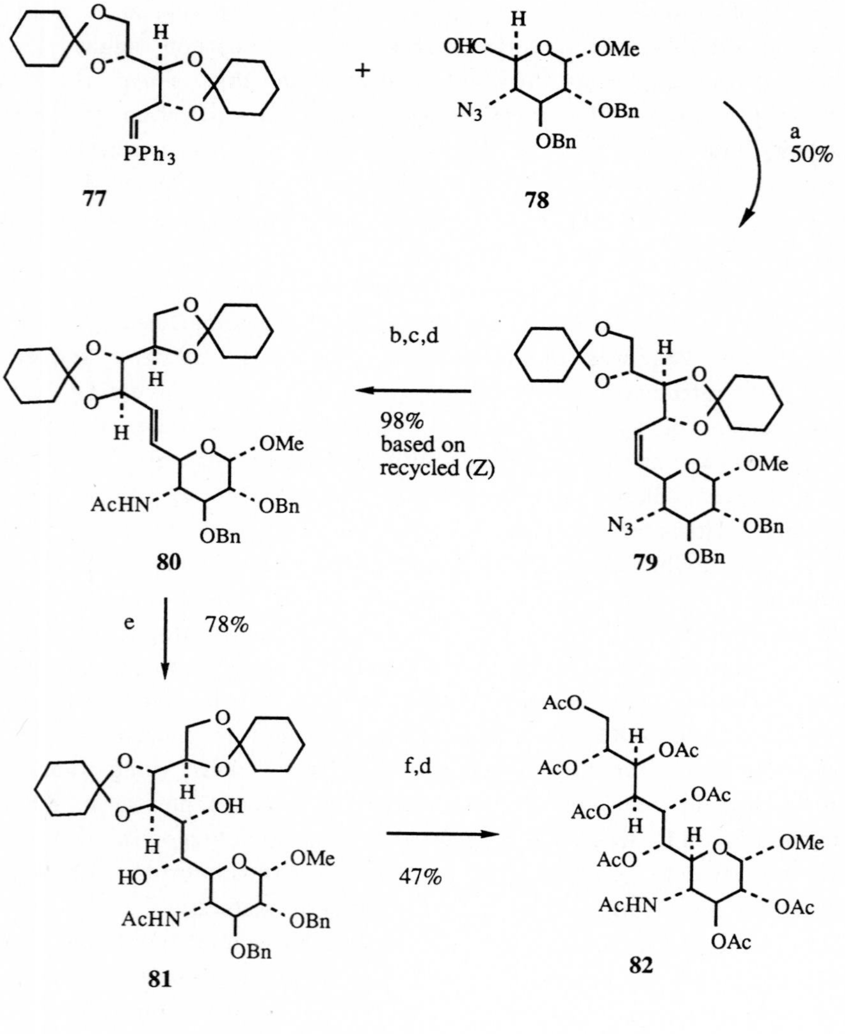

(a) tetrahydrofuran-hexamethylphosphoric triamide-65°->-10°C (b) lithium aluminum hydride (c) hv, C_6H_{12}, PhSSPh (d) acetic anhydride, C_5H_5N (e) cat. OsO_4, 4-methylmorpholine N-oxide (f) Pd on carbon/H_2, MeOH-H_2O-HCl.

Fig. 11. Secrist's synthesis of methyl peracetylhikosamine.

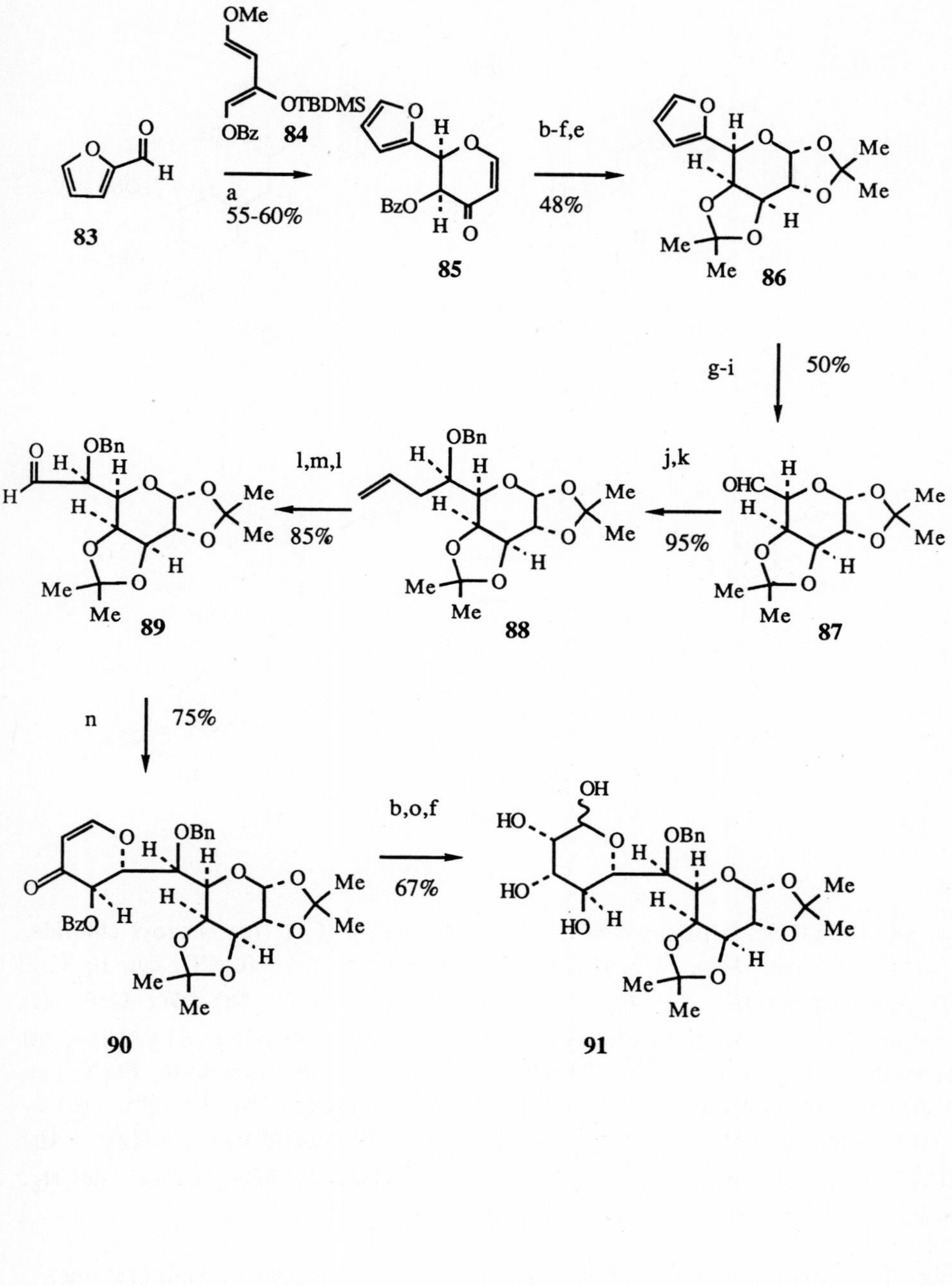

Fig. 12. Danishefsky's synthesis of methyl peracetylhikosamine.

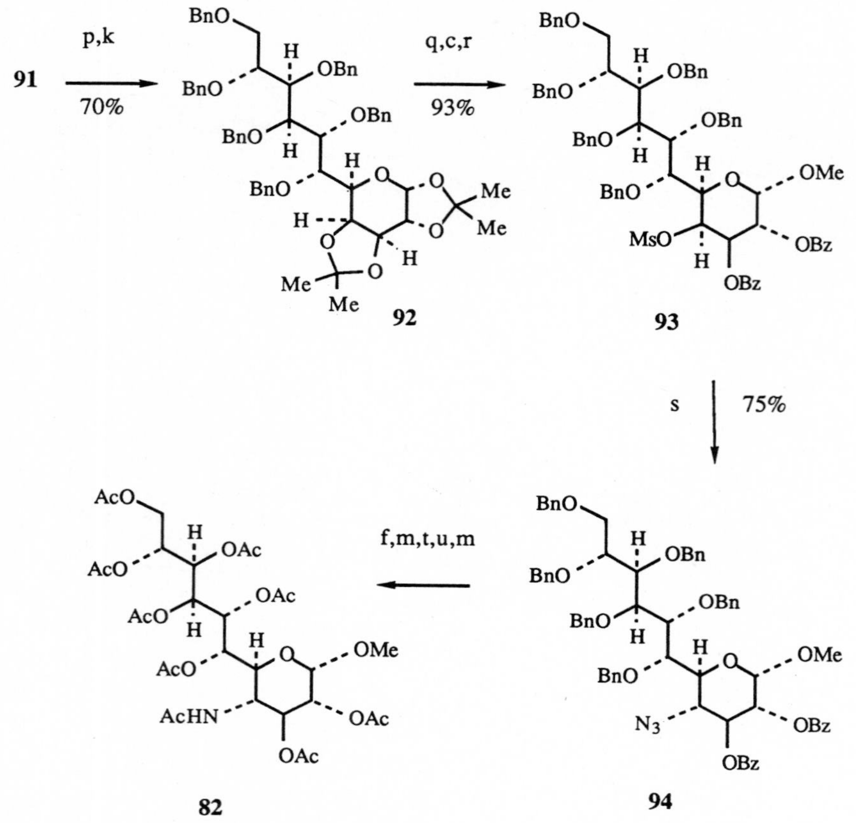

416

(a) **84**, EuFOD™; trifluoroacetic acid (b) $NaBH_4$-$CeCl_3$ (c) benzoyl chloride, C_5H_5N (d) cat. OsO_4, 4-methylmorpholine N-oxide (e) Me_2CO, cat. H_2SO_4 (f) K_2CO_3/MeOH (g) O_3/CH_2Cl_2-MeOH, -78°C (h) BH_3·THF (i) (unspecified) oxidation to aldehyde (j) CH_2=$CHCH_2SiMe_3$, BF_3·OEt_2 (k) benzylbromide, NaH (l) O_3/CH_2Cl_2, -78°C (m) acetic anhydride, Et_3N, cat. 4-dimethylaminopyridine (n) **84**, $MgBr_2$, CH_2Cl_2-$PhCH_3$, 0°C (o) 3-chloroperoxybenzoic acid (p) $LiBH_4$, tetrahydrofuran, reflux (q) HCl/MeOH (r) methanesulfonyl chloride, C_5H_5N (s) $^nBu_4N^+N_3^-$, $PhCH_3$, 85°C (t) PPh_3 ($-N_3 \rightarrow -NH_2$) (u) $Pd(OH)_2$/H_2.

Fig. 12. (cont'd) Danishefsky's synthesis of methyl peracetylhikosamine.

At this point a BF$_3$-catalyzed addition of 3-trimethylsilylpropene to **87** occurred in the Felkin-Anh sense to give the homoallylic ether **88** after benzylation. Stepwise degradation of the olefin afforded the α-benzyloxy aldehyde **89** which was the substrate for a second (MgBr$_2$-catalyzed) cyclocondensation reaction with the diene **84**. This time, however, the product **90** resulted from a chelation-controlled transition state as well as an *exo*-mode of addition. Once again ketone reduction using Luche's conditions gave an equatorial allylic alcohol which directed subsequent epoxidation by 3-chloroperoxybenzoic acid to the α-face and led to tetraol **91** after hydrolysis. Lactol reduction followed by perbenzylation afforded the undeca-pyranose **92** which has the side chain intact. The pyran moiety was then elaborated using procedures analogous to those employed for the synthesis of the 4-amino-4-deoxy-glucose moiety **78** in the Secrist synthesis mentioned above. Thus, transketalization with methanolic HCl followed by selective equatorial benzoylation and then axial mesylation resulted in the formation of **93**. Mesylate displacement with azide followed by deprotection and peracetylation gave (racemic) methyl peracetyl α-hikosaminide **82** whose ^{1}H NMR spectrum was identical to one obtained for Secrist's synthetic material (vide supra).

6 THE TUNICAMYCINS

The tunicamycins **6** are a family of closely related antibiotics isolated from *Streptomyces lysosuperficus* that differ from each other only with respect to the nature of the fatty acid components (ref. 30). A great deal of interest in these compounds has been generated by the wide range of biological properties which they exhibit, including antiviral and possibly even antitumor activity. Like the polyoxins, the tunicamycin antibiotics act by disrupting the metabolism of UDP-GlcNAc, this time however by inhibiting the uptake of N-acetylglucosamine which is required for glycolipid and glycoprotein biosynthesis in cells.

The major synthetic challenge in the tunicamycin series takes form in the unusual undeculose tunicamine residue. Suami reduced this problem to one of coupling the appropriate six carbon and five carbon sugar units to one another using the KF-catalyzed nitroaldol (Henry) reaction (refs. 31 & 32, Fig. 13). The D-galactosamine derived aldehyde **95** reacted cleanly with the nitrosugar **96** prepared from D-ribose in the presence of KF to give a *single* aldol product **97** of undetermined stereochemistry (compare **136** + **137** -> **138** in Fig. 19). This compound was converted to a nitroalkane using a dehydration/1,4-reduction sequence followed by oxidation of the resultant secondary nitroalkane to the ketone **98** with KMnO$_4$-NaO^{t}Bu. This ketone was reduced with NaBH$_4$ to give a (2:1)

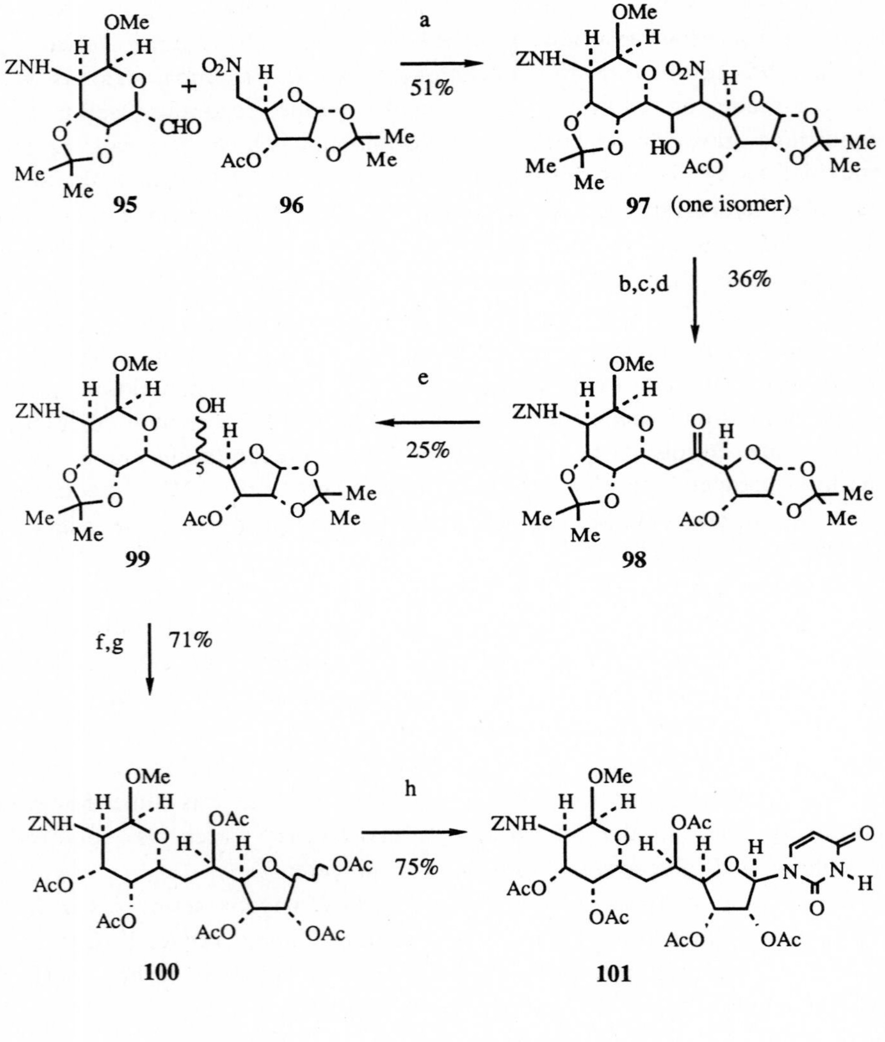

418

Fig. 13. Suami's tunicamycin synthesis.

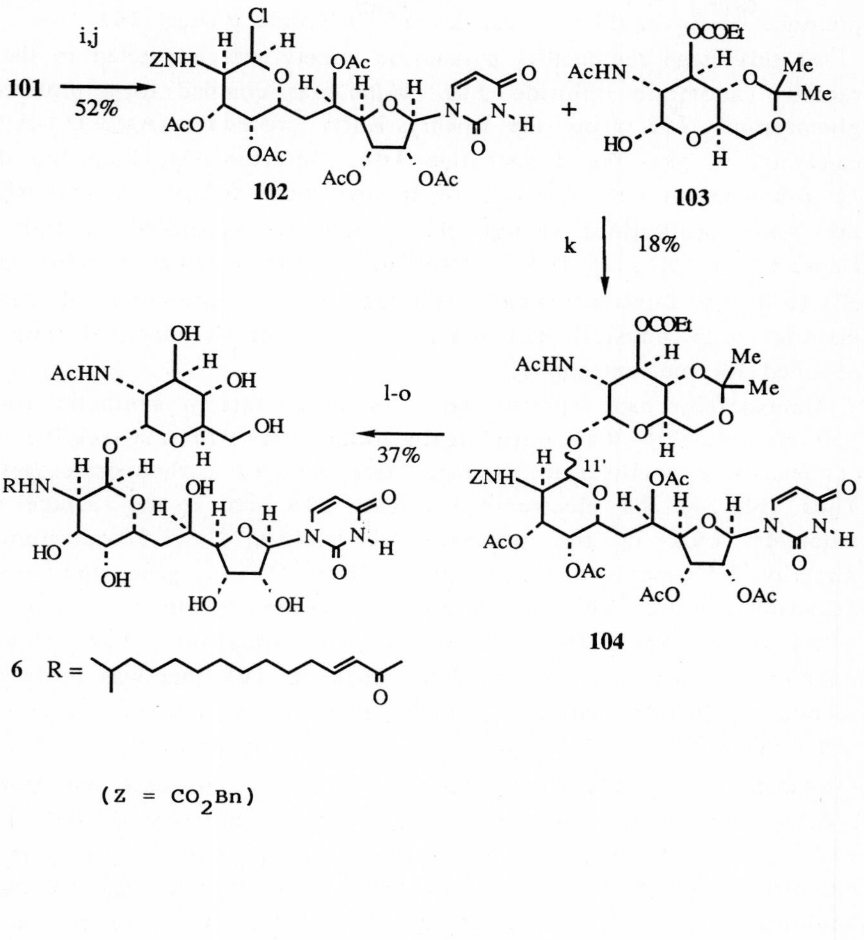

(a) KF, MeCN   (b) acetic anhydride, C_5H_5N, cat. 4-dimethylaminopyridine
(c) NaBH$_4$, MeOH, 0°C   (d) KMnO$_4$, NaO^{t}Bu   (e) NaBH$_4$, EtOH   (f) 60% acetic
acid   (g) acetic anhydride, C_5H_5N (h) 2,4-bis[(trimethylsilyl)oxy]pyridine,
SnCl$_4$, (CH$_2$Cl)$_2$ (i) acetic anhydride-H$_2$SO$_4$ (j) acetyl chloride, HCl   (k)
Ag$_2$CO$_3$-AgClO$_4$, molecular sieves   (l) NaOMe   (m) Pd on carbon/H$_2$ (n)
^{i}Pr(CH$_2$)$_9$CH=CHCO$_2$H, dicyclohexylcarbodiimide   (o) aq. acetic acid, 40°C.

Fig. 13. (cont'd)   Suami's tunicamycin synthesis.

mixture of alcohols **99** favoring the undesired (5S)-isomer. After chromatographic separation, the (5R)-isomer was carried forward to the anomeric acetates **100** which were coupled with disilylated uracil in the presence of SnCl4 to give the desired nucleoside product **101**.

At this stage the methyl pyranoside moiety was converted to the more reactive anomeric chloride **102** which was coupled to a protected D-glucosamine **103** using the Koenigs-Knorr procedure (Ag2CO3-AgClO4 catalysis) to give the disaccharide **104**. Unfortunately, both the desired 11'-β-anomer as well as the 11'-α-anomer were formed in approximately the same proportions though they could be separated at this stage. Deprotection of 11'-β-**104** followed by N-acylation with an active ester of (E)-13-methyl-2-tetradecenoate resulted in the formation of synthetic material with physical and biochemical properties identical with those obtained for tunicamycin V.

Danishefsky has reported an alternative, totally synthetic route to tunicaminyluracil that once again relies on two consecutive cyclo-condensation reactions to assemble this complex carbohydrate (ref. 33). Thus (Fig.14) the electron-rich diene **106** adds to the *re*-face of the aldehyde **105** in the presence of tris(6,6,7,7,8,8,8-heptafluoro-2,2-dimethyl-3,5-octanedionato)europium (EuFOD) to give the *erythro*-dihydropyranone **107**. Oxidative degradation led to formation of the benzyloxymethyl (BOM) protected β-hydroxyester **108** which was transformed into its corresponding aldehyde **109** that was ready for the second cyclocondensation. Aldehyde **109** now reacted with diene **84** under the influence of a mixed catalyst, Ce(OAc)3-BF3·OEt2, to give a new dihydropyranone **110** upon aqueous workup. In this case the developing (7R,8S) configuration appears to result from a chelation-controlled aldol-like addition followed by cyclization. Elaboration of the pyran ring commenced with CeCl3-catalyzed reduction (vide supra) followed by azidonitration with NaN3-(NH4)2Ce(NO3)6 (ref. 34) and anomeric substitution to give **111** which was further transformed into the protected amidodialdose **112** by deprotection and acetylation. Acetolysis of **112** afforded a mixture of four anomeric acetates **113** that could be separated by HPLC into equal proportions of the 11-β-OAc and 11-α-OAc isomers which were still anomeric mixtures at C-1. The former mixture reacted cleanly with disilylated uracil in the presence of trimethylsilyl triflate at the more reactive ribosyl center to give heptaacetyltunicaminyl uracil **114** which was identical with naturally derived material. Interestingly, the 11-α-OAc mixture resulted in competitive pyranosyl nucleoside formation under the same reaction conditions.

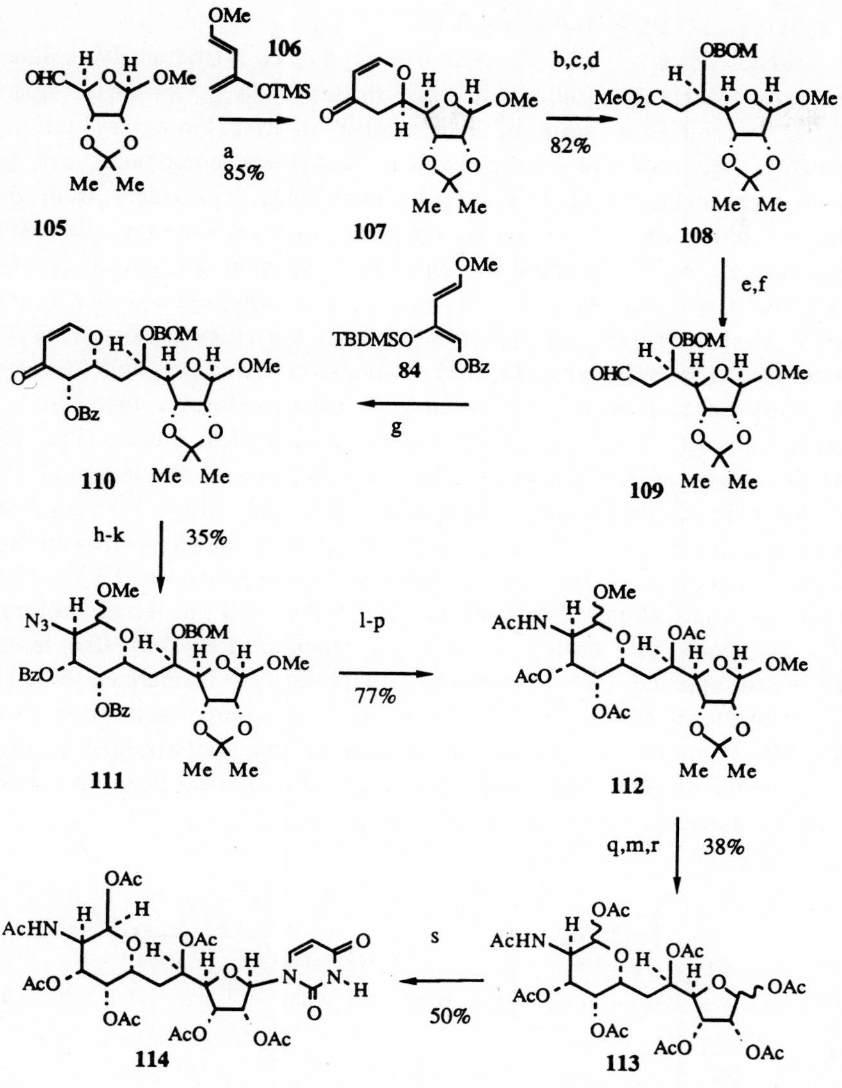

(a) **106**, EuFOD™ (b) O$_3$/CH$_2$Cl$_2$, -78°C; KOH, H$_2$O$_2$ (c) CH$_2$N$_2$ (d) benzyloxymethyl chloride, ^{i}Pr$_2$EtN (e) lithium-aluminum hydride (f) pyridinium chlorochromate, NaOAc (g) **84**, Ce(OAc)$_3$-BF$_3$·OEt$_2$, PhCH$_3$, -78°C (h) NaBH$_4$-CeCl$_3$ (i) benzoyl chloride (j) NaN$_3$, ceric ammonium nitrate, MeCN (k) silver triflate, (Me$_2$N)$_2$CO, MeOH (l) PPh$_3$ (-N$_3$→-NH$_2$) (m) acetic anhydride, C$_5$H$_5$N (n) Pd(OH)$_2$/H$_2$ (o) K$_2$CO$_3$/MeOH (p) acetic anhydride, Et$_3$N, cat. 4-dimethylaminopyridine (q) HCl/MeOH (r) acetic anhydride, acetic acid, H$_2$SO$_4$ (s) 2,4-bis[(trimethylsilyl)oxy]pyrimidine, trimethylsilyl triflate.

Fig. 14. Danishefsky's synthesis of tunicaminyluracil.

7 THE OCTOSYL ACIDS AND EZOMYCINS

The octosyl acids **7** and the ezomycins **8** are structurally related *Streptomyces* derived nucleoside antibiotics which possess quite challenging structures in spite of their rather modest biological activity (refs. 35-37 - see also: ref. 2, pp 23-30). That these compounds can be viewed as carboanalogues of a 3',5'-cyclic nucleosides is evident from the competitive inhibition of *cyclic*-adenosine monophosphate (cAMP) phosphodiesterase by the (chemically produced) adenine analog of octosyl acid A. Efforts directed towards the synthesis of the octosyl acids and ezomycins have focused on construction of the functionalized 3',7'-anhydrooctose (*trans*-dioxahydrindane) nucleus common to both targets. It is therefore appropriate to discuss both of these molecules together.

An early approach (ref. 38) to the octosyl acid nucleus began (Fig. 15) with the formation of the malonate derivative **116** via O-alkylation of the readily available D-allofuranose derivative **115** with ethyl bromoacetate followed by acylation. Selective deprotection of **116** followed by preferential tosylation of the primary alcohol gave substrate **117** which was ready for cyclization. Treatment of **117** with NaH in tetrahydrofuran not only produced the desired bicyclic compound **118** but also lesser amounts of an epoxide (not shown) that could not be isomerized to **118** under the reaction conditions. At this point anomeric activation was attempted as a prelude to nucleoside formation but the strained *trans*-dioxahydrindane system suffered spontaneous furan ring-opening under the required (acidic) conditions, terminating this effort.

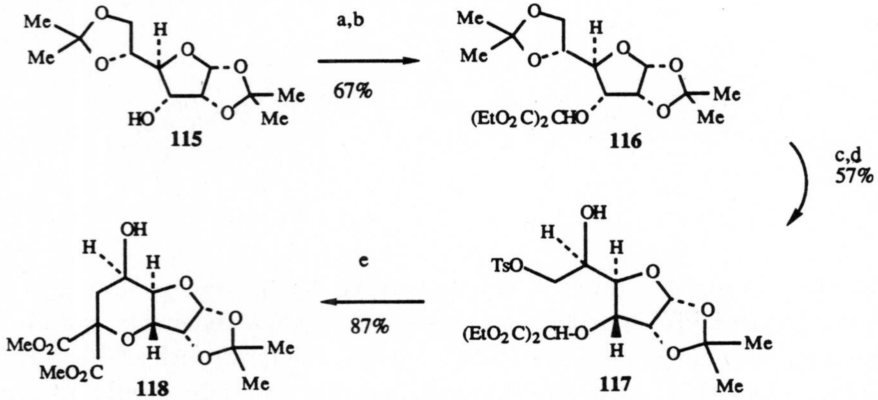

(a) EtO$_2$CCH$_2$Br, NaH (b) CO(OEt)$_2$ (c) 70% acetic acid (d) p-toluenesulfonyl chloride, C$_5$H$_5$N (e) NaH, tetrahydrofuran.

Fig. 15. Anzai and Saita's approach to the octosyl acid nucleus.

Szarek's synthesis of the *trans*-dioxahydrindane system also involved annulation of the pyran ring to a preformed furanose, but avoided the need for subsequent acidic reaction conditions by starting with an intact nucleoside (ref. 39, Fig. 16). Thus, reaction of the uridine-derived aldehyde **119** with a stabilized ylide **120** formed the octose **121** which was reduced stereorandomly to give the allylic alcohols **122**. Hydrogenation of this material followed tosylation and protecting group manipulation resulted in production of the diastereomeric tosylates **123**. This mixture was cyclized directly by treatment with base to give the corresponding 3',7'-anhydrooctose derivatives **124** which were now separated using column chromatography. Deprotection of the (7'S) isomer followed by catalytic oxidation of the primary alcohol with 10% Pt-C and O_2 afforded the octosyl acid analogue **125** after esterification.

Hanessian has reported a stereocontrolled synthesis octosyl acid A that also begins with a uridine derivative (ref. 40). The first step shown in Fig. 17 involves an *erythro*-selective addition of allylmagnesium bromide to **126** followed by benzyloxymethylation to give **127**. This stereochemical outcome is consistent with either a Felkin-Anh or an alternative β-chelated transition state model. After acetonide removal the key oxymercuration cyclization step yielded a fused bicyclic diol **129** possessing the correct configuration at both C-5' and C-7'. It should be noted that the transition state leading to the (7'R) diastereomer would entail the development of a severe 1,3-diaxial interaction with the O-benzyloxymethyl (BOM) group at C-5' (but see **117->118** in Fig. 15). With synthesis of the 3',7'-anhydrooctose skeleton complete, attention was turned to modification of the pyrimidine moiety so as to produce the 5-carboethoxy derivative **131** via a dihydrouridine analog **130**. Selective catalytic oxidation at C-8' followed by hydrolysis gave octosyl acid A (**7**) which was identified by comparison with an authentic sample.

In his own synthesis of octosyl acid A shown in Fig. 18, Danishefsky (ref. 41) began with the same pyranone **107** which he had employed in his synthesis of tunicaminyl uracil (vide supra). This time however, oxidative degradation of **107** was preceded by ketone reduction and protection (PMB = p-methoxybenzyl) so that γ-lactone **132** would be formed. This was transformed into the stereodefined (7R)-mesylate **133** using standard procedures that include the oxidative removal of the p-methoxybenzyl protecting group. In anticipation of nucleoside formation, the furanose residue was deprotected and then peracetylated to give **134** as an anomeric mixture. Reaction of **134** with disilylated 5-carbomethoxyuracil in the presence of trimethylsilyl triflate resulted in

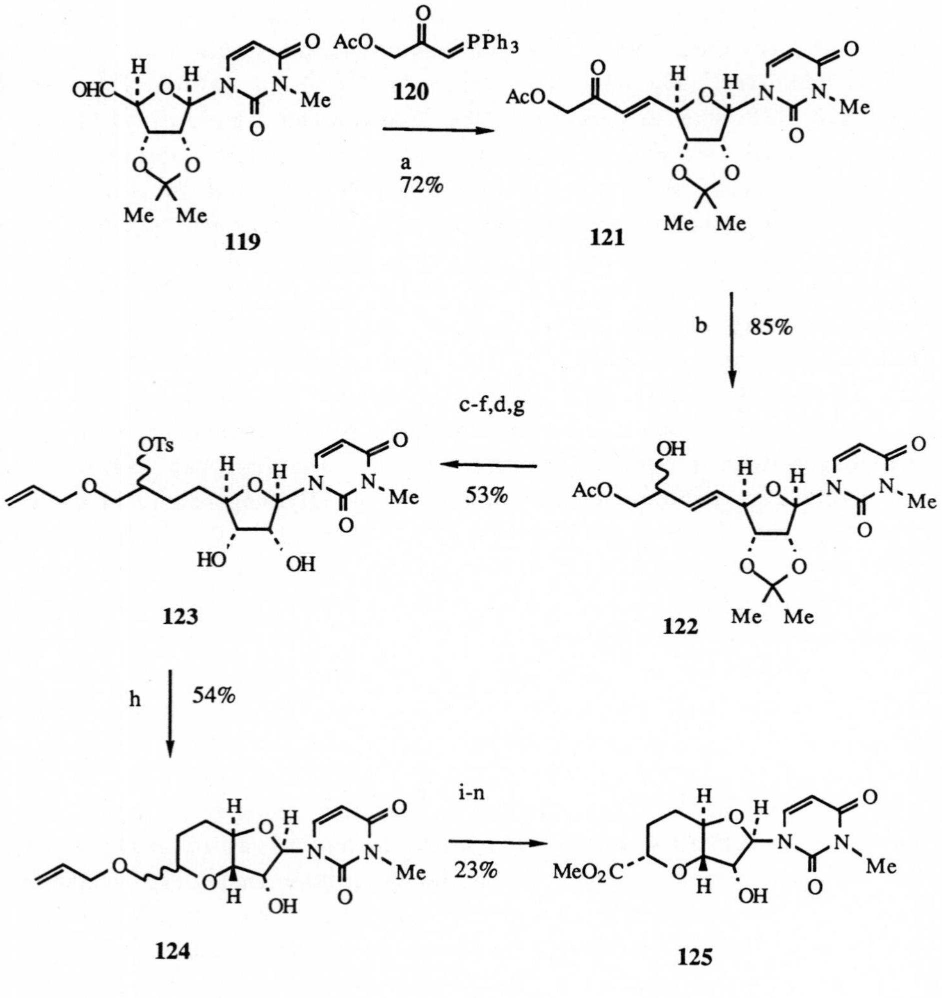

(a) **120**, dimethylsulfoxide   (b) $NaBH_4$, EtOH   (c) 10% Pd-C/H_2 (d) p-toluenesulfonyl chloride, C_5H_5N (e) NaOMe   (f) $CH_2=CHCH_2ONa$ (g) 90% HCO_2H, 0°C   (h) NaH   (i) RhCl(PPh$_3$)$_3$, 1,4-diazabicyclooctane   (j) benzoyl chloride, C_5H_5N (k) 0.1 N HCl, MeOH   (l) NaOMe/MeOH   (m) PtO_2/O_2, aq. $NaHCO_3$ (n) 5% HCl in MeOH.

Fig. 16.   Szarek's approach to the trans-dioxahydrindane system.

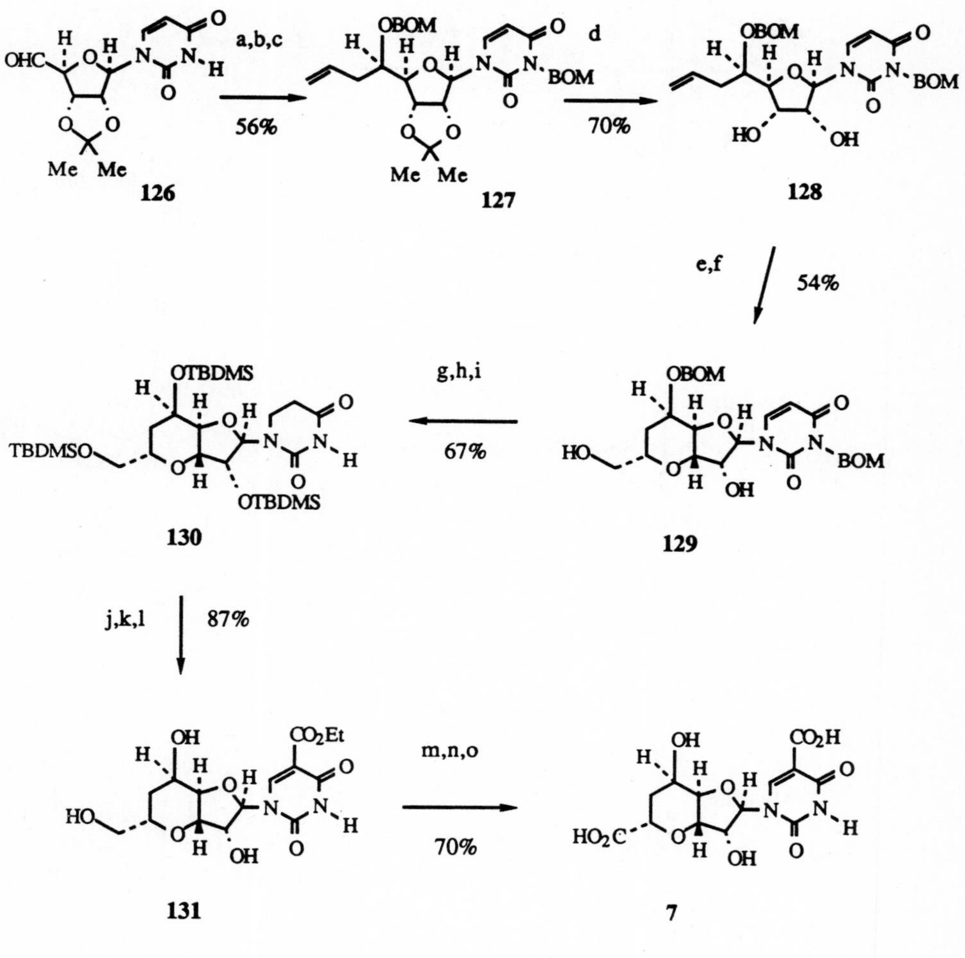

(a) $CH_2=CHCH_2MgBr$, tetrahydrofuran, -100°C (b) benzyloxymethyl chloride, 8-diazabicyclo-[5,4,0]undec-7-ene, dimethylformamide (c) benzyloxymethyl chloride, $^iPr_2EtN$, tetrahydrofuran (d) tetrahydrofuran-$MeCO_2H$-H_2O, 65°C (e) $Hg(OAc)_2$, tetrahydrofuran; NaBr (f) $NaBH_4$, O_2, dimethylformamide (g) 20% $Pd(OH)_2$-C/H_2, MeOH (h) 5% Rh/Al_2O_3, MeOH (i) $^tBuMe_2SiCl$, $^iPr_2EtN$, 4-dimethylaminopyridine (j) lithium diisopropylamide, $ClCO_2Et$, -78°C (k) PhSeCl, C_5H_5N; H_2O_2 (l) $^nBu_4N^+F^-$ (m) PtO_2/O_2, aq. $NaHCO_3$, 90°C (n) $H^+/EtOH$ (o) LiOH then Dowex-50(H^+) ion exchange resin .

Fig. 17. Hanessian's synthesis of octosyl acid A.

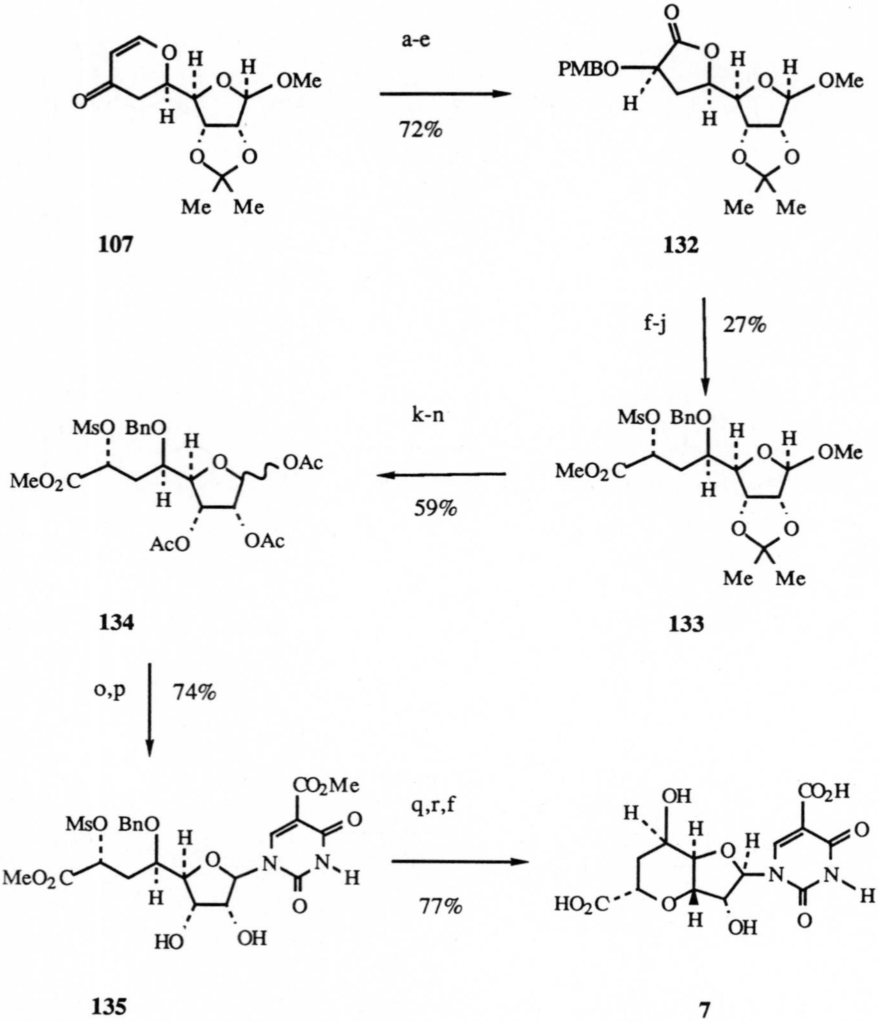

(a) NaBH$_4$-CeCl$_3$ (b) p-methoxybenzyl chloride, NaH (c) cat. OsO$_4$, NaIO$_4$
(d) K$_2$CO$_3$/MeOH (e) Ag$_2$CO$_3$-Celite, xylene reflux (f) LiOH (g) benzyl
bromide, NaH (h) CH$_2$N$_2$ (i) 2,3-dichloro-5,6-dicyano-1,4-benzoquinone
(j) methanesulfonyl chloride (k) HCl/MeOH (l) acetic anhydride (m)
acetic anhydride MeCO$_2$H-H$_2$SO$_4$ (n) 2,4-bis[(trimethylsilyl)oxy]-5-carbo-
methoxypyrimidine, trimethylsilyl triflate (o) NaOMe, MeOH (q) ^{n}Bu$_2$SnO,
MeOH; CsF, dimethyl formamide, 60°C (r) Pd(OH)$_2$/H$_2$.

Fig. 18. Danishefsky's synthesis of octosyl acid A.

formation of the desired nucleoside **135**. This material failed to cyclize cleanly under the usual base-catalyzed conditions. However, treatment of **135** with $^nBu_2SnO$ followed by CsF resulted in the formation of the desired trans-fused bicyclic system (presumably via a 2',3'-stannylene derivative) and this was then converted to synthetic octosyl acid A **7**.

Of course any approach to the ezomycin system must not only be capable of forming the 3',7'-anhydrooctose moiety but also selectively incorporate the required vicinal aminoalcohol at C-5' and C-6'. Suami (refs. 42 & 43) has attempted to address this problem by using a nitroaldol reaction analogous to that used in his tunicamycin synthesis. In this case (Fig. 19) reaction of aldehyde **136** with nitrofuranose **137** was not selective and produced a mixture three of the four possible aldol diastereomers **138** which were separated and characterized as the protected aminoalcohols **139**. As it turned out, the predominate diastereomer was shown to possess the desired (5R,6R) ezomycin stereochemistry.

Selective hydrolysis of the 7,8-O-isopropylidene group allowed further side chain manipulation that resulted in formation of the (7S) mesylate **140**. Removal of the methylthiomethyl (MTM) ether group yielded an alcohol (not shown) which refused to cyclize under basic conditions but gave instead the epimeric (7R) diol which presumably resulted from mesylate displacement by the neighboring acetamide group followed by hydrolysis of the imidate ether during workup. This competitive participation by the acetamide could be precluded by replacing it with a 2,4-dinitrophenylamine (DNP) group but this time unexpected E_2 elimination of methanesulfonic acid occurred across the 7-8 bond instead of the desired cyclization. Finally, mesylate **140** was converted to the (5R,6R,7R) diastereomer **141** using a sequence that involved acetamide-mediated hydrolysis followed by N-protection with 2,4-dinitro-fluorobenzene and treatment with methanesulfonyl chloride. This substrate did indeed cyclize to give the 3,7-anhydrooctose **142** in spite of the presence of four axial sustituents on the pyran ring. The fact that the configuration at C-8 must now be adjusted was not considered a major problem since the corresponding carboxylic acid (see structure **8**) should be readily epimerized to the more stable equatorial position.

In light of the subtle effects which can adversely influence the formation the highly substituted pyran ring present in the ezomycins, it is important to consider an alternative strategy that appends the furan moiety to an already functionalized pyran (cf. ref. 44). An example of this approach is provided by Hanessian's synthesis (ref. 45) of the *unnatural*

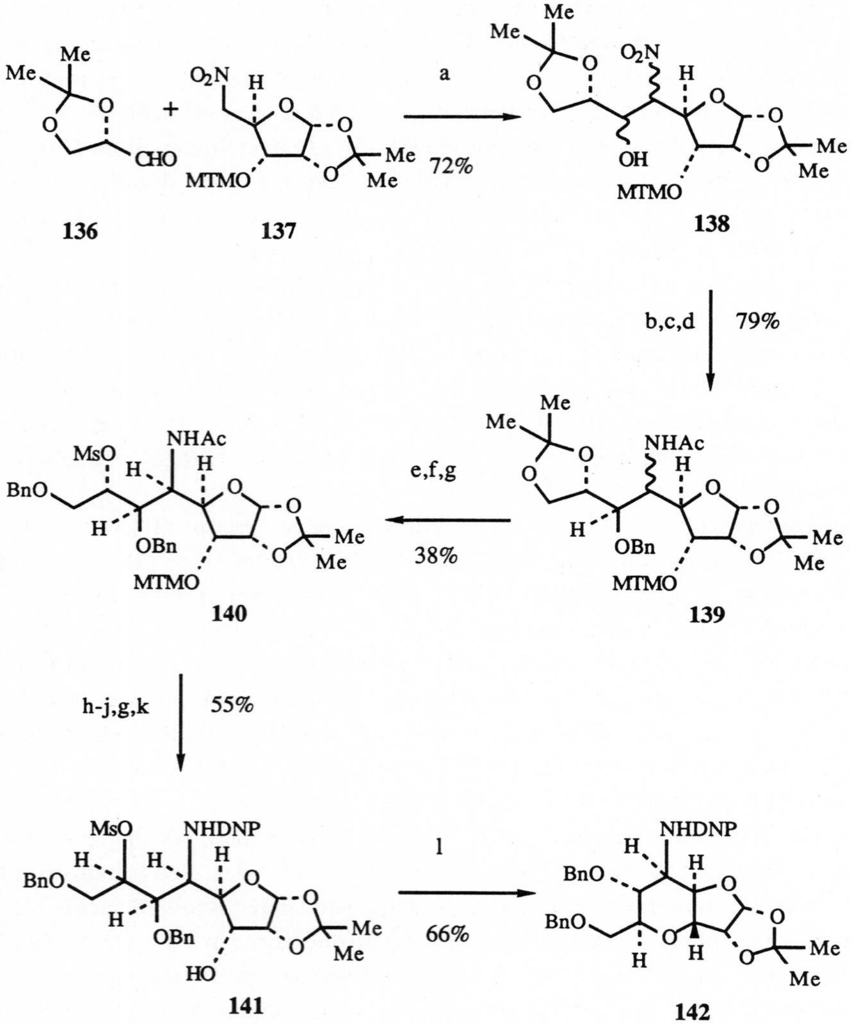

(a) KF, $^{n}Bu_4N^{+}I^{-}$ (b) Raney-Ni/H_2 (c) acetic anhydride, MeOH (d) benzyl bromide (e) aq. HOAc (f) benzyl bromide, NaH (g) methanesulfonyl chloride, Et_3N (h) NaH, dimethylformamide (i) NaOH (j) 2,4-dinitro-fluorobenzene (k) MeI, $NaHCO_3$ (l) NaH, dimethylsulfoxide.

Fig. 19. Suami's nitroaldol approach to the ezomycin nucleus.

nucleoside antibiotic quantamycin, as shown in Fig. 20. The sequence begins with the highly functionalized pyran **143** which is itself derived from the naturally-occurring antibiotic, lincomycin. Stereocontrolled C-glycosylation of **143** with vinylmagnesium bromide followed by benzoylation and oxidative degradation yielded the aldehyde **144**. The addition of $(PhS)_2CHLi$ to **144** was stereoselective (ratio = 7:1) in favor of the *erythro*-dithioacetal **145** which cyclized upon treatment with bromodimethylsulfonium bromide to give a mixture of anomeric thioglycosides **146** ($\alpha{:}\beta$ = 3:1). This mixture was reacted with N-benzoyladenine in the presence of Br_2 to give a (2:1) mixture of the β-nucleoside **147** and its α-isomer (which could be recycled to **147**). It should be noted that this sequence effectively overcomes the ring-opening problem encountered with the routes to the *trans*-dioxahydrindane system that did not begin with a preformed nucleoside (vide supra). Deprotection of **147** gave an amine which was coupled with ^{n}propylhygric acid **148** to give the target quantamycin **149**.

CONCLUSION

I have attempted to provide an overview of recent advances that have been made towards the asymmetric synthesis of complex nucleoside antibiotics **1-8**. A major tactical problem associated with these molecules involves the synthesis of unusual aminosugar/aminoacid moieties incorporated into the target structures. In particular, the controlled introduction of chirality has not always been successful. Our own work in this area has demonstrated the viability of using simple aminoacids to build up the aminosugar/aminoacid residues required for the polyoxins and amipurimycin. This aminoacid-based strategy (cf. ref. 46), or some variation thereof, should in principle be applicable to the synthesis of most if not all of the antibiotic targets discussed in this article.

Future work will undoubtedly be aimed not only at synthesis of naturally-occurring antibiotics but also "unnatural" antibiotics along the lines of quantamycin. The development of synthetic expertise for the assembly of these complex molecules may also prove quite valuable for probing actual biochemical interactions of antibiotics with their receptors (ie. molecular recognition) and lead to a new generation of rationally designed drugs. Thus it should be evident that natural product synthesis can provide a foundation for further interdisciplinary work that may ultimately result in a better understanding and control of important biological systems.

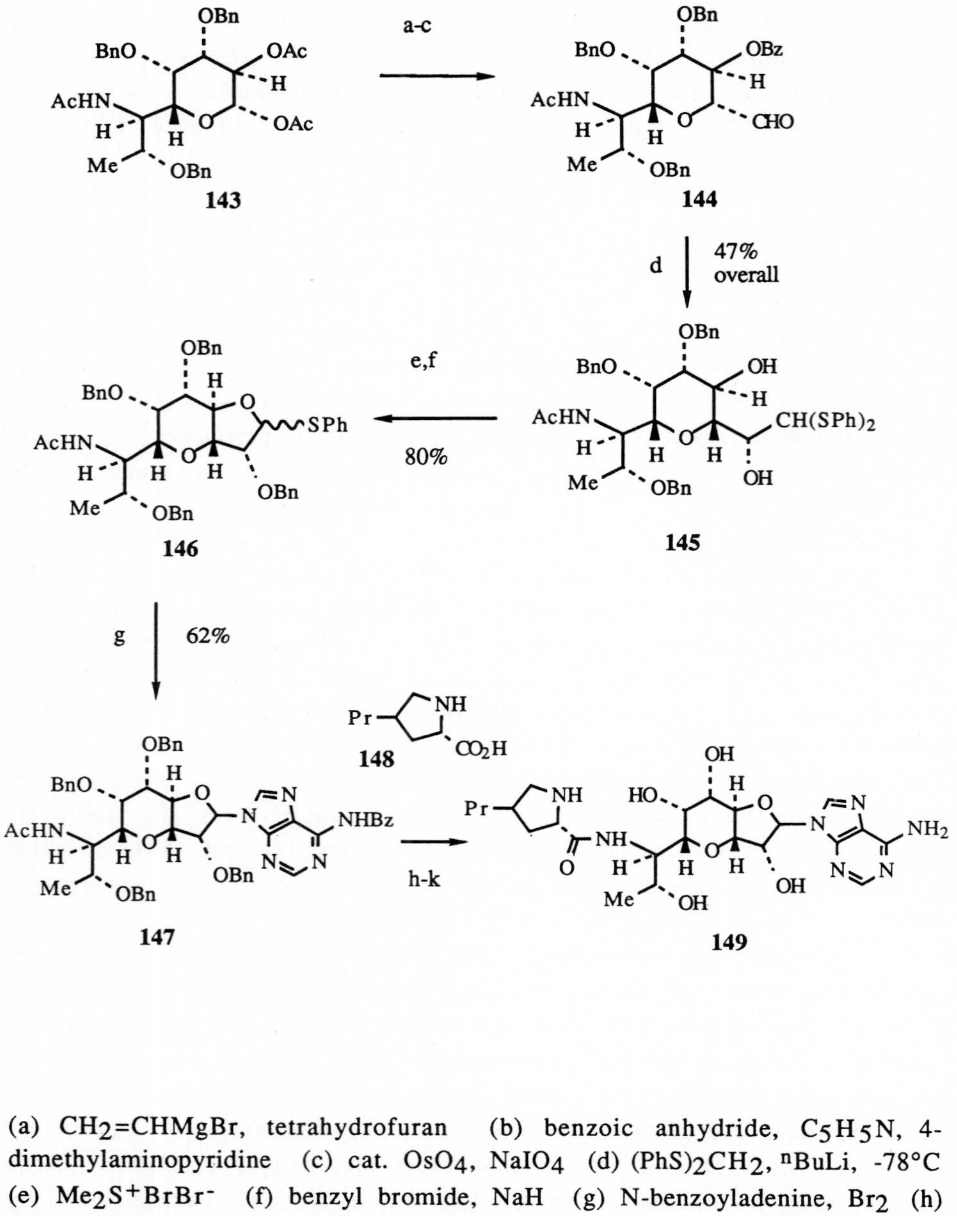

(a) $CH_2=CHMgBr$, tetrahydrofuran   (b) benzoic anhydride, C_5H_5N, 4-dimethylaminopyridine   (c) cat. OsO_4, $NaIO_4$ (d) $(PhS)_2CH_2$, ^nBuLi, -78°C (e) $Me_2S^+BrBr^-$ (f) benzyl bromide, NaH   (g) N-benzoyladenine, Br_2 (h) 20% $Pd(OH)_2/H_2$ (i) 0.5 N $Ba(OH)_2$, reflux   (j) hexamethyldisilazane, 120°C; **148**, BuOCOCl, N-methylmorpholine, -20°C   (k) $^nBu_4N^+F^-$.

Fig. 20.   Hanessian's quantamycin synthesis.

ACKNOWLEDGEMENT I would like to thank my coworkers, Dr. Sarabu Ramakanth and Ms. Jung Min Park, for their technical assistance and the National Institutes of General Medical Sciences for the continued support of our own research programs related to the synthesis of nucleoside antibiotics.

REFERENCES

1 R.J. Suhadolnik, Nucleosides as Biological Probes, Wiley-Interscience, New York, 1979.

2 J.B. Hobbs, Nucleosides, Nucleotides and Nucleic Acids, in: R.H. Thomson (Ed.), The Chemistry of Natural Products, Blackie & Son, Glasgow, 1985.

3 J. Goodchild, The Biochemistry of Nucleoside Antibiotics, in: P.G. Sammes (Ed.), Topics in Antibiotic Chemistry, Vol. 6, Halsted Press, New York, 1982.

4 J.G. Buchanan and R.H. Wightman, The Chemistry of Nucleoside Antibiotics, in: P.G. Sammes (Ed.), Topics in Antibiotic Chemistry, Vol. 6, Halsted Press, New York, 1982.

5 A. Zamojski and G. Grynkiewicz, The Total Synthesis of Carbohydrates, in: J. ApSimon (Ed.), The Total Synthesis of Natural Products, Vol. 6, Wiley-Interscience, New York, 1984, pp. 141-235.

6 K. Isono and S. Suzuki, The Polyoxins: Pyrimidine Nucleoside Peptide Antibiotics Inhibiting Fungal Cell Wall Biosynthesis, Heterocycles, 13(Spec. Issue) (1979) 333-351, and references cited therein.

7 H. Ohrui, H. Kuzuhara, and S. Emoto, Synthesis of Deoxypolyoxin C, "Thymine Polyoxin C", Tetrahedron Lett., (45) (1971) 4267-4270.

8 H. Kuzuhara and S. Emoto, Synthesis of 5-O-Carbamoylpolyoxamic Acid, Tetrahedron Lett., (50) (1973) 5051-5054.

9 H. Kuzuhara, H. Ohrui. and S. Emoto, Total Synthesis of Polyoxin J, Tetrahedron lett., (50) (1973) 5055-5058.

10 N. P. Damodaran, G. H. Jones, and J. G. Moffatt, Synthesis of the Basic Nucleoside Skeleton of the Polyoxin Complex, J. Am. Chem. Soc., 93(15) (1971) 3812-3813.

11 F. Tabusa, T. Yamada, K. Suzuki, and T. Mukaiyama, A Formal Total Synthesis of Polyoxin J Using 4-O-Benzyl-2,3-Isopropylidene-L-Threose As A Common Chiral Building Block, Chem. Lett., (1984) 405-408.

12 U. Niedballa and H. Vorbrueggen, A general Synthesis of N-Glycosides. I. Synthesis of Pyrimidine Nucleosides, J. Org. Chem., 39(25) (1974) 3654-3660.

13 N.T. Anh, Regio- and Stereo-Selectivities in Simple Nucleophilic Reactions, Top. Curr. Chem., 88 (1980) 145-162.

14 A.K. Saksena, R.G. Lovey, and A.T. McPhail, A Convenient Synthesis of Polyoxamic Acid, 5-O-Carbamoylpolyoxamic Acid, and Their Unnatural D Isomers, J. Org. Chem., 51(25) (1986) 5024-5028.

15 P. Garner and J. M. Park, An Asymmetric Synthesis of 5-O-Carbamoyl-polyoxamic Acid From D-Serine, manuscript in preparation.

16 P. Garner and J.M. Park, The Synthesis and Configurational Stability of Differentially Protected β-Hydroxy-α-amino Aldehydes, J. Org. Chem., 52(12) (1987) 2361-2364.

17 P. Garner, Stereocontrolled Addition to a Penaldic Acid Equivalent: An Asymmetric Synthesis of *threo*-β-Hydroxy-L-glutamic Acid, Tetrahedron Lett., 25(51) (1984) 5855-5858.

18 J.K. Cha, W.J. Christ, and Y. Kishi, On Stereochemistry of Osmium Tetraoxide Oxidation of Allylic Alcohol Systems: Empirical Rule, Tetrahedron, 40(12) (1984) 2247-2255.

19 T. Goto, Y. Toya, T. Ohgi, and T. Kondo, Structure of Amipurimycin, A Nucleoside Antibiotic Having a Novel Sugar Moiety, Tetrahedron Lett., 23(12) (1982) 1271-1274.

20 H. Seto, M. Koyama, H. Ogino, and T. Tsuruoka, The Structures of Novel Nucleoside Antibiotics, Miharamycin A and Miharamycin B, Tetrahedron Lett., 24(17) (1983) 1805-1808.

21 P. Garner and S. Ramakanth, Stereodivergent Syntheses of *threo*- and *erythro*-6-amino-6-deoxyheptosulose Derivatives via an Optically Active Oxazolidine Aldehyde, J. Org. Chem., 51(13) (1986) 2609-2612.

22 P. Garner and S. Ramakanth, The Stereoselective Synthesis of Acyclic and Exocyclic Trisubstituted Olefins via a Hydroxyl-Directed Wittig Reaction, J. Org. Chem., 52(12) (1987) 2629-2631.

23 P. Garner and S. Ramakanth, A Regiocontrolled Synthesis of N[7]- and N[9]-Guanine Nucleosides, submitted for publication.

24 G. A. Mock and J. G. Moffatt, An Approach to The Total Synthesis of Sinefungin, Nucleic Acids Res., 10(20) (1982) 6223-6234.

25 M. Geze, P. Blanchard, J. L. Fourrey, and M. Robert-Gero, Synthesis of Sinefungin and Its C-6' Epimer, J. Am. Chem. Soc., 105(26) (1983) 7638-7640.

26 J. W. Lyga and J. A. Secrist III, Synthesis of Chain-Extended and C-6' Functionalized Precursors of the Nucleoside Antibiotic Sinefungin, J. Org. Chem., 48(12) (1983) 1982-1988.

433

27   M. Vuilhorgne, S. Ennifar, B. C. Das, J. W. Paschal, R. Nagarajan, E. W. Hagaman, and E. Wenkert, Structure Analysis of the Nucleoside Disaccharide Antibiotic Anthelmycin by Carbon-13 Nuclear Magnetic Resonance Spectroscopy. A Structural Revision of Hikizimycin and Its Identity with Anthelmycin, J. Org. Chem., 42(20) (1977) 3289-3291.

28   J. A. Secrist III and K. D. Barnes, Synthesis of Methyl Peracetyl α-Hikosaminide, the Undecose Portion of the Nucleoside Antibiotic Hikizimycin, J. Org. Chem., 45(22) (1980) 4526-4528.

29 S. Danishefsky and C. Maring, A Fully Synthetic Route to Hikosamine, J. Am. Chem. Soc., 107(25) (1985) 7762-7764.

30 G. Tamura and T. Ito, Dicovery and Chemistry of Tunicamycin, in: G. Tamura (Ed.), Tunicamycin, Japan Scientific Societies Press, Tokyo, 1982, pp. 3-29.

31 T. Suami, H. Sasaki, and K. Matsuno, Synthesis of Methyl Hexaacetyl-Tunicaminyl Uracil, Chem. Lett., (1983) 819-822.

32   T. Suami, H. Sasaki, K. Matsuno, and N. Suzuki, Total Synthesis of Tunicamycin, Carbohydr. Res., 143 (1985) 85-96.

33   S. Danishefsky and M. Barbachyn, A Fully Synthetic Route to Tunicaminyluracil, J. Am. Chem. Soc., 107(25) (1985) 7761-7762.

34 R. U. Lemieux and R. M. Ratcliffe, The Azidonitration of Tri-O-acetyl-D-galactal, Can. J. Chem., 57(10) (1979) 1244-1251.

35   K. Isono, P. F. Crain, and J. A. McCloskey, J. Am. Chem. Soc., 97(4) (1975) 943-945.

36   K. Sakata, A. Sakurai, and S. Tamura, Structures of Ezomycins A_1 and A_2, Tetrahedron Lett., (49-50) (1974) 4327-4330.

37 K. Sakata, A. Sakurai, and S. Tamura, Structures of Ezomycins B_1, B_2, C_1, C_2, D_1, and D_2, Tetrahedron Lett., (37) (1975) 3191-3194.

38   K. Anzai and T. Saita, The Synthesis of Several Octose Derivatives Related to Octosyl Acids A and B, Bull. Chem. Soc. Jpn., 50(1) (1977) 169-174.

39 K. S. Kim and W. A. Szarek, Synthesis of 3',7'-Anhydrooctose Nucleosides Related to The Ezomycins and The Octosyl Acids, Can. J. Chem., 59(5) (1981) 878-888.

40 S. Hanessian, J. Kloss, and T. Sugawara, Stereocontrolled Access to the Octsyl Acids: Total Synthesis of Octosyl Acid A, J. Am. Chem. Soc. 108(10) (1986) 2758-2759.

41   S. Danishefsky and R. Hungate, Total Synthesis of Octosyl Acid A: A New Departure in Organostannylene Chemistry, J. Am. Chem. Soc.,108(9) (1986) 2486-2487.

42 O. Sakanaka, T. Ohmori, S. Kozaki, T. Suami, T. Ishii, S. Ohba, and Y. Saito, Synthetic Approach toward Antibiotic Ezomycins. I. Synthesis of 5-Amino-5-deoxyoctofuranose-(1,4) Derivatives by Henry Reaction and Their Stereochemistry, Bull. Chem. Soc. Jpn., 59(6) (1986) 1753-1759.

43 O. Sakanaka, T. Ohmuri, S. Kozaki, and T. Suami, Synthetic Approach toward Antibiotic Ezomycins. II. Synthesis of 5-Amino-3,7-anhydro-5-deoxyoctofuranose-(1,4) Derivatives, Bull. Chem. Soc. Jpn., 59(11) (1986) 3523-3528.

44 S. Hanessian, D. M. Dixit, and T. J. Liak, Studies Directed Towards the Total Synthesis of the Ezomycins, the Octosyl Acids and Related Substances, Pure & Appl. Chem., 53 (1981) 129-148.

45 S. Hanessian, K. Sato, T. J. Liak, N. Danh, and D. Dixit, "Quantamycin": A Computer-Simulated New-Generation Inhibitor of Bacterial Ribosomal Binding, J. Am. Chem. Soc., 106(20) (1984) 6114-6115.

46 G. M. Coppola and H. F. Schuster, Asymmetric Synthesis/Construction of Chiral Molecules Using Amino Acids, Wiley-Interscience, New York, 1987.

STUDIES DIRECTED TOWARD THE TOTAL SYNTHESIS OF THE MILBEMYCINS AND AVERMECTINS

MICHAEL T. CRIMMINS,* W. GARY HOLLIS, JR., and ROSEMARY O'MAHONY

1 INTRODUCTION

The milbemycins (**1a-t**, Fig. 1) and the avermectins (**2a-h**, Fig. 2) are related classes of macrocyclic lactone natural products with remarkable biological properties. The avermectins were isolated by Merck and Co. as a result of a broad screening program searching for naturally occurring anthelmintics.[1] Eight avermectins (**2a-h**) were extracted from the broth of a *Streptomyces avermitilis* culture in 1976 and several others (not shown) were subsequently isolated and identified.[2] The first milbemycins were isolated in 1975 by Mishima and coworkers,[3] and other milbemycins were later isolated bringing the total number to twenty.[4] Both the milbemycins and the avermectins contain a sixteen membered ring lactone incorporating a dioxaspiroundecane (6,6-spiroketal) moiety. A highly functionalized hexahydrobenzofuran unit is also common to both classes with the exception of the simpler milbemycins β3, β1, and β2 (**1a-c**) and the more recently isolated avermectins. The major difference between the two groups is the presence of a C13 hydroxyl group in the avermectins which bears an α-L-oleandrosyl-α-L-oleandrosyl disaccharide as well as the presence of a C22-C23 olefin in the avermectin 1-series. Additionally, the avermectins possess a C25 isopropyl or sec-butyl group while the milbemycins normally contain a C25 methyl, ethyl or isopropyl.

2 BIOLOGICAL ACTIVITY

The remarkable biological properties of the avermectins and the milbemycins have stirred an enormous interest in their medicinal application as well as their synthetic preparation and modification. The avermectins are active primarily against nematodes and arthropods (insects, ticks, mites) but not trematodes or cestodes.[5] Dihydro (Δ22,23) avermectin B$_{1a}$ (ivermectin) **2x** is marketed by Merck and Co. as a veterinary drug for the treatment of intestinal worms in livestock and heartworm in dogs.[6] Ivermectin has also been utilized successfully on an experimental basis to treat human onchocerciasis ("river blindness") in west Africa.[7] An interference in the GABA (γ-aminobutyric acid) regulated neurotransmission of nematodes and arthropods has been proposed as the mechanism of action of the avermectins. The milbemycins are generally less active than the avermectins, although milbemycin D (**1o**) is only slightly less active than the avermectins in its anthelmintic activity.[8] A detailed discussion of the biological activity, SAR studies and mechanism of action is beyond the scope of this review, and the reader is referred to an excellent review by Fisher.[9]

Figure 1: The Milbemycins

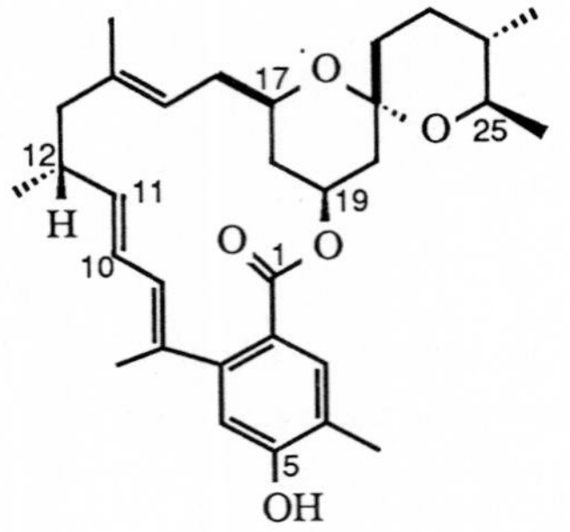

1a β_3

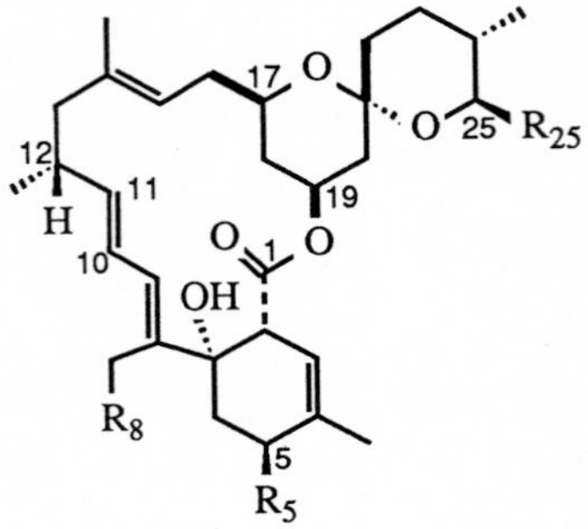

1b	β_1	R_5 = OCH$_3$, R_8 = OH, R_{25} = CH$_3$
1c	β_2	R_5 = OCH$_3$, R_8 = OH, R_{25} = C$_2$H$_5$
1d	E	R_5 = OCH$_3$, R_8 = OH, R_{25} = CH(CH$_3$)$_2$
1t	H	R_5 = =O, R_8 = H, R_{25} = CH(CH$_3$)$_2$

3 DEGRADATION STUDIES

In their efforts to effect the synthesis of milbemycins and avermectins, several groups have conducted research focusing on degradation of the natural materials. Successful degradation of these compounds provided intact fragments which were used to confirm the identity of intermediates generated in the course of their synthetic activities. In addition, compounds derived from natural material were used to confirm the viability of key steps in the synthetic sequence, particularly the later ones. Degradation products are attractive in such studies due to their availability in relatively substantial quantities from a few laboratory operations. The fragments which were produced in the degradation of natural material could also be subjected to biological testing, as could the hybrids and semi-synthetic compounds prepared from these fragments, in an effort to discover the identity of the structural components responsible for the activity of avermectins and milbemycins.

Smith's degradation[10] of avermectin B$_{1a}$ (**2f**) to yield northern fragment **3** and southern fragment **4** is shown in Scheme 1. Prior to oxidation and cleavage of the C8-C9 olefin, the secondary alcohols were selectively protected, with the β-hydroxyl at C5 exhibiting a higher reactivity than the free hydroxyl in the sugar, to generate macrocycle **5** in 90% yield. Regioselective epoxidation of the C8-C9 double bond affording epoxide **6** was achieved by using the C7-hydroxyl to direct the course of the reaction mediated by VO(acac)$_2$ and t-BuOOH. Methylation of the C7

Figure 1 (cont.): The Milbemycins

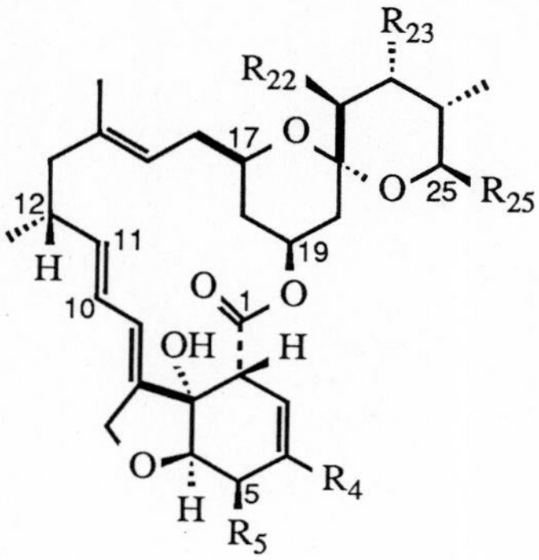

Milbemycin		R_4	R_5	R_{22}	R_{23}	R_{25}
1e	α_1	CH_3	OH	H	H	CH_3
1f	α_2	CH_3	OCH_3	H	H	CH_3
1g	α_3	CH_3	OH	H	H	C_2H_5
1h	α_4	CH_3	OCH_3	H	H	C_2H_5
1i	α_5	CH_3	OH	OH	OCO⟨⟩ 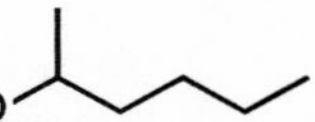	CH_3
1j	α_6	CH_3	OCH_3	OH	OCO⟨⟩ 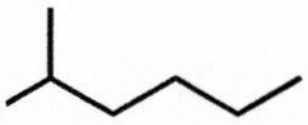	CH_3
1k	α_7	CH_3	OH	OH	OCO⟨⟩ 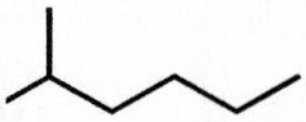	C_2H_5
1l	α_8	CH_3	OCH_3	OH	OCO⟨⟩ 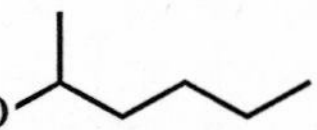	C_2H_5
1m	α_9	CH_2OCO-⟨pyrrole⟩ 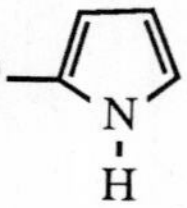	OH	H	H	CH_3
1n	α_{10}	CH_2OCO-⟨pyrrole⟩ 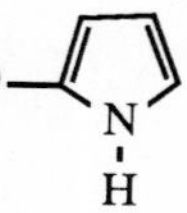	OH	H	H	C_2H_5
1o	D	CH_3	OH	H	H	$CH(CH_3)_2$
1p	F	CH_2OCO-⟨pyrrole⟩ 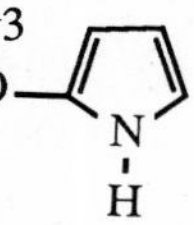	OH	H	H	$CH(CH_3)_2$
1q	G	CH_3	OCH_3	H	H	$CH(CH_3)_2$
1r	J	CH_3	=O	H	H	CH_3
1s	K	CH_3	=O	H	H	C_2H_5

438

Figure 2: The Avermectins

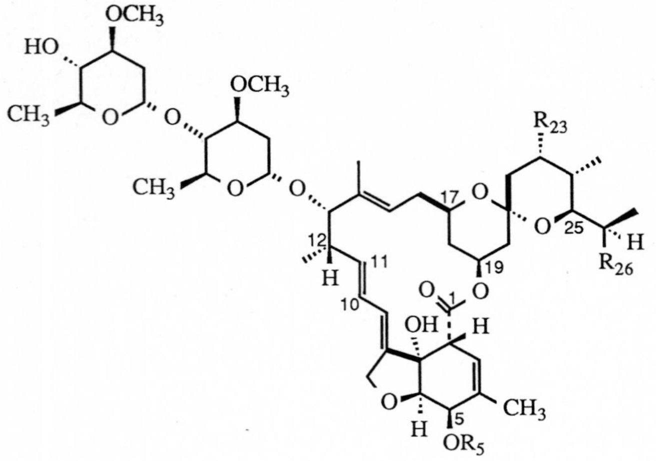

2a	A_{2a} $R_5 = CH_3$, $R_{26} = C_2H_5$, $R_{23} = OH$
2b	B_{2a} $R_5 = H$, $R_{26} = C_2H_5$, $R_{23} = OH$
2c	A_{2b} $R_5 = CH_3$, $R_{26} = CH_3$, $R_{23} = OH$
2d	B_{2b} $R_5 = H$, $R_{26} = CH_3$, $R_{23} = OH$
2x	ivermectin $R_5 = H$, $R_{26} = C_2H_5$, $R_{23} = H$

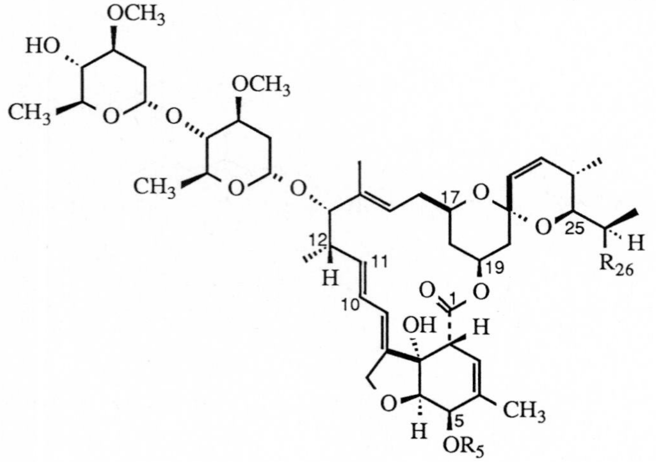

2e	A_{1a} $R_5 = CH_3$, $R_{26} = C_2H_5$
2f	B_{1a} $R_5 = H$, $R_{26} = C_2H_5$
2g	A_{1b} $R_5 = CH_3$, $R_{26} = CH_3$
2h	B_{1b} $R_5 = H$, $R_{26} = CH_3$

hydroxyl with CH_2N_2 occurred only after extended reaction but was necessary to insure the integrity of the southern subunit produced in the subsequent fragmentation. Diol **7** was provided as a mixture of diastereomers in 45% yield upon acid catalyzed epoxide opening. Reductive cleavage of the macrolactone was accomplished using $LiAlH_4$, and further oxidative cleavage of the 1,2-diol with $Pb(OAc)_4$ produced northern fragment **3** in 61% yield and southern fragment **4** in 40%.

Alternatively, the 2° alcohols could be protected as their acetates, the degradation sequence carried out, and the unprotected analogues of **3** and **4** generated, where the acetates were removed in the $LiAlH_4$ reduction. Smith also indicated that, based on research done by the Merck group, the saponification of the macrolactone is not feasible due to epimerization at C2 with subsequent aromatization and that ozonolysis of the macrolide results in oxidation of the C3-C4 olefin thereby prohibiting the isolation of an intact southern subunit such as **4.**

Scheme 1

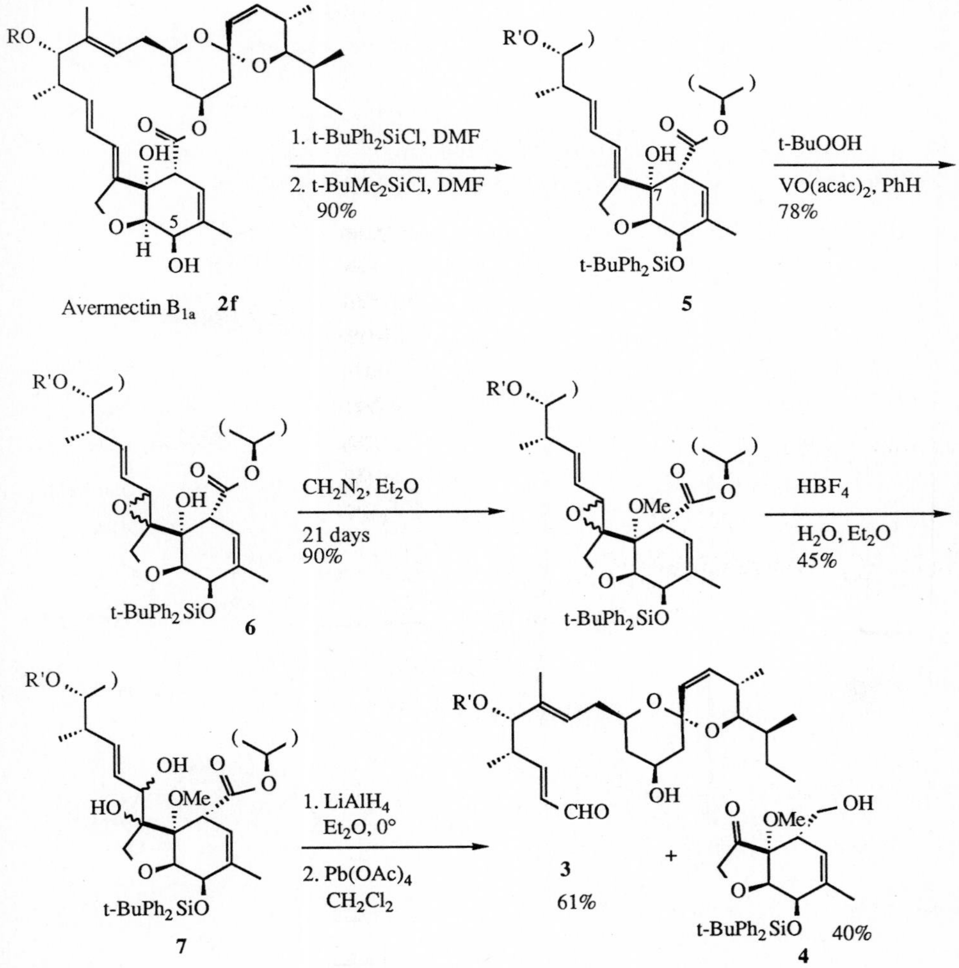

The degradation of avermectin B_{1a} (**2f**) by Hanessian[11] is depicted in Scheme 2. Exposure of B_{1a} (**2f**) to aqueous KOH / DME resulted in cleavage of the macrolide by saponification of the ester functionality with intentional conjugation of the C3-C4 olefin to minimize risks of aromatization of the hexahydrobenzofuran moiety. Methyl ester **8** was then generated in 82% yield using CH_2N_2. Protection of the 2° alcohols as their t-butyldimethylsilylethers followed by ozonolysis with a reductive work-up produced alcohol **9** in 92% yield and allylic alcohol **10** in 79% yield. Alcohol **9** could then be converted to α,β-unsaturated aldehyde **11** and disaccharide **12** by an oxidation-elimination sequence. Scheme 3 shows the products of the reductive fragmentation of olefin **13**, the agylcone analogue of **8**, where alcohol **14** resulted from cleavage of the C14-C15 double bond as well as the C10-C11 olefin.

Scheme 2

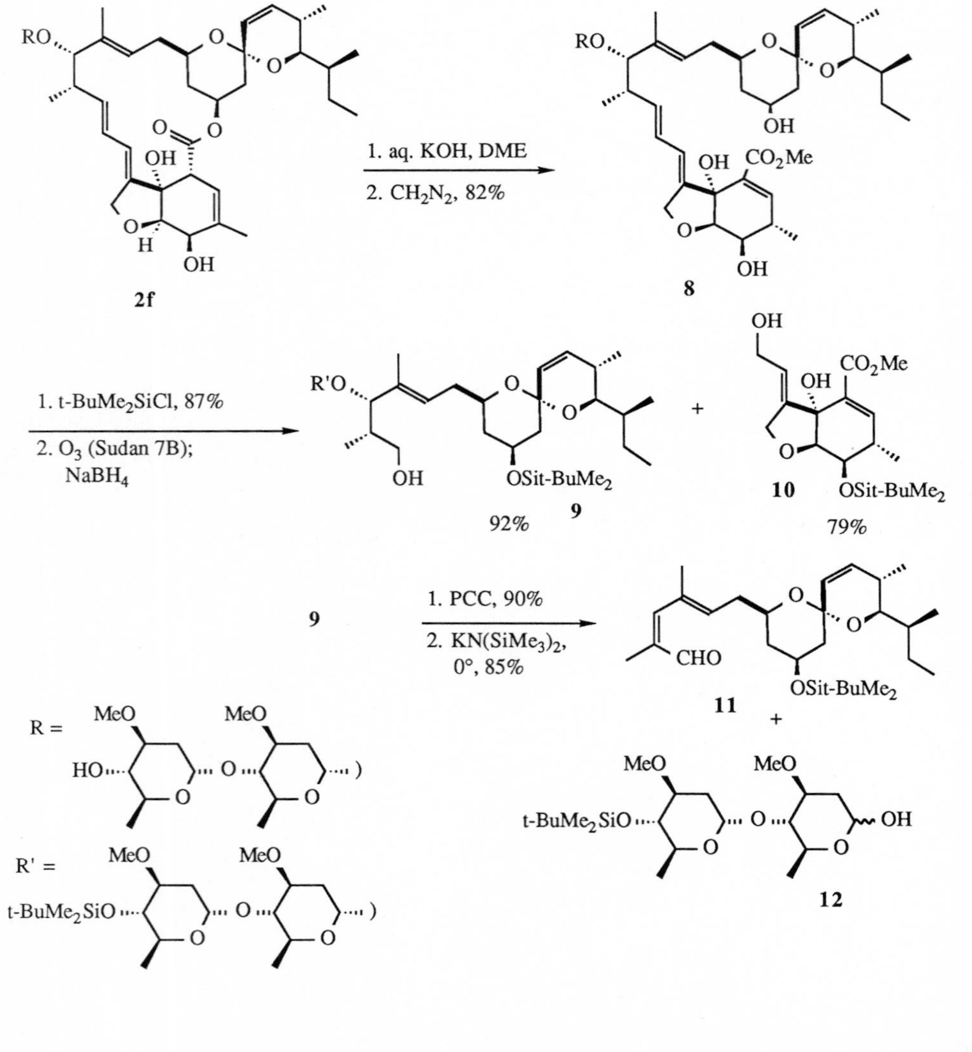

Scheme 3

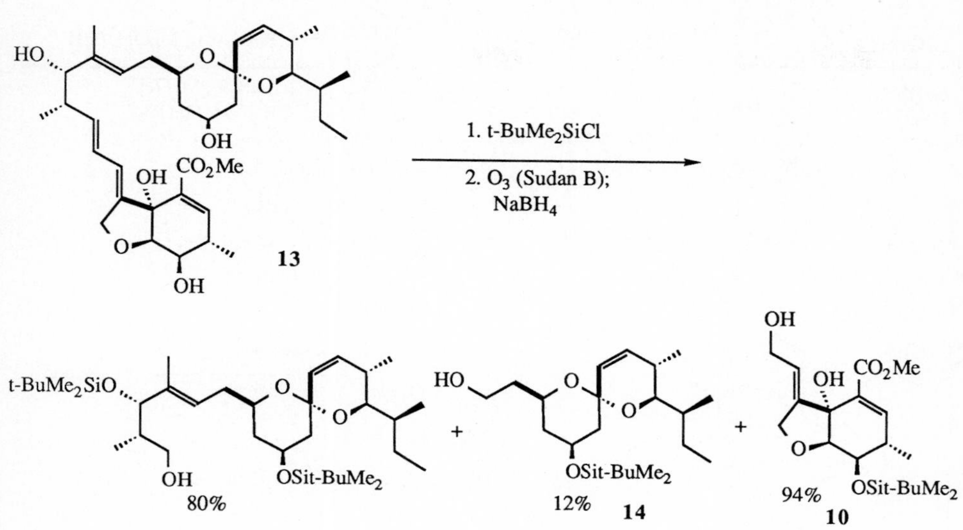
HO
OH
OH
CO2Me
OH
13
1. t-BuMe2SiCl
2. O3 (Sudan B);
NaBH4
t-BuMe2SiO
OH
OSit-BuMe2
80%
HO
+
OSit-BuMe2
12%
14
+
OH
OH
CO2Me
OSit-BuMe2
94%
10

Avermectin A_{1a} **(2e)** has been degraded by Danishefsky[12] to yield compounds **15** and **16** (see Scheme 4). Removal of the disaccharide moiety using aqueous acid and conjugation of the olefin with DBU (no opening of the macrocycle observed) afforded macrolide **17** in 93% yield. Upon treatment with osmium tetroxide, selective oxidation of the C14-C15 olefin was observed to produce tetraol **18** as a single isomer. Osmylation was believed to occur stereospecifically due to severe restrictions on the approach of the reagent to the β-face of the olefin. Exhaustive oxidation of the 1,2,3-triol, reduction with $NaBH_4$, and cleavage of the ester with K_2CO_3/MeOH gave diol **15** in 60% yield and α,β-unsaturated ester **16** in 83% yield. Attempts to oxidize **16** to the aldehyde were uniformly unsuccessful.

Scheme 4

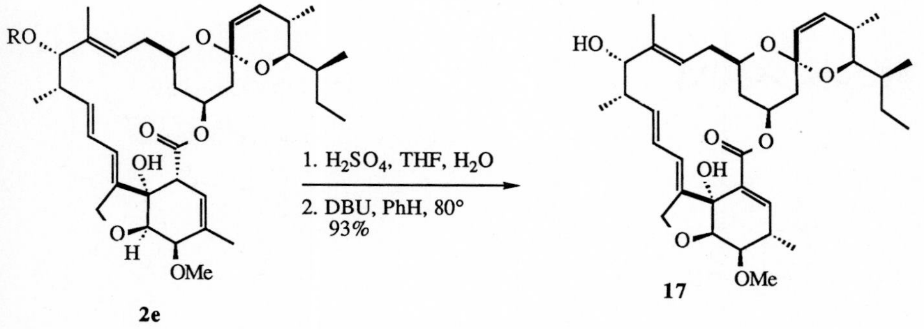
RO
O
O
OH
H
OMe
2e
1. H2SO4, THF, H2O
2. DBU, PhH, 80°
93%
HO
O
O
OH
OMe
17

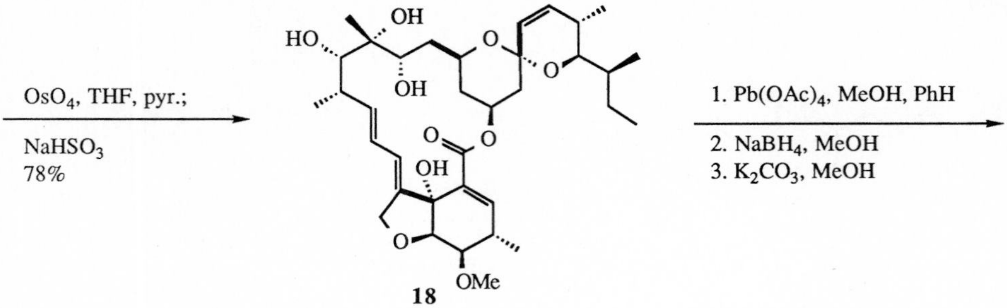

OsO$_4$, THF, pyr.;

NaHSO$_3$
78%

18

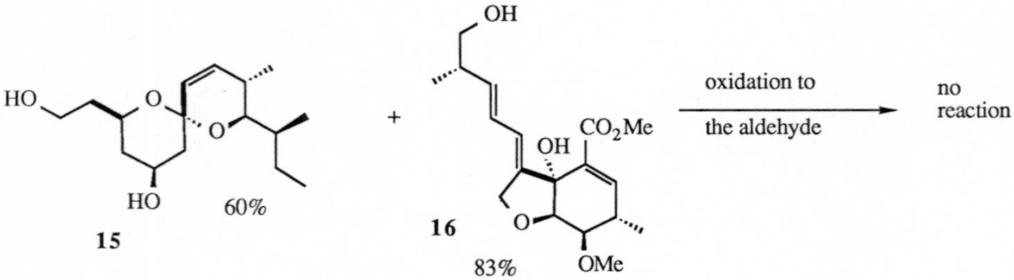

1. Pb(OAc)$_4$, MeOH, PhH

2. NaBH$_4$, MeOH
3. K$_2$CO$_3$, MeOH

HO

15   60%

+

16

83%   OMe

oxidation to
the aldehyde

no
reaction

4 SYNTHETIC MANIPULATIONS

In a study done at Merck,[13] Mrozik (Scheme 5) has been able to convert natural avermectin B$_{2a}$ (**2b**) to the more potent avermectin B$_{1a}$ (**2f**) and ivermectin (**2x**). To this end, **2b** was protected at the C5 and C4" hydroxyls, and the C23 hydroxyl was acylated with 4-O-methylphenylchlorothioformate to give **19**. Pyrolytic elimination of the thiocarbonate followed by removal of the protecting groups produced avermectin B$_{1a}$ (**2f**), while reduction of the thiocarbonate with Bu3SnH and deprotection gave ivermectin **2x**.

Scheme 5

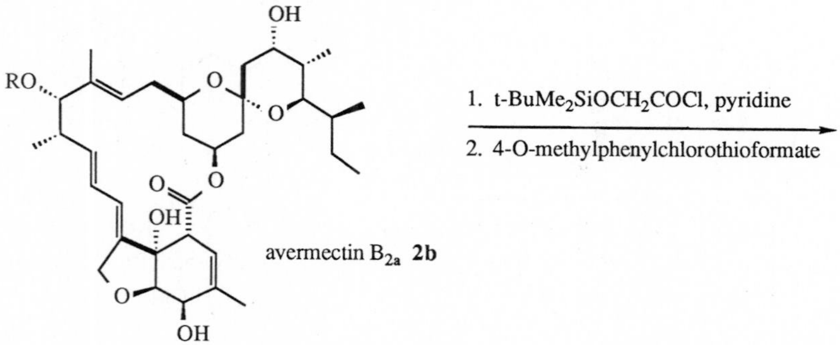

avermectin B$_{2a}$ **2b**

1. t-BuMe$_2$SiOCH$_2$COCl, pyridine

2. 4-O-methylphenylchlorothioformate

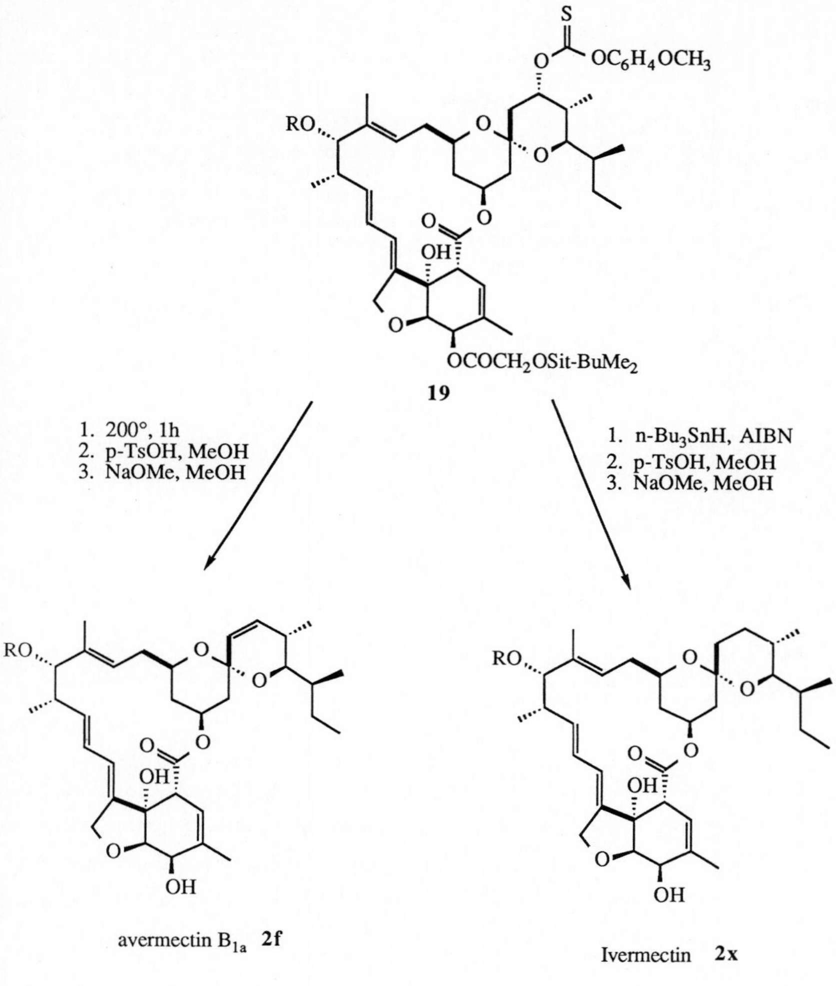
S
$OC_6H_4OCH_3$
RO
O
OH
$OCOCH_2OSit\text{-}BuMe_2$
19
1. 200°, 1h
2. p-TsOH, MeOH
3. NaOMe, MeOH
1. n-Bu_3SnH, AIBN
2. p-TsOH, MeOH
3. NaOMe, MeOH
avermectin B_{1a} 2f
Ivermectin 2x

Further studies at Merck[14] (Scheme 6) again by Mrozik provided for the preparation of α series milbemycins from 22-23-dihydroavermectin B_1 aglycones **20**. A mixture of **20a** and **20b** containing 85% **20a** (C25 sec-butyl) and 15% **20b** (C25 isopropyl) was first protected at the C5 hydroxyl. At this point the sec-butyl and isopropyl C25 isomers could be separated. The C13 chloride of either isomer could be prepared by reaction with 2-nitrobenzenesulfonyl chloride. Reduction mediated by Bu_3SnH followed by deprotection gave **21** or **1o**, completing the conversion of an avermectin to a milbemycin.

Mrozik (Scheme 7) was able to selectively cleave the methyl ether at C5 of avermectin A_{2a} **(2a)** to produce avermectin B_{2a} **(2b)**.[15] Addition of $Hg(OAc)_2$ to avermectin A_{2a} **(2a)** forms the labile methyl enol ether **22** which can be hydrolyzed to ketone **23** concurrent with elimination of the acetoxy group. Stereoselective reduction with $NaBH_4$ gave **2b**, identical to natural avermectin B_{2a}.

Scheme 6

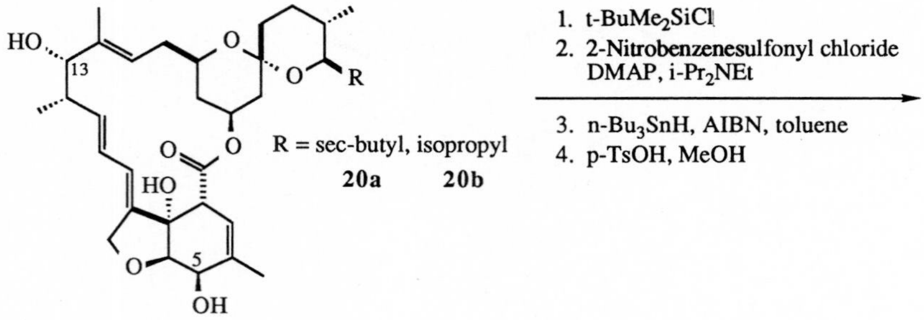
HO
13
O
O
R
O
O
HO
O
5
OH
1. t-BuMe$_2$SiCl
2. 2-Nitrobenzenesulfonyl chloride
DMAP, i-Pr$_2$NEt
3. n-Bu$_3$SnH, AIBN, toluene
4. p-TsOH, MeOH

R = sec-butyl, isopropyl
20a **20b**

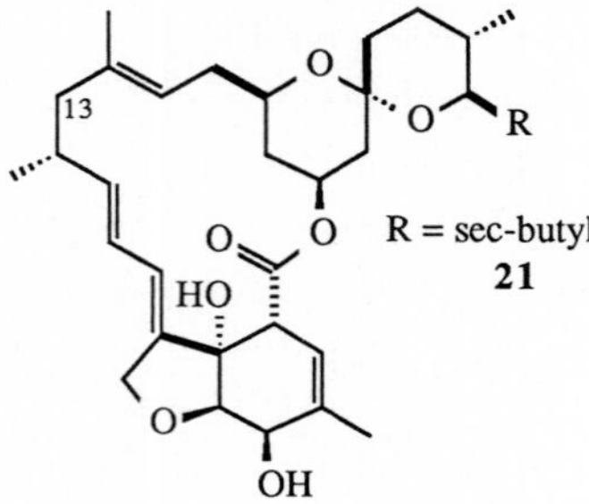
13
O
O
R
O
O
HO
O
OH

R = sec-butyl, isopropyl
21 **1o**

Scheme 7

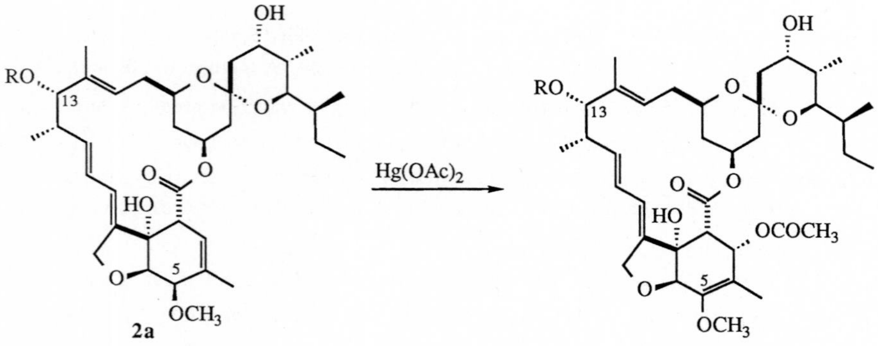
OH
RO
13
O
O
O
O
HO
O
5
OCH$_3$
Hg(OAc)$_2$
OH
RO
13
O
O
O
O
HO
OCOCH$_3$
O
5
OCH$_3$

2a

R = disaccharide

22

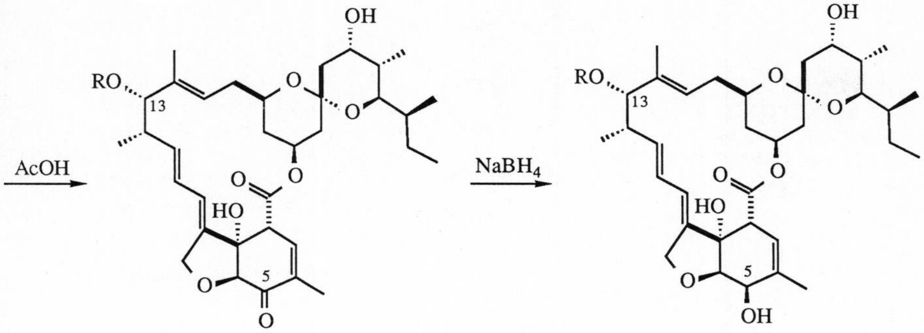
AcOH
NaBH4
RO
13
OH
HO
O
5

23 2b

An avermectin-milbemycin hybrid has been prepared by Smith[16] (Scheme 8) using aldehyde **24**, available in 5 steps from avermectin B_{1a} **(2f)** (see degradation studies). Protection of the C19 hydroxyl, reduction of the enal, and cleavage of the C10-11 olefin [$Mo(CO)_6$, t-BuOOH; then H_5IO_6] generated aldehyde **25** which could be coupled to synthetically prepared phosphine oxide **26** producing **27**. Removal of the protecting groups and macrolactonization completed the preparation of the hybrid **28**.

Scheme 8

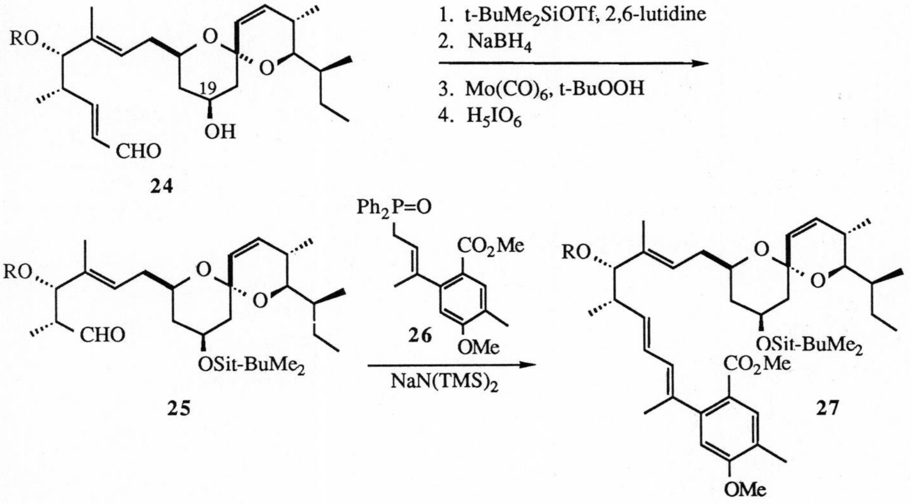
1. t-BuMe2SiOTf, 2,6-lutidine
2. NaBH4
3. Mo(CO)6, t-BuOOH
4. H5IO6
NaN(TMS)2
Ph2P=O
CO2Me
OMe
RO
CHO
OH
19
OSit-BuMe2
24
25
26
27

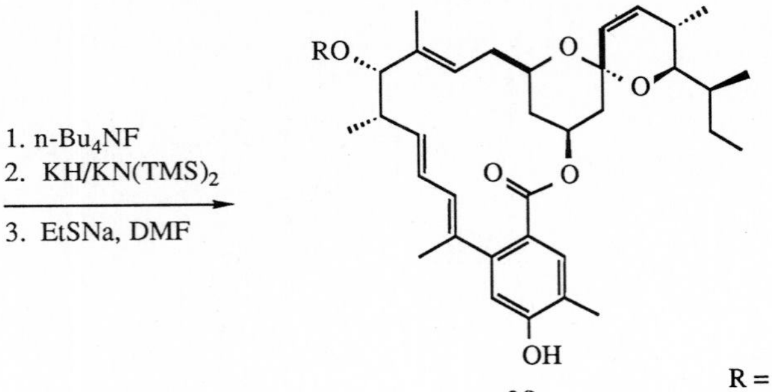

1. n-Bu$_4$NF
2. KH/KN(TMS)$_2$
―――――――――――
3. EtSNa, DMF

28

R = oleandrosyloleandrosyl

Thomas[17] (Scheme 9) has been able to synthesize a macrocyclic analogue of milbemycin β_1 by joining together a southern-like C1-C10 fragment and an analogue of the northern spiroketal moiety. Synthesis of the southern-like fragment begins with β-dicarbonyl **29**. A Robinson annelation with methyl propen-2-yl ketone followed by stereoselective reduction of the ketone and methylation of the alcohol gave **30**. At this point reaction of the furan with bromine in methanol, followed by acid, NBS and water gave the latent aldehyde **31**.

Preparation of the northern subunit analogue began with known lactone **32**. Hydrolysis of the lactone, protection of the alcohol and reduction of the ester to the aldehyde gave **33** which could be chain extended to ester **34** after addition of prop-2-enylmagnesium bromide and subsequent orthoester Claisen rearrangement. The unstable phosphonium salt **35** was readily formed from the corresponding iodide.

Scheme 9

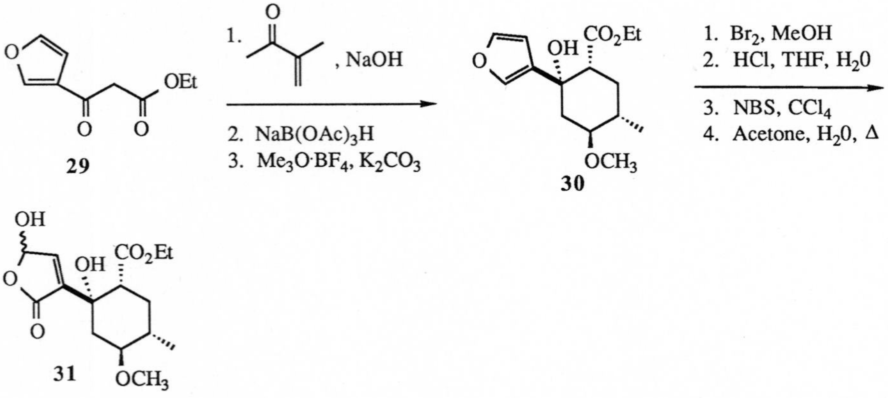

1.
, NaOH

2. NaB(OAc)$_3$H
3. Me$_3$O·BF$_4$, K$_2$CO$_3$

29

CO$_2$Et

1. Br$_2$, MeOH
2. HCl, THF, H$_2$O
―――――――――――
3. NBS, CCl$_4$
4. Acetone, H$_2$O, Δ

OCH$_3$
30

OH

OH CO$_2$Et

31 OCH$_3$

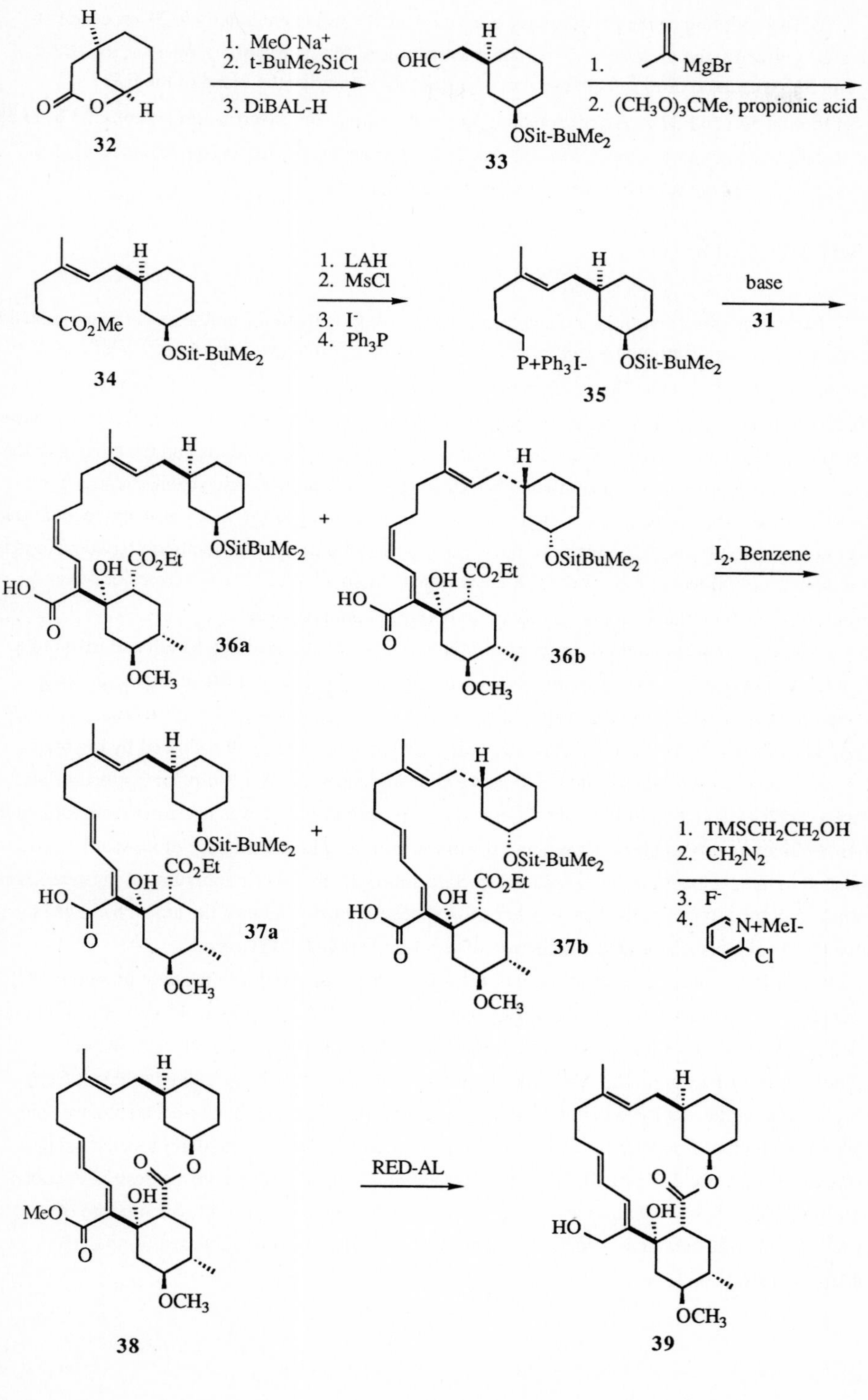

32

1. MeO⁻Na⁺
2. t-BuMe₂SiCl
3. DiBAL-H

33

1. (isopropenyl)MgBr
2. (CH₃O)₃CMe, propionic acid

34

1. LAH
2. MsCl
3. I⁻
4. Ph₃P

35

base

31

36a + **36b**

I₂, Benzene

37a + **37b**

1. TMSCH₂CH₂OH
2. CH₂N₂
3. F⁻
4. (pyridinium N⁺MeI⁻ Cl⁻)

38

RED-AL

39

Wittig coupling of the phosphorane derived from **35** and latent aldehyde **31** proceeded to give predominantly the undesired Z,Z diene diastereomers **36a** and **36b**. However, exposure of **36a** and **36b** to a trace of I_2 in benzene served to cleanly isomerize the diene to the E,Z diastereomers **37a** and **37b**. After protecting group manipulation, macrolactonization of **37a** to **38** was accomplished by exposure of **37a** to N-methyl-2-chloropyridinium iodide. Reduction of the ester to the alcohol completed the synthesis of the analogue **39**.

5 MILBEMYCIN β3

In their efforts to produce a viable program focusing on the total synthesis of the avermectins and milbemycins, most groups chose milbemycin β3 as their initial synthetic target. While attractive as an initial target due to its relatively simple aromatic southern zone, a successful approach to this molecule must address the challenges presented by the spiroketal subunit, the three olefinic moieties, the remote asymmetric center at C12, and finally the coupling of the subunits and the closing of the macrolide. It is worthy to note here that although a variety of strategies have been employed in the generation of the spiroketal subunit, most researchers have relied on the well precedented anomeric effect to establish the stereochemistry of the spirocyclic ketal moiety. The methodology developed while preparing milbemycin β3 was to be the cornerstone on which future syntheses of other members of this remarkably active family of compounds could be based.

The first total synthesis of milbemycin β3 was reported in 1982[18] by Smith and featured a Horner-Emmons route in the coupling of the two key subunits. Lactone **40**, available in both homochiral and racemic forms, served as the primary starting material for the northern, spiroketal-containing subunit (Scheme 10). Addition of allyl Grignard to lactone **40** followed by treatment with trimethyl orthoformate resulted in ketal **41** as a single isomer. A 1,3-dipolar cycloaddition between ketal **41** and the nitrile oxide generated *in situ* from acetal **42** was utilized to incorporate the requisite functionality to allow subsequent spiroketalization. The 2:1 mixture of isoxazolines **43** produced in the cycloaddition was reduced to yield amino alcohol **44** as a mixture of diastereomers. Protection of the alcohol as the benzyl ether and exhaustive methylation of the amine followed by exposure to aqueous acid afforded aldehyde **45** as a single crystalline isomer.

In order to establish the geometry of the C14-C15 olefin and the chirality of the remote C12 center, Ireland enolate Claisen methodology was employed. When aldehyde **45** was allowed to react with isopropenyl cuprate and the resulting mixture of allylic alcohols acylated, ester **46** was produced as a 7:1 mixture. The stereocontrol in the cuprate addition is postulated to result from chelation among the aldehyde oxygen, the spiroketal oxygens, and the metal which enhances the aldehydic stereofacial differences. The predominant ester was isolated and deprotonated using potassium hexamethyldisilylamide in THF/HMPA at low temperature to give the desired Z-enolate. After trapping with TMSCl and warming to facilitate rearrangement, acid **47** (6:1 mixture of diastereomers) was obtained. Aldehyde **48**, the desired northern subunit, was subsequently available after several standard manipulations.

Scheme 10

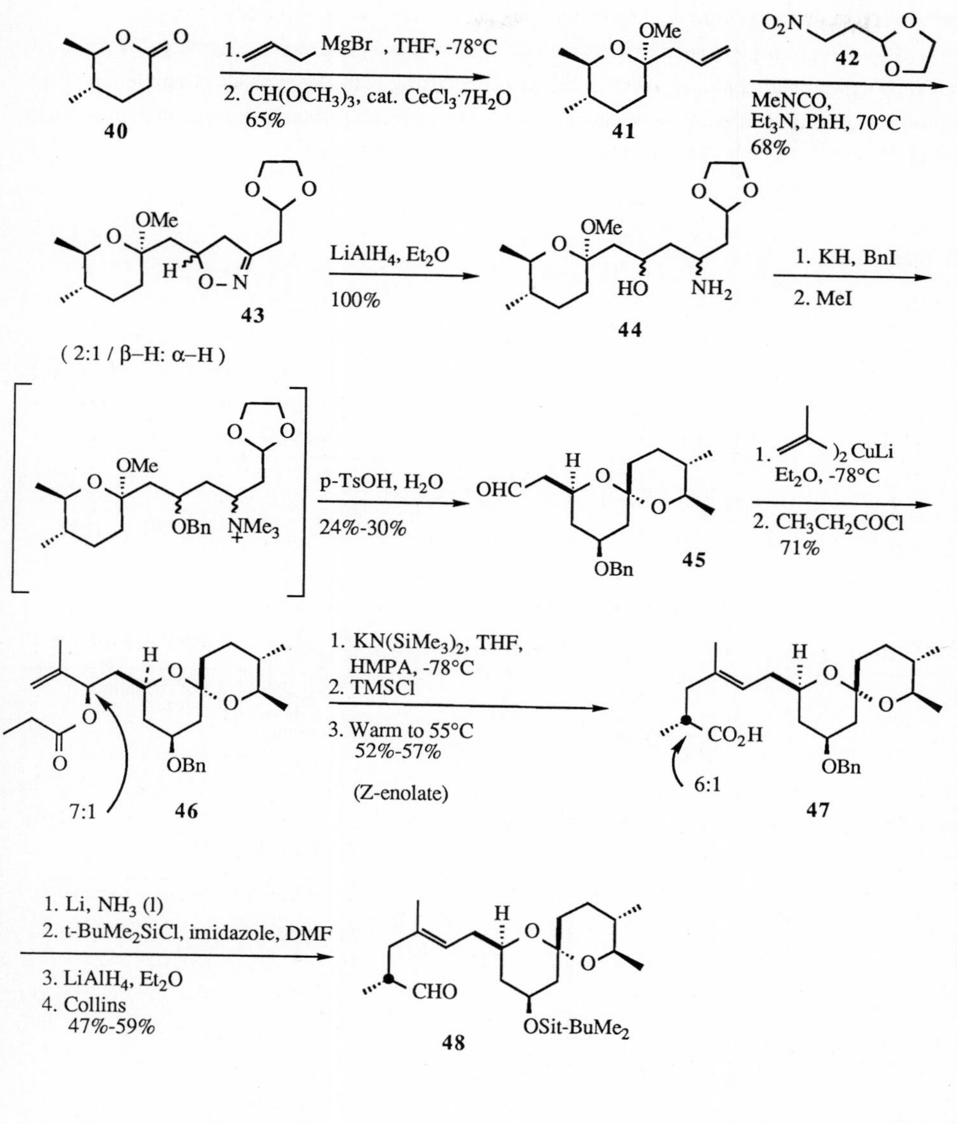

(2:1 / β–H: α–H)

Synthesis of the southern subunit began with anisic acid **49** (Scheme 11). Oxazoline **50** was prepared by converting anisic acid to the acid chloride followed by condensation with amino alcohol **51** and dehydration. Subsequent metalation and acylation yielded aromatic ketone **52** which was transformed into iminolactone **53** upon attack by vinyl Grignard. Phosphine oxide **54** was available from **53** as a mixture of olefin isomers *via* hydrolysis to the lactone, S$_N$2' attack by Ph$_2$PLi, and air oxidation. The 1:3 ratio of the desired E-olefin to the undesired Z could be compensated for to some extent by heating the mixture of the corresponding methyl esters **55** in ethylene glycol containing KOH. This protocol resulted in a 1:1 mixture of olefin isomers which were separable by flash chromatography allowing the Z-isomer to be recycled.

Coupling of the two subunits occurred in high yield and good stereoselectivity (7:1 / E:Z-olefin) by deprotonating phosphine oxide **55** and condensing with aldehyde **48**. Completion of the synthesis was accomplished by removing the silyl group, effecting macrolactonization in good yield using KH, and finally demethylating to produce milbemycin β_3 **(1a)**.

Scheme 11

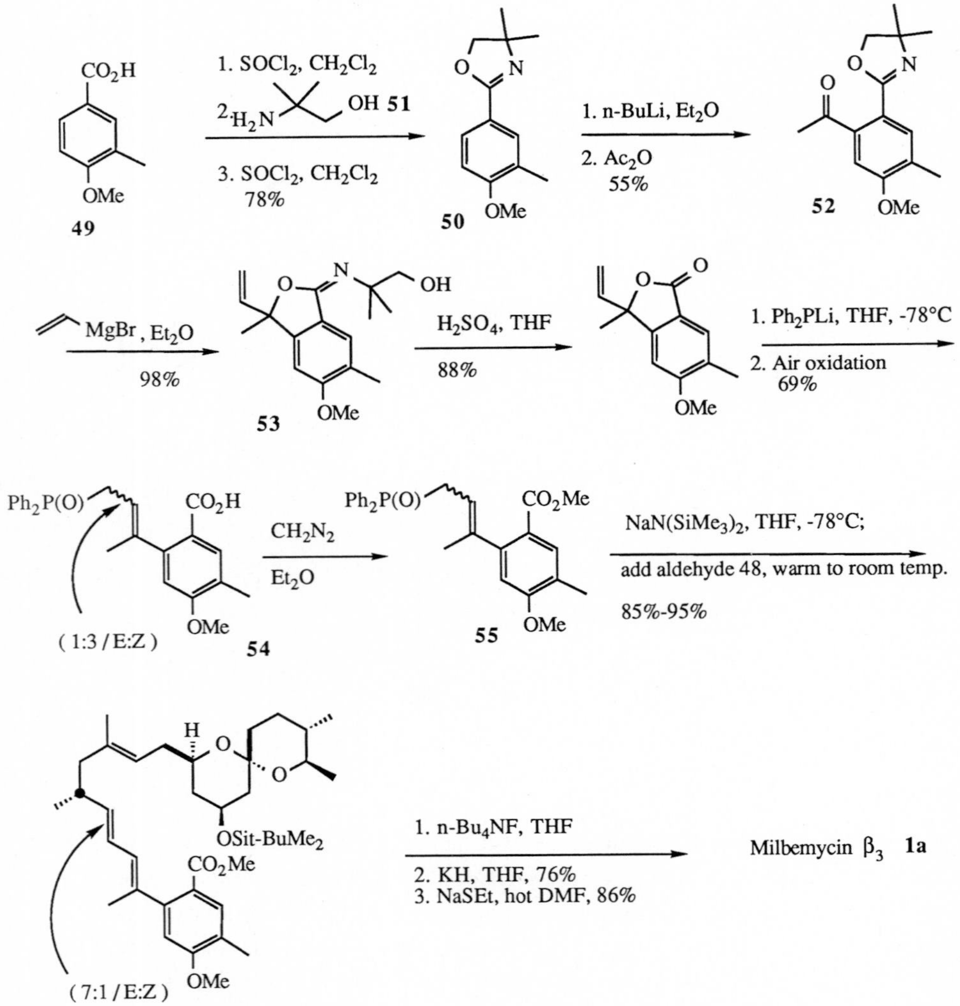

A homochiral synthesis of milbemycin β3 was also reported in 1982 by Williams.[19] Like Smith, Williams based the synthesis of the spiroketal subunit on lactone **40** which was prepared in five steps from (-)-(3S)-citronellol, **56**, in an overall 40% yield. Addition of lactone **40** (Scheme 12) to the sulfinyl carbanion derived from sulfoxide **57** gave ketone **58** as a mixture of diastereomers which was transformed to spiroketal **59** upon treatment with catalytic acid in moist benzene. Olefin **60** could then be obtained by protecting the primary alcohol as the benzoate and heating to facilitate elimination of the sulfoxide. Addition of t-BuOCl across the double bond yielded two diastereomeric chlorohydrins which were carried on to give a 5:1 mixture (α-OH:β-OH) of alcohols **61**. Although the predominant product was the α-alcohol, it was readily inverted *via* PCC oxidation and reduction with NaBH4. Aldehyde **62**, the chosen northern subunit, then resulted upon protection of the secondary alcohol, cleavage of the benzoate, and Swern oxidation.

Scheme 12

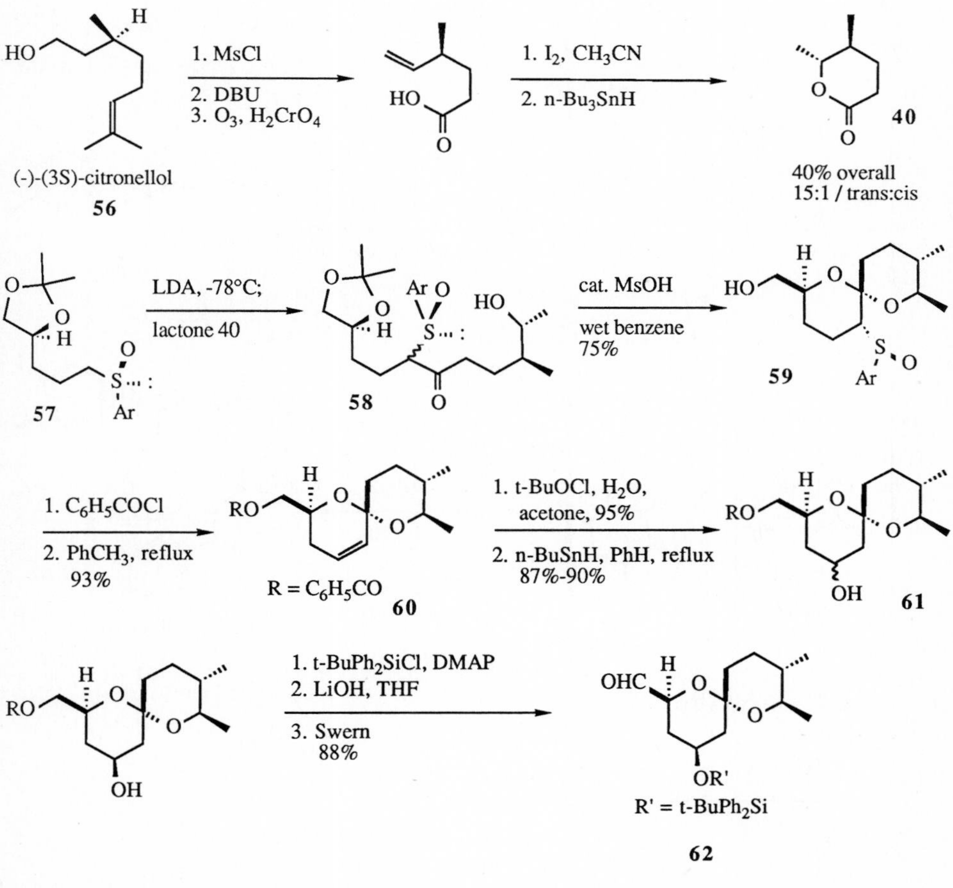

452

The requisite chirality at the remote C12 center was incorporated by beginning with (-)-(3S)-
citronellal, **63**, as shown in Scheme 13. Addition of lithiodibromomethane to **63** followed by
oxidative cleavage of the olefin and elimination *via* Zn/HOAc produced aldehyde **64** as a mixture of
olefin isomers. Formation of the enamine, selenenylation, reduction of the aldehyde, and elimination

Scheme 13

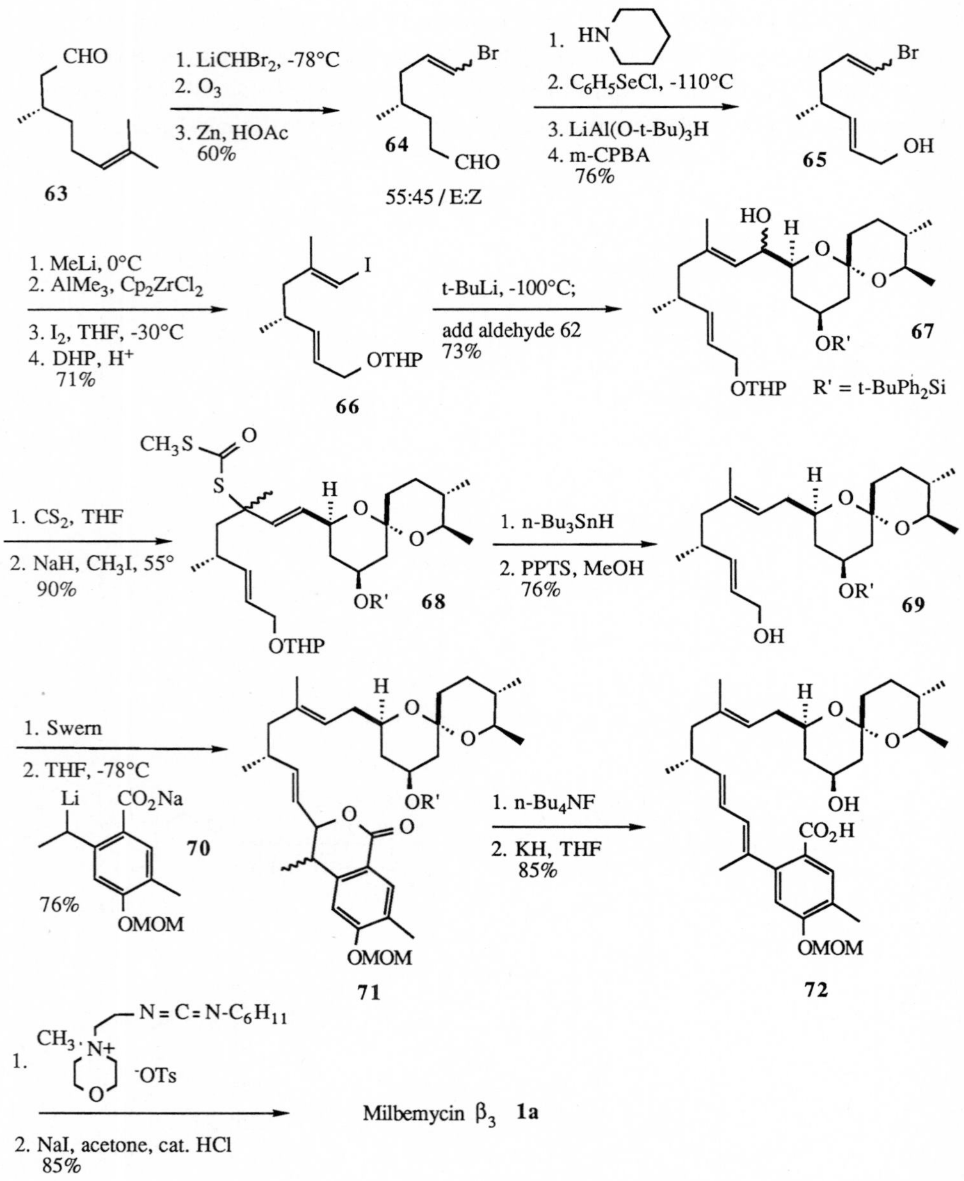

Milbemycin β₃ **1a**

of the selenoxide resulted in allylic alcohol **65** wherein the olefin geometry was set without compromising the integrity of the existing asymmetric center. Further elimination to the terminal alkyne with subsequent use of Negishi's protocol to methylate produced vinyl iodide **66** as a single isomer after protection of the alcohol as the THP ether.

Metalation of **66** to yield the vinyl anion and condensation with aldehyde **62** afforded spiroketal **67** as a mixture of diastereomers, epimeric at the secondary alcohol center. Removal of the undesired hydroxyl functionality was accomplished by conversion to the xanthate using CS_2 and rearrangement to dithiocarbonate **68**. Further transformation to allylic alcohol **69** was achieved *via* reduction with tributyltin hydride to give a single olefin isomer after deprotection. Coupling of the two fragments to give lactone **71** was facilitated by condensing the aldehyde derived from alcohol **69** with the aromatic dianion **70**, available from the neutral compound by sequential deprotonation with NaH and t-BuLi. Sequential exposure of the lactone **71** to n-Bu4NF to remove the silyl protecting group and KH to effect olefin formation yielded hydroxyacid **72**. Finally, macrolactonization followed by removal of the MOM group gave milbemycin β3 (**1a**).

The starting material used by Baker[20] in his homochiral synthesis of milbemycin β3 was laevoglucosan, **73** (see Scheme 14). Selective bistosylation of the hydroxyls at C2 and C4 of the sugar followed by reduction and etherification of the remaining alcohol with allyl bromide produced acetal **74** along with a minor amount of isomer **75**. Acetal exchange in acidic methanol and protection of the resulting primary alcohol as the benzyl ether gave glycoside **76**. Through the use of Rh(PPh3)3Cl, the allyl ether was isomerized to the vinyl ether which was cleaved *via*

Scheme 14

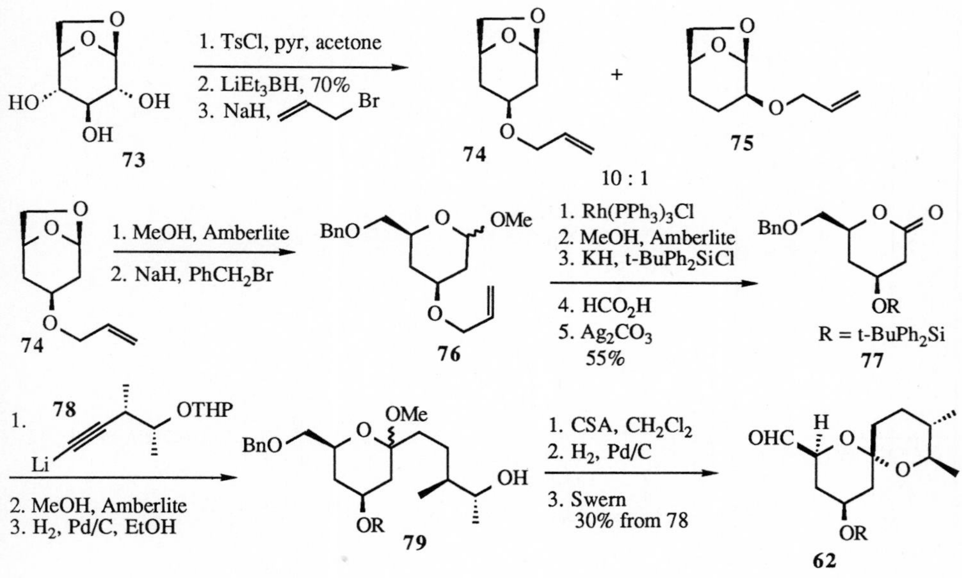

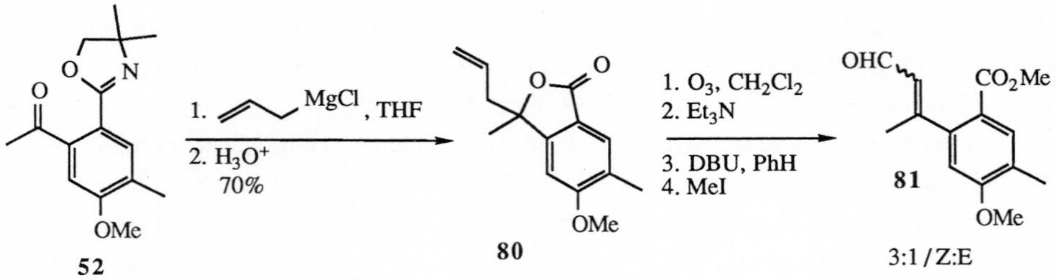

methanolysis and the product alcohol then protected as the t-butyldiphenylsilyl ether. Hydrolysis of the glycoside to the hemiacetal and subsequent oxidation with Ag_2CO_3 afforded the desired lactone **77**. Reaction of this lactone with lithium acetylide **78** provided the hemiketal which was converted to ketal **79** after methanolysis and hydrogenation. Exposure of ketal **79** to CSA in CH_2Cl_2 gave the appropriate spiroketal **62** after removal of the benzyl group and Swern oxidation.

Scheme 15

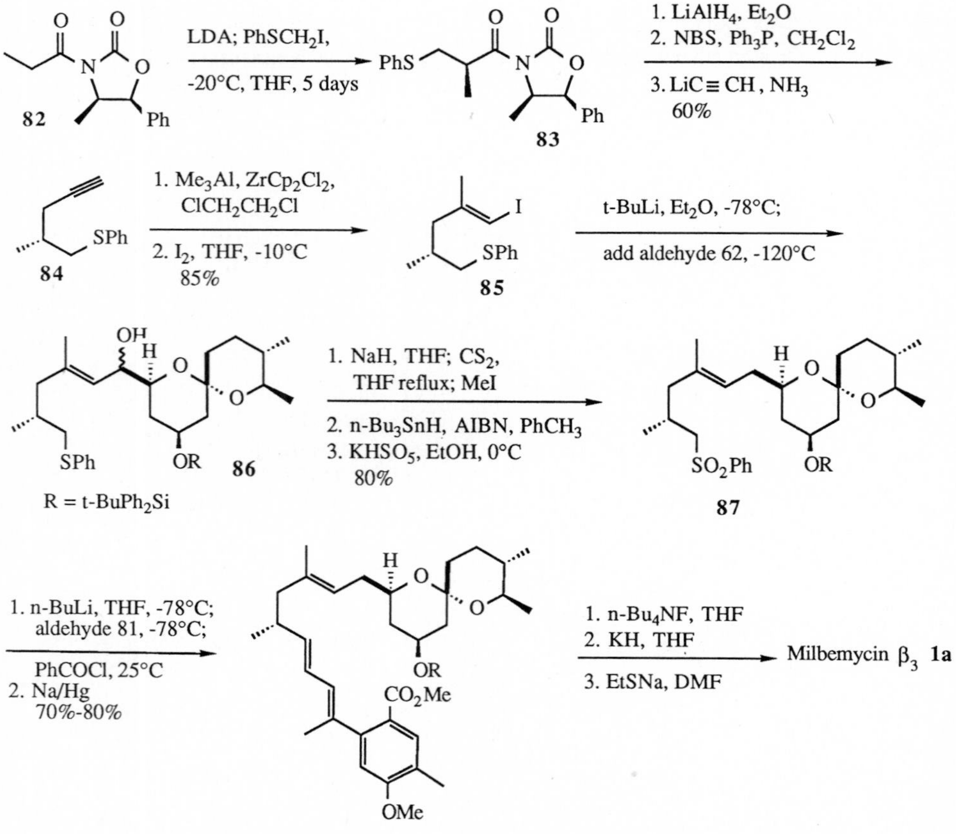

Aromatic ketone **52** (Scheme 14) was prepared as described previously and transformed to lactone **80** *via* treatment with allyl Grignard and hydrolysis of the intermediate iminolactone. Ozonolysis of the olefin to afford the aldehyde and subsequent elimination and esterification of the aromatic acid generated a 3:1 mixture of α,β-unsaturated aldehydes **81** in which the undesired Z-olefin predominated. After separating these isomers, the Z-olefin was isomerized to a 2:1 mixture (Z:E) under basic conditions.

The asymmetric center at C12 was introduced *via* an Evans asymmetric alkylation in which the norephedrine derived N-acyl oxazolidone **82** was utilized to provide phenyl sulfide **83** (Scheme 15). Reductive cleavage of the chiral auxiliary, bromination, and displacement of the resulting bromide with lithium acetylide gave alkyne **84** which could then be converted to vinyl iodide **85** by carboalumination and subsequent exposure of the intermediate alane to iodine. Condensation of aldehyde **62** with the vinyl lithium derivative of **85** provided spiroketal **86** as a mixture of epimers.

Scheme 16

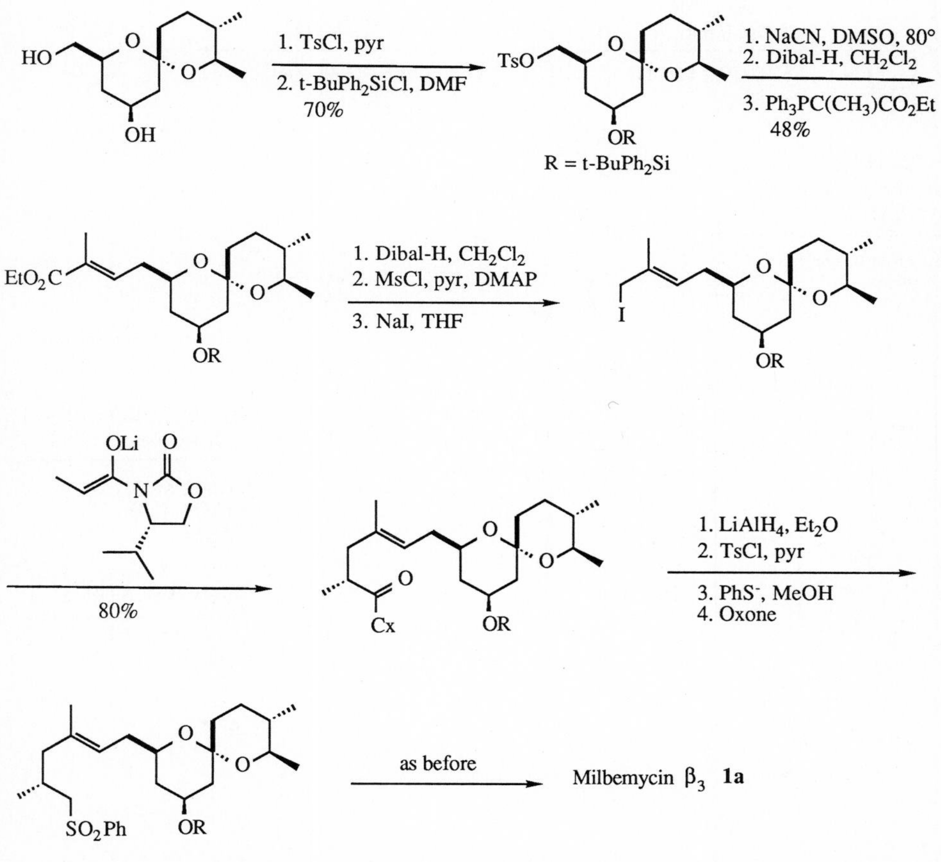

456

Removal of the hydroxyl group, accomplished *via* the xanthate as per Williams, and oxidation of the sulfide using Oxone generated sulfone **87**. Coupling of sulfone **87** to the aromatic aldehyde **81** was brought about by utilizing a modification of the Julia reaction, and milbemycin β3 was then available after several standard transformations.

Another approach by Baker to β3 is shown in Scheme 16.[21] This route used Wittig methodology to incorporate the desired C14-C15 olefin geometry and Evans' asymmetric alkylation protocol to set the desired chirality at C12.

Kocienski has reported two synthetic approaches to milbemycin β3.[22] In his "directed aldol" approach, key starting materials, diol **88** and ortholactone **89**, were prepared as shown in Scheme 17. Beginning with acetonide **90**, epoxide **91**, made *via* the tosylate, was subsequently opened with isopropenyl cuprate to provide alcohol **92**. Deprotection and inversion using the Mitsunobu protocol

Scheme 17

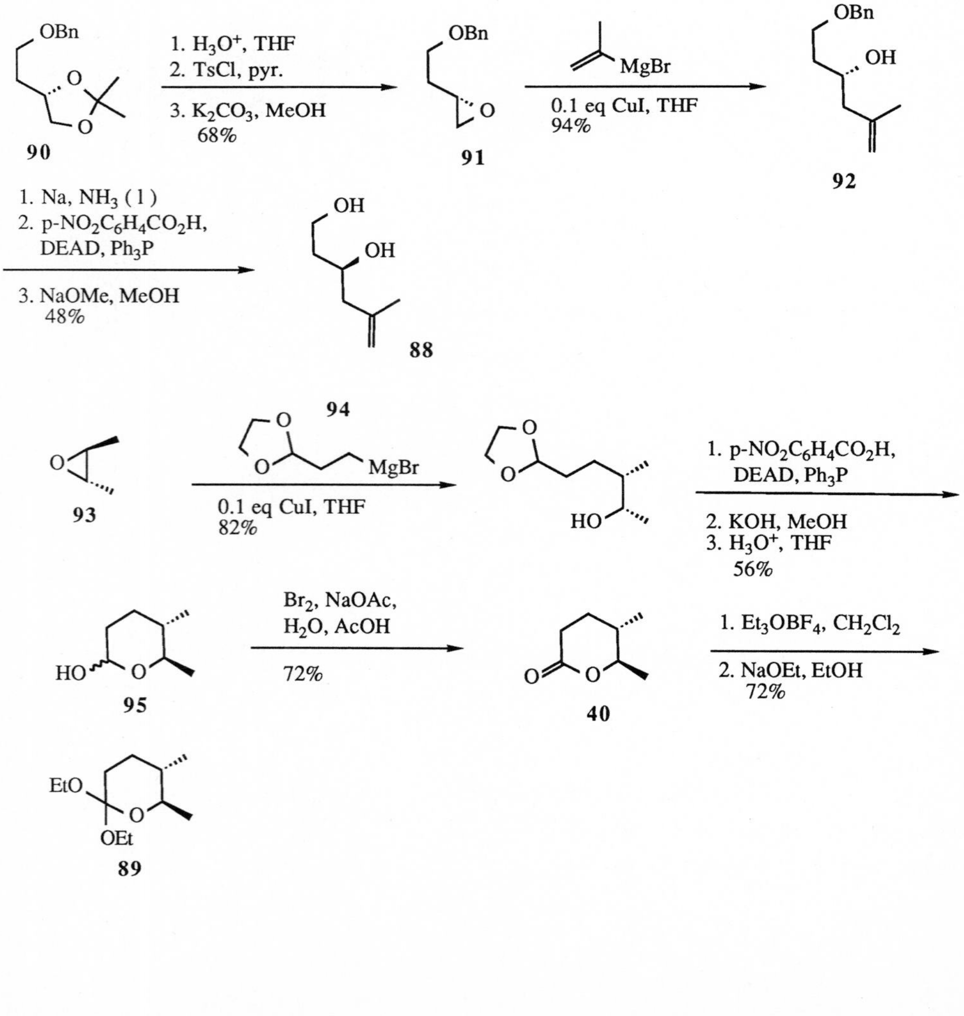

Scheme 18

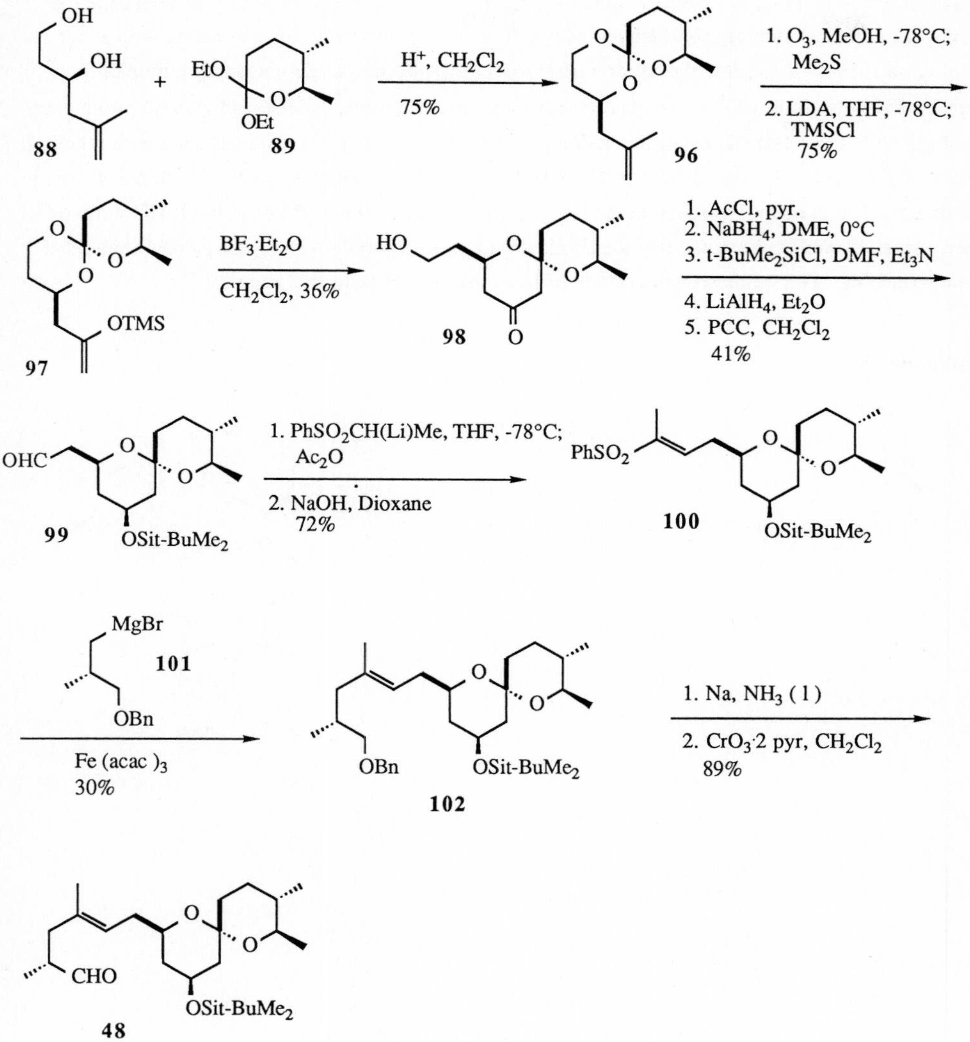

afforded the desired diol **88**. Optically active lactone **40** was generated from epoxide **93** by opening with cuprate **94**, inverting the alcohol center, hydrolyzing the dioxolane, and oxidizing intermediate hemiacetal **95**. Ortholactone **89** was then available after sequential treatment of the lactone with Et₃OBF₄ and NaOEt/EtOH.

Conversion to the spirocyclic ortholactone **96** was accomplished *via* an acid catalyzed transketalization reaction between diol **88** and ortholactone **89** (see Scheme 18) with the product being subsequently transformed to silyl enol ether **97** in a standard fashion. Generation of the

458

desired spiroketal **98** was achieved in moderate yield by subjecting **97** to a Lewis acid in the key intramolecular directed aldol reaction. Routine functional group manipulation then provided aldehyde **99** wherein it is worthy to note that the correct epimer at the secondary alcohol center is present. Condensation of this aldehyde with α-lithioethyl phenyl sulfone, quenching with acetic anhydride to give a mixture of diastereomeric β-acetoxysulfones and subsequently eliminating yielded vinyl sulfone **100** as a single olefin isomer. In the presence of Fe(acac)₃ union of the vinyl sulfone with Grignard **101** using methodology developed by Julia was accomplished with retention of olefin geometry to provide intermediate **102** which was carried on to aldehyde **48** in a standard manner. A Julia coupling (see Scheme 19) was chosen to join the aromatic subunit **103** with the spiroketal **48**, and it was noted that the elimination step using sodium amalgam proved somewhat unpredictable. The synthesis was completed in a precedented fashion.

Scheme 19

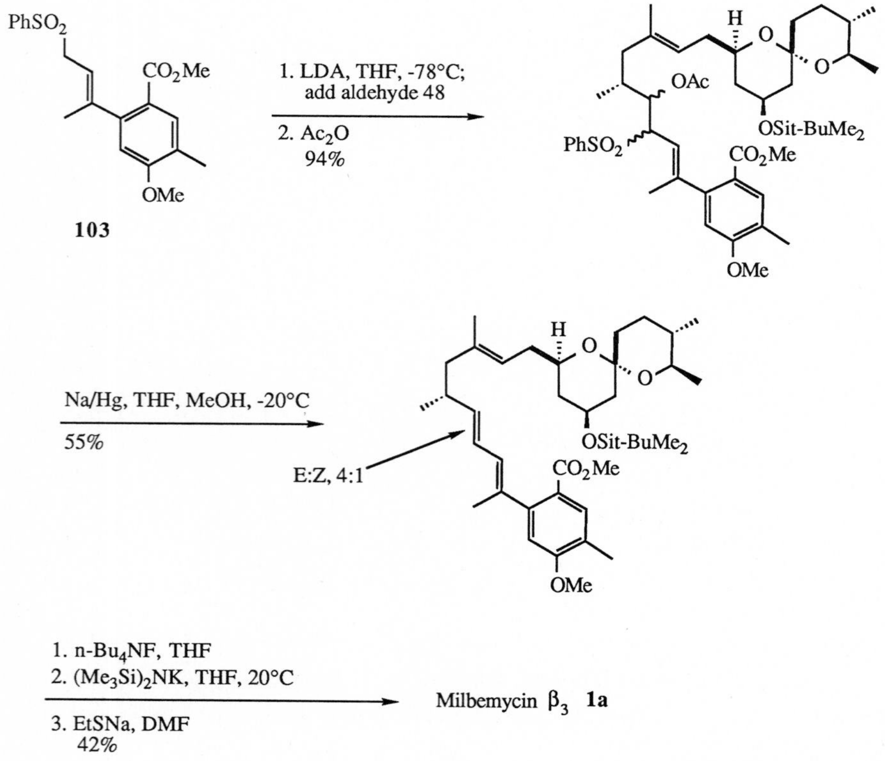

Kocienski's second approach[23] involved a different route to the spiroketal portion of the molecule (see Scheme 20). Epoxide **105** was formed as a mixture of diastereomers by condensing α-lithiomethyl phenyl sulfide with aldehyde **104** and then displacing the corresponding sulfonium functionality in an intramolecular fashion. Acetylene **107**, available from bromide **106**, was transformed *via* carboalumination methodology into the vinyl alane which was subsequently reacted with butyllithium to produce vinyl alanate **108**. This reactive intermediate was employed to open epoxide **105** and yield **109** as a mixture of epimeric alcohols which were readily separable by flash chromatography. The predominant isomer was the undesired one, but inversion of this center using Mitsunobu conditions proceeded smoothly. Transformation of the alcohol to epoxide **110** was readily achieved.

The 3,4-dihydro-2H-pyran **112**, prepared from hemiacetal **111**, was metalated and subsequently reacted with n-pentynylcopper to generate a mixed cuprate reagent. Exposure of epoxide **110** to this organocuprate afforded adduct **113** which was cyclized to spiroketal **114** under acidic conditions. Standard conversions provided aldehyde **48** which was an intermediate generated in his previous synthesis.

Scheme 20

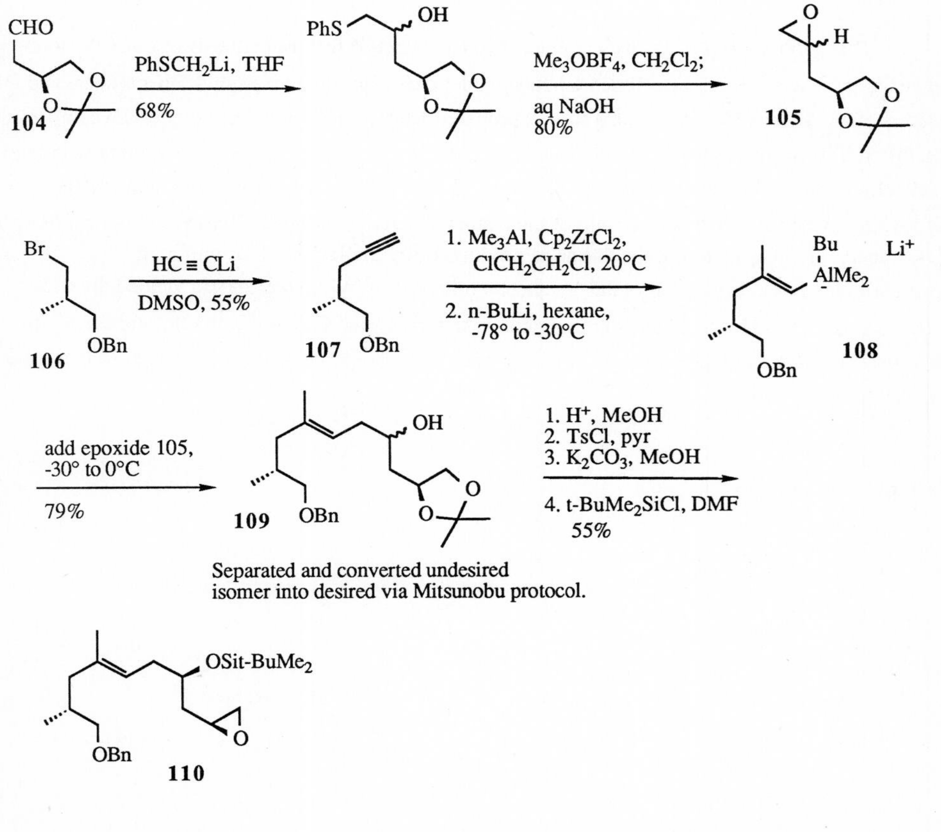

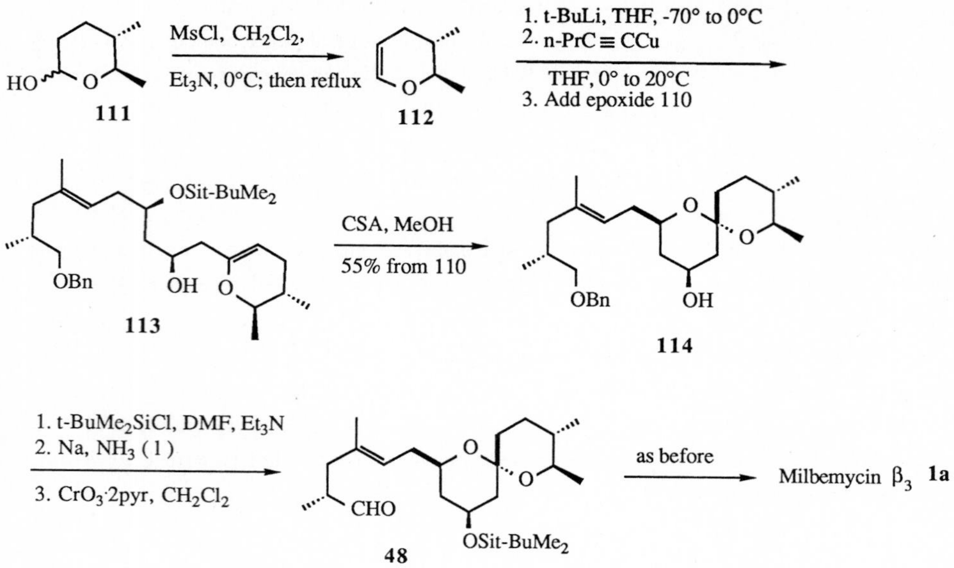

The synthesis of the aromatic subunit **116** by Barrett[24] for his synthesis of milbemycin β_3 is depicted in Scheme 21 and was based on Diels-Alder chemistry involving ethyl 2-pentynoate and the Danishefsky diene **115**. Furanone **120** was prepared from arylsulfonyl hydrazone **117** using a modified Shapiro reaction. The hydrazone was converted to trianion **118** upon treatment with base and subsequently to dianion **119** after warming. Smooth reaction of this dianion with (S)-(-) - propylene oxide followed by exposure of the adduct to acidic condtions afforded furanone **120** in good yield. Hydrogenation of the furanone produced dimethyllactone **121** with high diastereoselectivity. Reduction to the hemiacetal followed by Wittig olefination yielded the α,β- unsaturated ester **122** which was routinely converted to sulfone **123** in which C12 and the olefin possessed the desired geometries. Lactone **40** was synthesized in a manner analogous to that used by Williams.

Scheme 21

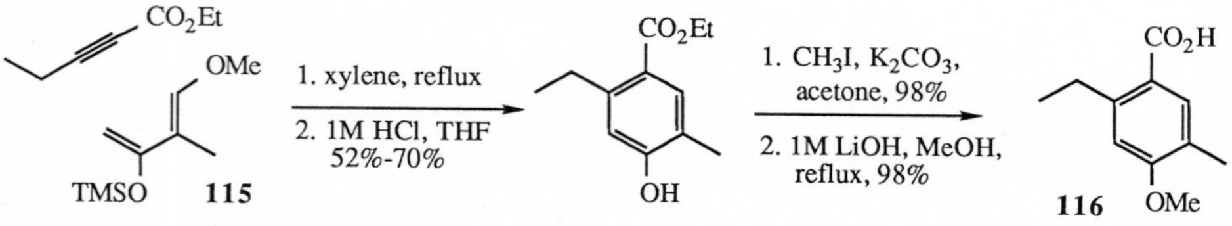

Scheme 21 (cont.)

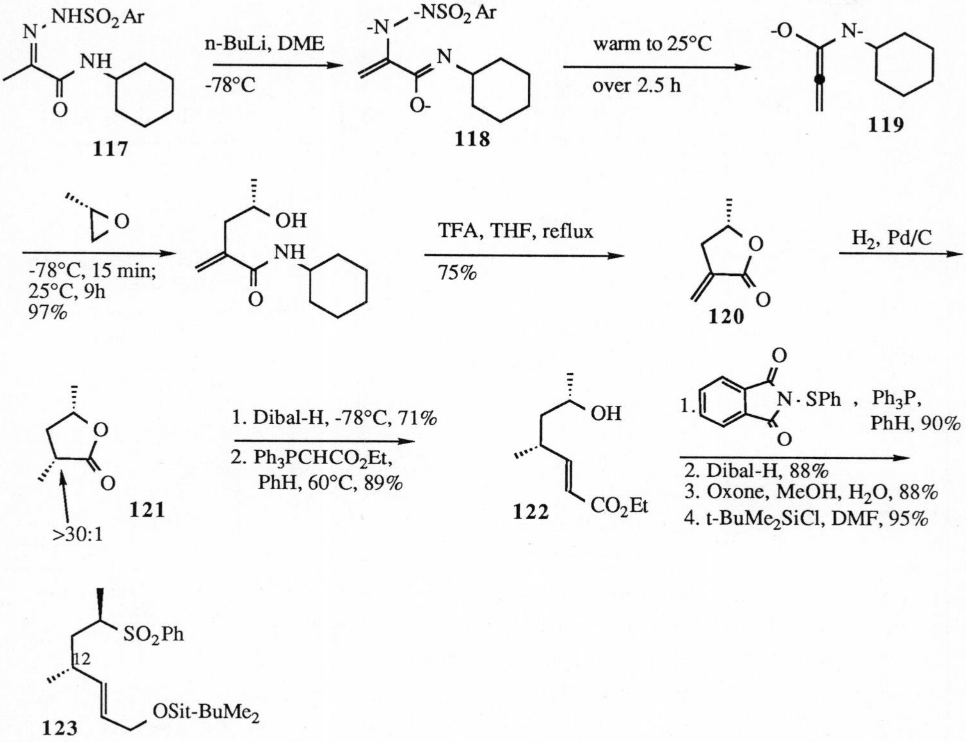

The dianion of β-diketone **124** when condensed with lactone **40** at low temperature yielded spiroenone **125** (84% yield) after exposure to acid (Scheme 22). Hydrogenation resulted in stereospecific saturation of the olefin and highly stereoselective reduction of the ketone in which the α-alcohol predominated over the β by 8:1. After protecting the alcohol, the C17 center was inverted through enolization with LDA followed by reprotonation at low temperature to provide spiroketal **126**. Homologation to aldehyde **127** proceeded in a straightforward fashion, and Julia coupling of this aldehyde to the α-lithiosulfone derived from **123** generated adduct **128** after reductive elimination of the intermediate diastereomeric β-acetoxysulfones. Only a modest degree of selectivity, 5:3/E:Z, was observed in the olefination.

Condensation of the aldehyde derived from **128** with aromatic dianion **70** afforded lactone **129** once the silyl group was removed. Elimination prompted by KH produced an intermediate hydroxyacid possessing the desired E,E-diene functionality. Macrolactonization with inversion was achieved using PPh$_3$ and DEAD and subsequent demethylation produced milbemycin β$_3$ in moderate yield.

462

Scheme 22

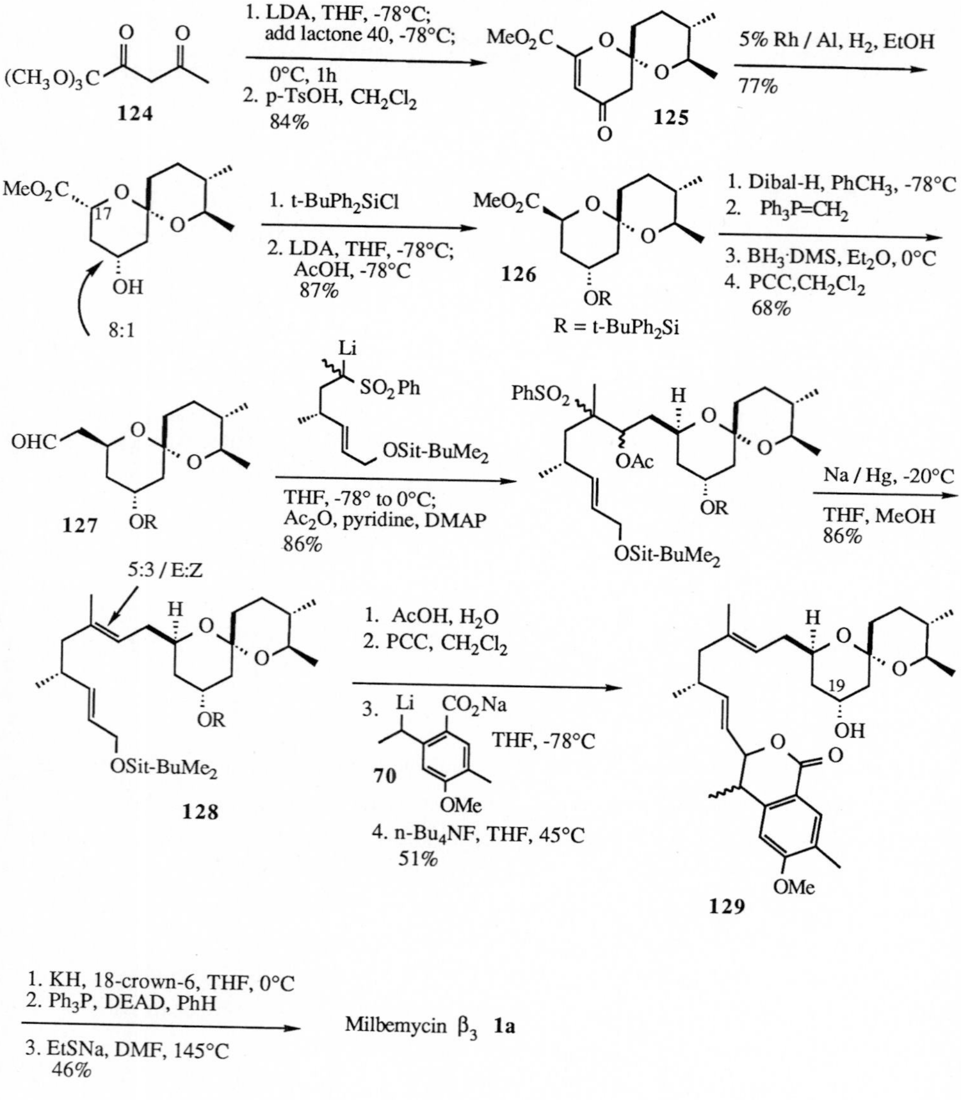

(CH₃O)₃C — structure **124**

1. LDA, THF, -78°C;
   add lactone 40, -78°C;
   0°C, 1h
2. p-TsOH, CH₂Cl₂
   84%

MeO₂C — structure **125**

5% Rh / Al, H₂, EtOH
77%

MeO₂C — structure (position 17, OH) 8:1

1. t-BuPh₂SiCl
2. LDA, THF, -78°C;
   AcOH, -78°C
   87%

MeO₂C — structure **126** OR
R = t-BuPh₂Si

1. Dibal-H, PhCH₃, -78°C
2. Ph₃P=CH₂
3. BH₃·DMS, Et₂O, 0°C
4. PCC, CH₂Cl₂
   68%

OHC — structure **127** OR

Li
SO₂Ph
OSit-BuMe₂

THF, -78° to 0°C;
Ac₂O, pyridine, DMAP
86%

PhSO₂ — structure (H, OAc, OR, OSit-BuMe₂)

Na / Hg, -20°C
THF, MeOH
86%

5:3 / E:Z

structure **128** (OR, OSit-BuMe₂)

1. AcOH, H₂O
2. PCC, CH₂Cl₂
3. Li CO₂Na structure **70** OMe
   THF, -78°C
4. n-Bu₄NF, THF, 45°C
   51%

structure **129** (position 19, OH, OMe)

1. KH, 18-crown-6, THF, 0°C
2. Ph₃P, DEAD, PhH
3. EtSNa, DMF, 145°C
   46%

Milbemycin β₃ **1a**

6 AVERMECTIN B₁ₐ

The only synthesis of an avermectin (B₁ₐ, **2f**) was reported in 1986 by Hanessian (Scheme 23).[25,26] In his approach, the northern C11-C28 fragment prepared from L-isoleucine, S-malic acid and (2S,3S)-2-hydroxy-3-methylsuccinic acid is coupled to the southern C1-C10 fragment, obtained from the natural material, utilizing a Julia sulfone strategy to connect the two subunits.

Scheme 23

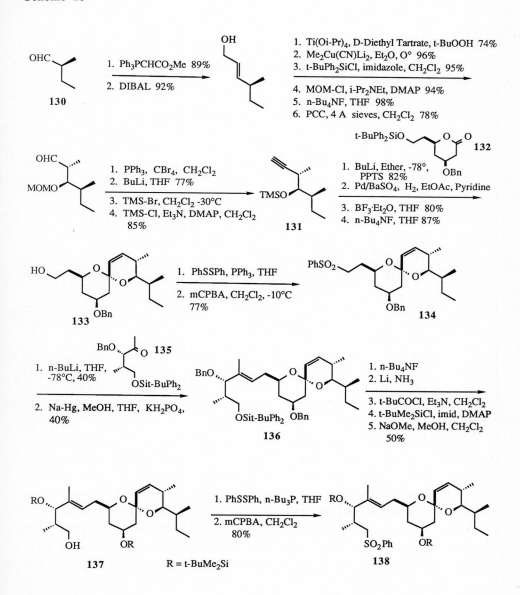

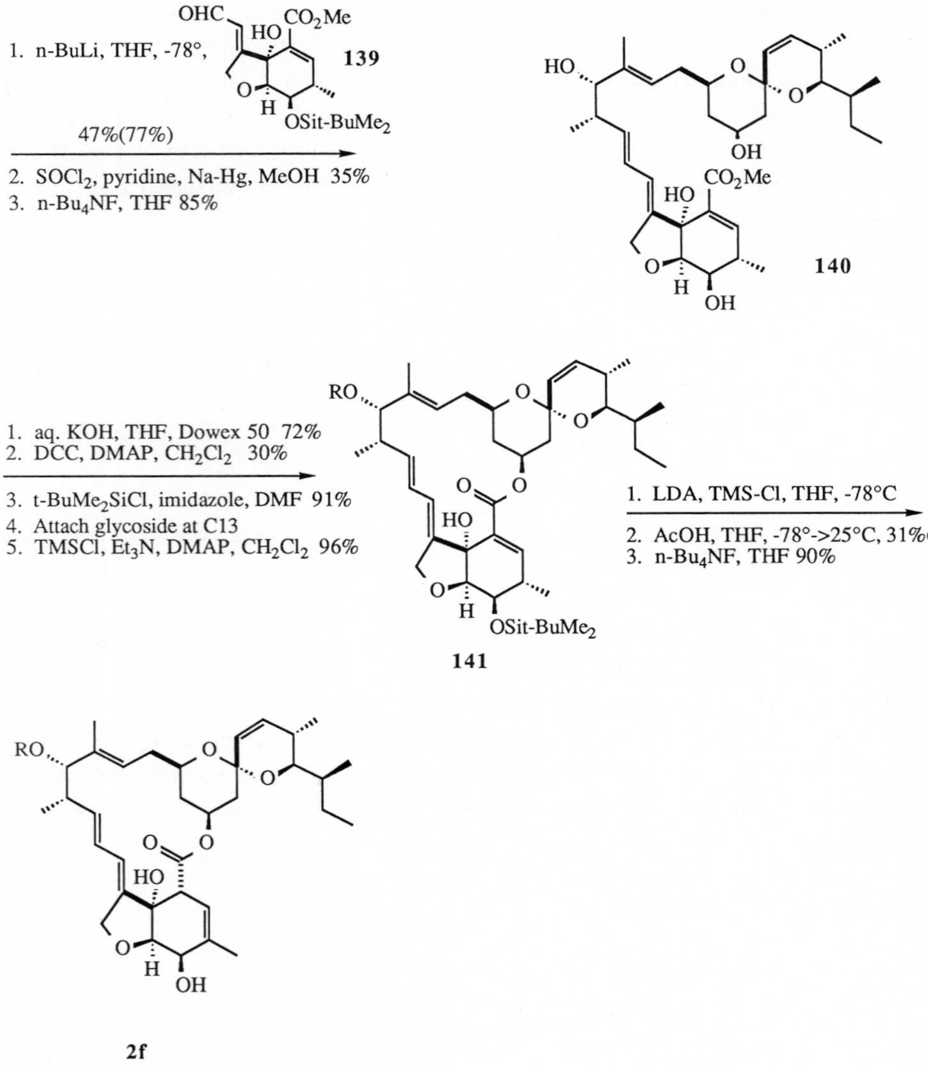

464

1. n-BuLi, THF, -78°, **139**

47%(77%)

2. SOCl$_2$, pyridine, Na-Hg, MeOH 35%
3. n-Bu$_4$NF, THF 85%

140

1. aq. KOH, THF, Dowex 50 72%
2. DCC, DMAP, CH$_2$Cl$_2$ 30%

3. t-BuMe$_2$SiCl, imidazole, DMF 91%
4. Attach glycoside at C13
5. TMSCl, Et$_3$N, DMAP, CH$_2$Cl$_2$ 96%

141

1. LDA, TMS-Cl, THF, -78°C

2. AcOH, THF, -78°->25°C, 31%
3. n-Bu$_4$NF, THF 90%

2f

Aldehyde **130**, derived from isoleucine, was converted to the alkyne **131** by a series of standard transformations. Addition of the acetylide of **131** to lactone **132** (similar to Baker's **77**, Scheme 14), available from either S-malic acid or D glucose, was followed by semihydrogenation to introduce the C22-C23 double bond with the necessary cis geometry. Exposure of the adduct to BF$_3$·Et$_2$O and removal of the silyl group produced the key spiroketal **133**. At this point a Julia coupling was utilized to extend the side chain. Consequently, the alcohol **133** was converted to the sulfide and oxidized to the sulfone **134**. Deprotonation of **134** using n-BuLi and exposure of the anion to ketone **135**, prepared from (2S, 3S)-2-hydroxy-3-methylsuccinic acid in 7 steps, gave the chain extended product **136** (with the required trans olefin geometry at C14-C15) after reductive elimination of the β–hydroxy sulfone. Spiroketal **136** was converted to the key intermediate **138** necessary for coupling to the southern fragment after manipulation of protecting groups to form **137** and formation of the sulfone at C11 to produce **138**. Deprotonation of **138** followed by addition of **139**, obtained from natural material, reductive cleavage of the resultant β-hydroxysulfone and removal of the protecting groups gave **140**, which contains the correct olefin geometry at C10-C11. Macrolactonization was effected by hydrolyzing the ester at C1 and treating with DCC/DMAP. The synthesis was completed by attaching the glycoside at C13 using AgOTf followed by deconjugation of the ester by treating **141** sequentially with LDA and TMSCl then AcOH. Removal of the silyl ether at C5 gave avermectin B$_{1a}$. It should be noted that Fraser-Reid has questioned the viability of approaches involving this deconjugation because of the propensity for the formation of the 2-epi-isomer upon deconjugation of **141** and similar compounds.[27]

7 SPIROKETALS

In his approach to the spiroketal subunit of avermectin A$_{2b}$, Barrett[28] (Scheme 24) begins with homoallylic alcohol **142**, available from isobutyraldehyde and crotonyl bromide. Protection of the alcohol and ozonolysis of the terminal olefin gave aldehyde **143**. Addition of the t-butyldimethylsilyl ketene acetal of t-butyl acetate followed by protecting group manipulation provided lactone **144** as the only diastereomer. Lactone **144** was converted to the spiroketal **145** by exposure to the dianion **146** followed by acid catalyzed cyclization as in his approach to milbemycin β$_3$. After removal of the silyl group, reduction proceeded rapidly to give 19.5:1 C19 diastereoselectivity and exclusively the C17 α isomer of spiroketal **147**. At this point, spiroketal **147** was resolved by formation of the bis-(S)-O-methylmandelate esters. After protecting the

Scheme 24

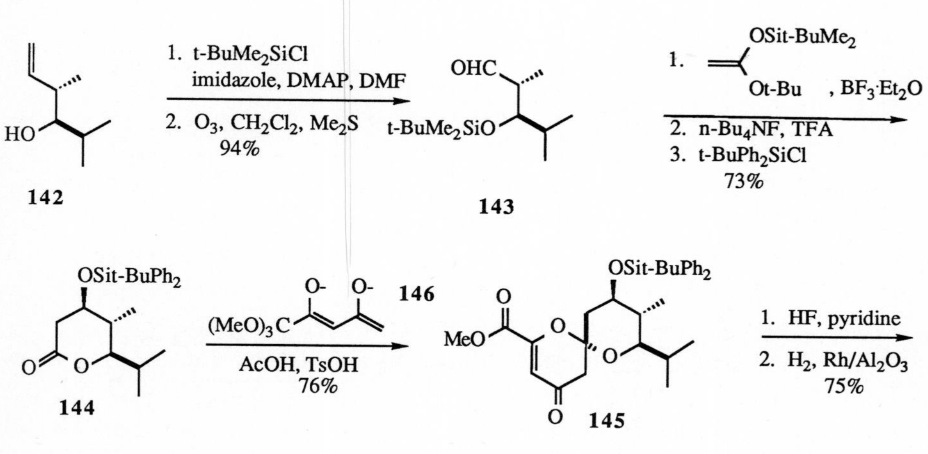

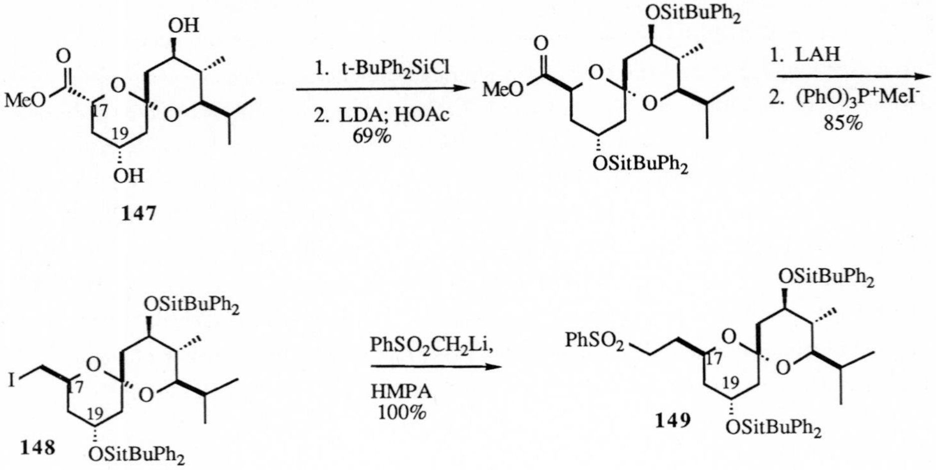

alcohols the C17 center was then epimerized to the desired stereochemistry by treatment with LDA then HOAc, and the C16 ester was reduced with LAH. The alcohol obtained from the reduction was converted to iodide **148** whose structure was confirmed by an X-ray crystallographic study. Nucleophilic displacement of iodide by $PhSO_2CH_2Li$ gave sulfone **149** suitably functionalized for further manipulation of the side chain.

Hirama[29] (Scheme 25) relies on the coupling of two optically pure fragments **154** and **155** followed by cyclization of the keto-diol in his approach to the spiroketal subunit of avermectin B_{1a}. To begin, the ester **151** was prepared from **150** by reaction with methyl acetate enolate. The ketone of **151** was then partially reduced to β-hydroxyester **152** with baker's yeast to give a 99:1 preference for the desired isomer at the new chiral center. Alkylation of **152** gave **153**

Scheme 25

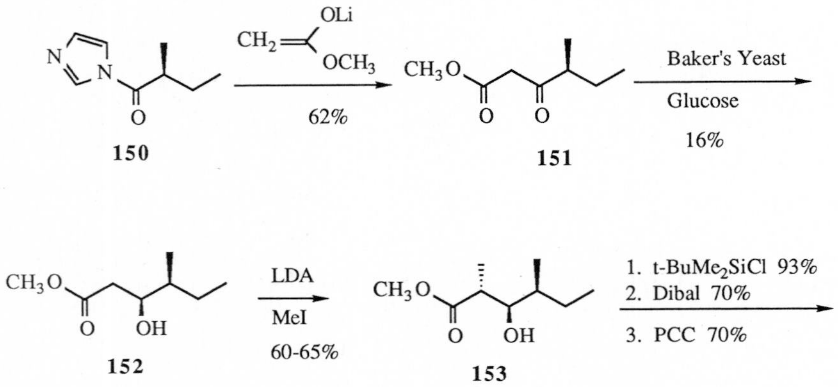

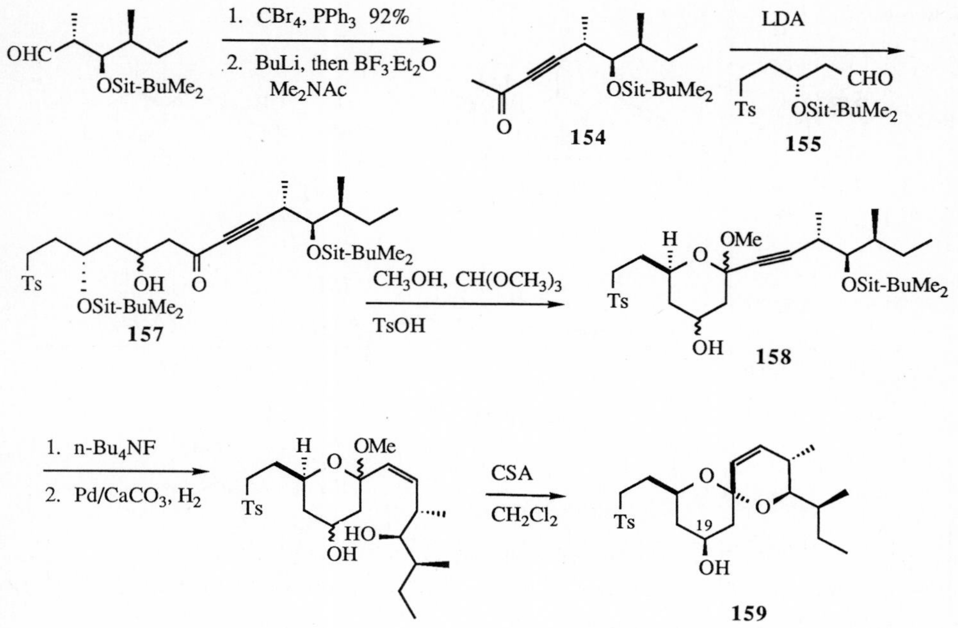

contaminated with 10% of the undesired diastereomer. Protection, reduction, oxidation, formation of the alkyne and acylation provided **154**, one of the two fragments necessary for construction of the spiroketal. The second fragment **155** was prepared in 4 steps (Scheme 26) from the triol **156**. The chirality of triol **156** was also introduced by a baker's yeast reduction. An aldol reaction served to couple the two segments providing a 1:1 mixture of alcohols at C19. Since both silyl groups could not be removed simultaneously, spiroketalization was accomplished by removing the protecting groups stepwise and closing each ring individually. Exposure of **157** to TsOH removed the silyl group at C17 and resulted in ketal **158**. Exposure of **158** to Bu$_4$NF removed the C25 silyl group, and semihydrogenation followed by treatment with CSA gave the desired spiroketal **159** as a 1.7:1 mixture at C19. The diastereomers could be separated at this point and an oxidation-reduction sequence allowed for recycling of the undesired isomer.

Scheme 26

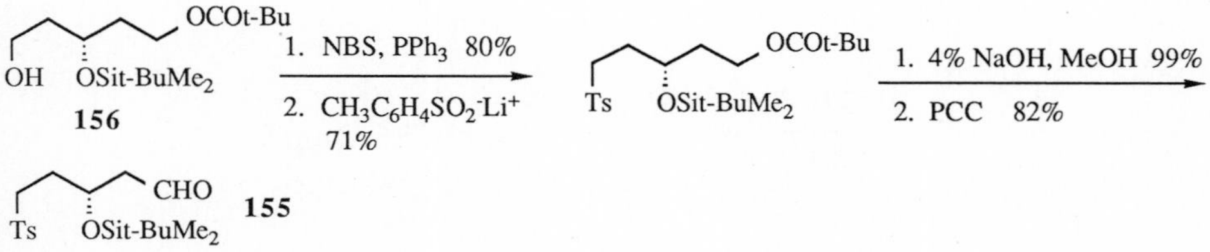

Scheme 27

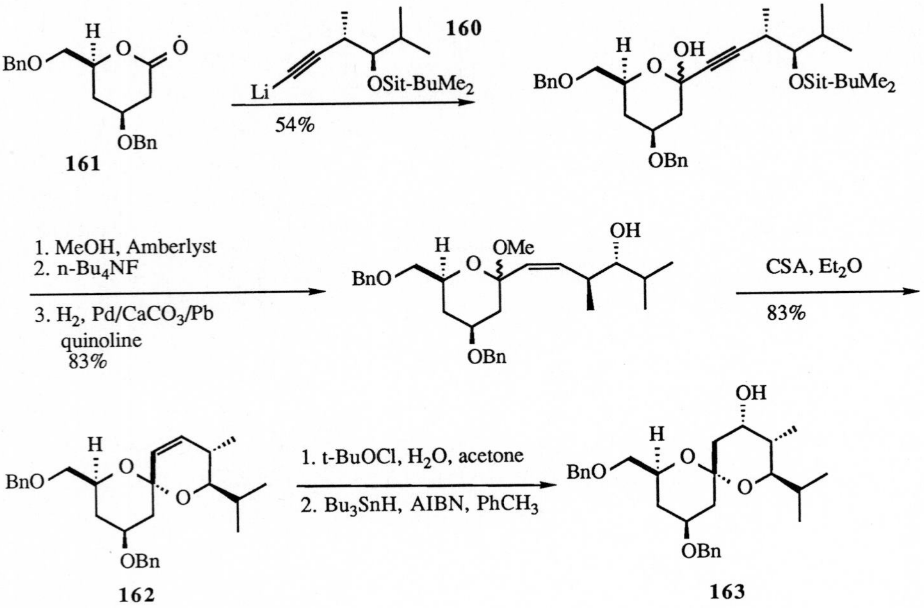

161

Baker has also reported an enantiospecific synthesis of the spiroketal moiety of avermectins B$_{1b}$ and B$_{2b}$ (Scheme 27).[30] In this approach the lithium acetylide **160**, readily available in optically pure form (see Scheme 28), is added to lactone **161** derived from laevoglucosan (see Scheme 14). Subsequent methanolysis, desilylation, semihydrogenation and exposure to acid resulted in

Scheme 28

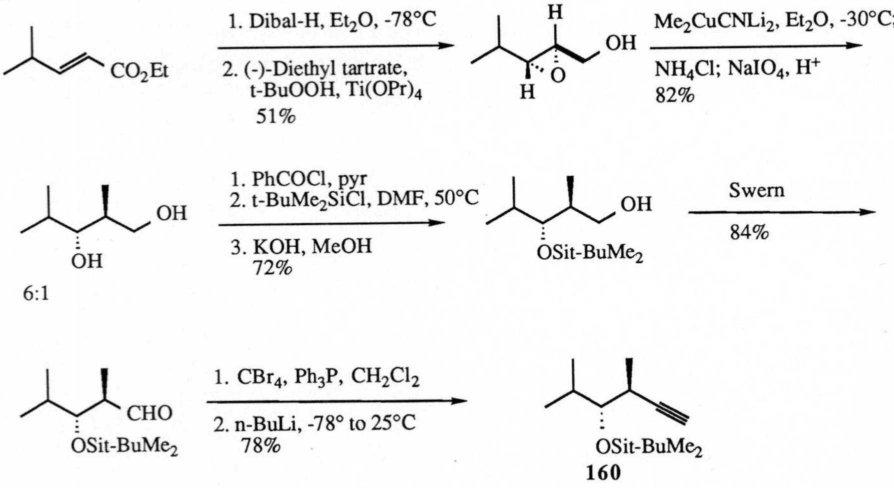

160

spiroketal **162**, required for avermectin B$_{1b}$. Unsaturated spiroketal **162** was converted to **163**, a subunit of avermectin B$_{2b}$, by treatment with t-BuOCl to give a 2:1 (desired:undesired) mixture of chlorohydrins which could be separated and the desired isomer reduced *via* Bu$_3$SnH. Regio- and stereocontrol were confirmed by NMR.

In his synthesis of the C11-C31 moiety of milbemycin D, Crimmins' (Scheme 29)[31] approach to the spiroketal subunit centers on lactone **165**, available in 10 steps from allyl alcohol **164**. Addition of 4-methoxy-3-butene-1-yne anion followed by exposure of the adduct to K$_2$CO$_3$ in methanol gave trimethoxy enone **166**. Treatment of **166** with aqueous acid produced a 4:1 mixture of the desired spiroketal enone **167** and its methanol adduct **168**. The undesired **168** could be converted to **167** upon exposure to Amberlyst 15. Elaboration of the side chain began with the conjugate addition of vinyl cuprate to produce **169** as a 25:1 (desired:undesired) mixture at C17. Reduction of the C19 carbonyl yielded a 2:1 mixture of alcohols of which the desired isomer was the major product **170**. An oxidation-reduction sequence allows for recycling of the undesired isomer **171**. After protection of the C19 hydroxyl the side chain was extended in 5 steps to yield bromide **172**. An Evans asymmetric alkylation served to set the stereochemistry of the C12 methyl, and reductive removal of the chiral auxiliary completed the synthesis of the C11-C31 fragment **173**.

Scheme 29

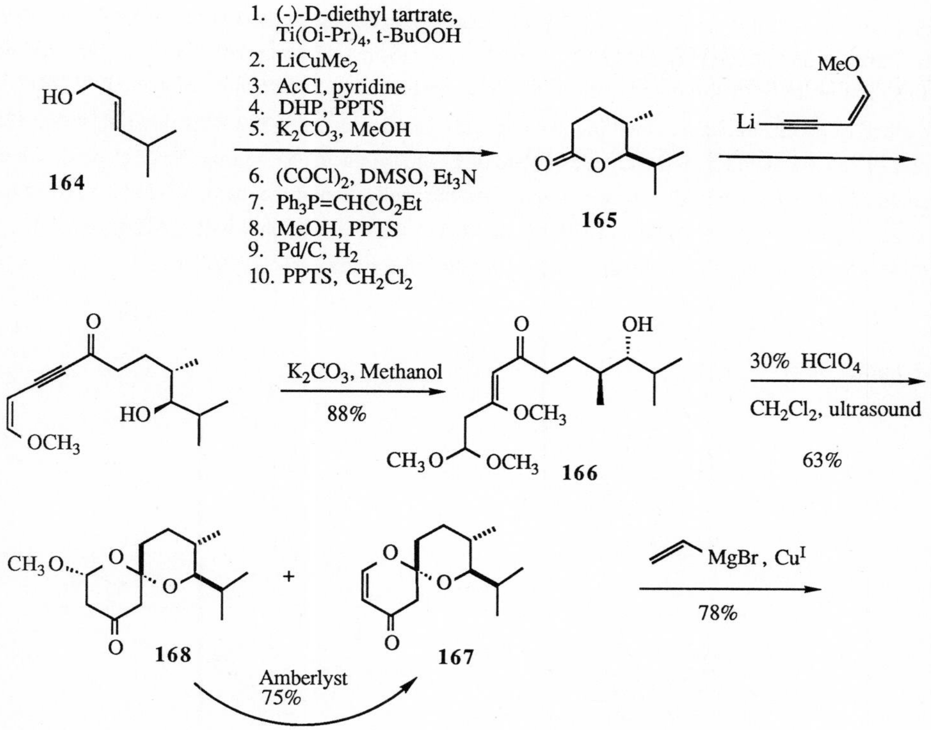

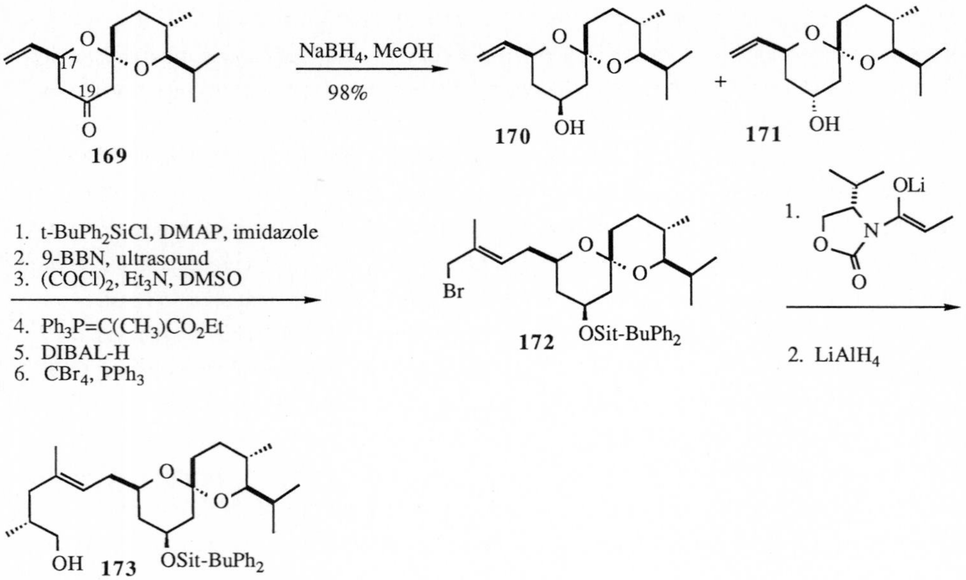

NaBH$_4$, MeOH

98%

169

170 OH

171 OH

+

1. t-BuPh$_2$SiCl, DMAP, imidazole
2. 9-BBN, ultrasound
3. (COCl)$_2$, Et$_3$N, DMSO

4. Ph$_3$P=C(CH$_3$)CO$_2$Et
5. DIBAL-H
6. CBr$_4$, PPh$_3$

172 OSit-BuPh$_2$

1.

2. LiAlH$_4$

OH **173** OSit-BuPh$_2$

Julia[32] (Scheme 30) was able to execute a synthesis of the spiroketal portion of C22-C23 dihydroavermectin B$_{1b}$, relying on a stepwise elaboration of both ends of pentane-2,4-dione. The first alkylation involved bromide **174**, available in 5 steps from isobutyraldehyde, and the dianion **175**. Exposure of the product β–diketone **176** to acid resulted in formation of ketal **177**. The method of choice to chain extend involved the aldol condensation of **177** with either aldehyde **178** or **179**. Each reacted with the kinetic enolate of **177** to produce a 55:45 mixture of alcohols which could be separated via flash chromatography. Acidic cyclization of the major alcohol in each case led to the desired spiroketals, **180** and **181**. Reduction of the C19 carbonyl with LAH gave a 4:1 (desired:undesired) ratio of alcohols **182** to complete the study.

Scheme 30

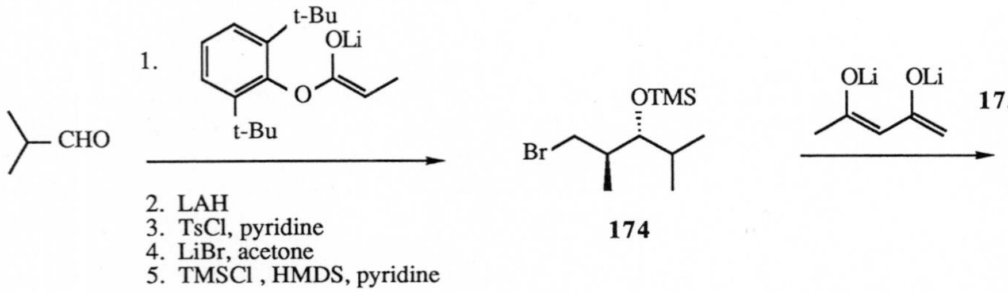

1.

t-Bu
OLi

CHO

t-Bu

2. LAH
3. TsCl, pyridine
4. LiBr, acetone
5. TMSCl, HMDS, pyridine

OTMS

Br

174

OLi OLi **175**

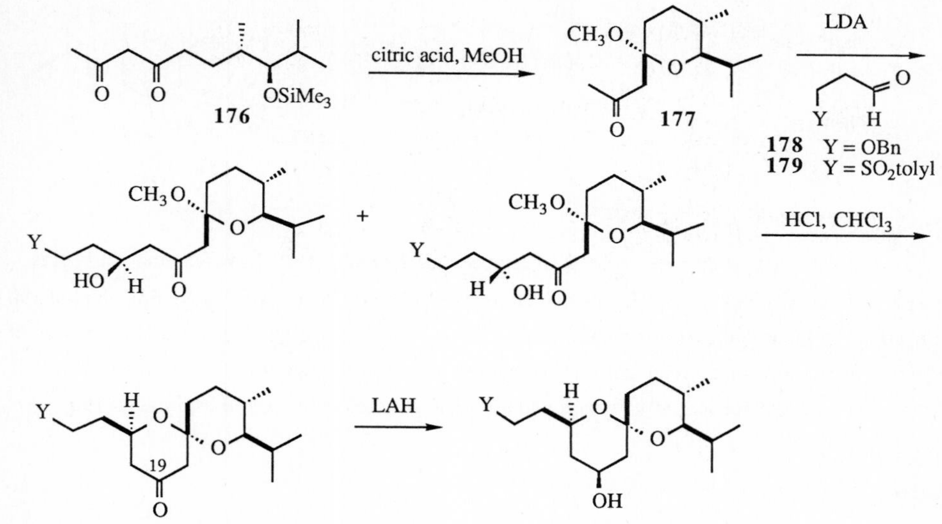

180 Y = OBn
181 Y = SO_2tolyl

182 Y = SO_2tolyl

In his most recently published work in this area, Ley[33] (Scheme 31) prepared the C11 to C25 fragment of the milbemycins by the addition of sulfone anion **183** to either epoxide **184** or **185** followed by acidic work-up. The spiroketal **186** produced in each case contains the correct chirality at the C19 hydroxyl. Epoxides **184** and **185** were prepared from reaction of the bis-epoxide **187** with cuprate **188**. Cuprate **188** is available in 5 steps from the Roche acid [(S)-β–hydroxyisobutyric acid].

Scheme 31

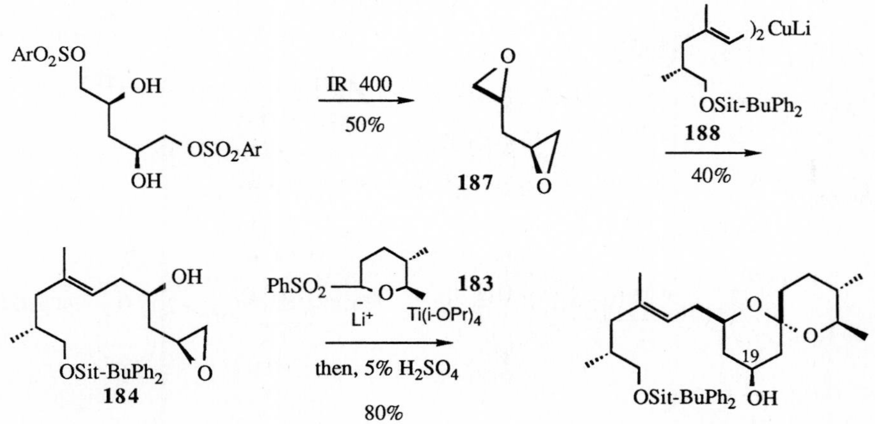

186

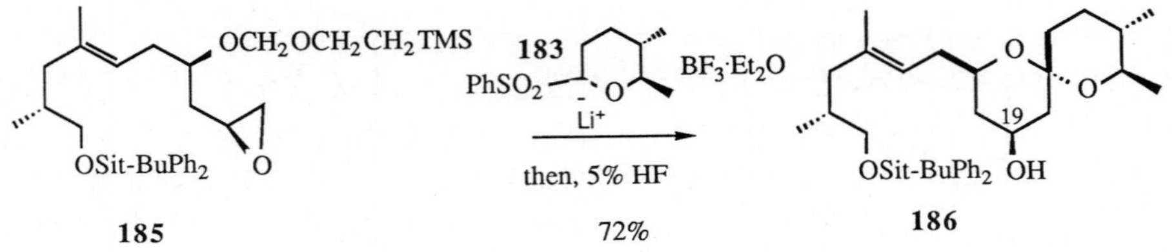

In another approach to the spiroketal segment of milbemycin α_1, Ley (Scheme 32) begins by opening hemiacetal **189** with ethane dithiol to give dithiane **190**.[34] After acylating alcohol **190** the resulting dithiane was treated with thallium trifluoroacetate providing aldehyde **191**. Exposure of the aldehyde to the ylide derived from **192** (prepared in two steps from the known alcohol **194**, Scheme 33) followed sequentially by base and then acid resulted in the desired spiroketal **193**.

Scheme 32

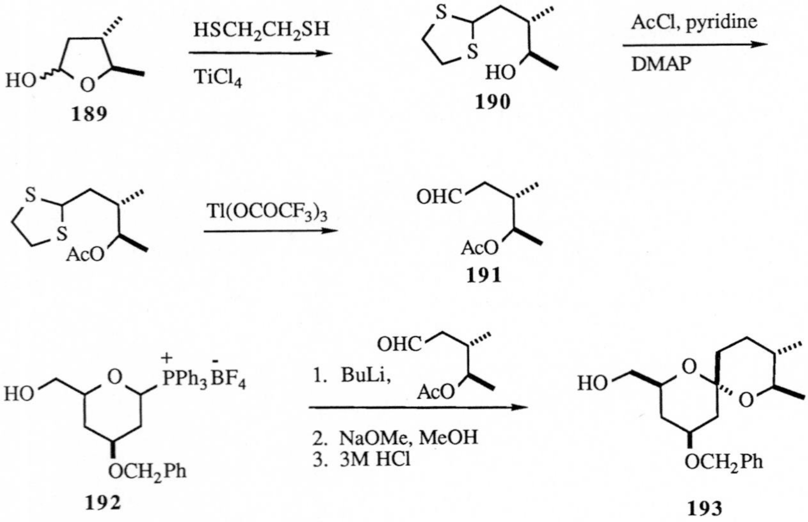

Scheme 33

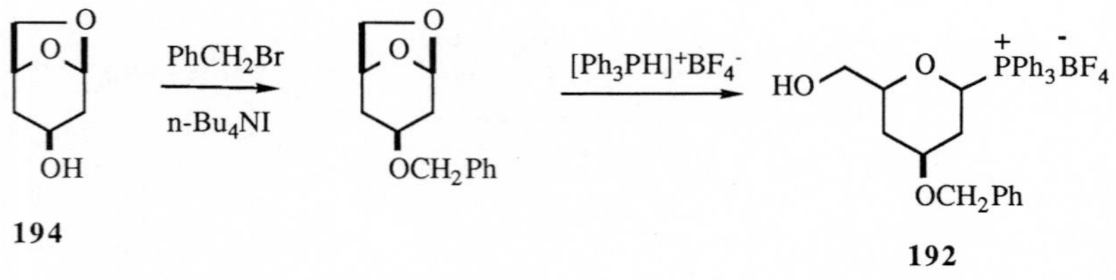

A third approach explored by Ley (Scheme 34) involves the same Wittig methodology, but in this study the ylide contains a larger segment of the bridging side chain.[35] Beginning with (S)-(+)-methyl-3-hydroxy-2-methylpropionate the iodide **195** was prepared in five steps. Formation of the cuprate followed by the addition of epoxide **196** (readily available in three steps from **194**) resulted in alcohol **197**. At this juncture the phosphonium salt was prepared by hydrolysis of the dithiane, treatment of the aldehyde with dimethoxypropane and exposure of the resultant ketal to triphenylphosphoniumtetrafluoroborate. The Wittig reaction and cyclization proceeded under the previously described conditions to give spiroketal **198**.

Scheme 34

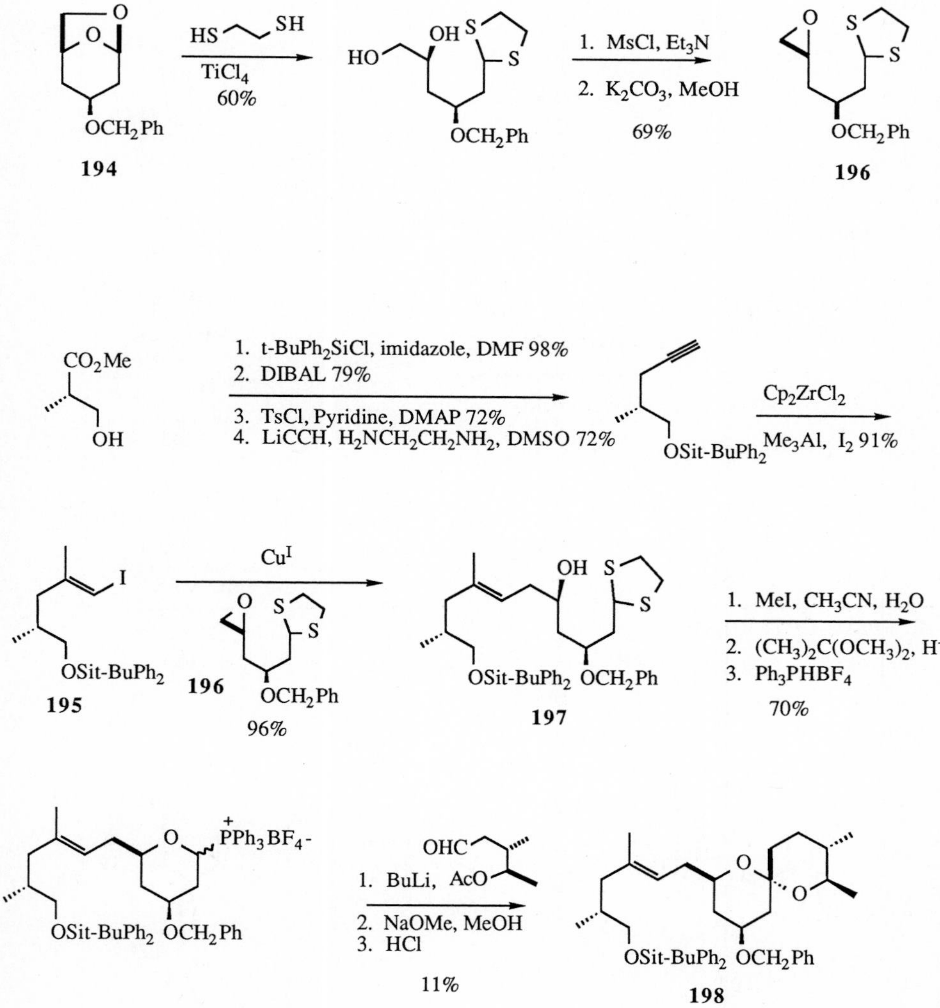

474

In a model study directed toward the synthesis of the spiroketal portion of the avermectins, Danishefsky (Scheme 35) employs a hetero Diels-Alder reaction and an oxidative cyclization of a suitably functionalized dihydropyran.[36] Beginning with triacetate 199, reaction with allyl trimethylsilane and osmylation provided aldehyde 200, the necessary heterodienophile. Cycloaddition of 200 with diene 201 (Danishefsky's diene) mediated by MgBr$_2$ gave dihydropyrone 202 as a 5:1 mixture at C19. The stereochemistry shown is that of the major isomer, which was confirmed by X-ray crystallography. Subsequent Luche reduction and protection of the alcohol gave 203 with the chirality at C17 in place. Cleavage of the enol ether and further reduction led to the dihydropyran 204, the envisioned spiroketal precursor. Deprotection of the C17 hydroxyl followed by oxidative cyclization upon exposure to HgO and I$_2$ gave spiroketal 205.

Scheme 35

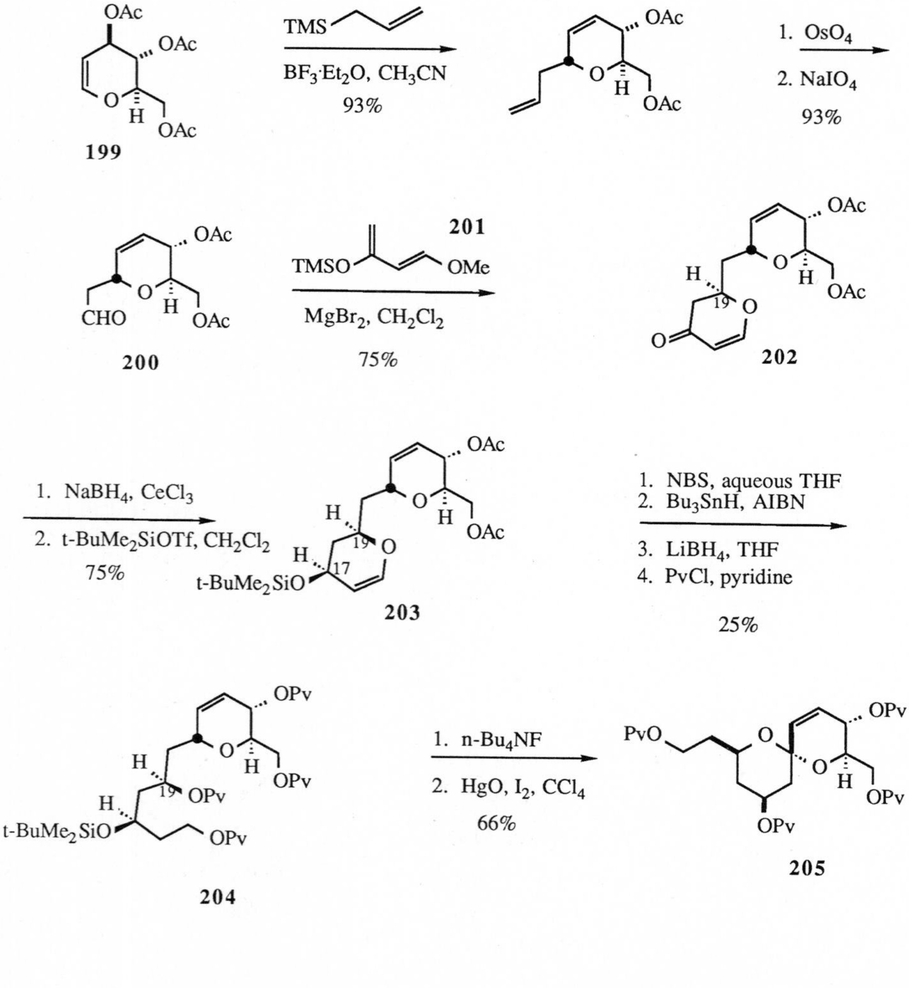

White[37] (Scheme 36) prepared a spiroketal model beginning with lactone **206**. Hydrolysis of the lactone and protection of the secondary hydroxyl, followed by formation of the acid chloride, provides key intermediate **207**. Racemic tin acetylide **208**, available in six steps (Scheme 37), is then acylated with **207** in a palladium mediated reaction. Semihydrogenation to the alkene produced **209** which unfortunately could not be cyclized. Further reduction provided **210** which closed to the spiroketal **211** upon exposure to aqueous HF.

Scheme 36

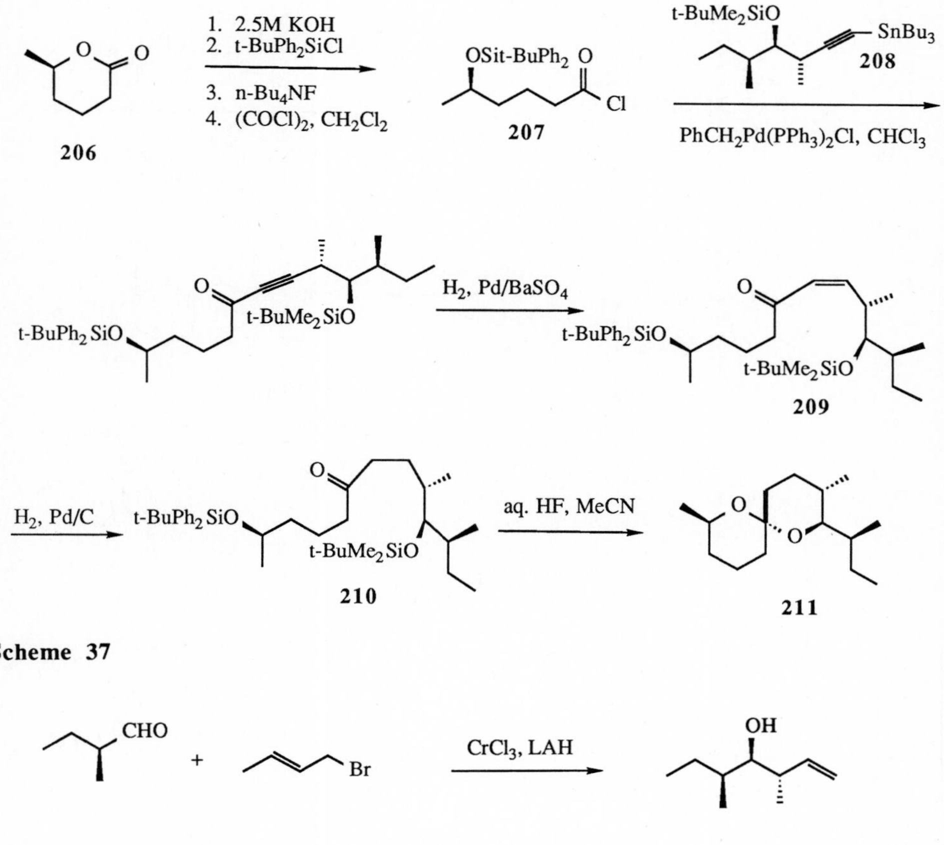

Scheme 37

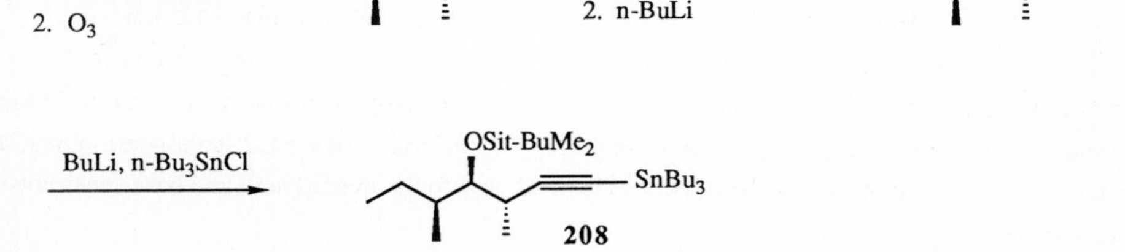

In a second approach to the spiroketal moiety of milbemycin β3, Williams[38] (Scheme 38) began by opening lactone **40** with methyl lithium followed by protection of the alcohol which gave **212**. Acylation with enone **213** followed by exposure to fluoroboric acid resulted in spiroketals **214** and **215** as a 60/40 mixture at C17. After removal of the protecting group, the isomers **214** and **215** could be separated. Treatment of the undesired isomer **215** with LiOH allowed for equilibration of the C17 center providing **216** and **217**.

Scheme 38

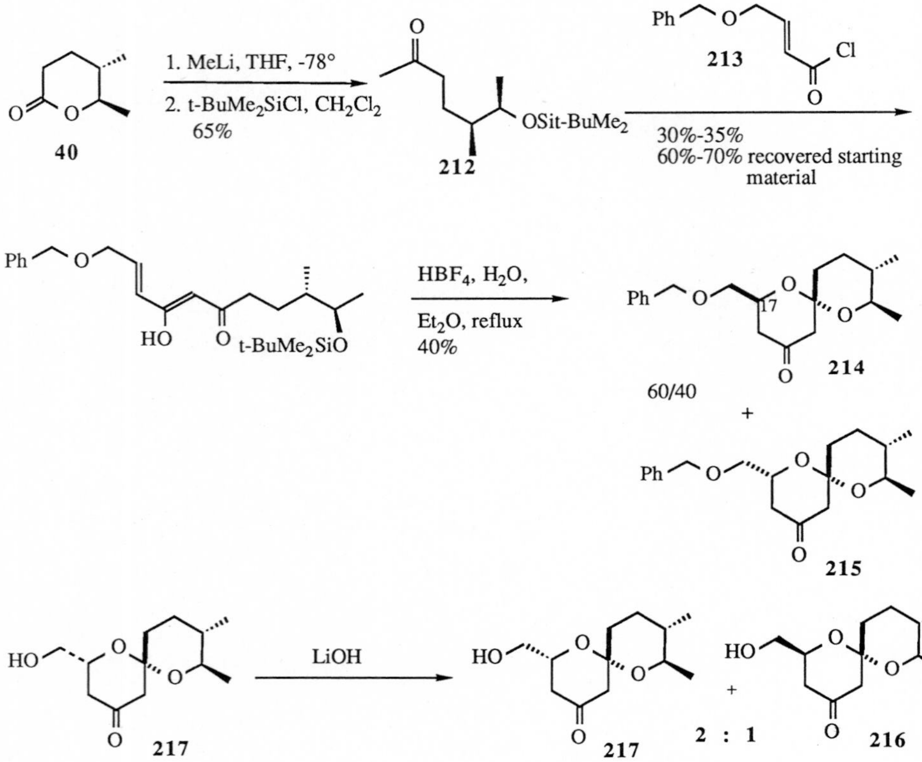

8 THE HEXAHYDROBENZOFURAN SUBUNIT

The large number of successful approaches to the spiroketal fragment of the milbemycins and the avermectins has been due largely to the known thermodynamic preference for the diaxial configuration of spiroketals. In contrast, several factors have contributed to the difficulty of the preparation of the hexahydrobenzofuran subunit: 1) the high density of oxygen functionality, 2) four contiguous stereogenic centers on a cyclohexenyl ring, 3) the lability of the C2 stereogenic center, 4) the propensity for conjugation of the Δ3,4 double bond, and 5) the overall acid and base sensitivity of the system.

Fraser-Reid[39] reported the first successful synthesis of the hexahydrobenzofuran subunit in 1985 utilizing a carbohydrate based approach with an intramolecular nitrile oxide cycloaddition as a key step (Scheme 39). Allyl alcohol **218** which was readily available from diacetone glucose was converted to alcohol **219** by a series of standard transformations. Oxidation of **219** followed by condensation with nitromethane gave α,β-unsaturated nitro compound **220** which upon conversion to the nitrile oxide underwent spontaneous cyclization to produce a single diastereomer **221**. Reduction of the isoxazoline and the resulting ketone gave mesylate **222** after tritylation and mesylation. The critical mesylate elimination in **222** could only be regioselectively accomplished if

Scheme 39

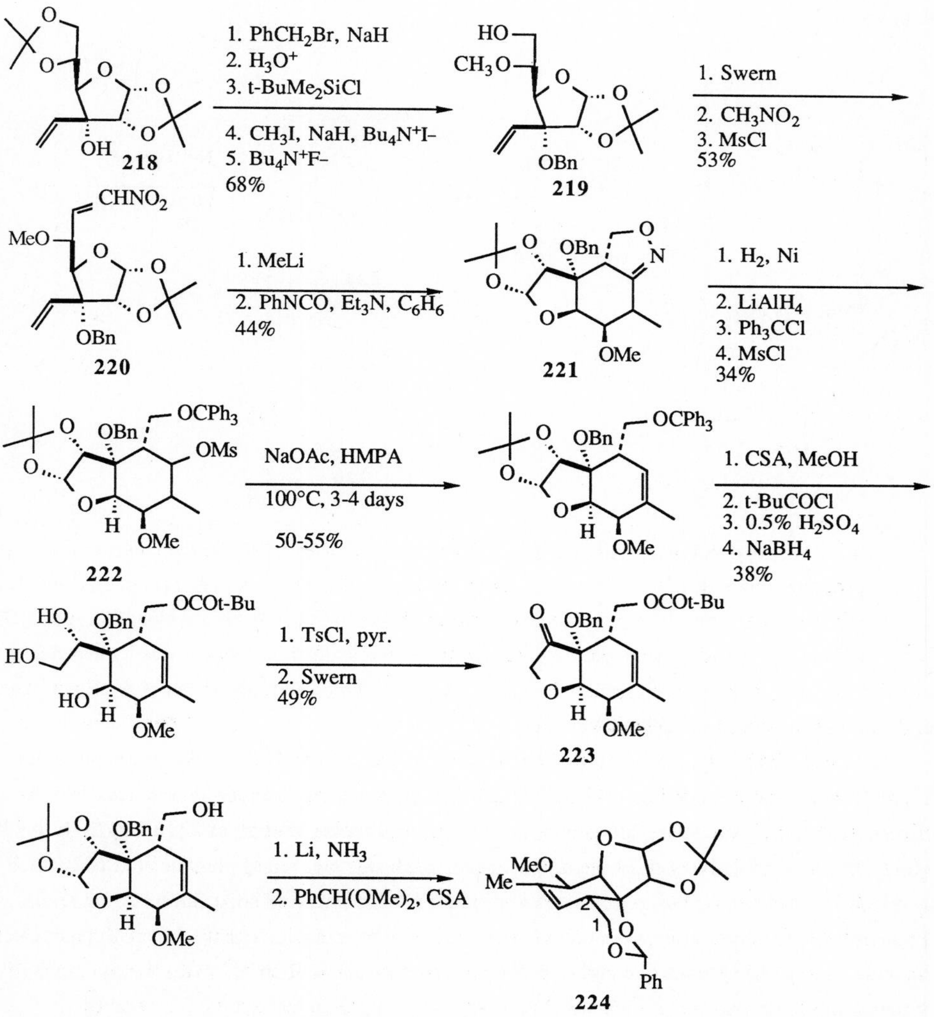

carefully prescribed conditions were followed (NaOAc, HMPA, 100°C, 3-4 days). The resulting Δ3,4 olefin was then converted into the desired hexahydrobenzofuran **223** by refunctionalization of the furan ring. That the nitrile oxide cyclization had occurred with the desired selectively was determined by examination of the C1-C2 ^{1}H NMR coupling constants in benzylidine **224**.

Prior to Fraser-Reid's report there had been two preliminary reports on studies toward the "southern" fragment of the milbemycins and the avermectins. The first of these was a report by Turnbull on an approach to the cyclohexenyl portion of milbemycin β1.[40] In his report Turnbull describes a clever, short approach (Scheme 40) centered around a Robinson annelation on methyl 4,4-dimethoxy-3-oxopentanoate to prepare cyclohexanone **225**. The ketone was subsequently stereoselectively reduced to give **226** after hydrolysis of the ketal.

Scheme 40

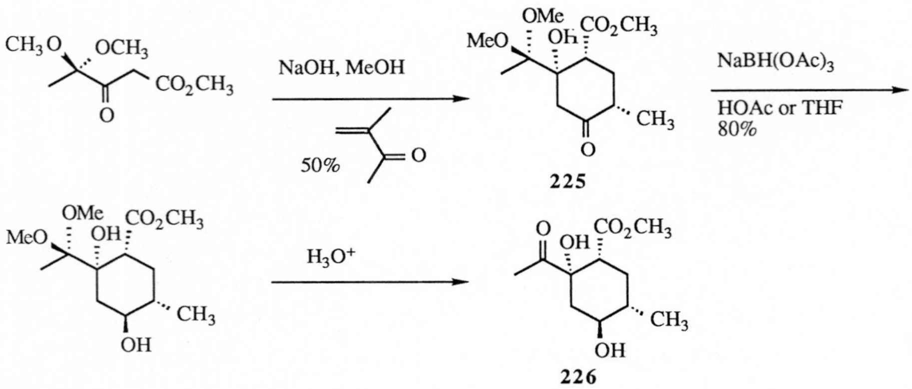

Jung has reported two separate Diels-Alder approaches directed at the synthesis of the hexahydrobenzofuran unit.[41,42] One involves an intramolecular Diels-Alder of furan **227** which was readily prepared from 2-acetyl-5-methylfuran as shown in Scheme 41. Cycloaddition of **227** proceeded cleanly in quantitative yield to produce the tricyclic ether **228** which was hydroxylated and subsequently cleaved with sodium in THF to give **229** which contains much of the functionality and stereochemistry of the desired subunit.

Alternatively, Jung[43] has carried out an intermolecular furan Diels-Alder wherein a furan serves as the dienophile (Scheme 42). Here 3,4-dibenzyloxyfuran is exposed to an excess of 2-trimethylsilylethyl coumalate **230** generating a nearly equimolar mixture of **231a** and **231b** in 88% yield. The trimethylsilylester of the minor isomer **231b** was cleaved to give the acid **232** which was reduced to a methyl group in 4 steps to produce a 4:1 mixture of **233a:233b**. At this point Jung was able to demonstrate an important point: the stereochemical integrity of the C2 center can be maintained in some systems in contrast to the rapid epimerization observed in the methyl esters of some seco acid derivatives of the natural products. Hydrolysis of the lactone followed by

Scheme 41

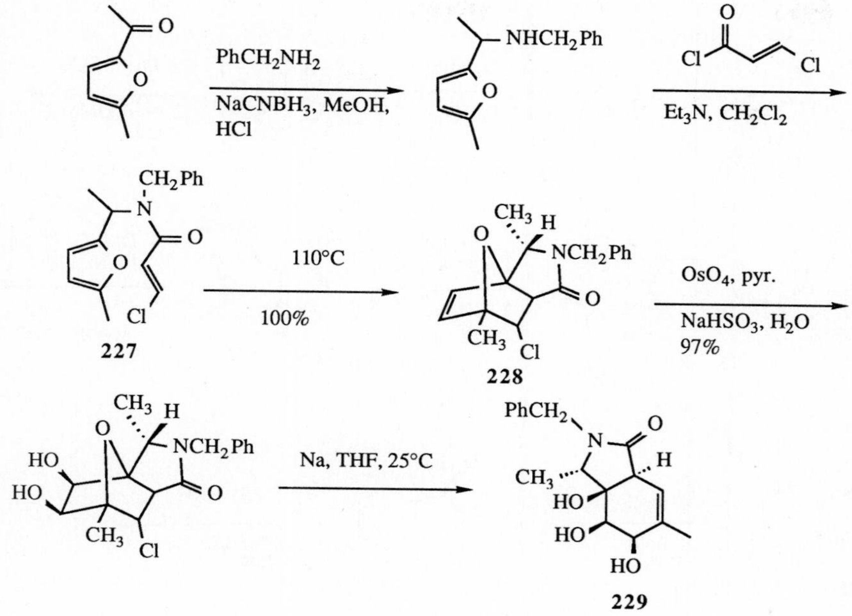

esterification of the acid gave **234**, a stable C1 ester, which required only inversion at C5 and removal of protecting groups. An oxidation-reduction sequence inverted the C5 hydroxyl (9:1 β:α) to produce **235** and sequential hydrogenolysis of the benzyl ethers provided the racemic southern fragment **236** in 12 steps in 9% overall yield.

Scheme 42

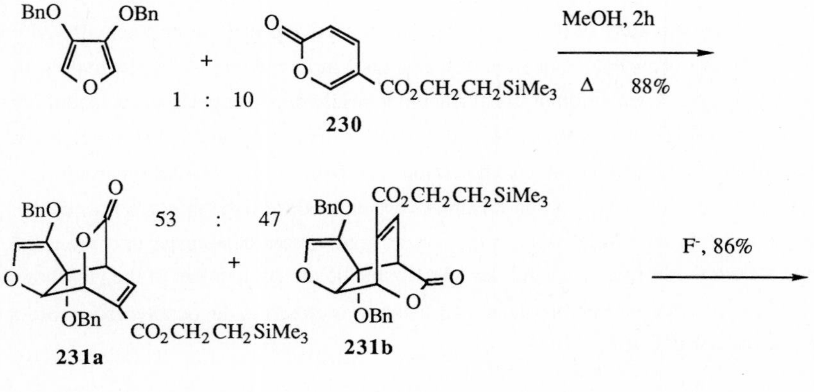

Kozikowski has utilized an intramolecular nitrile oxide approach as illustrated in Scheme 43.[44] The nitrile oxide precursor was prepared from aldehyde **237** by condensation with the anion of **238** to produce a 16:1 mixture of diastereomers **239**. Selective protection of the hydroxyl groups gave **240**. Removal of the pivalate, oxidation of the alcohol and treatment of the aldehyde with hydroxylamine gave **241**. Generation of the nitrile oxide resulted in spontaneous cycloaddition to produce four diastereomeric isoxazolines **242a-d** in yields of 37,48,2.5 and 2.5% respectively. The INOC reaction proceeds through a chair-like transition state with the phenylsulfonyl group occupying an equatorial site. The two minor isomers are derived from the minor isomer of the sulfone condensation. Isomer **242a** contains the correct relative stereochemistry at carbons 2,5, and 6 lacking only the tetrahydrofuran ring and the Δ3,4 double bond. Attachment of the vinylbromide and cleavage of the isoxazoline ring produced **243** which was closed to the octahydrobenzofuran **244** by exposure to n-Bu₂CuLi.

Scheme 43

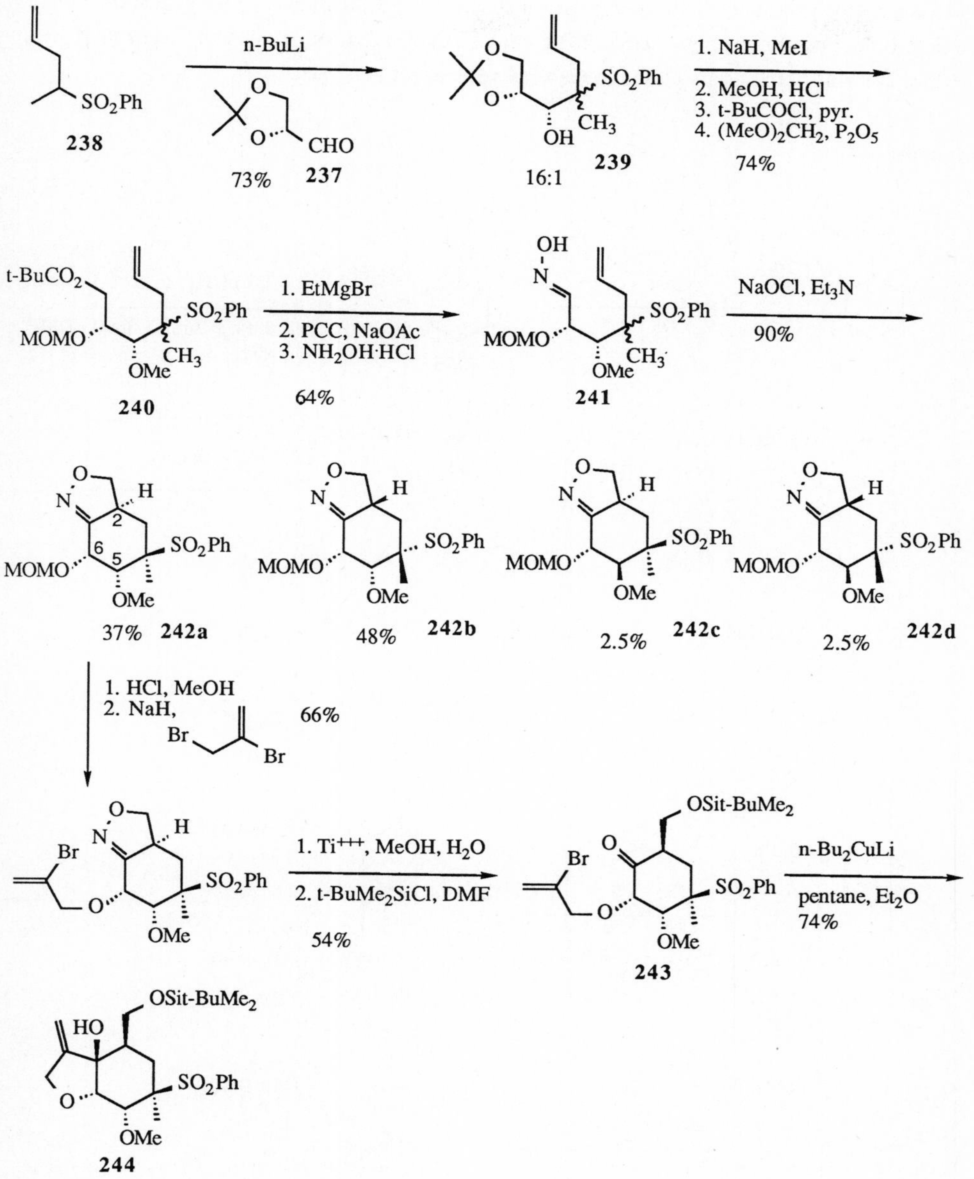
SO2Ph
238
n-BuLi
CHO
73% 237
SO2Ph
OH CH3
16:1 239
1. NaH, MeI
2. MeOH, HCl
3. t-BuCOCl, pyr.
4. (MeO)2CH2, P2O5
74%
t-BuCO2
SO2Ph
MOMO
OMe CH3
240
1. EtMgBr
2. PCC, NaOAc
3. NH2OH·HCl
64%
OH
N
SO2Ph
MOMO
OMe CH3
241
NaOCl, Et3N
90%
O N H SO2Ph MOMO OMe
37% 242a
48% 242b
2.5% 242c
2.5% 242d
1. HCl, MeOH
2. NaH,
Br Br
66%
Br
SO2Ph
OMe
1. Ti+++, MeOH, H2O
2. t-BuMe2SiCl, DMF
54%
OSit-BuMe2
SO2Ph
OMe
243
n-Bu2CuLi
pentane, Et2O
74%
OSit-BuMe2
HO
SO2Ph
OMe
244

In a recent model study Barrett (Scheme 44) has taken advantage of a yeast reduction of carboethoxycyclohexanone to prepare homochiral hydroxyester **245**.[45] The hydroxyester was conveniently transformed into enone **246** which gave a 1:6 ratio of epoxides **247a:247b** on exposure to alkaline hydrogen peroxide. Alternatively, reduction of the ketone followed by epoxidation and reoxidation to the ketone provided a 55:25 mixture of the two epoxyketones with the desired isomer predominating. Ketone **247a** was then condensed with dithianedianion **248** to yield 83% of **249** which was easily closed to the octahydrobenzofuran **250**.

Scheme 44

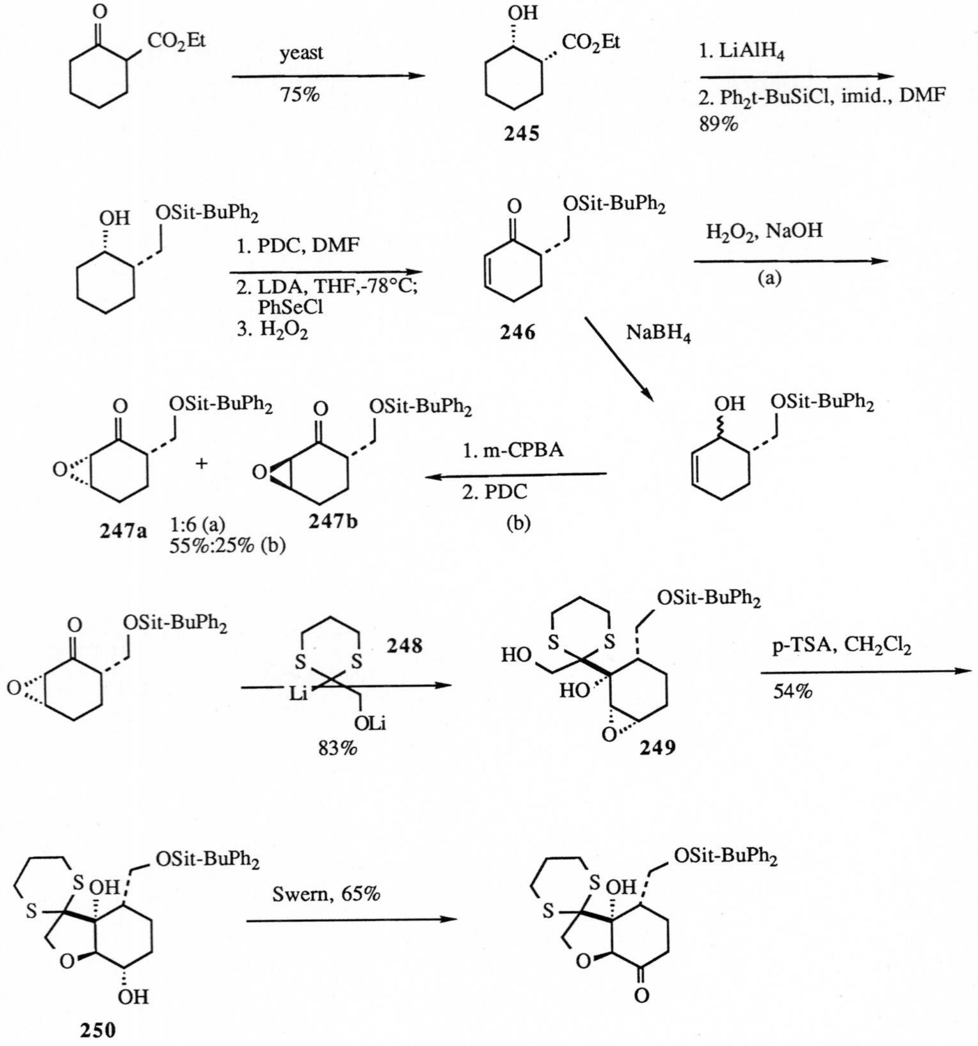

Hanessian has accomplished a synthesis of an octahydrobenzofuran and a desoxy "southern" subunit in both racemic and homochiral form as shown in Scheme 45.[46,26] The racemic synthesis begins with maleic anhydride and 2-acetoxyfuran to produce the Diels-Alder adduct **251**. Hydroxylation and hydrolysis of **251** provided the tetrasubstituted cyclohexanone **252** which was readily converted to **253**. Selective addition of methylmagnesium chloride to the ketone carbonyl followed by treatment with sodium methoxide gave the triol **254**. Alkylation of the allylic alcohol and acetylation of the free hydroxyls yielded vinyl bromide **255** which underwent tin mediated radical cyclization to the octahydrobenzofuran **256**. Cleavage of the exocyclic olefin followed by

Scheme 45

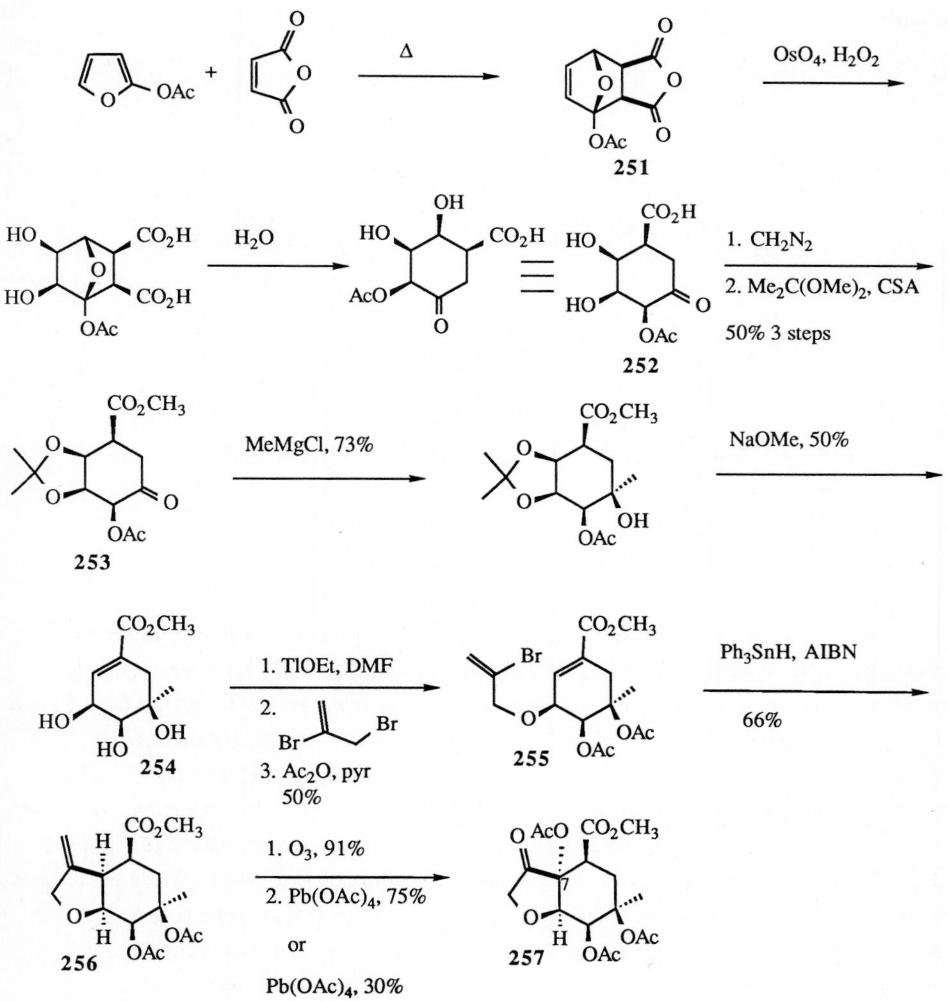

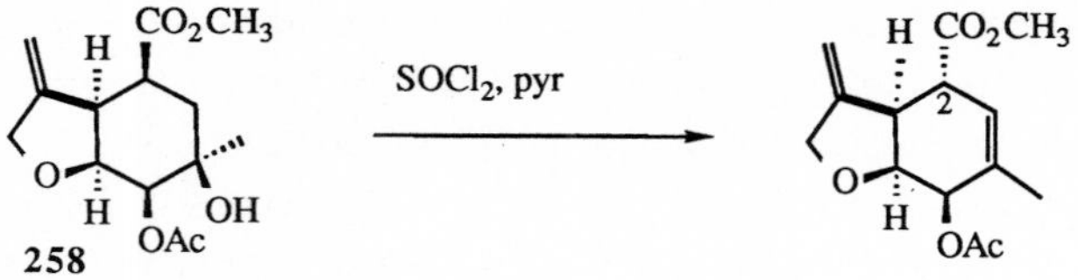

treatment with lead tetraacetate allowed introduction of the C7 hydroxyl in **257**. Another intermediate **258** has been dehydrated to give the Δ3,4 olefin with the correct stereochemistry at C2!

Triol **254** has also been prepared in homochiral form from (-)-quinic acid as shown in Scheme 46.

Scheme 46

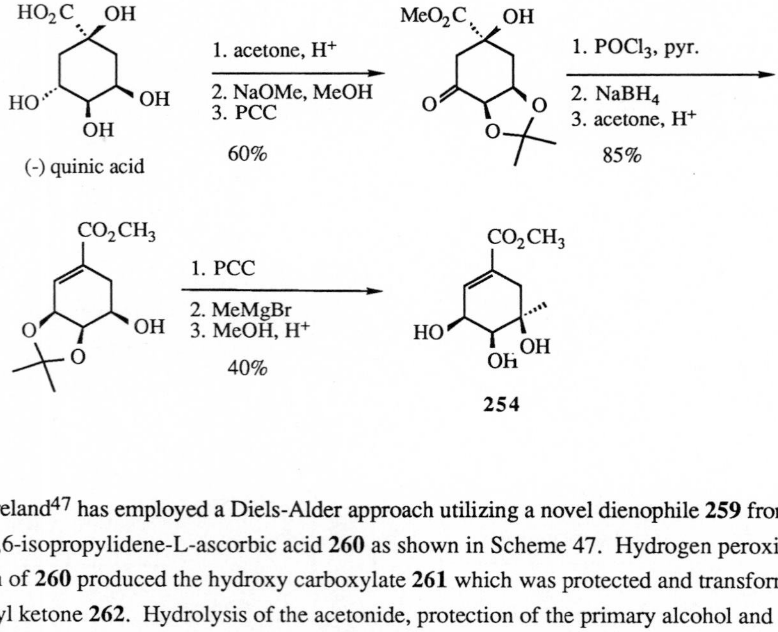

Ireland[47] has employed a Diels-Alder approach utilizing a novel dienophile **259** from the known 5,6-isopropylidene-L-ascorbic acid **260** as shown in Scheme 47. Hydrogen peroxide oxidation of **260** produced the hydroxy carboxylate **261** which was protected and transformed into the methyl ketone **262**. Hydrolysis of the acetonide, protection of the primary alcohol and mesylation of the secondary alcohol gave mesylate **263**. Treatment of **263** with DBU in THF provided enone **264** which was converted to the dienophile **259** by mCPBA oxidation of the silyl enol ether of **264**. When dienophile **259** and diene **265** were heated in benzene at 125° C for 2 days a 74% yield of the cyclohexene **266** resulted. Exposure of **266** to BF₃·Et₂0 resulted in clean transformation to the acyloxy enone **267** which underwent cyclization to the bicyclic ketone **268** upon acid treatment. This ketone has been transformed into its tosylhydrazone **269**, but final conversion to the desired hexahydrobenzofuran subunit **270** remains to be completed.

Scheme 47

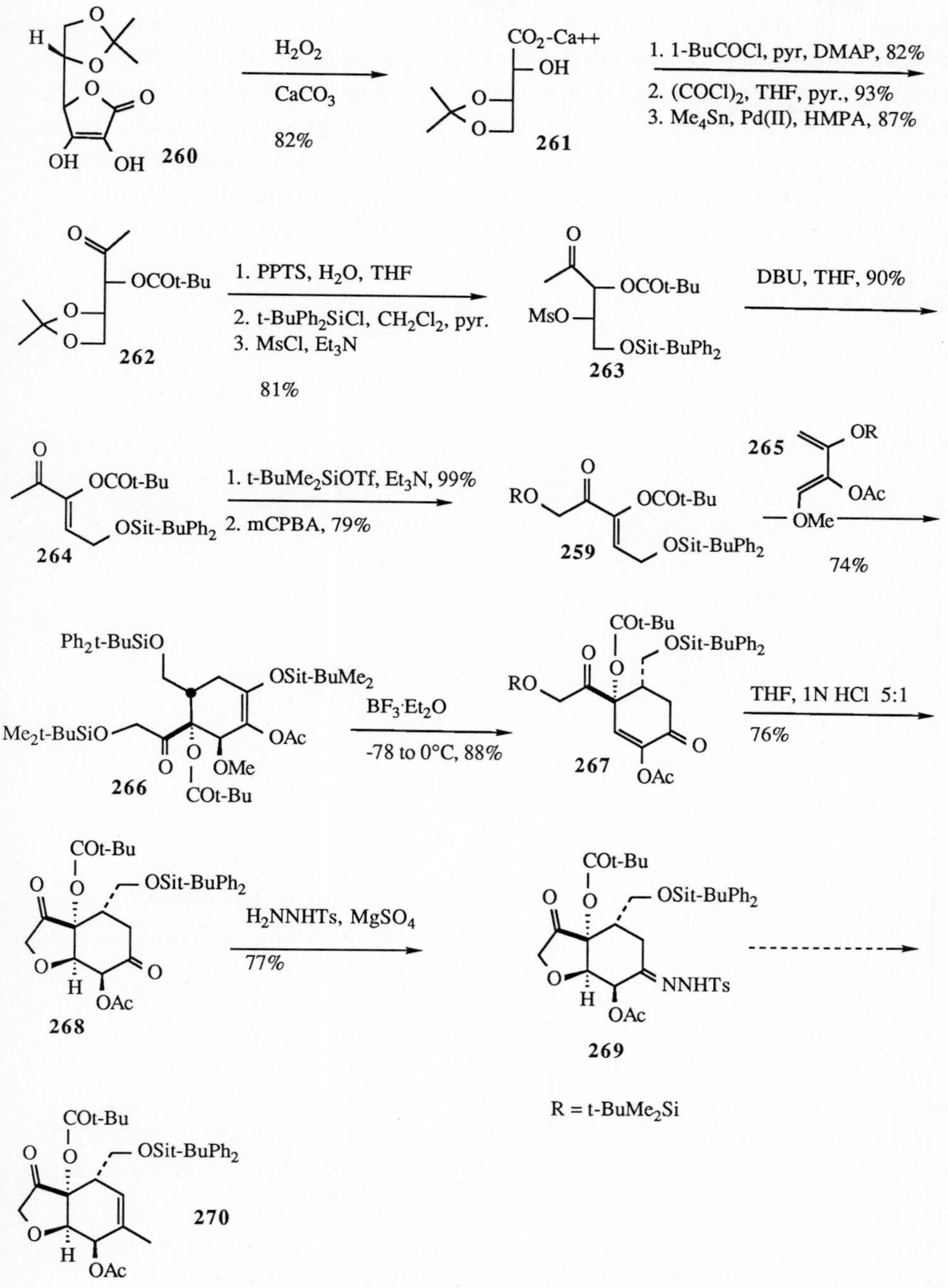
260
H2O2
CaCO3
82%
CO2-Ca++
OH
261
1. 1-BuCOCl, pyr, DMAP, 82%
2. (COCl)2, THF, pyr., 93%
3. Me4Sn, Pd(II), HMPA, 87%
OCOt-Bu
262
1. PPTS, H2O, THF
2. t-BuPh2SiCl, CH2Cl2, pyr.
3. MsCl, Et3N
81%
MsO
OSit-BuPh2
263
DBU, THF, 90%
264
1. t-BuMe2SiOTf, Et3N, 99%
2. mCPBA, 79%
RO
259
265
OR
OAc
OMe
74%
Ph2t-BuSiO
OSit-BuMe2
Me2t-BuSiO
COt-Bu
266
BF3·Et2O
-78 to 0°C, 88%
267
THF, 1N HCl 5:1
76%
268
H2NNHTs, MgSO4
77%
NNHTs
269
R = t-BuMe2Si
270

Crimmins and coworkers[48] have carried out extensive studies on the preparation of the hexahydrobenzofuran subunit and have completed an efficient preparation of a fragment which includes C9 and C10 of the bridging chain appropriately functionalized for further manipulation. In early studies the diketone **271** was prepared according to Scheme 48. A Diels-Alder reaction of propiolaldehyde with diene **272** gave the 1,4 diene **273** which was treated with methylmagnesium iodide to generate a 1.8:1 mixture of diastereomers **274a:274b** in 85% overall yield. Epoxidation of the mixture gave the corresponding epoxyalcohols **275a** and **275b** which were readily separated by flash chromatography. The major isomer was exposed to Swern conditions and yielded 95% of ketone **276**. Formation of the t-butyldimethylsilylenol ether and rearrangement of the epoxide produced **277** and **278** in 30 and 54% yields respectively. The ketone **278** which resulted from

Scheme 48

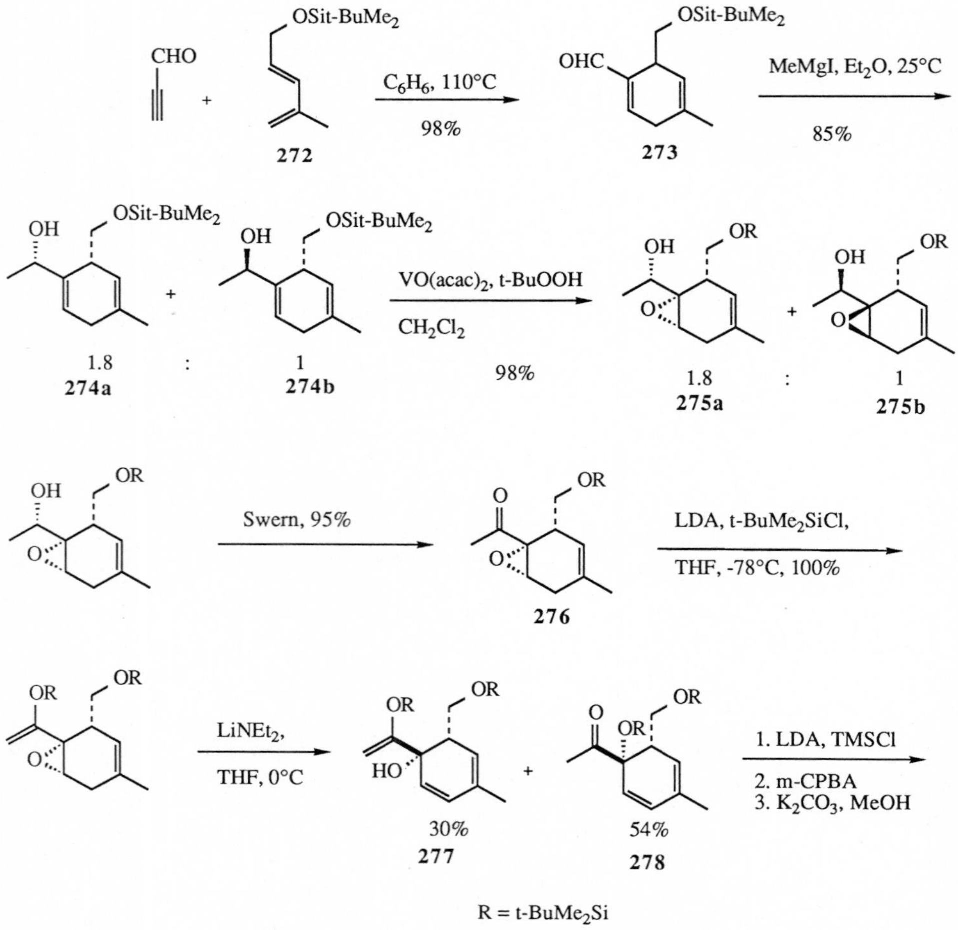

migration of the t-butyldimethylsilyl group was α-hydroxylated to give **279** which underwent smooth electrophilic cyclization on treatment with NBS to produce the 1,4 adduct **280**. Selective hydrolysis of the primary t-butyldimethylsilyl ether and oxidation of the resultant alcohol produced the aldehyde **281**. The allylic bromide was solvolyzed in acetic acid with the assistance of silver acetate resulting in production of the tricyclic acetate **282** which upon reduction with LiAlH4 gave triol **283**. The primary alcohol was selectively protected, and the remaining secondary hydroxyls were oxidized to provide **271**.

In a more recent improved route[49] to homochiral material (Scheme 49) the Diels-Alder adduct **273** was reduced with lithium aluminum hydride, and the allylic alcohol was exposed to Sharpless conditions for asymmetric epoxidation producing alcohols **284a** and **284b**. The diastereomers were separated after conversion to the corresponding methyl ketones providing the desired enantiomer and diastereomer **285a** in >95% e.e. Exposure of **285a** to excess LDA and excess trimethylsilyl chloride followed immediately by oxidation with m-CPBA yielded 50% of the hydroxyketone **286** directly from the ketoepoxide. When this material was treated with phenylselenenyl chloride followed by oxidation with hydrogen peroxide, allylic alcohol **287** was isolated in 70% yield. This critical step proceeds by a selenium mediated 1,4-electrophilic addition to the diene followed by oxidation to the selenoxide which spontaneously undergoes 2,3-sigmatropic rearrangement to the selenenate which eventually collapses to the allylic alcohol on further oxidation. The ketone was then condensed with cyanomethylenetriphenylphosphorane to produce a 15:1 mixture of E:Z olefin isomers **288** in 85% yield. This is an important development because many workers in the field felt that a Wittig type olefination at C8 might be problematic due to the steric hindrance at the carbonyl and the acidity of the α-protons. Hydrolysis of the silyl ethers followed by

488

formation of the benzylidene gave a 5:1 mixture of the β:α isomers **289**. The major isomer was inverted at C5 by an oxidation- reduction sequence to produce the β-alcohol **290** in 79% yield. Protection of the alcohol and subsequent Dibal-H reduction of the nitrile gave the desired aldehyde **291** in 50% yield.

Scheme 49

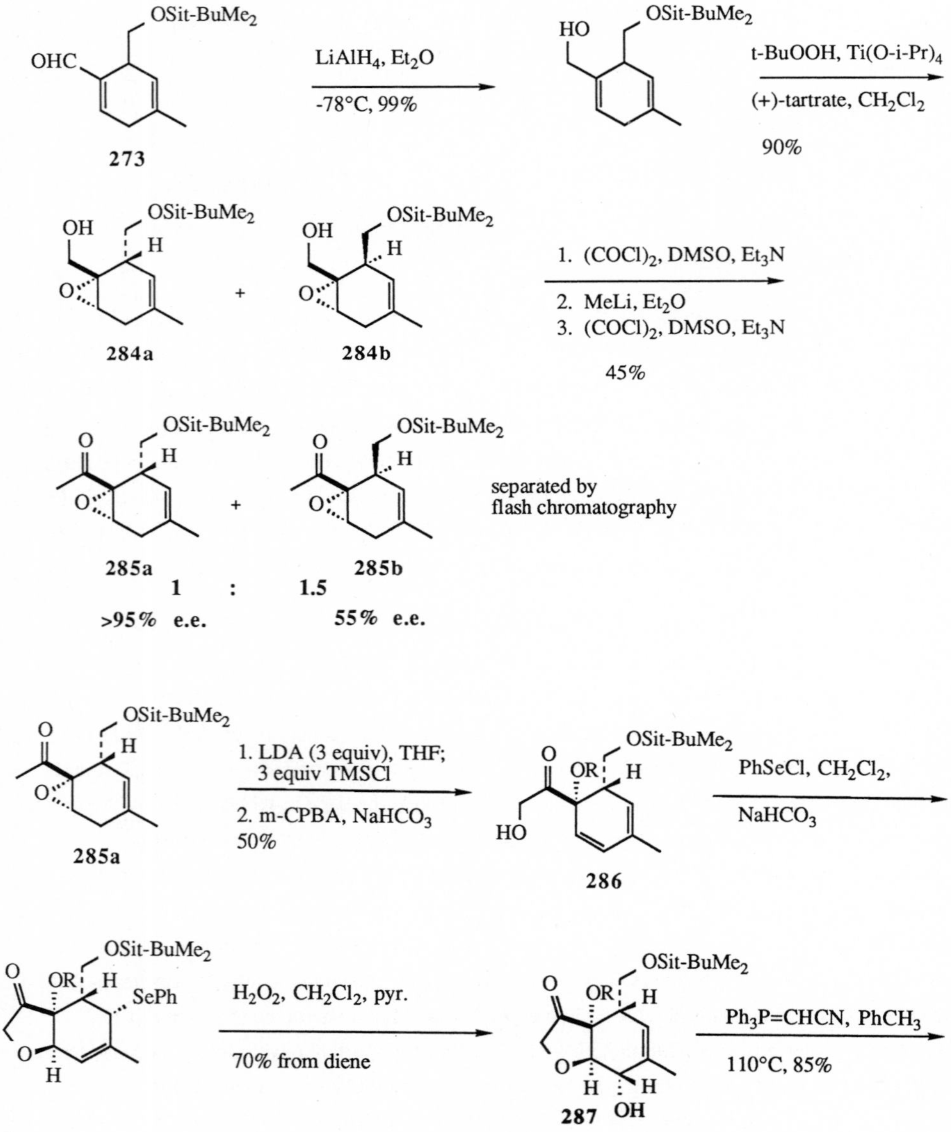

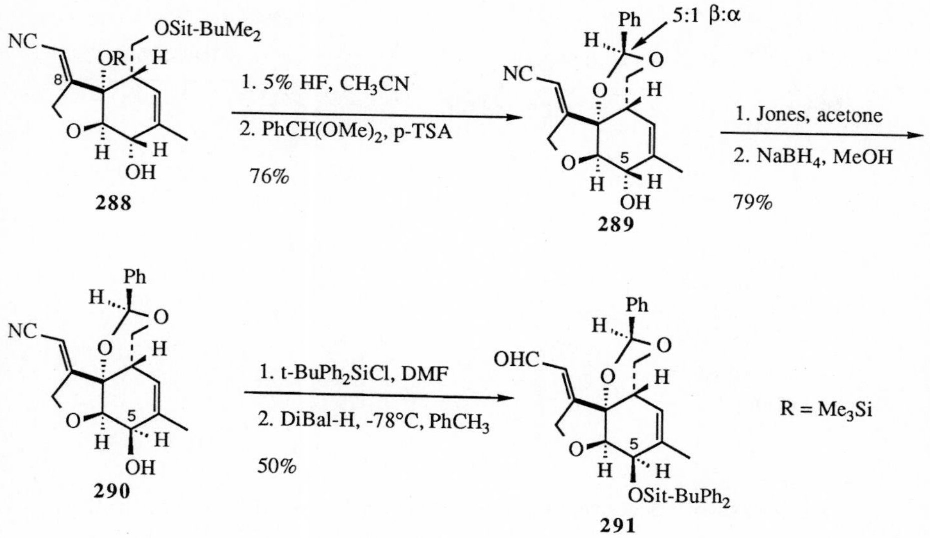

Danishefsky has completed a model study for the construction of the oxahydrindene subunit from **292** which is available from ribose.[50] Condensation of **292** with allyltrimethylsilane (Scheme 50) followed by alkylation of the resultant alcohol produced methyl ether **293**. Two deoxygenations were performed on **293** to provide olefin **294** which was readily converted to ester **295**. The key transformation is the highly stereoselective Michael addition of phenylthiolate to the unsaturated ester **295** to close the six membered ring and provide **296** which contains a handle for the introduction of the unsaturation. Oxidation of **296** gave the target **297** upon elimination of phenylsulfenic acid.

Scheme 50

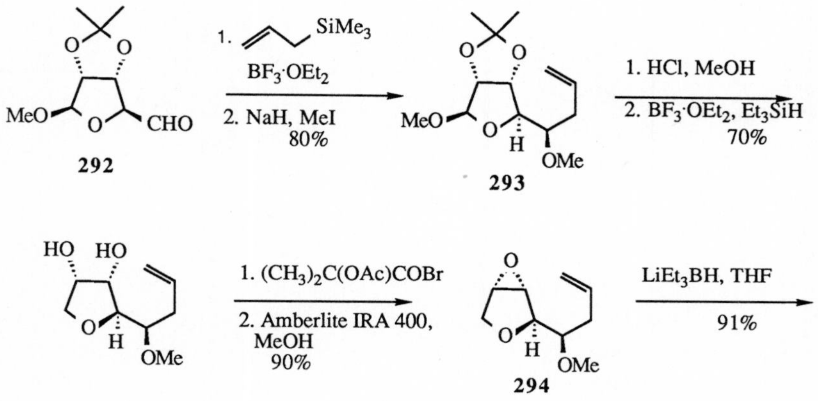

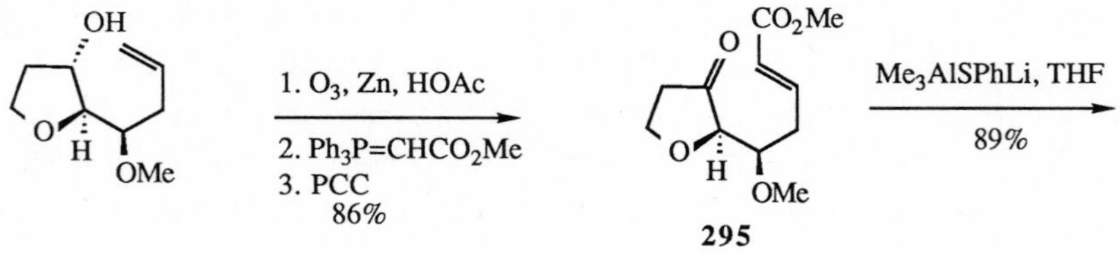

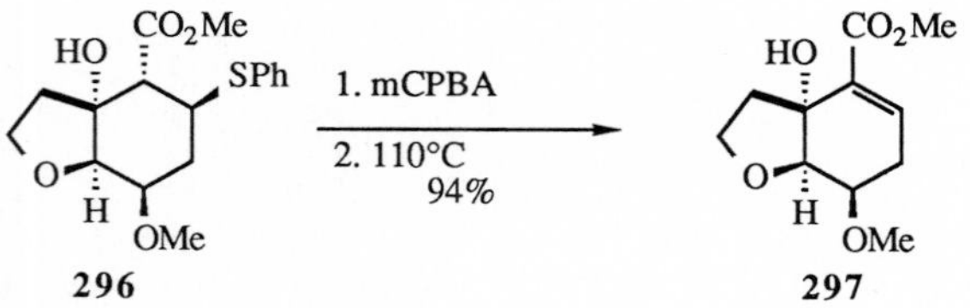

Julia and coworkers[51] have executed an interesting radical cyclization to form the five membered ring of the southern subunit. Propargyl ethers **298a** and **298b** were readily prepared from dimethyl 2-hydroxypimelate in two steps (Scheme 51). Addition of tributyltin radical to the acetylene gave the vinyl stannane **299** which was protodestannated to produce the olefin **300**.

Scheme 51

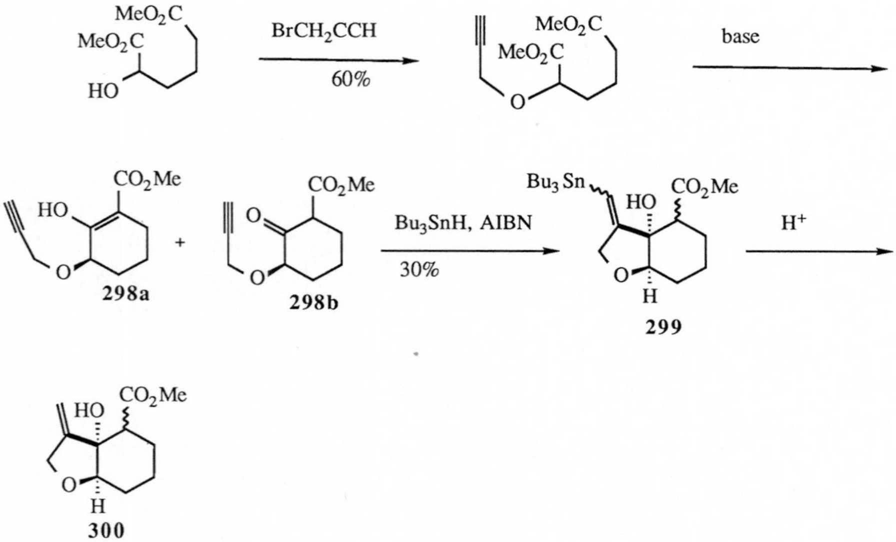

A stereoselective intramolecular Claisen condensation is the central reaction in the approach taken by Williams (Scheme 52).[52] Acetonide **301** which is readily prepared from L-rhamnose was treated with methylmagnesium chloride, and the resultant primary alcohol was selectively protected. The secondary alcohol was oxidized to the ketone which was treated with butenyl magnesium bromide to form tertiary alcohol **302** as the sole product. The olefin was cleaved with ozone, and the resultant hemiacetal was oxidized to the lactone to give **303** after removal of the silyl ether and oxidation to the aldehyde. The aldehyde **303** was then transformed to the 3-furyl ketone **304** which smoothly underwent an intramolecular Claisen condensation upon exposure to LDA at -78°C.

Scheme 52

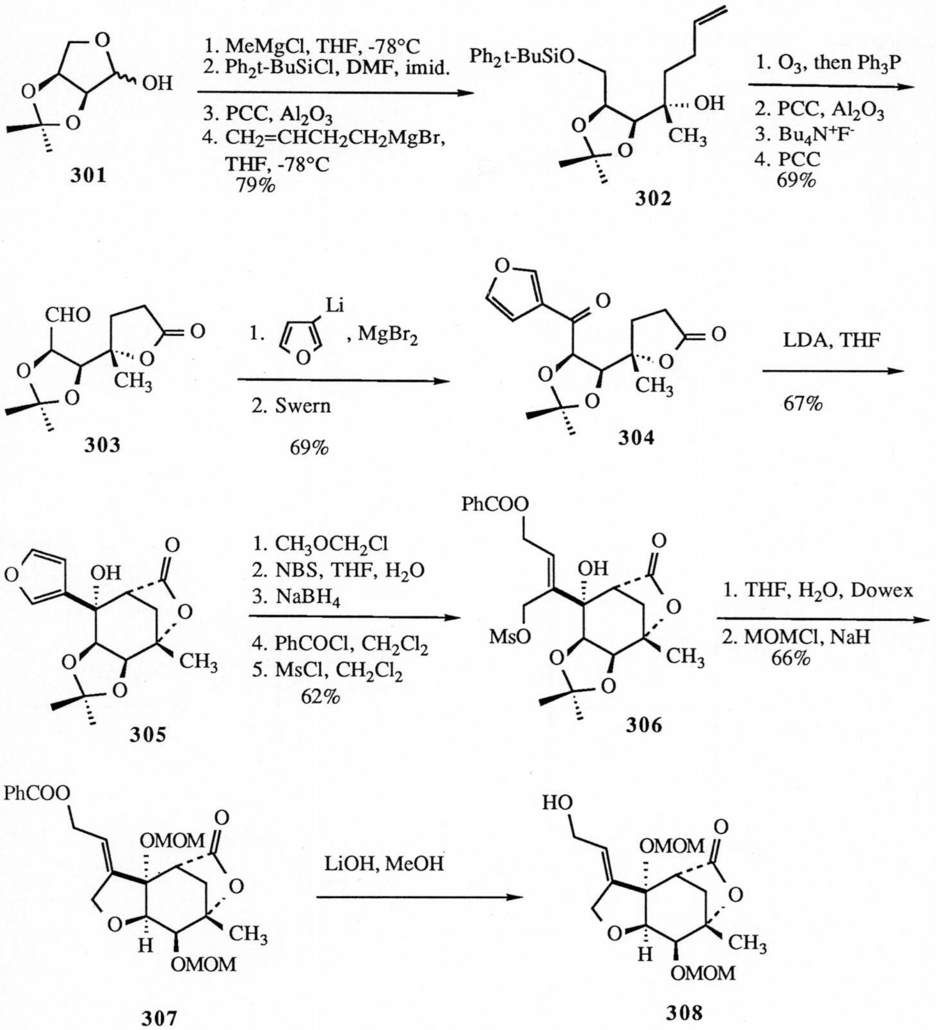

Cyclohexane **305** was formed in a 4:1 preference over the alternative diastereomer. After protection of the tertiary alcohol, the furan was cleaved to give the bis-allyl alcohol which was differentially protected to provide the mesylate **306**. Hydrolysis of the acetonide led to spontaneous closure of the tetrahydrofuran ring to give **307** after replacement of the MOM ethers. The benzoate was then readily hydrolyzed to unveil the allylic alcohol in **308** ready for further manipulation. One of the major advantages of this approach is the ability to tie back the C1 ester and prevent the epimerization at C2 which has plagued many approaches.

In an initial study by White[53] (Scheme 53), the Diels-Alder adduct **309** of maleic anhydride and 1-trimethylsilyloxy-3-methyl butadiene was converted to the bicyclic lactonic acid **310** which was transformed into the diazoketone **311**. In a unique approach to the tetrahydrofuran ring, **311** was exposed to sulfuric acid to effect the cyclization, and the resultant acid was esterified to provide ester **312**. While further transformations met with difficulties, the C7 hydroxyl could be incorporated to give low yields of **313**.

Scheme 53

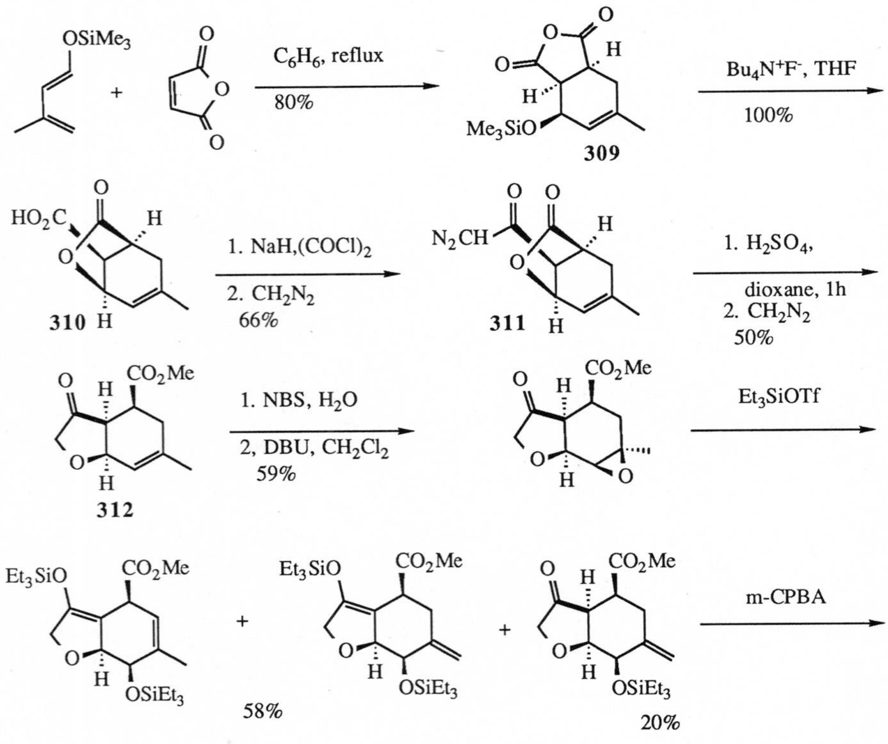

313 28%   10%

References

1. Albers-Schonberg, G.; Arison, B. H.; Chabala, J.C.; Douglas, A.W.; Eskola, P.; Fisher, M.H.; Lusi, A.; Mrozik, H.; Smith, L.J.; Tolman, R.L. *J. Am. Chem. Soc.* **1981**, *103*, 4216. Springer, J.P.; Arison, J.P.; Hirschfield, J.M.; Hoogsten, K. *J. Am. Chem. Soc.* **1981**, *103*, 4221.

2. Arison, B.H.; Goegelman, R.T.; Gullo, V.P. U.S. Patent 4,285,963 **1981**. Goegelman, R.T.; Gullo, V.P.; Kaplan, L. U.S. Patent 4,378,353 **1983.**

3. Mishima, H.; Kurabayashi, M.; Tamura, C.; Sato, S.; Kuwano, H.; Saito, A. *Tetrahedron Lett.* **1975**, 711.

4. Mishima, H.; Ide, J.; Muramatsu, S.; Ono, M. *J. Antibiotics* **1983**, *36*, 980. Ono, M.; Mishima, H.; Takiguchi, Y.; Terao, M.; Kobayashi, H.; Iwasaki, S.; Okuda, S. *J. Antibiotics* **1983**, *36*, 991. Mishima, H. 5th International Congress of Pesticide Chemistry, Kyoto, Japan 1982.

5. Putter, I.; MacConnell, J.G.; Preiser, F.A.; Haidri, A.A. Ristich, S.S.; Dybas, R.A. *Experientia* **1981**, *37*, 963.

6. Chabala, J.C.; Mrozik, H.; Tolman, R.L.; Eskola, P.; Lusi, A.; Peterson, L.H.; Woods, M.F.; Fisher, M.H.; Campbell, W.C.; Egerton, J.R.; Ostlind, D.A. *J. Med. Chem.* **1980**, *23*, 1134.

7. Cupp, E.M.; Bernardo, M.J.; Kiszewski, A.E.; Collins, R.C.; Taylor, H.R.; Aziz, M.A.; Greene, B.M. *Science* **1986**, *231*, 740.

8. Wang, C.C.; Pong, S.S. "Progress in Clinical and Biological Research," **1981**, *97*, 373.

9. Fisher, M.H. "The Avermectins" in "Recent Advances in the Chemistry of Insect Control," Janes, N.F., Ed.,, The Royal Society of Chemistry Special Publication no. 53, London, p.53. (1985)

10. Smith, A.B., III; Thompson, A.S. *Tetrahedron Lett.* **1985**, *26*, 4279.

11. Hanessian, S.; Ugolini, A.; Hodges, P.J.; Dube', D. *Tetrahedron Lett.* **1986**, *27*, 2699.

12. Selnick, H.G.; Danishefsky, S.J. submitted.

13. Mrozik, H.; Eskola, P.; Fisher, M.H. *Tetrahedron Lett.* **1982**, *23*, 2377.

14. Mrozik, H.; Chabala, J.C.; Eskola, P.; Matzuk, A.; Waksmunski, F.; Woods, M.; Fisher, M.H. *Tetrahedron Lett.* **1983**, *24*, 5333.

15. Mrozik, H.; Eskola, P.; Fisher, M.H. *J. Org. Chem.* **1986**, *51*, 3058.

494

16. Smith, A.B., III; Thompson, A.S. *Tetrahedron Lett.* **1985**, *26*, 4283.

17. Hughes, M.J.; Thomas, E.J.; Turnbull, M.D.; Jones, R.H.; Warner, R.E. *J. Chem. Soc. Chem. Commun.* **1985**, 755.

18. Smith, A.B., III; Schow, S.R.; Bloom, J.D.; Thompson, A.S.; Winzenberg, K.N. *J. Am. Chem. Soc.* **1982**, *104*, 4015.   Schow, S.R.; Bloom, J.D.; Thompson, A.S.; Winzenberg, K.N.; Smith, A.B., III *J. Am. Chem. Soc.* **1986**, *108*, 2662.

19. Williams, D.R.; Barner, B.A.; Nishitani, K.; Phillips, J.G. *J. Am. Chem. Soc.* **1982**, *104*, 4708.

20. Baker, R.; Boyes, R.H.O.; Broom, D.M.P.; Devlin, J.A.; Swain, C.J. *J. Chem. Soc. Chem. Commun.* **1983**, 829. Baker, R.; O'Mahony, M.J.; Swain, C.J. *J. Chem. Soc. Chem. Commun.* **1985**, 1326.

21. Baker, R.; O'Mahony, M.J.; Swain, C.J. *Tetrahedron Lett.* **1986**, *27*, 3059.

22. Kocienski, P.; Street, S.D.A. *J. Chem. Soc. Chem. Commun.* **1984**, 571. Street, S.D.A.; Yeates, C.; Kocienski, P.; Campbell, S.F. *J. Chem. Soc. Chem. Commun.* **1985**, 1386.

23. Yeates, C.; Street, S.D.A.; Kocienski, P.; Campbell, S.F.   *J. Chem. Soc. Chem. Commun.* **1985**, 1388.

24. Attwood, S.V.; Barrett, A.G.M.; Carr, R.A.E.; Richardson, G. *J. Chem. Soc. Chem. Commun.* **1986**, 479.   Barrett, A.G.M.; Carr, R.A.E.; Attwood, S.V.; Richardson, G.; Walshe, N.D.A. *J. Org. Chem.* **1986**, *51*, 4840.

25. Hanessian, S.; Ugolini, A.; Dube, D.; Hodges, P.J.; Andre, C. *J. Am. Chem. Soc.* **1986**, *108*, 2776.

26. Hanessian, S.; Ugolini, A.; Hodges, P.J.; Beaulieu, P.; Dube, D.; Andre, C. *Pure and Applied Chem.* **1987**, *59*, 299.

27. Fraser-Reid, B.; Wolleb, H.; Faghih, R.; Barchi, J. *J. Am. Chem. Soc.* **1987**, *109*, 933.

28. Barrett, A.G.M.; Raynham, T.M. submitted.

29. Hirama, M.; Nakamine, T.; Ito, S. *Tetrahedron Lett.* **1986**, *27*, 5281.

30. Baker, R.; Swain, C.J.; Head, J.C. *J. Chem. Soc. Chem. Commun.* **1985**, 309.

31. Crimmins, M.T.; Hollis, W.G., Jr.; Bankaitis-Davis, D.M. *Tetrahedron Lett.*, **1987**, *28*, 3651.

32. Ardisson, J.; Ferezou, J.P.; Julia, M.; Lenglet, L.; Pancrazi, A. *Tetrahedron Lett.* **1987**, *28*, 1997.

33. Godoy, J.; Ley, S.V.; Lygo, B. *J. Chem. Soc. Chem. Commun.* **1984**, 1381.

34. Culshaw, D.; Grice, P.; Ley, S.V.; Strange, G.A. *Tetrahedron Lett.* **1985**, *26*, 5837.

35. Greck, C.; Grice, P.; Ley, S.V.; Wonnacot, A. *Tetrahedron Lett.* **1986**, *27*, 5277.

36. Wincott, F.; Danishefsky, S.J.; Schulte, G. submitted.

37. White, J.D. unpublished results.

38. Williams, D.R.; Barner, B.A. *Tetrahedron Lett.* **1983**, *24*, 427.

39. Prashad, M.; Fraser-Reid, B. *J. Org. Chem.* **1985,** *50,* 1564.

40. Turnbull, M.D.; Hatter, G.; Ledgerwood, D.E. *Tetrahedron Lett.* **1984,** *25,* 5449.

41. Jung, M.E.; Street, L.J. *J. Am. Chem. Soc.* **1984,** *106,* 8327.

42. Jung, M.E.; Street, L.J.; Usui, Y. *J. Am. Chem. Soc.* **1986,** *108,* 6810.

43. Jung, M.E.; Usui, Y.; Vu, C.T. submitted.

44. Kozikowski, A.P.; Maloney-Huss, K.E. *Tetrahedron Lett.* **1985,** *26,* 5759.

45. Barrett, A.G.M.; Capps, N.K. *Tetrahedron Lett.* **1986,** *27,* 5571.

46. Hanessian, S.; Beaulieu, P.; Dube', D. *Tetrahedron Lett.* **1986,** *27,* 5071.

47. Ireland, R.E.; Obrecht, D.M. *Helv. Chim. Acta* **1986,** *69,* 1273.

48. Crimmins, M.T.; Lever, J.G. *Tetrahedron Lett.* **1986,** *27,* 291.

49. Crimmins, M.T.; Hollis, W.G., Jr.; Lever, J.G. *Tetrahedron Lett.* **1987,** *28,* 3647.

50. Armistead, D.M.; Danishefsky, S.J. submitted.

51. Ardisson, J., Ferezou, J.P.; Julia, M.; Pancrazi, A. *Tetrahedron Lett.* **1987,** *28,* 2001.

52. Williams, D.R.; Klingler, F.D. submitted.

53. White, J.D.; Dantanarayana, A.P. unpublished results.

STEREOSELECTIVE SYNTHESIS OF BIOLOGICALLY ACTIVE COMPOUNDS

A V RAMA RAO

1 INTRODUCTION

In recent years, natural products with significant biological activity are increasingly being identified and coming into prominence. Their isolation and identification, however complex, have been simplified by the availability of various modern separation techniques augmented by sophisticated instrumental facilities. As the presence of these products in natural sources is in very low concentration, need has been felt for obtaining such products in large quantities in order to assess their physiological properties and understand the mechanism of their activity which in turn may help in designing products with better therapeutic properties. Although, some of the products like antibiotics (e.g.β-lactam antibiotics) could be manufactured in abundant quantities by fermentation, their synthesis either partial or total, has often helped in better understanding of the mechanism of their action in the human body. All these factors coupled with the availability of new tools and synthetic methods involving new reactions and reagents have enabled synthetic organic chemists to take up more challenging structures for total synthesis.

The crucial aspect which needs to be addressed, in planning the total synthesis of any compound of natural origin, is the problem of selectivity (ref. 1). Selectivity can be broadly classified as (a) regioselectivity, related to the concept of orientational control, (b) chemoselectivity, which deals with functional group differentiation and (c) stereoselectivity. The main problem is to obtain enantiomerically pure compounds. The earlier practitioners of organic synthesis achieved this objective by generally effecting the resolution of a racemic mixture at a convenient stage via a diastereomeric intermediate and separating the desired enantiomer for further elaboration and completing the synthesis. The main disadvantage in this approach is the loss of more than 50% of the unwanted stereoisomer. While this approach even now continues to be practised in many large scale operations, other methods for controlling chirality in total synthesis have attained prominence in recent years. Most of these approaches can be broadly classified into two categories: (i) asymmetric synthesis, in which the enantiomerically enriched product is generated by carrying out the reaction in a chiral environment with the use of suitable reagents, catalysts etc. (ref. 2), and (ii) the chiron approach, in which the introduction of chirality into a target molecule is achieved by employing a chiral starting material (ref. 3). These two methods are complementary to each other and the final choice is dictated by the structure of the target molecule. These two methods are becoming common for devising the synthesis of enantiomerically pure compounds of diverse structures. In this chapter, the author makes a presentation of these strategies taking the synthesis of anti-tumour antibiotics, rifamycin-S and hydroxy octadecadienoic and trienoic acids (octadecanoids) as typical examples.

2. ANTI-TUMOUR ANTIBIOTICS

2.1 Anthracycline antibiotics

The usefulness of certain anthracycline antibiotics as antineoplastic agents is now widely accepted. Daunomycin (1) (ref. 4), and adriamycin (2) (ref. 5) have become established among the most useful weapons in the oncologists' armoury and their targets include lymphomas, osteogenic and soft tissue carcinomas and solid tumours (ref. 6). Their potent anticancer activity has made them the subject of interest of many synthetic organic chemists all over the world and extensive work has been carried out on their isolation, structure determination and large scale preparation. Consequently other closely related anthracyclines like carcinomycin (3) (ref. 7), 11-deoxydaunomycin (4)

1 : R = OMe ; R' = H ; X = OH
2 : R = OMe ; R' = X = OH
3 : R = X = OH ; R' = H
4 : R = OMe ; R' = X = H
5 : R = OMe ; R' = OH ; X = H
6 : R = R' = H ; X = OH

7 : R = OMe ; X = OH
8 : R = OMe ; X = H
9 : R = H ; X = OH
10 : R = X = H

and 11-deoxyadriamycin (5) (ref. 8), have been isolated. A number of synthetic analogues have also been prepared and screened for their antitumor activity. Among them, the most promising is 4-demethoxydaunomycin (6), which is 8 to 10 times more active than daunomycin (ref. 9).

2.1.1 Synthesis of anthracyclinones

The best possible way of elaborating the synthesis of anthracyclines is illustrated by Smith et al (ref. 10), where the degradation of daunomycin (1) to tetracyclic ketone and functionalization of A ring of daunomycin and adriamycin were carried out as depicted in Scheme 1. As several elegant syntheses of L-daunosamine, including two best approaches starting either from D-glucose (ref. 11), or from D-glucosamine (ref. 12), and its coupling to daunomycinone have been accomplished, most of the efforts were directed towards the synthesis of the aglycone moieties (anthracyclinones). A comprehensive review (ref. 13) and a book on anthracyclines (ref. 14), covering the work upto 1980 had earlier appeared.

Scheme-1

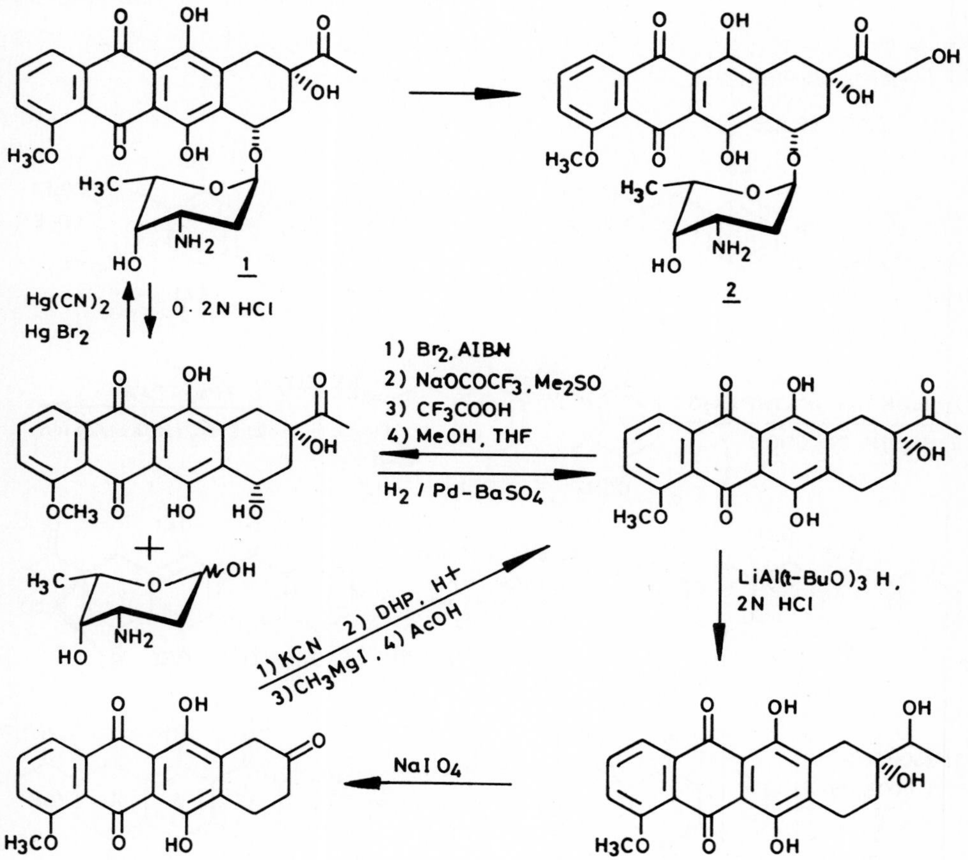

Hg(CN)₂ 0·2N HCl
Hg Br₂

1) Br₂, AIBN
2) NaOCOCF₃, Me₂SO
3) CF₃COOH
4) MeOH, THF

H₂ / Pd–BaSO₄

LiAl(t-BuO)₃ H,
2N HCl

1) KCN 2) DHP. H⁺
3) CH₃MgI . 4) AcOH

NaIO₄

In evolving a strategy for the synthesis of anthracyclinones, due consideration should be given to the presence of OMe substituent at C-4 in ring D, in addition to the stereochemical control of the substituent at C-7 and C-9 in ring A. To meet this objective in the synthesis, a number of classical reactions have been tried, viz. Friedel-Crafts reactions, Diels-Alder reactions, nucleophilic reactions and other miscellaneous reactions. A few specific and selected examples are highlighted in the following pages.

(i) <u>Friedel-Crafts reactions.</u> Friedel-Crafts reactions including photo-Fries rearrangements have played an important role in the regiospecific synthesis of many anthracyclinones. In the versatile method first employed by Wong et al (ref. 15), for

the assemblage of tetracyclic systems, AB+CD coupling seems to be most appropriate. It involves the preparation of a tetralin derivative, representing final AB rings
and its condensation by cyclization with phthalic acid derivative to CD rings in tetracyclic system. Later, Arcamone et al (ref. 16), developed a one step method for
obtaining tetracyclic system by fusion of tetralin derivative with phthalic anhydride.

A new type of Friedel-Crafts alkylation directly on the ring B of anthracyclinones was reported by Johnson and his colleagues (ref. 17), for the synthesis
of daunomycinone (7) (Scheme 2). They have extended this approach to the synthesis
of 11-deoxydaunomycinone (8) (ref. 18).

Scheme - 2

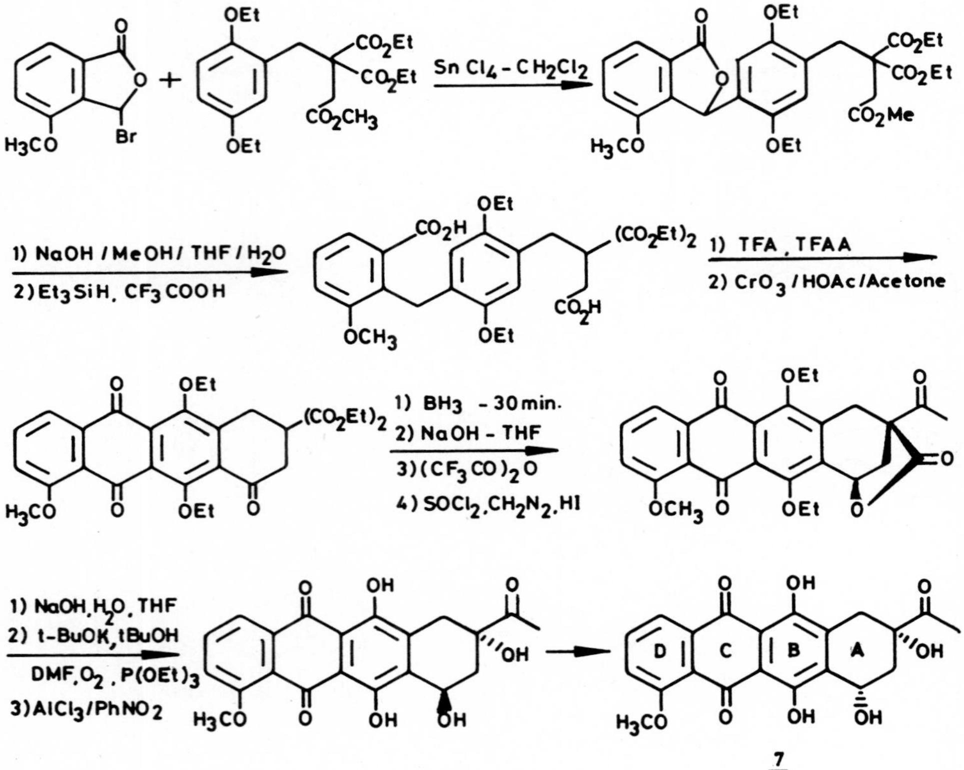

Rama Rao et al (ref. 19), have developed a conceptually different strategy
(Scheme 3) which is effective for the synthesis of both (±) daunomycinone and (±)
11-deoxydaunomycinone simultaneously. The key intermediate (13), obtained by condensation of 2-acetyl-5-hydroxy-8-bromo-1,2,3,4-tetrahydronaphthalene (11) with 3-bromo-
4-methoxy-phthalide (12), was reduced with zinc dust in ethanolic KOH to give 14
which was then converted to daunomycinone as shown by the sequence of reactions.
Alternatively, heating 14 with zinc dust in aqueous KOH gave 15, which was finally

transformed into 11-deoxydaunomycinone (**8**). This approach was further extended to the synthesis of 4-demethoxydaunomycinone (**9**) and 11-deoxy-4-demethoxydauno-mycinone (**10**) (ref. 20).

Scheme 3

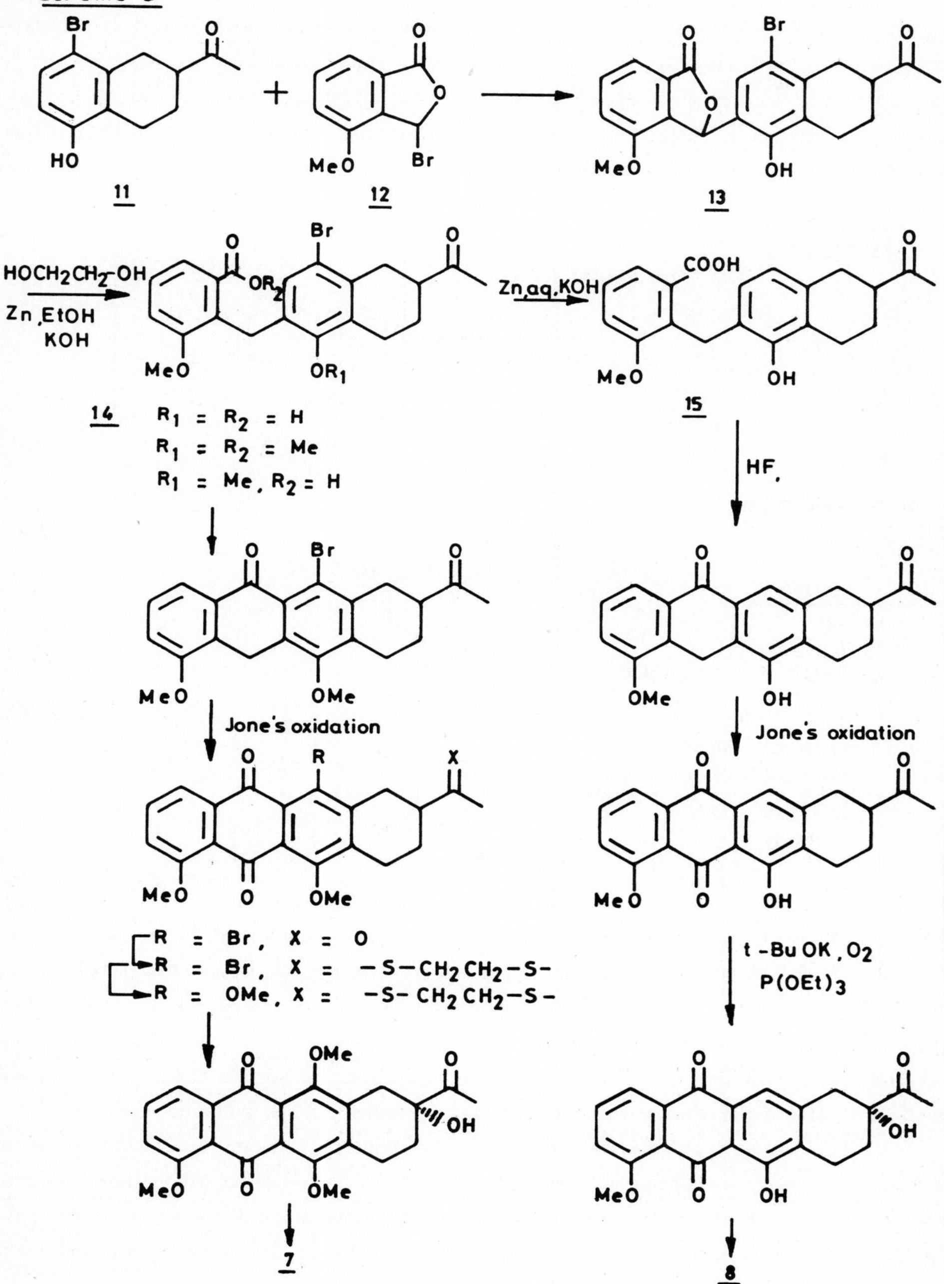

(ii) <u>Diels-Alder reaction</u>. The utility of Diels-Alder reaction in the synthesis of anthracyclinones was initially fraught with many difficulties and disadvantages, the first and foremost being the lack of regiospecificity commonly observed in earlier synthesis. A general strategy, however, appeared from Kelly's group (ref. 21), wherein, regiocontrolled Diels-Alder reaction of substituted napthaquinone (17) and 1-methoxy-cyclohexa-1,3-diene (16) furnished an adduct (18). The high degree of regioselectivity was attributed to the electron deficient character of the carbonyl group at C-4 involved in hydrogen bonding interaction with <u>peri</u>-hydroxy group. The second Diels-Alder reaction with 1,3-bistrimethylsiloxy-1,3-butadiene followed by thermolysis, dehydration etc. afforded (±) daunomycinone (7) (Scheme 4).

Scheme – 4

Substitution of halogen atom in naphthaquinone ring in order to induce inductive effect and direct the course of Diels-Alder reaction has been exploited by many groups (refs. 22,23). 2,7-Dichloronaphthaquinone (19) and 20 undergo Diels-Alder reaction to furnish 21. Elimination of HCl gave the ABC synthon. Chloronaphthaquinone segment present in the molecule controls the orientation of second Diels-Alder reaction with 22 to afford the 4-substituted tetracyclic system (ref. 23), (Scheme 5).

Tamura <u>et al</u> (ref. 24), reported the enolate induced cycloaddition of 4-acetoxy-8-methoxyhomophthalic anhydride (23) with halo-1,4-naphthaquinone derivative to provide short, convergent synthesis of daunomycinone (Scheme 6). They have applied the same strategy for the regiospecific synthesis of 11-deoxydaunomycinone by using the anhydride in which ring A of anthracyclinone was functionalized (ref. 25).

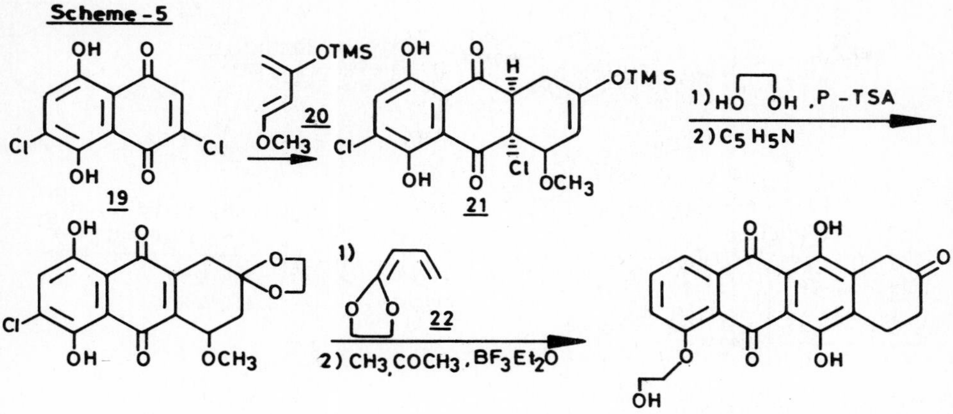

Scheme - 5

(iii) <u>Nucleophilic reactions.</u>   The search for effective solutions to obtain aglycone regiospecifically has led to the development of a number of new methods involving anionic species for the construction of anthracyclinones. A regioselective

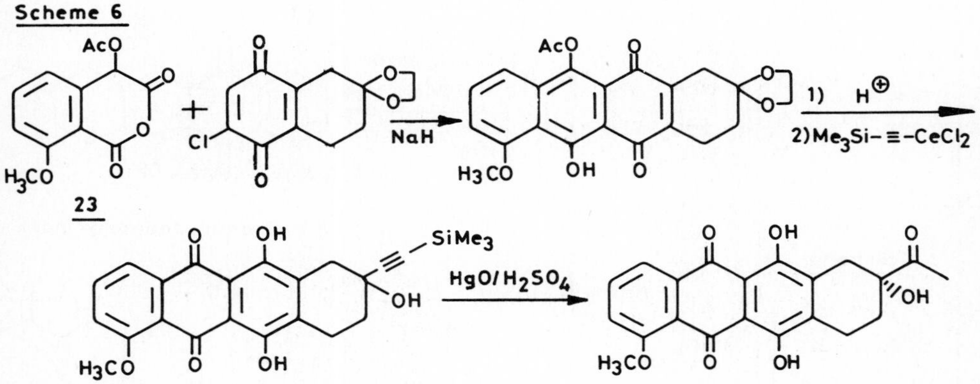

Scheme 6

approach for the synthesis of daunomycinone was applied by Kende (ref. 26), Parker (ref. 27), Braun (ref. 28), Broadhust (ref. 29), Hauser (ref. 30), Furukawa (ref. 31), Swenton (ref. 32), Russel (ref. 33), and their associates. Sih and coworkers (ref. 34), were the first to report the synthesis of 7,11-dideoxydaunomycinone involving the nucleophilic condensation of the CD ring synthon to a suitably functionalized A-ring intermediate, followed by subsequent cyclization of the B-ring to complete the tetra-cyclic aglycone skeleton (Scheme 7).

Hauser (ref. 35), synthesised 7,11-dideoxydaunomycinone starting from a bicyclooctanol which was converted to a suitably functionalised AB ring synthon. Condensation of this synthon with the anion of methoxy-phenylsulphonylisobenzofura-none furnished the tetracyclic product, which was further transformed into 7,11-di-deoxydaunomycinone in single step (Scheme 8).

Scheme - 7

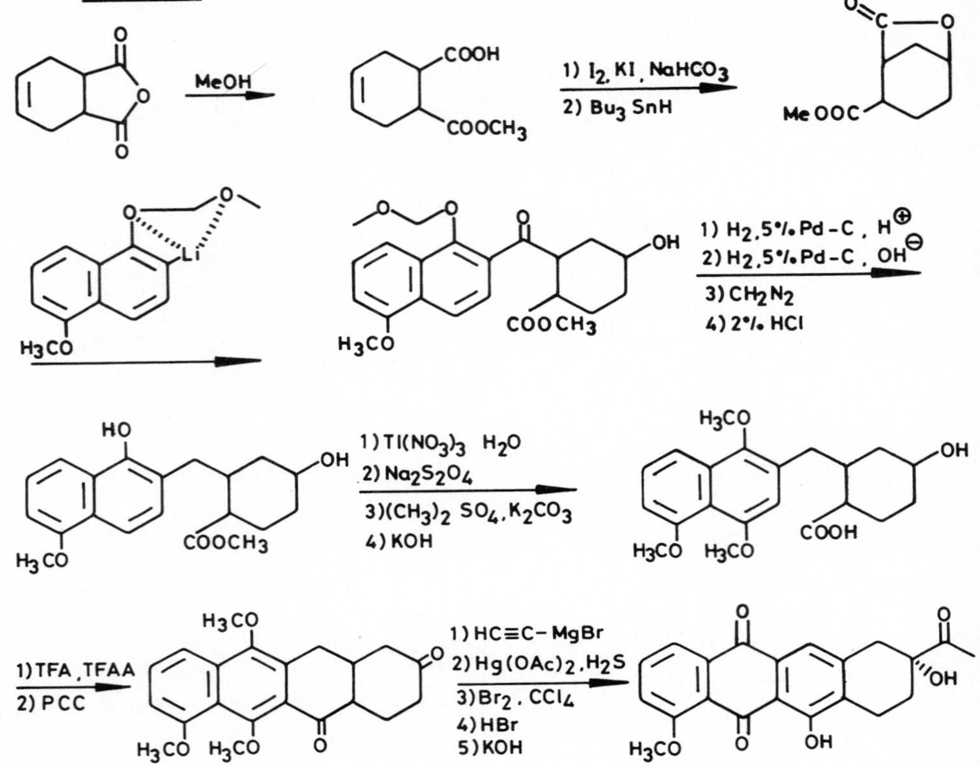
MeOH
COOH
COOCH3
1) I2, KI, NaHCO3
2) Bu3 SnH
MeOOC
Li
H3CO
OH
COOCH3
1) H2, 5% Pd-C, H⊕
2) H2, 5% Pd-C, OH⊖
3) CH2N2
4) 2% HCl
HO
1) Tl(NO3)3 H2O
2) Na2S2O4
3) (CH3)2 SO4, K2CO3
4) KOH
COOH
1) TFA, TFAA
2) PCC
1) HC≡C-MgBr
2) Hg(OAc)2, H2S
3) Br2, CCl4
4) HBr
5) KOH

7, 11-Dideoxydaunomycinone

Scheme - 8

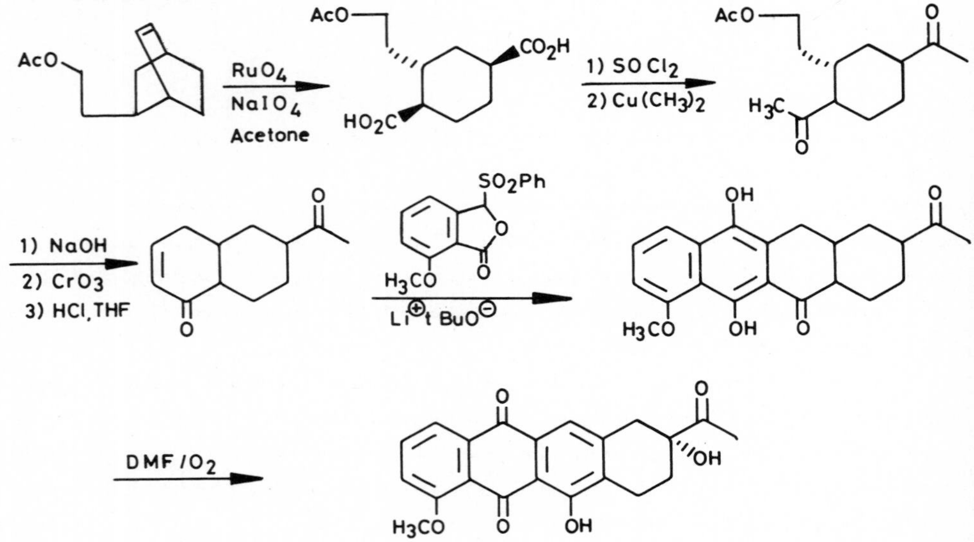
AcO
RuO4
NaIO4
Acetone
CO2H
HO2C
1) SO Cl2
2) Cu(CH3)2
H3C
1) NaOH
2) CrO3
3) HCl, THF
SO2Ph
H3CO
Li⊕ t BuO⊖
OH
DMF / O2

(iv) <u>Miscellaneous reactions</u>. Various other types of chemical reactions have also been used to achieve the regioselectivity in the synthesis of anthracyclinones. Sutherland <u>et al</u> (ref. 36), reported a regiospecific synthesis of 7,9-dideoxydaunomycinone (Scheme 9) from 5-hydroxyquinizarin. Their strategy was to form the initial

Scheme – 9

C-C bond by regiospecific nitronate addition to C-2 of 5-hydroxyquinizarin and the second C-C bond by Marschalk condensation. Later, they (ref. 37), reported the same synthesis by reverse strategy using regioselective piperdinium acetate catalysed Marschalk-Lewis condensation to form the C-C bond, followed by nitronate cyclisation.

Very recently, Wulff and Tang (ref. 38), demonstrated the importance of Fischer carbene complexes in the regiospecific synthesis of anthracyclines (Scheme 10). Thus, the benz-annulation reaction of acetylene derivative and the chromium complex gave the lactone regiospecifically. Protection of phenolic hydroxyl and reductive cleavage of lactone resulted in the formation of acid which was cyclized, oxidized and demethylated to give the known tetracyclic ketone.

2.1.2   Synthesis of 4-demethoxydaunomycine

4-Demethoxydaunomycin (**6**) was the first synthetic analogue of daunomycin which was proved to be clinically useful and eight times more effective compared to daunomycin or adriamycin. Further, it was found to be orally active. As there was no possibility of obtaining **6** by fermentation, many groups had undertaken its synthesis. This involved the synthesis of the aglycone, (+)-4-demethoxydaunomycinone, the amino sugar, L-daunosamine and finally building (+)-4-demethoxydaunomycin. Most of the efforts were directed towards the synthesis of 4-demethoxy-7-deoxydaunomycinone (**24**), as this product could be processed to 4-demethoxydaunomycin by the same sequence of reactions depicted in Scheme 1.

Scheme – 10

The synthesis of **9** represents a simplified model of **1** owing to the absence of regioselectivity problems encountered in the lattr. As a result, innumerable approaches have appeared in literature displaying original features in the choice of the reactants and in the solution of important details relating to functionalisation and stereochemistry. In this chapter, the main emphasis is focussed on the stereoselective synthesis of (+)-4-demethoxydaunomycinone.

(+)-4-Demethoxydaunomycinone can be obtained either by resolution of the racemic mixture with an appropriate resolving agent or by asymmetric synthesis. This can be achieved either by obtaining the optically pure key intermediates and later building up the tetracyclic ring system, or preparing optically pure tetracyclic ring system itself. Both these approaches have been made for obtaining (+) **9**.

Arcamone and his co-workers (ref. 39), resolved the key intermediate (±)-2-acetyl-2-hydroxy-5,8-dimethoxy-1,2,3,4-tetrahydronaphthalene (**25**) by reacting with (-) α-phenylethylamine followed by several crystallisations. By this method, they could obtain optically pure R(-)-(**25**) which was condensed with phthalic anhydride in one step (ref. 16), to give (-)-4-demethoxy-7-deoxydaunomycinone (**24**). The conversion of (-) **24** to (+)-4-demethoxydaunomycinone (**9**) and ultimately to (+)-4-demethoxy-daunomycin (**6**) was achieved by established methods (Scheme 1).

Many reports have appeared on the resolution of either the key intermediate (**25**) (ref. 40), or the tetracyclic 4-demethoxy-7-deoxydaunomycinone (ref. 41), for the preparation of (+)-4-demethoxydaunomycinone. It has become apparent that the intermediate (±)-**25** is indeed a valuable precursor for the resolution of the dl-mixture. As a consequence, innumerable approaches towards preparation of **25** have appeared. Rama Rao et al (ref. 42), described a number of approaches for (±) **25** with a view to prepare optically active (+) **9**. Since most of the existing approaches were found not suitable for large scale preparation of (+) **9**, they adopted a conceptually different stereoconvergent approach towards (+) **9**; (-)-**25** was initially synthesised by employing Sharpless asymmetric epoxidation (ref. 43), (Kinetic resolution) of allylic alcohol (**26**) followed by acylation reaction (ref. 44), (Scheme 11).

Scheme 11

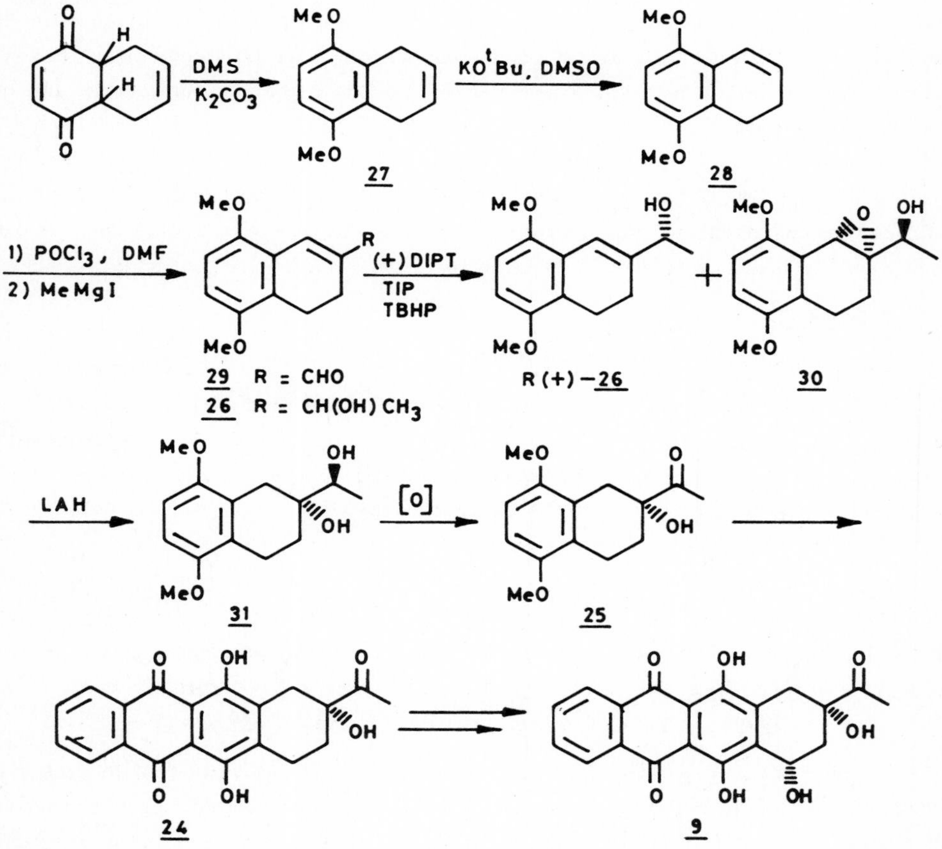

Diels-Alder reaction of benzoquinone with butadiene afforded an adduct, which on methylation with dimethyl sulphate, and K_2CO_3 in acetone gave 5,8-dimethoxy-1,4-dihydronaphthalene (27) and base catalysed isomerisation of 27 resulted in the formation of 5,8-dimethoxy-3,4-dihydronaphthalene (28). Vilsmeir formylation of 28 gave the aldehyde (29) which on treatment with Grignard reagent afforded the (±)-2-(1-hydroxyethyl)-5,8-dimethoxy-3,4-dihydronaphthalene (26).

Kinetic resolution of (±) 26 was carried out at -50° in CH_2Cl_2 by using titanium tetraisopropoxide (TIP), (+)-diisopropyltartrate [L(+)DIPT] and t-butylhydroperoxide (TBHP) in 1:1:0.6 molar ratio. After the completion of the reaction, the mixture of two products (30 and (+)-26) was directly subjected to LAH reduction and chromatographic separation to give 38% yield of R-(+)-26; $[\alpha]_D$+20.3°) and R-(-)-[2-(S)-1-hydroxyethyl]-2-hydroxy-5,8-dimethoxy-1,2,3,4-tetrahydronaphthalene (31) (40% yield; $[\alpha]_D$-49.4°). The undesirable antipode (+)-26 was inverted (ref. 45), by reacting with triphenylphosphine, diethylazodicarboxylate and benzoic acid in THF to give the benzoate of (-)-26. Alkaline hydrolysis afforded (-) 26 which was subjected to Sharpless epoxidation followed by reduction to give (-)-31. Oxidation of (-) 31 with Fetizone reagent gave R-(-)-25, which on fusion with phthalic anhydride in the presence of an intimate mixture of $AlCl_3$-NaCl (5:1) at 180° and usual working gave (-)-4-demethoxy-7-deoxydaunomycinone (24). This was converted to (+)-4-demethoxydaunomycinone (9) by established methods.

Cava et al (ref. 46), reported the synthesis of R-(-)-4-demethoxy-7-deoxydaunomycinone (24) from (±)-4-demethoxy-7,9-dideoxydaunomycinone. The diacetate of 32a was subjected to Sharpless epoxidation to obtain the epoxy alcohol (-) 32b which was further converted to (-) 24 (Scheme 12).

Scheme – 12

Monneret et al (ref. 47), synthesised optically active 4-demethoxyanthracyclinone (+) 9 in several steps from lactose as a chiral precursor of ring A and leucoquinizarin as that of BCD rings. Thus lactose was converted by a series of reactions to the aldehyde (33) which on condensation with leucoquinizarin in presence of piperidine acetate followed by acid treatment gave 34. This was further elaborated to (+) 9 (Scheme 13). An enantio-controlled synthesis of (+) 9 (Scheme 14) was reported

Scheme -13

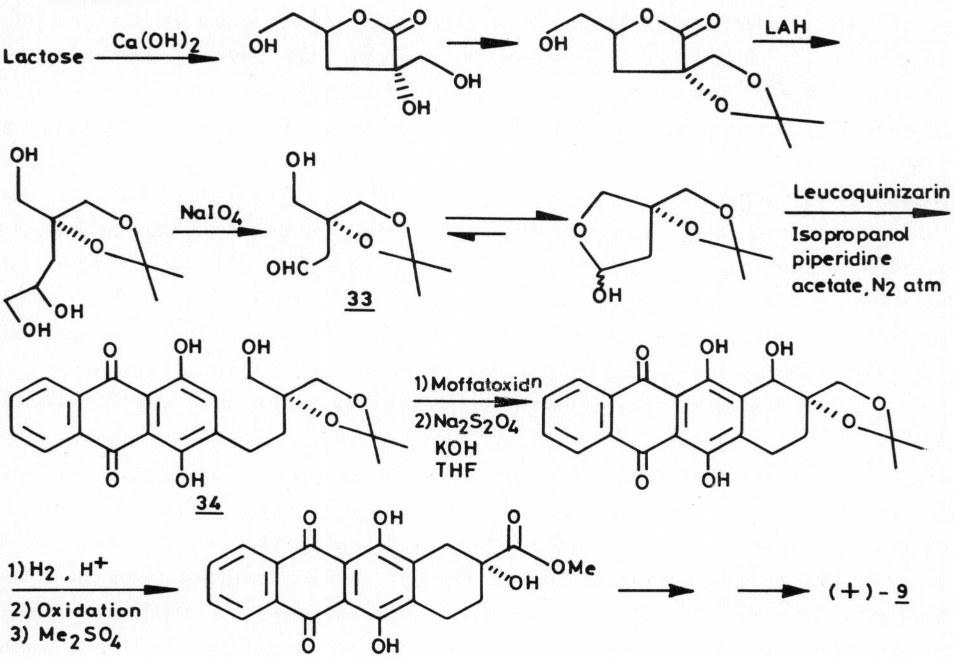

Scheme -14

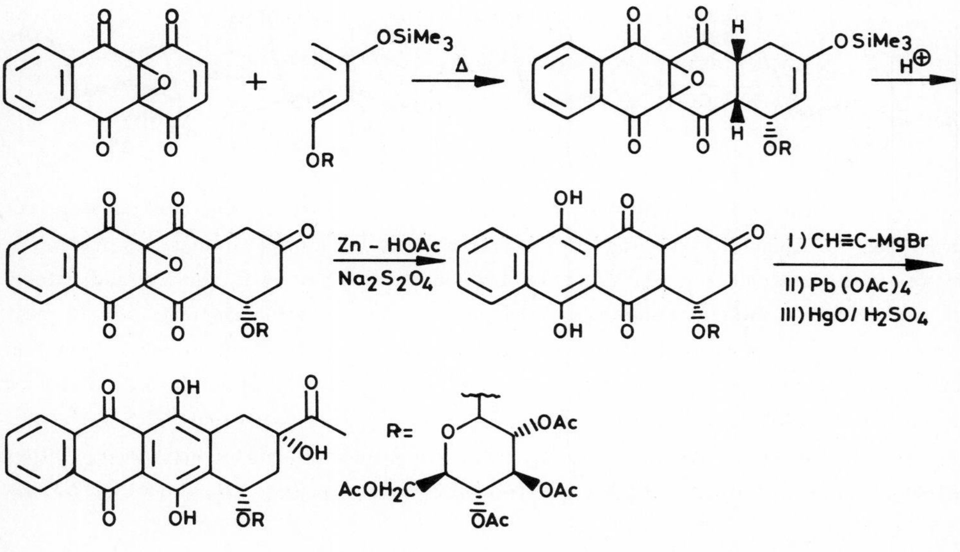

510

by Stoodley et al (ref. 48), wherein Diels-Alder reaction of (E)-3-trimethylsilyloxybuta-1,3-dienyl 2,3,4,6-tetra-O-acetyl-α -D-glucopyranoside with 4a,9a-epoxy-4a,9a-dihydro-anthracene-1,4,9,10-tetrone resulted in a cycloadduct which was converted to (+) **9** in six steps.

Much effort has gone in for the synthesis of optically active anthracycli-nones (ref. 49). Among all the approaches, Sharpless epoxidation method is by far the most superior approach for the synthesis of optically active anthracyclinones. Sharpless epoxidation was also exploited by Shibasaki et al (ref. 50), for the synthesis of R-(-)-**25**.

2.1.3   L-Daunosamine

L-Daunosamine, the sugar component present in almost all the anthracycli-ne antibiotics, is systematically known as 3-amino-2,3,6-trideoxy-L-lyxo-hexo-pyranose (**35**). Among the trideoxyaminohexoses, the isomeric L-daunosamine by far is the most celebrated sugar. This is testified by numerous examples of its synthesis both from carbohydrate and non-carbohydrate precursors. Two review articles (refs. 51, 52), on **35** have already appeared and therefore, in this chapter only a brief overview is given.

The first total synthesis of **35** from L-rhamnose by Goodman et al (ref. 53), appeared in 1965. The derived L-rhamnal (**36**) was subjected to methoxy-mercura-tion-demercuration reaction to afford the 2,6-dideoxyproduct whose 3,4-dihydroxy function was converted into the epoxide (**37**) via the corresponding 3-tosylate. The nucleophilic epoxide opening with azide ion occurred at position 3 to afford the azido product (**38**). The subsequent epimerisation at C-4 was effected by OMs - OBz trans-formation followed by catalytic hydrogenation in the presence of an acid giving rise to methyl glycoside of L-daunosamine hydrochloride (**35**) (Scheme 15).

Scheme –15

In principle, naturally occurring L-rhamnose is by far the most appropriate precursor to L-daunosamine, as evident from their structural similarities (ref. 54). However, because of the high cost of L-rhamnose, its use was limited and efforts to use cheaply available D-hexoses received considerable attention (ref. 55). It must be pointed out that in achieving L-stereochemistry to be acquired in the preparation of L-daunosamine from D-hexose, the inversion of configuration at C-5 becomes one of the crucial steps in the synthesis.

Horton and Weckerle (ref. 56), employed D-mannose in an impressive synthe-sis of **35** wherein the modified Klemer-Rhodomeyer reaction (ref. 57), of methyl 2,3;4,6-

di-O-benzylidene-α-D-mannopyranoside (39) afforded the key intermediate (40). The derived oxime (41) was reduced with lithium aluminium hydride and acetylated to give a mixture of ribo/arabino amines (88:12). The major isomer (42) was treated with NBS to cleave 1,3-dioxane ring in accordance with Hanessian reaction (ref. 58). The resulting 6-bromide (43) was dehydrobrominated with AgF-pyridine to give the 5,6-ene (44). Removal of the benzoyl group by hydrolysis and catalytic hydrogenation gave exclusively the lyxo-isomer (45). The regiospecific reduction of 44 occurred due to the presence of α-glycosidic linkage which directs the delivery of hydrogen from above the ring and also the ring flips from $^4C_1$ to $^1C_4$ conformation. Hydrolysis of N-acetyl group with Ba(OH)$_2$ and methyl glycoside with HCl gave 35 as a hydrochloride (Scheme 16).

Scheme –16

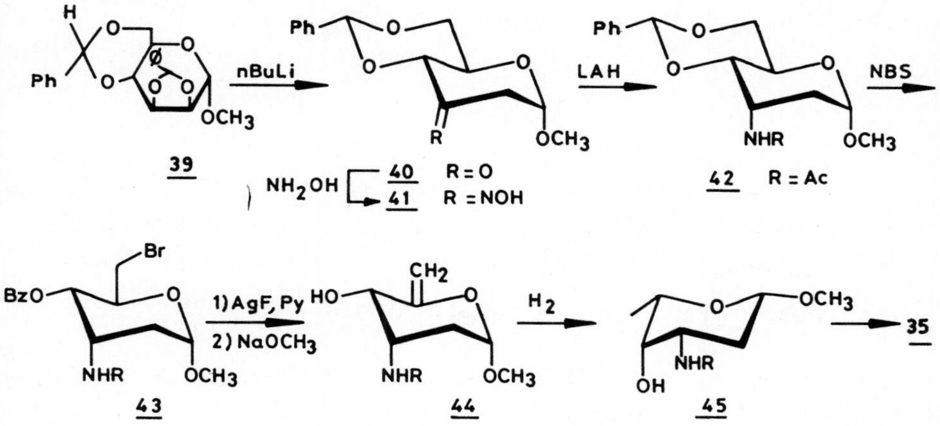

The key intermediate (40) has also been synthesised by LAH reduction (ref. 59), of the corresponding allo-epoxide (46) or by Bu$_3$SnCl-NaBH$_4$ reduction of the known 2-iodoaltropyranoside (47) followed by oxidation (ref. 60), with PDC (Scheme 17).

Scheme – 17

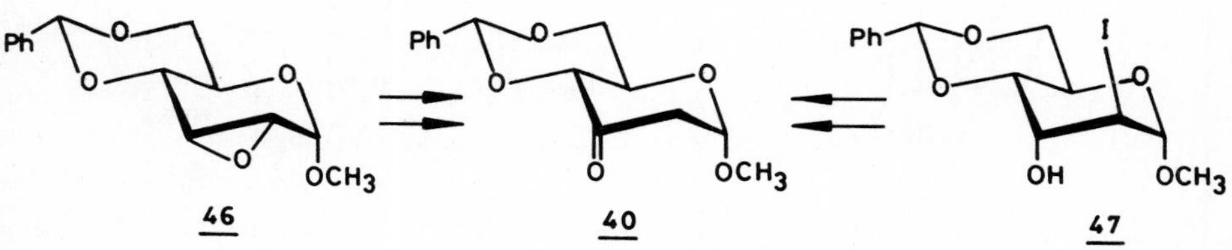

Arcamone (ref. 61), and Terashima (ref. 62), and their coworkers reported the utility of 40 in the synthesis of 35. Both the approaches are sequentially identical to the route by Horton (ref. 56), however, the reagents employed have certain specific selectivities. For example, the stereoselective reduction of the oxime (41) with either aluminium amalgam or vitride furnished the required ribo-isomer (42) as an exclusive product. In addition, the dehydrobromination of 43 (R=COCF$_3$) was effected with DBU which is reasonably inexpensive as compared to AgF used by Horton.

Rama Rao et al (ref. 12), accomplished the synthesis of L-daunosamine from cheaply available 2-amino-2-deoxy-D-glucose hydrochloride. The salient features of the synthesis (Scheme 18) involve the transformation of amino function of 2-amino-2-

Scheme -18

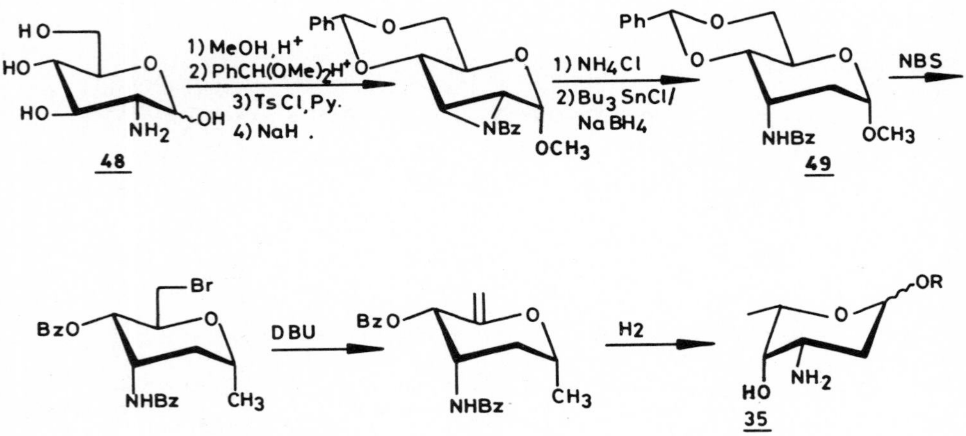

deoxy-glucose (48) from C-2 to C-3 via 2,3-epimine in order to yield the 3-ribo isomer (49) with right stereochemistry. In order to introduce methyl group at C-5 with L-configuration, Hanessian reaction, dehydrobromination and reduction were performed to afford 35 (as a tribenzoate). This particular approach came handy to prepare (refs. 63-67), the 2-fluoro-L-daunosaminide (50) (Scheme 19). There is considerable amount

Scheme -19

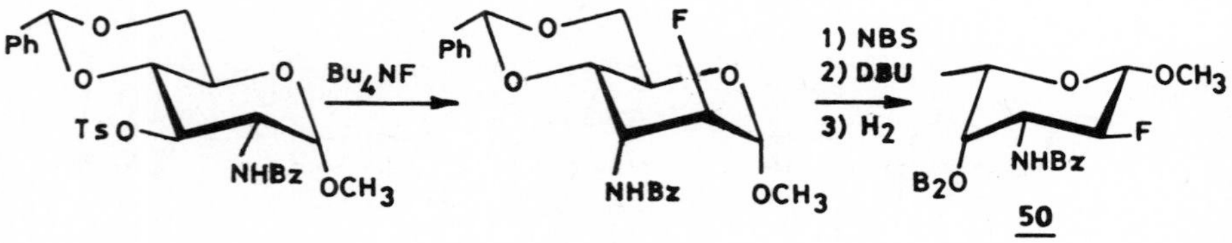

of work being done to introduce halogen substituent (ref. 68), at position 2. The inductive effect of halogen at C-2 indirectly retards the hydrolysis of the glycosidic bond in vivo and allows the antibiotic to remain at the target site for longer period.

Ironically, in all the approaches of the synthesis of 35, carbon atoms 1/6 of the starting material are transformed into carbon atoms 1/6 of L-daunosamine.

It can, however, be pointed out that if carbon atoms 1/6 of D-glucose are transformed into carbon atoms 6/1 of **35** then the crucial step of epimerisation at C-5 can be curtailed. This approach was adopted by Gurjar and Pawar (ref. 69). The 1,2-O-iso-propylidene derivative (**51**) was reduced via the corresponding chloride to give **52** which could then be converted to diethylmercaptan (**53**) by conventional methods. Further transformation involves reduction, desulfurisation and nucleophilic displacement reactions to give **54** from which **35** can be synthesised (ref. 70), (Scheme 20).

Scheme –20

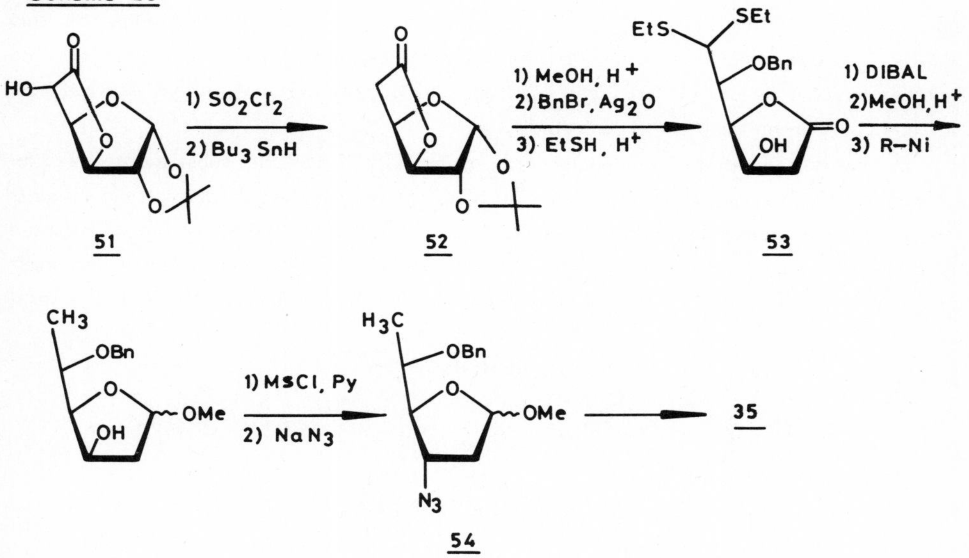

2.1.4   Glycosylation of anthracyclinones

The coupling reaction between aglycone and L-daunosamine is one of the crucial aspects in the synthesis of anthracycline antibiotics. By and large, the glycosidic bond is formed by treating anthracyclinone with a glycosyl halide (**55**) in the presence of a catalyst (refs. 71-74). The reaction normally affords the α-glycoside as a major product if not as an exclusive product. Several groups utilised this approach to separate the diastereomeric mixture of anthracyclines. Because of the difficulty

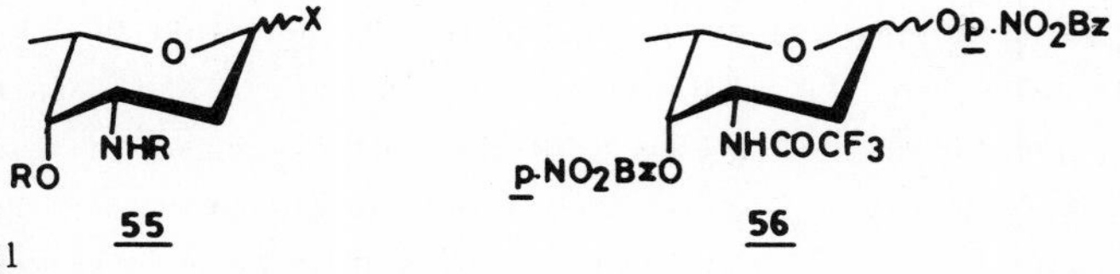

Scheme 21

in the preparation of the glycosyl halide of L-daunosamine coupled with its unstability towards storage, the corresponding 1-acyl derivatives found considerable utility (refs. 75,76). The latter reaction is normally carried out in presence of a Lewis acid. Tera-

514

shima et al (ref. 77), demonstrated that trimethylsilyl trifluorosulfonate (TMS triflate) was by far the most superior catalyst and gave high yield of the glycoside with 98% anomeric purity. For instance, 4-demethoxydaunomycin (6) was synthesised in good yield by the glycosilation of (+)-demethoxydaunomycinone (9) with N-trifluoroacetyl-1,4-di-O-p-nitrobenzoyl-L-daunosamine (56) in the presence of TMS triflate.

No single area attracted the attention of chemists and biologists all over the world as that of anthracyclines during 70s and the early half of 80s. Several thousands of scientific publications have appeared and quantum wise, probably it may occupy the third largest after β-lactam antibiotics and prostaglandins. In this chapter, various aspects relating to their synthhesis, especially daunomycin (1) and 4-demethoxydaunomycin (6) are dealt with special attention on regio- and stereospecific selectivity in building these molecules

Various synthetic analogues are made with different substituents on the aglycone and amino sugar moiety. Some of them showed promise in animal experiments and a few are undergoing clinical trials. Taking the lead, several simple substituted anthraquinones have been made and evaluated for the anti-tumour activity. Amongst these, mitoxanthrone (57) (ref. 78), has been introduced for therapeutic use in several

57

countries as an antineoplastic agent. This drug, being simple in its structure and easy to manufacture, is regarded as "Poor man's adriamycin".

2.2   Sesbanimide A

In recent years, two other natural products, fredericamycin A (58), an anti-tumour antibiotic isolated from Streptomyces griseus (refs. 79-82), and sesbanimide A (59), a plant product isolated from the seed of Sesbania drummondii (Rydb) Cory (ref. 83), have shown pronounced activity in experimental animals. Both these compounds, having novel structural features, attracted the attention of many synthetic organic chemists. However, this chapter is confined only to the various strategies that were developed by different groups for the total synthesis of sesbanimide A.

The isolation and structural elucidation of sesbanimide A is a result of joint efforts by different groups (ref. 84). Only a few milligrams of pure sesbanimide A could be obtained from thousand pounds of Sesbania drummandii seeds, a poisonous plant also known as coffee bean and rattle box. The structure and relative stereochemistry of sesbanimide A (59) was established by X-ray crystallographic and NMR spectral analysis, as having a tricyclic structure in which the three rings are linked by single bonds. Its absolute configuration, however, was not derived. Two more struc-

Scheme 22

59	$R_1 = CH_3$, $R_2 = H$, $X = \beta - OH$
61	$R_1 = H$, $R_2 = CH_3$, $X = \beta - OH$
62	$R_1 = H$, $R_2 = CH_3$, $X = \measuredangle - OH$

turally related active principles were later isolated from the same source and identified as sesbanimide B1 (**61**) and sesbanimide B2 (**62**) (ref. 85). Because of its unique structural features coupled with its remarkable anti-tumour activity, compound **59** distinguishes itself as a very interesting target for its synthesis. For convenience, the three rings in **59** have been designated as rings A, B and C as represented in the structure **59**.

In focussing the attention on sesbanimide A (**59**) or its antipode (**60**), the commonly employed retro-synthetic analysis divides the molecule into three segments viz. glutarimide ring (**63**), heavily substituted and chemically differentiated 1,3-dioxane-triol derivative (**64**), and the tetra hydrofuran unit (**65**), which respectively

63 **64** **65** **Scheme** 23

correspond to rings A, B and C of sesbanimide A. Appropriately, various approaches were developed for the construction of the individual rings of **59** which are briefly described below.

(i) <u>Construction of ring A.</u> Since early 1983, attention of various groups in the field has been directed towards the synthesis of A and B rings of **59** (refs. 86-91). The most logical approach for the synthesis of glutarimide ring (ring A) would be to prepare glutaric acid or the diester and react it with a suitable amine. The corresponding glutaric acid unit is generally prepared by a conjugate addition of a metal salt of dimethylmalonate (ref. 92), or Meldrum acid (ref. 93), to an α,β-unsaturated ester to obtain the triester derivative (66) which, after decarboxylation, and thermal reaction with ammonia, urea or benzylamine is converted into glutaramide derivative (63) as exemplified in Scheme 24. Alternatively, ring A is also prepared

Scheme 24

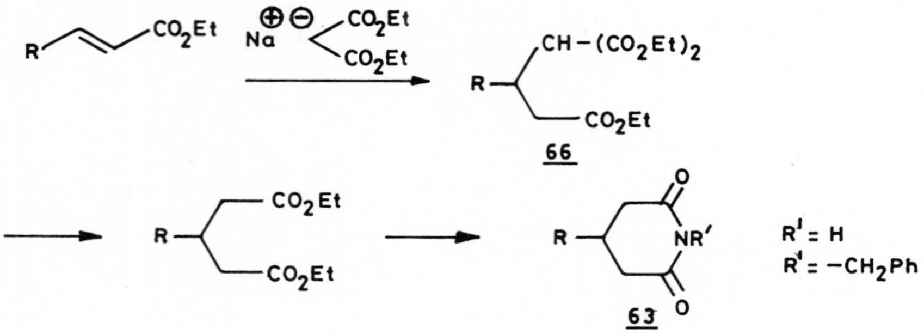

by 1,4-conjugate addition of a derived free radical of iodoacetamide or cyanotrimethyl-silyl methyl lithium to α,β-unsaturated ester (ref. 94).

(ii) <u>Construction of ring B.</u> Examination of the structure of sesbanimide A indicated that a compound of type **64**, having actual or potential aldehyde groups at both ends of the molecule could be an ideal synthon for ring B as it would provide room to manoeuver at both ends of the molecule. The utility of the masked aldehyde groups can be exploited to build ring A or C at any of the two available positions. The fragment **64** can be derived from easily accessible carbohydrate precursors, as the elements of chirality at C-7, C-8 and C-9 of **59** could be correlated with C-2, C-3 and C-4 carbon atoms of several commonly available sugars (Scheme 25). D-glu-

Scheme 25

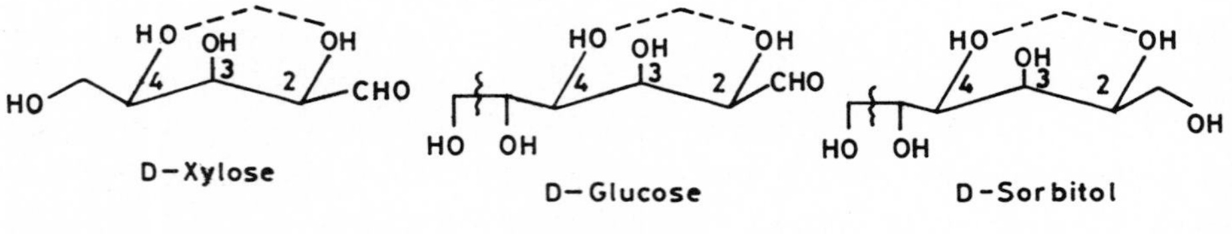

cose, D-sorbitol and D-xylose, have been utilized for the construction of B-ring of sesbanimide A (refs. 86-91).

Scheme 25, no doubt, suggested that the choice of making D-xylose as a precursor for building ring-B would be most appropriate as other sugars such as D-glucose or sorbital having an extra carbon atom would be required to undergo chopping at appropriate stage of the sequence. D-xylose was indeed employed, first by Terashima (ref. 89), and later by Pandit (ref. 88), and Rama Rao (ref. 91). The general approach adopted by Rama Rao et al (ref. 91), for the construction of AB rings of Sesbanimide A is given in Scheme 26.

Scheme 26

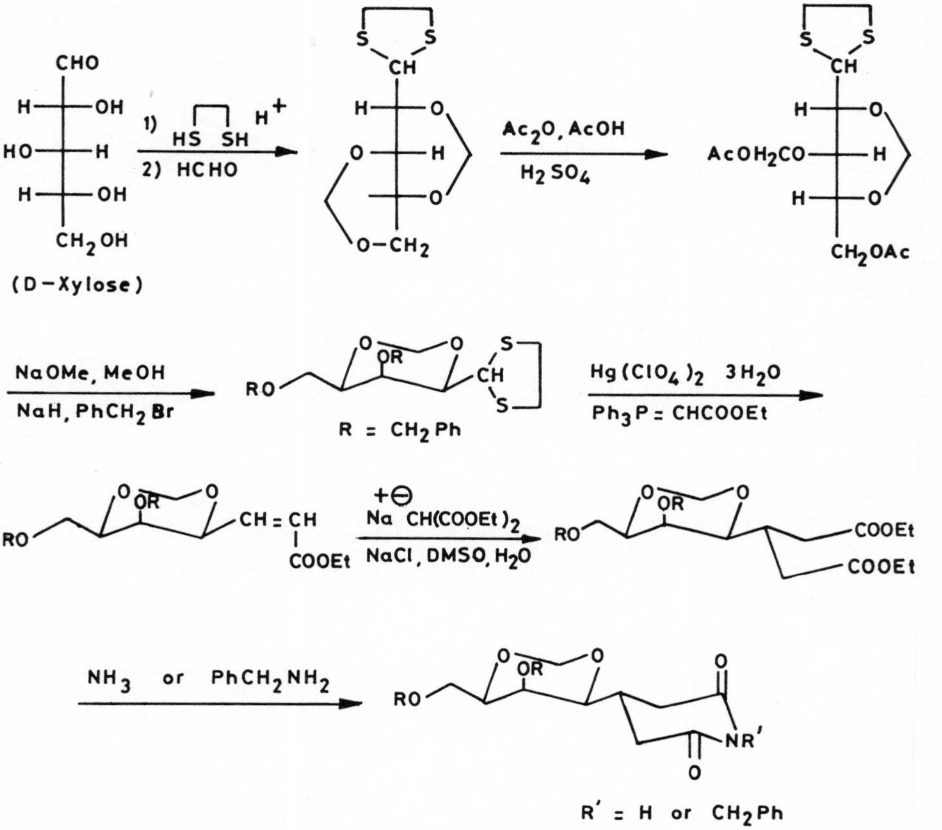

(iii) Construction of ring C. Rama Rao et al (ref. 95), were the first to devise a simple methodology for the construction of C-ring of Sesbanimide A. Their approach is depicted in Scheme 27. Condensation of the bromide (67) with 2-(2-methyl-2-dioxolanyl)-ethanal (68) was initially attempted in the presence of magnesium or lithium metal but the overall yield of the desired product was very poor, the main product being the dimer arising from the bromide. However, this problem was circumvented by adopting Zaitzev reaction in which the bromide (67) was reacted with the aldehyde (68) in the presence of activated zinc and iodine as a catalyst to give 69.

518

Although this reaction was not stereoselective as judged by NMR spectrum, a high degree of regioselectivity was undoubtedly observed. Since the chirality of the carbon bearing the hydroxyl group was destroyed at a later stage, the non-stereoselectivity of the Zaitzev reaction was not of any consequence and would not effect the overall synthetic plan.

Scheme 27

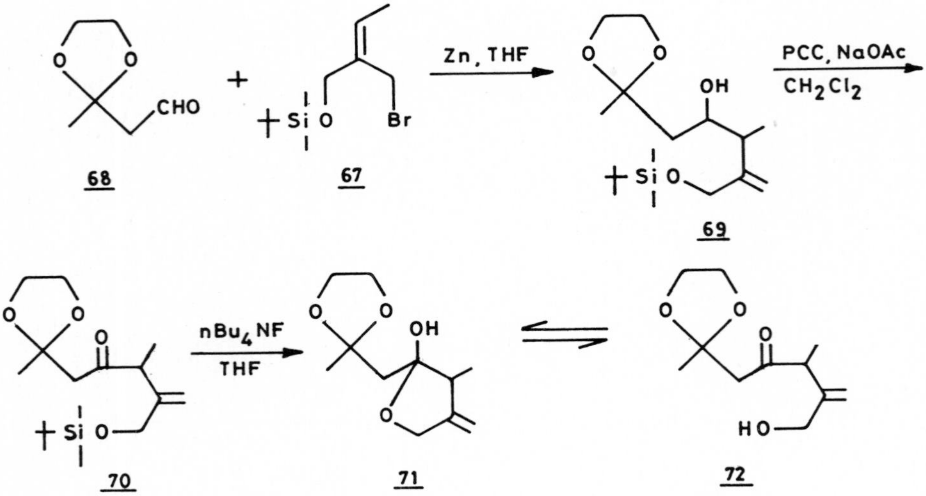

Oxidation of **69** with PCC-NaOAc in CH_2Cl_2 gave the ketone (**70**). Deprotection of the silyl group of the primary hydroxyl group resulted in formation of the desired hemiketal (**71**). The PMR spectrum of **71** showed the existence of the product in both forms, closed (**71**) and open (**72**) in the ratio of 4:1. A similar approach for the construction of ring C was adopted by Pandit (ref. 96), Terashima (ref. 97), and Schlessinger (ref. 98), for the total synthesis of Sesbanimide A.

2.2.1   Synthesis of Sesbanimide A

Pandit and his colleagues (ref. 96), were the first to report the total synthesis of the antipode [(7R, 9R, 10S, 11S) - **59**] of natural Sesbanimide A, therefore, the absolute stereochemistry of the natural alkaloid as (7S, 8R, 9S, 10R, 11R) - **60** was eventually established.

Their strategy was to build ring B from the appropriate carbohydrate unit and to construct the remaining rings in the sequence B→AB→ABC (Scheme 28). Initiating the synthesis with the aldehyde **73** (obtained from sorbitol), they constructed the glutaramide ring (A) in three steps. The biacetal product (**74**) was first converted to the diol (**75**) by heating with acetic anhydride - acetic acid - sulphuric acid mixture, followed by deacetylation. **75** was converted into 3,4-dimethoxybenzylidene derivative (**76**) which was then selectively opened under reductive conditions (Et$_2$AlCl - CH$_2$Cl$_2$. Et$_3$SiH -55°) to the primary alcohol (**77**). It was then oxidized to give the aldehyde (**78**). Coupling of aldehyde (**78**) with the allylsilane (**79**) in presence of BF$_3$-

Scheme - 28

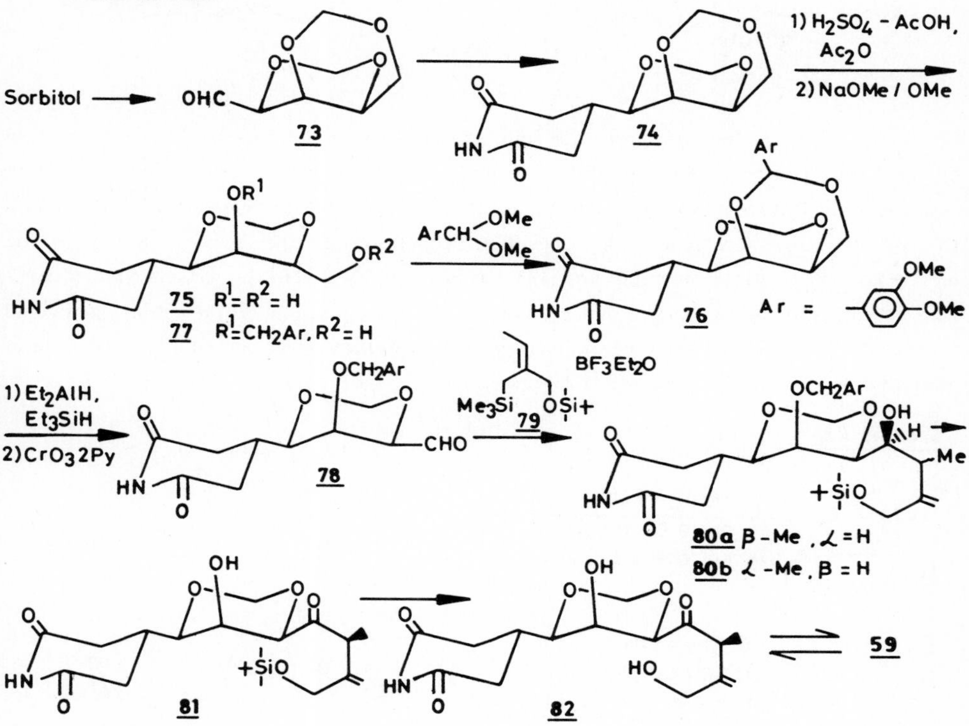

etherate led to the formation of two diastereomers (**80a, 80b**) (1.7:1) which were separated chromatographically and the major alcohol (**80a**) was oxidized to the ketone (**81**). The protective groups of the two hydroxyls were removed by sequential reactions with DDQ/CH$_2$Cl$_2$ and AcOH/THF/H$_2$O to give the hydroxy ketone (**82**) which forms crystalline hemiacetal (**59**). The NMR spectrum of **59** is identical to that of natural sesbanimide A, but the optical rotation of **59** had a value of -52° exactly opposite to that observed for natural sesbanimide A. These results led to the conclusion that **59** is an antipode of natural sesbanimide, whose absolute structure should be repre- sented as given in structure **60**.

Later, Pandit and his colleagues (ref. 96), accomplished the synthesis of natural sesbanimide A (**60**) from D-xylose. Thus D-xylose was converted to compound **83** (Scheme 28a). Since **83** is the antipode of diol (**75**), the same sequence of chemical transformations earlier adopted by them for the synthesis of **59** led to the final synthe- sis of Sesbanimide A (**60**) which was identical in all respects to the natural product.

Subsequently, three other groups (refs. 97-99), had also completed the synthesis of Sesbanimide A (60). Matsuda and Terashima (ref. 97), accomplished its synthesis

Scheme 28a

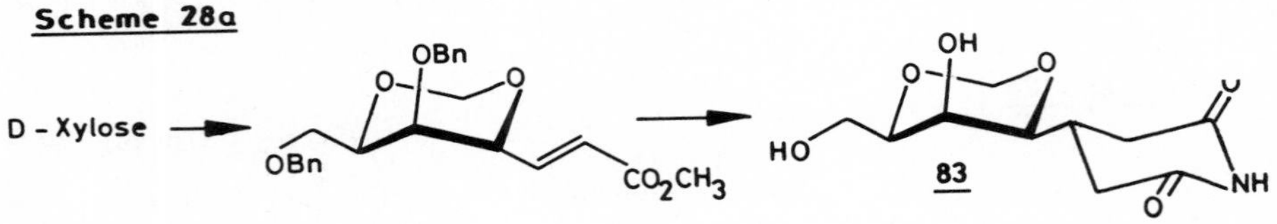

by first obtaining the (-)-glutarimide diol (83), corresponding to the AB ring system of 60, from D-xylose. 83 was then converted to the aldehyde (84), which was then subjected to Reformatsky reaction by employing (2)-ethyl 2-(bromoethyl)-2-butanoate to give the _exo_-methylene γ-lactone (85). Reduction of the lactone (85) gave the diol (85a) which was converted into 1:1 mixture of 60 and 61 by a) selective protection of the primary hydroxy group of 87a, b) Collins oxidation of the resulting alcohol (85b) and c) removal of the two silyl groups. The mixture of 60 and 61 gave the natural Sesbanimide A and B respectively (Scheme 29).

Scheme 29

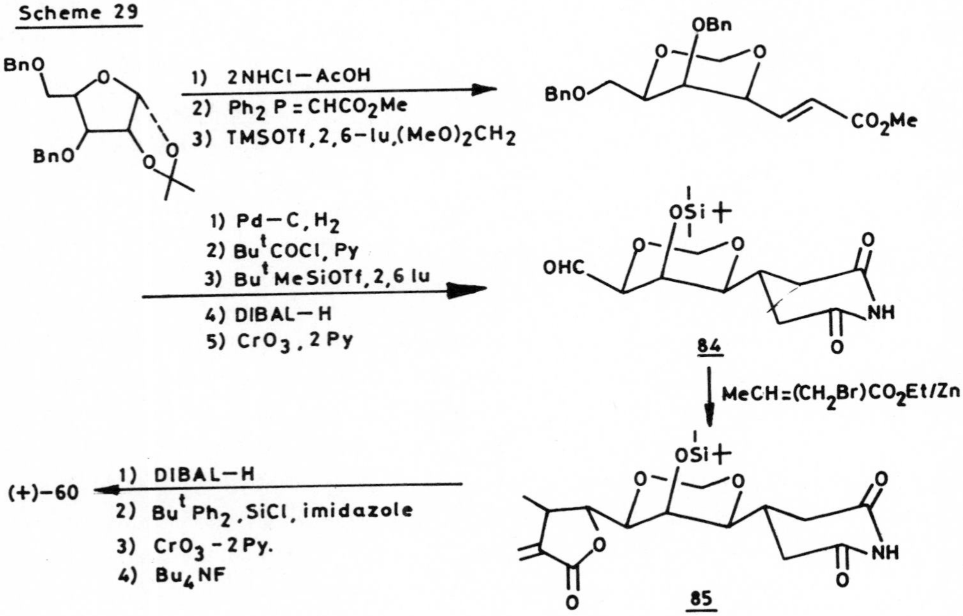

Rama Rao _et al_ (ref.99-100) have recently completed the synthesis of sesbanimide A, by making the AB synthon (83) starting from D-glucose (ref. 101), or D-xylose (ref. 96). The diol (83) was then converted to the aldehyde (78) (ref. 96). Ring C of 60 was built on 78a as reported by them (ref. 95), to give 81a, which was then transformed to sesbanimide A (60), by the removal of the two protective groups in the former compound (81a) (Scheme 30).

Scheme-30

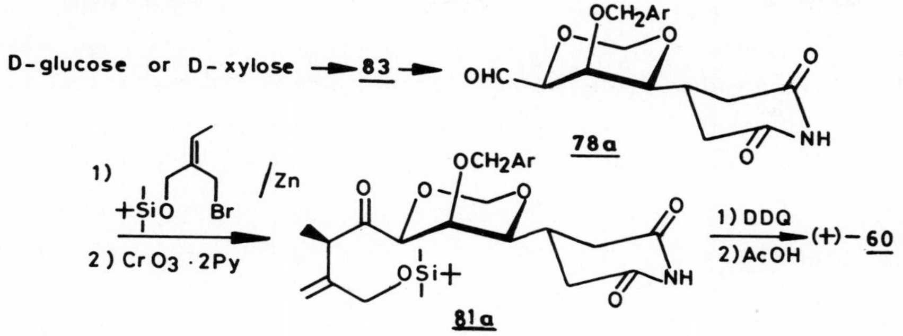

Synthetic studies on sesbanimide A elegantly demonstrated how the existing asymmetric centres of sugars can be employed for the synthesis of enantiomerically pure compounds.

3   Studies directed towards the synthesis of rifamycin-S

3.1   Rifamycin-S

Rifamycin-S (86) (refs. 101-2), belonging to the novel ansamycin family of antibiotics has been a target compound for many synthetic organic chemists. The first and still the only total synthesis of this antibiotic was reported by Kishi and coworkers (ref. 103). In their approach, rifamycin-S was disconnected at the carbon-hetero atom bonds to give an aromatic moiety (87) and the ansa chain (88) having eight contiguous asymmetric centres had been the focus of considerable interest.

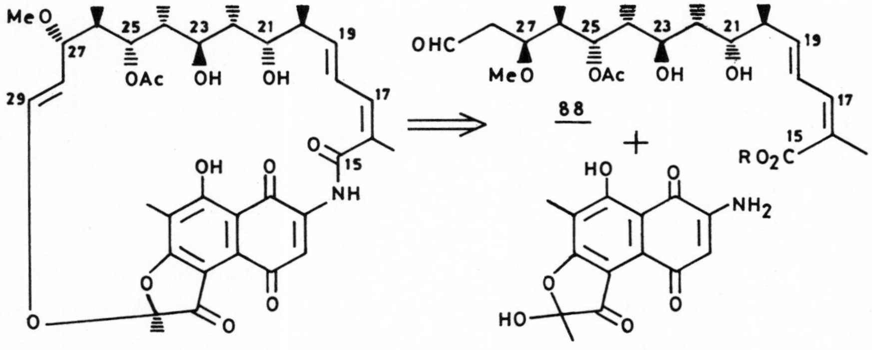

The acyclic stereocontrol approaches involving chiral sequences were described first by Kishi (ref. 104), and later by Masamune (ref. 105), Still (ref. 106), and Corey (ref. 107), and their coworkers while Hanessian (ref. 108), Kinoshita (ref. 109) and Fraser-Reid (ref. 110), and their coworkers reported chiron approach using carbohydrate precursors. Their work was extensively covered in a recent review by Paterson and Mansuri (ref. 111).

3.2 Synthesis of ansa bridge

In all the syntheses reported todate, acyclic stereocontrol is the method of choice. Rama Rao et al (ref. 112), report the synthesis of ansa chain, which is different from the earlier reports, in that it employs a cyclic building block for achieving an efficient stereocontrol functionalization of the ansa chain. Retrosynthetic analysis of the ansa chain (88) by Rama Rao et al revealed three fragments A, B and C (Scheme 31).

Scheme 31

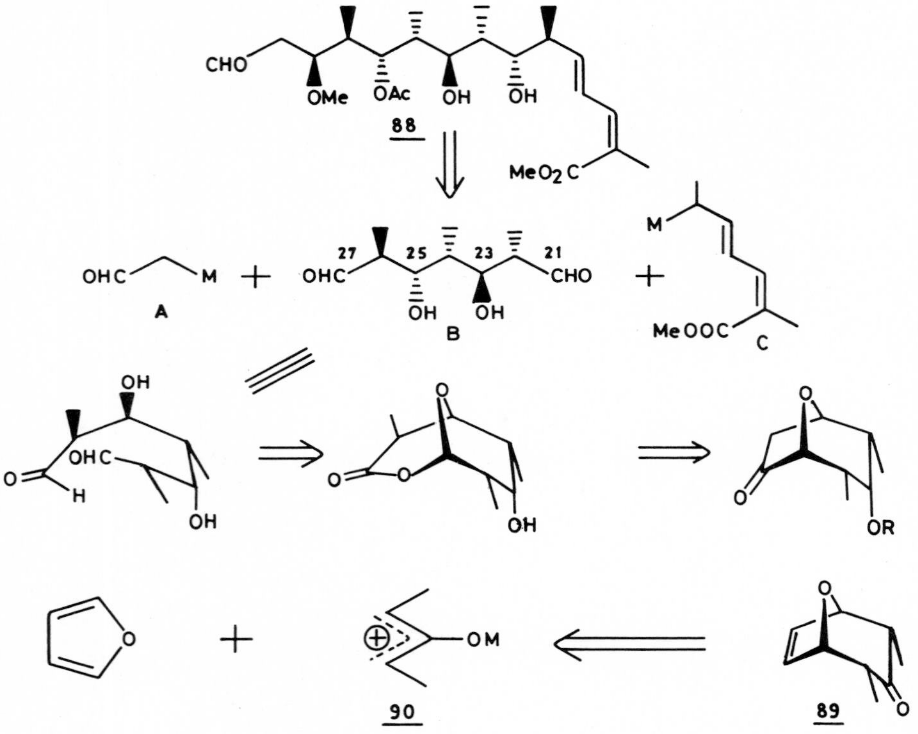

The fragment B constitute the C-21 to C-27 carbon chain, with terminal aldehydic functionalities which could be easily elaborated by an aldol condensation with suitable intermediates (A and C). Therefore, the synthesis of the ansa chain centers around the formation of the key intermediate B, with all the asymmetric centres having required stereochemistry. This fragment B can be derived from the bicyclic intermediate (89) (Scheme 31) which may be prepared from [3+4] cycloaddition of furan and the oxyallyl cation (90) derived from 2,4-dibromo-3-pentanone. The choice of making use of a bicyclic molecule is two fold. In the first place, the inherent rigidity of bicyclic intermediate facilities functionalization in a stereocontrol fashion and secondly the characterization of the intermediate is rather easy and can be done by simple spectroscopic techniques.

The bicyclic ketone (89) was obtained along with other isomers (20%) by employing [3+4] cycloaddition of oxyallyl cation (90) with furan by the method of Hoffman and his coworkers (ref. 113) (Scheme 32). The separation of isomeric ketones proved to be futile and the total cycloadduct (89 + isomers) was reduced with diisobutylaluminium hydride which provided a mixture of alcohols. The required alcohol (91) was found to be least polar and obtained by column chromatography over silica gel (50% overall yield).

Scheme 32

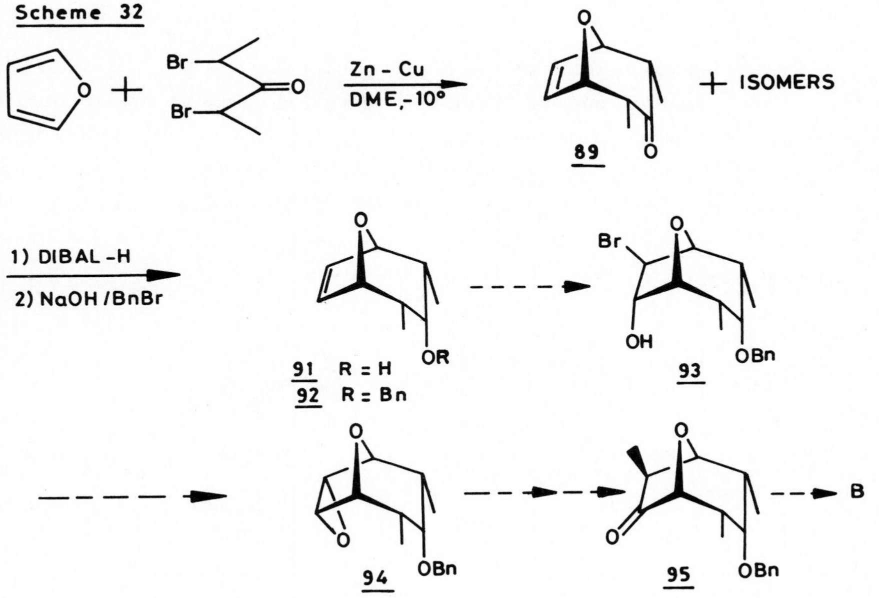

With the preparation of synthetic intermediate 92 with three carbons stereo-selectively functionalised to serve as the C-22 to C-24 carbons of the ansa chain, the next task was to introduce an exo methyl group at C-6 to serve as the C-26 carbon, with a convenient functionality at C-7 to serve as the 'key' to cleave the ring to obtain the acyclic moiety. For the introduction of the exo methyl group at C-6, it was envisaged that the most logical method would be to construct an endo epoxide (94) at C-6, C-7 followed by opening with dimethyl cuprate to afford the required exo methyl product (95). Since it is known that the direct epoxidation of olefinic bicyclic compounds invariably ends up with an exo epoxide, an indirect method for endo epoxide (94) formation was employed. The reaction of hypobromous acid with a bicyclic compound would lead to the addition of the electrophilic bromonium ion from the exo face, to give rise to the bromohydrin (93). Base treatment of the bromohydrin (**93**) should lead to the formation of the endo epoxide (**94**). Thus, when compound (**92**) was treated with N-bromosuccinimide in presence of calcium carbonate under aqueous conditions, it resulted in a colourless solid m.p. 60° which was found to be a cage molecule (**96**), instead of the bromohydrin (**93**). The structure of **96** was confirmed by chemical as well spectroscopic methods.

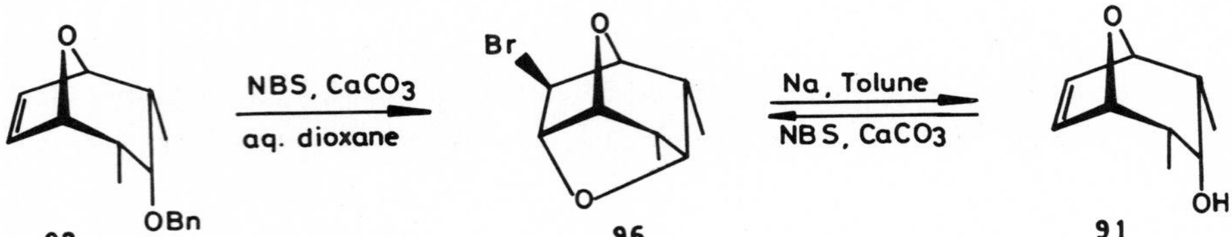

In another approach (Scheme 33), the methyl group at C-6 was introduced by an enolate alkylation reaction on ketone (**97**) which was prepared by the stepwise oxidation of **92**. The alkylation on the ketone (**97**) was carried out with methyl iodide in presence of lithium diisopropyl amide (LDA) at -78° giving rise to a single product which was confirmed as exo 6-methyl derivative (**98**) on the basis of PMR spectrum.

Scheme 33

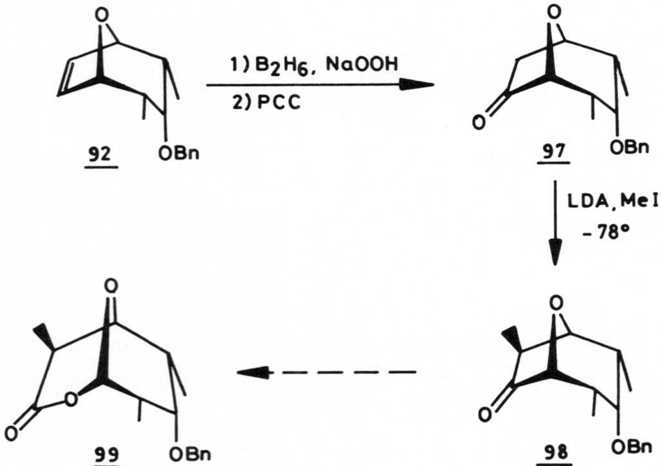

The most logical method to open the bicyclic derivative (**98**) would be to prepare a lactone (**99**) by Baeyer Villiger reaction. Accordingly, the C-methylated ketone (**98**) was treated with various peracids but this proved to be a futile.
With the failure to obtain the lactone (**99**) from the alkyl ketone (**98**), it was felt worthwhile to look into the Baeyer-Villiger reaction with the parent ketone (**97**) (Scheme 34). Then the derived lactone could be used to introduce the exo methyl at C-8 (lactone numbering). Thus, the ketone (**97**) was subjected to Baeyer-Villiger oxidation

reaction with trifluoroperoxyacetic acid in presence of disodium hydrogen phosphate as a buffer to give **100** as the sole product. Alkylation of the lactone (**100**) with LDA and methyl iodide at -78° furnished a solid (**101**) which was found to have C-8 methyl in <u>exo</u> position as evidenced by X-ray studies.

Scheme 34

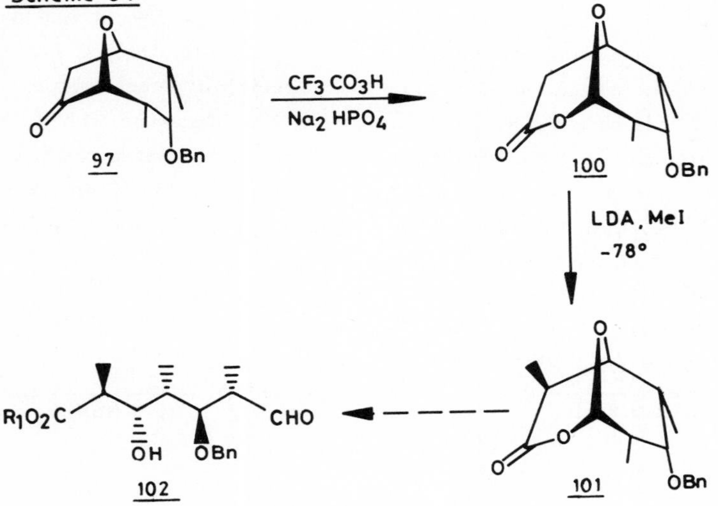

Having obtained the bicyclic derivative (**101**) with all the required functionalities, attention was directed to open the bicyclic ring to **102**. Initially, hydrolysis under variety of conditions with aqueous bases met with failure. Because of the fact that mild base and mild reaction conditions did not open the ring, while stronger bases and drastic conditions resulted in side products, it was felt that reduction of the lactone (**101**) would be the ideal choice. Accordingly (Scheme 35), the lactone

Scheme 35

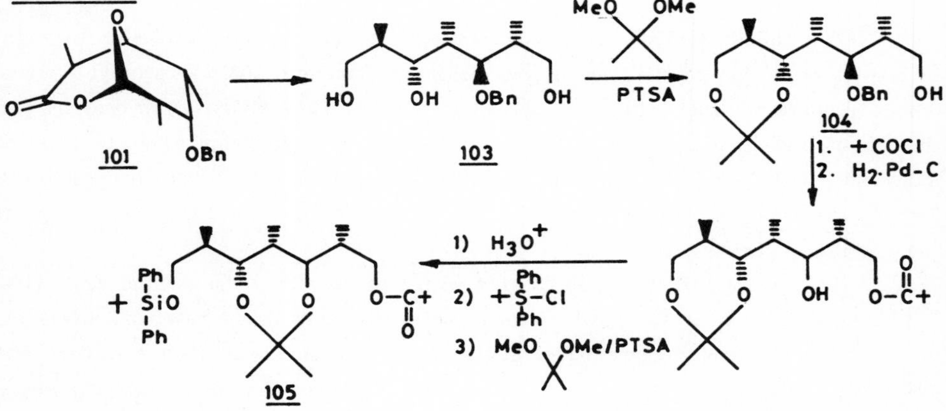

(**101**) on treatment with LAH in ether at -10° gave the expected triol (**103**).

In order to differentiate between the two primary hydroxyls, the triol (**103**) was subjected to acetalation with dimethoxypropane and p-toluenesulfonic acid to give the monoacetonide (**104**) with a primary alcohol left out which acts as a handle for further elaboration. Incidentally, **104** was conveniently converted to the Kishi intermediate (**105**) (ref. 103), which was earlier elaborated to the ansa chain (**88**).

To complete the construction of the ansa chain from **104**, it was desirable to introduce the 5-substituted-2-methyl-2Z,4E-pentadienoic acid (**110**; C-15 to C-20) in the correct stereochemical arrangement. The stereoselective construction of **110** was accomplished independently by Corey (ref. 115), and Masamune (ref. 105), from cyclic intermediates and by Kishi (ref. 104), by making use of Wittig reaction. Rama Rao et al (ref. 116), developed a simple and stereoselective methodology for the construction of 5-substituted-2-methyl-2Z,4E-pentadienoic acid present in the ansa chain of rifamycin-S (Scheme 36).

Scheme 36

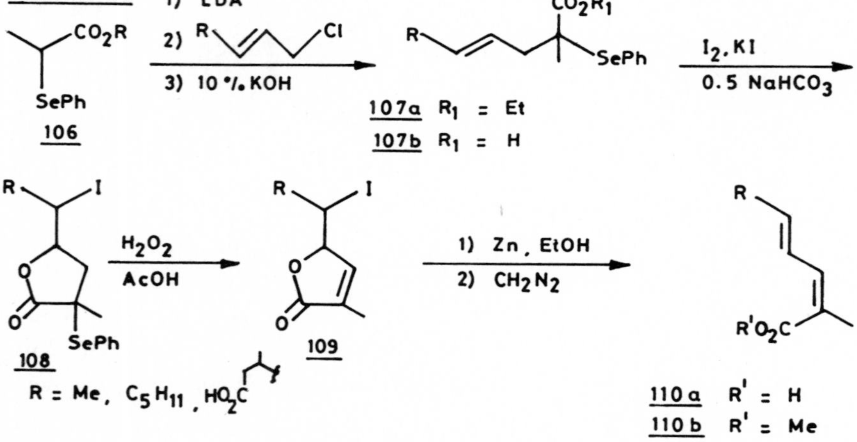

1) LDA
2) R ~ Cl
3) 10 % KOH
CO2R
SePh
106
CO2R1
SePh
107a R1 = Et
107b R1 = H
I2, KI
0.5 NaHCO3
H2O2
AcOH
108
R = Me, C5H11, HO2C
109
1) Zn, EtOH
2) CH2N2
R'O2C
110a R' = H
110b R' = Me

The salient feature for the construction of **110** involves exclusive formation of the cis olefin via deselenation of the cyclic intermediate (**108**). Thus ethyl 2-selenophenyl propionate (**106**) was first alkylated to give **107a** which on hydrolysis gave the corresponding acid (**107b**). This was subjected to iodolactonization to give the iodolactone (**108**). Treatment of **108** with hydrogen peroxide furnished the unsaturated lactone (**109**) as an exclusive product. The cis-double bond in **109** would constitute the Z olefin of the target molecule.

Compound (**109**) was then treated with zinc to afford the dienoic acid (**110a**). The E configuration of the newly formed olefin was established based on ample evidences available in literature (refs. 104-5, 115). The acid was then esterified with CH_2N_2 to give the methyl dienoate (**110b**). Further work for the total synthesis of ansa chain is being actively pursued.

Recently, three more reports appeared (refs. 117-9), on the synthesis of the ansa segment of rifamycin-S. Roush and Palkowitz (ref. 117), made use of tartrate ester of modified allylic boronate to achieve the stereoselective synthesis of the C-19 to C-29 ansa segment. Their synthesis (Scheme 37) commenced with the reaction of (S,S)-111 and chiral aldehyde, [(S)-113]. This transformation is a mismatched

Scheme-37

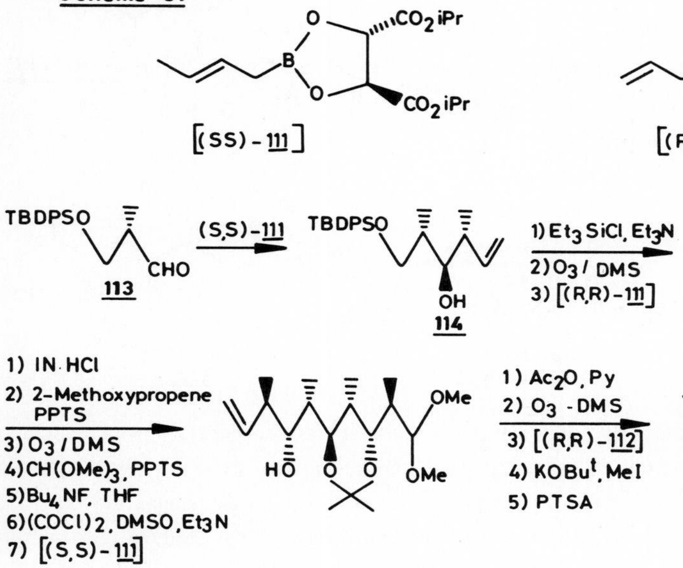

double asymmetric reaction (ref. 120), and the major component **114** serves as the substrate for the second crotyl boronate addition reaction to give a matched pair **115**. Conversion of **115** by the repeated sequence of reaction as depicted in Scheme 37 resulted in the synthesis of **116** which was earlier elaborated by Kishi and his colleagues for the synthesis of the ansa chain (ref. 104).

Danishefsky et al (ref. 118), took advantage of Lewis acid catalysed cyclocondensation reactions of activated dienes with α-alkylated aldehydes in the synthesis of the C-19 to C-29 polypropionate segment of rifamycin-S. Thus, cyclocondensation of the diene (**117**) with the aldehyde (**118**) led to **119**, which on reduction with LAH followed by treatment with methanol and PTS-acid provided **120** (Scheme 38). Hydroboration on **120** and subsequent oxidation followed by reduction resulted in the α-alcohol (**121**). The pyran (**121**) was then converted to the aldehyde (**122**). Cyclocondensation of **122** with the diene (**123**) in the presence of BF_3-etherate gave the pyrone (**124**), which was further elaborated by a series of reactions to the racemic **125**, earlier reported by Kishi et al (ref. 103).

Scheme 38

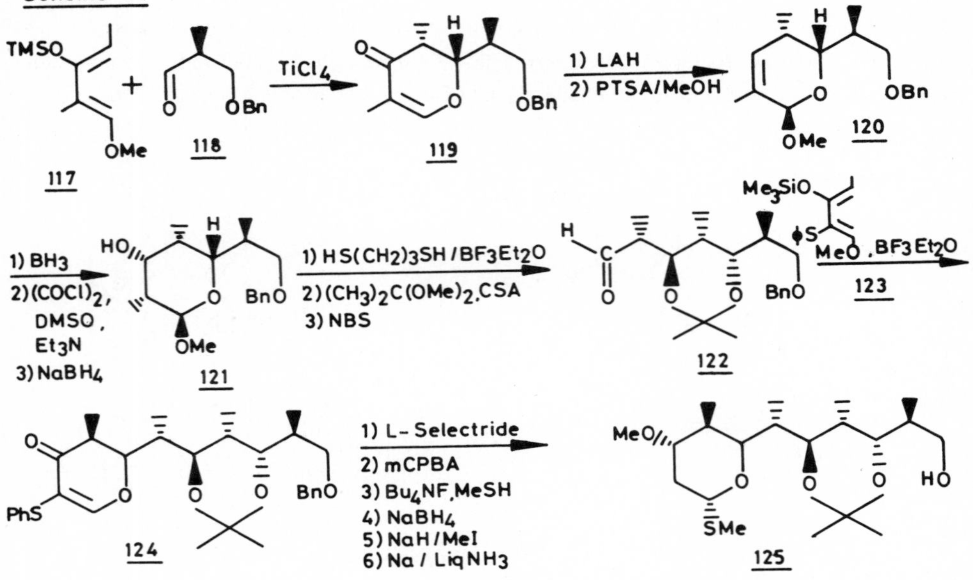

Ziegler et al (ref. 119), employed non-stereoselective reactions which include alkylation of 4-alkyl-3-methyl-γ-butyrolactone and reduction of 2-methyl-3-hydroxy ketone to prepare Kishi's intermediate corresponding to C 19 - C 27 fragment.

Synthetic studies carried out on rifamycin-S have thus contributed to the better understanding of various factors in controlling the stereochemistry of esta-blished reaction in particular the aldol condensation (ref. 120).

4   Synthesis of unsaturated hydroxy C-18 fatty acids

Polyunsaturated fatty acids play an important role in the biological systems either being involved directly in various physiological functions or as precursor mole-cules which are transformed into potent mediators with far ranging effects. In the recent years, fatty acids such as leukotrienes (eicosanoids) and unsaturated hydroxy C-18 fatty acids (octadecanoids) have been the subject of great interest owing to their wide range of biological activity. While leukotrienes are associated with phar-macological activities related to hypertensive reactions, such as, asthama and inflam-mation (ref. 121), these octadecanoids have an altogether different biological roles such as ionophoric activity and self-defensive properties in plants.

Coriolic acid (126) and dimorphecolic acid (127) were the first two octa-decanoids isolated from beef heart mitochondria by Blondin (ref. 122), and shown to possess unique divalent cation ionophoric activity. 126 and 127 are also potent inducers of mitochondrial swelling in the presence of Mg^{+2} ions and are capable of restoring the transport of Ca^{+2} ions even in the presence of ruthenium red, a calcium transport inhibitor. In addition, coriolic acid and dimorphecolic acid along with other octadecanoids (128, 129 and 130) have also been isolated from the resistant cultivar of rice plant <u>FUKUYUKI</u> (<u>Oryza</u> <u>sative</u> L) and shown to be active as self-defensive substances against rice blast disease (ref. 123). These acids seem to inhibit the spore

germination and germ tube growth of <u>Conidia</u> of rice blast fungus and thus playing a self-defensive role in the rice plant without being infected from the disease. Further, the susceptible variety of rice <u>SASANISHIKI</u> becomes resistant to the fungus when the rice plant is cultured for 10 days in aqueous solution containing these fatty acids.

Recently, Samuelsson et al (ref. 124), established the presence of coriolic acid and dimorphecolic acid in the sera of patients with Familial Mediterranean Fever (FMF), an autosomal recessive disorder. The presence of these fatty acids in the lipid extracts of sera of the patients with FMF suggests that the defects formation of/or elimination of these compounds might play a role in the pathogenesis of FMF.

4.1   Synthesis of coriolic acid

Rama Rao et al (ref. 125), were the first to describe the synthesis of (±)-coriolic acid (126). The strategy involved in their approach centers round (E)-pent-2-en-4-yn-1-ol (131) which turned out to be a convenient precursor for such type of molecules. The availability of two functional moieties in 131 allowed the extension of the side chain and at the same time the acetylenic group served as a precursor for introducing cis double bond (Scheme 39). Thus, alkylation of 131 with the bromide (132) followed by oxidation and Grignard reaction with n-pentylmagnesium bromide furnished compound **133**. The secondary hydroxyl group in **133** was protected, the derived primary hydroxyl group was stepwise oxidised and esterified to afford **134**, which on partial hydrogenation and hydrolysis furnished (±)-**126**.

Scheme-39

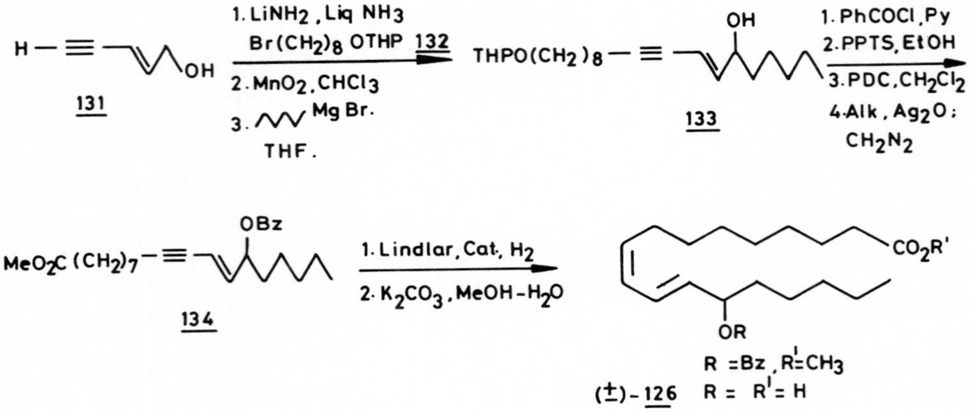

Although the above approach is the first report on the synthesis of coriolic acid (**126**), the desire to obtain multigram quantities of **126** by this approach seems to be unattainable because of the lengthy sequence of synthesis. The quest in this direction culminated into yet two other short and efficient approaches for (±)-**126** (ref. 126).

The first alternative approach was a straight forward sequence which cut short the number of steps (Scheme 40). Logically, alkylation of **131** with 8-bromo-

Scheme-40

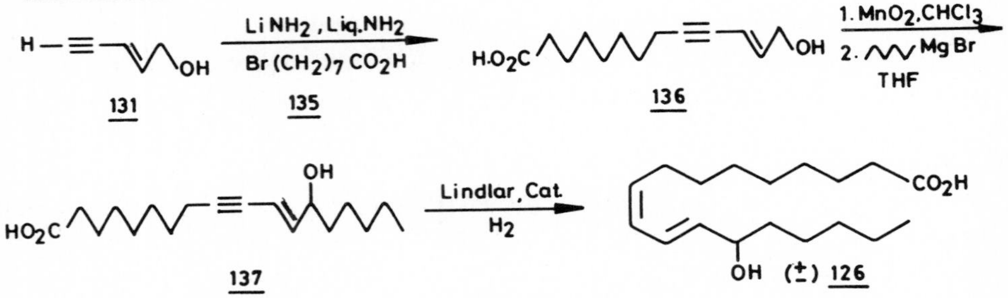

octanoic acid (**135**) gave the hydroxy acid (**136**). This was oxidised with active manganese dioxide and further Grignard reaction with excess of n-pentylmagnesium bromide afforded the carbinol (**137**). Compound **137** on partial hydrogenation using Lindlar catalyst furnished (±)-**126**.

The salient feature of their other approach involved, in situ generation of 1,3-butadiyne for further elaboration of appropriate side chains at each end of the intermediate (Scheme 41). 1,4-Dichloro-but-2-yne (138) and capronaldehyde were treated in the presence of sodium amide in liquid ammonia to furnish 139. Due to the presence of a hydroxyl group in 139, the adjacent acetylenic group was selectively reduced to the trans double bond to give 140. Subsequent alkylation of 140 with the bromo acid (135) followed by catalytic hydrogenation gave (±)-126.

Scheme -41

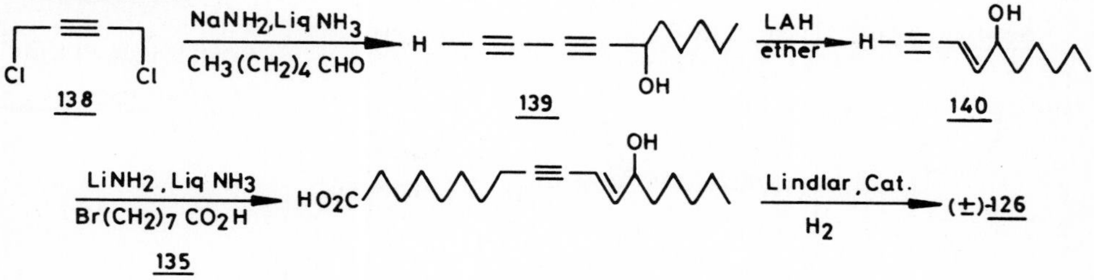

The synthesis reported by Gunn (ref. 127), involves the utilisation of 1,4-di-carbonyl compound (Scheme 42). Thus 2-pentylfuran (141) on oxidative cleavage followed by treatment with PTS acid gave the ene-dione (142). Compound 142 on Wittig reaction with phosphonium salt (143) gave the keto ester (144). The keto functionality in 144 on reduction with $NaBH_4$ followed by ester hydrolysis afforded (±)-126.

Scheme - 42

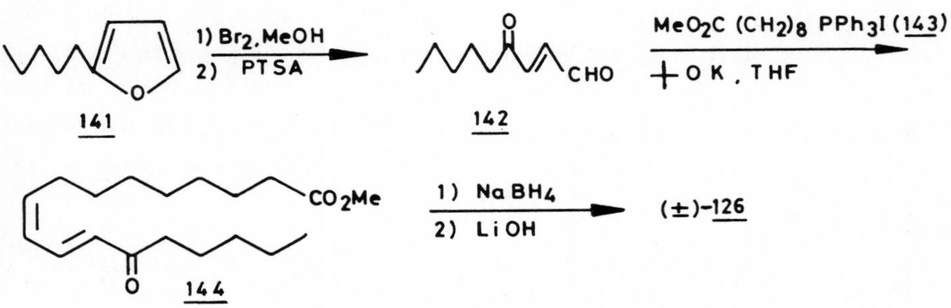

Later workers diverted the attention to provide a synthesis sequence for optical-ly active coriolic acid. The first to report (Scheme 43) asymmetric synthesis of

Scheme – 43

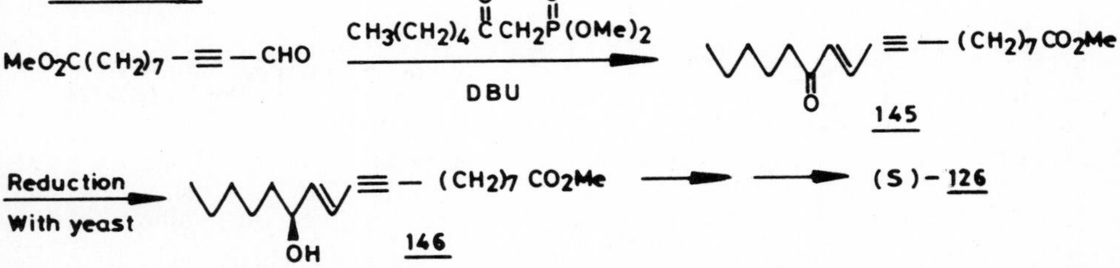

(S)-126 was Sakai et al (ref. 128), by involving enzymatic reduction. In their synthetic

sequence, the intermediate **145** was reduced with a yeast to furnish specifically the S-isomer (**146**) which was taken to S-coriolic acid (**126**).

Falck et al (ref. 129), described the synthesis of (S)-**126** starting from D-glucose (Scheme 44). The derived aldehyde (**147**) was transformed into the product **148** whose anomeric centre was further elaborated to **149**. The resulting diol inter-mediate **149** was converted into an epoxy acid (**150**) whose isomerisation in the pre-sence of methyl magnesium derivative of N-cyclohexylisopropylamine (MMA) gave (S)-**126**.

Scheme – 44

Recently, Sato et al (ref. 130), reported a generalised technique to prepare all isomers of **126**. The approach (Scheme 45) consists of palladium catalysed [PdCl$_2$(PPh$_3$)-CuI] coupling reaction of optically active (S)-acetylenic alcohol (**151**) with vinyl bromide (**152**) to give **153**, which on straight forward reactions furnished (S)-**126**. In a similar fashion the corresponding R-alcohol [(R)-**151**] gave rise the (R)-**126**. The group also studied the correlation of the stereochemistry of **126** with physio-logical activity and observed that both the isomers (R) and (S) are equally active.

Scheme –45

Very recently, Rama Rao et al (ref. 131), accomplished the synthesis of (R)-**126** by adopting Sharpless asymmetric epoxidation method (Scheme 46). Thus, the (R)-allylic alcohol (**154**) obtained in the Sharpless asymmetric epoxidation (kinetic resolution) step during their studies on the synthesis of 14,15-leukotriene A_4 (ref. 132), was benzoylated and cleaved to give aldehyde (155). Compound 155 on successive Wittig reactions and hydrolysis afforded (R)-126.

Scheme-46

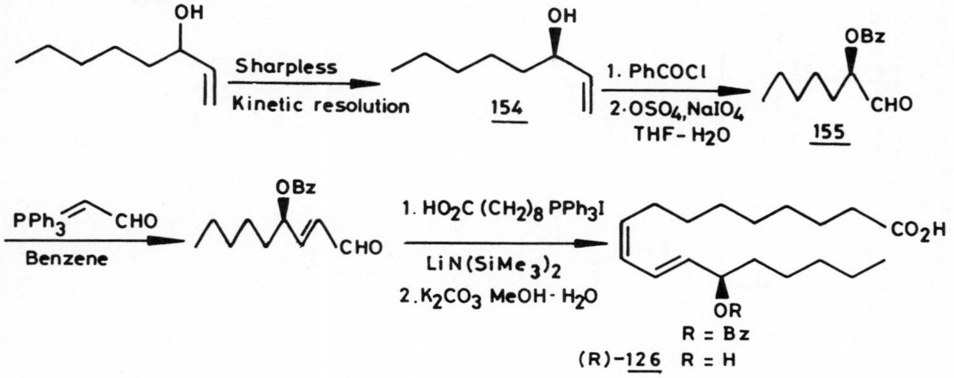

Sharpless Kinetic resolution → 154

1. PhCOCl
2. OsO4, NaIO4 THF-H2O → 155 (OBz, CHO)

PPh3 CHO Benzene → (OBz, CHO)

1. HO2C (CH2)8 PPh3I, LiN(SiMe3)2
2. K2CO3 MeOH-H2O → CO2H, OR

R = Bz
(R)-126   R = H

4.2   Synthesis of dimorphecolic acid

The first synthesis of dimorphecolic acid (127) which is having similar properties to those of coriolic acid (126) also appeared from Rama Rao's group (refs. 125, 133). Due to the anology of 127 and 126, with regard to functionalities, a similar strategy (Scheme 39) was logically developed for 127 (Scheme 47). Thus, (E)-pent-2-

Scheme-47

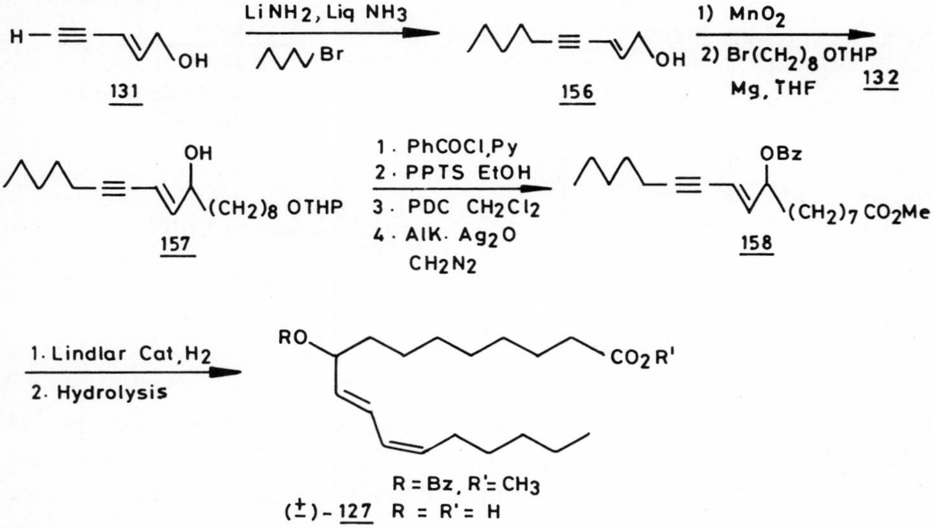

H—≡—OH   131   LiNH2, Liq NH3 / Br →   ≡—OH   156

1) MnO2
2) Br(CH2)8 OTHP Mg, THF   132

OH (CH2)8 OTHP   157

1. PhCOCl, Py
2. PPTS EtOH
3. PDC CH2Cl2
4. Alk. Ag2O CH2N2 →

OBz (CH2)7 CO2Me   158

1. Lindlar Cat, H2
2. Hydrolysis →

RO, CO2R'

R = Bz, R'= CH3
(±)-127   R = R'= H

en-4-yn-1-ol (131) on alkylation with n-pentyl bromide gave 156 which on oxidation and Grignard reaction with 132 afforded the alcohol (157). Compound 157 was transformed into **158** by sequential benzoylation, depyranilation, oxidation and esterification. **158** on partial hydrogenation and hydrolysis afforded (±)-**127**.

Sakai's approach (ref. 128), is similar to the one earlier adopted for the synthesis of coriolic acid (**126**) (Scheme 48). It is interesting to note that the microbiol reduction of keto ester (**159**) to give optically active **127**, did not work. However, compound **159** was reduced with sodium borohydride and further elaborated to (±)-**127**.

Scheme –48

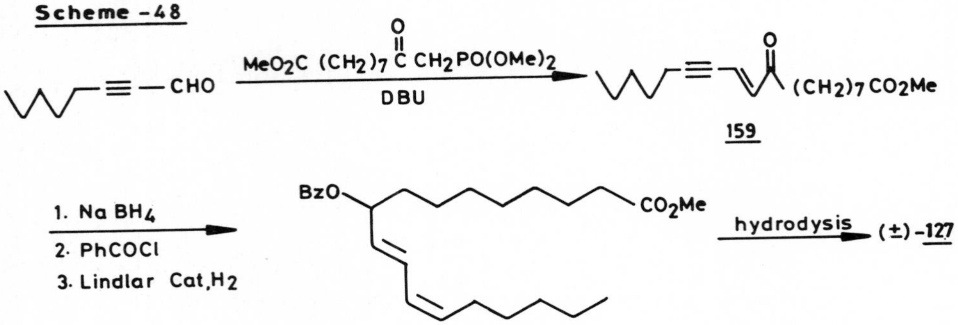

Takeda _et al_ (ref. 134), utilised the allenic ester (**160**) to introduce cis-trans diene system (Scheme 49). Thus, the ester was isomerised over alumina to furnish **161** which was converted to the aldehyde and subjected to Grignard reaction with 8-bromooctanol to give **162**. Compound **162** was exhaustively oxidised and then reduced with sodium borohydride to afford (±)-**127**.

Scheme – 49

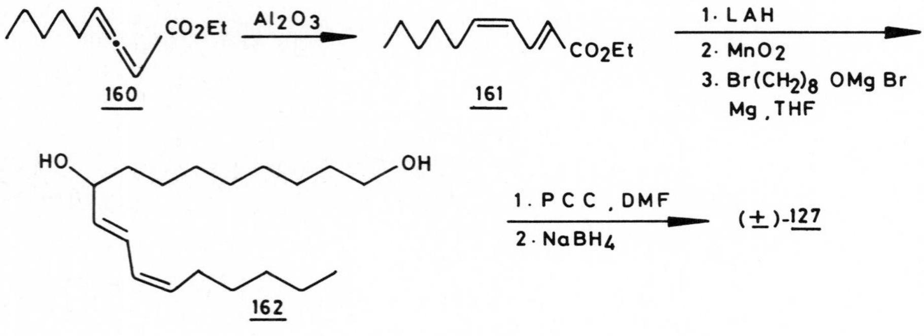

4.3 Synthesis of hydroxy octadecatrienoic acids

The other two hydroxy octadecatrienoic acids namely 16-hydroxy-9Z,12Z,14E-octadecatrienoic acid (**128**) and 9-hydroxy-10E,12Z,15Z-octadecatrienoic acid (**129**) have been demonstrated to have self-defensive activity against rice blast diseases. Rama Rao _et al_ (ref. 135), developed synthetic sequences for these molecules. Their strategy for compound **128** consists of utilising 1,3-butadiyne for elaboration purpose either by alkynylation or by alkylation reactions and also for stereoselective introduction of double bonds (Scheme 50). Thus, the reaction of 1,4-dichloro-but-2-yn (138) with propionaldehyde in presence of sodamide in liquid ammonia gave hept-5,6-diyn-3-ol (163) which on subsequent reduction with lithium aluminium hydride in ether afforded the trans olefinic alcohol (164). Compound 164 on Grignard coupling reaction

Scheme – 50

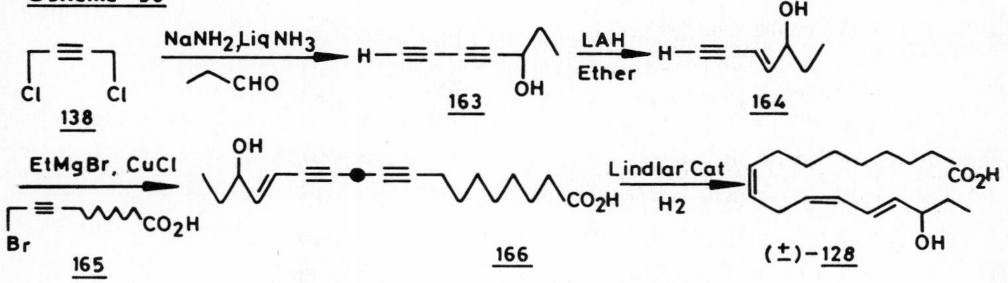

with the bromo acid (165) in the presence of CuCl afforded the hydroxy acid (165). Finally, the required cis double bonds were introduced by partial hydrogenation of the triple bonds in 166 using Lindlar catalyst to afford (±)-128.

The strategy for the construction (ref. 135, 136), involves the introduction of allylic hydroxyl group by selenium dioxide catalysed tert-butyl hydroperoxide on olifinic compound, followed by protection and cleavage of double bond to give α-hydroxy (protected) aldehyde which was elaborated by successive Wittig reactions (Scheme 51).

Scheme – 51

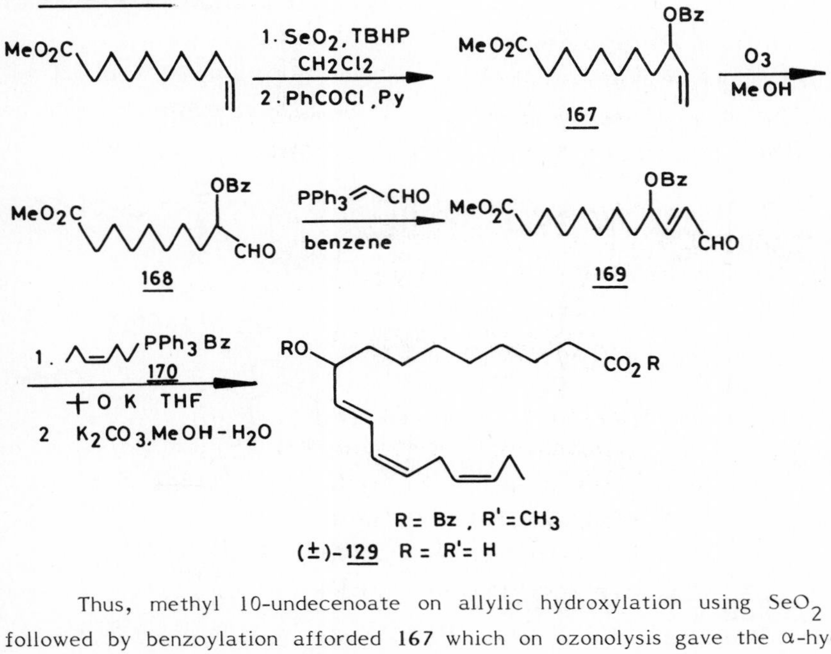

Thus, methyl 10-undecenoate on allylic hydroxylation using SeO$_2$ and TBHP followed by benzoylation afforded 167 which on ozonolysis gave the α-hydroxy (protected) aldehyde (**168**), a key synthon for the preparation of **129**. **168** on treatment with formylmethylene triphenylphosphorane gave the α,β-unsaturated aldehyde **169** which on Wittig reaction with phosphonium bromide (**170**) in presence of potassium tert-butoxide in THF, followed by hydrolysis afforded (±)-**129** (ref. 136).

The strategies developed for the synthesis of **128** and **129** are conceptually simple, elegant and novel. The yields in the synthesis are commendable and may have even commercial possibilities.

5   R (+)-α-lipoic acid

R (+)-α-lipoic acid (**171**), an important protein bound coenzyme and a prosthetic group is a substrate in plants, animals and micro-organisms. It was first isolated by Reed et al (ref. 137), and has been identified as a vital cofactor in the multi-enzyme complexes that catalyze the oxidative decarboxylation of α-keto acids (e.g. pyruvate, α-ketoglutarate and branched chain α-keto-acids) (ref. 138). Recently, it has also shown beneficial effects on diabetic rabbits during glucose tolerance test and curative effect in heavy metal poisoning (ref. 139). In addition, derivatives of lipoic acid have acquired importance in consemetic preparations (ref. 140).

Since its isolation, α-lipoic acid has been a subject of synthetic endeavour and this can be testified by number of syntheses and patents appeared in literature (ref. 141), to find out its therapeutic values. Available evidences suggest that the biological activities of α-lipoic acid is confined to the 'R' isomer which is also the naturally occurring compound (ref. 142). The 'R' isomer of **171** in the past was obtained by the resolution of the racemate. However, in recent years, interest in an asymmetric synthesis of **171** has been aroused.

The first optically active synthesis of unnatural S (-)-α-lipoic acid was reported by Golding et al (ref. 143) (Scheme 52), based on organo cuprate mediated

Scheme – 52

opening of (S)-2-benzyloxyethyl oxirane (172) (obtained from S-malic acid) with but-3-enylmagnesium chloride to afford the eight carbon 1,3-diol intermediate (173). This diol system turned out to be a useful precursor for the total synthesis of α-lipoic acid. The transformation of the 1,3-diol to 1,3-dithialane system was realised by Golding et al by the treatment of the derived dimesylate with sodium sulfide-sulfur in DMF. It is pertinent to mention that this approach for building up of 1,3-dithialane system of 171 was adopted by almost all the research groups.

Scheme - 53

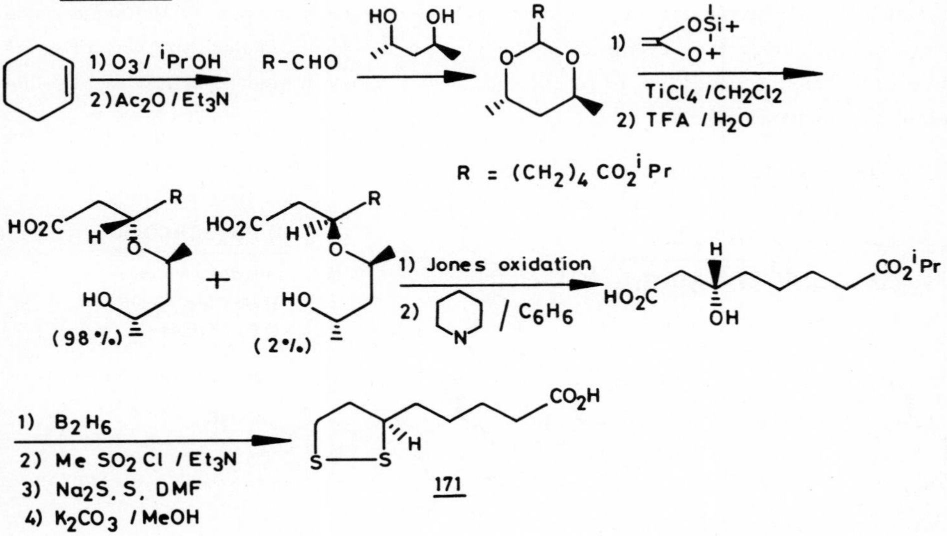

R = $(CH_2)_4 CO_2^i Pr$

Scheme - 54

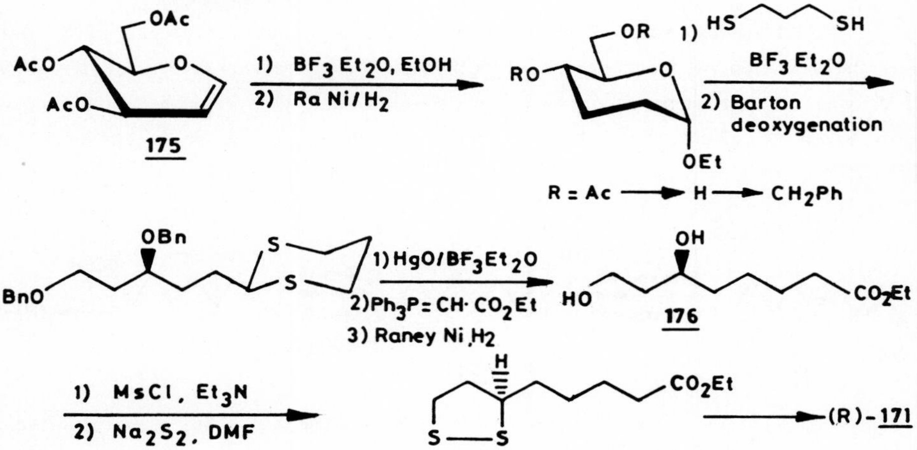

Elliott et al (ref. 144), were the first to synthesise R(+)-α-lipoic acid based on titanium tetrachloride mediated aldol type of condensation between the chiral acetal and subsequent reduction afforded the 1,3-diol which was converted into **171** by adopting Goldings approach (Scheme 53).

The enantiospecific synthesis of R (+)-**167** has been developed by Rama Rao et al (ref. 145), from D-glucose (Scheme 54). The commercially available 3,4,6-tri-O-acetyl-D-glucal (**175**) was transformed into the six-carbon aldehyde whose C_2 homologation by Wittig reaction afforded the 1,3-diol (**176**) which was converted into **171** by the known approach (Scheme 54). The second approach by the same group (ref. 146), involves the reverse strategy wherein the C-2 homologation was effected first and then the undesired 7-hydroxyl group was deoxygenated to afford the 1,3-diol derivative (Scheme 55).

Scheme – 55

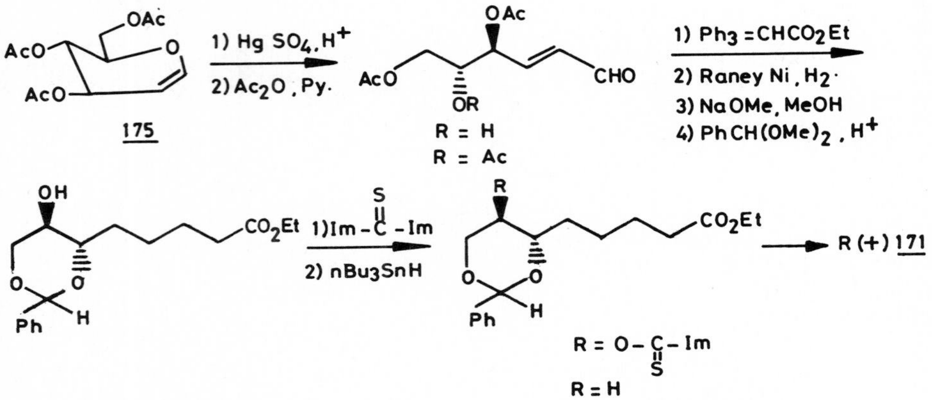

The utility of Sharpless epoxidation for the asymmetric synthesis of R (+)-171 was the key step of the approach of Sutherland et al (ref. 147). The allylic alcohol (**177**) was converted into the chiral epoxide (**178**) and then regiospecifically opened with Redal to afford the 1,3-diol derivative (**179**) which was later converted into 171 (Scheme 56).

Scheme –56

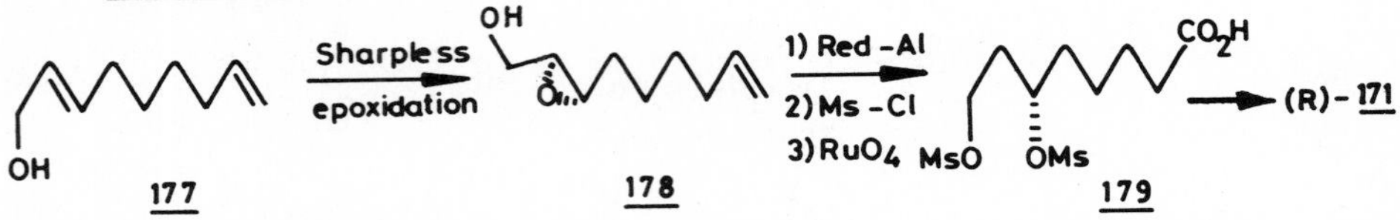

The presence of C-2 symmetry in R,R-divinylglycol (**180**) was exploited by Rama Rao et al (ref. 148), for the synthesis of R (+)-**171**. Compound **180**, conveniently obtained from D-mannitol, was functionalised to afford the 1,3-diol derivative which was transformed into R (+)-171 by Goldings approach (Scheme 57).

Scheme 57

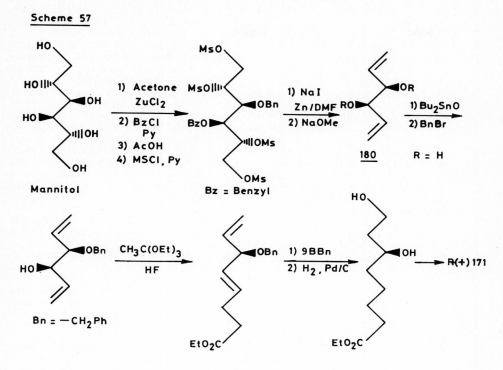

Mannitol

Bz = Benzyl

180   R = H

Bn = —CH$_2$Ph

R(+) 171

CONCLUSION

The major problem encountered in the synthesis of complex natural products is to achieve the desired selectivity viz., regioselectivity, chemoselectivity and stereoselectivity. In this chapter, the author presented some examples of such compounds of topical interest and having significant biological properties with emphasis on his own contributions in the area of anti-tumour agents (anthracyclines and sesbanimide A), macrolides (rifamycin-S), and some lipids (hydroxy fatty acids and lipoic acid). The total synthesis/construction of the complex natural products depends on the availability of new reagents and methodologies. It has to be admitted that while considerable progress has been made in the recent past, much remains to be done to obtain the type of selectivity that is achieved by enzymes in the biological systems.

ACKNOWLEDGEMENTS

I am indebted to my colleagues, especially to Drs J S Yadav, M K Gurjar and E Rajarathnam Reddy for their invaluable help in preparing this manuscript.

540

REFERENCES

1   a) B.M. Trost, Science, **219**, 245, 1983; b) B.M. Trost, Chemistry in Briton, 315 (1984).
2   J.D. Morrison, "Asymmetric Synthesis", Vol. 1-5, Academic Press (1984).
3   a) S. Hanessian, The Total Synthesis of Natural Products, The "Chiron approach". Organic Chem. Series, Vol. 3, Pergamon Press, Oxford (1983); b) D. Seebach and A. Vasella, Chiral building blocks in enantiomers synthesis in "Modern Synthetic Methods", 1980, Vol. 2, R. Scheffold (ed.), Frankfurt, 1980.
4   a) F. Arcamone, G. Franceschi, P. Orezzi, G. Cassinelli, W. Barbieri and R. Mondelli, J. Am. Chem. Soc., **86**, 5334 (1964); b) J. Bernand, R. Paul, M. Boiron, C. Jaquillat and R. Miral, Ed. "Rubidomycin", Springer Verlag, New York, 1969; c) R.B. Livingston and S.K. Carter, "Daunomycin", Chemotherapy Fact Sheet, National Cancer Institute, Bethesda, Md., 1970.
5   a) F. Arcamone, G. Franceschi, S. Penco and A. Selva, Tetrahedron Lett., 1007 (1969); b) R.H. Blum and S.K. Carter, Ann. Intern. Med., **86**, 249 (1974).
6   a) M.G. Brazhnikov, V.B. Zbarsky, V.I. Ponomarenko and N.P. Potapova, J. Antibiot., **27**, 254, 1974; b) G.F. Gause, M.G. Brazhnikov and V.A. Shorin, Cancer Chemother. Rep. Part I, **58**, 255 (1974).
7   a) M.C. Wani, H.L. Taylor, M.E. Wall, A.T. McPhail and K.D. Onan, J. Am. Chem. Soc., **97**, 5955 (1975); b) G.R. Pettit, J.J. Einck, C.L. Herald, R.H. Ode, R.B. Vondreele, P. Brown, M.G. Brazhnikova and G.J.F. Gause, J. Am. Chem. Soc., **97**, 7387 (1975).
8   F. Arcamone, G. Cassinelli, F. DiMatteo, S. Forenza, M.C. Ripamonti, G. Rivola, A. Vigevani, J. Clardy and T. McCabe, J. Am. Chem. Soc., **102**, 1462, (1980).
9   a) F. Arcamone, L. Bernardi, P. Giardino, B. Patelli, A. DiMarco, A.M. Casazza, G. Pratesi and P. Reggiani, Cancer Treat. Rep., **60**, 829 (1976); b) S. Neidel, Nature, **268**, 195 (1977); c) F. Arcamone, Lloydia, **40**, 45 (1977).
10   T.H. Smith, A.N. Fujiwara, W.W. Lee, H.Y. Wu and D.W. Henry, J. Org. Chem., **42**, 3653 (1977).
11   M.K. Gurjar, J.S. Yadav and A.V. Rama Rao, Indian J. Chem., **22B**, 1139(1983).
12   M.K. Gurjar, V.J. Patil, J.S. Yadav and A.V. Rama Rao, Carbohydrate Research, **129**, 264 (1984).
13   T.R. Kelly, Ann. Rep. Med. Chem., **14**, 288 (1979).
14   F. Arcamone, "Doxyrubicin" Anticancer antibiotics, Medicinal Chemistry, Vol. 17, Academic Press, New York (1981).
15   C.M. Wong, D. Popien, R. Schwenk and J. Te Raa, Can. J. Chem., **49**, 2712 (1971).
16   F. Arcamone, L. Bernardi, B. Patelli, P. Giardino, A. DiMarco, M.A. Casazza, C. Soranzo and G. Pratesi, Experientia, **34**, 1255 (1978).
17   K.S. Kin, E. Vanotti, A. Suarato, F. Johnson, J. Am. Chem. Soc., **101**, 2483 (1979).
18   S.D. Kimball, K.S. Kim, D.K. Mohanty, E. Vanotti and F. Johnson, Tetrahedron Lett., 3871 (1982).
19   A.V. Rama Rao, K. Bal Reddy and A.R. Mehendale, J. Chem. Soc. Chem. Commun., 565 (1983).
20   A.V. Rama Rao, A.R. Mehendale and K. Bal Reddy, Tetrahedron Lett., **24**, 1093 (1983).
21   Y. Bessieri and P. Vogel, Helv. Chim. Acta., **63**, 232 (1980).
22   a) J.P. Gesson, J.C. Jacquesy and M. Mondon, Tetrahedron Lett., **21**, 3351 (1980); b) J.G. Bauman, R.B. Barber, R.D. Gless and H. Rapoport, Tetrahedron Lett., **21**, 4777 (1980).
23   A. Echavarpen, P. Prados and F. Farina, Tetrahedron, **40**, 4561 (1984).
24   Y. Tamura, M. Sasho, S. Akai, H. Kishimoto, J. Sekihachi and Y. Kita, Tetrahedron Lett., **27**, 195 (1986).
25   Y. Tamura, M. Sasho, H. Ohe, S. Akai and Y. Kita, Tetrahedron Lett., **26**, 1549 (1985).
26   A.S. Kende, J. Rizzi and J. Riemer, Tetrahedron Lett., 1201 (1979).
27   K.A. Parker and J. Kallmerten, J. Am. Chem. Soc., **102**, 5881 (1984).

28   M. Braun, Tetrahedron, **40**, 4585 (1984).

29   M.J. Broadhurst and C.H. Hassall, J. Chem. Soc., (Perkin Trans-1) 2227, (1982).

30   F.M. Hauser and S. Prasanna, J. Am. Chem. Soc., **103**, 6378 (1981).

31   M. Watanabe, H. Maenosono and S. Furukawa, Chem. Pharm. Bull., **31**, 2662 (1983).

32   M.G. Dolson, B.L. Chenard and J.S. Swenton, J. Am. Chem. Soc., **103**, 5263 (1981).

33   R.A. Russel, A.S. Krauss, R.N. Warrener and R.W. Irvine, Tetrahedron Lett., **25**, 1517 (1984).

34   J.S. Yadav, P. Corey, C.T. Hsu, K. Perlman and C.J. Sih, Tetrahedron Lett., **22**, 811 (1981).

35   F.M. Hauser, J. Am. Chem. Soc., **105**, 5688, (1983).

36   J.K. Sutherland and A.F. Asheroft, J. Chem. Soc. Chem. Commun., 1075(1981).

37   J.K. Sutherland, P.S. Jones, D.T. Davies, Tetrahedron Lett., 519, 1983.

38   W.D. Wulff and Peng-cho Tang, J. Am. Chem. Soc., **106**, 434 (1984).

39   F. Arcamone, L. Bernardi, B. Patelli and A. DiMarca, Ger. Offen., **2**, 601, 785 (July 29, 1976) C.A. **85**, 142918 (1976).

40   D. Dominguez, R.J. Ardecky and M.P. Cava, J. Am. Chem. Soc., **105**, 1608 (1983).

41   a) S. Terashima, K. Tamato and Sulgimori, Tetrahedron Lett., 4507 (1982). b) M.J. Broadhurst, C.H. Hassall and G.J. Thomas, J. Chem. Soc., Perkin, Trans.I, 2249 (1982).

42   a) A.V. Rama Rao, V.H. Deshpande and N. Lakshma Reddy, Tetrahedron Lett., **23**, 775 (1982); b) A.V. Rama Rao, V.H. Deshpande, B. Ramamohan Rao and K. Ravichandran, Tetrahedron Lett., **23**, 1115 (1982); c) A.V. Rama Rao, B. Chanda and H.B. Borate, Tetrahedron, 3555 (1982); d) A.V.Rama Rao, J.S.Yadav, K. Bal Reddy and A.V. Mehendale, J. Chem. Soc. Chem. Commun., 453 (1984); e) A.V. Rama Rao, V.H. Deshpande, K. Ravichandran and B. Ramamohan Rao, Synthetic Commun., **13**, 1219 (1983); f) A.V. Rama Rao, V.H. Deshpande, K.M. Sathaye and S.M. Jaweed, Ind. J. Chem., **24B**, 697 (1985); g) V.H. Deshpande, K. Ravichandran and B. Ramamohan Rao, Synthetic Commun., **14**, 477, (1984).

43   V.S. Martin, S.S. Woodard T. Katsuk, Y. Yamada, M.I. Keda and K.B. Sharpless, J. Am. Chem. Soc., **103**, 6237 (1981).

44   A.V. Rama Rao, J.S. Yadav, K. Bal Reddy and A.V. Mehendale, Tetrahedron, **40**, (22), 4643 (1984).

45   O. Mitsunobu, Synthesis 1 (1981).

46   D. Dominguez and M.P. Cava, J. Org. Chem., **48**, 2820 (1983).

47   F. Bennani, J.C. Florent, M. Kochi and C. Monneret, Tetrahedron, **40**, 4669 (1984).

48   R.C. Gupta, P.A. Harland and R.J. Stoodley, Tetrahedron, **40**, 4657 (1984).

49   a) S. Terashima and K. Tamato, Chem. Pharm. Bull., **32**, 4328, (1984); b) S. Terashima, S.S. Jew and K. Koga, Tetrahedron Lett., 4939 (1978); c) R.A. Russell, A.S. Krans, R.N. Warrener and R.W. Irvine, Tetrahedron Lett., **25**, 1517 (1984); d) S. Terashima, N. Tanno and K. Koga, Tetrahedron Lett., **21**, 2753 (1980); e) F. Arcamone et al., Chemistry and Industry, 810 (1981).

50   M. Sodeoka, T. Limori and M. Shibasaki, Tetrahedron Lett., **26**, 6497 (1985).

51   F.M. Hauser and S.R. Ellenberger, Chem. Rev., **86**, 35 (1986).

52   M.K. Gurjar, V.J. Patil and S.M. Pawar, J. Sci. Ind. Res., **45**, 433 (1986).

53   J.P. Marsh (Jr.), C.W. Mosher, E.M. Acton and L. Goodman, Chem. Commun., 973 (1967).

54   a) B. Frasa Reid and H.W. Pauls, J. Chem. Soc. Chem. Commun., 1031 (1983); b) J.S. Brimacombi, R. Hanna, L.C.N. Tucker, Carbohydrate Res., **136**, 419 (1985).

55   a) S. Tahaka, Japan Kokal Tokky Kobo, 7914913 (1979), <u>Chem. Abstr.,</u> **91**, 5437 (1979); b) G. Medgyes and J. Kuszmann, Carbohydrate Res., **92**, 225 (1981).

56   D. Horton and W. Weckerle, Carbohydrate Res., **44**, 227 (1975).

57   A. Klemer and G. Rhodemeyer, Chem. Ber., **107**, 2612 (1974).

58   S. Hanessian, Carbohydrate Res., **2**, 86 (1966).

59   N.K. Richtmyer, "Methods in Carbohydrate Chemistry", R.L. Whister and M.L. Wolfrom, (Eds.), Academic Press, London, 1962, Vol. 1, p.108.

60   A.V. Rama Rao, T.R. Ingle, V.R. Kulakarni and V.V. Dhekne, Indian J. Chem. 22B, 69 (1983).

61   A. Crugnola, P. Lombardi, C. Gandolfi and F. Arcamone, Gazz, Chim. Italiana, 111, 395 (1981).

62   Y. Kimura, T. Matsumoto, M. Suzuki and S. Terashima, Bull. Chem. Soc. Japan, 59, 663 (1986).

63   M.K. Gurjar, V.J. Patil, J.S. Yadav and A.V. Rama Rao, Carbohydrate Res., 135, 174 (1984).

64   D. Picq and D. Anker, J. Carbohydrate Chem., 4, 113 (1985).

65   G. Luckas et al., Carbohydrate Res., 140, 51 (1985).

66   H.H. Baer and A.J. Sobiesiak, Carbohydrate Res., 140, 201 (1985).

67   G. Luckas et al., J. Org. Chem., 50, 4913 (1985).

68   D. Harton and W. Priebe, Carbohydrate Res., 136, 391 (1985).

69   M.K. Gurjar and S.M. Pawar, Tetrahedron Lett., 1327 (1987).

70   T. Yamaguchi and M. Kojima, Carbohydrate Res., 59, 343 (1977).

71   E.M. Acton, A.N. Fujiwara and D.W. Henry, J. Med. Chem., 17, 659 (1974).

72   F. Arcamone et al., J. Med. Chem., 19, 703 (1975).

73   H.S. El Khadem and D.L. Swartz, Carbohydrate Res., 65, C-1 (1978).

74   E.F. Fuchs, D. Harton and W. Weckerle, Carbohydrate Res., 57, C-36 (1977).

75   H.S. El Khadem, A. Liav and D.L. Swartz, Carbohydrate Res., 74, 345 (1979).

76   J. Boivin, C. Monneret and M. Pais, Tetrahedron Lett., 1111 (1978).

77   Y. Kimura, M. Suzuki, T. Matsumoto, R. Abe and S. Terashima, Chem. Lett., 501 (1984).

78   R.K.Y. Zee-cheng, C.C. Cheng, J. Med. Chem., 21, 291 (1978).

79   R. Misra and R.C. Pandey, J. Am. Chem. Soc., 104, 4478 (1982).

80   a) R.G. Powell, C.R. Smith Jr, R.V. Madrigal, Planta Med., 30, 1 (1976); b) R.G. Powell, C.R. Smith Jr, D. Weisleder, D.A. Muthard and J. Clandy, J. Am. Chem. Soc., 101, 2784 (1979).

81   T.R. Kelly, N. Ohashi, R.T. Armstrong-chong and S.M. Bell, J. Am. Chem. oc., 108, 7100 (1986).

82   a) A.V. Rama Rao, D. Reddeppa Reddy and V.H. Deshpande, J.C.S. Chem. Comm., 1119 (1984); b) K.A. Parker, K.A. Koziski and G. Breautt, Tetrahedron Letters, 26, 2181 (1985); c) A.S. Kende, F.H. Ebetino and T. Ohta ibid, 26, 2063 (1985); d) G. Eek, M. Julia, B. Pfeiffer and C. Rolando, ibid, 26, 4723, 4725 (1985); e) M. Braun and R. Veith ibid, 27, 179 (1986); f) R.D. Bach and R.C. Flix, J. Org. Chem., 51, 749 (1986); g) R.D. Bach and R.C. Flix, Tetrahedron Letters, 27, 1983 (1986); h) S.M. Bennett and D.L.J. Clive, J. Chem. oc. Chem. Commun., 878 (1986); i) M.A. Ciufolini and M.E. Browne, Tetrahedron Letters, 28, 171 (1987); j) G. Mehta and D. Subrahmanyam, Tetrahedron Letters, 28, 479 (1987).

83   a) A.V. Rama Rao and D.R. Reddy, J. Chem. Soc. Chem. Comm., 574 (1987); b) A.V. Rama Rao, D.R. Reddy, G.S. Annapurna and V.H. Deshpande, Tetrahedron Letters, 28, 451 (1987); c) A.V. Rama Rao, N. Srinivasan, D.R. Reddy and V.H. Deshpande, ibid 28, 455 (1987); d) K.A. Parker, M. Spero and K.A. Koziski, J. Org. chem., 52, 183 (1987); e) D.L.J. Clive and J. Sedgeworth, Heterocycles, 24, 509

84   a) R.G. Powell, C.R. Smith Jr, D. Weisleder, G.K. Matsumoto, J. Clardy and J. Kozlowski, J. Am. Chem. Soc., 105, 3739 (1983); b) C.P. Gorst-Allman, P.S. Steyn, R. Vleggar and N. Grobbelaar, J.C.S. Perkin I, 1311 (1984).

85   R.G. Powell, C.R. Smith Jr and D. Weisleder, Phytochemistry, 23, 2789 (1984).

86   G.W.J. Fleet and T.K.M. Shing, J.C.S. Chem. Commun., 835 (1984).

87   M. Shibuya, Heterocycles, 23, 61 (1985).

88   M.J. Wanner, G-J. Koomen and U.K. Pandit ibid 22, 1483 (1984).

89   F. Matsuda, M. Kawasaki and S. Terashima, Tetrahedron Letters, 26, 4639 (1985).

90   W.R. Roush and M.R. Machaelides, Tetrahedron Letters, 27, 3353 (1986).

91   A.V. Rama Rao, J.S. Yadav, A.M. Naik and A.G. Chaudhary, Ind. J. Chem., 25B, 579 (1986).

543

92   T. Kinoshita, K. Okamoto and J. Clardy, Synthesis, 402 (1985).
93   G. Sacripante, C. Tan and G. Jurt, Tetrahedron Letters, **26**, 5643 (1985).
94   K. Tomioka and K. Koga, Tetrahedron Letters, **25**, 1599 (1984).
95   A.V. Rama Rao, J.S. Yadav, A.M. Naik and A.G. Chaudhary, Tetrahedron Letters, **27**, 993 (1986).
96   M.J. Wanner, N.P. Willard, G-J. Koomen and U.K. Pandit, J.C.S.Chem.Commun. 396 (1986); Ibid , Tetrahedron, **43**, 2549 (1987).
97   F. Matusuda and S. Terashima, Tetrahedron Letters, **27**, 3407 (1986).
98   R.H. Schlessinger and J.L. Wood, J. Org. Chem., **51**, 2621 (1986).
99   A.V. Rama Rao, J.S. Yadav, A.G. Chaudhary and A.M. Naik, Unpublished results.
100   A.M. Naik, "Synthesis of Biologically Active Compounds" Ph.D. Thesis, Bombay University, Bombay, 1986.
101   P. Sensi, A.M. Greco and R. Ballotta, Antibiot. Ann., 262 (1960).
102   V. Prelog, Pure Appl. Chem., **7**, 551 (1963).
103   H. Nagaoka, W. Rutsch, G. Schmid, H. Iio, M.R. Johnson and Y. Kishi, J. Am. Chem. Soc., **102**, 7962 (1980); H. Iio, H. Nagaoka and Y. Kishi, J. Am. Chem. Soc., **102**, 7965 (1980).
104   H. Nagaoka and Y. Kishi, Tetrahedron, **37**, 3873 (1981).
105   S. Masamune, B. Imperiali and D.S. Carvey, J. Am. Chem. Soc., **104**, 5528 (1982).
106   W.C. Still and J.C. Barrish, J. Am. Chem. Soc., **105**, 2487 (1983).
107   E.J. Corey and T. Hase, Tetrahedron Letters, 335 (1979).
108   S. Hanessian, J-R. Pougny and I. Boessenkool, J. Am. Chem. Soc., **104**, 6164 (1982); S. Hanessian, J-R. Pougny and I. Buessenkool, Tetrahedron, **40**, 1289 (1984).
109   M. Nakata, M. Takao, Y. Ikeyama, T. Sakai, K. Tatsuka and M. Kinoshita, Bull. Chem. Soc. Japan, **54**, 1749 (1981); M. Nakata, T. Sakai, K. Tatsuka and M. Kinoshita, Ibid **54**, 1743 (1981); M. Nakata, Y. Ikeyama, H. Takao and M. Kinoshita, Ibid **53**, 3252 (1980).
110   B. Fraser-Reid, L. Magdizinski and B. Molino, J. Am. Chem. Soc., **106**, 731 (1984).
111   I. Paterson and M.M. Mansuri, Tetrahedron, **41**, 3569 (1985).
112   A.V. Rama Rao, J.S. Yadav and V. Vidyasagar, J. Chem. Soc. Chem. Comm., 55 (1985).
113   H.M.R. Hoffmann, K.E. Clemens and R.H. Smithers, J. Am. Chem. Soc., **94**, 3940 (1972).
114   T.N. Guru Row and S.S. Tawale, Unpublished work.
115   E.J. Corey and G. Schmidt, Tetrahedron Letters, 2317 (1979).
116   A.V. Rama Rao, J.S. Yadav and C.S. Rao, Tetrahedron Letters, **27**, 3297 (1986).
117   W.R. Roush and A.D. Palkowitz, J. Am. Chem. Soc., **109**, 953 (1987).
118   S.J. Danishefsky, D.C. Myles and D.F. Harvey, J. Am. Chem. Soc., **109**, 862 (1987).
119   F.E. Ziegler and A. Kneisley, Tetrahedron Lett., **28**, 1725 (1987).
120   S. Masamune, W. Choy, J.S. Paterson and L.R. Sita, Angw. Chem. Int., Ed. **24**, 1 (1985) and references cited therein.
121   a) R.H. Green and P.F. Lambeth, Tetrahedron, **39**, 1687 (1987); b) Georgiev, In Survey of drug research in immunological diseases, Aliphatic derivatives-1, S. Karger Basel. London, p.182 (1984); c) R.P. Evstignecva and G.I. Myagkova, Russian Chemical Reviews, **55**, 455 (1986).
122   G.A. Blondin, Ann. N.Y. Acad. Sci., **264**, 98 (1975).
123   T. Kato, Y. Yamaguchi, T. Hirano, T. Yokoyama, T. Uyehara, T. Namai, S. Yamanaka and N. Harada, Chemistry Lett., 409 (1984).
124   P.S. Aisen, K.A. Haines, W.Given, S.B. Aramson, M. Pras, C. Serhan, M. Hamberg, B. Samuelsson and W. Weissmann, Proc. Natl. Acad. Sci., **82**, 1232 (1985).
125   A.V. Rama Rao, E. Rajarathnam Reddy, G.V.M. Sharma, P. Yadagiri and J.S. Yadav, Tetrahedron Lett., **26**, 465 (1985).
126   A.V. Rama Rao, S. Pulla Reddy and E. Rajarathnam Reddy, J. Org. Chem., **51**, 4158 (1986).
127   B. Gunn, Heterocycles, **23**, 3061 (1985).

544

128   H. Suemune, N. Hayashi, K. Funakoshi, H. Akita, T. Oishi and K. Sakai, Chem. Pharm. Bull., **33**, 2168 (1985).

129   C.A. Moustakis, D.K. Weerasinghe, P. Mosset and J.R. Falck, Tetrahedron Lett., **27**, 303 (1986).

130   Y. Kobayashi, S. Okamoto, T. Shimazaki, Y. Ochiai and F. Sato, Tetrahedron Lett., **28**, 3959 (1987).

131   A.V. Rama Rao, A.V. Purandare and E. Rajarathnam Reddy (Unpublished results).

132   A.V. Rama Rao, A.V. Purandare and E. Rajarathnam Reddy (Unpublished results).

133   A.V. Rama Rao, E. Rajarathnam Reddy, G.V.M. Sharma, P. Yadagiri and J.S. Yadav, Tetrahedron, **42**, 4523 (1986).

134   S. Tsuboi, S. Maeda and A. Takeda, Bull. Chem. Soc (Japan), **59**, 2050(1986).

135   A.V. Rama Rao and E. Rajarathnam Reddy, Tetrahedron Lett., **27**, 2279 (1986).

136   A.V. Rama Rao, E. Rajarathnam Reddy, A.V. Purandare and Ch.V.N.S. Varaprasad, Tetrahedron, **43**, 0000 (1987).

137   L.J. Reed, I.C. Counsalus, B.G. DeBusk, and C.S. Homberger Jr., Science, **114**, 93 (1951).

138   H. Sigel, Angew. Chem. Int. Edn. Engl., **21**, 389 (1982).

139   C.V. Natraj, V.M. Gandhi and K.K.G. Menon, J. Biosci., **6**, 37 (1984).

140   Pola Chemical Industries Inc. Jpn. Kokai Tokkyo Koho J.P., 81, 120, 611; Chem. Abstr., **96**, p.11517g (1982).

141   L.G. Chebotoreva, A.M. Yurkevich, Khim. Farm.Zh., **14**, 92 (1980); Chem. Abstr. **94**, 103722g (1981) Ibid USSR 667, 556 (1979); Chem. Abstr. 91, p.123732b (1979).

142   I.C. Gunsalus, M.I. Dolin and L. Struglia, J. Biol. Chem., **194**, 849 (1952).

143   M.H. Brookcs, B.T. Golding, D.A. Howes and A.T. Hudson, J. Chem. Soc. Chem. Commun., 1051 (1981).

144   J.D. Elliott, J. Steele and W.S. Johnson, Tetrahedron Lett., **26**, 2535 (1985).

145   A.V. Rama Rao, K. Garyali, M.K. Gurjar and T. Ravindranathan, Carbohydrate Res., **51**, 148 (1986).

146   A.V. Rama Rao, A.V. Purandare, E. Rajarathnam Reddy and M.K. Gurjar, Synthetic Commun., **17**, 1095 (1987).

147   P.C.B. Page, C.M. Rayner and I.O. Sutherland, J.Chem. Soc. Chem. Commun., 1408 (1986).

148   A.V. Rama Rao, S.V. Mysorekar, M.K. Gurjar and J.S. Yadav, Tetrahedron Lett., **28**, 2183 (1987).

TROPONE: VERSATILE BUILDING BLOCK FOR NATURAL PRODUCT SYNTHESIS

J. H. RIGBY

1. INTRODUCTION

Nature abounds with compounds which display a seven-membered carbocycle as a prominent architectural feature. The carbon-skeleta of these natural products can range in complexity from simple bicyclic systems to exceedingly intricate polycyclic and multifunctional arrays. The multifarious structural types and potent biological activity exhibited by many of these materials has prompted an intense international total synthetic effort in recent years.

In this account we will attempt to briefly outline the current state of affairs in the synthesis of natural products containing seven-membered rings and place within the context of this burgeoning field the advances made in the use of tropone-based chemistry, with particular emphasis on contributions from our laboratory.

The perhydroazulene ring system is perhaps the most commonly encountered of the cycloheptane derived structural categories in nature and is characterized by the fusion of a seven-membered ring to a five-membered carbocycle. The bicarbocyclic guaiazulene sesquiterpenes such as bulnesol (1) and the tricyclic terpene aromadendrene (2) are examples of relatively simple hydroazulenic compounds and were among the earliest cycloheptane-derived species to yield to total synthesis. (ref. 1).

1

2

More recently, the guaianolide and pseudoguaianolide sesquiterpene lactones have captured the attention of the synthetic community (refs. 2,3). These species are intriguing from a number of perspectives. In addition to the obvious synthetic challenges that these compounds present, many members of this class of natural products also possess important biological activity. These include significant tumor inhibitory (refs 4a,b), schistosomicidal (ref 4c) and plant growth regulatory activity (ref 4d).

Damsin (3) and estafiatin (4) are typical examples of the pseudoguaianolide and guaianolide families, respectively. It is interesting to note, at this point, that pseudoguaianes

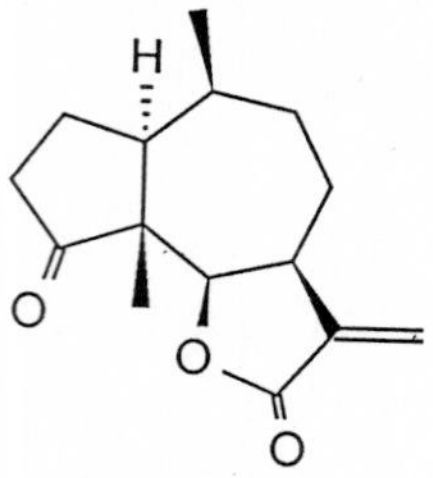

3

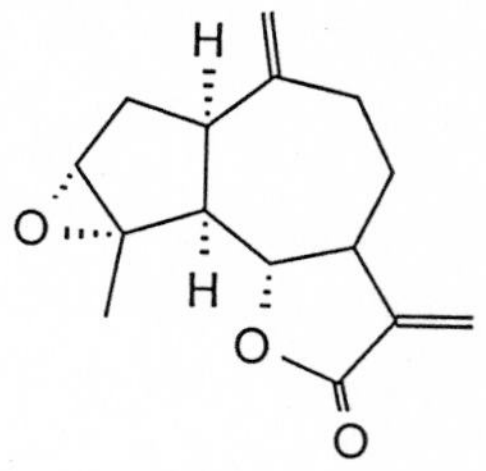

4

have attracted considerably more attention, in terms of total synthesis, than have the related guaianolides (ref. 3). Despite the apparent structural similarities between these two classes of plant derived materials, the synthetic challenges inherent in each differ dramatically. Much of this difference stems from the absence of the angular methyl group in the guaianolide series. This substituent serves as a potent stereochemical control element in many of the

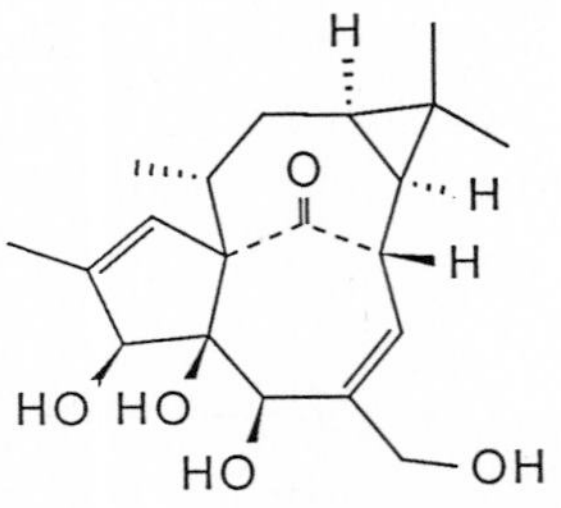

5

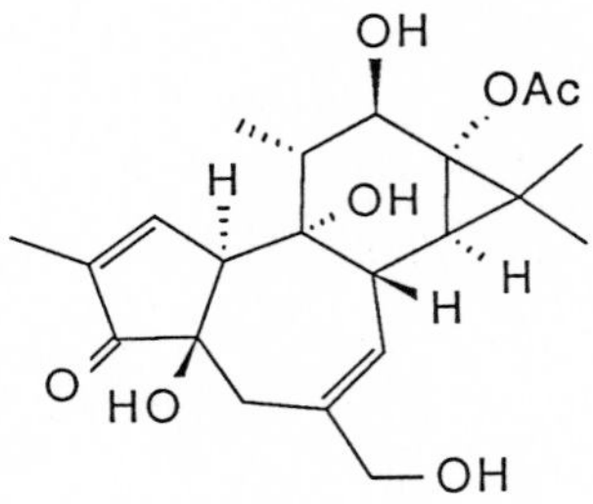

6

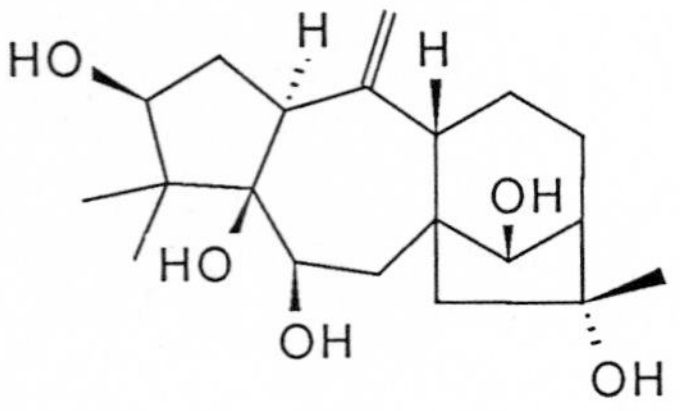

7

pseudoguaianolide syntheses that have been reported.

Even more structurally elaborate compounds have come to the fore recently. Among the most exciting of these from both a synthetic and medicinal standpoint are the cocarcinogenic diterpenes, ingenol (5) (refs. 5,6) and phorbol (6) (refs. 5,7) as well as the broadly active grayanotoxins (7) (refs. 8,9). The level of molecular complexity displayed by these natural products has added impetus to developing more efficient methodology for the construction of complicated seven-membered ring carbon frameworks.

Among the most interesting and powerful approaches into the perhydroazulene system is the stereocontrolled rearrangement of an appropriately substituted decalin derivative into the fused 7,5-system (ref. 10). This strategy has found considerable utility in relay syntheses which start from decalin based natural products (refs. 10e,f). The strengths of this synthetic design are quite apparent. Conformational analysis of both the *cis*- and the *trans*-fused decalin ring systems

is well established and stereorational substituent introduction can often be achieved with a minimum of difficulty. Furthermore, the stereochemical relationships set in the decalin species can be transferred to the hydroazulene with complete integrity through the aegis of a stereoselective rearrangement protocol. A typical example of this approach can be seen in Heathcock's synthesis of the pseudoguaianolide, confertin (8) (ref. 10c).

The decalin-based "ring expansion-contraction" route continues to play a major role in the construction of complex natural products. Paquette's pioneering investigations on the extremely demanding ingenane ring system is a case in point (ref. 6a). The hydroazulene portion of the tricycle was generated by a photochemically induced rearrangement of a decalinic epoxyketone to provide an intermediate which could be carried on to ingenol (5).

Other bicyclic starting materials that are well-behaved conformationally or that display significant stereochemical biases in their structures are also well represented in the chemical literature as solutions to the hydroazulene problem (ref. 11). Biomimetic type cyclizations of 10-membered rings have received considerable attention in this context as well (ref. 12). Numerous successful perhydroazulene syntheses have commenced with an appropriately functionalized cyclopentane ring, followed by elaboration of the elements of the seven-membered ring (ref. 13). This strategy has been among the most popular for assembling pseudoguaianolides. Vandewalle's clever entry into the hydroazulene skeleton through a photochemical [2+2] cycloaddition of two cyclopentene derivatives is an intriguing example (ref. 13a) which has been parlayed into the syntheses of a number of guaiane and pseudoguaiane species.

In spite of the growing interest in the synthesis of hydroazulenic natural products, there remains a conspicuous absence of approaches which introduce the elements of the five-membered ring onto a pre-existing cycloheptane template (ref. 14). This initially surprising situation becomes less so when examined in the light of several potential limitations on a cycloheptane-based synthetic strategy.

A principal drawback to the successful implementation of this conceptually attractive approach is the paucity of commercially available candidates for starting materials which are sufficiently functionalized to be useful for initiating a complex total synthesis. This is in stark contrast to the multitude of highly functionalized and inexpensive six-membered carbocyclic starting materials that are routinely available.

The well known conformational ambiguities associated with the cycloheptane and hydroazulene ring systems is a second point of concern when planning a synthesis (ref. 15). The ability to predict with confidence the stereochemical outcome of a particular transformation

when performed on these flexible rings can be difficult. Recently, this obstacle has been ameliorated somewhat by the advent of several systematic methods for evaluating the various conformations of these systems (ref. 16). Nonetheless, conformational analysis of cycloheptane derivatives is still not as accessible on a routine basis as it is for the cyclohexane series.

With these considerations in mind, we embarked on a program to evaluate the utility of 2,4,6-cycloheptatrienone (tropone) as a starting material for the synthesis of hydroazulenic sesquiterpenes. We were attracted by the notion that selective manipulation of the considerable reactivity displayed by the tropone nucleus would provide ample opportunity for controlling or circumventing the problems alluded to above.

2. TROPONE AS A STARTING MATERIAL FOR PERHYDROAZULENE SYNTHESIS

Tropone (9) displays many of the characteristics one normally associates with a useful starting material for natural product synthesis (ref. 17). It possesses a high level of functionalization which can be selectively manipulated in a number of ways and undergoes a remarkably rich variety of carbon-carbon bond forming reactions which are particularly suited to elaborating multiple ring systems. In recent years we have had the opportunity in our laboratory to exploit a number of these fascinating processes for the construction of interesting natural products. It is somewhat surprising that prior to 1982 relatively little activity had been reported concerning the use of troponoid derivatives for synthetic applications (ref. 18), although the groundwork had been laid early on by a number of classic studies on the reactions of this interesting group of compounds (ref. 17,19). Examples include Crabbe's clever use of the well-

known photochemically induced electrocyclic closure of tropolone to prepare prostanoid precursors (ref. 18a) as well as Garst and Ito who have examined in some detail the cycloaddition of dienes to tropone to prepare a number of interesting substrates (ref. 18b, 18d).

A major concern which may have discouraged the use of tropone in complex synthesis in the past was the perceived difficulties with the large-scale accessibility of this compound and its derivatives. The parent tropone can be prepared by a number of different routes, many of which are somewhat cumbersome and not amenable to large scale manipulation (ref. 20). However, the selenium dioxide mediated allylic oxidation of cycloheptatriene has proven to be a most reliable method in our hands for generating tropone on a reasonably large scale. Yields in the 50% range are now routinely obtained from this reaction by carefully controlling conditions (ref. 21). Cycloheptatriene is easily the most inexpensive, functionalized seven-membered ring compound available commercially, making this convenient procedure relatively economical.

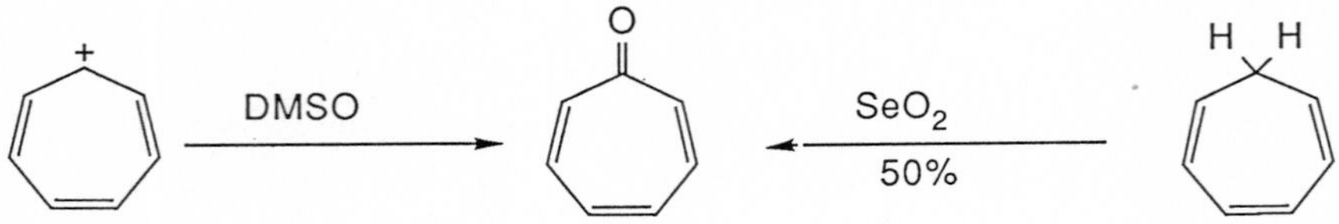

9

With a reliable supply of the starting material available, our fundamental plan of entry into the perhydroazulene manifold rested on the facile attachment of a terminally functionalized three-carbon unit onto the tropone nucleus. Subsequent bond formation at an adjacent carbon center would complete the assembly of the requisite hydroazulene with ample functionality suitably disposed around the ring system periphery for use in attacking a number of target natural products. At the outset of this investigation, the generalized hydroazulene species depicted in Scheme I exemplified what was viewed as an ideal intermediate which could be obtained by successful implementation of this strategy. Convenient manipulation of both the five- and seven-membered rings was anticipated by virtue of the oxygenation at C_4 and C_{10} and

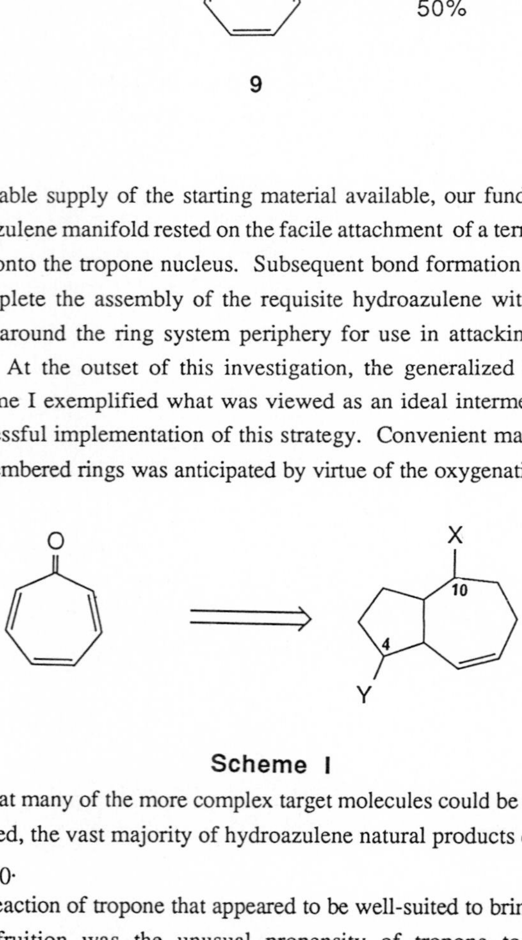

Scheme I

it was envisaged that many of the more complex target molecules could be directly accessed from this product. Indeed, the vast majority of hydroazulene natural products display substituents at C_4, C_6, C_7 and C_{10}.

A unique reaction of tropone that appeared to be well-suited to bringing the initial step of this protocol to fruition was the unusual propensity of tropone to add Grignard and organolithium reagents to the trienone system exclusively in a 1,8-fashion (refs. 19,22). This process was attractive principally because the elements of the five-membered ring could be conveniently introduced while retaining substantial functional group handles for subsequent bond formation in the resulting dihydrotropone product (Eq. 1).

A three-carbon unit was required at this stage which could effectively participate both in the initial 1,8-addition to tropone and subsequently permit the maximum flexibility for five-

9 $\xrightarrow{\text{RM}}$ Eq. 1

membered ring formation. At the time of our original efforts in the area, a limited selection of candidates which satisfied the aforementioned criteria were available. Readily accessible 2-(2-bromoethyl)-1,3-dioxolane, which has enjoyed considerable success as a cyclopentane building

9

10

block, emerged as the obvious first choice for our purposes (refs. 23,24). The normally unreactive Grignard derivative of this compound was found to add smoothly to tropone to provide the key dihydrotropone adduct **10** in a yield in excess of 80%. The resultant intermediate could, in principle, be induced to undergo cyclization in either of two complimentary directions to yield hydroazulene products. Species which retained the oxygenation at C_{10} were adjudged to provide access to a greater range of natural product targets and, therefore, these routes were vigorously pursued.

10

As a prelude to the proposed cyclopentannulation, the seven-membered ring ketone in **10** was reduced and the resulting mixture of alcohol epimers protected to insure the regiochemical integrity of the diene system. Although a number of alcohol protection schemes were examined, a compound bearing a methyl ether, **11a** and one displaying an acetoxy group, **11b** emerged as the most useful. These two manipulations were deemed necessary to preclude the migration of the unsaturation into conjugation with the carbonyl group, a phenomenon known to proceed with unusual facility in the dihydrotropone series (ref. 25).

10   ⟶

11a, R=Me
11b, R=Ac

A surprising characteristic of dihydrotropones which came to light during this investigation was their reluctance to undergo nucleophilic addition at the carbonyl carbon with most carbanionic reagents. Attempted additions of Grignard and organolithium reagents to a series of dihydrotropone derivatives resulted in only recovered starting material, suggesting that enolization had occurred in preference to addition in these cases. This result is not entirely surprising since the carbon atoms adjacent to the ketone in these compounds are flanked on both sides by acidifying substituents. This carbonyl addition problem can be alleviated in a quite satisfactory manner by employing the corresponding "cerium carbanion" which appears to behave as a much less basic, more carbonylophilic species (ref. 26). The contrast in the behavior of compound 12 toward methyllithium and its reaction with methyllithium in the presence of $CeCl_3$ is a dramatic example of this situation.

S.M. $\xleftarrow{CH_3Li}$

12    $\xrightarrow[\substack{CeCl_3 \\ 88\%}]{CH_3Li}$

13

The continuing goal of this synthetic program was the efficient construction of a hydroazulene intermediate which would afford the maximum flexibility for building a wide range of target molecules. Toward this end, construction of a functionalized *cis*-fused hydroazulene was the pivotal event for successfully implementing this broad based strategy. The *trans*-locked isomer in most perhydroazulenes is favored relative to the corresponding *cis*-fused species in an equilibrium process (ref. 27). Thus it was envisaged that both ring fusions could be obtained from a base catalyzed equilibration of a common *cis*-fused intermediate. Furthermore, in the absence of other stereochemical factors, such as the angular methyl group present in the pseudoguaianolide series, the facial bias inherent in the *cis*-fused species could contribute some guidance to the approach of external reagents in subsequent operations.

With compounds **14a,b** available in large quantities from the acid hydrolysis of **11a,b**, the stage was set for examining the means for effecting a stereoselective closure to provide the annulated five-membered ring. It was initially anticipated that a mixture of hydroazulenic

14a, R=Me
14b, R=Ac

15a, R=Me
15b, R=Ac

diene alcohols would be the principle product resulting from an acid-mediated cyclization of 14 (ref. 28). However, aldehyde 14 was rapidly converted into the corresponding cyclic ether 15 in yields of 88 - 95% on treatment with 2 equiv. of BF$_3$•Et$_2$O at 0°C. This remarkably facile transformation marked the first documented example of an *intra*molecular Lewis acid catalyzed hetero Diels-Alder reaction (ref. 29-31) and provided a particularly efficient entry into the requisite hydroazulene skeleton. This powerful process produces, in a single step, a molecule rich in stereochemical information and the requisite *cis*-ring fusion was assured by virtue of the rigid geometrical constraints placed on the approach of the tethered heterodienophile toward the diene partner. Aldehyde 14b was converted into 15b with only slightly less efficiency.

A number of fascinating aspects of this reaction came to our attention during the course of our investigations. For example, the facility of this cyclization in the dihydrotropone series was demonstrated in dramatic fashion, when acetal 11a provided cyclic ether 15a directly in 77% yield upon exposure to a solution of 50% aqueous trifluoroacetic acid in THF. This remarkable development permitted the hydrolysis of the acetal and the cyclization to be performed in a single operation without sacrificing overall yield. A further observation with

16a, R=CH$_3$
16b, R=CH(CH$_3$)$_2$

17a R=CH$_3$ 73%
17b , R=CH(CH$_3$)$_2$ 58%

profound synthetic implications was the surprising ease with which a ketone dienophilic partner underwent cycloaddition when tethered to the dihydrotropone ring.

Normally ketones are much less useful than aldehydes as participants in acid catalyzed olefin-cyclization reactions and we were gratified to find that tethered ketone dienophiles participated in this cyclization protocol with only a small decrease in efficacy relative to the aldehyde examples (ref. 32). The success of this "cycloaddition" process in the isopropyl series (16b to 17b) was particularly exciting (ref. 33).

Unfortunately, not all substrates examined were as amenable to cycloaddition under these conditions. Extending the side-chain in aldehyde 14a by one methylene unit completely suppressed the cycloaddition mode of reaction. A myriad of products were produced when compound 18 was exposed to a variety of Lewis acid catalysts. Apparently, the additional

OMe

decomposition products

18

methylene group in the side-chain raises the energy level of the conformation required for cycloaddition through unfavorable nonbonding interactions.

An alternative stereoselective entry into the cis-perhydroazulene skeleton, developed recently in our laboratory, is also based on the concept of attaching a 3-carbon unit onto the tropone nucleus. In this example, lithio *tert*-butylacetate is added in 1,8-fashion to tropone to provide keto-ester 19 in excellent yield. Interestingly, this was the first reported example of an enolate addition to tropone or a tropone derivative of which we are aware. We have thoroughly examined this transformation and it appears to be general for a range of enolate species. Selected examples are shown in Table I (ref. 34). With a functionalized two carbon side-chain appended to the cycloheptadienone ring, methods for cyclization to the five-membered carbocycle were

9 ⟶ t-BuO

OCOt-Bu

19   **20**

considered. A diazoketone insertion into one of the adjacent diene double bonds appeared viable in this situation and the resulting cyclopropyl ketone could provide ample reactivity from which both carbon and heteroatom substituents could be introduced onto the hydroazulene system.

Table I. Addition of Enolates to Tropone

Enolate	Product	Yield
		68%
		64%
		95%
		90%[a]
		84%[b]

[a] Addition performed on 2-methyltropone. [b] Addition performed on 2-chlorotropone.

Prior to converting the side-chain carboxylate into the corresponding diazoketone, the cycloheptanone ketone was selectively reduced and protected. Copper mediated insertion of the carbenoid carbon into the proximate unsaturation of the cycloheptadiene system yielded a conformationally rigid and facially biased hydroazulenic intermediate in the form of cyclopropyl ketone 20 (ref. 35). Once again the constraints placed on the cyclization step insure the formation of the desired cis-fusion between the newly created five-membered ring and the existing cycloheptane moiety.

Comparison of the structures of compounds 15 and 20 reveals that both species harbor reactive functionality at key positions in the perhydroazulene skeleton and possess significant stereochemical information which should be readily exploitable for construction of a variety of target molecules. Indeed, both of these intermediates offer a number of sites for stereocontrolled elaboration of substituents. The arrows are suggestive of locations that may be particularly susceptible to selective manipulation.

15b   20

2.1   SYNTHESIS OF (±)-DEHYDROCOSTUS LACTONE

The intriguing paucity of synthetic successes in the guaianolide area stimulated our interest in utilizing tropone-based chemistry as a possible starting point for the construction of some members of this large group of sesquiterpene lactones. Dehydrocostus lactone (21) and grosshemin (22) are typical examples of this family. Brief examination of the structural features of these materials reveal several distinctions which set them apart from the related pseudoguaianolides. The presence of a *cis*-fusion in the hydroazulene moiety and the absence of

21   22

a methyl group at C_5 are initially the most obvious differences which would be expected to have an impact on the planning of a synthesis. One of the crucial questions that emerged at the outset of our investigations in this area was whether substituents could be effectively appended to intermediates such as 15 or 20 with an appropriately high degree of stereoselectivity. Hydroazulene precursor 15 was viewed as a particularly attractive starting point from which to assemble guaianolides since the contrathermodynamic *cis*-ring fusion was already intact. There is some evidence, however, to suggest that once a *trans*-fused butyrolactone is in place the *cis*-ring fusion in the hydroazulene moiety becomes thermodynamically preferred (ref. 3b and references cited therein). Furthermore, useful functionality in this compound was disposed at four key positions around the bicyclo[5.3.0]decane skeleton. The key to realizing the utility of this substance for perhydroazulene synthesis remained the facility with which the allylically activated carbon-oxygen bond could be severed in a synthetically viable fashion.

Compound 15a was selected for initial examination because the robust methyl ether substituent at C_{10} was expected to survive a wide range of reaction conditions intact and our plans included postponing the C_{10} methylene group introduction until late in the synthesis. Thus this substituent was expected to play a relatively benign role in subsequent transformations. Unfortunately, the well-known difficulties in removing methyl ethers could eventually prove to be a liability in the latter stages of this synthesis.

The relatively simple guaianolide, dehydrocostus lactone (21) was selected as the first target of this study because its synthesis would test the viability of the salient features of our synthetic planning without necessitating cumbersome functional group manipulations which can often be the most stubborn obstacle to completing a total synthesis. In addition, this lactone presented a number of interesting opportunities for conversion into some more highly oxygenated guaianolides at a future stage (ref. 36).

As alluded to earlier, the utility of compound 15a in hydroazulene synthesis depended critically on the ability to selectively break the carbon-oxygen bond at C_8 in the presence of three other carbon-oxygen bonds. Selective reductive cleavage of the C_8-bond employing dissolving metal conditions was an attractive option, particularly in the context of the synthesis of simple guaianolides which have a minimum of substituents in the seven-membered ring. This type of transformation has ample precedence in the literature (ref. 37), and exposing compound 15a to lithium metal in refluxing monomethylamine provided hydroazulene **23** in a straightforward fashion. Remarkably, only a single double-bond regioisomer was detected in the product isolated from this reaction. This fortuitous result greatly simplified the plans for subsequent elaboration of the *trans*-γ-butyrolactone unit and may represent a thermodynamic preference for the double bond isomer proximate to the ring fusion in the hydroazulene system. In every case examined to date the 6,7-double bond isomer was the exclusive isomer observed from reductive cleavage of the C_8 carbon-oxygen bond.

Protection of the C_4 alcohol as a MEM-ether and stereoselective epoxidation set the stage for elaborating the elements of the *trans*-butyrolactone ring. Regioselective oxirane opening with dilithioacetate provided the key lactone intermediate **24**. The attack of the nucleophile in this process proved to be very regioselective, yielding less than 5% of the unwanted isomer resulting from attack at C_6. Simultaneous deprotection of the C_4 and C_{10} alcohols with the mild TMSCI/NaI reagent (ref. 38) followed by oxidation and careful *bis*-methylenation (ref. 39) gave the penultimate target **25**. During this phase of the total synthesis, we had occasion to examine in some detail the TMSCI/NaI reagent for cleaving ether linkages. This *in situ* source of TMSI was very effective in removing MEM-ethers in the presence of other functions. Returning to the synthesis, routine introduction of the α-methylene carbon onto the lactone in compound **25** completed the first total synthesis of dehydrocostus lactone in only 12 steps from tropone. As suggested at the outset of our efforts in this area, dehydrocostus lactone serves admirably as a point of embarkation for making more functionalized guaianolides. Selective equilibration of the C_4 exocyclic methylene double bond to the corresponding trisubstituted olefin followed by stereoselective epoxidation gave (±)-estafiatin (**4**) (ref. 36 and references cited therein).

Two principal observations about the differences between the behavior of intermediates in the guaianolide series and pseudoguaianolide series were made during this synthetic study. Not surprisingly, transannular interactions appear to be more pronounced in the flexible *cis*-fused hydroazulene relative to the somewhat more rigid *trans*-locked system. This distinction manifested itself in the epoxidation step of the dehydrocostus lactone synthesis. Substantial quantities of cyclic ether **26** accompanied the formation of the desired epoxide **27**. After an exhaustive survey of other protection schemes for the C_{10} position and epoxidation reaction conditions, the methyl ether proved to be the least susceptible protecting group to neighboring group interaction with the C_6 position (ref. 40).

26

27

A second and somewhat more pernicious difference between these closely related skeleta which has had a major impact on our synthetic planning is the location of the C_{15} carbon. In the pseudoguaianolide family this substituent resides at the ring fusion and plays a well-recognized role as a stereochemical control element which has been exploited in a number of syntheses. However, this carbon, in perhaps a less well appreciated role, also serves as an "insulator" between the functionality located on the five-membered ring and the lactone unit attached to the cycloheptane. This phenomenon made its presence felt when we attempted to introduce the exocyclic methylene carbon at the C_4 ketone. Most conventional reagents for methylenation

28

gave only the elimination product **28**. Vandewalle and coworkers may also have observed a similar response to attempted methylene group introduction at the C_4 position in the guaianolide series (ref. 46). When the relatively non-basic, nucleophilic reagent P-(lithiomethyl)-N,N-dimethyl-P-phenyl phosphinothioic amide, which has met with good success with notoriously enolizible ketones, was employed modest yields of addition products were obtained but substantial amounts of **28** were also formed (ref. 39). Approaches to solving the problems arising from the 1,3-disposition of the C_4 ketone and the butyrolactone oxygen have met with only limited success to date. One solution which has been effective and will be explored in the next section is simply not to have a carbonyl group located at this position.

2.2 SYNTHESIS OF (±)-GROSSHEMIN

The hydroazulene intermediate **15** is a veritable storehouse of stereochemical information. Unfortunately, the reductive fission protocol used to break the allylic carbon-oxygen bond in the dehydrocostus lactone synthesis squanders, in a sense, the potentially useful stereogenic center at C_8. In view of our desire to develop a general strategy into the guaianolide family from a common hydroazulene intermediate, it was envisaged that the activated carbon-oxygen bond at C_8 in **15** could, under certain circumstances, be severed with concomitant formation of a new heteroatom-carbon bond with inverted stereochemistry. Oxygen would be the heteroatom ultimately required at this position, thus permitting a rapid, stereoselective entry into the series of guaianolides which exhibit an α-oriented hydroxyl group at C_8. Grosshemin

560

(22) is a typical example of the group of compounds which could be accessed by this strategy. The well-established reluctance of alkoxides to participate as leaving groups in bimolecular nucleophilic substitution reactions would necessitate employing a nucleophilic process assisted by the intervention of a Lewis acid, but early efforts at direct introduction of an oxygen using Lewis acid mediated conditions failed in this system (ref. 41).

15　　　　　　　　　　　　　　　**22**

After considerable investigation, the most reproducible results were obtained using a sulfur nucleophile in the presence of a Lewis acid (ref. 42). The sulfur could be viewed as an equivalent to the requisite oxygen at this position by virtue of their interconvertibility *via* a [2,3]-sigmatropic rearrangement of the corresponding sulfoxide. Exposure of **15b** to excess $BF_3 \cdot Et_2O$ in thiophenol as solvent provided an easily separable 2:1 mixture of sulfides **29** and

29　　　　　　　　　　　　　　　**30**

30, respectively, in 62% yield. Single crystal x-ray analysis established the stereo- and regio-chemistry of the phenylthio substituent in each case. Efforts to identify reaction conditions which were more regioselective were generally unsuccessful. Conversion of the resultant sulfide regioisomers into the corresponding allylically transposed alcohols *via* sulfoxide rearrangement followed by several routine manipulations to convert the C_{10} acetoxy group into the exocyclic methylene group required for grosshemin gave alcohols **31** and **32**. With the formation of substantial quantities of two regioisomeric series of compounds, the ultimate success of this synthetic endeavor was predicated on the utility of both regioisomeric alcohols as productive precursors to the grosshemin target molecule. The key to this notion was an observation made by Danishefsky which suggested that dilithioacetate can attack *syn*-epoxy alcohols preferentially at the oxirane carbon adjacent to the hydroxyl substituent in a number of circumstances (ref. 45). Therefore, if this phenomenon was general in scope, there was reason to expect that epoxides derived from both 31 and 32 would lead to the correct lactone when treated with the acetic acid dianion nucleophile.

31   **32**

Directed epoxidation of each allylic alcohol followed by exposure to excess dilithioacetate gave lactone 33 as the only isolable product in each case. This type of "regioconvergent" process greatly facilitates the general entry into the entire genre of natural products which display a hydroxy-lactone in their structures. Introduction of the C_4 methyl group and transposition of the C_4 ketone to the C_3 carbon were the only operations remaining in this synthesis of grosshemin. Unfortunately, ketone 34 refused to undergo nucleophilic addition under any conditions examined. As in earlier examples, enolate induced β-elimination appeared to be the principal reaction pathway; indeed, in this particular substrate elimination proved to be the only pathway followed.

33   **34**

At this junction, intermediate 20, prepared as outlined in Section 2, appeared to offer a number of advantages as a potential solution to the problems encountered when employing synthetic intermediates which possess a ketone at C_4 (ref. 47). Crucial to the success of compound 20 as an intermediate in the construction of grosshemin would be the ability to regio- and stereoselectively introduce the methyl group at C_4 and elaborate an appropriate oxygen substituent at C_6 or C_8. In principle, the cyclopropyl ketone portion of this species can provide reactivity commensurate with the task at hand. (Scheme II). In this context, several interesting reports dealing with acid-mediated heteroatom attack at the γ-carbon of cyclopropyl ketones have been forthcoming recently (ref. 47) and a variation on this theme would be particularly

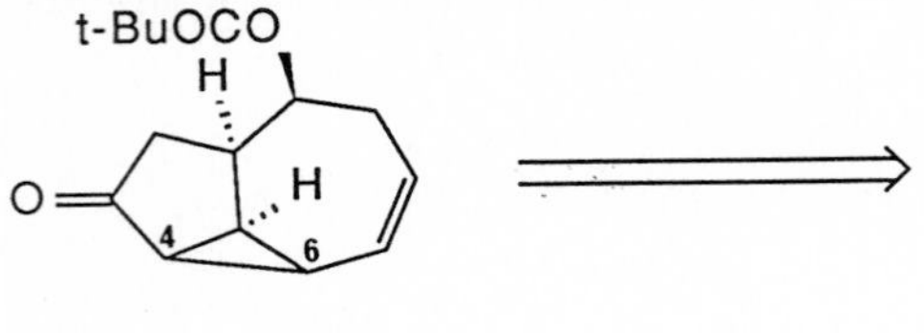

applicable to the grosshemin problem. Incumbent upon any methodology selected would be the requirement that the C$_4$ position would be ultimately susceptible to selective alkylation. While an enol or enolate species is a putative intermediate in all cyclopropyl ketone opening reactions,

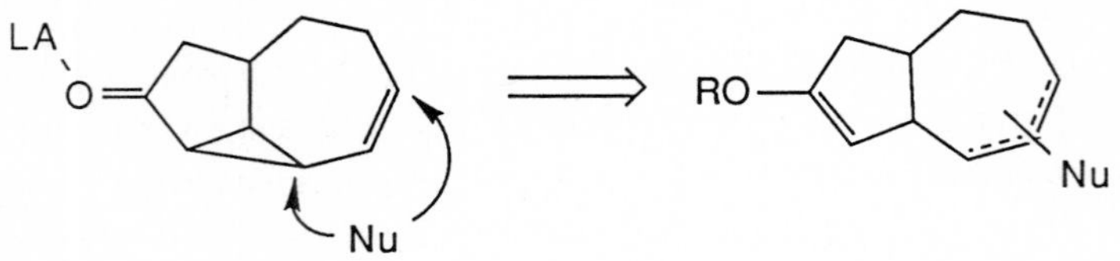

Scheme II

little success had been reported in efforts to trap this intermediate in a usable form. In response to this situation, we developed an alternative technology which has proven to be exceptionally useful in this regard. Treatment of 20 with excess BF$_3$·Et$_2$O in acetic anhydride at -40°C gave bis-acetate 35 in 93% yield (ref. 49). The vinyl acetoxy group in this compound can be viewed

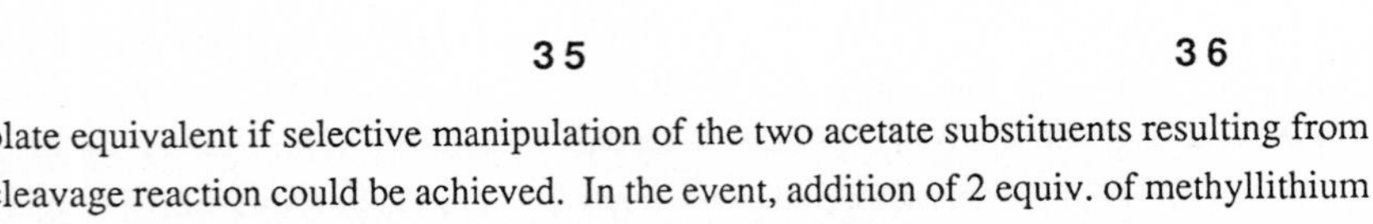

as an enolate equivalent if selective manipulation of the two acetate substituents resulting from the ring cleavage reaction could be achieved. In the event, addition of 2 equiv. of methyllithium to 35 at -78°C followed by quenching with excess methyl iodide led to acetoxy-ketone 36 in good yield. Complete selectivity was observed in the attack of the organolithium reagent on the vinyl acetate to yield the required regiospecifically generated enolate. After careful protection of the cyclopentanone, the C_{10} methylene group was elaborated in conventional fashion and the requisite trans-butyrolactone was again introduced by regioselective attack on the *syn*-epoxy-alcohol 37 culminating in the first total synthesis of the C_8-oxygenated guaianolide, grosshemin.

3 7

2.3 CONSTRUCTION OF THE HYDRODICYCLOPENTA[a,d]CYCLOOCTANE RING SYSTEM.

In recent years the ophiobolane sesquiterpenes, such as ceroplasteric acid (38) and ophiobolin F (39) have attracted considerable attention in the synthetic community (ref. 50). The unusual carbon skeleton with its attendant stereochemical intricacies have made these C_{25} terpenoids particularly challenging synthetic targets. To date, no member of this relatively small group of compounds has yielded to total synthesis.

3 8   **3 9**

While a variety of ring expansion strategies have already been examined as entries into the 5-8-5 tricycle, we envisioned pursuing a novel one-carbon ring enlargement protocol based on the stereoselective elaboration of the hydroazulene precursor 15 into the A and B rings of the hydrodicyclopenta[a,d]cyclooctane target (ref. 51). The introduction of the elements of the C-ring was based on transferring the easily established chirality of the C_6 carbon-oxygen bond in compound 40 into the allylically transposed carbon-carbon bond at the C_{10} (ophiobolane numbering) position through a silyl enolate Claisen rearrangement (ref. 52). Compound 40 was prepared by Lewis-acid assisted nucleophilic opening of compound 15 with thiophenol followed by allylic sulfoxide rearrangement in a manner analogous to that described in the previous section. In the initial model study, the Claisen technology was performed on the propionate ester of **40**, providing after rearrangement, compound **41** as an easily separable 85:15 mixture of isomers. The cycloheptanone ring in the major isomer of **41** was subjected to a completely regiospecific one-carbon ring expansion using trimethylsilyldiazomethane in the presence of $BF_3 \cdot Et_2O$ (ref. 53). The direction of ring homologation and the relative

1 5

configurations at the newly created stereogenic centers were established by single-crystal x-ray analysis on compound **42**. Of particular significance was the formation of the correct

4 0   **4 1**

configuration at C_{14} for accessing ceroplasteric acid. After some routine oxidation level adjustments, the side-chain was cyclized *via* a copper-mediated diazoketone insertion into the proximate double bond in the eight-membered ring to give **43** as a single compound. Reductive

4 2   **4 3**

fission of the cyclopropylketone using dissolving metal conditions provided the tricyclic compound **44** as the only isolable product.

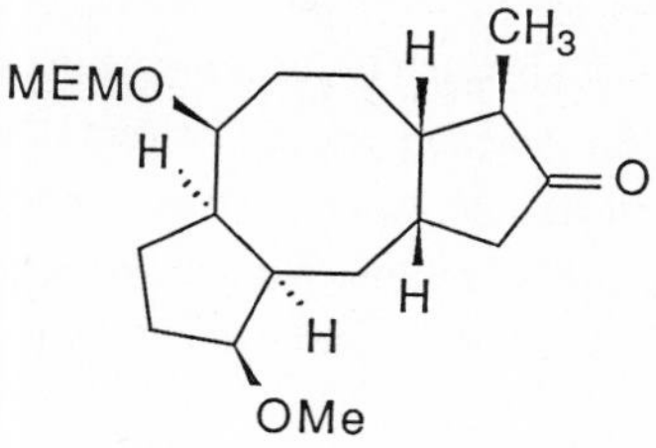

4 4

This model study has clearly provided further validation for the notion of building complex multi-ring systems directly from the functionality and stereochemistry contained in the readily available cyclic ether **15**. Further work is now underway to utilize the basic tenets of this strategy in the asymmetric synthesis of ceroplasteric acid.

3.　TROPONE CYCLOADDITION REACTIONS IN SYNTHESIS

The varied nature of tropone cycloaddition chemistry has stimulated the development of numerous fascinating strategies designed for constructing the carbon framework of complex

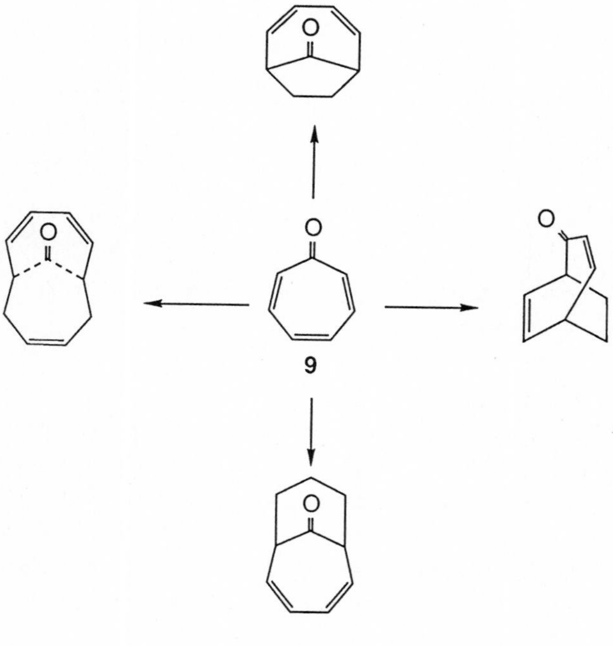

Scheme III

natural products. Scheme III depicts several of the more synthetically interesting modes of tropone "cycloaddition" which have been examined in recent years.

While the diversity of cycloaddition chemistry exhibited by tropone and its derivatives can present difficulties in rational synthetic design, a rapidly growing data base in this area affords the synthetic chemist ample opportunity for constructing complex ring systems with a high degree of selectivity when reaction conditions are chosen judiciously. Several cycloaddition pathways available to tropone which can, under appropriate circumstances, display synthetically useful periselectivity will be examined in subsequent sections. Examples have been selected for their utility or potential utility in natural product synthesis.

3.1 TROPONE AS THE DIENE PARTICIPANT: THE [4π+2π] MODE OF CYCLOADDITION

It has been known for some time that tropone and its derivatives can participate as the 4-electron partner in [4+2] cycloadditions in the presence of a wide variety of dienophiles (ref. 54). Both electron-rich and electron-deficient double bonds appear to cycloadd with facility to the tropone nucleus.

The regioselectivity of reactions involving 2-substituted tropones is not particularly high in many cases (ref. 54d) but unsymmetrically substituted dienophiles tend to add to tropone with quite high selectivity. This is particularly true in "inverse electron demand" cycloadditions which

employ electron rich olefins such as vinyl ethers and vinyl sulfides as dienophilic partners with the electron-poor diene tropone(ref. 55). The high regioselectivity observed in these cases is consistent with predictions based on the currently accepted FMO treatment of tropone (ref. 22). Compound 45, the reaction product of tropone and butyl vinyl ether, has been converted into compound 46, a potential precursor to α-himachalene (47), via an alkoxide accelerated Cope rearrangement (ref. 55,56). This strategy was designed to ensure exclusive formation of the requisite *cis*-fusion between the six- and seven-membered rings. A short route into bicyclo[3.2.2]nona-3,6,8-trien-2-one(48), an important precursor to a number of theoretically

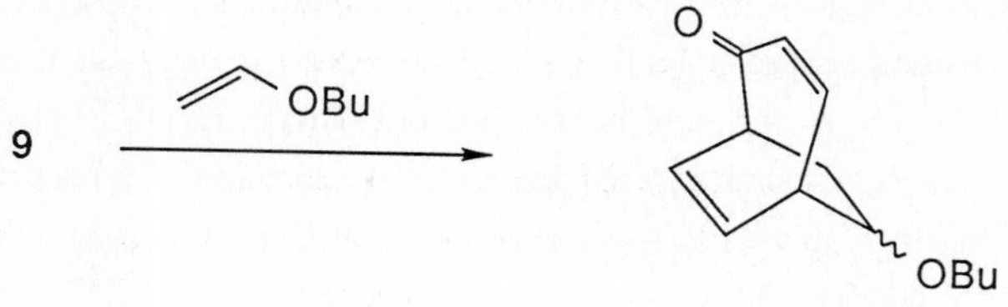

45

interesting molecules, has as the crucial step a regioselective cycloaddition of vinyl phenyl sulfide to tropc :e (ref. 57). Other work dealing with the regioselectivity of dienophile

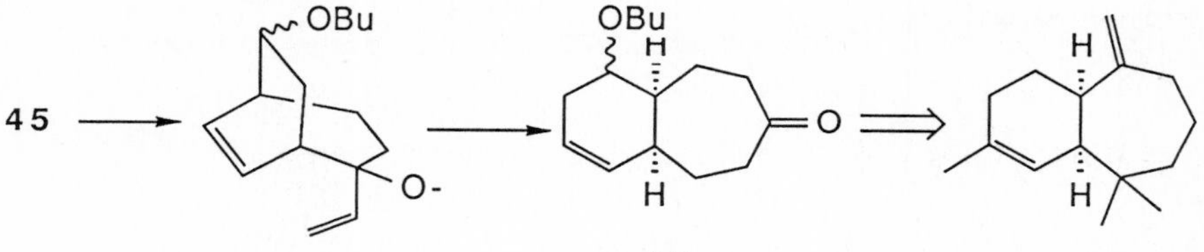

46 **47**

cycloaddition to troponoids have been reported (ref. 58) and the adducts obtained from these reactions are capable of further undergoing some interesting transformations.

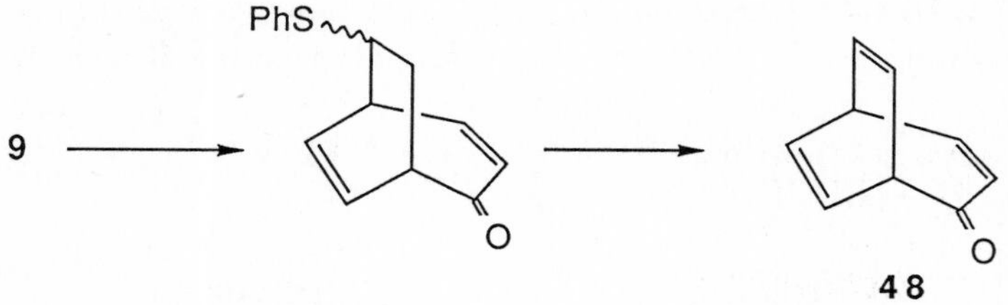

48

Uyehara has reported a fascinating photo-induced conversion of bicyclo[3.2.2] nonadienone, derived from the addition of ethylene to tropone, into a norcarene derivative (ref.

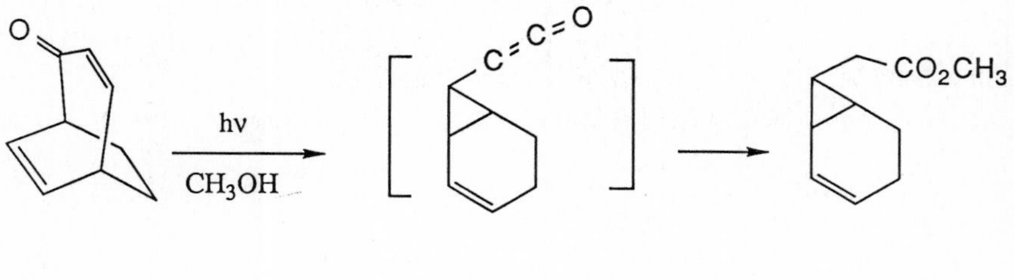

59).   He and his coworkers have utilized this process to good advantage for the synthesis of several natural products (ref. 59b).

Not surprisingly, the intramolecular variant of the [4+2] tropone-olefin cycloaddition technology has emerged recently.This protocol offers promise as a powerful route into a number of polycyclic natural products. Funk has disclosed a new strategy into the cedrane ring system with the well-conceived total synthesis of pipitzol (49)(refs. 6b,60). Typically, these reactions proceed with high endo selectivity and apparently the cycloaddition is not particularly susceptible to steric inhibition. Oxidative ring cleavage followed by a facile intramolecular aldol reaction completed the synthesis.

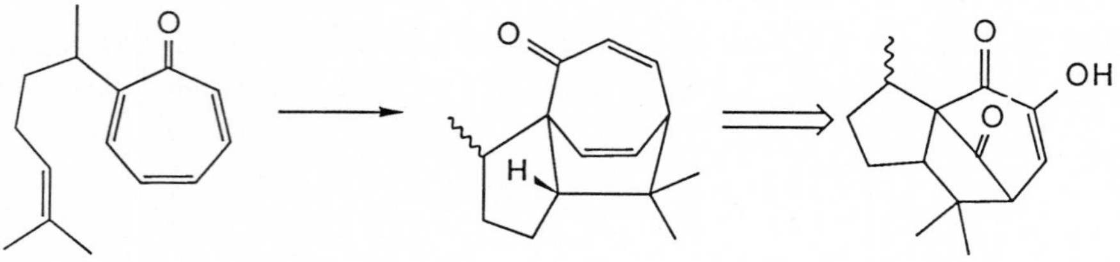

4 9

3.2   INTRAMOLECULAR [6π+2π] PHOTOCYCLOADDITION OF ALKENYLTROPONES

Feldman has recently reported an exciting new tropone cycloaddition process which quickly assembles potential precursors to the important bicyclo[6.3.0]undecane carbon skeleton (ref. 61). Compounds 50 are produced in a [6π+2π] cycloaddition of the olefin across the triene of tropone. Tricycle 51 apparently arises from a competitive [8+2] addition process (ref. 62). Interestingly, the reaction appears to proceed *via* the hydroxytropylium ion 52 and generally

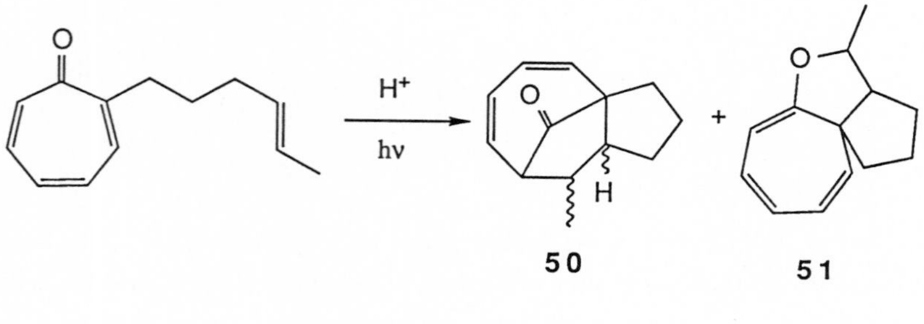

5 0   **5 1**

5 2

provided mixtures of diastereomeric products. This important cycloaddition may offer a particularly efficient entry into synthetically useful bicyclo[6.3.0]undecane intermediates and may find applications in the total synthesis of natural products such as the ophiobolins and dactylol.

3.3   [6+3] CYCLOADDITION OF TRIMETHYLENEMETHANE PRECURSORS WITH TROPONE

Bicyclo[4.3.1]decanone species, such as **53**, can be prepared with a high degree of efficiency by the Pd(0) catalyzed addition of trimethylenemethane precursor **54** across the triene system of tropone (ref. 63).

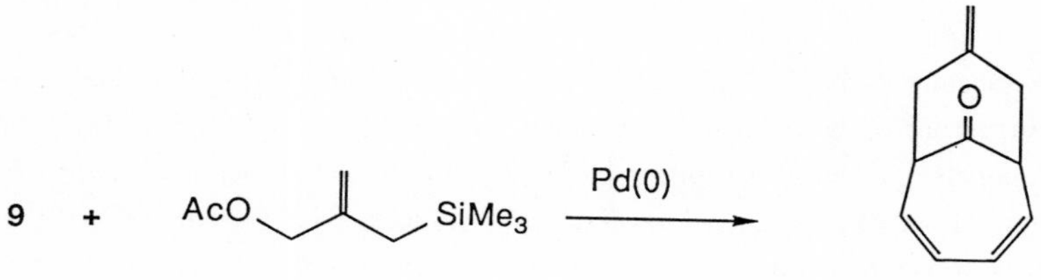

9 + AcO⟍⤴⟋SiMe₃ —Pd(0)→ **53**

5 4   **5 3**

This formal [6+3] cycloaddition process apparently proceeds through an initial nucleophilic addition to the C_2 position of tropone in a manner analogous to the addition of other carbanions. This is followed by attack of the resultant enolate onto the electrophilic π-allyl palladium center as depicted in Scheme IV.

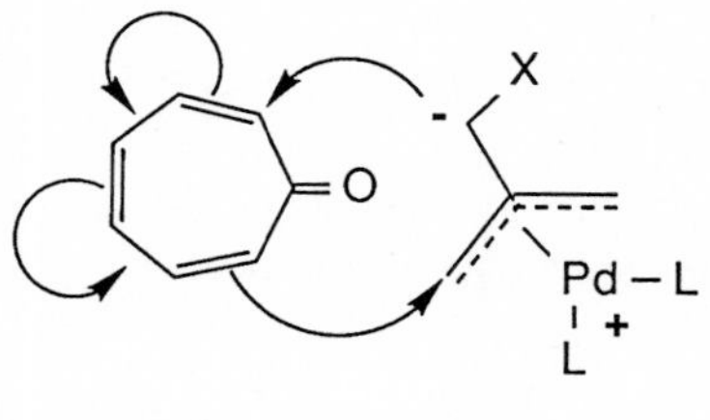

Scheme IV

The regiochemistry of this addition is another manifestation of the propensity for reaction at C_2 and C_7 displayed by cycloheptatrienone (tropone).

3.4   [6+4] CYCLOADDITION OF DIENES TO TROPONE

Tropone is known to readily undergo a thermally induced [6+4] cycloaddition with a wide range of diene partners, including cyclopentadiene, (ref 64a,b), cyclohexadiene (ref. 64c), butadiene (ref. 64d) as well as a host of other compounds. The periselectively of the reaction of a given diene with tropone is critically dependent on the temperature and duration of the reaction as well as the structures of the two participants. Heating tropone and cyclopentadiene at 80°C yields the [6+4] adduct 55, but performing the reaction at 145°C yields the presumed

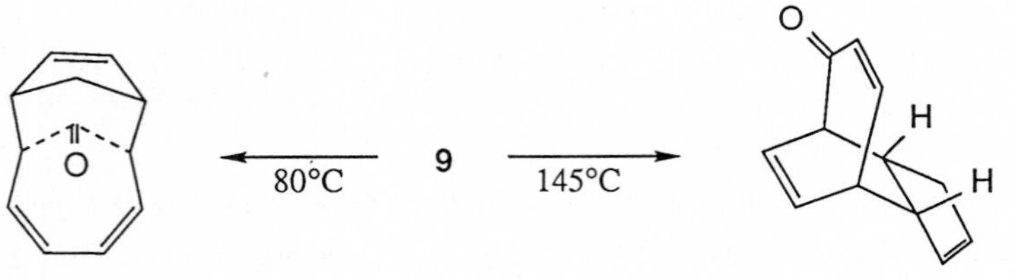

5 5   9   **5 6**

thermodynamically favored [4+2] adduct 56 (ref. 64). The substitution pattern on either reaction partner and electron-density alteration in the tropone nucleus can also play a major role in determining the product distribution in these transformations (ref. 65). For example, we have observed that 1-acetoxy butadiene provided only products arising from the [4+2] mode of cycloaddition when reacted with a 2-alkyl tropone partner (ref. 66).

An early effort to use the [6+4] cycloaddition of cycloheptatrienone for synthesis was reported by Garst (ref. 18b). The relatively high level of exo-selectivity displayed by this

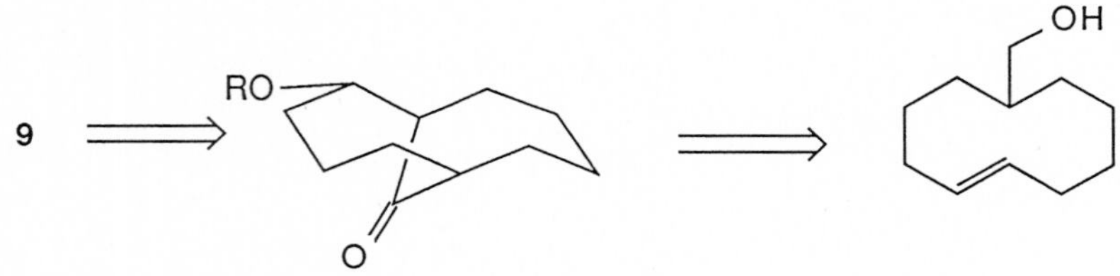

process and the rigidity of the resultant bicyclo[4.4.1]undecanone molecule was exploited in a well-conceived entry into substituted 10-membered rings.

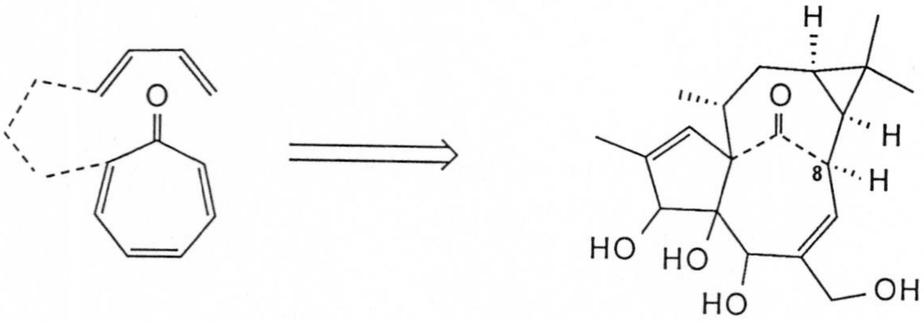

5 7

A particularly exciting application of the [6+4] cycloaddition process is in the direct assembly of the A,B and C rings of the cocarcinogenic diterpene, ingenol (5). This type of cycloaddition in the tropone series is eminently suitable for construction of the bicyclo[4.4.1]undecanone nucleus which is characteristic of the ingenane class of natural products. Immediate access to the carbon framework of so-called isoingenol (57) (ref. 6a) which is epimeric with the natural product only at C8 can be achieved in a highly efficient fashion using this protocol.

Indeed, both an *inter-* and *intra*molecular approach into basic tricyclic building blocks for the ingenol diterpenes have been developed in our laboratory (ref. 66). Cycloaddition of tropone (9) with 1-acetoxybutadiene gave a single cycloadduct 58 which was converted into enone 59

by a series of routine oxidation level adjustments. Efficient alkylation of a regiospecifically generated bridgehead enolate with 2-methoxyallyl bromide followed by stereoselective 1,4-addition of a methyl substituent gave after hydrolysis and aldol condensation tricycle 60. The key to the success of this strategy was the general efficiency of the alkylation of the bridgehead enolate, a reaction with relatively little precedence in the literature (ref. 67). Intermediate 60 has functionality well-positioned for use in elaborating the remaining substituent of the ingenane system.

During the course of this study, we also developed the first example of an *intra*molecular [6+4] cycloaddition of a diene to tropone. Tethered diene 61 led to tricycle 62 in 82% yield when heated for several hours. This reaction is quite general and has been effected with a number of different substituents located at various positions on the side-chain (ref. 69). Subsequent to our report, Funk also disclosed several examples of the intramolecular cycloaddition of dienes to tropone (ref. 6b).

4.   ORGANOMETALLIC DERIVATIVES OF TROPONE: ALTERATIONS IN
      NORMAL REACTIVITY PATTERNS

The normal reaction patterns displayed by tropone can be altered in rather dramatic ways
by pre-complexation with transition metals such as iron. Franck-Neumann and coworkers have
revealed several fascinating examples of how the tropone-iron tricarbonyl complex (63)

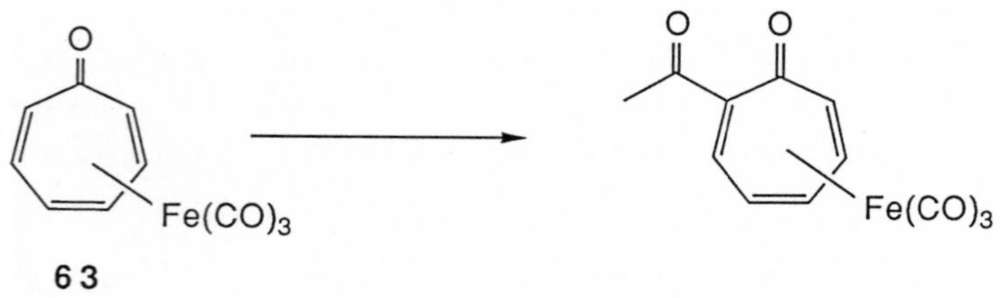

63

undergoes reactions that are not normally accessible to tropone itself (ref. 65e, 69). Compound
63 provides the 2-acylated species on exposure to conventional Friedel-Crafts conditions. This
intermediate also displayed reactivity not customary for tropone by becoming susceptible to bond
formation at the 3-position in the tropone ring (ref. 70). The resulting products from these
reactions were then efficiently converted into the terpenes, β-thujaplicin (64) and β-dolabrin
(65).

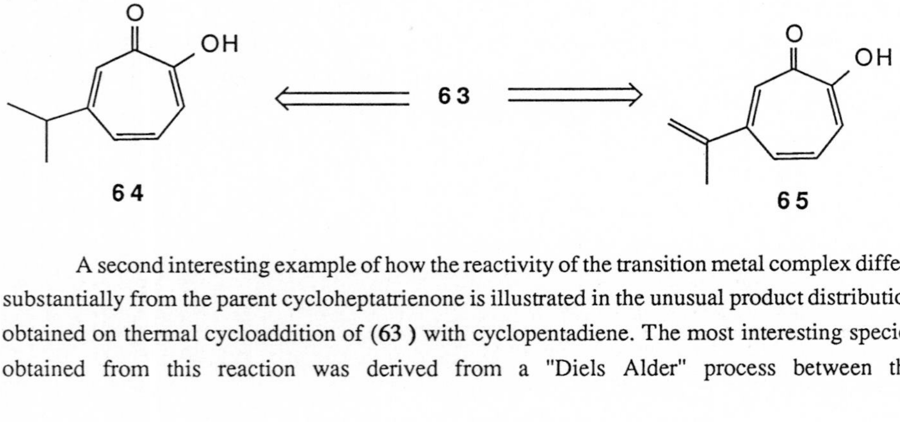

6 4   63   6 5

A second interesting example of how the reactivity of the transition metal complex differs
substantially from the parent cycloheptatrienone is illustrated in the unusual product distribution
obtained on thermal cycloaddition of (63) with cyclopentadiene. The most interesting species
obtained from this reaction was derived from a "Diels Alder" process between the

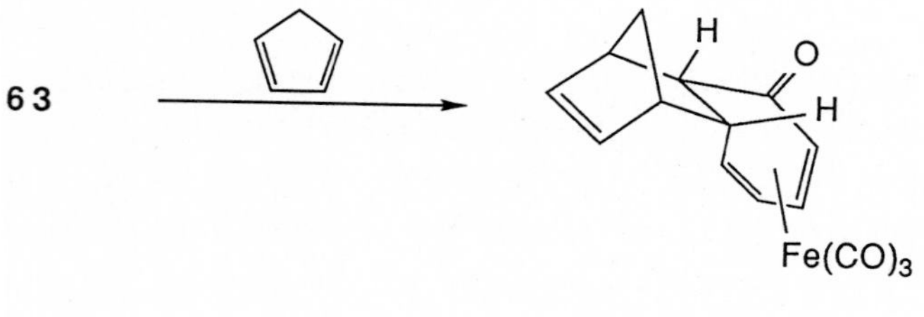

63

complex (63) which behaved as the 2π component and the diene. This result nicely complements the normal mode of the reactivity of tropone itself which rarely participates as the 2π moiety in a Diels-Alder type reaction.

Yet another fascinating change in the course of a normal tropone reaction has been reported by Brookhart and coworkers (ref. 71). Addition of Grignard or organolithium reagents to complex 63 resulted in a 1,2-addition of the carbanion directly to the carbonyl carbon. In the absence of complexation, this addition occurs exclusively in a 1,8-fashion (see Section 2). This remarkable transformation greatly expands the range of positions in the tropone nucleus which can undergo carbon-carbon bond forming reactions.

5.   SUMMARY

The multifarious reactivities exhibited by tropone and its derivatives which have been discussed in this review, can provide the synthetic chemist with ample opportunities to construct highly-functionalized polycyclic carbon skeleta. Many of the systems accessible by these routes may be difficult or impossible to assemble with equal efficacy in other ways. It is clear that continuing efforts in this direction will reap substantial benefits.

ACKNOWLEDGEMENT

I wish to thank the following co-workers whose hard work and enthusiasm have played a major role in bringing our tropone work to fruition: Jean-Marc Sage, Dr. JoAnn Zbur Wilson, Dr. Neel Balasubramanian, Dr. Chrisantha Senanayake, Terry Moore, Sushil Rege, Dr. David Head and Pawel Kierkus. I would like to thank Helen Brewster for technical assistance. I would also like to thank the National Institutes of Health for their generous support of our research programs through grants GM 30771 and CA36543.

REFERENCES

1. C. H. Heathcock, Total Synthesis of Sesquiterpenes, in: J. ApSimon (Ed.), The Total Synthesis of Natural Products, Vol. 2, Wiley-Interscience, New York, 1973, pp. 395-428.
2. N. H. Fischer, E. J. Olivier, H. D. Fischer, Fortschr. Chem. Org. Naturst., 38 (1979) 47.
3. For some synthetic approaches to these sesquiterpenes, see:
    a) C. H. Heathcock, S. L. Graham, M. C. Pirrung,F. Plavac, C. T. White, The Total Synthesis of Sesquiterpenes, in: J. ApSimon (Ed.)., The Total Synthesis of Natural Products, Vol. 5, Wiley-Interscience, New York (1983).
    b) M. Vandewalle, P. J. DeClercq, Tetrahedron, 41 (1985) 1767.
    c) P. A. Jacobi, H. G. Selnick, J. Am. Chem. Soc., 106 (1984) 3041.

4. a) S. D. Jolad, R. M. Wiedhopf, J. R. Cole, J. Pharm. Sci., 63 (1974), 1321.
   b) S. M. Kupchan, M. A. Eakin, A. M. Thomas, J. Med. Chem., 14 (1971) 1147.
   c) W. Vichnewski, S. J. Sarti, B. Gilbert, W. Herz. Phytochemistry, 15 (1976) 191.
   d) E. Rodriguez, G. H. Towers, J. C. Mitchell, Ibid., 15 (1976) 1573.
5. a) F. J. Evans; A. D. Kinghorn, Phytochemistry, 13 (1974) 1011.
   b) E. Hecker, Pure Appl. Chem., 49 (1977) 1423.
   c) F. J. Evans, C. J. Soper, Lloydia, 41 (1978), 193.
6. For synthetic work on the ingenanes, see:
   a) L. A. Paquette, T. J. Nitz, R. J. Ross, J. P. Springer, J. Am. Chem. Soc., 106 (1984) 1446.
   b) R. L. Funk, G. L. Bolton, Ibid., 108 (1986) 4655.
   c) J. D. Winkler, K. E. Henegar, P. G. Willard, Ibid., 109 (1987) 2850.
7. For synthetic work on the tigliane ring system, see:
   a) P. A. Wender, C. L. Hillemann, M. J. Szynonifka, Tetrahedron Lett., (1980) 2205.
   b) P. A. Wender, K. Koehler, R. Wilhelm, P. Williams, P. Keenan, H. Y. Lee, in "New Synthetic Methodology and Functionally Interesting Compounds," A. Yoshida, Ed. (1986) p. 163.
   c) P. A. Wender, R. M. Keenan, H. Y. Lee, J. Am. Chem. Soc., 109 (1987) 4390.
8. K. Nakanishi, T. Goto, S. Ito, S. Natori, S. Nozoe, "Natural Products Chemistry" Vol. 1, Academic Press, New York, pp. 285-289.
9. For recent synthetic work, see:
   a) T. Kametani, H. Matsumoto, T. Honda, K. Fukuroto, Tetrahedron Lett., 22 (1981) 2379.
   b) S. Gasa, N. Hamanaka, S. Matsunaga, T. Okunu, N. Takeda, T. Matsumoto, Ibid., (1976) 553.
10. For some examples of the decalone rearrangement strategy, see:
    a) G. Buechi, W. Hofheinz, J. V. Paukstelis, J. Am. Chem. Soc., 91 (1969) 6473.
    b) C. H. Heathcock, R. Ratcliffe, Ibid., 93 (1971) 1746.
    c) C. H. Heathcock, E. G. Delmar, S. L. Graham, Ibid. 104 (1982) 1907.
    d) J. B. P. A. Wijnberg, A. deGroot, Tetrahedron Lett., 28 (1987) 3007.
    e) M. Ando, A. Ono, K. Takase, Chem. Lett. (1984) 493.
    f) J. B. Hendrickson, C. Ganter, D. Dorman, H. Link, Tetrahedron Lett. (1968) 2235.
    g) G. H. Posner, G. L. Loomis, R. D. Mithal, W. J. Frazee, D. M. Schmit, K. A. Babiak, Ibid., (1978) 4213.
11. For the use of other bicyclic systems in hydroazulene synthesis, see:
    a) J. A. Marshall, J. J. Partridge, J. Amer. Chem. Soc., 90 (1968) 1090.
    b) P. A. Grieco, Y. Ohfune, G. Majetich, Ibid., 99 (1977) 7393.
    c) G. J. Quallich, R. H. Schlessinger, Ibid., 101 (1979) 7627.
12. a) J. A. Marshall, W. R. Snyder, J. Org. Chem., 40 (1975) 1656.
    b) R. A. Kretchner, W. S. Thompson, J. Am. Chem. Soc., 98 (1976) 3379.
13. a) P. J. DeClercq, M. Vandewalle, J. Org. Chem., 42 (1977) 3447.
    b) P. T. Lansburgy, D. Hanguaer, J. P. Vacca, J. Am. Chem. Soc., 102 (1980) 3964.
14. a) C. H. Heathcock, C. M. Tice, T. C. Germroth, J. Am. Chem. Soc., 104 (1982) 6081.
    b) A. E. Greene, J. P. Depres, J. Am. Chem. Soc., 101 (1980) 4003.
15. a) J. B. Hendrickson, Tetrahedron, 19 (1963), 1387.
    b) J. B. Hendrickson, J. Am. Chem. Soc., 83 (1961) 4537.
16. a) P. J. DeClercq, Tetrahedron, 40 (1984) 3717.
    b) P. J. DeClercq, Ibid., 37 (1981) 4277.
    c) P. J. DeClercq, J. Org. Chem., 46 (1981) 667.
    d) H. O. House, P. C. Gaa, D. VanDerveer, Ibid., 48 (1983) 1661.
    e) E. Toromanoff, Tetrahedron, 36 (1980) 2809.
17. For a review of the reactions and properties of troponoids, see: F. Pietra, Chem. Rev., 73 (1973) 293.
18. a) A. E. Greene, M. A. Teixeira, E. Barreiro, A. Cruz, P. Crabbe, J. Org. Chem., 47 (1982) 2553.
    b) M. E. Garst, V. A. Roberts, C. Prussin, Ibid., 47 (1982) 3969.
    c) C. A. Cupas, L. Hodakowski, J. Am. Chem. Soc., 96 (1974) 4668.
    d) S. Ito, K. Sakan, Y. Fujise, Tetrahedron Lett., (1970) 2873.
19. a) T. Nozoe, T. Mukai, J. Tezuka, Bull. Chem. Soc. Jpn., 34 (1961) 619.
    b) O. L. Chapman, D. J. Pasto, A. A. Griswold, J. Am. Chem. Soc., 84 (1962) 1213.

20. a) E. Garfunkel, I. D. Reingold, J. Org. Chem., 44 (1979) 3725.
    b) L. N. McCullagh, D. S. Wulfman, Synthesis, 8 (1972) 422.
    c) G. Jones, J. Chem. Soc. C., (1970) 1230.
    d) K. M. Harmon, A. B. Harmon, T. T. Coburn, J. M. Fisk, J. Org. Chem., 33 (1968) 2567.
21. a) P. Radlick, J. Org. Chem., 29 (1964) 960.
    b) P. K. Freeman, D. M. Balls, D. J. Brown, Ibid., 33 (1968) 2211.
22. The unusual regiochemistry of this addition as well as many other reactions of tropone addition can be rationalized, in part, by considering the large coefficients at C_2 and C_7 in the LUMO of tropone:
    a) L. Salem, J. Am. Chem. Soc., 90 (1968) 543,553.
    b) G. Biggi, F. DelCima, F. Pietra, Ibid., 95 (1973) 7101.
23. G. Buechi, H. Wueest, J. Org. Chem., 34 (1969) 1122.
24. A. Marfat, P. Helquist Tetrahedron Lett., (1978) 4217.
25. A. P. TerBorg, H. Kloosteszicl, Recl. Trav. Chim. Pays-Bas, 82 (1963) 741.
26. T. Imamoto, N. Takiyama, K. Nakamura, Tetrahedron Lett., 26 (1985) 4763.
27. a) P. W. Concannon, J. Ciabattoni, J. Am. Chem. Soc., 95 (1973) 3284.
    b) J. A. Marshall, W. F. Huffman, Ibid., 92 (1970) 6358.
    c) P. J. DeClercq, M. Vandewalle, J. Org. Chem., 42 (1977) 3447.
28. For a related example, see: J. A. Marshall, P. G. Wuts, J. Org. Chem., 42 (1977) 1794.
29. J. H. Rigby, Tetrahedron Lett., 253 (1982) 1863.
30. For examples of the intermolecular version using electron-rich diene partners, see: S. Danishefsky, N. Kato, D. Askin, J. F. Kerwin, J. Am. Chem. Soc., 104 (1982) 360.
31. For another example of an intramolecular process, see: B. M. Trost, M. Lautens, M. -H. Hung, C. S. Carmichael, J. Am. Chem. Soc., 106 (1984) 7641.
32. J. H. Rigby, J. Z. Wilson, C. Senanayake, Tetrahedron Lett., 27 (1986) 3329.
33. The use of the term "cycloaddition" for this process does not necessarily have any mechanistic implications.
34. J. H. Rigby, C. Senanayake, S. Rege, manuscript in preparation.
35. For other diazoketone mediated cyclopropanations, see:
    a) A. B. Smith, R. K. Dieter, Tetrahedron, 37 (1981) 2407.
    b) S. D. Burke, P. A. Grieco, Organic Reactions, 26 (1979) 361.
36. J. H. Rigby, J. Z. Wilson, J. Am. Chem. Soc., 106 (1984) 8217.
37. a) M. Mongrain, J. Lafontaine, A. Belanger, P. Delongchamps, Can. J. Chem., 48 (1970) 3273.
    b) W. I. Fanta, W. F. Erman, J. Org. Chem., 33 (1968) 1656.
38. J. H. Rigby, J. Z. Wilson, Tetrahedron Lett., 25 (1984) 1429.
39. C. R. Johnson, R. C. Elliott, J. Am. Chem. Soc., 104 (1982) 7041.
40. Other examples of this process have been reported:
    a) T. Hudlicky, S. V. Govindan, J. O. Frazier, J. Org. Chem., 50 (1985) 4166.
    b) K. K. Weinhardt, Tetrahedron Lett., 25 (1984) 1761.
41. B. Ganem, V. R. Small, J. Org. Chem., 39 (1974) 3728.
42. J. H. Rigby, J. Z. Wilson, J. Org. Chem., 52 (1987) 34.
43. The more labile acetyl protecting group could be used without difficulty when the C_8, carbon-oxygen bond was not cleaved reductively.
44. D. A. Evans, G. C. Andrews, Acc. Chem. Res., 7 (1974) 147.
45. a) S. Danishefsky, P. F. Schuda, T. Kitahara, S. J. Etheredge, J. Am. Chem. Soc., 99 (1979) 6066.
    b) S. Danishefsky, M. Y. Tsai, T. Kitahara, J. Org. Chem., 42 (1977) 394.
46. A. A. Devreese, M. Demuynck, P. J. DeClercq, M. Vandewalle, Tetrahedron, 39 (1983) 3049.
47. J. H. Rigby, C. Senanayake, J. Am. Chem. Soc., 109 (1987) 3147.
48. a) M. Demuth, P. R. Rayhanan, Helv. Chim. Acta., 62 (1979) 2338.
    b) R. K. Dieter, S. Pounds, J. Org. Chem., 47, (1982) 3174.
    c) M. Demuth, G. Mikhail, Tetrahedron, 39 (1983) 991.
49. J. H. Rigby, C. Senanayake, J. Org. Chem., 52 (1987) 0000, in press.
50. a) T. K. Das, A. D. Gupta, P. K. Ghosal, P. C. Dutta, Indian J. Chem., Sect. B 14B (1976) 238.
    b) T. K. Das, P. C. Dutta, Synth. Commun. 6 (1976) 253.

32

    c) T. K. Das,P. C. Dutta,G. Kartha, J. M. Bermassan, J. Chem. Soc., Perkin Trans., 1 (1977) 1287.

    d) R. K. Boeckman, Jr., J. P. Bershas, J. Clardy, B. J. Solheim, J. Org. Chem., 42 (1977) 3630.

    e) W. G. Dauben,D. J. Hart, J. Org. Chem., 42 (1977), 922.

    f) W. R. Baker, P. D. Senter, R. M. Coates, J. Chem. Soc., Chem. Commun., (1980) 1011.

    g) R. M. Coates,P. D. Senter,W. R. Baker, J. Org. Chem., 47 (1982) 3597.

    h) L. A. Paquette, D. R. Andrews, J. P. Springer, Ibid., 48 (1983) 1148.

    i) H. Takeshita,N. Kato,K. Nakanishi, Chem. Lett. (1984) 1495.

    j) R. M. Coates, J. W. Muskopf, P. A. Senter, J. Org. Chem., 50 (1985) 3541.

    k) L. A. Paquette,J. A. Colapret, D. R. Andrews, Ibid., 50 (1985) 201.

    l) G. Metha,N. Krishnamurthy, J. Chem. Soc., Chem. Commun. (1986) 1319.

51. J. H. Rigby, C. Senanayake, J. Org. Chem., 52 (1987) 0000, in press.
52. R. E. Ireland, R. H. Mueller, A. K. Willard, J. Am. Chem. Soc., 98 (1976) 2868.
53. S. Mori, I. Sai, T. Aoyama, T. Shioiri, Chem. Pharm. Bull., 30 (1982) 3380.
54. a) S. Ito, H. Takeshita, Y. Shoji, Tetrahedron Lett., (1969) 1815.

    b) T. Uyehara, Y. Kitahara, Chem. Ind. (1971) 354.

    c) S. Ito, A. Mori, S. Yoshikazu, H. Takeshita, Tetrahedron Lett., (1972) 2865.

    d) T. Uyehara, Y. Kitahara, Bull. Chem. Soc. Jpn., 52 (1979) 3355.

55. J. H. Rigby, J.-M. Sage, J. Raggon, J. Org. Chem., 47 (1982) 4815.
56. D. A. Evans, A. M. Golob, J. Am. Chem. Soc., 97 (1975) 4765.
57. J. H. Rigby, J. -M. Sage, J. Org. Chem., 48 (1983) 3591.
58. a) H. Tanida, T. Yano, M. Ueyama, Bull. Chem. Soc. Jpn., 45 (1972) 946.

    b) Y. Kashman, O. Awerbrouch, Tetrahedron, 29 (1973) 191.

    c) Y. Fujise, Y. Chonan, H. Sakurai, S. Ito, Tetrahedron Lett., (1974) 1585.

59. a) T. Uyehara, J. Yamada, K. Ogata, T. Kato, Bull. Chem. Soc., Jpn., 58 (1983) 211.

    b) T. Uyehara, J. Yamada, T. Kato, F. Bohlmann, Ibid., 58 (1985) 861.

60. R. L. Funk, G. L. Bolton, J. Org. Chem., 52 (1987) 3173.
61. K. S. Feldman, J. H. Come, A. J. Freyer, B. J. Kosmider, C. M. Smith, J. Am. Chem. Soc., 108 (1986) 1327.
62. For other examples of [8+2] cycloadditions in the tropone series, see:

    a) T. S. Cantrell, Tetrahedron Lett., (1975) 907.

    b) T. Sasaki, K. Kanematsu, K. Iizuka, J. Org. Chem., 41 (1976) 1105.

63. B. M. Trost, P. R. Seone, J. Am. Chem. Soc., 109 (1987) 615.
64. a) R. C. Cookson, B. V. Drake, J. Hudec, A. Morrison, J. Chem. Soc. Chem. Commun., (1966) 15.

    b) S. Ito, Y. Fujise, T. Okuda, Y. Inoue, Bull. Chem. Soc., Jpn., 39 (1966) 1351.

    c) T. Mukai, Y. Akasaki, T. Hagiwara, J. Am. Chem. Soc., 94 (1972) 675.

    d) S. Ito, H. Ohtani, S. Narita, H. Honma, Tetrahedron Lett., (1972) 2223.

65. For excellent discussions on the mechanistic details of the [6+4] tropone-diene cycloaddition, see:

    a) L. A. Paquette, S. J. Hathaway, P. F. T. Schirch, J. Org. Chem., 50 (1985) 4199.

    b) M. E. Garst, V. A. Roberts, K. N. Houk, N. G. Rondan, J. Am. Chem. Soc., 106, (1984) 3882.

    c) M. E. Garst, V. A. Roberts, C. Prussin, Tetrahedron, 39 (1983) 581.

    d) D. Mukherjee, C. R. Watts, K. N. Houk, J. Org. Chem., 43 (1978) 817.

    e) M. Frank-Neumann, D. Martina, Tetrahedron Lett., (1977) 2293.

66. J. H. Rigby, T. L. Moore, S. Rege, J. Org. Chem., 51 (1986) 2398.
67. See, for example: R. M. Williams, R. W. Armstrong, J. -S., Dung, J. Am. Chem. Soc., 106, (1984) 5748.
68 S. Rege, Wayne State University, unpublished results.
69. M. Franck-Neumann, F. Brion, D. Martina, Tetrahedron Lett., (1978) 5033.
70. 3-substituted tropones can be difficult to prepare, see: M. E. Garst, V. A. Roberts, J. Org. Chem., 47 (1982) 2188.
71. C. P. Lewis, W. Kitching, A. Eisenstadt, M. Brookhart, J. Am. Chem. Soc., 101 (1979) 4896.

HOMOCHIRAL KETÁLS AND ACETALS IN ORGANIC SYNTHESIS*

EUGENE A. MASH

1. INTRODUCTION

The development of contemporary synthetic organic methodology for the production of homochiral substances from achiral or heterochiral starting materials is a most exciting and important area of research. For economic and aesthetic reasons, the modern synthetic chemist often plans an asymmetric synthesis of a complex target rather than a synthesis of a dl mixture. With this advance, the availability of homochiral molecules has become a pressing problem. Sugars, amino acids, and terpenes are three natural sources of such molecules (ref. 1). Unfortunately, these materials are sometimes expensive and often require extensive "tailoring" to remove and/or introduce functionality. Many other molecules are available by traditional methods of optical resolution (ref. 2). However, optical resolution methods are frequently inapplicable to a new substance despite encouraging literature precedent. Additionally, half of the material being resolved is generally discarded as unusable, and assignment of configuration to the separated enantiomers is not always straightforward. Microorganisms can sometimes be employed to perform selective chemical transformations, though the general applicability of these biological methods to a variety of substrates has been demonstrated in only a few instances (ref. 3). Thus, the development of synthetic methodology for the construction of molecules unavailable from the "chiral pool" has drawn much attention (ref. 4).

Synthetic chemists have long sought to mimic the regiochemical and absolute stereochemical selectivity observed for many enzymatic processes. Enzymatic chemoselectivity is thought to result from well-defined, obligate alignments for homochiral enzyme·substrate complexes during enzymatic catalysis. Although regiocontrol and relative stereocontrol for a number of non-enzymatic chemical transformations have been achievable for some time

---

*The terms "homochiral" and "heterochiral" are used as contracted forms of "homogeneously chiral" and "heterogeneously chiral" in this article.

578

(e.g., eqns. 1 and 2), only recently have substantial advances in absolute

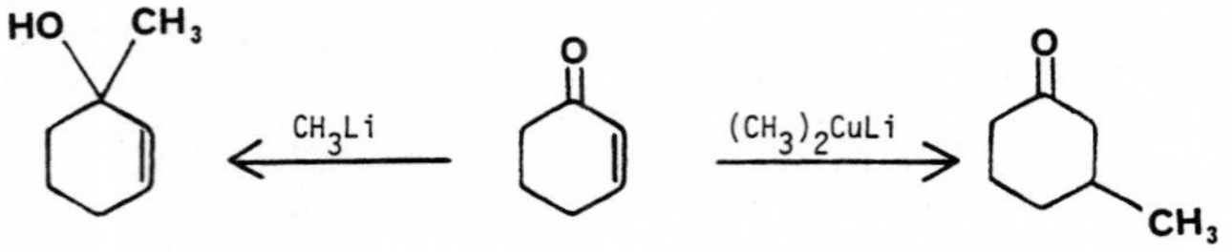

(Eqn. 1)

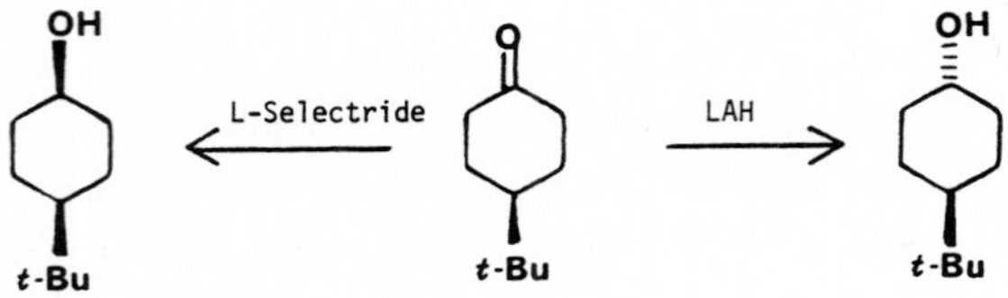

(Eqn. 2)

stereocontrol been made. Given the premise that chirality begets chirality, absolute stereocontrol can be achieved by one of the following methods:

(1) Utilization of achiral reagents for diastereoselective functionalization of homochiral compounds.

(2) Utilization of homochiral reagents for diastereoselctive functionalization of achiral compounds (ref. 5).

(3) Utilization of "matched" homochiral reagents for diastereoselective functionalization of homochiral compounds (ref. 6).

(4) Utilization of achiral reagents for functionalization of achiral compounds in chiral media.

Some achiral compounds can be rendered homochiral by attachment of a homochiral auxiliary (e.g., eqn. 3). Absolute stereocontrol via (1) or (3),

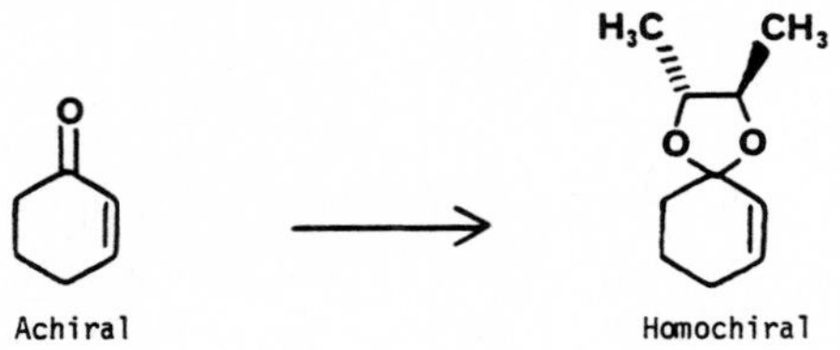

(Eqn. 3)

Achiral   Homochiral

followed by removal of the auxiliary, can provide homochiral products. Where reactions produce mixtures of separable diastereoisomers, their separation prior to removal of the homochiral auxiliary can provide products of 100% homochirality, regardless of the observed diastereoselectivity.

Homochiral auxiliaries may also serve as "traditional" protecting groups
before, during, and after their use in establishing absolute stereochemistry.*

Although achiral ketals and acetals of many kinds have been employed as
protecting agents for carbonyl groups and for 1,2- and 1,3-diols (ref.
7), studies describing diastereoselective reactions of homochiral acetals
and ketals have only recently appeared. This article will review uses of
homochiral 1,3-dioxolane and 1,3-dioxane acetals and ketals in synthesis
appearing prior to July, 1987. Particular attention will be drawn to cases
where diastereoselective functionalization of one component of the acetal
or ketal occurs as a result of asymmetry in the second component.

1.1 Homochiral Diol and Ketone Auxiliaries

A partial list of the homochiral diols and dithiols available for preparation of homochiral ketals and acetals is presented in Table 1. A similar list of homochiral ketones is presented in Table 2. Except for
derivatives of L-tartaric acid, most of the commercially available homochiral
diols are relatively expensive. However, many of these auxiliaries can be
prepared efficiently and in quantity from less expensive starting materials
(see references in Table 1). The most useful ketone auxiliaries, 8-
substituted menthone derivatives, are prepared synthetically from pulegone
(ref. 30b, 35). (R)-(+)-Pulegone is commercially available, while (S)-(-)-
pulegone can be prepared from commercially available (S)-(-)-citronellol
(ref. 35c).

1.2 Methods of Acetalization and Ketalization

Protic or Lewis acid-catalyzed dehydrative or exchange acetalization
and ketalization (ref. 7) are most commonly employed for preparations of
homochiral acetals and ketals (e.g., eqn. 4).

(Eqn. 4)

---

*It is interesting to note that, historically, asymmetry in a protecting
group was shunned as undesirable, in part because separable diastereoisomers were sometimes encountered and because they occasionally displayed
different chemical reactivity.

TABLE 1
Available Homochiral Diols and Dithiols (Ref. 8)

Entry	Compound[a]	CAS Registry No.	Cost, $/gram[b]	References for Preparation
1	H_3C CH_3 HO OH (S,S)-1	19243-06-0 24347-58-8	60 40	9,10,11,12
2	H_3C CH_3 HO OH (S,S)-2	72345-23-4 42075-32-1	50 25	13, 14
3	H_3C CH_3 HS SH (S,S)-3	68170-33-2 69307-86-4	NA NA	15
4	H_3C CH_3 HS SH (S,S)-4	84799-98-4	NA	16
5	Ph Ph HO OH (R,R)-5	52340-78-0 2325-10-2	100 100	17,18
6	H_3OOC $COOCH_3$ HO OH (R,R)-6	608-68-4 5057-96-5	0.15 3	12

TABLE 1., (continued)
Available Homochiral Diols and Dithiols (ref. 8)

Entry	Compound[a]	CAS Registry No.	Cost, $/gram[b]	References for Preparation
7	EtOCO COOEt HO OH (R,R)-7	87-91-2 13811-71-7	0.25 2	19,20,21
8	Me_2N CO $CONMe_2$ HO OH (R,R)-8	26549-65-5 63126-52-3	NA NA	20,21
9	CH_3O— —OCH_3 HO OH (R,R)-9	50622-10-1 33507-82-3	NA NA	21,22
10	$PhCH_2O$— —OCH_2Ph HO OH (R,R)-10	17401-06-8 91604-41-0	NA NA	23,24
11	4-Cl-$PhCH_2O$— —OCH_2-4-Cl-Ph HO OH (R,R)-11	--- ---	NA NA	25

582

TABLE 1., (continued)
Available Homochiral Diols and Dithiols (ref. 8)

Entry	Compound[a]	CAS Registry No.	Cost, $/gram[b]	References for Preparation
12	(S,S)-12		NA NA	26
13	(S)-13	4254-15-3 4254-14-2	11 NA	27
14	(S)-14	24621-61-2 6290-03-5	20 20	28
15	(S)-15	17199-29-0 611-71-2	1 1	-- --
16	(S)-16	6168-83-8 625-72-9	NA NA	29 29

(a) The enantiomer shown corresponds to the first CAS registry number given. The second CAS registry number belongs to the mirror image of the enantiomer shown.
(b) 1987 U.S. dollars; NA = not commercially available.

TABLE 2.
Available Homochiral Ketones (ref. 30)

Entry	Compound[a]	CAS Registry No.	Cost, $/gram[b]	References for Preparation
1	(R)-17	6672-30-6	18	31
		6672-24-8	NA	
2	(R)-18	13368-65-5	3	32
		24965-87-5	NA	
3	(2S,5R)-19	14073-97-3	0.25	33,34
		3391-87-5	NA	
4	(2R,5R)-20	57707-92-3	NA	35
		97371-54-5	NA	

(a) The enantiomer shown corresponds to the first CAS registry number.
(b) 1987 U.S. dollars; NA = not commercially available.

A recent report by Noyori (ref. 36) that bis(trimethylsilyl)ethylene glycol reacts with aldehydes and ketones under catalysis by trimethylsilyl-trifluoromethanesulfonate to produce 1,3-dioxolanes in high yield without double bond migration followed a report by Evans (ref. 37) which described similar behavior for bis(trimethylsilyl)-1,2-ethanedithiol under catalysis by anhydrous zinc iodide (e.g., eqns. 5 and 6). The higher yields and

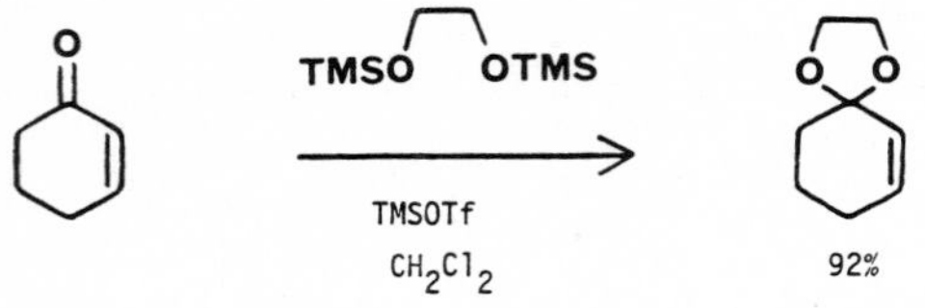

(Eqn. 5)

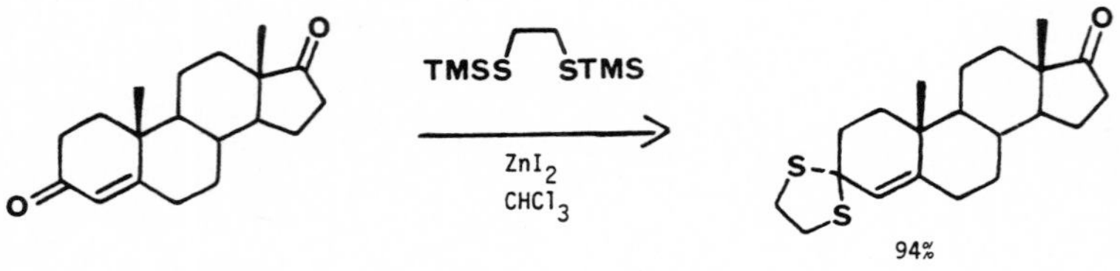

TMSOTf
CH_2Cl_2

92%

(Eqn. 6)

ZnI_2
$CHCl_3$

94%

greater selectivities offered by these methodologies (ref. 38) should increase their use (vide infra) in preparations of homochiral ketals and acetals.

Other routes to homochiral acetals, by palladium catalyzed acetalization of terminal olefins (ref. 39), are also available (e.g., eqns. 7 and 8).

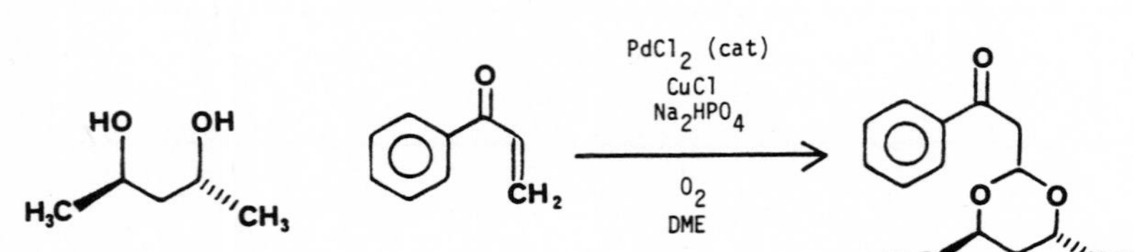

(Eqn. 7)

$PdCl_2$ (cat)
CuCl
O_2
DME

61% (20% d.e.)

(Eqn. 8)

$PdCl_2$ (cat)
CuCl
Na_2HPO_4
O_2
DME

75%

1.3 Diastereomeric Resolution

The earliest uses in synthesis of homochiral diol and dithiol auxiliaries involved preparation and separation of diastereomeric ketals. Such
use continues to the present day. Presented in Table 3 are representative
resolutions accomplished by preparation and separation of diastereomeric
ketals.

TABLE 3.
Diastereomeric Resolutions Via Preparation and Separation of Homochiral
Ketones.

Entry	Auxiliary	Substance Resolved[a]	Method of Separation	Ref.
1	H₃C CH₃ HO OH (R,R)-1	21	GLC	40
2	(R,R)-1	22	GLC	10
3	(R,R)-1	23	GLC	41
4	(R,R)-1	24	GLC	42

TABLE 3., (continued)
Diastereomeric Resolutions Via Preparation and Separation of Homochiral
Ketones.

Entry	Auxiliary	Substance Resolved[a]	Method of Separation	Ref.
5	(R,R)-1	**25**	GLC	42
6	(R,R)-1	**26**	HPLC	43
7	(R,R)-1	**27**	HPLC	44
8	(R,R)-1	**28**	GLC	45
9	(R,R)-1	**29**	LC	46

TABLE 3., (continued)
Diastereomeric Resolutions Via Preparation and Separation of Homochiral
Ketones.

Entry	Auxiliary	Substance Resolved[a]	Method of Separation	Ref.
10	(R,R)-1	30	HPLC	47
11	(S,S)-3	31	LC	48
12	(S,S)-3	32	Recryst.	15
13	(S,S)-3	33	LC	49
14	(R,R)-7	34	GLC	50

588

TABLE 3., (continued)
Diastereomeric Resolutions Via Preparation and Separation of Homochiral
Ketones.

Entry	Auxiliary	Substance Resolved[a]	Method of Separation	Ref.
15	(R,R)-7	35	MPLC	51
16	(R,R)-7	36	MPLC	52,53
17	(R,R)-10	37	LC	54
18	(R,R)-10	38	LC	54
19	(R,R)-11	39	Recryst.	55

(a) Only one diastereoisomer is shown.

1.4 Kinetic Resolution

Resolution based upon differences in rates of formation for diastereomeric acetals or ketals has not yet proven to be generally useful (ref. 56). Limited success was reported when dimethyl tartrate and hydrobenzoin were used as homochiral diol components (e.g., eqn. 9) (ref. 57). Greater success was reported for diastereoselective monoketalization

of 9-methyl-cis-decalin-1,8-dione 40 using (2R,3R)-2,3-butanediol as the homochiral diol component (eqn. 10). Attempts to similarly resolve 9-methyl-trans-decalin-1,8-dione were unsuccessful (refs. 58 and 59). Extensions of these results to other rigid polycycles bearing diastereotopic carbonyls have not been disclosed.

2. REACTIONS AT THE ACETAL OR KETAL CARBON

2.1 Diastereoselective Intramolecular Cyclizations

In 1968, W. S. Johnson reported that stannic chloride-catalyzed cation-olefin cyclization of trans dienic (2R,3R)-2,3-butanediol acetal **41** produced four bicyclic diastereoisomers in unequal amounts (eqn. 11) (refs. 60 and 61).

(Eqn. 11)

Fifteen years later, Bartlett and Johnson explained the observed product distribution as the result of a combination of

(1) Diastereoselective (~9:1) acetal bond cleavage.

(2) Synchronous attack by the C(5)-C(6) π-bond on the incipient cation (diastereofacial selectivity ~2:1; assumes inversion of configuration of the acetal carbon).

(3) Diastereoselective cation-olefin cyclization (ref. 62).

They rationalized the remarkably selective pro-R oxygen-acetal carbon bond cleavage on steric grounds. Of the four diastereomeric transition states depicted (eqn. 12), the least sterically congested transition state **D** should predominate, making the pro-R oxygen of the acetal the preferred leaving group. Johnson's early observations paved the way for his

(Eqn. 12)

later synthesis (Fig. 1) of the Inhoffen-Lythgoe diol **42**, a versatile precursor to Vitamin D metabolites (ref. 63), and may lead to constructions
of more complex polycyclic systems under absolute stereocontrol. As importantly, Johnson's early observations led to exploration of intermolecular
reactions of homochiral acetals and ketals by his group and other research
groups.

2.2 Intermolecular Reactions
2.2.1 Diastereoselective Reductive Cleavage

Richter's examination of the Lewis acid-catalyzed reductive cleavage of
homochiral (2R,3R)-2,3-butanediol acetals and ketals **43** (Table 4) was the
first reported study of the intermolecular reactivity of such molecules
(ref. 64). He observed remarkable diastereoselectivity for reductions of
acetals and lesser diastereoselectivity for reductions of ketals, but
apparently did not deduce the absolute stereochemistries of the butanediol
monoether products **44**. Later work by Yamamoto (refs. 65-68) established a
synthetic utility for the reductive cleavage of related (2R,4R)-2,4-pentane-
diol ketals **45** (Table 5). Diastereoselectivity observed for the 1,3-dioxane
system was generally greater than that observed for the 1,3-dioxolane
system (e.g., compare Table 4, entry 9, with Table 5, entries 4 and 5) and
gave pentanediol monoether products **46** of predictable chirality. The
pentanediol auxiliary was also more easily removed, via oxidation and
β-elimination, from the monoether intermediates **46** than was the
butanediol auxiliary (ref. 62).

The stereochemical results reported by Yamamoto are consistent with the
proposed mechanism depicted in Figure 2 (refs. 68 and 82). Complexation of
the Lewis acid occurs at the pro-R oxygen atom of the more stable 1,3-
dioxane pseudochair conformation, lengthening, but not breaking, its bond
to the ketal carbon while shortening the bond between this carbon and the
uncomplexed dioxane pro-S oxygen atom. Lengthening of this bond relieves
the 1,3-diaxial interaction between the smaller appendage "S" and the axial
methyl group. Nucleophile-dependent attack then occurs by either Sn2
(inversion) or Sni (retention) pathways.

Figure 1. Johnson's Synthesis of the Inhoffen-Lythgoe Diol (42)

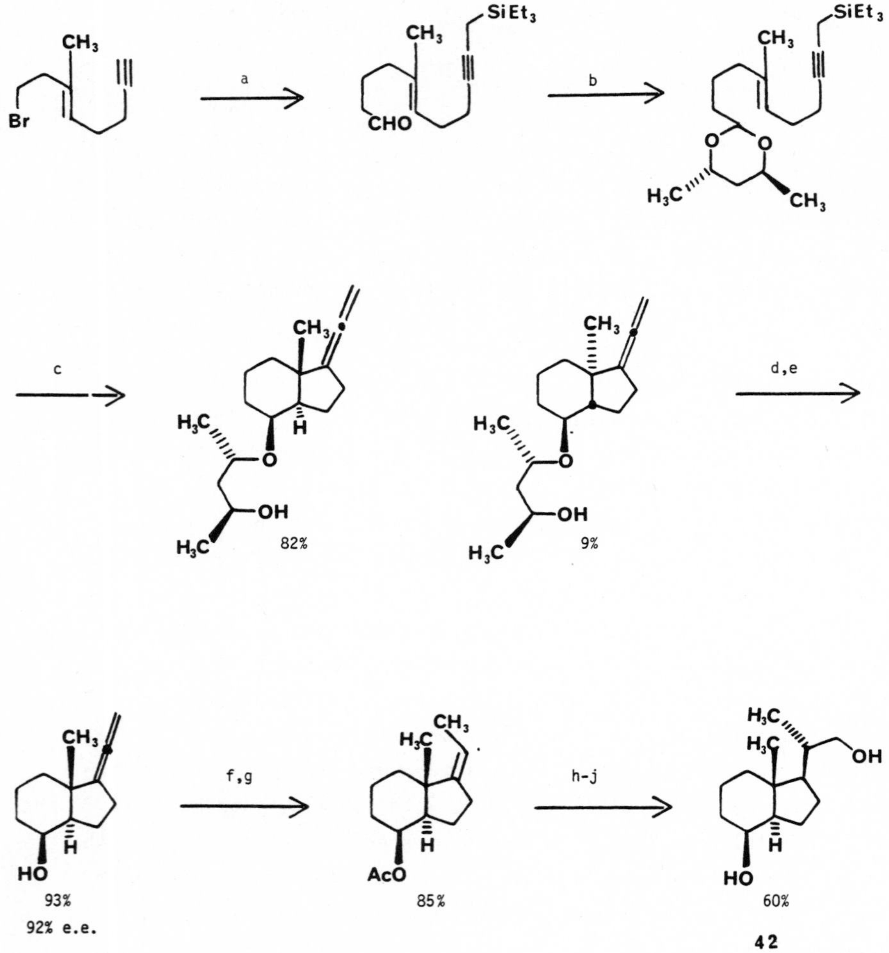

Reagents:

(a) See Ref. 63

(b) (2S,4S)-Pentanediol, H^+

(c) $TiCl_4$, 2,4,6-trimethylpyridine, CH_2Cl_2

(d) PCC, CH_2Cl_2

(e) Aq. KOH, THF, CH_3OH

(f) Ac_2O, DMAP, pyr.

(g) H_2, Lindlar Cat.

(h) Formaldehyde, $BF_3 \cdot Et_2O$

(i) H_2, PtO_2

(j) Aq. KOH, THF, CH_3OH

TABLE 4.
Diastereoselective Reductive Cleavage of (2R,3R)-2,3-Butanediol Acetals and
Ketals **43** (Ref. 64).

Entry	R_1	R_2	R_3	Diastereoisomer Ratio
1	H	H	Me	49:1
2	H	H	Et	99:1
3	H	H	CH_2Ph	10:1
4	H	H	Ph	6:1
5	H	H	Cyclohexyl	9:1
6	H	H	iPr	7:2
7	H	Me	$C{\equiv}CH$	1:1
8	H	Me	$CH=CH_2$	1:1
9	H	Me	Ph	1:1
10	H	Me	Et	3:1
11	H	Me	nPr	2:1
12	H	Me	CH_2Ph	3:2
13	H	Me	iBu	3:1
14	H	Me	CH_2tBu	2:1
15	H	Me	iPr	4:1
16	H	Me	Cyclohexyl	3:1
17	H	Me	tBu	7:1
18	H	Me	1-Adamantyl	8:1
19	H	Me	CH_2OMe	1:1
20	OMe	Me	Et	1:1
21	OMe	Me	iPr	1:1

TABLE 5.
Diastereoselective Reductive Cleavage of (2R,4R)-2,4-Pentanediol Ketals 45
(Refs. 65-68)

Entry	R1	R2	Hydride[a] Reagent	Yield,%	Ratio 46a:b
1	Cyclohexyl	Me	Br_2AlH	99	1:23
2	Cyclohexyl	Me	DIBAL	88	1:13
3	Cyclohexyl	Me	$TiCl_4+Et_3SiH$	85	49:1
4	Ph	Me	Br_2AlH	94	1:57
5	Ph	Me	DIBAL	88	1:28
6	n-Hexyl	Me	Br_2AlH	64	1:8
7	n-Hexyl	Me	DIBAL	58	2:7
8	C≡C-nBu	Me	Br_2AlH	100	13:1
9	C≡C-nBu	Me	DIBAL	90	24:1
10	C≡C-nBu	Me	$TiCl_4+Et_3SiH$	58	1:24
11	C≡C-nBu	Et	Br_2AlH	98	49:1
12	C≡C-nBu	Et	DIBAL	93	32:1
13	C≡C-Me	iBu	Br_2AlH	99	49:1
14	C≡C-Ph	Me	Br_2AlH	92	9:1
15	C≡C-Ph	Me	DIBAL	86	9:1
16	C≡C-Ph	Et	Br_2AlH	99	19:1
17	C≡C-Me	Cyclohexyl	Br_2AlH	98	99:1

(a) Solvent for Br_2AlH reactions, ether; solvent for DIBAL and Et_3SiH
reactions, CH_2Cl_2.

FIGURE 2.
Johnson-Yamamoto Mechanistic Rationale Applied to the Diastereoselective
Reductive Cleavage of (2R,4R)-2,4-Pentanediol Ketals **45**. (S = small group,
L = large group)

For Br$_2$AlH and DIBAL:

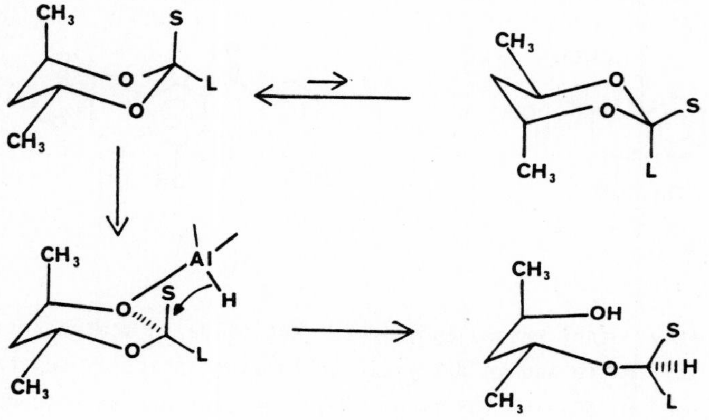

For TiCl$_4$ + Et$_3$SiH:

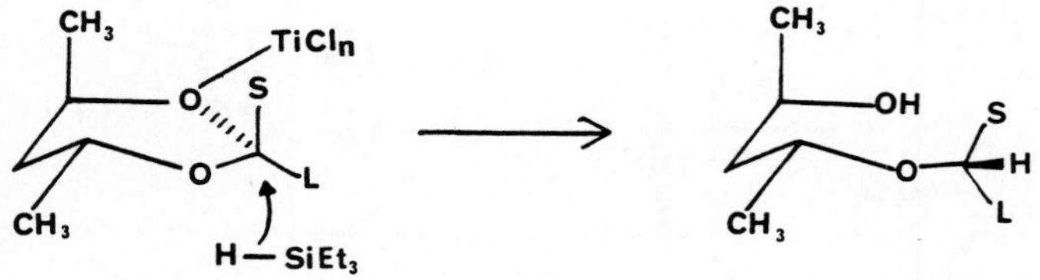

2.2.2   Diastereoselective Crossed-Aldol Condensations

The first reported intermolecular use of homochiral acetals in diastereoselective carbon-carbon bond formation was due to Kishi, who developed a Lewis acid-catalyzed crossed-aldol process for use in his enantioselective syntheses of aklavinone 47 and 11-deoxydaunomycinone 48 (Fig. 3 (refs. 69-71). Following unsuccessful attempts involving dimenthoxy acetals, he found that the treatment of (2R,3R)-2,3-butanediol acetal 49 with

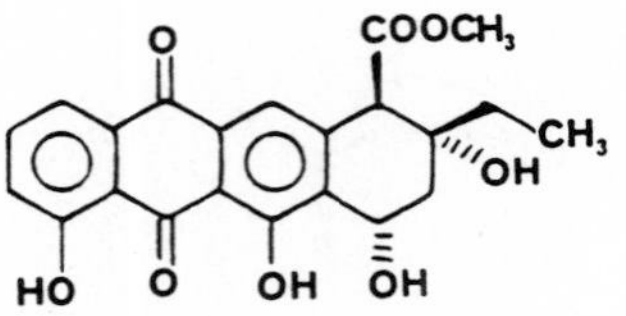

47

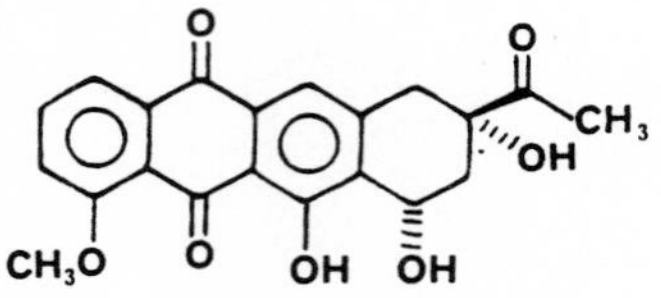

48

trimethylsilylmethyl ethyl ketone and tin tetrachloride in acetonitrile produced, in a 10:1 ratio and in 83% yield, chromatographically separable crossed-aldol products 50a and 50b (eqn. 13). The absolute stereochemistry of the major diastereoisomer 50a was established by its conversion to (+)-alkavinone of known absolute configuration . Predominant formation of 50a is in accord with the Johnson-Bartlett mechanistic model (ref. 62).

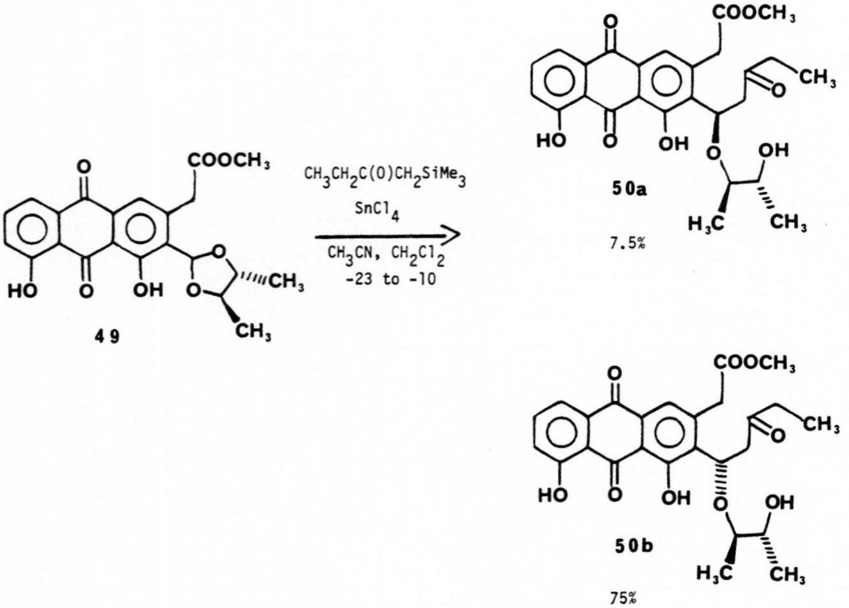

(Eqn. 13)

Figure 3.

Enantioselective Synthesis of Aklavinone (47).

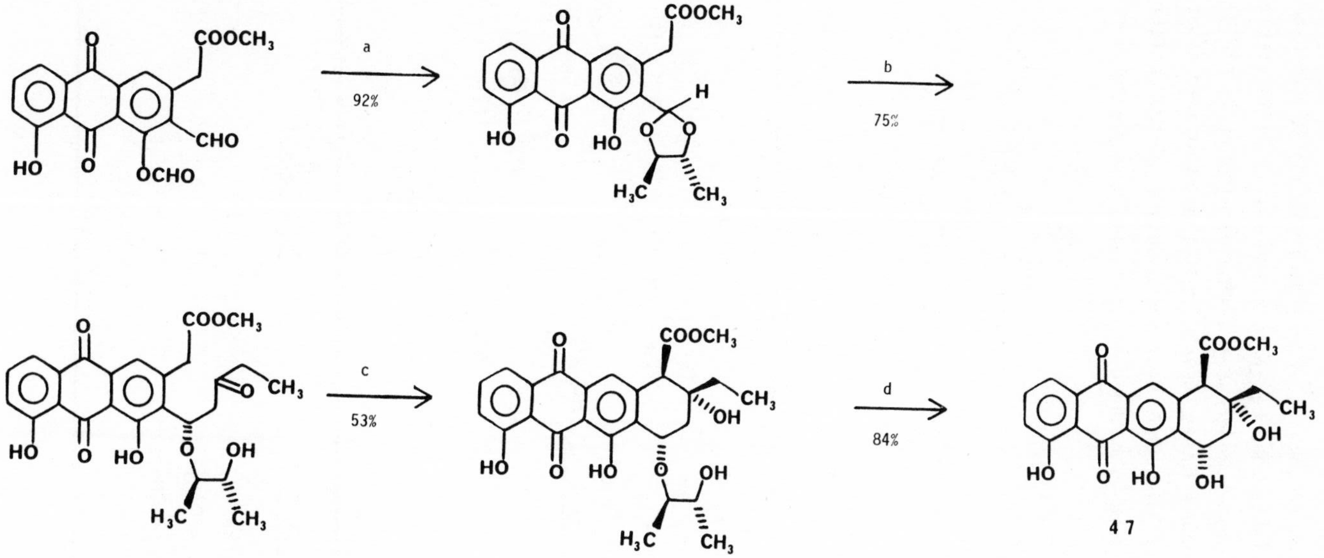

Reagents:

(a) (2R,3R)-Butanediol, PPTS, C_6H_6, heat

(b) 1-Trimethylsilyl-2-butanone, $SnCl_4$, CH_3CN

(c) K_2CO_3, CH_3OH

(d) CF_3COOH, -78° to RT

Kishi reported that the Lewis acid-catalyzed crossed-aldol condensations of simple aliphatic (2R,3R)-2,3-butanediol acetals 51 exhibited
much less diastereoselectivity than benzylic acetals such as 49 (Table 6)
ref. 69). Johnson found that much greater diastereoselectivity could be
obtained for aliphatic systems by employing (2R,4R)-2,4-pentanediol acetals
53 (Table 7, refs. 72 and 73). The invariant absolute stereochemistry
observed for 54 was in accord with the Johnson-Yamamoto mechanistic model
outlined previously (Fig. 2).

TABLE 6.
Diastereoselective Crossed-Aldol Condensations of (2R,3R)-2,3-Butanediol
Acetals 51 (Ref. 69).

Entry	R	Ratio 52a:b
1	Ph	16:1
2	n-Octyl	3:1
3	CH=CMe2	2:1
4	iPr	1:1

TABLE 7.
Diastereoselective Crossed Aldol Condensations of (2R,4R)-2,4-Pentanediol
Acetals **53** (Refs. 72 and 73).

Entry	R	Nucleophile	Yield, %	Ratio 54a:b
1	n-Octyl	$CH_3COCH_2SiMe_3$	92	32:1
2	n-Octyl	CH_2=C($OSiMe_3$)Et	95	32:1
3	n-Octyl	CH_2=C($OSiMe_3$)tBu	96	24:1
4	n-Octyl	CH_2=C($OSiMe_2tBu$)OtBu	98	32:1
5	CH_2=CH(CH_2)$_2$	$CH_3COCH_2SiMe_3$	93	32:1
6	CH_2=CH(CH_2)$_2$	CH_2=C($OSiMe_3$)Me	89	32:1
7	Cyclohexyl	$CH_3COCH_2SiMe_3$	93	49:1
8	Cyclohexyl	CH_2=C($OSiMe_2tBu$)OtBu	98	49:1
9	iPr	$CH_3CH_2COCH_2SiMe_3$	48	19:1
10	iPr	CH_2=C($OSiMe_3$)Et	95	32:1
11	$(CH_2)_4CO_2iPr$	CH_2=C($OSiMe_2tBu$)OtBu	93	49:1

Johnson also studied the Lewis acid-catalyzed crossed-aldol condensations
of aliphatic (3S)-1,3-butanediol acetals **56** (Table 8) (ref. 74). Acetalization of aldehydes **55** under equilibrating conditions gave the expected
2,4-cis-disubstituted-1,3-dioxanes **56**. High diastereoselectivity was
then observed for the crossed-aldol condensations of acetals **56** with trimethylsilyl 2-propenyl ether. Formation of the predominant diastereoisomers
57a can be rationalized as depicted in Figure 4. Coordination of the
Lewis acid occurs at the less hindered oxygen atom of the thermodynamically
preferred conformation of the 1,3-dioxane ring. Synchronous dioxane
cleavage and Sn2 attack by the nucleophile leads to diastereoselective product formation.

TABLE 8.
Diastereoselective Crossed-Aldol Condensations of (3S)-1,3-Butanediol
Acetals 56 (Ref. 74)

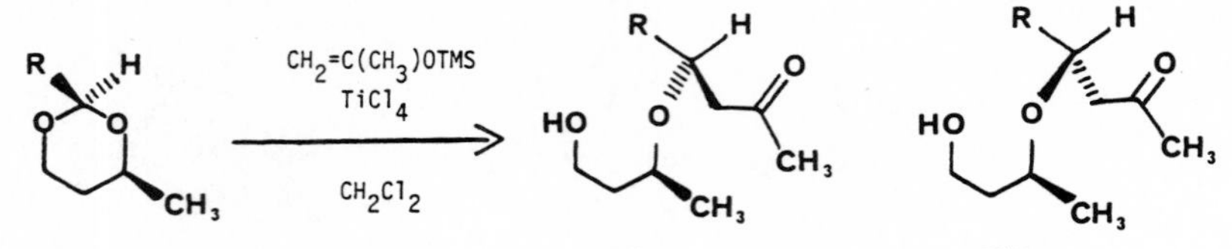

Entry	R	Yield, %	Ratio 57a:b
1	CH2=CHCH2CH2	93	24:1
2	Cyclohexyl	90	32:1
3	n-Octyl	84	99:1

Figure 4. Johnson-Yamamoto Mechanism for the Diastereoselective Crossed-Aldol Condensation of (3S)-1,3-Butanediol Acetals 56.

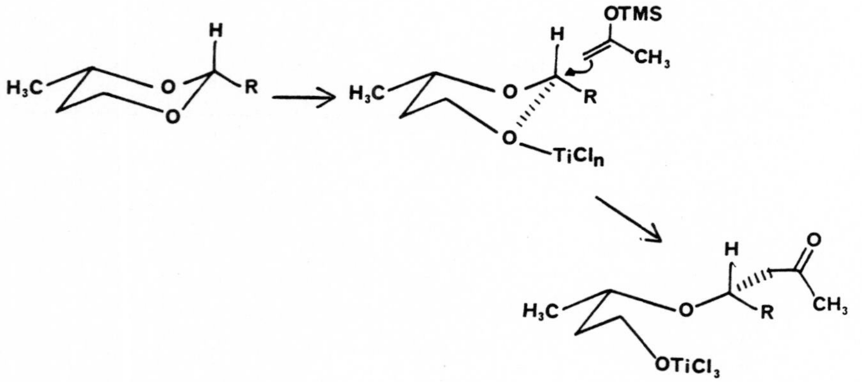

Johnson and coworkers have employed this methodology in syntheses of nonactic acid intermediate 58 (refs. 72 and 74) (Fig. 5) and of (R)-(+)-lipoic acid 59 (ref. 73) (Fig. 6).

Figure 5. Enantioselective Synthesis of Nonactic Acid Intermediate 58.

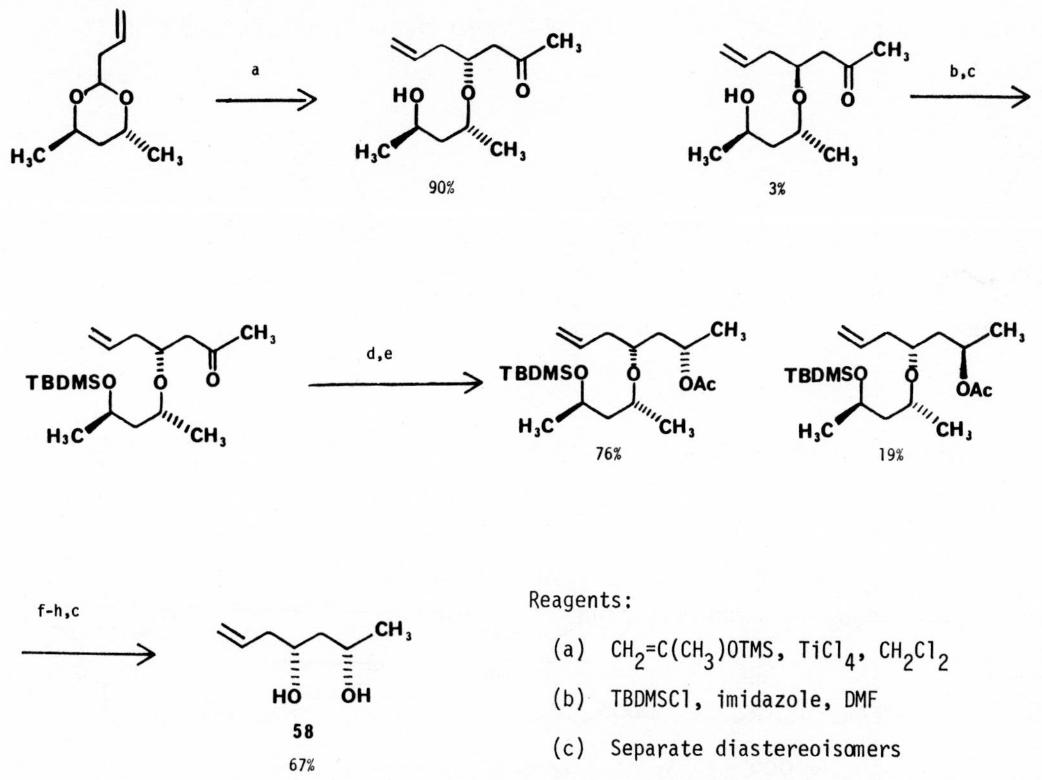

Reagents:

(a) CH$_2$=C(CH$_3$)OTMS, TiCl$_4$, CH$_2$Cl$_2$

(b) TBDMSCl, imidazole, DMF

(c) Separate diastereoisomers

(d) L-Selectride, THF, -78°

(e) Ac$_2$O, DMAP, pyr

(f) nBu$_4$N$^+$ F$^-$, THF

(g) PCC, CH$_2$Cl$_2$

(h) Aq. KOH, THF, CH$_3$OH

602

Figure 6. Enantioselective Synthesis of (R)-(+)-Lipoic Acid 59.

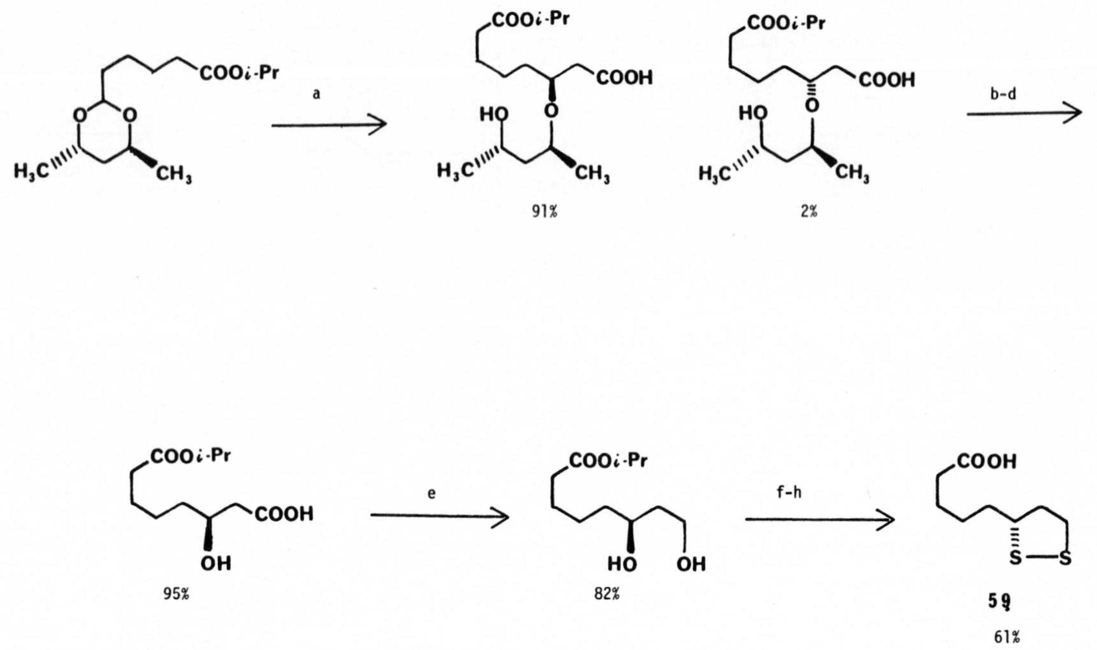

Reagents:

(a) CH₂=C(OtBu)(OTBDMS), TiCl₄, CH₂Cl₂; Aq. TFA

(b) Separate Diastereoisomers

(c) Jones' Oxidation

(d) Piperidinium Acetate, C₆H₆, heat

(e) BH₃·THF; Aq. KOH

(f) CH₃SO₂Cl, Et₃N

(g) Na₂S, S, DMF

(h) Aq K₂CO₃, CH₃OH

Kellogg described a related aldol process for 1,3-dioxolan-4-ones 60 (Table 9) in which (R)- and (S)-mandelic acids, both of which are commercially available and inexpensive, function as homochiral auxiliaries (ref. 75). The observed diastereoselectivity is in keeping with the Bartlett-Johnson mechanistic model (eqn. 14) (ref. 62).

TABLE 9.
Diastereoselective Aldol Condensations of (S)-Mandelic Acid Acetals and Ketals 60 (Ref. 75).

Entry	R_1	R_2	Nucleophile	Yield, %	Ratio 61a:b
1	Ph	H	$CH_2=C(C_6H_5)(OTMS)$	98	14:1
2	Ph	H	$(CH_3)_2C=CH(OTMS)$	95	10:1
3	Ph	Me	$CH_2=C(C_6H_5)(OTMS)$	65	2:1
4	2-Furyl	H	$CH_2=C(C_6H_5)(OTMS)$	76	3:1
5	2-Furyl	H	$(CH_3)_2C=CH(OTMS)$	90	5:1

(Eqn. 14)

2.2.3   Diastereoselective Allylations

Following Kishi's report of the intermolecular allylation of (2R,3R)-2,3-butanediol acetal 62 (eqn. 15) (ref. 69), a more complete

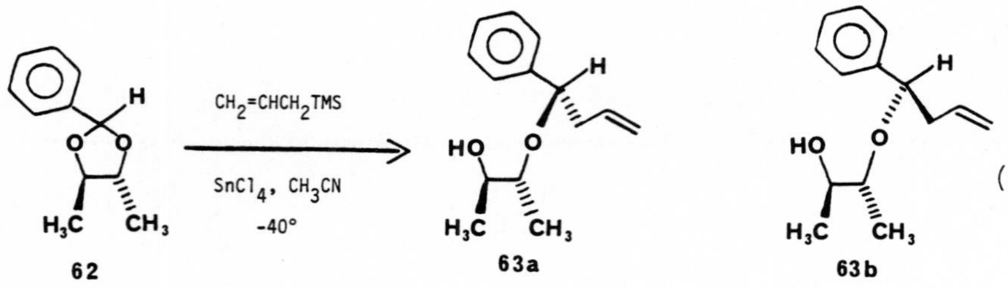

(Eqn. 15)

ratio 63a: 63b   5:1

study of the allylation reactions of a series of (2R,4R)-2,4-pentanediol acetals 64 was reported by Bartlett and Johnson (ref. 62) and by Johnson (refs. 76 and 77) (Table 10). Their study demonstrated that dramatic improvement in the observed diastereoselectivity could be obtained by "fine-tuning" the strength of the Lewis acid employed (Table 10, entries 2 and 3, 4 and 5, 6 and 7). Presumably, enhanced diastereoselectivity is due to an increase in the synchronousness of bond breaking and bond making in the transition state. Johnson and coworkers have employed their allylation reaction in syntheses of (-)-dihydromyoporone 66 (ref. 76) (Fig. 7) and calcitriol lactone 67 (ref. 77) (Fig. 8).

TABLE 10.
Diastereoselecive Allylations of (2R,4R)-2,4-Pentanediol Acetals 64
(refs. 62, 76 and 77).

Entry	n	R_1	R_2	Lewis Acid	Yield	Ratio 65a:b
1	0	n-Octyl	H	$TiCl_4$	90	7:1
2	1	n-Octyl	H	$TiCl_4$	98	7:1
3	1	n-Octyl	H	$TiCl_2(OiPr)_2$	98	49:1
4	1	$CH_2=CHCH_2CH_2$	H	$TiCl_4$	92	5:1
5	1	$CH_2=CHCH_2CH_2$	H	$TiCl_2(OiPr)_2$	92	49:1
6	1	Cyclohexyl	H	$TiCl_4$	96	5:1
7	1	Cyclohexyl	H	$TiCl_2(OiPr)_2$	98	19:1
8	1	n-Octyl	Me	$TiCl_2(OiPr)_2$	98	99:1
9	1	cyclohexyl	Me	$TiCl_2(OiPr)_2$	90	99:1
10	1	i-Bu	Me	$TiCl_2(OiPr)_2$	96	99:1

Figure 7. Enantioselective Synthesis of (-)-Dihydromyoporone 66.

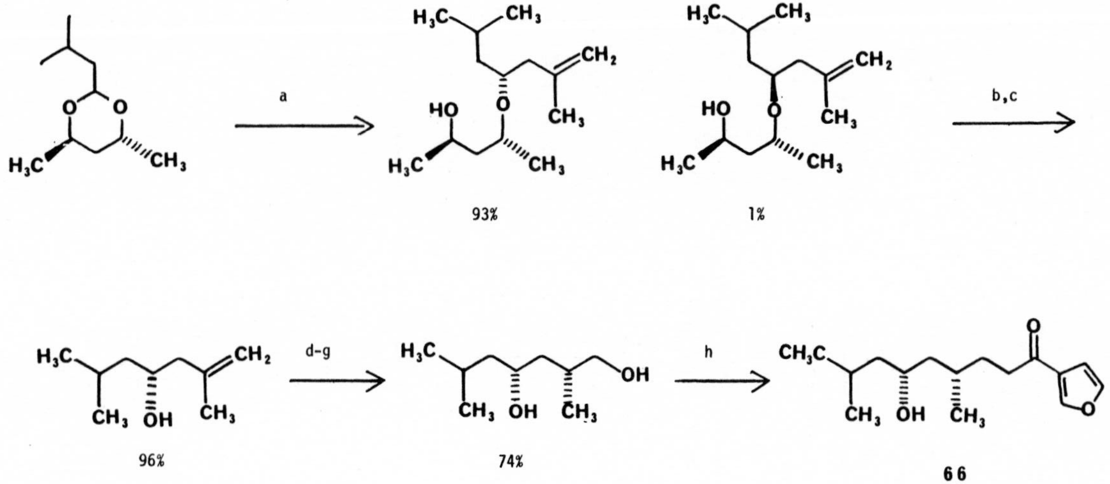

Reagents:

(a) CH₂=C(CH₃)CH₂TMS, TiCl₄

(b) PCC, CH₂Cl₂

(c) Aq. KOH, CH₃OH

(d) (CF₃CO)₂O, pyr.

(e) Thexylborane, THF, -78° to -15°

(f) H₂O₂, EtOH, H₂O

(g) Separate Diastereoisomers

(h) See JACS, 1980, 102, 7385-7387

Figure 8. Enantioselective Synthesis of Calcitriol Lactone 67.

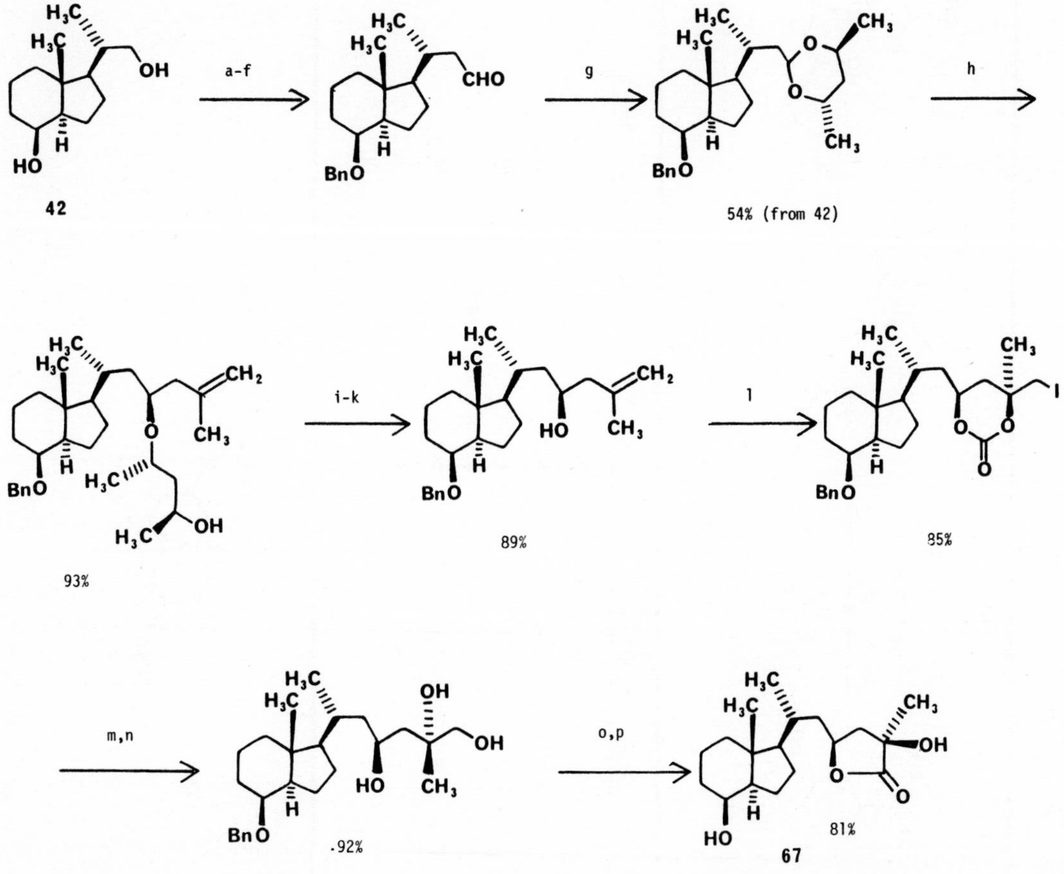

Reagents:

(a) TBDMS-Cl, imidazole

(b) NaH, DMF; BnBr

(c) nBu$_4$NF, THF

(d) TsCl, pyridine

(e) KCN, DMF

(f) DIBAL, CH$_2$Cl$_2$

(g) (2S,4S)-Pentanediol, PPTS

(h) 6TiCl$_4$·5Ti(OiPr)$_4$, CH$_2$Cl$_2$

(i) PCC, CH$_2$Cl$_2$

(j) Aq. KOH, CH$_3$OH

(k) nBuLi; BOC-ON (JOC 1982, 47, 4013)

(l) I$_2$, EtCN, -40°

(m) Aq. KOH, DME

(n) Aq. H$_2$SO$_4$, THF

(o) H$_2$, Pd·C, CH$_3$OH

(p) O$_2$, Pt, H$_2$O, diglyme, sodium lauryl sulfate

54% (from 42)

85%

93%

89%

.92%

81%

Yamamoto examined the effect of the nucleophilicity of the allylating agent on diastereoselectivity (Table 11) (ref. 78). For allylating reagents of higher nucleophilicity, bond breaking and bond making can be made sufficiently synchronous by judicious choice of the Lewis acid so that "template selectivity" (see Path a, Fig. 9) will predominate over "Cram rule" selectivity. However, where bond breaking substantially precedes nucleophilic attack (see Path B, Fig. 9), either due to low nucleophilicity of the allylating agent or to excessive Lewis acidity of the catalyst, "template selectivity" will be diminished.

TABLE 11.
Effect of Nucleophilicity of the Allylating Agent on Diastereoselectivity (Ref 78).[a]

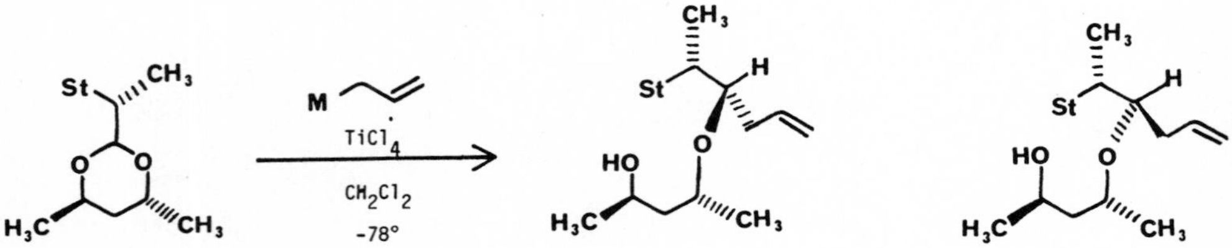

Entry	Dioxolane Configuration	M	Yield	Ratio 69a:b
1	R,R	TMS	93	99:1
2	S,S	TMS	93	9:1
3	R,R	9-BBN	90	99:1
4	S,S	9-BBN	88	7:1
5	R,R	SnBu$_3$	85	24:1
6	S,S	SnBu$_3$	84	1:2
7	S,S	SnPh$_3$	80	3:1

a

St =

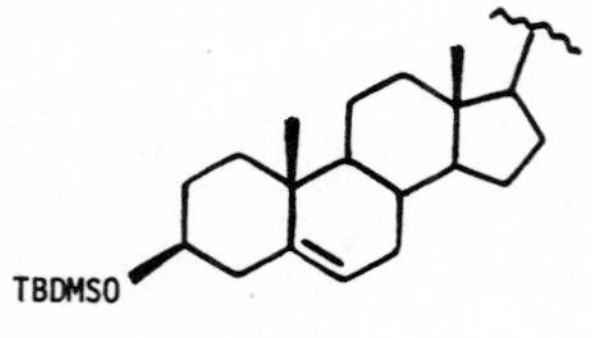

Figure 9. The Effect of Timing of Bond-Breaking and Bond-Making on
Diastereoselectivity.

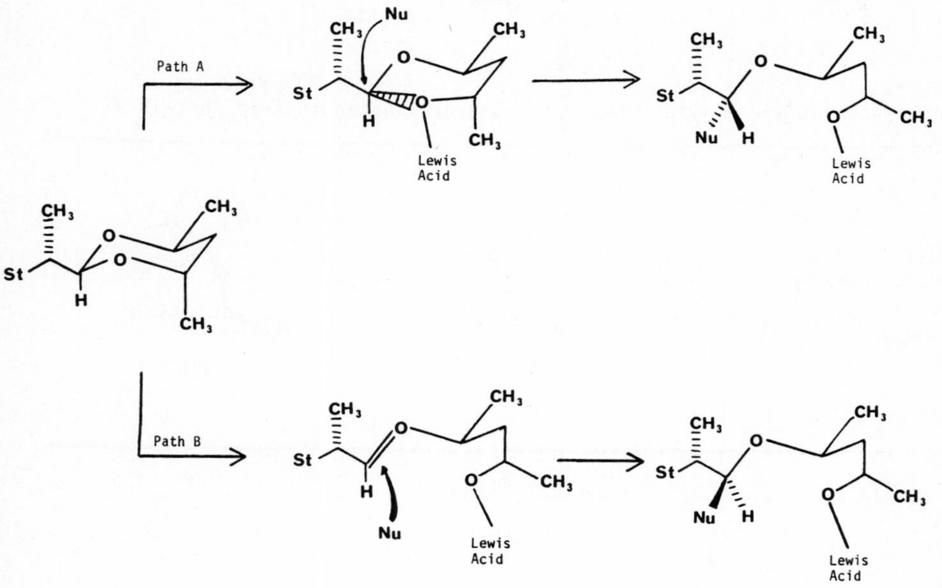

Kellogg described allylations of 1,3-dioxolan-4-ones 70 prepared
from optically active mandelic acids (ref. 75). The yields and
diastereoselectivities reported thus far are somewhat lower than those
reported from the pentanediol auxiliary (e.g., eqn. 16).

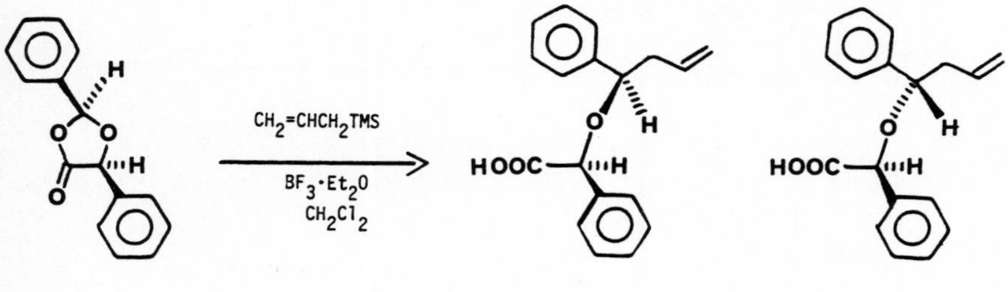

(Eqn. 16)

70   71a   71b

Ratio 71a:b 1:2

Seebach described allylations of 1,3-dioxan-4-ones **72** prepared from homochiral 3-hydroxybutanoic acids (Table 12) (ref. 79). He also observed increased diastereoselectivity when a less potent Lewis acid was used.

TABLE 12.
Diastereoselective Allylations of (<u>R</u>)-3-Hydroxybutanoic Acid Acetals **72** (Ref. 79).

Entry	R	Lewis Acid	Yield	Ratio 73a:b
1	n-Octyl	$TiCl_3(OiPr)$	99	24:1
2	CH_2CH_2Ph	$TiCl_4$	95	7:1
3	CH_2CH_2Ph	$TiCl_3(OiPr)$	95	32:1

2.2.4 Diastereoselective Alkynylations

Johnson discovered that silylacetylenes could be coupled diastereoselectively with (<u>2R</u>,<u>4R</u>)-2,4-pentanediol acetals **74** under Lewis acid catalysis (Table 13) (ref. 80). Seebach reported similar results for homochiral 1,3-dioxan-4-ones **76** derived from (<u>R</u>)-3-hydroxybutyric acid (Table 14) (ref. 79). The predominance of diastereoisomers **75a** and **77a** is in keeping with previously discussed mechanistic logic.

TABLE 13.
Diastereoselective Acetylenations of (2R,4R)-2,4-Pentanediol Acetals **74**
(Ref. 80).

Entry	R^1	R^2	Yield	Ratio 75a:b
1	n-Octyl	Me	81	13:1
2	n-Octyl	TMS	98	28:1
3	i-Bu	Me	94	6:1
4	i-Bu	TMS	85	16:1
5	n-pentyl	TMS	91	28:1

TABLE 14.
Diastereoselective Acetylenations of (R)-3-Hydroxybutanoic Acid Acetals **76**
(Ref. 79)

Entry	R^1	R^2	Yield	Ratio 77a:b
1	CH$_2$CH$_2$Ph	TMS	87	65:1
2	CH$_2$CH$_2$Ph	Me	50	49:1

Yamamoto examined the effect of the nucleophilicity of the acetylinic
component on diastereoselectivity (Table 15) (ref. 78). As observed for
allylation, optimization of "template control" requires a balance between

the Lewis acidity of the catalyst and acetylene nucleophilicity, so that bond breaking and nucleophilic attack are sufficiently synchronous to afford high diastereoselectivity.

TABLE 15.
Effect of Nucleophilicity of the Acetylene on Diastereoselectivity (ref. 78)[a]

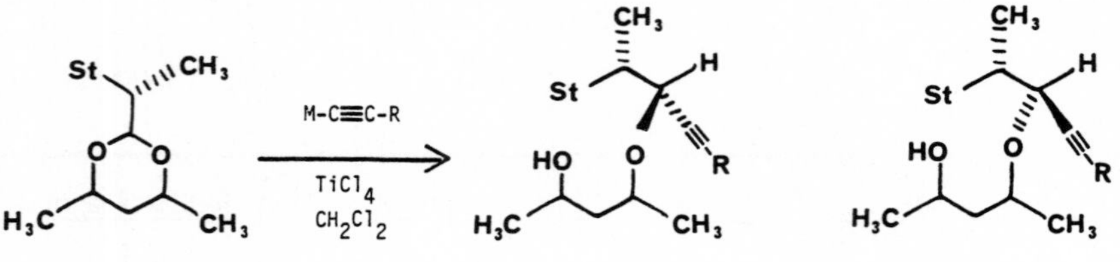

Entry	Dioxane Configuration	M	R	Yield %	Ratio 79a:b
1	R,R	SnBu3	i-Pr	80	19:1
2	S,S	SnBu3	i-Pr	78	1:9
3	R,R	SnBu3	n-Bu	82	19:1
4	S,S	SnBu3	n-Bu	78	1:11
5	R,R	SiMe3	n-Bu	72	49:1
6	S,S	SiMe3	n-Bu	82	11:1

(a)

St =

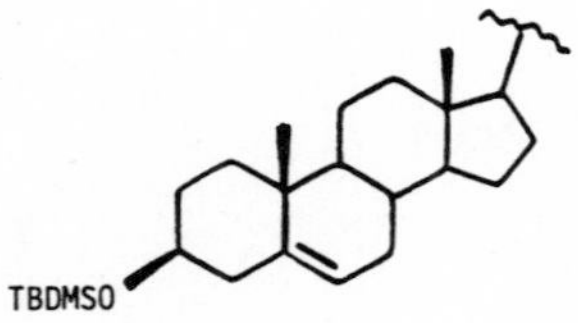

2.2.5   Diastereoselective Cyanations

Johnson described the reactions of 2,4-pentanediol and 1,3-butanediol acetals 80 with trimethylsilylcyanide under Lewis acid catalysis (Table 16) (refs. 81 and 82). The predominance of 81a is as expected.

Seebach demonstrated that homochiral 1,3-dioxan-4-ones derived from (R)-3-hydroxybutanoic acid undergo a similar diastereoselective cyanation (Table 16, entries 7 and 8) (ref. 79).

TABLE 16.
Diastereoselective Cyanations of 2,4-Pentanediol and 1,3-Butanediol Acetals 80 (refs. 79, 81, and 82)

Entry	R$_1$	R$_2$	Configuration	Yield,%	Ratio 81a:b
1	Me	n-Undecyl	R,R	98	19:1
2	Me	n-Octyl	R,R	100	24:1
3	Me	Ph	R,R	97	28:1
4	Me	m-(Ph)C$_6$H$_4$	S,S	100	1:39
5	H	m-(Ph)C$_6$H$_4$	S	92	1:15
6	H	n-Octyl	S	~90	1:9
7	=O	Me	R	65	65:1
8	=O	CH$_2$CH$_2$Ph	R	99	49:1

2.2.6 Diastereoselective Alkylations with Organometallic Reagents

Alexakis and Normant reported the first Lewis acid-catalyzed intermolecular alkylations of homochiral acetals using organocopper reagents (Table 17) (ref. 83). The (2R,4R)-2,4-pentanediol auxiliary was superior to both the (2R,3R)-2,3-butanediol auxiliary and the (2S,5S)-2,5-hexanediol auxiliary in achieving high diastereoselectivity (compare Table 17, entries 1,4, and 5).

More recently, Schrieber reported high diastereoselectivity for additions of organocopper reagents to homochiral 1,3-dioxan-4-ones derived from various aldehydes and (R)-3-hydroxybutanoic acid (Table 18) (ref. 84).

TABLE 17.
Diastereoselective Alkylations of Homochiral Acetals 82 Using Organocopper
Reagents (Ref. 83)

Entry	n	Config.	R_1	R_2Cu	Lewis Acid	Yield	Ratio 83a:b
1	0	R,R	Ph	Me$_2$CuLi	BF$_3$	95	5:1
2	0	R,R	Ph	MeCu	BF$_3$	50	6:1
3	0	R,R	Ph	Me$_2$CuLi	TiCl$_4$	48	15:1
4	1	R,R	Ph	Me$_2$CuLi	BF$_3$	93	21:1
5	2	S,S	Ph	Me$_2$CuLi	BF$_3$	92	1:13
6	0	R,R	n-Hexyl	Me$_2$CuLi	BF$_3$	96	>99:1
7	0	R,R	Me	(Ph)$_2$CuLi	BF$_3$	94	>99:1
8	0	R,R	Me	(n-Hex)$_2$CuLi	BF$_3$	89	> 99:1

TABLE 18.
Diastereoselective Alkylations of (R)-3-Hydroxybutanoic Acid Acetals 84
Using Organocopper Reagents (Ref. 84).

Entry	R	Yield,%	Ratio 85a:b
1	Me	58	33:1
2	n-Bu	68	70:1
3	Cyclohexyl	90	66:1
4	CH$_2$=CH	56	4.8:1

Johnson has examined the titanium tetrachloride-catalyzed reactions of a number of organometallic reagents with several (2R,4R)-2,4-pentanediol acetals and ketals 86 (Table 19) (Ref. 85). In a related study, Yamamoto has studied alkylations of acetals 88 using preformed alkyltitanium halides (Table 20) (ref. 86). In all cases, the observed diastereoselectivity is in accord with the Bartlett-Johnson or Johnson-Yamamoto mechanistic models. The variable diastereoselectivities observed for reactions of a particular acetal with different nucleophiles (e.g., compare Table 19, entries 3-5) reflect small differences in energies in the competing transition states and underscore the need to closely match the Lewis acidity of the catalyst with the nucleophilicity of the organometallic reagent in order to maximize diastereoselectivity.

TABLE 19.
Diastereoselective Alkylations of (2R,4R)-2,4-Pentanediol Acetals and Ketals 86 Using Organometallic Reagents (Ref. 85).

Entry	R_1	R_2	R_3M	Yield,%	Ratio 87a:b
1	m-(Ph)C$_6$H$_4$	H	MeMgCl	97	49:1
2	m-(Ph)C$_6$H$_4$	H	MeLi	83	49:1
3	n-Octyl	H	MeMgBr	90	24:1
4	n-Octyl	H	CH$_2$=CHCH$_2$MgBr	93	32:1
5	n-Octyl	H	n-BuLi	82	7:1
6	Cyclohexyl	H	n-BuMgCl	36	12:1
7	Cyclohexyl	H	n-BuLi	74	7:1
8	(CH$_3$)$_2$C=CHCH$_2$CH$_2$	Me	Et$_2$MgBr	46	2:1
9	(CH$_3$)$_2$C=CHCH$_2$Ch$_2$	Me	Et$_2$CuMg	80	9:1

TABLE 20.
Diastereoselective Alkylations of (2R,4R)-2,4-Pentanediol Acetals 88 Using
Organotitanium Halides (Ref. 86).

Entry	R_1	R_2M	Yield	Ratio 89a:b
1	Cyclohexyl	MeTiCl$_3$	100	24:1
2	Cyclohexyl	Me$_2$TiCl$_2$	100	24:1
3	Cyclohexyl	MeTiBr$_3$	91	32:1
4	Cyclohexyl	TiCl$_4$; n-BuLi	92	8:1
5	n-Hexyl	MeTiCl$_3$	93	16:1
6	n-Hexyl	MeTiBr$_3$	77	32:1
7	n-Hexyl	TiCl$_4$; Et$_2$Zn	100	4:1
8	n-Hexyl	TiBr$_4$; Et$_2$Zn	100	4:1
9	n-Hexyl	TiCl$_4$; nBuLi	47	10:1
10	n-Butyl	MeTiCl$_3$	81	32:1
11	n-Butyl	TiCl$_4$; Et$_2$Zn	86	4:1

2.2.7   Other Reactions at Acetal Carbon

Snider reported limited success in attempts at inducing diastereoselectivity in ene-type reactions of alkenes with homochiral acetals (ref. 87).
Some of his results are summarized in equations 17 and 18.

(Eqn. 17)

n	yield	diastereomer ratio
0	61	2:1
1	53	2:1

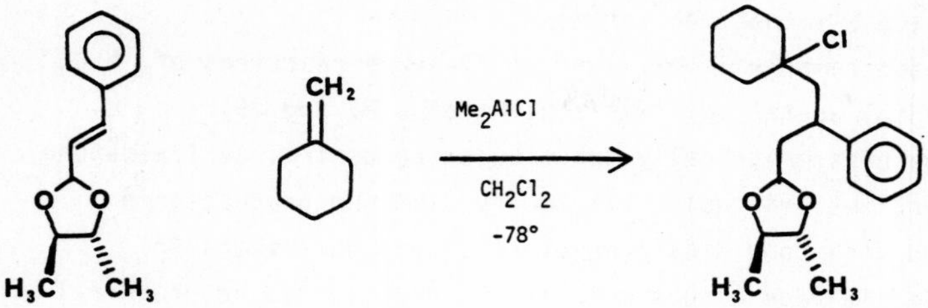

(Eqn. 18)

92

93

70% yield

diastereomer ratio 4:1

3. REACTIONS AT CARBON ALPHA TO ACETAL OR KETAL CARBON

3.1 Diastereoselective Eliminative Cleavage

Yamamoto described remarkable eliminative cleavage reactions of (2R,4R)-2,4-pentanediol acetals 94 (Table 21) (refs. 88 and 89). Eliminative cleavage of symmetrically substituted homochiral cycloalkanone ketals (Table 21, entries 1-4) using triisobutylaluminum produces, in excellent yields and with good diastereoselectivity, enol ethers 95. Incomplete reductive cleavage of unsymmetrically substituted heterochiral

TABLE 21.
Diastereoselective Eliminative Cleavages of (2R,4R)-2,4-Pentanediol Cycloalkanone Ketals 94 (Refs. 88 and 89).

Entry	n	R(Position)	Aluminum Reagent (equiv.)	Yields, % 94	Yields, % 95	Diastereomer Ratios 94	Diastereomer Ratios 95
1	1	Me(4)	TIBA(4)	--	99	7:1	--
2	1	Et(4)	TIBA(4)	--	99	7:1	--
3	1	t-Bu(4)	TIBA(4)	--	99	7:1	--
4	1	Me,Me(3,5)	TIBA(4)	--	99	10:1	--
5	1	Me(2)	TIBA(2)	--	33	--	32:1
6	1	Me(2)	DIBAH(2)	42	--	99:1	--
7	1	CH$_2$CH=CH$_2$(2)	TIBA(2)	35	33	99:1	32:1
8	1	Cyclohexyl(2)	TIBA(4)	25	--	99:1	--
9	1	Me(3)	TIBA(4)	49	--	2:1	--
10	2	Me(2)	TIBA(4)	27	--	49:1	--
11	0	n-Undecyl(2)	DIBAH(1.5)	37	--	49:1	--

cycloalkanone ketals (Table 21, entries 5-11) produces, via kinetic resolution, a mixture of resolved ketals 94 and enol ethers 95. Assuming that the Johnson-Yamamoto mechanism remains operative, deprotonation, presumably by a group attached to aluminum, must occur concomitantly with diastereoselective cleavage of the dioxane bond that relieves the worst 1,3-diaxial interaction(s). If the symmetrically substituted cycloalkanone ketals are anchored by their substituents, then two chair-chair conformations are available (eqn. 19). Given the preponderant formation of $\underline{S}$ enol ether products, conformer A must be more reactive and/or be present in a higher concentration than conformer B. For unsymmetrically substituted cycloalkanone

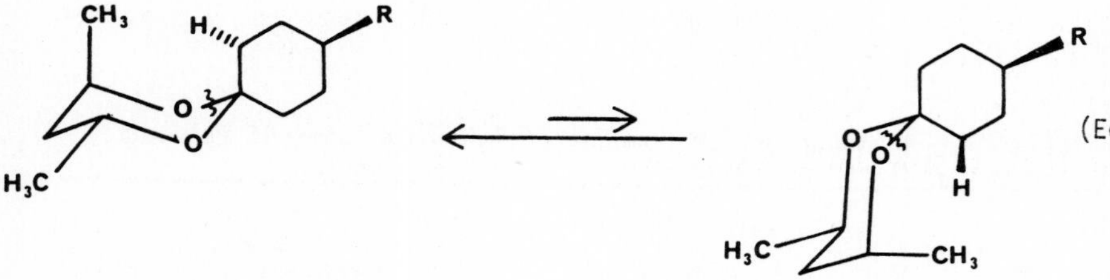

(Eqn. 19)

A

B

ketals two chair-chair conformers must be considered for each diastereoisomer, assuming the cycloalkane substituents occupy equitorial positions (eqns. 20 and 21). On steric grounds conformers C and E should predominate. Proton H_C is more accessible than H_E; thus, conformer C should be (and apparently is) more reactive than E.

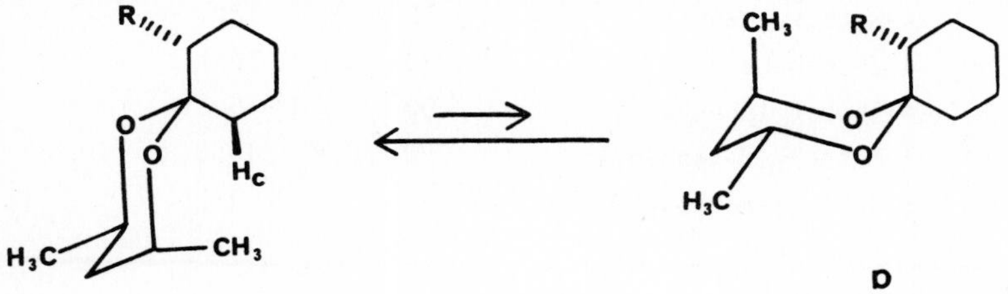

(Eqn. 20)

C

D

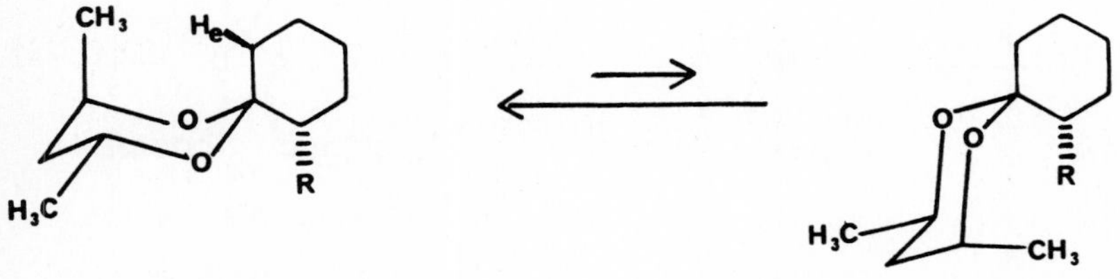

(Eqn. 21)

E

F

3.2 Diastereoselective Halogenations

Castaldi and Giordano have described the diastereoselective bromination of homochiral 2-aryl-2-ethyl-1,3-dioxolanes **96** derived from (R,R)-dimethyl tartrate (Table 22) (ref. 90). This reaction may represent a case where diastereoselectivity results during dioxolane ring reformation (eqn. 22). Mechanistic studies and extensions to other homochiral ketals or halogens have not been reported.

TABLE 22.
Diastereoselective Brominations of Homochiral 1,3-Dioxolanes 96 Derived from (R,R)-Dimethyl Tartrate (Ref. 90).

Entry	Ar	Yield	Ratio 97a:b
1	Phenyl	94	13:1
2	4-Methoxyphenyl	95	12:1
3	4-Chlorophenyl	93	16:1
4	4-Isobutylphenyl	90	5:1
5	6-Methoxy-2-naphthyl	98	10:1
6	5-Bromo-6-methoxy-2-naphthyl	98	9:1

(Eqn. 22)

3.3 Diastereoselective Additions

Tamura described diastereoselective additions of Grignard reagents to homochiral monoketals of 1,2-cycloalkanediones 98 (Table 23) (ref. 91). Much greater diastereoselectivity was observed using 1,4-di-O-methyl-L-threitol as the auxiliary than for (2R,3R)-2,3-butanediol. Interestingly, the major diastereoisomer was the same for both auxiliaries. These observations are consistent with reagent delivery under chelatory control for the threitiol auxiliary and under steric approach control for the butanediol auxiliary (eqns. 23 and 24).

TABLE 23.
Diastereoselective Additions of Grignard Reagents to Homochiral Monoketals of 1,2-Cycloalkanediones 98 (Ref. 91)

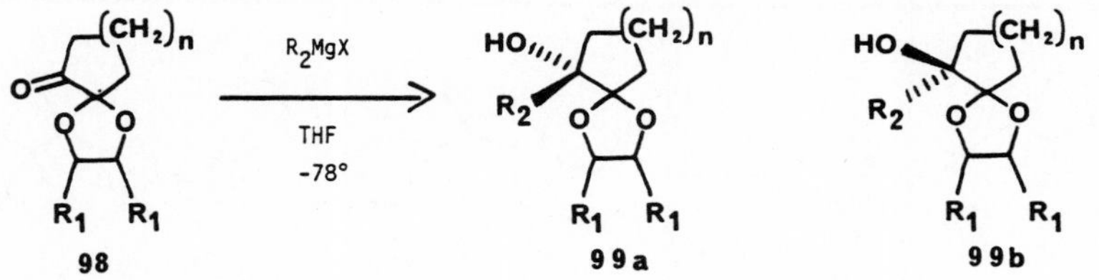

Entry	R_1	Configuration	n	R_2MgX	Yield	Ratio 99a:b
1	CH_2OMe	(S,S)	1	MeMgBr	91	49:1
2	CH_2OMe	(S,S)	2	MeMgBr	93	99:1
3	CH_2OMe	(S,S)	1	EtMgCl	95	99:1
4	CH_2OMe	(S,S)	2	EtMgCl	95	99:1
5	CH_2OMe	(S,S)	2	$CH_2=CHMgBr$	95	32:1
6	CH_2OMe	(S,S)	2	PhMgBr	85	19:1
7	Me	(R,R)	1	MeMgBr	95	2:1
8	Me	(R,R)	2	MeMgBr	91	2:1

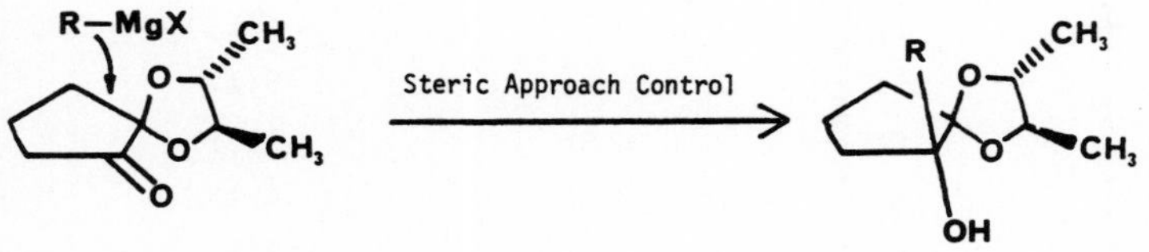

Steric Approach Control

(Eqn. 23)

(Eqn. 24)

4. REACTIONS AT CARBON BETA TO THE ACETAL OR KETAL CARBON

4.1 Diastereoselective Carbonyl Reductions

Matsumoto examined the diastereoselective reductions of the ketone moiety in homochiral acetals **100** (Table 24) for use in his synthesis of (+)-pedamide **102** (Fig. 10) (refs. 92, 93 and 94). The best

TABLE 24.
Diastereoselective Reductions of β-Ketoacetals **100** (Refs. 92-94).

Entry	R_1	R_2	MX	Ratio 101a:b
1	H	Et	none	4:1
2	H	Et	LiBr	5:1
3	H	Et	MgBr$_2$	1:4
4	Ph	Et	none	2:1
5	Ph	Et	LiBr	2:1
6	Ph	Et	MgBr$_2$	1:4
7	OPh	Et	none	2:1
8	OPh	Et	LiBr	3:1
9	OPh	Et	MgBr$_2$	1:10
10	OCH$_2$Ph	Et	none	3:1
11	OCH$_2$Ph	Et	LiBr	5:1
12	OCH$_2$Ph	Et	MgBr$_2$	4:1
13	OMe	Et	none	5:1
14	OMe	Et	LiBr	11:1
15	OMe	Et	MgBr$_2$	4:1
16	Me	CH=CH$_2$	none	6:1

Figure 10. Enantioselective Synthesis of Pedamide (102).

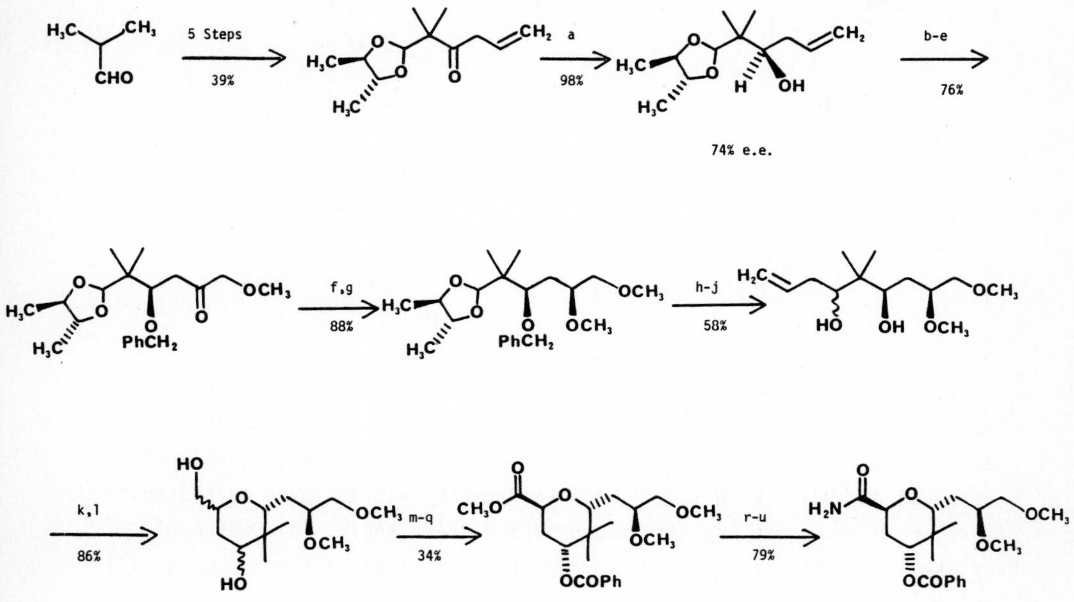

Reagents:

(a) LAH, ether, toluene, -123°

(b) BnCl, tert-AmylONa, DMSO, RT

(c) MCPBA, CH$_2$Cl$_2$

(d) NaOCH$_3$, CH$_3$OH

(e) Collins Oxidation

(f) LiAlH(OtBu)$_3$, ether, -78°

(g) CH$_3$I, NaH, C$_6$H$_6$, heat

(h) Aq. HCl, acetone, heat

(i) CH$_2$=CHCH$_2$MgBr, ether

(j) Na, liq. NH$_3$

(k) MCPBA, CH$_2$Cl$_2$

(l) TsOH, C$_6$H$_6$, heat

(m) Jones' Oxidation

(n) CH$_2$N$_2$, ether

(o) NaBH$_4$, EtOH

(p) PhCOCl, pyridine

(q) LDA, THF, -78°; HOAc

(r) Recryst. to optical purity

(s) Et$_3$N, H$_2$O, CH$_3$OH

(t) SOCl$_2$, DMF, CH$_2$Cl$_2$, heat

(u) NH$_3$, CH$_2$Cl$_2$, 0°

selectivity was observed using 1,4-di-<u>O</u>-methyl-D-threitol as the auxiliary in the presence of lithium cation (Table 24, entry 14). The authors ascribe these results to chelation-induced biasing of the conformations available to the ketoacetal (eqn. 25) (ref. 94).

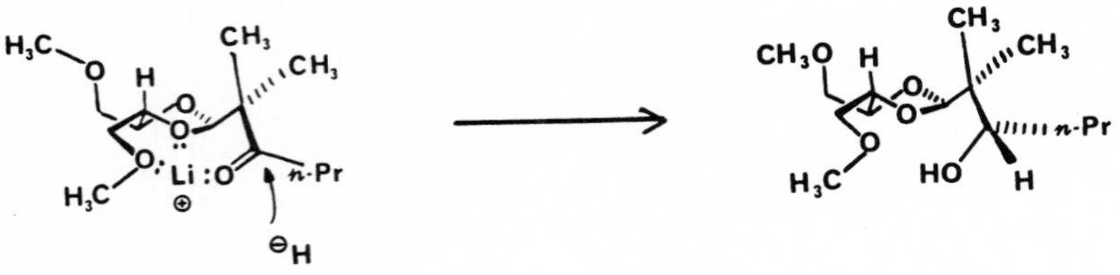

(Eqn. 25)

4.2 Additions to α,β-Unsaturated Homochiral Acetals and Ketals

Yamamoto described regiocontrolled diastereoselective alkyl transfer from trialkylaluminum reagents to homochiral 2-vinyl-1,3-dioxolanes **103** derived from (S,S)-N,N,N',N'-tetramethyltartaric acid diamide (Table 25) (refs. 95 and 96). When (<u>2R</u>,<u>4R</u>)-2,4-pentanediol was employed as the auxiliary, no diastereoselectivity was observed (ref. 68b). This methodology was used to prepare alcohol **106**, a component of Vitamins E and K (Fig. 11) (ref. 95).

Figure 11. Enantioselective Synthesis of Alcohol 106.

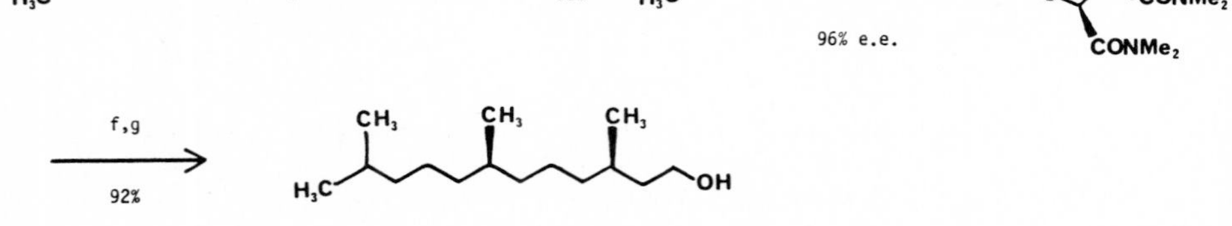

Reagents:

(a) (EtO)$_2$POCH=CHNLi(tBu)

(b) H$_2$, Pd·C, THF

(c) HC(OEt)$_3$, NH$_4$NO$_3$, EtOH

(d) (S,S)-Tetramethyltartramide, PPTS

(e) (CH$_3$)$_3$Al, toluene

(f) Aq. HCl, dioxane

(g) NaBH$_4$

TABLE 25.
Regioselective and Diastereoselective Alkyl Transfer from Trialkylaluminum
Reagents to Tetramethyltartramide Acetals and Ketals 103 (Refs. 95 and
96).[a]

Entry	R_1	R_2	R_3Al	Solvent(Temp)	Yields (diastereoselectivity) 104	105
1	H	nPr	Me$_3$Al	ClCH$_2$CH$_2$Cl$_2$(RT)	84(15:1)	13
2	H	nPr	Me$_3$Al	CHCl$_3$(RT)	0	85(15:1)
3	H	nPr	Me$_3$Al	CH$_2$Cl$_2$(RT)	70(19:1)	21
4	H	nPr	Me$_3$Al	Toluene (RT)	48(19:1)	31
5	H	Me	nPr$_3$Al	CH$_2$Cl$_2$(0°C)	65(14:1)	31
6	H	Me	nPr$_3$Al	Toluene (0°C)	41(27:1)	41
7	H	Ph	Me$_3$Al	CH$_2$Cl$_2$(0°C)	62(32:1)	0
8	H	Ph	Me$_3$Al	Toluene (0°C)	72(49:1)	19
9	Me	Ph	Me$_3$Al	CH$_2$Cl$_2$(RT)	83(3:1)	0
10	Me	Ph	Me$_3$Al	Toluene(RT)	81(5:1)	0
11	Me	Ph	Et$_3$Al	Toluene(RT)	74(6:1)	0
12	-CH$_2$CH$_2$CH$_2$-		Me$_3$Al	CH$_2$Cl$_2$ (RT)	94(6:1)	0
13	-CH$_2$CH$_2$CH$_2$-		Me$_3$Al	Toluene (RT)	91(8:1)	0

(a) Predominant product diastereoisomers shown.

Alexakis and Normant have examined the boron trifluoride-catalyzed additions of organocopper reagents to homochiral α,β-unsaturated acetals and ketals 107 (Table 26) (refs. 83, 97 and 98).

TABLE 26.
Diastereoselective Boron Trifluoride-Catalyzed Additions of Organocopper Reagents to Homochiral α,β-Unsaturated Acetals and Ketals 107 (Refs. 83, 97 and 98).

Entry	R_1	R_2	$R_3Cu \cdot L$	Yield,%	Ratio 108a:b
1	$-CH_2CH_2CH_2-$		Me_2CuLi	71	2:1
2	H	Me	PhCu	70	6:1
3	H	Me	$PhCu \cdot PBu_3$	75	19:1
4	H	nPr	$PhCu \cdot PBu_3$	71	10:1
5	H	Me	$(CH_3)_2C=CHCu \cdot Me_2S$	72	5:1
6	H	Me	$(CH_3)_2C=CHCu \cdot PBu_3$	70	12:1
7	H	Me	$CH_3(CH_2)_5CH=CHCu \cdot Me_2S$	70	6:1
8	H	Me	$CH_3(CH_2)_5CH=CHCu \cdot PBu_3$	68	12:1
9	H	$CH_2=CH(CH_2)_2$	$CH_2=C(CH_3)Cu \cdot PBu_3$	>50	12:1

These same authors have described additions of organocopper reagents to homochiral α,β-alkynyl acetals 109 (Table 27) (ref. 99). Under appropriate conditions, the intermediate α-metallated alkenyl acetals ring open to provide alkoxyallenes 110 diastereoselectively.

TABLE 27.
Diastereoselective Synthesis of Alkoxyallenes by Addition of Organocopper
Reagents to α,β-Alkynyl Acetals 109 (ref. 99).

Entry	n	R	L	Yield	Ratio 110a:b
1	1	Me	--	100	2:1
2	0	Me	--	100	4:1
3	0	Me	P(OEt)₃	100	9:1
4	0	nBu	--	100	6:1
5	0	tBu	--	100	99:1
6	0	Ph	--	100	3:1

5. REACTIONS INVOLVING CARBONS ALPHA AND BETA TO THE ACETAL OR KETAL CARBON

5.1 Diastereoselective Halolactonizations

Terashima has reported a novel diastereoselective bromolactonization
of cycloalkene ketals 111 derived from homochiral tartramides (ref. 100)
(eqn. 26). Bromolactonization of related ketals 113 (Fig. 12) provided,

(Eqn. 26)

in ~80% yields, 6:1 mixtures of seven and six-membered bromolactones 114
and 115 with complete diastereoselectivity (refs. 101 and 102). These
could be converted, via epoxide 116, into anthracyclinone intermediates 117

Figure 12. Synthesis of Anthracyclinone Intermediates 117.

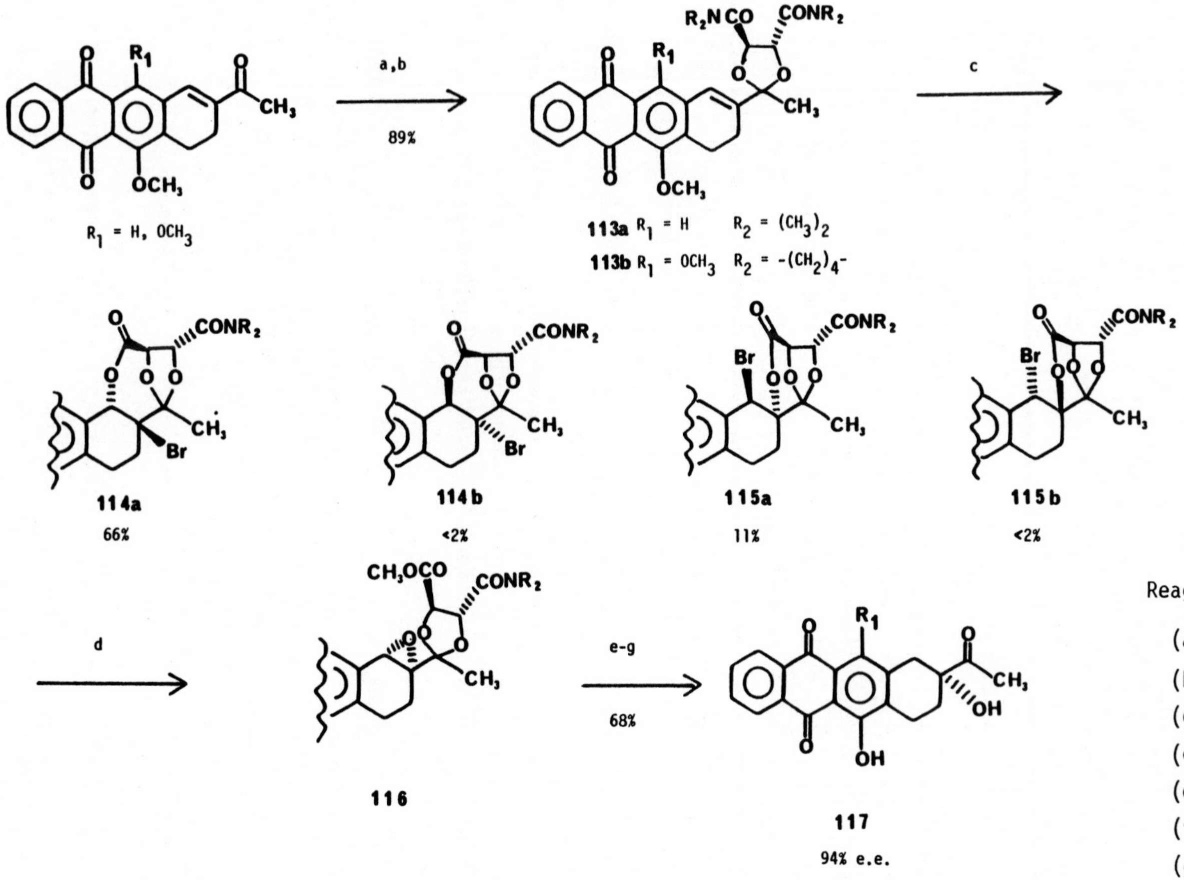

Reagents:

(a) $(CH_3O)_3CH$, PPTS

(b) Tartramide, PPTS

(c) $CH_3CONHBr$, Aq. DMF

(d) K_2CO_3, CH_3OH

(e) H_2, Pd·C, THF

(f) HCl, EtOH, heat

(g) $AlCl_3$, C_6H_6

(Fig. 12). The authors suggest that steric repulsion between the methyl attached to the acetal carbon and the bromine atom of the appendage-stabilized bromium ion is responsible for the observed diastereoselectivity (eqn. 27). Studies examining the generality of this bromolactonization process have not been reported.

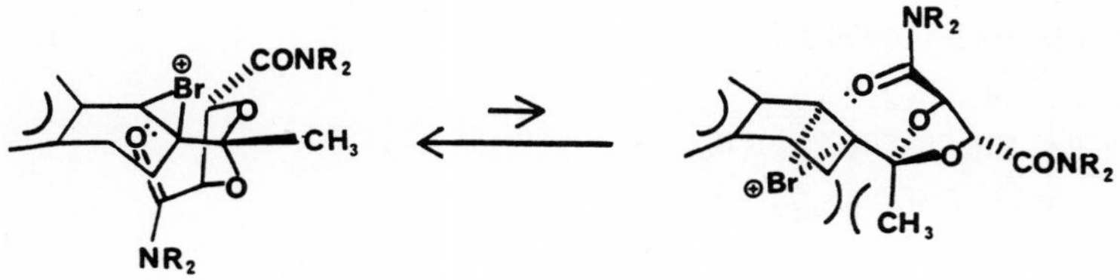

(Eqn. 27)

5.2 Diastereoselective Cyclopropanations

Yamamoto reported that homochiral α,β-unsaturated acetals 118 derived from tartaric acid esters could be cyclopropanated diastereoselectively under alkylidine transfer conditions (Table 28) (refs. 103 and 104). Inferior diastereoselectivities were observed when (2R,4R)-2,4-pentanediol was employed as the auxiliary. This methodology was used to construct 5,6-methanoleukotriene A₄ 120 enantioselectively (Fig. 13) (ref. 104).

Figure 13. Enantioselective Synthesis of 5,6-Methanoleukotriene A₄ (120).

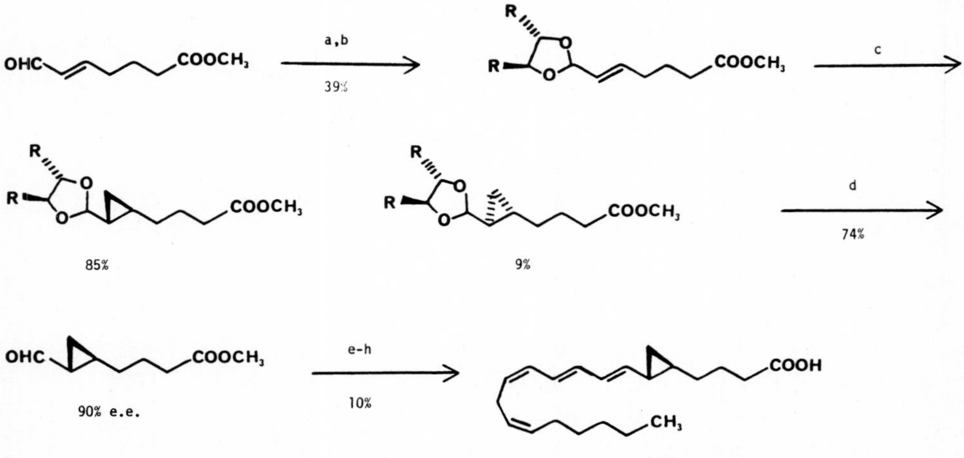

630

Legend for Figure 13.

Reagents:

      (a) $(CH_3O)_3CH$, NH_4NO_3, EtOH

      (b) $(+)$-Diisopropyltartrate, PPTS, C_6H_6

      (c) CH_2I_2, Et_2Zn, hexanes, $-20°$

      (d) TsOH, CH_3OH, H_2O

      (e) $nBu_3SnCH=CH-CH=CHOEt$, nBuLi, THF

      (f) TsOH, THF, H_2O

      (g) $Ph_3P=CHCH_2CH=CH(CH_2)_4CH_3$, THF

      (h) Aq. NaOH, THF, CH_3OH

TABLE 28.
Diastereoselective Cyclopropanations of Homochiral Acetals 118 Under Alkylidene Transfer Conditions (Refs. 103 and 104).

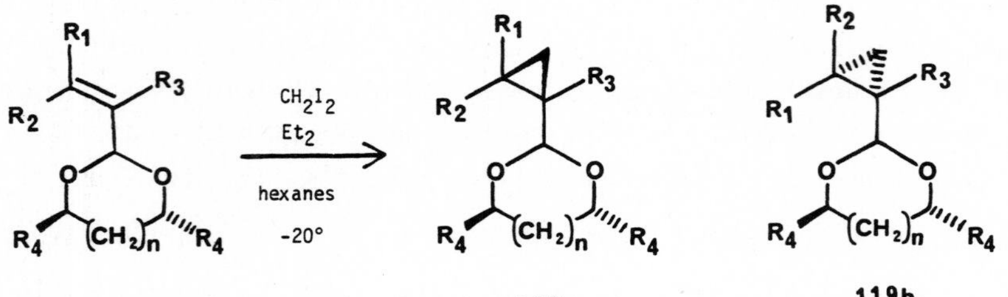

Entry	R_1	R_2	R_3	R_4	n	Yield, %	Ratio 119a:b
1	Me	H	H	CO_2iPr	0	90	32:1
2	Me	H	H	Me	1	74	5:1
3	nPr	H	H	CO_2Et	0	95	16:1
4	nPr	H	H	CO_2iPr	0	80	21:1
5	nPr	H	H	Me	1	95	6:1
6	Ph	H	H	CO_2Et	0	82	14:1
7	Ph	H	H	CO_2iPr	0	92	21:1
8	Ph	H	H	Me	1	85	5:1
9	Et	H	Me	CO_2iPr	0	81	17:1
10	Et	H	Me	Me	1	69	7:1
11	Me	Me	H	Me	1	81	1:2

Mash reported that homochiral cycloalkenone ketals 121 derived from 1,4-di-$\underline{O}$-benzyl-L-threitol undergo diastereoselective cyclopropanation when treated with the Simmons-Smith reagent (Tables 29 and 30) (refs. 105-108). Acetals 123 displayed little or no diastereoselectivity under these conditions (ref. 109).

TABLE 29.
Diastereoselective Cyclopropanations of Homochiral 2-Cycloalken-1-one Ketals 121 Using Simmons-Smith Reagent (Refs. 105-108).

Entry	n	R_1	R_2	R_3	Yield, %	Ratio 122a:b
1	1	CH_2OCH_2Ph	H	H	72	9:1
2	1	CH_2OCH_2Ph	Me	H	88	9:1
3	1	CH_2OMe	Me	H	88	8:1
4	1	CO_2iPr	Me	H	36	4:1
5	1	CH_2OCH_2Ph	H	iPr	54	9:1
6	1	CH_2OCH_2Ph	$-(CH_2)_3-$		78	9:1
7	1	CH_2OCH_2Ph	$-(CH_2)_4-$		90	7:1
8	1	CH_2OCH_2Ph	$-(CH_2)_5-$		84	9:1
9	2	CH_2OCH_2Ph	H	H	98	9:1
10	2	CH_2OMe	H	H	86	5:1
11	2	CH_2OH	H	H	50	1:2
12	2	CO_2Me	H	H	37	2:1
13	2	$C(CH_3)_2OMe$	H	H	91	4:1
14	2	$(CH_2)_3Ph$	H	H	92	9:1
15	2	Me	H	H	86	9:1
16	2	CH_2OCH_2Ph	Me	H	99	20:1
17	2	CH_2OCH_2Ph	$-(CH_2)_4-$		92	7:1
18	3	CH_2OCH_2Ph	H	H	90	8:1

TABLE 30.
Diastereoselective Cyclopropanations of Acyclic Homochiral Acetals and
Ketals **123** Using Simmons-Smith Reagent (Refs. 105-108).

Entry	R_1	R_2	R_3	R_4	Yield, %	Diastereomer Ratio 124
1	Me	$-(CH_2)_4-$		H	88	14:1
2	H	$-(CH_2)_4-$		H	70	2:1
3	Me	H	Me	Me	58	2:1
4	H	H	Me	Me	94	1:1
5	H	H	$(CH_2)_3CO_2Me$	H	62	1:1

This methodology has been used in syntheses of (<u>R</u>)-muscone **125** (Fig. 14)
(ref. 110), (+)-β-eudesmol **126** (Fig. 15) (ref. 111), (-)-chokol A **127** (Fig.
16) (ref. 112), and in an enantioselective approach to modhephene **128** (Fig.
17), (ref. 113).

125   **126**   **127**   **128**

Figure 14. Enantioselective Synthesis of (R)-Muscone (125).

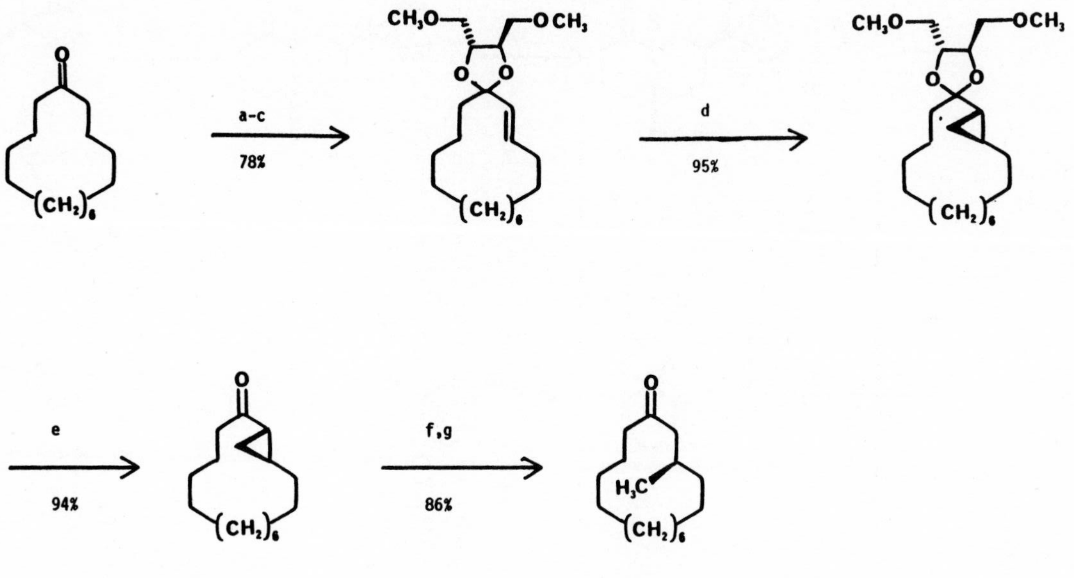

Reagents:

(a) Br_2, CH_3OH

(b) 1,4-Di-0-methyl-L-threitol, TsOH, C_6H_6, heat

(c) $NaOCH_3$, DMSO

(d) CH_2I_2, $Zn(Cu)$, I_2, Et_2O, heat

(e) Aq. HCl, CH_3OH

(f) Li, tBuOH, liquid NH_3

(g) PDC, CH_2Cl_2

634

Figure 15. Enantioselective Synthesis of (+)-β-Eudesmol (126).

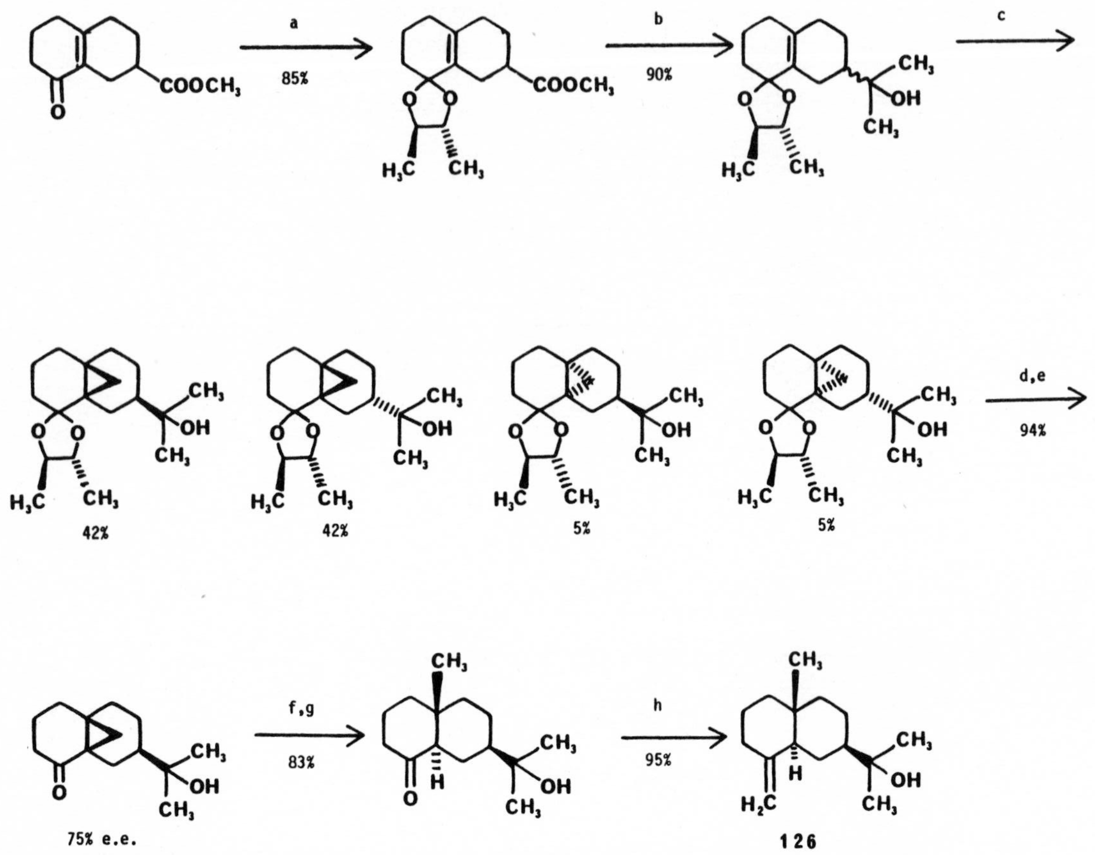

Reagents:

(a) (2S,3S)-Butanediol
    bis-TMS Ether, TMSOTf

(b) CH$_3$MgBr, THF

(c) CH$_2$I$_2$, Zn(Cu), I$_2$,
    Et$_2$O, heat

(d) Aq. HCl, CH$_3$OH

(e) HOCH$_2$CH$_2$OH, PPTS,
    C$_6$H$_6$, heat

(f) Li, tBuOH, liq. NH$_3$

(g) PDC, CH$_2$Cl$_2$

Figure 16. Enantioselective Synthesis of (-)-Chokol A (127).

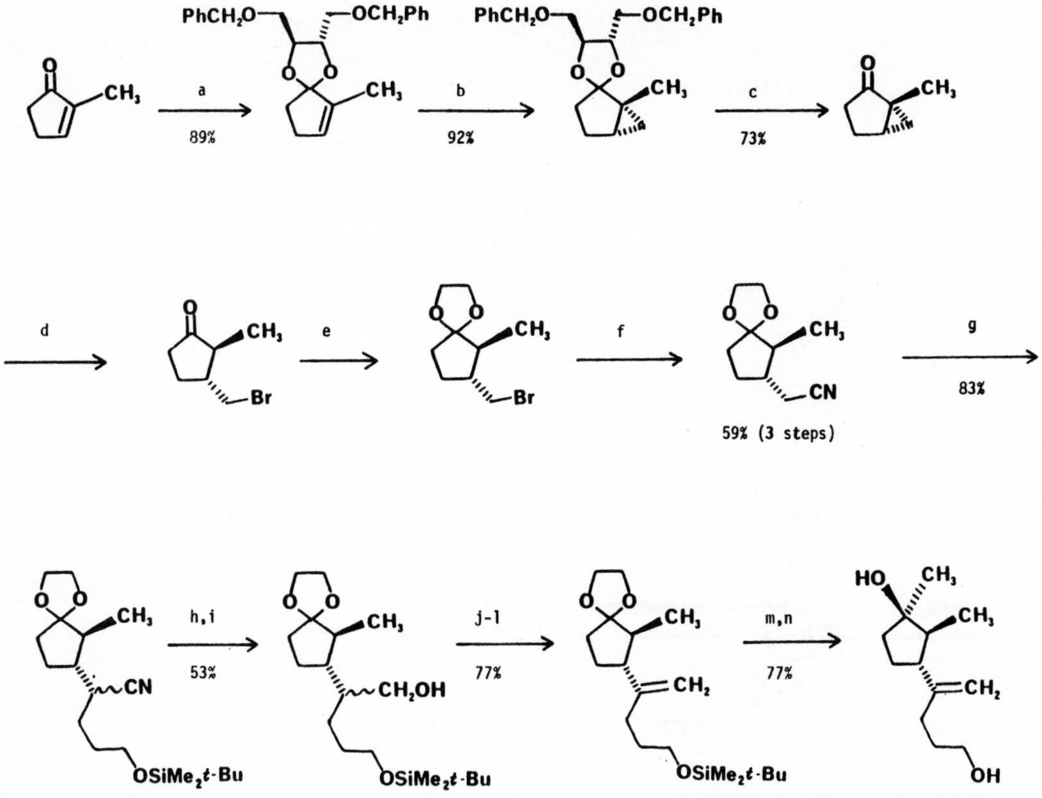

Reagents:

(a) 1,4-Di-O-benzyl-L-threitol, PPTS, C_6H_6, heat

(b) CH_2I_2, Zn(Cu), I_2, ether, heat

(c) Aq. HCl, CH_3OH

(d) TMS-Br, CH_2Cl_2, -10°

(e) $HOCH_2CH_2OH$, PPTS, C_6H_6, heat

(f) KCN, NaI, DMSO

(g) $LiN(SiMe_3)_2$, THF, -78°; $ICH_2CH_2CH_2OTBDMS$

(h) DIBAL, CH_2Cl_2

(i) $NaBH_4$, EtOH

(j) TsCl, pyridine, 0°

(k) $Na^+ ~^-SePh-o-NO_2$, EtOH

(l) H_2O_2, THF, EtOH

(m) Aq. HCl, CH_3OH

(n) $(CH_3)CeCl_2$, THF, -78°

Figure 17. An Enantioselective Approach to Modhephene (128).

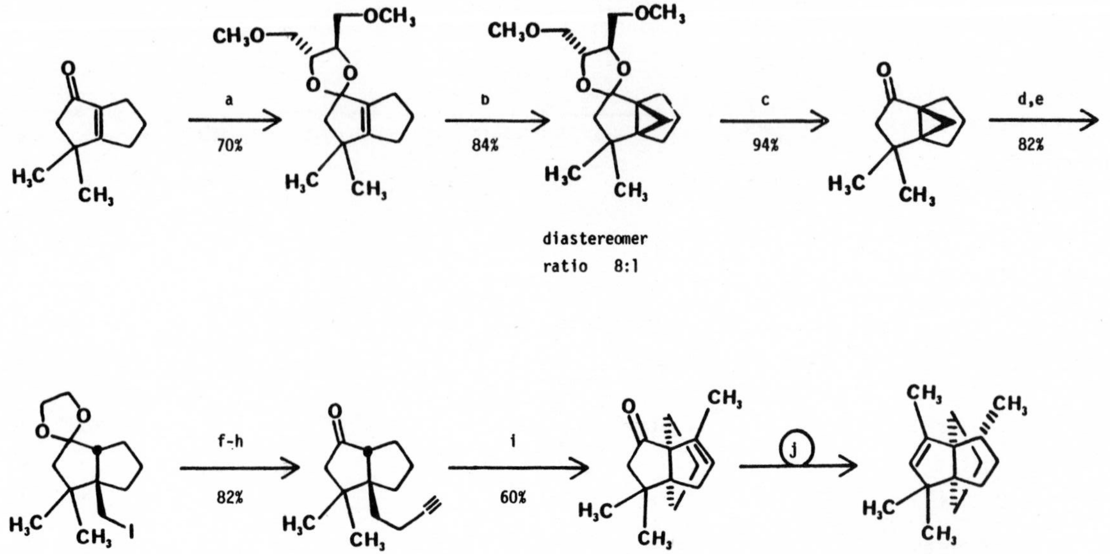

diastereomer
ratio 8:1

128

Reagents:

(a) 1,4-Di-0-methyl-2,3-di-0-trimethylsilyl-D-threitol, TMSOTf, CH_2Cl_2

(b) CH_2I_2, Zn(Cu), I_2, Et_2O, heat

(c) Aq. HCl, CH_3OH

(d) TMS-I, CCl_4

(e) $TMSOCH_2CH_2OTMS$, TMSOTf, CH_2Cl_2

(f) Li^+ $^-C{\equiv}C$-TMS, TMED, HMPA, Et_2O

(g) Aq. HCl, CH_3OH

(h) nBu_4NF, THF

(i) Decalin, 380°

(j) See JACS 1981, <u>103</u>, 722.

Chelation-controlled delivery of the Simmons-Smith reagent by allylic oxygen is well-known. The diastereoselectivity observed for these cyclopropanations may be due to preferential chelation of the reagent by one of the two dioxolane or dioxane oxygen atoms, resulting from conformational or rotational isomerism. Facts from which a secure mechanistic picture can be assembled are not yet available.

6. REACTIONS WHICH INCORPORATE THE AUXILIARY

Masaki and coworkers have reported an enantioselective approach to tetrahydropyrans in which the carbon skeleton of a homochiral acetal derived from tartaric acid is incorporated into the THP molecule (refs. 114-116). Molecules so synthesized include exo-brevicomin 129 (Fig. 18) (ref. 114); acid 130, a component of civet (Fig. 19) (ref. 115); and (-)-pestalotin 131 (Fig. 20) (ref. 116).

Figure 18.
Enantioselective Synthesis of Exo-brevicomin (129)

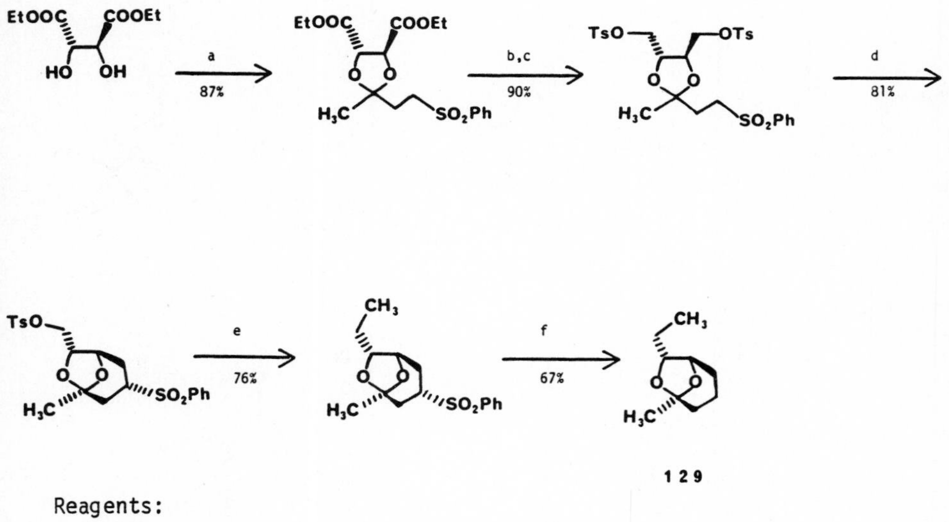

Reagents:

(a) $CH_3C(OCH_3)_2CH_2CH_2SO_2Ph$, TsOH, C_6H_6
(b) $NaBH_4$, EtOH
(c) TsCl, pyridine
(d) nBuLi, THF, -20°
(e) $(CH_3)_2CuLi$, Et_2O, $(CH_3)_2S$
(f) Na, THF, EtOH

638

Figure 19.
Enantioselective Synthesis of Civet Acid 130.

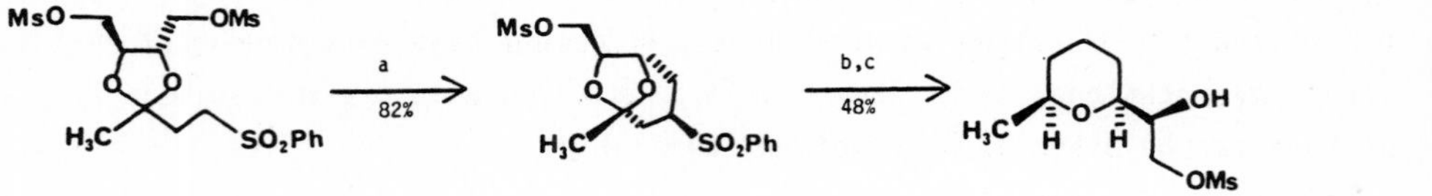

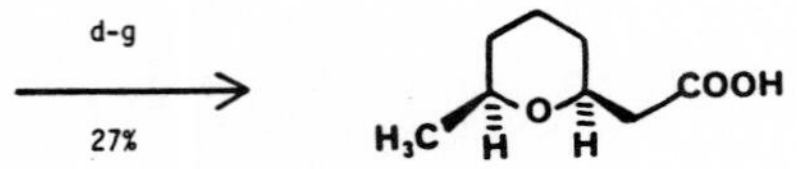

Reagents:

(a) nBuLi, THF, -20°
(b) Na, EtOH, THF, -20°
(c) LAH, AlCl$_3$, Et$_2$O
(d) K$_2$CO$_3$, CH$_3$OH
(e) KSeCN, H$_2$O, CH$_3$OH
(f) BH$_3$·THF; H$_2$O$_2$, NaOH
(g) Jones' Oxidation

Figure 20.
Enantioselective Synthesis of (-)-Pestalotin (131).

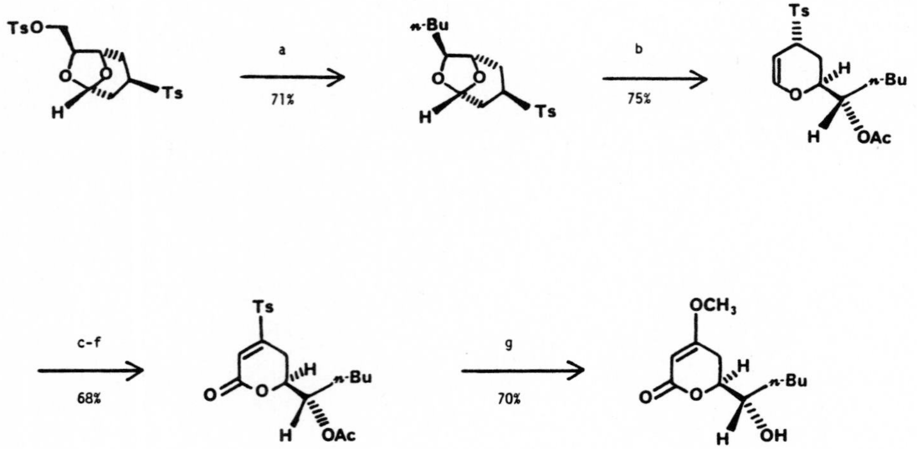

Reagents:
(a) (nPr)$_2$CuLi, Et$_2$O, (CH$_3$)$_2$S
(b) Ac$_2$O, BF$_3$·Et$_2$O, CH$_2$Cl$_2$
(c) Br$_2$, CH$_2$Cl$_2$
(d) Aq. K$_2$CO$_3$, THF

(e) NaOCH$_3$, THF
(f) Jones' Oxidation
(g) K$_2$CO$_3$, CH$_3$OH

7. MISCELLANEOUS USES OF HOMOCHIRAL DIOLS

Truesdale employed (2R,3R)-2,3-butanediol as a protecting group in an enantioselective approach to 11-deoxyprostanoids **132** (Fig. 21) (ref. 117). The homochiral auxiliary apparently did not influence the subsequent chemistry.

Ikegami used the same auxiliary in an enantioselective approach to iso-carbacyclin **133** (Fig. 22) (ref. 118). The bulk of a dioxolane methyl appendage inhibits insertion at the carbon alpha to the acetal (compare eqns. 28 and 29).

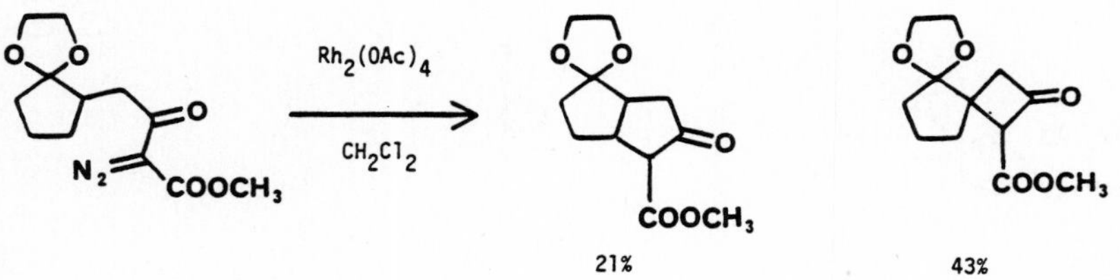

(Eqn. 28)

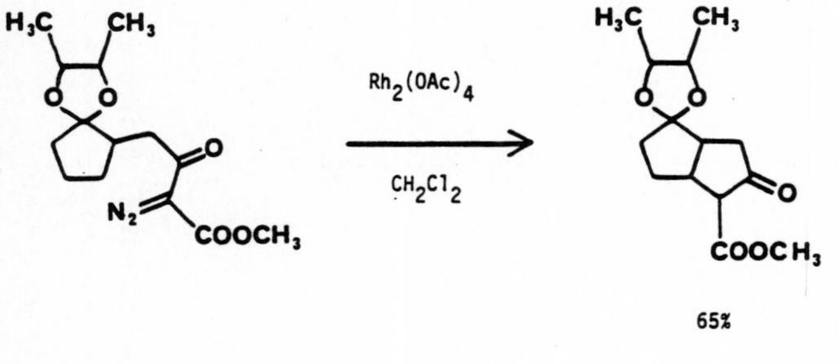

21%   43%

(Eqn. 29)

65%

Figure 21. Enantioselective Synthesis of 11-Deoxyprostanoid 132.

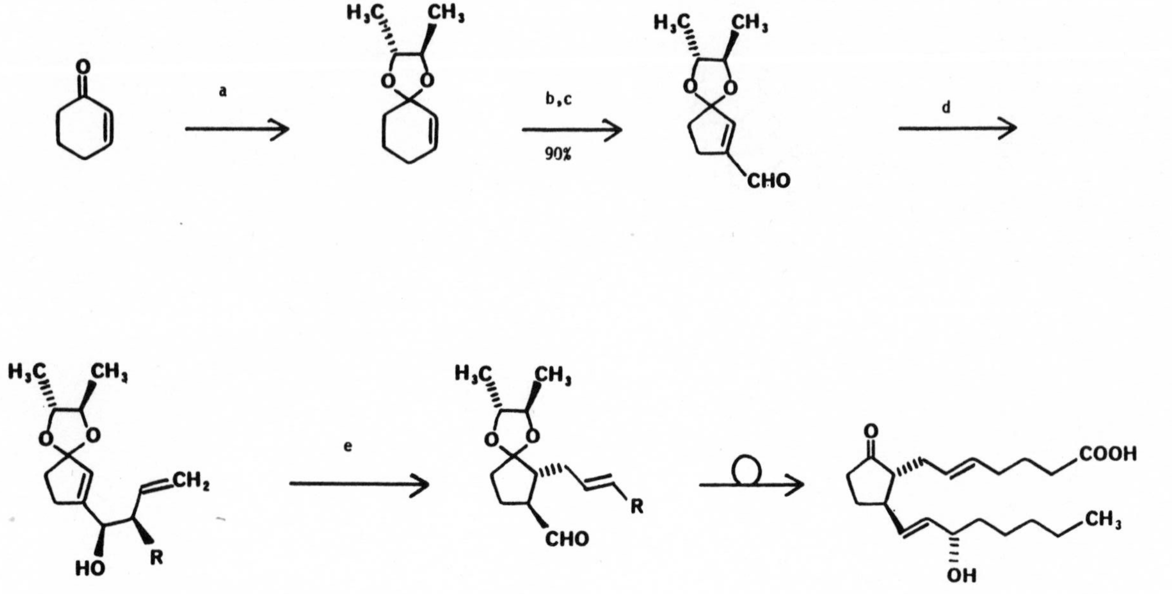

R = H, Ph, $(CH_2)_3$COBO

diastereomer ratio 4:1

132

Reagents:

(a) (2R,3R)-Butanediol, oxalic acid
(b) O_3, CH_2Cl_2; $(CH_3)_2S$
(c) Piperidinium acetate
(d) (1)-B-Allyldiisopinocampheylborane
(e) KH, glyme, heat

Figure 22. Enantioselective Synthesis of Isocarbacyclin (133).

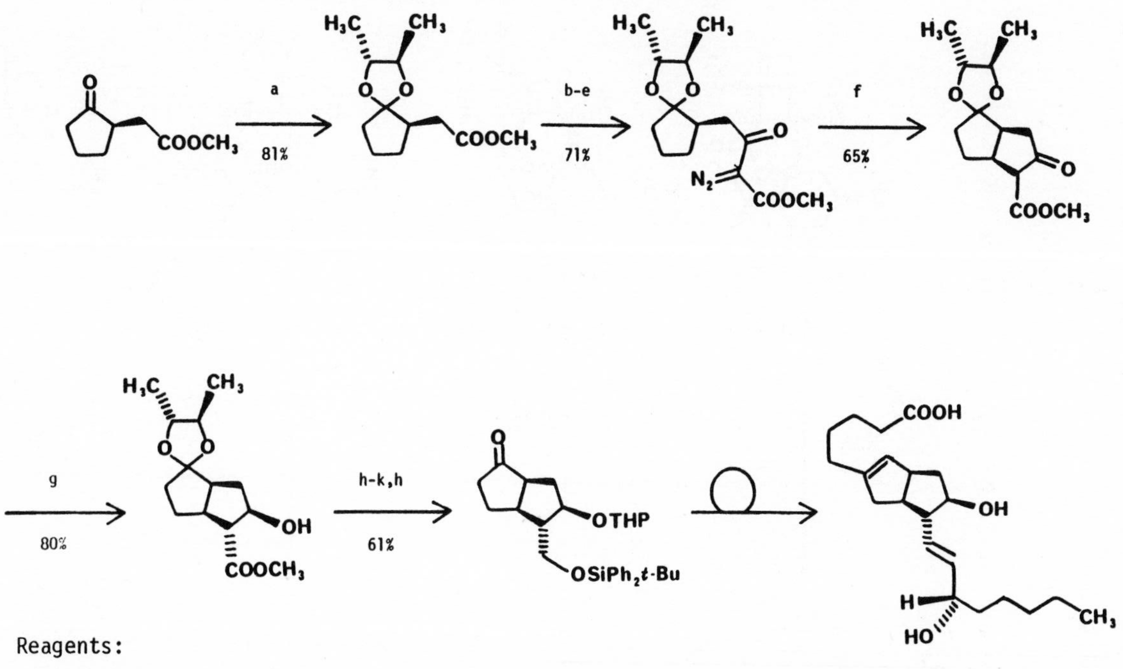

Reagents:

(a) (2R,3R)-Butanediol, $BF_3 \cdot Et_2O$, CH_2Cl_2

(b) KOH, CH_3OH

(c) Carbonyldiimidazole, THF

(d) $CH_3O_2CCHCO_2Mg$, THF

(e) TsN_3, Et_3N

(f) $Rh_2(OAc)_4$, CH_2Cl_2

(g) $NaBH_4$, EtOH, $-50°$

(h) DHP, PPTS, CH_2Cl_2

(i) LAH, Et_2O, $-20°$

(j) TBDPS-Cl, DMAP, $(iPr)_2NEt$

(k) Aq. HCl, THF

8. USES OF HOMOCHIRAL KETONE AUXILIARIES

Relatively few synthetic uses for ketals derived from homochiral ketone auxiliaries have been described. Demuth prepared and separated dioxacyclo-hexenones **134** and **135** and studied their photochemical cycloadditions with olefins (Table 31) (ref. 119). This methodology was used to synthesize (+)-grandisol **136** (Fig. 23).

TABLE 31.
Photochemical Additions of Dioxacyclohexenones **134** and **135** to Alkenes (Ref. 119)).

Figure 23. Enantioselective Synthesis of (+)-Grandisol (136).

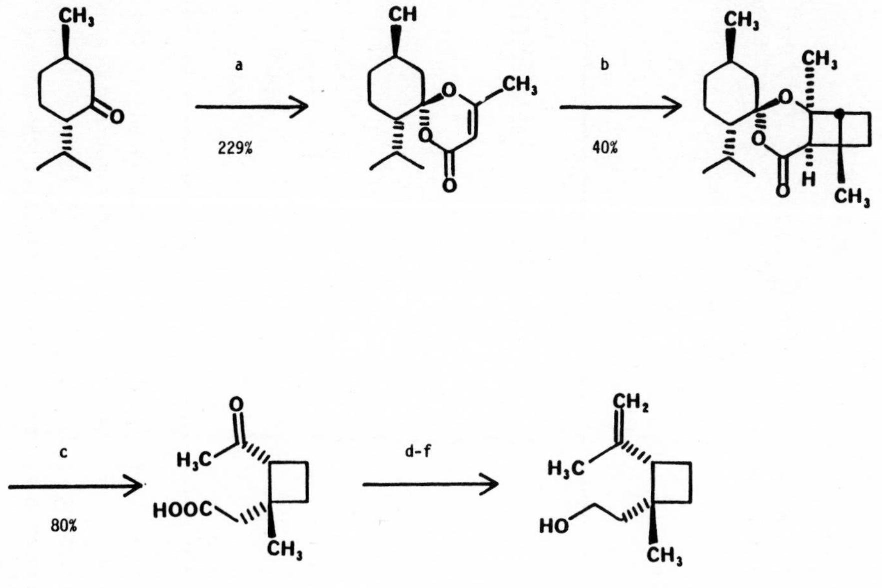

Reagents:

(a) $CH_3COCH_2CO_2tBu$, Ac_2O, H_2SO_4

(b) 1-Methylcyclobutene, light

(c) HCOOH, H_2O, acetone

(d) $TMSCH_2MgCl$, THF

(e) SO_2Cl_2, rt

(f) LAH, Et_2O

Pearson prepared and separated dioxolanes **137** and **138** and studied the alkylations of the corresponding enolates (Table 32) (ref. 120).

TABLE 32.
Alkylations of Dioxolanones **137** and **138** (Ref. 120).

Entry	RX	Yield	Ratio 139a:b	Ratio 140a:b
1	MeI	99	24:1	
2	CH$_2$=CH$_2$CH$_2$I	96	123:1	
3	PhCH$_2$Br	92	123:1	
4	n-BuI	82	28:1	
5	MeI	97		14:1
6	CH$_2$=CHCH$_2$I	97		42:1
7	PhCH$_2$Br	96		58:1
8	n-BuI	85		24:1

These methods are handicapped by the required separations of the mixtures 134/135 and 137/138. Unfortunately, nature does not produce useful homochiral ketones possessing C$_2$ symmetry!

Finally, Oku has reported asymmetric preparations of the chroman ring **141** and sidechain **142** of α-tocopherol based upon enantioselective functionalization of prochiral diols via cleavage of homochiral spiroketals (Figs. 24 and 25) (ref. 121).

Figure 24. Enantioselective Synthesis of Chroman 141.

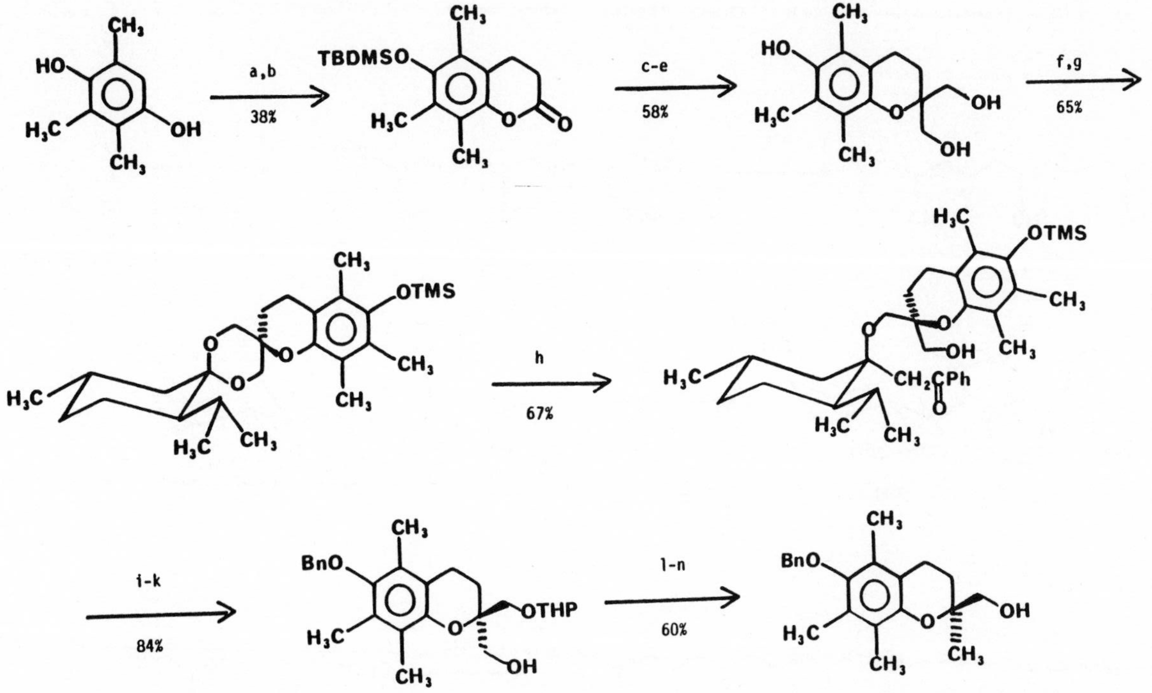

Reagents: (a) $CH_2=CHCN$, $AlCl_3$, HCl, (b) TBDMS-Cl, imidazole, DMF, (c) $LiCH_2OCH(CH_3)OEt$, THF, (d) Aq. HCl, CH_3OH, (e) TsOH, C_6H_6, heat (f) TMS-Cl, $(Me_3Si)_2NH$ (g) d-Menthone, TMSOTf, (h) $CH_2=C(Ph)OTMS$, $TiCl_4$, CH_2Cl_2, (i) nBu_4NF, BnBr, (j) DHP, TsOH, (k) Aq. KOH, THF, CH_3OH, (l) $P(NMe_2)_3$, CCl_4, THF, (m) $LiEt_3BH$, (n) TsOH, CH_3OH.

Figure 25. Enantioselective Synthesis of Alcohol 142.

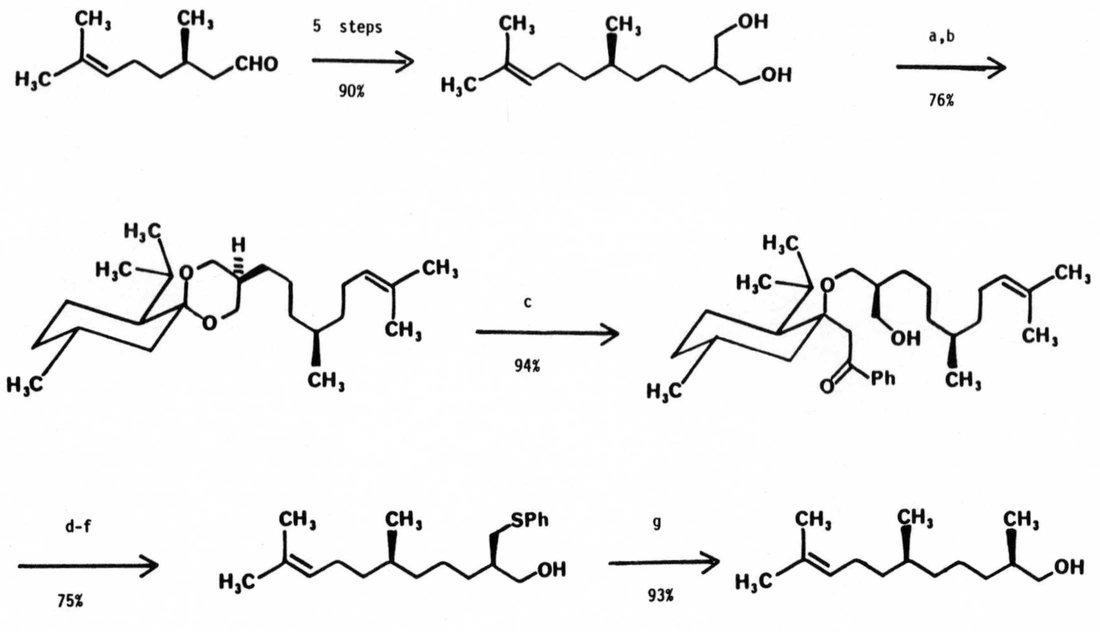

Reagents: (a) TMS-Cl, (Me₃Si)₂NH, l-Menthone, TMSOf, CH₂Cl₂, (c) CH₂=C(Ph)OTMS, TiCl₄, CH₂Cl₂, (d) MsCl, Pyridine, (e) PhSNa, THF, EtOH, (f) Aq. KOH, THF, CH₃OH, (g) Li⁺ [naphth]⁻, THF.

REFERENCES

1. Sugars: S. Hanessian, Total Synthesis of Natural Products: The 'Chiron' Approach; Pergamon Press, New York, 1983. Amino Acids: (a) A. Fischli, in Modern Synthetic Methods, R. Scheffold, Ed., Otto Salle Verlag, Frankfurt, 1980, p. 269. (b) K. Drauz, A. Kleeman, J. Martens, Angew. Chem. Int. Ed. Eng., 21 (1982) 584-608. (c) L. R. Smith, H. J. Williams, J. Chem. Ed. 56 (1979) 696-698. (d) G. M. Coppola, H. F. Schuster, Asymmetric Synthesis: Construction of Chiral Molecules Using Amino Acids, Wiley-Interscience, New York, 1987. Terpenes: (a) W. A. Szabo, H. T. Lee, Aldrichimica Acta 13, (1980) 13. (b) K. Mori in The Total Synthesis of Natural Products, Vol. 4, J. ApSimon, Ed., Wiley-Interscience, New York, 1981, p. 1.

2. J. Jaques, A. Collet, S. H. Wilen, Enantiomers, Racemates and Resolutions, J. Wiley and Sons, New York, 1981.

3. (a) G. M. Whitesides, C. H. Wong, Aldrichimica Acta, 16, (1983) 37-34. (b) J. B. Jones, Tetrahedron, 42 (1986) 3351-3403.

4. Reviews: (a) Asymmetric Synthesis, Vol. 1-5, J. D. Morrison, Ed., Academic Press, Orlando. (b) D. Valentine, J. W. Scott, Synthesis, (1978) 329-356.(c) J. W. ApSimon, R. P. Seguin, Tetrahedron, 35 (1979), 2797-2842.

5. Utilization of homochiral reagents for diastereoselctive functionalization of heterochiral compounds can be useful if the heterochiral forms are equilibratable or if the heterochiral forms display significantly different reactivity.

6. For a review, see: S. Masamune, W. Choy, J. S. Petersen, L. R. Sita, Angew. Chem. Int. Ed. Eng., 24 (1985) 1-30.

7. T. W. Greene, "Protecting Groups in Organic Synthesis, Wiley, New York, 1981, pp. 72-82 and 114-140.

8. See also: S. H. Wilen, Tables of Resolving Agents and Optical Resolutions, University of Notre Dame Press, Notre Dame, Indiana, 1972, pp. 214-217.

9. L. J. Rubin, H. A. Lardy, H. O. L. Fischer, J. Am. Chem. Soc., 74, (1952) 425-428.

10. J. J. Plattner, H. Rapoport, J. Am. Chem. Soc., 93 (1971) 1758-1761.

11. S. Ui, H. Masuda, H. Muraki, J. Ferment. Technol., 61 (1983) 253-259.

12. P. Kocienski, S. D. A. Street, Synthetic Communications, 14 (1984) 1087-1092.

13. K. Ito, T. Harada, A. Tai, Y. Izumi, Chem. Lett. (1979) 1049-1050.

14. J. Bakos, I. Toth, L. Marko, J. Org. Chem., 46 (1981) 5427-5428.

15. E. J. Corey, R. B. Mitra, J. Am. Chem. Soc. 84 (1962) 2938-2941.

16. D. Danneels, M. Anteunis, L. van Acker, D. Tavernier, Tetrahedron, 31 (1975) 327-331.

648

17. (a) L. F. Fieser, Organic Experiments, D. C. Heath and Co., Boston, 1964, 233-235. (b) F. Dietl, J. Haunschild, A. Merz, Tetrahedron, 41 (1985) 1193-1197.

18. J. Brugidou, H. Christol, R. Sales, Bull. Soc. Chim. Fr., 1974, 2033-2034.

19. T. M. Lowry, J. O. Cutter, J. Chem. Soc. 121 (1922) 532-544.

20. D. Seebach, H.-O. Kalinowski, W. Langer, G. Crass, E.-M. Wilka, Organic Synthesis, 61 (1983) 24-34.

21. D. Seebach, H.-O. Kalinowski, B. Bastani, G. Crass, H. Daum, H. Dorr, N. P. DuPreez, U. Ehrig, W. Langer, C. Nussler, H.-A. Oei, M. Schmidt, Helv. Chim. Acta, 60 (1977) 301-325.

22. M. Schmidt, R. Amstutz, G. Crass, D. Seebach, Chem. Ber. 113 (1980) 1691-1707.

23. N. Ando, Y. Yamamoto, J. Oda, Y. Inouye, Synthesis (1978) 688-690.

24. E. A. Mash, K. A. Nelson, S. Van Deusen, S. A. Hemperly, Org. Syn., 66 (1988) 000-000.

25. K. Tamoto, M. Sugimori, S. Terashima, Tetrahedron, 40 (1984) 4617-4623.

26. D. S. Matteson, personal communication.

27. M. D. Fryzuk, B. Bosnich, J. Am. Chem. Soc., 100 (1978) 5491-5494.

28. (a) P. A. MacNeil, N. K. Roberts, B. Bosnich, J. Am. Chem. Soc., 103 (1981) 2273-2280.
(b) H. B. Kagan, J.C. Fiaud, C. Hoornaert, D. Meyer, J. C. Poulin, Bull. Soc. Chim. Belg., 88 (1979) 923-931.

29. (a) D. Seebach, M. Zuger, Helv. Chim. Acta, 65 (1982) 495-503.
(b) D. Seebach, M. F. Zuger, Tetrahedron Lett., 25 (1984) 2747-2750.
(c) D. Seebach, M. A. Sutter, R. H. Weber, M. F. Zuger, Org. Syn., 63 (1984) 1-9.

30. See also (a) Ref. 8, pp. 226-235. (b) J. K. Whitesell, R. M. Lawrence, H.-H. Chen., J. Org. Chem. 51 (1986) 4779-4784.

31. W. C. M. C. Kokke, F. A. Varkevisser, J. Org. Chem., 39 (1974) 1535-1539.

32. R. Adams, C. M. Smith, S. Loewe, J. Am. Chem. Soc., 64 (1942) 2087-2089.

33. A. S. Hussey, R. H. Baker, J. Org. Chem, 25 (1960) 1434.

34. H. C. Brown, C. P. Garg, K.-T. Liu, J. Org. Chem., 36 (1971) 387-390.

35. (a) O. Ort, Org. Syn., 65 (1987) 203-214. (b) E. J. Corey, H. E. Ensley, J. Am. Chem. Soc., 97 (1975) 6908-6909. (c) E. J. Corey; H. E. Ensley, J. W. Suggs, J. Org. Chem., 41 (1976) 380-381.

36. T. Tsunoda, M. Suzuki, R. Noyori, Tetrahedron Lett., 21 (1980) 1357-1358.

37. D. A. Evans, L. K. Truesdale, K. G. Grimm, S. L. Nesbitt, J. Am. Chem. Soc., 99 (1977) 5009-5017.

38. (a) J. R. Hwu, L.-C. Leu, J. A. Robl, D. A. Anderson, J. M. Wetzel, J. Org. Chem., 52 (1987) 188-191. (b) J. R. Hwu, J. M. Wetzel, J. Org. Chem., 50 (1985) 3946-3948.

39. T. Hosokawa, T. Ohta, S. Kanayama, S.-I. Murahashi, J. Org. Chem., 52 (1987) 1758-1764 and references cited therein.

40. J. Casanova, E. J. Corey, Chem. Ind., (1961) 1664-1665.

41. J. D. White, D. N. Gupta, J. Am. Chem. Soc., 90 (1968) 6171-6177.

42. M. Sanz-Burata, J. Irurre-Perez, S. Julia-Arechaga, Afinidad, 281 (1970) 698-704.

43. J. D. White, M. A. Avery, S. C. Choudhry, O. P. Dhingra, M. Kang, A. J. Whittle, J. Am. Chem. Soc., 105 (1983) 6517-6518.

44. H. Shibuya, S. Tsujii, Y. Yamamoto, H. Miura, I. Kitagawa, Chem. Pharm. Bull., 32 (1984) 3417-3427.

45. G. Dana, F. Weisbuch, J.-M. Drancourt, Tetrahedron, 41 (1985) 1233-1239.

46. R. Wiskin, J. Oren, B. Fuchs, Tetrahedron Lett., 26 (1985) 2365-2368.

47. R. Zibuck, N. J. Liverton, A. B. Smith, J. Am. Chem. Soc., 108 (1986) 2451-2453.

48. E. J. Corey, M. Ohno, R. B. Mitra, P. A. Vatakencherry, J. Am. Chem. Soc., 86 (1964) 478-485.

49. H. Buding, B. Deppisch, H. Musso, G. Snatzke, Chem. Ber., 118 (1985) 4597-4612.

50. J.-C. Milhavet, C. Sablayrolles, J.-P. Girard, J.-P. Chapat, J. Chem. Res. (M), (1980) 1901-1913.

51. M. Demuth, S. Chandrasekhar, K. Schaffner, J. Am. Chem. Soc., 106 (1984) 1092-1095.

52. R. K. Hill, G. H. Morton, J. R. Peterson, J. A. Walsh, L. A. Paquette, J. Org. Chem., 50 (1985) 5528-5533.

53. M. Demuth, P. Ritterskamp, E. Weigt, K. Schaffner, J. Am. Chem. Soc., 108 (1986) 4149-4154.

54. C. J. Flann, E. A. Mash, Synthetic Communic., 18 (1988) 0000-0000.

55. K. Tamoto, M. Sugimori, S. Terashima, Tetrahedron, 40 (1984) 4617-4623.

56. H. Wynberg, J. P. Lorand, J. Org. Chem., 46 (1981) 2538-2542.

650

57. J. Brugidou, H. Christol, R. Sales, Bull. Soc. Chim. Fr., (1979) 40-47.

58. R. O. Duthaler, P. Maienfisch, Helv. Chim. Acta, 65 (1982) 635-653.

59. R. O. Duthaler, P. Maienfisch, Helv. Chim. Acta, 67 (1984) 832-844.

60. W. S. Johnson, C. A. Harbert, R. D. Stipanovic, J. Am. Chem. Soc., 90 (1968) 5279-5280.

61. W. S. Johnson, C. A. Harbert, B. E. Ratcliffe, R. D. Stipanovic, J. Am. Chem. Soc., 98 (1976) 6188-6193.

62. P. A. Bartlett, W. S. Johnson, J. D. Elliott, J. Am. Chem. Soc., 105 (1983) 2088-2089.

63. W. S. Johnson, J. D. Elliott, G. J. Hanson, J. Am. Chem. Soc., 106 (1984) 1138-1139.

64. W. J. Richter, J. Org. Chem., 46 (1981) 5119-5124.

65. A. Mori, J. Fujiwara, K. Maruoka, H. Yamamoto, Tetrahedron Lett., 24 (1983) 4581-4584.
66. K. Ishihara, A. Mori, I. Arai, H. Yamamoto, Tetrahedron Lett., 26 (1986) 983-986.

67. A. Mori, K. Ishihara, H. Yamamoto, Tetrahedron Lett., 26 (1986) 987-990.

68. (a) A. Mori, J. Fujiwara, K. Maruoka, H. Yamamoto, J. Organomet. Chem., 285 (1985) 83-94. (b) K. Maruoka, H. Yamamoto, Angew. Chem. Int. Ed. Engl., 24 (1985) 668-682.

69. J. M. McNamara, Y. Kishi, J. Am. Chem. Soc., 104 (1982) 7371-7372.

70. H. Sekizaki, M. Jung, J. M. McNamara, Y. Kishi, J. Am. Chem. Soc., 104 (1982) 7372-7374.

71. J. M. McNamara, Y. Kishi, Tetrahedron, 40 (1984) 4685-4691.

72. W. S. Johnson, C. Edington, J. D. Elliott, I. R. Silverman, J. Am. Chem. Soc., 106 (1984) 7588-7591.

73. J. D. Elliott, J. Steele, W. S. Johnson, Tetrahedron Lett., 26 (1985) 2535-2538.

74. I. R. Silverman, C. Edington, J. D. Elliott, W. S. Johnson, J. Org. Chem., 52 (1987) 180-183.

75. S. H. Mashraqui, R. M. Kellogg, J. Org. Chem., 49 (1984) 2513-2516.

76. W. S. Johnson, P. H. Crackett, J. D. Elliott, J. J. Jagodzinski, S. D. Lindell, S. Natarajan, Tetrahedron Lett., 25 (1984) 3951-3954.

77. W. S. Johnson, M. F. Chan, J. Org. Chem., 50 (1985) 2598-2600.

78. Y. Yamamoto, S. Nishii, J. Yamada, J. Am. Chem. Soc., 108 (1986) 7116-7117.

79. D. Seebach, R. Imwinkelried, G. Stucky, Angew. Chem. Int. Ed. Eng., 25
     (1986) 178-180.

80. W. S. Johnson, R. Elliott, J. D. Elliott, J. Am. Chem. Soc., 105 (1983)
     2904-2905.

81. J. D. Elliott, V. M. F. Choi, W. S. Johnson, J. Org. Chem., 48 (1983)
     2294-2295.

82. V. M. F. Choi, J. D. Elliott, W. S. Johnson, Tetrahedron Lett., 25
     (1984) 591-594.

83. A. Ghribi, A. Alexakis, J. F. Normant, Tetrahedron Lett., 25 (1984)
     3083-3086.

84. S. D. Lindell, J. D. Elliott, W. S. Johnson, Tetrahedron Lett., 25
     (1984) 3947-3950.

85. A. Mori, K. Maruoka, H. Yamamoto, Tetrahedron Lett., 25 (1984)
     4421-4424.

 6. S. L. Schreiber, J. Reagan, Tetrahedron Lett., 27 (1986) 2945-2948.

87. B. B. Snider, B. W. Burbaum, Synthetic Commun., 16 (1986) 1451-1460.

88. A. Mori, H. Yamamoto, J. Org. Chem., 50 (1985) 5444-5446.

89. Y. Naruse, H. Yamamoto, Tetrahedron Lett., 27 (1986) 1363-1366.

90. (a) G. Castaldi, S. Cavicchioli, C. Giordano, F. Uggeri, Angew. Chem.
     Int. Ed. Engl. 25 (1986) 259-260. (b) G. Castaldi, S. Cavicchiali,
     C. Giordano, F. Ugerri, J. Org. Chem., 52 (1987) 3018-3027.

91. Y. Tamura, H. Kondo, H. Annoura, R. Takeuchi, H. Fujioka, Tetrahedron
     Lett., 27 (1986) 81-82.

92. M. Yanagiya, F. Matsuda, K. Hasegawa, T. Matsumoto, Tetrahedron Lett.,
     23 (1982) 4039-4042.

93. T. Matsumoto, F. Matsuda, K. Hasegawa, M. Yanagiya, Tetrahedron, 40
     (1984) 2337-2343.

94. K. Hasegawa, F. Matsuda, M. Yanagiya, T. Matsumoto, Tetrahedron Lett.,
     28 (1987) 1671-1672.

95. J. Fujiwara, Y. Fukutani, M. Hasegawa, K. Maruoka, H. Yamamoto, J. Am.
     Chem. Soc., 106 (1984) 5004-5005.

96. Y. Fukutani, K. Maruoka, H. Yamamoto, Tetrahedron Lett., 25 (1984)
     5911-5912.

97. P. Mangeney, A. Alexakis, J. F. Normant, Tetrahedron Lett., 27 (1986)
     3143-3146.

98. P. Mangeney, A. Alexakis, J. F. Normant, Tetrahedron Lett., 28 (1987)
     2363-2366.

99. A. Alexakis, P. Mangeney, J. F. Normant, Tetrahedron Lett., 26 (1985) 4197-4200.

100. M. Suzuki, Y. Kimura, S. Terashima, Chem. Lett., (1985) 367-370.

101. M. Suzuki, Y. Kimura, S. Terashima, Tetrahedron Lett., 26 (1985) 6481-6484.

102. M. Suzuki, Y. Kimura, S. Terashima, Bull. Chem. Soc. Jpn., 59 (1986) 3559-3572.

103. I. Arai, A. Mori, H. Yamamoto, J. Am. Chem. Soc., 107 (1985) 8254-8256.

104. A. Mori, I. Arai, H. Yamamoto, Tetrahedron, 42 (1986) 6447-6458.

105. E. A. Mash, K. A. Nelson, J. Am. Chem. Soc., 107 (1985) 8256-8258.

106. E. A. Mash, K. A. Nelson, Tetrahedron Lett., 27 (1986) 1441-1444.

107. E. A. Mash, K. A. Nelson, Tetrahedron, 43 (1987) 679-692.

108. E. A. Mash, K. A. Nelson, P. C. Heidt, Tetrahedron Lett., 28 (1987) 1865-1868.

109. Cyclopropanations of homochiral acetals are best carried out under alkylidine transfer conditions. However, for the few cycloalkenone ketals that have been examined so far, Simmons-Smith cyclopropanation has given better diastereoselectivity than has low temperature alkylidine transfer cyclopropanation; E. A. Mash, unpublished results.

110. K. A. Nelson, E. A. Mash, J. Org. Chem., 51 (1986) 2721-2724.

111. E. A. Mash, J. A. Fryling, J. Org. Chem., 52 (1987) 3000-3003.

112. E. A. Mash, J. Org. Chem., 52 (1987) 4142-4143.

113. E. A. Mash, C. J. Flann, S. K. Math, unpublished results.

114. Y. Masaki, K. Nagata, Y. Serizawa, K. Kaji, Tetrahedron Lett., 23 (1982) 5553-5554.

115. Y. Masaki, Y. Serizawa, K. Nagata, K. Kaji, Chem. Lett., (1983) 1601-1602.

116. Y. Masaki, K. Nagata, Y. Serizawa, K. Kaji, Tetrahedron Lett., 25 (1984) 95-96.

117. L. K. Truesdale, D. Swanson, R. C. Sun, Tetrahedron Lett., 26 (1985) 5009-5012.

118. S. Hashimoto, T. Shinoda, Y. Shimada, T. Honda, S. Ikegami, Tetrahedron Lett., 28 (1987) 637-640.

119. M. Demuth, A. Palomer, H.-D. Sluma, A. K. Dey, C. Kruger, Y.-H. Tsay, Angew. Chem. Int. Ed. Engl., 25 (1986) 1117-1119.

120. (a) W. H. Pearson, M.-C. Cheng, J. Org. Chem., 51 (1986) 3746-3748.
     (b) W. H. Pearson, M.-C. Cheng, J. Org. Chem., 52 (1987) 3176-3178.

121. T. Harada, T. Hayashiya, I. Wada, N. Iwa-ake, A. Oku, J. Am. Chem.
     Soc., 109 (1987) 527-532.

RECENT ADVANCES IN BIOMIMETIC OLEFIN CYCLIZATION USING MERCURY(II) TRIFLATE/AMINE COMPLEX[+]

MUGIO NISHIZAWA

1 INTRODUCTION

Olefin cyclization is beautifully controlled in nature to give a variety of polycyclic compounds with many asymmetric centers (refs. 1,2). This elegant biosynthetic process, namely squalene cyclization, was first pointed out by Stork and Eschenmoser in 1955 (ref.3,4), and this hypothesis stimulated a number of studies dealing with biomimetic olefin cyclization (refs.5,6).

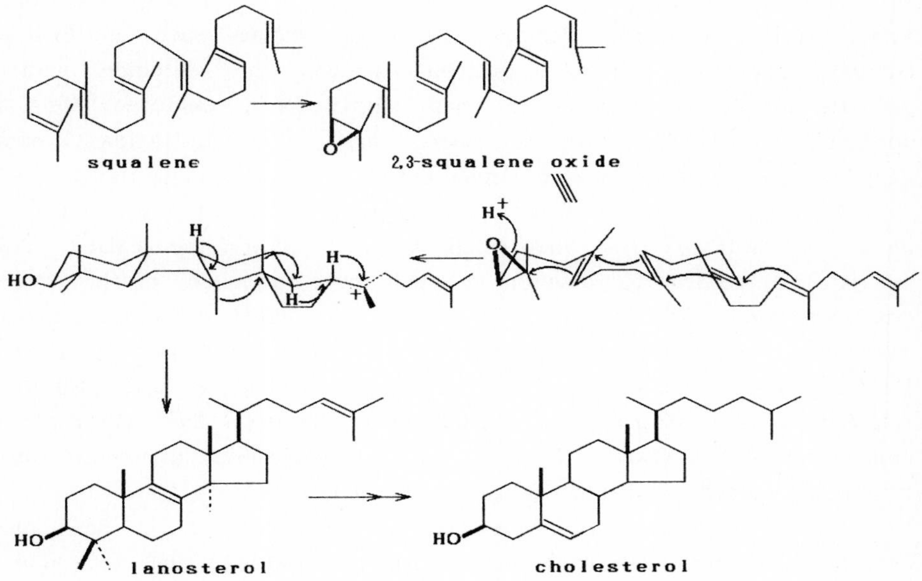

squalene   2,3-squalene oxide

lanosterol   cholesterol

[+]Dedicated to Professor Emeritus Takeo Sakan of Osaka City University on the occasion of his 77th birthday.

van Tamelen's approach (ref.7) induced by an oxirane ring opening and Johnson's approach (ref.8) via an acid catalyzed acetal ring cleavage must be the most important contributions to this field of chemistry. Alternatively, a variety of electrophiles such as protons (refs.9-11), bromonium ions (refs.12-17), acyl cations (refs.18-19), boron trifluoride (refs.20-23), stannic chloride (refs.24-27), selenium salts (refs.28,29), or mercuric salts (refs.30-32), have been employed as the initiators of cationic polyene cyclization.

$$E = H^+, Br^+, CH_3CO^+, BF_3, SnCl_4, PhSeX, HgX_2 \text{ etc.}$$

Among these electrophiles, mercuric salt is the most versatile reagent since a mercury group introduced at the initiated position is easily transformed into various functional groups such as hydrogen (ref.33), hydroxyl (ref.34), halogen (ref.35), or selenide (refs.36, 37). Mercury(II) trifluoroacetate has mainly been employed for this purpose (refs.38-43). However, there are some limitations, namely poor selectivity and lack of generality.

According to our synthetic program of duku metabolites 6 - 11, which we isolated from Indonesian tropical fruits, so called "duku" (ref.44), we needed to develop a much more efficient olefin cyclization reagent. We tried to prepare a super cationic species, mercury(II) trifluoromethanesulfonate (hereafter triflate). Although the attempted cyclization of E,E-farnesol derivative with mercury(II) triflate itself afforded a disappointing result, the corresponding amine complex prepared in situ showed remarkable efficiency as a biomimetic olefin cyclization agent (ref.45).

Here we wish to review the development of a new olefin cyclization agent, mercury(II) triflate/amine complex, and its application to the total synthesis of a variety of polycyclic terpenoids. In the last stage we will discuss mechanistic aspects of a biomimetic olefin cyclization on the basis of our experimental evidence. Our conclusion is that "biomimetic olefin cyclization proceeds through a stepwise mechanism via conformationally flexible cationic intermediates" (ref.46).

2 DEVELOPMENT OF MERCURY(II) TRIFLATE/N,N-DIMETHYLANILINE COMPLEX.

Mercury(II) triflate is prepared from mercury(II) oxide yellow and
an equimolar amount of triflic anhydride in nitromethane. The
resulting creamy white suspension is not effective enough for olefin
cyclization probably due to some side reactions by the liberated
triflic acid. The addition of one equivalent of amine is expected to
quench the liberated super acid. When a variety of amines are added
to the nitromethane solution of mercury(II) triflate, the creamy white
suspension turns into a pale yellow clear solution instantaneously, in
which the mercury(II) triflate/amine complex is formed (refs. 45,47).
This solution is stable under 0℃ in a sealed tube for several months.

$$HgO + (CF_3SO_2)_2O \longrightarrow Hg(OSO_2CF_3)_2$$

$$\xrightarrow{C_6H_5NMe_2} Hg(OSO_2CF_3)_2 \cdot C_6H_5NMe_2$$

1

Mercury(II) triflate/amine complex, thus prepared, was first
applied to the cyclization of (E,E)-farnesylsulfone (2) by using a
variety of amines as ligands. In every experiment, organomercuric
product 3 was obtained as the major product. Although, tertiary
amines generally afforded satisfactory yields of 3, secondary or
primary amines decreased the yields. Therefore, we decided to employ
mercury(II) triflate/N,N-dimethylaniline complex (1) as our standard
reagent for our biomimetic olefin cyclization studies.

Table I Cyclization of 2 with Hg(OTf)₂/amine complex.

amine	PhNMe$_2$	Ph$_3$N	(pyridine)	i-Pr$_2$NEt$_2$	i-Pr$_2$NH	n-BuNH$_2$
yield (%)	74	70	70	54	45	45

By using 1, cyclization of a variety of (E,E)-farnesol derivatives
4 was examined and 3β-mercurio-A/B-trans-Δ^{7,8}-decalines 5 were the
major products in each case. Thus the reagent 1 was shown to be the
best olefin cyclization reagent due to the yield and selectivity
(ref.45).

Table II Cyclization of (E, E)-farnesol derivatives with 1.

R	$CH_2SO_2C_6H_4CH_3$	CH_2OH	$CH_2OCH_2OCH_3$	CO_2CH_3	$CH_2OCOCCl_3$
yield (%)	74	51	50	51	51

3 TOTAL SYNTHESIS OF UNSYMMETRICAL ONOCERANE TRITERPENOIDS.

When we extracted duku peels (Lansium domesticum, Meliaceae) a number of unique unsymmetrical onoceranoids, named α, γ-onoceradiene-dione (6), lansic acid (7), lansiolic acid (8), and lansioside A, B, C (9 - 11), were isolated and characterized (refs.44,48-50). Not only the structural interest but also the important physiological activities prompted us to promote synthetic studies towards this class of compounds. Although a symmetrical onoceranoid, α-onocerin, was already synthesized by Stork in 1959 (refs.51,52), nothing had been done towards the synthesis of unsymmetrical onoceranoids. As seen from our synthetic plan, the key reaction is the biomimetic olefin

8 R=H
9 R=N-acetyl-β-D-glucosamine
10 R=β-D-glucose
11 R=β-D-xylose

cyclization of (E,E)-farnesol derivatives.

The mercury group of the cyclization product **3** was readily replaced by the hydroxyl group according to Whitesides' procedure (ref. 34), and following oxidation and reduction sequence afforded the 3β-alcohol **12**. Double bond migration from $\Delta^{7,8}$ to $\Delta^{8,12}$ was simply achieved by two steps, HOCl promoted chlorination (ref.53) followed by reduction with zinc afforded **14** in 62% yield. After the introduction of a second farnesyl residue, the sulfonyl group along with the benzyl moiety were smoothly removed by lithium/liquid ammonia reduction. The acetate **16** was subjected to the second cyclization with mercury(II) triflate/N,N-dimethylaniline complex (1) to give a diastereomeric mixture of an unsymmetrical onoceranoid skeleton. The organomercuric compound **17** was transformed into ($\pm$)-α,γ-onoceradiene-dione (**6**) via a three step sequence and identified with the natural product. This is the first total synthesis of the unsymmetrical onoceranoid triterpene (ref. 54).

Transformation of the natural diketone **6** to lansic acid (**7**) was simply accomplished via three steps, 1) synthesis of a dioxime **18** by NH_2OH/C_2H_5OH, 2) Beckmann fragmentation with CH_3SO_2Cl/pyridine to give a seco-nitrile **19**, and 3) hydrolysis of nitrile **19** with KOH/CH_3OH, in

660

11% overall yield (ref.54).

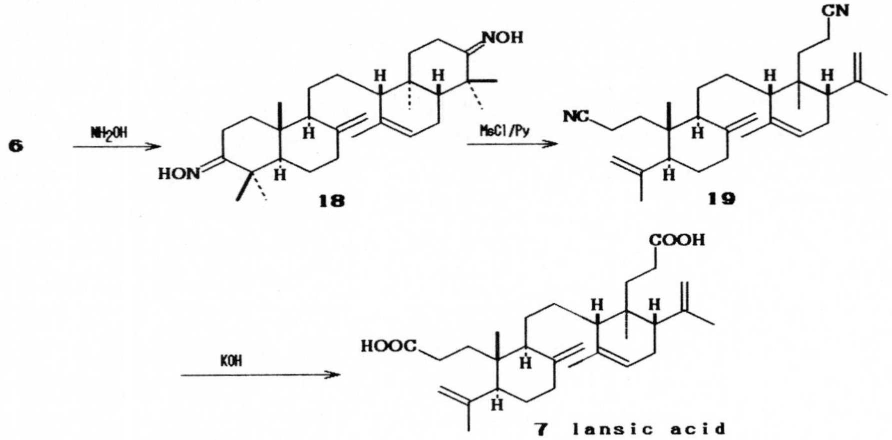

6 → (NH$_2$OH) → **18** → (MsCl/Py) → **19**

18 → (KOH) → **7 lansic acid**

Two kinds of carbonyl moieties of α,γ-onoceradienedione (6) are distinguishable under reduction conditions using NaBH$_4$ in isopropyl alcohol at -78°C to give a keto alcohol **20** as the major product. The resulting ketone **20** was converted to lansiolic acid (8) via the three steps sequence described above (ref.55).

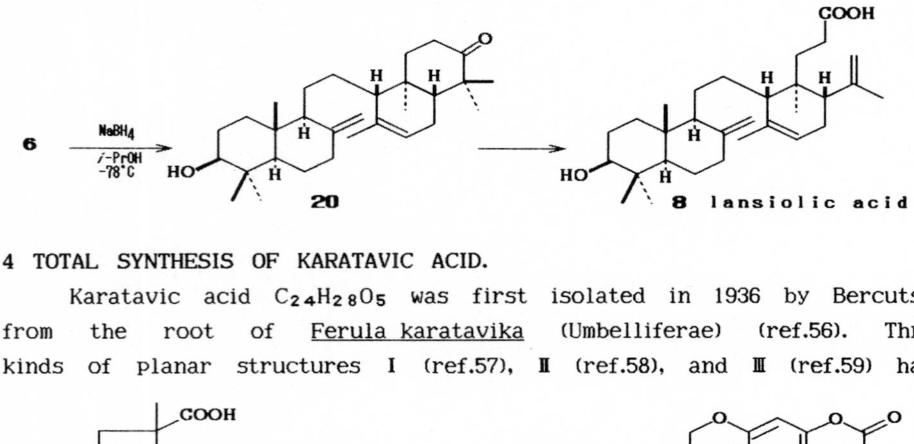

6 → (NaBH$_4$, i-PrOH, -78°C) → **20** → **8 lansiolic acid**

4 TOTAL SYNTHESIS OF KARATAVIC ACID.

Karatavic acid C$_{24}$H$_{28}$O$_5$ was first isolated in 1936 by Bercutsky from the root of <u>Ferula karatavika</u> (Umbelliferae) (ref.56). Three kinds of planar structures I (ref.57), II (ref.58), and III (ref.59) have

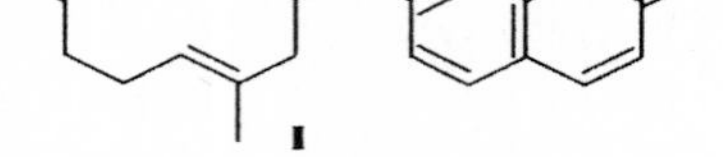

I

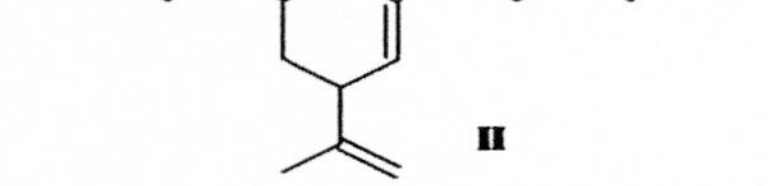

II

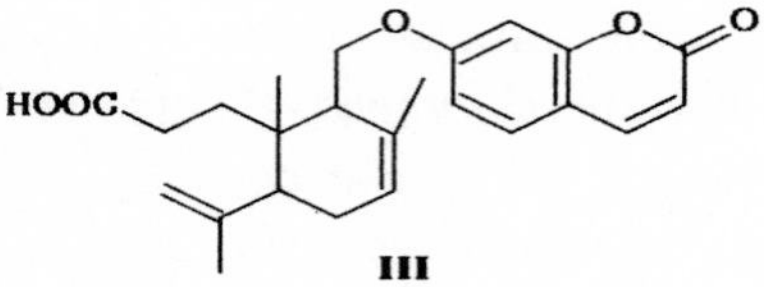

III

been proposed without any clear experimental evidence. Among these three, Paknikar's speculation (formula Ⅲ) seems most reasonable from its spectral data and consideration of sesquiterpene biogenesis. Although many coumarin-containing sesquiterpenoids with a drimane skeleton have been reported, effective cyclization of umbelliprenin (21), a possible biogenetic precursor, has not yet been recorded.

In 1966, van Tamelen and Coates reported the Lewis acid catalyzed cyclization of umbelliprenin terminal epoxide (ref.60). However, the reaction proceeds in a non-selective manner yielding a variety of cyclization products. Bicyclic products were obtained only 9% yield as a mixture of double bond regio isomers.

When 21 was treated with 1 in nitromethane at -20℃, very clean cyclization occurred giving a bicyclic product 22 in 75% yield along with a sole side product 26 (6% yield). The stereochemistry at C9 of 22 and 26 has been established by ^{13}C NMR chemical shift analogy. The chemical shift of the C10 methyl group of 4,4,10-trimethyldecalin system reflects the stereochemistry at C9 and a characteristic low field shift is observed when the C-9 alkyl substituent is located on the α side (ref.47).

Transformation of the organomercuric compound 22 into the carboxylic acid 25a has been achieved by an analogous procedure with the total synthesis of the α,γ-onoceradienedione (6) and lansic acid (7) via a ketone 23 and a seco-nitrile 24. The resulting acid 25a was converted into methyl ester 25b using diazomethane and its ^{1}H NMR (100 MHz) was identified by comparison with the reported spectral data (ref.58). Thus the structure and relative stereochemistry of karatavic acid has been established to be 25a. This is the first example of a 3,4-seco-drimane type sesquiterpenoid (ref.61).

5 CYCLIZATION OF (E,E,E)-GERANYLGERANIOL DERIVATIVES.

We have been interested in the cyclization of (E,E,E)-geranyl-geraniol derivatives, since only a few cyclization studies have been recorded on geranylgeraniol, probably due to its increased complexity (refs.62-64). We examined the cyclization of (E,E,E)-geranylgeranyl p-nitrobenzoate (31) with the reagent 1 at -20°C in nitromethane. After successive treatment of the reaction mixture with aqueous sodium chloride and sodium borohydride, three kinds of cyclization products were isolated and characterized. A tricyclic product 32 (22% yield) was identified with the authentic material which had been reported to be an intermediate towards the total synthesis of spongian diterpe-noids such as isoagatholactone (35) and 12α-hydroxyspongia-13(16),14-diene (36) by Nakano (ref.65). Therefore a stereoselective one-step formation of the three-ring system was realized in a moderate yield. Two kinds of by-products 33 and 34 were also isolated in 9% and 19% yields, respectively. Both of these products were identified as the labdane diterpenoids isolated from Nicotiana setchelli (ref.66). Thus, the simulation of the biogenesis of spongian and labdane diterpenoids was easily accomplished under our conditions (ref.67).

When (E,E)-farnesyl acetate (37) was subjected to the cyclization with 1, a tertiary alcohol 39 was obtained as the major product in 45% yield along with 20% of the usual endocyclic olefin 40. The stereochemistry of 39 was established by its conversion to a tobacco leaf sesquiterpenoid, driman-8,11-diol (41) (ref.68). The formation of 39 is recognized to occur by the intramolecular participation of the carbonyl group which stabilizes the resulting tertiary cation as seen in 38. Thus, we have developed an efficient procedure for the stereospecific introduction of a hydroxyl moiety to the C8 position of the decalin skeleton (ref.45).

If an analogous reaction takes place with (E,E,E)-geranylgeranyl acetate (43), the total synthesis of a bromine containing tricyclic diterpenoid, (±)-isoaplysin-20 (reported structure 42) (refs.69,70), can be accomplished in only two steps. Therefore, we attempted the reaction of 43 with 1 under the same reaction conditions as above. By the treatment of the reaction mixture with aqueous KBr solution, two kinds of products 44 and 45 were easily isolated by column chromatography, followed by recrystallizations, in 16% and 17% yield, respectively. The structures of these products was confirmed by their chemical transformation to the known demercuration products **46** and **47**, respectively (ref.70). Therefore, compound **44** appeared to have the stereochemistry corresponding to (±)-isoaplysin-20. The bromination of **44** was simply achieved by applying Hoye's procedure (LiBr/Br$_2$/ Pyridine/O$_2$) (ref.35). The stereochemistry of the C3 bromine of the product **48** was clearly shown to be β-equatorial on the basis of the ^{1}H NMR data of the C3 proton (δ 3.94, dd, J = 12 and 6 Hz). However the whole spectrum of **48** was not identical with that of natural isoaplysin-20 acetate. Thus, we concluded that the structure of isoaplysin-20 is not **48**, which had been proposed by Yamamura (ref.69) and Rúveda (ref.70).

The cyclization of **43** was not entirely stereospecific, and the contamination of some minor stereoisomeric products was detected in the recrystallization mother liquor of **44**. This mixture was subjected to bromination as mentioned before and to exhaustive purification by HPLC. Two kinds of tricyclic compounds **49** (mp 178℃, 1.8% yield from **43**) and **50** (mp 147℃, 1.4% yield from **43**) were isolated as crystals. The ^{1}H NMR spectrum of **49** was entirely superimposable with that of natural isoaplysin-20 acetate. Hydrolysis of **49** afforded the diol **52**, which was also identified with natural isoaplysin-20. Thus, we have accomplished the first total synthesis of isoaplysin-20 without any information about the structure. The structure **49** was definitely established by the single-crystal X-ray diffraction study. The perhydrophenanthrene skeleton of **49** confirmed by X-ray analysis, involves an anti/syn/anti ring juncture which forces the ring systems to take the chair/boat/chair conformation. Now the structure of isoaplysin-20 must be revised from the all chair structure **51** to a chair/boat/chair structure **52**. Thus, this result is a very novel example in which the perhydrophenanthrene derivative with a boat form B-ring is prepared by means of a biomimetic olefin cyclization (ref.47). It is very interesting to note that the biosynthetic squalene oxide cyclization takes place through a boat form B-ring intermediate (refs.1-4).

The structure of the other minor product **50** was also established by X-ray analysis. The formation of a novel anti/syn/syn ring system, structure **53**, by a biomimetic olefin cyclization is also very interesting from the mechanistic standpoint (ref.47).

6 CYCLIZATION OF AMBLIOFURAN ANALOGS: TOTAL SYNTHESIS OF BAIYUNOSIDE.

Sweet taste is part of the fascinating fields of organic chemistry based on a social requirements of low calorie and high quality taste. We have been interested to develop a synthetic approach to the sweet tasting diterpene glycoside, baiyunoside **54**, isolated from a Chinese drug Bai-Yun-Shen (Phlomis betonicoides, Labiatae) by Tanaka and co-workers in 1983 (refs.70,71). We planned a retrosynthetic scheme shown below in order to prepare not only the natural product itself but also a wide variety of artificial analogs in order to find much more efficient sweeteners. To this end, we needed to prepare the aglycon, (±)-baiyunol **(56)**, in a highly efficient way in a large quantity by means of the biomimetic olefin cyclization of 1. Thus, we started this synthetic study from the cyclization of ambliofuran **(57)**, a possible biogenetic precursor.

When a nitromethane solution of mercury(II) triflate/N,N-dimethyl-aniline complex (1) was slowly added to ambliofuran **(57)** at -20℃, we found that the cyclization was mainly initiated from an internal double bond (Δ^{10}) to give **58** in 38% yield along with a doubly cyclized product **59** (15% yield) and the normal cyclization product **60** (13% yield). The structures of these cyclization products were confirmed by the spectral analysis of derived demercuration products **61a~d**, **62**, and **63**, respectively. Formation of **58** and **59** in this study is important since this is the first example of the biomimetic olefin cyclization initiated from an internal double bond selectively.

Attempted cyclization of dendrolasin **(64)** with 1 under the same conditions was also initiated from the internal double bond (Δ^6) to give **65** in 53% yield along with a tricyclic **66** in 10% yield.

We found that no bicyclic product was available by the cyclization of ambliofuran, itself. We then investigated the cyclization of
the sulfone analog **67**, a synthetic precursor of ambliofuran (**57**).
Although the steric bulk of the sulfonyl moiety at C13 was efficient
to control the initiation point to the terminal double bond (Δ^2), the
major product was the tetracyclic product **68** (42% yield) along with
the bicyclic **69** in 19% yield. The latter product was a mixture of
regioisomers around the double bond ($\Delta^{7,8}:\Delta^{8,9}:\Delta^{8,17}$ = 9:4:3) (ref.
73).

In order to achieve the efficient total synthesis of (±)-baiyunol
(**56**), not only the initiation point but also the termination mode of
the cyclization must be controlled. We expected the electron
withdrawing effect of the conjugated carbonyl group to control both
factors, and prepared 13-oxoambliofuran (**70**) from 3-furancarbaldehyde.
When the reaction of **70** with 1 was quenched with aq sodium chloride
solution at -20℃, a keto alcohol **73** was obtained in 33% yield along
with the desired ketone **71** in 31% yield. This result suggested that
the cationic intermediate **74**, stabilized by intramolecular participation of the carbonyl moiety, was stable enough to exist under the

reaction conditions. Thus, the reaction mixture was warmed to 5℃ prior to quenching in order to promote the proton elimination. By the latter procedure, the desired product **71** was obtained in 68% yield along with an isomeric **72** (8% yield). Although the separation of these isomers at this stage was not easy on a large scale, it was achieved simply in a later step by recrystallization. Neither the exocyclic double bond isomer nor the internal double bond cyclization product or the fully cyclized products were detected. Thus, the cyclization was efficiently controlled by the introduction of the C13 carbonyl group. The reason for the selective elimination of the C9 proton from the intermediate **74** is not yet clear.

The organomercuric ketone **71** was subjected to Whiteside's hydroxylation ($NaBH_4/O_2/DMF$) (ref.34) to give the 3α- and 3β-hydroxy-lated ketone **75**. The carbonyl moiety at C12 was entirely inert under these conditions. The mixture **75** was oxidized with Jones reagent to give rise to a single diketone **76** in 86% yield. Recrystallization at this stage eliminated the contaminating $\Delta^{7,8}$ isomer. The diketone **76** was then transformed to a diastereomeric mixture of diacetate **78** in 98% yield through lithium aluminumhydride reduction followed by acetylation. Finally the mixture of diacetate was subjected to a lithium/ammonia reduction and alcohol **56** was obtained via a simultane-ous fragmentation of the acetoxyl moiety from C12 and the acetyl group from C3 in 58% yield. The spectral properties of (±)-**56** were indistinguishable from those of (+)-baiyunol derived from baiyunoside (**54**). The sole by-product of the final reaction was diol **77**, which was easily separated from **56** and employed for recycling. Therefore we have developed a very simple and efficient procedure for the synthesis of (±)-baiyunol (**56**) by means of the biomimetic olefin cyclization with mercury(II) triflate/N,N-dimethylaniline complex (1) (ref.74).

This method allowed us to obtain sufficient quantities of **56** in order to prepare a variety of artificial analogs of the sweet taste glycoside.

In order to complete the total synthesis of baiyunoside (**54**) from (±)-baiyunol (**56**), we needed to introduce glycosyl moiety to baiyunol in a 2' position discriminated manner. Thus we employed the 2' free glycosyl halide **79** in the presence of silver triflate and tetramethyl-urea (TMU). The condensation took place smoothly but in a non-stereo-selective manner to give the desired 2' discriminated glycoside (**55**) and its diastereomer (**80**) along with the corresponding α-glycosides **81** and **82** in a 10:5:8:6 ratio in 60% yield. Four isomers were separated by HPLC and fully characterized by spectral analysis. For example, the NMR spectra of **55** showed characteristic β-glycosyl features (δ 4.44, d, J = 7.8 Hz for 1' proton and δ 104.9, d, J = 159.9 Hz for 1' carbon in CDCl$_3$) without acetyl moiety at 2' (δ 3.62, 1H, br, t, J = 7.8 Hz, 2'H; δ 5.16, t, J = 9.2 Hz, 3'H; δ 5.04, t, J = 9.2 Hz, 4'H). The absolute configuration of **55** was determined by ^{13}C NMR chemical shift analogy with a related class of natural products, namely by glycosidation shift. The absolute configuration of α-glycosides were definitively established by converting **82** to (-)-baiyunol via basic hydrolysis (CH$_3$ONa/CH$_3$OH) followed by an enzymic hydrolysis (α-glyco-sidase/pH 5 phosphate buffer solution). Thus the glycoside **81** has the normal-type and **82** has the ent-type diterpenoid moieties. In this novel 2' discriminated glycosidation no trace of oligo-saccharide formation was detected presumably due to the increased steric bulk of generated 2' hydroxyl moieties of mono-glycosides. Actually glycoside **55** does not react with glycosyl halide under a variety of Koenigs-Knorr type reaction condition in order to introduce second sugar moiety. This is the reason that the novel 2' discriminated glycosidation works well.

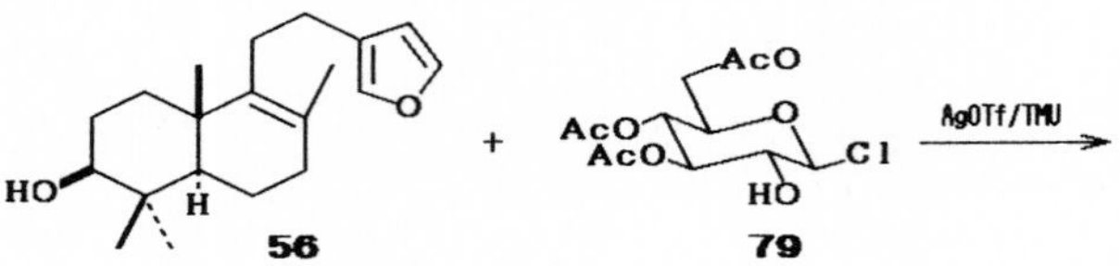

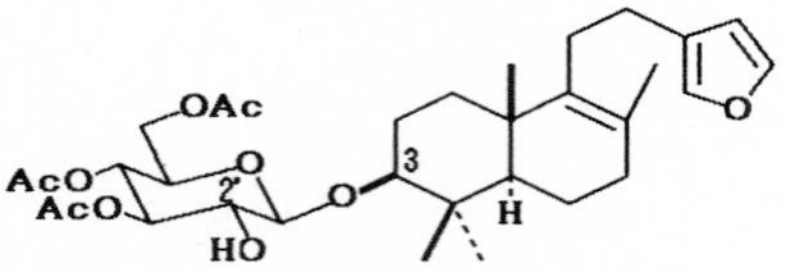

55 ¹H-NMR
1' δ 4.44 J = 7.8 Hz d
2' δ 3.64 J = 7.8 Hz br t
3' δ 5.04 J = 9.2 Hz t
4' δ 5.16 J = 9.0 Hz t
¹³C-NMR
3 δ 90.4
1' δ 104.9 J = 160 Hz

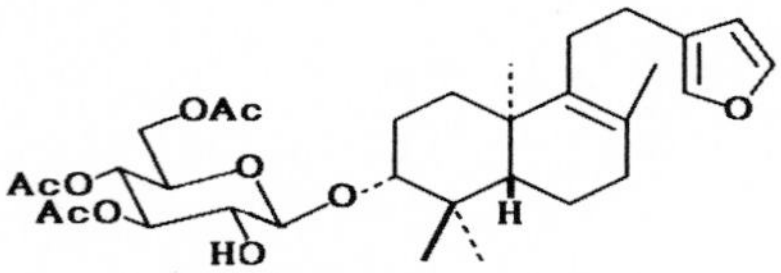

80 ¹H-NMR
1' δ 4.42 J = 7.8 Hz d
2' δ 3.55 J = 8.4 Hz br t
3' δ 5.03 J = 9.6 Hz t
4' δ 5.14 J = 9.4 Hz t
¹³C-NMR
3 δ 85.5
1' δ 100.2 J = 157 Hz

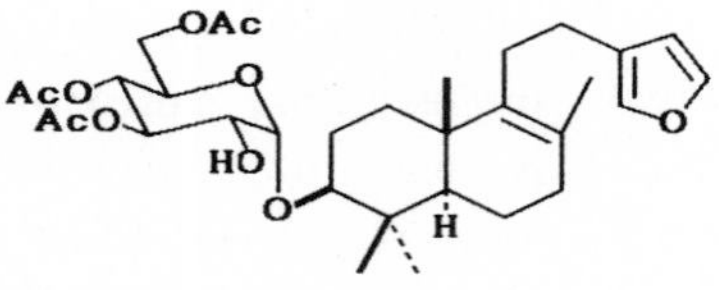

81 ¹H-NMR
1' δ 5.08 J = 4.1 Hz d
2' δ 3.71 J = 9.6, 3.9 Hz dt
3' δ 5.03 J = 9.1 Hz t
4' δ 5.17 J = 9.6 Hz t
¹³C-NMR
3 δ 85.4
1' δ 85.5 J = 172 Hz

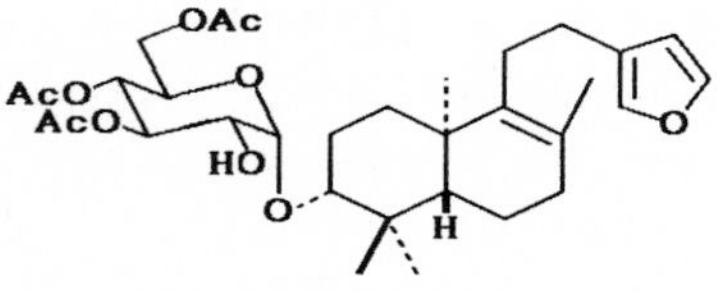

82 ¹H-NMR
1' δ 5.03 J = 3.9 Hz d
2' δ 3.69 J = 9.6, 3.9 Hz dt
3' δ 5.01 J = 9.5 Hz t
4' δ 5.22 J = 9.5 Hz t
¹³C-NMR
3 δ 89.4
1' δ 100.4 J = 169 Hz

The difficulty of the second glycosidation was overcome by employing Noyori's glycosidation based upon the strong affinity between silicon and fluorine. Trimethylsilyl triflate catalyzed glycosidation of **83** with xylopyranosyl fluoride **84** in toluene at 0℃ underwent smoothly to give **85** and its α-xyloside in 58% yield in 62:38 ratio. HPLC separation afforded pure **85**, and successive deprotection with lithium/ammonia followed by CH₃OLi/CH₃OH afforded baiyunoside (**54**). The synthetic baiyunoside was also very sweet. Therefore, we have established a general way to prepare a large numbers of baiyunoside analogs in order to evaluate the taste (ref.76).

7 MECHANISTIC ASPECTS OF A BIOMIMETIC OLEFIN CYCLIZATION

Little direct evidence has been reported to allow one to decide whether the cationic polyene cyclization proceeds by a "stepwise" or "concerted" mechanism (refs.5,6). The pioneering workers in this field explained their stereospecific results by a concerted mechanism (refs.3,4). According to Johnson, the balance of the evidence is somewhat in favor of a concerted process (ref.8). van Tamelen (ref. 78) and Sharpless (ref.79) have indicated a stepwise mechanism which involves a series of conformationally rigid cationic intermediates. On the theoretical point of view, Dewar recently reported a stepwise mechanism via an intermediate of olefin-carbenium ion π-complex on the basis of MINDO/3 caluculation (ref.80). In 1974, Gleiter also discussed the mechanism of olefin cyclization on a theoretical basis (CNDO/2 and INDO) to be a non-concerted reaction (ref.81).

We have observed that the cyclization agent 1 is stable but still reactive enough in the presence of water. When the cyclization of (E,E,E)-geranylgeranyl acetate (43) with 1 was conducted in nitro-methane at -20°C in the presence of water (12 equiv), five kinds of tertiary alcohols 86, 87, 88, 89, and 44 were isolated in 4.9%, 0.8%, 9.0%, 2.9%, and 8.8% yield, respectively, along with 40% recovery of the starting material after the treatment with aqueous KBr solution. The tricyclic 44 was the same compound as that obtained by the cyclization under anhydrous conditions. The structure of the bicyclic product 88 was established by its conversion to (±)-(13E)-13-labdene-8,15-diol (90), a diterpenoid isolated from Nicotiana tabacum (ref.82), by means of the demercuration with sodium borohydride. The structure of 89 was confirmed by the completion of the total synthesis of (±)-aplysin-20 (91) (refs.83,84), a marine diterpenoid from Aplysia kurodai (ref.85), using Hoye's bromination. The structure of monocyclic products 86 and 87 was determined by spectral analysis especially the ^{13}C NMR spectra. Thus, the major hydroxylations was found to occur from the α-side and minor hydroxylations proceeded from the β-side.

Therefore, this result is the first clear experimental evidence that the biomimetic olefin cyclization proceeds through a stepwise mechanism, and suggests the existence of cationic intermediates such as 92, 93, and 94. These intermediates should retain sufficient stability via solvation, and they should be slowly converted to each of the hydroxylated products in competition with the subsequent ring closure. The predominant formation of α-equatorial alcohols (86:87 = 6:1, 88:89 = 3:1) is important. If the conformation of the alkyl side chain of the cationic intermediate is fixed via π-complexation by the next olefinic bond as seen in 95, as stated by van Tamelen (ref.78) or Dewar (ref.80), the α-side of the cation should be entirely shielded and the nucleophile should attack predominantly from the β-side. Thus, the conformation of the polyenyl side chain should be flexible, and the cationic center should be stabilized by solvation.

92   **93**   **94**   **95**

The following experiments support our conclusion: 1) Not only water, but also methanol is effective in trapping the cationic intermediates giving rise to only α-methoxylated products **96** and **97** along with the tricyclic carbinol **46** after reductive workup with NaBH₄. 2) The ratio of mono-, bi-, and tricyclic products was proportionally changed according to the quantity of water employed. A large quantity of water favours the formation of the mono-cyclic carbinol, while a limited amount of water favours the formation of the tricyclic product, as seen in Table Ⅲ.

43   1. 1/MeOH   2. KBr   3. NaBH₄   **96**   +   **97**   +   **46**

Table III   Isolation yield (%) of mono-, bi-, and tricyclic alcohols.

H₂O (equiv)	86 + 87	88 + 89	44
12	5. 7	11. 9	8. 8
1. 2	1. 3	5. 4	11. 2

Thus, we propose that "THE BIOMIMETIC OLEFIN CYCLIZATION PROCEEDS BY A STEPWISE MECHANISM VIA CONFORMATIONALLY FLEXIBLE CATIONIC INTERMEDIATES"   (refs.46,47).

674

Acknowledgement: The author is deeply indebted to the following co-workers, Professor Yuji Hayashi, Dr. Hisaya Nishide, Mr. Hideyuki Takenaka, of Osaka City University and Mr. Hidetoshi Yamada of Tokushima Bunri University for kind concern throughout this study.

REFERENCES
1 K. Bloch, Science, 150 (1965) 19-28.
2 R. B. Clayton, Quart. Rev., (London), 19 (1965) 168-200.
3 G. Stork, A. W. Burgstahler, J. Am. Chem. Soc., 77 (1955) 5068-5055.
4 A. Eshenmoser, L. Ruzicka, O. Jeger, D. Arigoni, Helv. Chim. Acta, 38 (1955) 1890.
5 D. Goldsmith, Prog. Chem. Org. Nat. Prod., 29 (1971) 363-394.
6 P. A. Bartlett, Olefin cyclization processes that form carbon-carbon bonds, in J. D. Morrison (Ed.), Asymmetric Synthesis, Vol. 3, Academic Press, Orlando, 1984, pp 341-409.
7 E. E. van Tamelen, Acc. Chem. Res., 1 (1968) 111-120. 8 (1975) 152-158.
8 W. S. Johnson, Acc. Chem. Res., 1 (1968) 1-8. Angew. Chem., Int. Ed. Engl., 15 (1976) 9-17. Bioorg. Chem., 5 (1976) 51-.
9 A. Eschenmoser, D. Felix, M. Gut, J. Meier, P. Stadler, in Ciba foundation symposium on the biosynthesis of terpenes and sterols, G. E. W. Wolstenholme and M. O'Connor (Ed.) J. and J. Churchill, London, 1959.
10 P. A. Stadler, A. Nechvatal, A. J. Frey, A. Eschenmoser, Helv. Chim. Acta, 40 (1957) 1373-1409.
11 G. E. Muntyan, M. Kurbanov, V. A. Smit, A. V. Semenovski, V. F. Kucherov, Kucherov, Izu. Akad. Nauk. SSSR Ser. Khim., (1973) 633-.
12 T. Kato, I. Ichinose, S. Kumazawa, Y. Kitahara, Bioorg. Chem., 4 (1975) 188-192.
13 T. Kato, I. Ichinose, A. Komoshida, Y. Kitahara, J. Chem. Soc., Chem. Commun., (1976) 518-519.
14 T. Kato, K. Ishii, I. Ichinose, Y. Nakai, T. Kumazawa, J. Chem. Soc., Chem. Commun., (1980) 1106-1108.
15 L. E. Wolinsky, D. J. Faulkner, J. Org. Chem., 41 (1976) 597-600.
16 T. R. Hoye, M. J. Kurth, J. Org. Chem., 43 (1978) 3693-3697.
17 H. M. Shieh, G. D. Prestwich, Tetrahedron Lett., 23 (1982) 4643-4646.
18 S. Kumazawa, Y. Nakano, T. Kato, Y. Kitahara, Tetrahedron Lett., (1974) 1757-1760.
19 T. Kato, S. Kumazawa, C. Kabuto, T. Honda, Y. Kitahara, Tetrahedron Lett., (1975) 2319-2322.
20 Y. Kitahara, T. Kato, T. Suzuki, S. Kanno, M. Tanemura, Chem. Commun., (1969) 342-343.
21 K. Ima-ye, H. Kakizawa, J. Chem. Soc., Perkin I, (1973) 2591-2595.
22 G. L. Trammel, Tetrahedron Lett., (1978) 1525-1528.
23 D. Nashipuri, G. Das, J. Chem. Soc., Perkin I, (1979) 2776-2778.
24 T. Kato, M. Maenuma, S. Konno, T. Suzuki, Y. Kitahara, Bioorg. Chem., 1 (1971) 84-91.
25 R. W. Skeean, G. L. Trammel, J. D. White, Tetrahedron Lett., (1976) 525-528.
26 S. Torii, K. Uneyama, I. Kawahara, M. Kuyama, Chem. Lett., (1978) 455-456.
27 A. Saito, H. Matsushita, Y. Fujino, T. Kisaki, K. Kato, M. Noguchi, Chem. Lett., (1978) 1065-1068.
28 W. P. Jackson, S. V. Ley, A. J, Whittle, J. Chem. Soc., Chem. Commun., (1980) 1173-1174.
29 F. Rouessac, H. Zamarlik, N. Gnonlonfoun, Tetrahedron Lett., 24 (1983) 2247-2250.
30 M. Julia, E. Colomer, S. Julia, Bull. Soc. Chim. Fr., (1966) 2397.
31 M. Kurbanov, A. V. Semenovsky, W. S. Smit, L. V. Shmelev, V. F.
32 Kucherov, Tetrahedron Lett., (1972) 2175-2178.
33 T. R. Hoye, A. J. Caruso, M. J. Kurth, J. Org. Chem., 46 (1981) 3550-

675

3552.
34 C. H. Hill, G. M. Whitesides, J. Am. Chem. Soc., 96, (1974) 870-877.
35 T. R. Hoye, M. J. Kurth, J. Org. Chem., 44 (1979) 3461-3467.
36 R. V. Stevens, K. F. Albizati, J. Org. Chem., 50 (1985) 632-640.
37 M. D. Erion, J. E. McMurry, Tetrahedron Lett., 26 (1985) 559-562.
38 T. R. Hoye, A. J. Caruso, J. F. Dellaria, Jr., M. J. Kurth, J. Am. Chem. Soc., 104 (1982) 6704-6709.
39 E. J. Corey, J. J. Das, J. Am. Chem. Soc., 104 (1982) 5551-5553.
40 C. Sato, S. Ikeda, H. Shirahama, T. Matsumoto, Tetrahedron Lett., 23, (1982) 2099-2102.
41 J. D. White, T. Nishiguchi, R. W. Skeean, J. Am. Chem. Soc., 104 (1982) 3980-3987.
42 R. J. Armstrong, F. L. Harris, L. Weiler, Can. J. Chem., 60 (1982) 673-675.
43 J. E. McMurry, M. D. Erion, J. Am. Chem. Soc., 107 (1985) 2712-2720.
44 M. Nishizawa, H. Nishide, Y. Hayashi, S. Kosela, Tetrahedron Lett., 23 (1982) 1349-50.
45 M. Nishizawa, H. Takenaka, H. Nishide, Y. Hayashi, Tetrahedron Lett., 24 (1983) 2581-2584.
46 M. Nishizawa, H. Takenaka, Y. Hayashi, J. Am. Chem. Soc., 107 (1985) 522-523.
47 M. Nishizawa, H. Takenaka, Y. Hayashi, J. Org. Chem., 51 (1986) 806-813.
48 M. Nishizawa, H. Nishide, S. Kosela, Y. Hayashi, J. Org. Chem., 48 (1983) 4462-4466.
49 A. K. Kiang, E. L. Tan, F. Y. Lim, K. Habaguchi, M. Watanabe, Y. Nakada-ira, K. Nakanishi, L. Fachan, G. Ourisson, Tetrahedron Lett., (1967) 3571-3574.
50 K. Habaguchi, M. Watanabe, Y. Nakadaira, K. Nakanishi, A. K. Kiang, F. Y. Lim, Tetrahedron Lett., (1968) 3731-3734.
51 G. Stork, J. E. Davies, A. Meisels, J. Am. Chem. Soc., 81 (1959) 5516-5517..
52 G. Stork, A. Meisels, J. E. Davies, J. Am. Chem. Soc., 85 (1963) 3419-3425.
53 S. G. Hedge, M. K. Vogel, J. Saddler, T. Hrinyo, N. Rockwell, R. Haynes, M. Oliver, J. Wolinsky, Tetrahedron Lett., 21 (1980) 441-444.
54 M. Nishizawa, H. Nishide, Y. Hayashi, Tetrahedron Lett., 25 (1984) 5071-5074.
55 M. Nishizawa, H. Nishide, K. Kuriyama, Y. Hayashi, Chem. Pharm. Bull., 34 (1986) 4443-4447.
56 V. P. Vercutsky, Tradi. Sredniaz gosainversiteta, Tashkent,(1936) 3.
57 N. P. Kiryalov, V. Yu. Bagirov, Khim. Prir. Soedin, (1968) 283.
58 V. Yu. Bagirov, V. I. Sheichenko, Khim. Prir. Soedin, (1975) 700-703.
59 S. K. Paknikar, J. Veervalli, Chem. Ind., (1978) 431-432.
60 E. E. van Tamelen, R. M. Coates, Chem. Commun., (1966) 413-414.
61 M. Nishizawa, H. Takenaka, Y. Hayashi, Tetrahedron Lett., 25 (1984) 437-440.
62 A. G. Gonzalez, J. D. Martin, M. L. Rodoriguez, Tetrahedron Lett., (1973) 3657-3660.
63 S. M. Kumanireng, T. Kato, Y. Kitahara, Chem. Lett., (1973) 1045-1048.
64 E. E. van Tamelen, S. A. Marson, J. Am. Chem. Soc., 97 (1975) 5614-5616.
65 T. Nakano, M. I. Hernandez, J. Chem. Soc., Perkin I, (1983) 135-139.
66 H. Suzuki, M. Noma, N. Kawashima, Phytochemistry, 22 (1983) 1294-1295.
67 M. Nishizawa, H. Takenaka, Y. Hayashi, Chem. Lett., (1983) 1459-1460.
68 J. R. Hlubucek, A. J. Aasen, S. O. Almqvist, C. R. Enzell, Acta Chem. Scand., B28 (1974) 289-294.
69 S. Yamamura, Y. Terada, Tetrahedron Lett., (1977), 2171-2172.
70 P. M. Iwamura, E. A. Ruveda, J. Org. Chem., 45 (1980) 510-515.
71 T. Tanaka, O. Tanaka, Z. W. Lin, J. Zhou, H. Ageta, Chem. Pharm. Bull., 31 (1983) 780-783.
72 T. Tanaka, O. Tanaka, Z. W. Lin, J. Zhou, Chem. Pharm. Bull., 33 (1985) 4275-4280.

73 M. Nishizawa, H. Yamada, Y. Hayashi, Tetrahedron Lett., 27 (1986) 187-190.
74 M. Nishizawa, H. Yamada, Y. Hatashi, Tetrahedron Lett., 27 (1986) 3255-3256.
75 S. Hashimoto, M. Hayashi, R. Noyori, Tetrahedron Lett., 25 (1984) 1379-1382.
76 M. Hayashi, S. Hashimoto, R. Noyori, Chem. Lett., 1747-1750.
77 H. Yamada, M. Nishizawa, Tetrahedron Lett., in press.
78 E. E. van Tamelen, D. R. James, J. Am. Chem. Soc., 99 (1977) 950-952.
79 K. B. Sharpless, Pd.D. Thesis, Stanford University, July 1968.
80 M. J. S. Dewar, C. H. Reynols, J. Am. Chem. Soc., 106 (1984) 1744-1750.
81 R. Gleiter, K. Mullen, Helv. Chim. Acta, 57 (1974) 823-831.
82 I Wahlberg, I. Wallin, K. Nordfors, T. Nishida, C. R. Enzell, W. Reid, Acta Chem. Scand., B33 (1979) 541-548.
83 A. Murai, A. Abiko, T. Masamuna, Tetrahedron Lett., 25 (1984) 4955-4958.
84 Y. Yamaguchi, T. Uyehara, T. Kato, Tetrahedron Lett., 26 (1985) 343-346.
85 H. Matsuda, Y. Tomiie, S. Yamamura, Y. Hirata, Chem. Commun., (1967) 898-899.

CHIRAL SYNTHESIS OF BIOACTIVE NATURAL PRODUCTS EMPLOYING THE BUILDING BLOCKS OF MICROBIAL ORIGIN

KENJI MORI

1   INTRODUCTION

The recent trend in the synthesis of natural products is to prepare the natural enantiomer itself in an efficient manner. The enantioselective synthesis is especially important in the field of bioactive natural products, because in general the bioactivity depends on the absolute configuration of the molecule. This dependence is evidently the reflexion of the chiral nature of the receptor sites in biological systems. It then follows that we can take advantage of this enantioselectivity of the biological reactions for the preparation of chiral building blocks useful in organic synthesis.

Optically active compounds are prepared by one of the following three methods: a) derivation from chiral starting materials such as terpenes or other natural products, b) optical resolution of an intermediate or the final product, and c) asymmetric synthesis. Microbes and enzymes can carry out asymmetric transformations which lead to either optical resolution or asymmetric synthesis as first discovered by Louis Pasteur in 1858 (ref.1). The microbial or enzymatic products thus obtained can serve as chiral building blocks and enrich the existing chiral carbon pool (refs.2 and 3). By virtue of the ability of microorganisms to transform even unnatural substrates, we can obtain the building blocks otherwise quite difficult to secure. It should be emphasized that the chiral natural products are those generated according to the biogenetic rule, while the products of microbial transformations sometimes possess quite exotic structures which can never be seen among natural products. Preparation of chiral building blocks by means of biochemical transformation is therefore the combination of man's endeavor to design and prepare the substrate for that transformation and the capacity of Mother Nature to modify the substrate.

As there are so many excellent general reviews on the use of biological systems in organic synthesis (refs.4-16), the present article will include only those works which were done in my own laboratories with emphasis on the experimental aspects of biochemical hydrolysis and reduction. The standard monograph in this area is that by Jones et al. (ref.4).

2   α-AMINO ACIDS AS CHIRAL BUILDING BLOCKS

2.1   Amino acylase as the versatile tool for the preparation of optically active α-amino acids

Amino acylase was first discovered by Greenstein et al. from hog kidney source, and found to catalyze the hydrolysis of N-acetyl group of (S)-α-amino acids (ref.17). Subsequent extensive works by Chibata and his co-workers proved that the amino acylase of Aspergillus is a highly enantioselective biocatalyst to effect the asymmetric hydrolysis of racemic N-acetyl α-amino acids, giving (S)-α-amino acids and the unchanged (R)-N-acetyl-α-amino acids (ref.18). By employing immobilized Aspergillus acylase, this asymmetric hydrolysis is now being industrialized on a large scale as shown in Fig.1 in combination with the racemization of the recovered (R)-N-acetyl-α-amino acid to its racemate (ref.18).

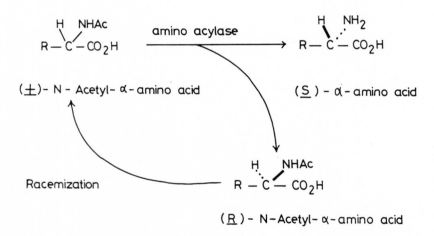

Fig.1.   Asymmetric hydrolysis with amino acylase.

I became interested in investigating the scope of the substrate specificity of Aspergillus amino acylase. If this enzyme can hydrolyze even unnatural (S)-N-acetyl-α-amino acids with various different alkyl groups, then it becomes possible to obtain the enantiomers of new unnatural amino acids. So far we have found that enzyme to be of broad substrate specificity tolerating several alkyl groups which do not exist among naturally occurring α-amino acids.

α-Amino acids are useful chiral building blocks as such, or as the corresponding α-hydroxy acids which can be converted into epoxides as detailed below.

2.2 α-Amino acid itself as a building block -- synthesis of gizzerosine

A serious disease named "black vomit" is a big problem in poultry production all over the world. The disease is accompanied by gizzard erosion or ulceration in chicks, and is known to be caused by brown fish meal in the diet. In 1983 Okazaki et al. isolated 2 mg of a toxic compound from 10 kg of heated mackerel meal. The toxin caused severe gizzard erosion in chicks within a week when fed to them at the level of 2.2 ppm in the diet (or ca. 50 µg/day) (ref.19). This toxic compound was named gizzerosine and assigned structure **1** with unknown absolute configuration (ref.19). Our synthesis of (±)-**1** by the reductive cou-

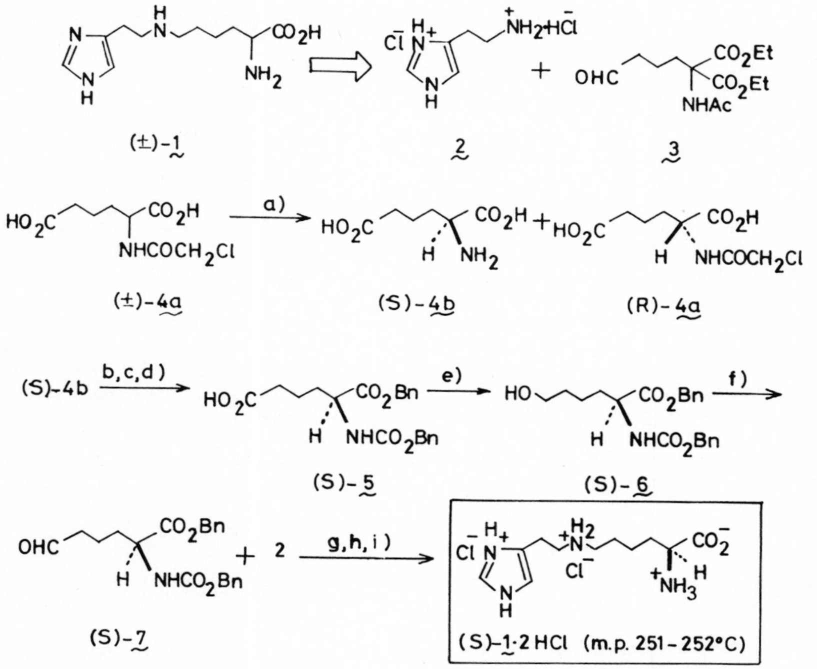

Reagents: a) Aspergillus amino acylase, 37°C, 48 h; b) CbzCl/toluene, NaOH aq-An (92%); c) Cs_2CO_3/DMF; d) BnBr/DMF (45%); e) $BH_3 \cdot$THF (83%); f) $(COCl)_2$/DMSO-CH_2Cl_2, Et_3N (quant.); g) $NaBH_3CN$, MS 3Å/MeOH (69%); h) H_2/Pd-C/EtOH; i) dil HCl, Recrystallization from MeOH (33%).

Fig.2. Synthesis of gizzerosine.

pling of histamine dihydrochloride (2) and an aldehyde 3 confirmed the correctness of the proposed structure (ref.20).

To clarify the absolute configuration of gizzerosine, we carried out the synthesis of both the enantiomers of 1 as shown in Fig.2 (ref.21). For the synthesis of (S)-gizzerosine (1), the (S)-aldehyde 7 was necessary. To prepare 7, (±)-2-chloroacetaminoadipic acid (4a) was resolved by employing Aspergillus amino acylase to give (S)-4b and (R)-4a. Conversion of (S)-4b to (S)-7 via (S)-5 and (S)-6 was followed by the coupling of (S)-7 with 2 to give the desired product (S)-1 after deprotection and neutralization with hydrochloric acid. In the same manner, (R)-gizzerosine was synthesized from (R)-4a. Only (S)-1 was found to be toxic when fed to chicks. The absolute configuration of the toxin obtained from heated mackerel meal was therefore determined to be S. Our synthetic gizzerosine served as a standard in developing the microanalytical method of 1, by which the content of 1 in fish meal can be measured accurately. Owing to the development of this analytical method (ref.22), people in poultry industry will no longer suffer from the "black vomit".

Experimental. Enzymatic resolution of (±)-4a (ref.21).

A soln of (±)-4a (159.8 g) in deionized water (1500 ml) was adjusted to pH 7.0 by the addition of 2M LiOH (325 ml) and the mixture was filled up to 3300 ml using deionized water. To this was added Aspergillus amino acylase (Tokyo Kasei Co., 6.5 g) and CoCl$_2$·6H$_2$O (10 mg) and the mixture was left to stand at 37°C for 24 h. The mixture was adjusted to pH 5 by the addition of AcOH, then filtered through Norit. The filtrate was concentrated in vacuo to a volume of 370 ml. The residual soln was adjusted to pH 3.2 by the addition of conc HCl, diluted with EtOH (740 ml) and then left to stand at 5°C overnight. The precipitates were recovered by filtration, washed with EtOH and dried to give (S)-4b (52.0 g, 96%), m.p. 189-190°C, [α]$_D^{20}$ +23.1° (c=1.08, 5N HCl); νmax (KBr) 3270 (s), 1715 (s), 1640 (s), 1260 (s) cm^{-1}.

The filtrate and washings after removing (S)-4b were combined and concentrated in vacuo. The residue was adjusted to pH 0.5 by the addition of conc HCl and extracted several times with EtOAc. The EtOAc soln was dried (Na$_2$SO$_4$) and concentrated in vacuo. The residue was azeotropically dried with benzene. The residue was dissolved in EtOAc and filtered to remove insoluble materials. The filtrate was concentrated in vacuo and the residual solid was recrystallized from EtOAc to give (R)-4a (59.1 g, 79%), m.p. 105-106°C, [α]$_D^{21}$ -9.2° (c=0.58, acetone); νmax (KBr) 3270 (s), 1715 (s), 1640 (s), 1260 (s) cm^{-1}. This was hydrolyzed with dil HCl in a usual manner to give (R)-4b (37.7 g, 96%), m.p. 187-188°C, [α]$_D^{21.5}$ -21.5° (c=1.08, 5N HCl). Its IR spectrum was identical with that of (S)-4b.

2.3 α-Hydroxy acid as a building block -- synthesis of a bioactive cerebroside

Fruiting body formation in Basidiomycetes is indeed a spectacular phenomenon especially to those who love to taste mushrooms. Its mechanism, however, is still a mystery in spite of the tremendous efforts to clarify it. In 1982 Kawai and Ikeda found that the fruiting body formation of Schizophyllum commune can be stimulated by some cerebrosides in its mycelia (ref.23). They then clarified the structure of one of the active substances as depicted in 8 (Fig.3) (ref.24). We synthesized 8

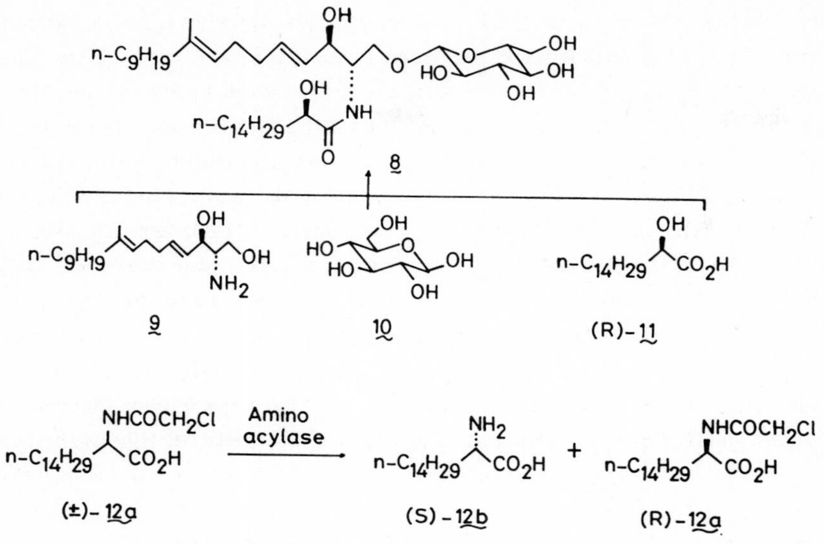

Fig.3. Synthesis of the fruiting-inducing cerebroside in a Basidiomycete *Schizophyllum commune*.

by combining the three chiral building blocks, the sphingadienine **9**, D-glucose **10** and (R)-2-hydroxyhexadecanoic acid **11** (refs.25 and 26). Our synthetic cerebroside **8** was as active as the natural one when assayed against *Schizophyllum commune*.

(R)-2-Aminohexadecanoic acid (**12b**) was the ideal intermediate for the preparation of (R)-**11**, because deamination of (R)-**12b** with nitrous acid would give (R)-**11** with retention of configuration. Even in the present case of an α-amino acid with a lengthy alkyl group, *Aspergillus* amino acylase was a good biocatalyst to give (S)-**12b** and (R)-**12a** starting from (±)-**12a**.

Experimental. Enzymatic resolution of (±)-12a (ref.25).
Aspergillus amino acylase (Tokyo Kasei Co., 5 g) and $CoCl_2 \cdot 6H_2O$ (10 mg) were added to a soln of (±)-**12a** (36.0 g) in water (4000 ml) adjusted to pH 7.3 by the addition of NaOH. The soln was left to stand for 44 h at 37°C. The precipitated crystalline (S)-**12b** was collected on a filter, washed with MeOH and ether, and dried over P_2O_5 to give (S)-**12b** (14.0 g, quantitative), m.p. 234-236°C, $[\alpha]_D^{26}$ +21.8° (c=0.1, AcOH); νmax (nujol) 1575 (s), 1510 (s) cm^{-1}. The filtrate obtained after removal of (S)-**12b** was acidified with 3M HCl. The precipitated solid was collected on a filter, and dissolved in EtOAc (1000 ml). The insoluble material was filtered off and the filtrate was concentrated in vacuo. The residue was recrystallized from n-hexane to give (R)-**12a** (15.5 g, 86.1%), m.p. 87.0-88.0°C, $[\alpha]_D^{21}$ -28.0° (c=0.5, $CHCl_3$). The IR spectrum of (R)-**12a** was identical with that of (±)-**12a**.

2.4 Epoxides as building blocks

(i) Synthesis of (3S,4S)-4-methyl-3-heptanol and invictolide.

(3$\underline{S}$,4$\underline{S}$)-4-Methyl-3-heptanol (**18**) is a component of the pheromone responsible for the aggregation of the smaller European elm bark beetle,
$\underline{Scolytus}$ $\underline{multistriatus}$ (ref.27).   Chiral epoxide **17** was required for its
synthesis, and prepared from amino acid (2$\underline{R}$,3$\underline{S}$)-**13b** as shown in Fig.4
(ref.28).   Asymmetric hydrolysis of (±)-**13a** with a bulky $\underline{sec}$-alkyl group
proceeded rather slowly none the less with good enantioselectivity to
give (2$\underline{S}$,3$\underline{R}$)-**13b** and (2$\underline{R}$,3$\underline{S}$)-**13a**.   Hydrolysis of (2$\underline{R}$,3$\underline{S}$)-**13a** with acid
to (2$\underline{R}$,3$\underline{S}$)-**13b** was followed by a sequence of reactions to give (2$\underline{R}$,3$\underline{S}$)-
17 via **14**, **15** and **16+16'**.   Treatment of (2$\underline{R}$,3$\underline{S}$)-**17** with lithium dimethylcuprate yielded the pheromone component **18**.

   The antipodal epoxide (2$\underline{S}$,3$\underline{R}$)-**17** was employed in the synthesis of
(−)-invictolide, the bioactive enantiomer of the component of the queen
recognition pheromone of the red imported fire ant, $\underline{Solenopsis}$ $\underline{invicta}$

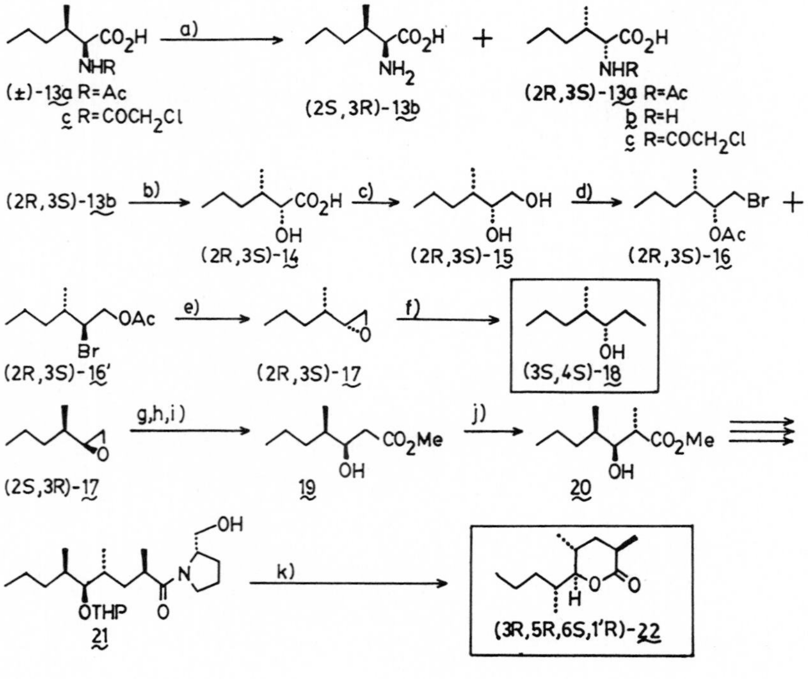

Reagents: a) Amino acylase; b) HNO_2 (66–68%); c) LAH/ether (75%); d)
HBr/AcOH (quant.); e) NaOMe/MeOH (47–52%); f) Me_2CuLi/ether (75%); g)
NaCN/40% EtOH aq; h) dil HCl; i) CH_2N_2/ether (56%); j) LDA/THF-HMPA;
MeI (65%); k) dil HCl, Δ (55%).

Fig.4.   Synthesis of (3S,4S)-4-methyl-3-heptanol (**18**)
and (−)-invictolide (**22**).

(ref.29). In this case N-chloroacetyl derivative (±)-**13c** was chosen as the substrate for the enzymatic resolution in order to accelerate the hydrolysis. Replacement of N-acetyl group with N-chloroacetyl group generally accelerates the enzymatic hydrolysis with amino acylase. The epoxide **17** was converted to hydroxy ester **20** via **19**. The alkylating agent derived from **20** was used for the preparation of **21** by the Evans asymmetric alkylation. The amide **21** yielded (-)-invictolide **22**.

Experimental. Enzymatic resolution of (±)-**13c** (ref.29).

(±)-**13c** (95 g) was dissolved in deionized water (4000 ml) by adding 2M NaOH up to pH 7.1. Aspergillus amino acylase (Amano Pharmaceutical Co., 20 g) and CoCl$_2$·6H$_2$O (40 mg) were added to the soln, and the mixture was left to stand at 37°C for 2.5 days. This was decolorized with Norit, filtered and concentrated in vacuo to a volume of ca. 150 ml. The separated crystals were collected on a filter, which were dried to give (2S,3R)-**13b** (31.8 g, 51.2%). An analytical sample was obtained by recrystallization from water, m.p. 237-241°C, [α]$_D^{20}$ +42.1° (c=0.275, 5M HCl); νmax (nujol) 3050 (m,br), 2700 (m), 2602 (m), 1580 (s), 1510 (s) cm^{-1}. The filtrate after removing **13b** was then acidified with 5M HCl to pH 2. The precipitates were collected on a filter and dried to give (2R,3S)-**13c** (40.6 g, 42.3%). An analytical sample was obtained by recrystallization from acetone, m.p. 129-130°C; [α]$_D^{20}$ -30.7° (c=1.020, EtOH); νmax (nujol) 3420 (m), 2670 (m), 2630 (m), 1715 (m), 1630 (s), 1540 (m) cm^{-1}; ^{13}C NMR δ (CDCl$_3$) 14.1, 15.1, 19.8, 34.5, 35.2, 42.4, 55.6, 166.4, 172.9.

(ii) Synthesis of 8-hydroxyhexadecanoic acid and 4-dodecanolide.

(+)-8-Hydroxyhexadecanoic acid (**26**) was isolated by Yamane et al as an endogenous inhibitor for spore germination in Lygodium japonicum

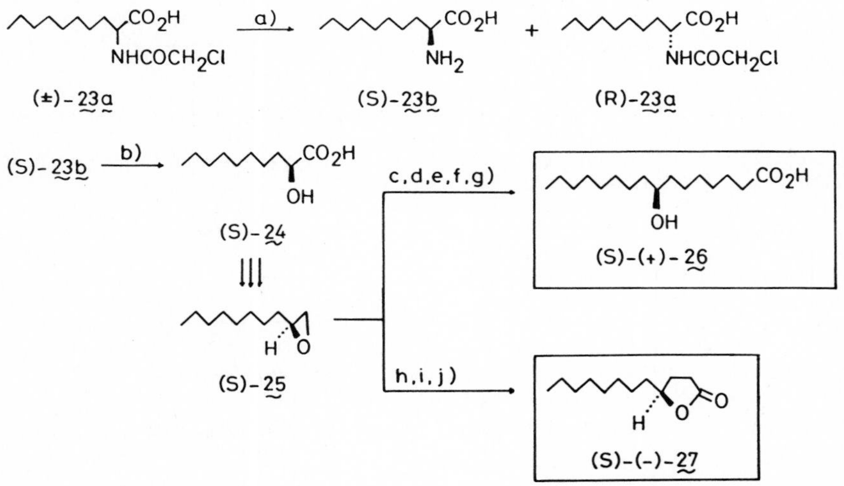

Reagents: a) Amino acylase; b) HNO$_2$ (58%); c) BrMg(CH$_2$)$_6$OTHP, CuI/THF (86%); d) Ac$_2$O, DMAP/C$_5$H$_5$N (quant.); e) TsOH/MeOH (86%); f) Jones CrO$_3$/An (quant.); g) KOH/aq MeOH; dil HCl (60%); h) CH$_2$(CO$_2$Et)$_2$, NaOEt/EtOH; i) NaOH-H$_2$O; j) dil H$_2$SO$_4$ (70% from **24**).

Fig.5. Synthesis of (S)-(+)-8-hydroxyhexadecanoic acid and (S)-4-dodecanolide.

(ref.30). In order to determine the absolute configuration of the natural (+)-**26** and also to know the stereochemistry-bioactivity relationship, we synthesized both the enantiomers of **26** starting from the enantiomers of 2-aminodecanoic acid (**23b**) as shown in Fig.5 (refs.31 and 32). The enzymatic resolution of (±)-N-chloroacetyl-α-amino acid **23a** proceeded smoothly to give (S)-**23b** and (R)-**23a**. Deamination of **23b** gave **24**, which was converted to epoxide **25**. Coupling of **25** with 6-tetrahydropyranyloxyhexylmagnesium bromide provided the C_{16}-skeleton of **26** with retention of configuration at C-8. Subsequent transformation of the product yielded **26**. Because (S)-**26** was dextrorotatory, the natural germination inhibitor was shown to be with S-configuration (ref.31). Both the enantiomers were bioactive as the germination inhibitor (ref.32).

Starting from the enantiomers of epoxide **25**, both the enantiomers of 4-dodecanolide (**27**), a defense secretion of rove beetles (Bredius mandebularis and B. spectabilis), were synthesized as also shown in Fig.5 (ref.33).

Experimental. Enzymatic resolution of (±)-**23a** (ref.31).
a) (±)-2-Chloroacetaminodecanoic acid (±)-**23a**: Chloroacetyl chloride (67.8 g) and 2M NaOH (525 ml) was added simultaneously during 1 h to an ice-cooled and vigorously stirred soln of (±)-**23b** (56.1 g) in 2M NaOH (150 ml). Subsequently the mixture was left to stand for 2 h at room temp and acidified with conc HCl to pH 1.7. The precipitated solid was collected on a filter and dissolved in EtOAc. The soln was washed with water, decolorized with Norit, dried (MgSO4) and concentrated in vacuo. The residue was recrystallized from acetone-pet. ether to give 51.7 g (61.5 %) of (±)-**23a**, m.p. 86-87°C; νmax (KBr) 3375 (s), 3400-2100 (br.s), 1720 (s), 1620 (s), 1530 (s) cm^{-1}.
b) Enzymatic resolution: A soln of (±)-**23a** (28.0 g) in water (4000 ml) was prepared by adjusting its pH to 7.3 with NaOH (ca. 4 g). To this soln were added Aspergillus amino acylase (Tokyo Kasei Co., 5 g) and CoCl2·6H2O (10 mg). The mixture was left to stand at 37°C for 2 days. The precipitate was collected on a filter, washed with MeOH and ether, dried and recrystallized from AcOH to give (S)-**23b** (8.6 g, 92%), m.p. 220-222°C (dec), $[\alpha]_D^{20}$ +29.3° (c=0.49, AcOH); νmax (KBr) 3200-2400 (br.s), 1600 (sh.s), 1580 (s), 1510 (s), 1460 (s), 1440 (s), 1400 (s) cm^{-1}. The filtrate after removing (S)-**23b** was acidified with 6M HCl to pH 1.7. The precipitate was collected on a filter and dissolved in acetone. The acetone soln was filtered to remove insoluble material and the filtrate was concentrated in vacuo. The residue was recrystallized to give (R)-**23a** (11.9 g, 85%) with a constant optical rotation <$[\alpha]_D^{21}$ -35.1° (c=0.93, CHCl3)>. This was mixed with 4M HCl (125 ml) and the mixture was stirred and heated under reflux for 3 h. After cooling, the mixture was neutralized with aq. NH3 and the precipitate was collected on a filter. This was washed with MeOH and ether, dried and recrystallized from AcOH to give (R)-**23b** (7.6 g, 81%), m.p. 220-222°C (dec), $[\alpha]_D^{21}$ -29.9° (c=0.48, AcOH). Its IR spectrum was identical with that of (S)-**23b**.

Synthesis of the enantiomers of 5-hexadecanolide, the pheromone of the queen of the oriental hornet (Vespa orientalis) was also achieved by employing the enzymatic resolution of (±)-2-aminotridecanoic acid as the key-step (ref.34).

3   ALCOHOLS AS CHIRAL BUILDING BLOCKS

3.1 Lipases as tools for the preparation of optically active alcohols

Lipases and other esterases achieve asymmetric hydrolysis of esters to provide optically active alcohols. Lipase became a popular tool of organic chemists to be used for the preparation of optically active alcohols because of its stability and reasonable price. It now can be used even in organic media to effect enantioselective transesterification (ref.35). In Fig.6 are shown the works dealing with the use of esterase and lipases published in the first half of 1987. It can be seen from Fig.6 that the scope of lipase-catalyzed asymmetric hydrolysis is broad enough to be used in organic synthesis.

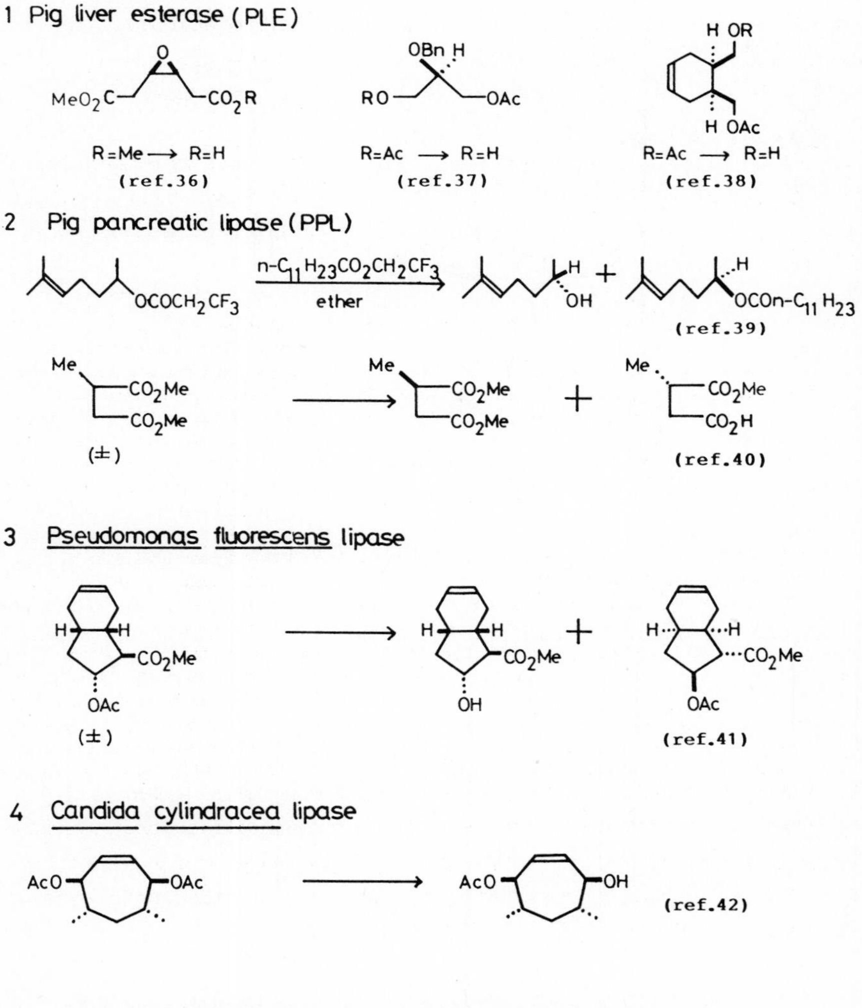

Fig.6. Recent examples of asymmetric hydrolysis with esterase and lipases.

3.2 (1S,4R)-4-Acetoxy-2-cyclopenten-1-ol as a building block in prostaglandin synthesis

(1S,4R)-(+)-4-Acetoxy-2-cyclopenten-1-ol (**29b**) is an important chiral building block for the synthesis of prostaglandins. Enzymatic preparation of this compound or its antipode was reported by several groups (refs.43-48). We recently reported an efficient preparative method of **29b** employing pig pancreatic lipase (PPL) as shown in Fig.7 (ref.49). A mixture of trans-**28a** and cis-**29a** was used as the substrate for the hydrolysis with PPL. The hydrolysis was both enantioselective and

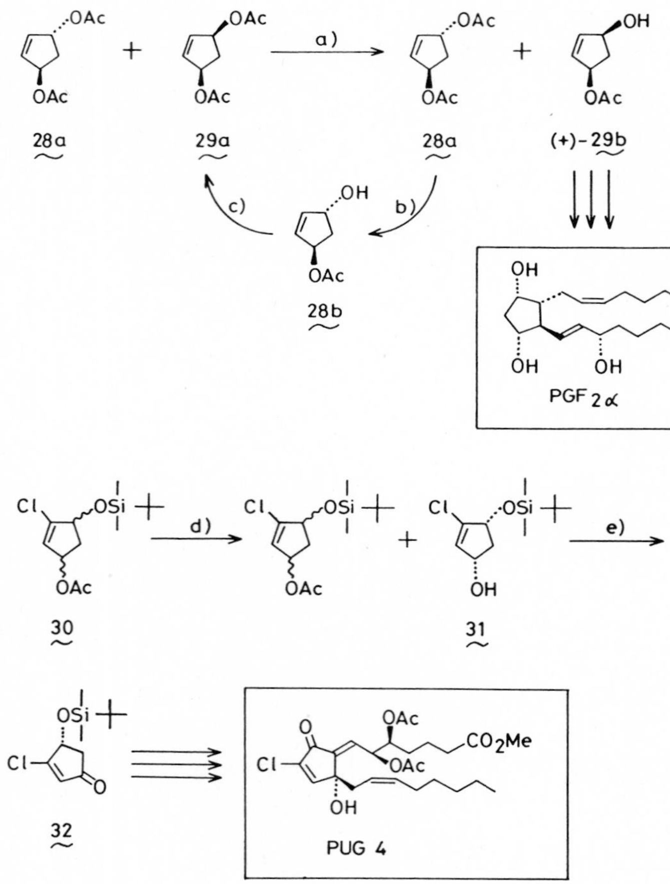

Reagents: a) PPL (75% from **29a**); b) lipase P (80%); c) Ph₃P, AcOH, EtO₂CN=NCO₂Et/THF, MS 3Å (90%); d) PPL (25%); e) PDC/DMF (91%).

Fig.7. Synthesis of (1S,4R)-4-Acetoxy-2-cyclopenten-1-ol and (1S,4R)-4-t-butyldimethylsilyloxy-3-chloro-2-cyclopenten-1-ol.

substrate-selective, yielding mainly (+)-**29b** and the recovered **28a.** The recovered **28a** was hydrolyzed with lipase P from <u>Pseudomonas</u> (Amano Pharmaceutical Co.) to give **28b**, which was treated with acetic acid under Mitsunobu conditions to give <u>cis</u>-**29a** with inversion of configuration. The regenerated **29a** was again submitted to PPL hydrolysis giving an additional amount of (+)-**29b.** Enzymatic conversion of prochiral **29a** to chiral **29b** was thus proved to be a simple way to furnish a prostaglandin intermediate.

Experimental. <u>Enzymatic hydrolysis of</u> **29a** <u>to</u> (+)-**29b** (ref.49).

PPL (Sigma Chemical Co., L-3126, 5 g) and **28a+29a** (10 g) was added to 0.1M phosphate buffer (pH 7, 50 ml) and the mixture was vigorously stirred at 37°C. Its pH was maintained by the continuous addition of 1M KOH using pH controller. After 8 h, the mixture was saturated with NaCl and extracted with EtOAc (5 x 50 ml). The organic soln was washed with brine (200 ml), dried (MgSO$_4$) and concentrated <u>in vacuo</u> to give a crude oil (8.1 g). GLC [5% PEG-20M (2 m, 90°C+2°C/min), N$_2$ (1.0 kg/cm^2)] Rt 15.4 min (**28a**, 46.2%); 20.1 min (**29b**, 47.3%); 22.3 min (**28b**, 6.5%). This was chromatographed over SiO$_2$ (40 g). Elution with <u>n</u>-hexane-EtOAc yielded **28a** (3.6 g, 36%) and a mixture of (+)-**29b** and **28b** (4.0 g, 52%). The latter was recrystallized from <u>n</u>-pentane-ether (1:1, 20 ml) to give crystalline (+)-**29b** (3.2 g, 75% from **29a** contained in the starting material), m.p. 47.5-48°C, $[\alpha]_D^{22}$ = +75.0° (c=1.16, CHCl$_3$); νmax (film) 3450, 1725 cm^{-1}.

3.3 (1S,4R)-4-t-Butyldimethylsilyloxy-3-chloro-2-cyclopenten-1-ol as a building block in punaglandin synthesis

Punaglandin 4 (PUG 4, Fig.7) is one of the chlorinated marine prostanoids isolated from the Hawaiian octocoral <u>Telesto riisei</u> by Scheuer et al. (ref.50). We recently synthesized PUG 4 (ref.51). The key chiral building block **31** was prepared by treating a stereoisomeric mixture of **30** with PPL in 25% yield. Fortunately, the desired stereoisomer **31** of 100% c.e. was the only product of PPL-catalyzed hydrolysis. PDC oxidation of **31** gave **32**, which was finally converted to PUG 4.

Experimental. <u>PPL-catalyzed hydrolysis of</u> **30** <u>to</u> **31** (ref.51).

To a soln of **30** (30.0 g) in MeOH (600 ml) was added 0.1M phosphate buffer (pH 7, 1800 ml) and PPL (Sigma Chemical Co., L-3126, 15.0 g). The mixture was stirred for 12 h at 15°C. Then the mixture was extracted with ether. The ether soln was dried (MgSO$_4$) and concentrated <u>in vacuo.</u> The residue was chromatographed over SiO$_2$ (450 g). Elution with <u>n</u>-hexane-EtOAc (15:1) gave a mixture of <u>trans</u>-**30** and <u>cis</u>-(3<u>R</u>,5<u>S</u>)-**30** (21.2 g). Further elution with <u>n</u>-hexane-EtOAc (3:1) gave **31** (6.4 g, 25%) as a colorless oil, n$_D^{25}$ 1.4711, $[\alpha]_D^{25}$ -31.7° (c=0.75, MeOH); νmax (film) 3370 (s), 1630 (m) cm^{-1}.

3.4 (R)-2-Acetoxymethyl-3-phenyl-1-propanol as a building block for A-factor

Khokhlov et al. discovered A-factor (**40**, Fig.8) as the inducer of the biosynthesis of streptomycin in inactive mutants of <u>Streptomyces griseus</u> (ref.52). It also induces the formation of spores in asporophological modifications of <u>S</u>. <u>griseus</u> (ref.52). We became interested in synthesizing this microbial bioregulator **40**, and completed its chiral synthesis in 1982 (ref.53) employing (<u>S</u>)-(-)-paraconic acid (**37**) as the start-

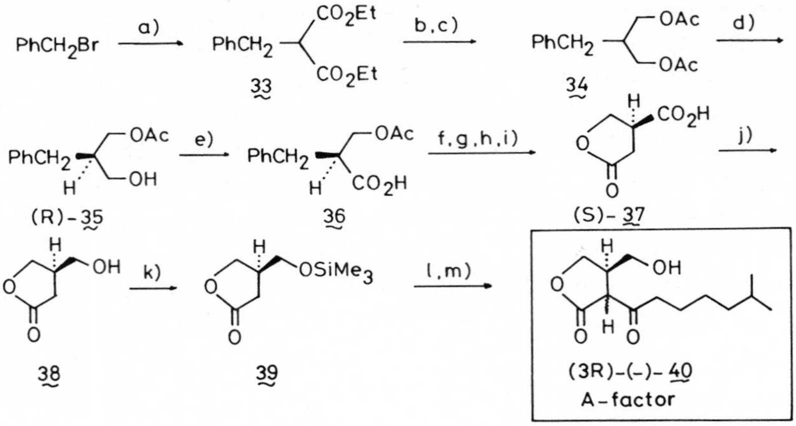

Reagents: a) $CH_2(CO_2Et)_2$, NaOEt/EtOH (58%); b) LAH/ether (78%); c) Ac_2O/C_5H_5N (66%); d) PPL/An-H_2O (56%); e) CrO_3 (98%); f) O_3; g) H_2O_2; h) H^+ (83%); i) purification via (R)-α-phenethylamine salt; j) $BH_3 \cdot THF$ (92%); k) Me_3SiCl, $(Me_3Si)_2NH$ (52%); l) LDA, $Me_2CH(CH_2)_4COCl$; m) aq EtOH, Δ (30%).

Fig.8.   Synthesis of (3R)-(-)-A-factor.

ing material, which was obtained by the optical resolution of (±)-**37** (refs.53 and 54). Very recently, we developed a route in which (S)-**37** was prepared by an enzymatic process (ref.55). Treatment of diacetate **34** with PPL gave (+)-**35** of 86% e.e. Its absolute configuration was determined as R by converting (+)-**35** into (S)-(-)-**37**. Tombo et al. reported that the PPL-catalyzed hydrolysis of **34** yielded (+)-**35** (ref.56). Our configurational assignment was based on direct conversion of (+)-**35** to a known compound as mentioned above, and was therefore unambiguous. After purifying (S)-(-)-**37** as an amine salt, (S)-**37** was converted into (3R)-(-)-A-factor (**40**) via **38** and **39**. There are some other reports on the asymmetric hydrolysis of prochiral diacetates of derivatives of 1,3-propanediol (refs.37,56-58).

Experimental.   **PPL-catalyzed hydrolysis of 34 to (R)-35 (ref.55).**

A soln of **34** (35 g) in acetone (600 ml) was added to a soln of Triton X-100 (3 g) in deionized water (1400 ml). The suspension was emulsified by ultrasonification for 1 h. To the stirred and cooled emulsion of **34** was added PPL (Sigma Chemical Co., L-3126, 70 g) at -5°C. The reaction mixture was stirred vigorously with automatic neutralization of the liberated AcOH with 1M NaOH to keep the mixture at pH 7.0. After 2 h at -5°C, 70 ml of 1M NaOH was consumed. At this point, the reaction was quenched by the addition of EtOAc, NaCl and Celite. The mixture was filtered and EtOAc soln was separated from the aq layer. The aq layer was then extracted with EtOAc (x 3). The EtOAc soln was dried ($MgSO_4$) and concentrated in vacuo to give an oil (34.5 g). This was chromatographed over SiO_2 (350 g). Elution with n-hexane-EtOAc (8:1) gave back the starting material **34** (13.5 g, 39 %). Further elution with n-hexane-EtOAc (3:1) gave (R)-**35** (16.2 g, 56%), n_D^{24} 1.5046, $[\alpha]_D^{24}$ +24.2° (c=1.03, $CHCl_3$); νmax (film) 3460 (s), 1735 (s), 1605 (m) cm^{-1}.

4   β-HYDROXY ESTERS AS CHIRAL BUILDING BLOCKS

4.1 Yeast as a versatile biocatalyst for asymmetric reduction

Thanks to the works by Prelog and Mosher in 1960's, reduction of prochiral ketones with yeasts is known to yield predominantly alcohols of S-configuration (p.71 of ref.4). As will be discussed in this and the next sections, yeasts are now known to be the most popular biocatalysts for organic chemists in preparing optically active alcohols from prochiral ketones. Baker's yeast (Saccharomyces cerevisiae) is the commonest reducing agent which should be tried first. If that yeast does not give a good result, other yeasts should be screened so as to obtain a better result with regard to yield and enantioselectivity.

4.2 Ethyl 3-hydroxybutanoate as a building block

(i)   Preparation of the enantiomers of ethyl 3-hydroxybutanoate. Reduction of ethyl acetoacetate with baker's yeast was first studied by Ridley et al. (ref.59). The reduction furnished ethyl (S)-3-hydroxybutanoate (41) of 83-92% e.e. (refs.59-64). If the reduction is carried out under dilute concentration of the substrate (≤1g/1000 ml), the product of 95-97% e.e. will be obtained (ref.62). We prefer the use of a thermophilic yeast Saccharomyces bailii KI 0116 to baker's yeast, because S. bailii KI 0116 gives (S)-41 of 96% e.e. as shown in Fig.9 (ref.64). The hydroxy ester (S)-41 can be purified by recrystallizing the corresponding 3,5-dinitrobenzoate. Hydrolysis of the pure 3,5-dinitrobenzoate gives (S)-41 of 100% e.e. (refs.60 and 64).

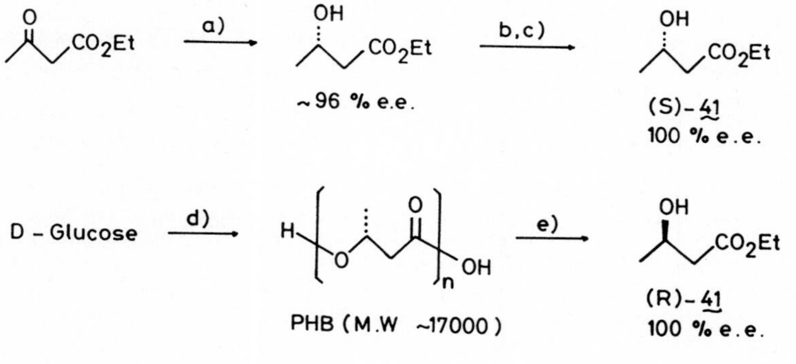

~96 % e.e.

(S)-41
100 % e.e.

PHB (M.W ~17000)

(R)-41
100 % e.e.

Reagents: a) Saccharomyces bailii KI 0116 (80%); b) 3,5-(O₂N)₂C₆H₃CO₂H, DMAP, DCC/CH₂Cl₂; Recrystallization; c) KOH/THF-EtOH-H₂O (45%); d) Zoogloea ramigera; e) EtOH-H₂SO₄/CH₂Cl₂ [33 g of (R)-41 from 50 g of Z. ramigera cells].

Fig.9. Preparation of the enantiomers of
ethyl 3-hydroxybutanoate.

Ethyl (R)-3-hydroxybutanoate of 100% e.e. is readily obtainable by ethanolysis of PHB (poly-3-hydroxybutanoate) (ref.65; see also refs.64 and 66). A wide variety of microorganisms accumulates granules composed of the polymeric ester of (R)-3-hydroxybutanoic acid as their intracellular reserve of organic carbon and/or chemical energy. Seebach et al. used PHB obtained from Alcaligenes eutrophus (pp.85-126 of ref.10 and ref.65), while we employed Zoogloea ramigera I-16-M for the production of PHB (refs.64 and 66). Because of the usefulness of (R)- and (S)-**41** in chiral syntheses, their preparative procedure is recorded below.

Experimental. Preparation of (R)- and (S)-**41** (refs.64 and 66).

a) Poly-3-hydroxybutanoate (PHB): Stock cultures of Zoogloea ramigera I-16-M (ATCC 1923) were maintained on agar slants containing 1.5% Trypticase soy agar (BBL Co.) and 1.0% agar (Difco Co.) at 4°C after incubating at 30°C for 1 day. The stock cultures were transferred every two weeks in order to obtain optimal growth. The basal glucose-starved medium for the dispersed growth of the cells was prepared as described by Tomita et al. (ref.67) and contained 0.5% casamino acid (Difco Co.), 0.5% yeast extract (Oriental Yeast Co.), 0.2% K_2HPO_4 and 0.1% KH_2PO_4 in deionized water. The basal medium (50 ml each) was put into two 500 ml-shaking flasks and inoculated. The flasks were then incubated at 30°C for 24 h on a reciprocal shaker (130 cpm). The combined media in the two flasks were transferred to a 5000 ml-Erlenmeyer cultivating flask with two internal projections for aeration, containing 2000 ml of the medium. Antifoam (Nakarai Co., Antifoam AF emulsion, 10% soln, 4 ml) was added to the culture broth and then incubated at 30°C for 24 h on a gyrorotary shaker (150 rpm). The above process was carried out under sterilized conditions. The culture broth was then transferred to a 5000 ml four-necked flask provided with a glass tube for aeration, a thermometer, air-tight sealed mechanical stirrer and a glass tube for evacuation. Glucose (100 g) and deionized water (1000 ml) were added to the medium and the mixture was stirred (360-400 rpm) for 24 h at 30°C with aeration (16000 ml/min) by evacuation with an aspirator. During that period Z. ramigera cells flocculated. The culture broth was centrifuged (3000 rpm, 20 min) to collect the flocculated cells. The cells were suspended in cold water and the centrifugation was repeated as above. The washed cells were suspended in acetone (1000 ml) and filtered. The collected cells were washed with acetone and dried in vacuo for 1 h at 60°C to give 31.5 g (14.3 g/1000 ml of medium) of dry cells of Z. ramigera. PHB could be extracted from the cells with chloroform (ref.67).

b) Ethyl (R)-3-hydroxybutanoate (R)-**41**:

(1) From PHB. Ethanolysis of PHB was carried out essentially according to Seebach (ref.65). PHB (4.0 g) was suspended in dry EtOH (29 ml) and $ClCH_2CH_2Cl$ (29 ml). Ultrasonic treatment of the mixture effected the swelling of PHB. To this was added conc H_2SO_4 (0.9 ml) and the mixture was stirred and heated under reflux for 37 h. After cooling, it was diluted with 18 % NaCl soln (14 ml) and extracted with ether. The ether soln was washed with $NaHCO_3$ soln and brine, dried ($MgSO_4$) and concentrated in vacuo. The residue was distilled to give (R)-**41** (4.7 g, 80%), b.p. 84-85°C/20 Torr, n_D^{24} 1.4158, $[\alpha]_D^{24}$ -43.6° (c=1.34, $CHCl_3$); νmax (film) 3450 (m), 1740 (s), 1185 (s) cm^{-1}.

(2) Directly from Z. ramigera cells. A suspension of dry cells of Z. ramigera (5.0 g) in dry EtOH (30 ml) and $ClCH_2CH_2Cl$ (36 ml) containing conc H_2SO_4 (1.1 ml) was stirred and heated under reflux for 57 h. After cooling, the mixture was diluted with brine and filtered through Celite. The filtrate was extracted with ether (1 x 70 ml, 3 x 20 ml). The residue was stirred with ether (100 ml) for 30 min, filtered through Celite and the ether layer was separated. The combined ether soln was washed with sat $NaHCO_3$ soln and brine, dried ($MgSO_4$) and concentrated in vacuo. The residue was distilled to give (R)-**41** (3.3 g), b.p. 86.0-86.5°C/22 Torr, n_D^{21} 1.4164, $[\alpha]_D^{21}$ -43.4° (c=1.36, $CHCl_3$).

(3) HPLC analysis of the MTPA ester of (R)-**41**. Column, Partisil 5, 4.6 mm x 250 mm; eluent, n-hexane-THF-MeOH (6000:100:1), 0.95 ml/min; detected at 254 nm: Rt 51.0 min (single peak). Therefore, the optical purity of (R)-**41** was determined to be 100% e.e.

c) Cultivation of thermophilic yeasts: Stock cultures of Saccharomyces bailii KI 0116 or Pichia terricola KI 0117 were maintained on a usual malt extract-agar slants at 4°C after incubating at 37°C for 1-1.5 days. Two loops of S. bailii KI 0116 or P. terricola KI 0117 were inoculated into a 100 ml-portion of the medium containing malt extract (20 g), peptone (1 g), glucose (20 g) in deionized water (1000 ml) placed in a 500 ml-shaking flask. This was cultivated at 37°C for 36 h on a reciprocal shaker (100 cpm). Two batches of the seed culture were added into a 1800 ml of medium with the same ingredient in a 5000 ml-Erlenmeyer cultivating flask with two internal projections for aeration. After cultivating at 37°C for 36 h on a gyrorotary shaker (100 rpm), the wet cells of yeast (ca. 10 g of S. bailii KI 0116 or ca. 22 g of P. terricola KI 0117) were harvested by centrifugation (3000 rpm) for 5 min. The cells of S. bailii KI 0116 were further cultivated by recycling the procedure (cultivation and harvest). After five times of recycling this procedure, 33 g of wet cells were obtained. These cells were as active as those obtained from the initial cultivation. Moreover, after the incubation of the substrate, the cells could be reactivated by recycling the cultivation procedure to give further amount of cells. Recycling of the cultivation of P. terricola KI 0117 gave further amount of cells. However, these cells were far less active than those obtained originally.

d) Preservation of cells of S. bailii KI 0116: This procedure must be done under sterilized condition as carefully as possible. After having been reactivated, the cells (ca. 400 g) were suspended into a medium containing KH_2PO_4 (43.6 g), $Na_2HPO_4 \cdot 12H_2O$ (20.8 g), L-glutamic acid mono-sodium salt (200 g), lactose monohydrate (400 g), and inositol (400 g) in deionized water (2000 ml). After stirring for several minutes, the cells were harvested by centrifugation (3000 rpm) for 5 min. The recovered cells were treated in the same manner once again. The cells were put into a polypropylene vessel which was previously sterilized with steam (120°C, 20 min). Then the vessel was cooled to -70°C with an acetone-dry ice bath. The frozen cells could be stored at -20°C for several months. The cells could be used after defreezing at room temp and activating under the cultivation condition.

e) Ethyl (S)-3-hydroxybutanoate (S)-**41**:

(1) Yeast reduction. The cells of S. bailii KI 0116 (33 g) were suspended into a soln of glucose (60 g) in water (600 ml) and the mixture was stirred at 40°C. Ethyl acetoacetate (5 g) was added to it and the stirring was continued for 4 h. Glucose (30 g) was added to the mixture and the stirring was further continued for 4.5 h. A small portion was extracted with ether, and GLC analysis of the extract revealed that 90 % of the substrate was reduced. Then the broth was centrifuged, and the supernatant was saturated with NaCl and extracted with ether (x 3). The extract was washed with brine, dried ($MgSO_4$) and concentrated in vacuo. The residue was chromatographed over SiO_2 followed by distillation to give (S)-**41** (3.07 g, 61%), b.p. 70-71°C/12 Torr, $n_D^{21.5}$ 1.4173; $[\alpha]_D^{21.5}$ +43.9° (c=0.998, CHCl$_3$). Its IR spectrum was identical with that of (R)-**41**.

(2) HPLC analysis of the MTPA ester of (S)-**41**. HPLC (under the same condition as above): Rt 42.0 min (2.0 %), 51.0 min (98.0 %). Therefore, the optical purity of (S)-**41** was determined to be 96.0% e.e.

(3) Purification via 3,5-dinitrobenzoate. This can be done according to ref.60.

(ii) Synthesis of sulcatol and pityol. Sulcatol (**43**) is the aggre-

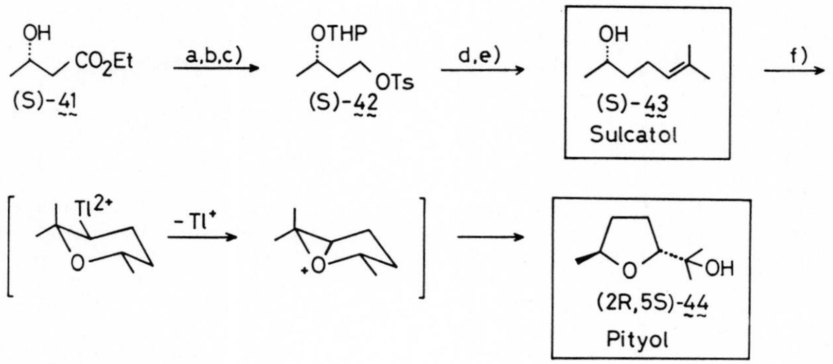

Reagents: a) DHP, TsOH (quant.); b) LAH/ether (89%); c) TsCl/C$_5$H$_5$N (quant.); d) Me$_2$C=CHMgBr, CuI/THF (quant.); e) AcOH-THF-H$_2$O, Δ (82%); f) Tl(OAc)$_3$/HBF$_4$-An-H$_2$O (99%; 12% after MPLC purification).

Fig.10. Synthesis of (S)-sulcatol and (2R,5S)-pityol.

gation pheromone first isolated from the boring dust of <u>Gnathotrichus</u> <u>sulcatus</u> (ref.68). Its identification as a 65:35 mixture of (<u>R</u>)- and (<u>S</u>)-**43** evoked interest in its chiral synthesis. My first synthesis of sulcatol enantiomers from glutamic acid (ref.69) led to the discovery of quite unexpected stereochemistry-pheromone activity relationship: <u>G.</u> <u>sulcatus</u> responded to sulcatol only when both the enantiomers were present (ref.70). This meant a synergistic response of the insect to the enantiomeric mixture. We later synthesized (<u>R</u>)- and (<u>S</u>)-**43** starting from the enantiomers of ethyl 3-hydroxybutanoate (**41**) as shown in Fig.10 (refs.61 and 71). Conversion of **41** to **42** was followed by its coupling with 2-methyl-1-propenylmagnesium bromide to give sulcatol THP ether. Removal of the THP protective group furnished sulcatol in 73% overall yield from **41**.

The enantiomers of sulcatol (**43**) were converted by treatment with thallium (III) triacetate into the enantiomers of pityol (**44**), which is a male-specific attractant of the bark beetle <u>Pityophthorus pityographus</u> (ref.71).

(iii) <u>Synthesis of 2,8-dimethyl-1,7-dioxaspiro[5.5]undecane.</u> The main component of the mandibular gland secretion of bees <u>Andrena</u> <u>wilkella</u> was identified by Francke <u>et al</u>. as 2,8-dimethyl-1,7-dioxa-spiro[5.5]undecane **48** (ref.72). We synthesized the spiroacetal **48** as shown in Fig.11 starting from ethyl 3-hydroxybutanoate **41** (refs.73-75). Alkylation of the dianion of ethyl acetoacetate with **45** to give **46** was followed by another alkylation of **46** with **45** to give **47**. The keto ester **47** furnished (2<u>S</u>,6<u>R</u>,8<u>S</u>)-**48**.

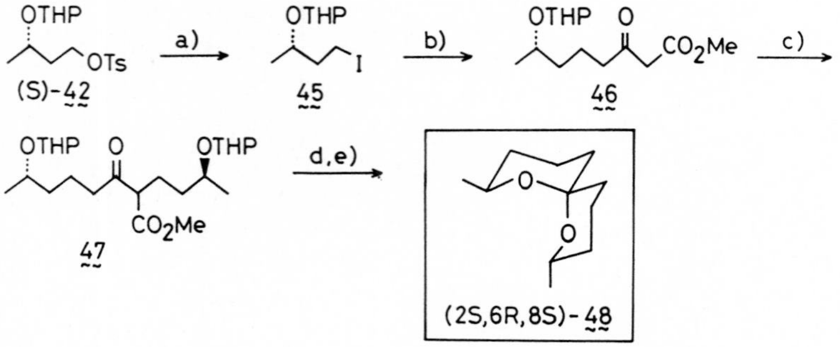

Reagents: a) NaI, NaHCO$_3$/An (80%); b) MeCOCH$_2$CO$_2$Me, NaH, <u>n</u>-BuLi/THF (75%); c) (<u>S</u>)-**45**, K$_2$CO$_3$/An-DMF (89%); d) KOH/aq MeOH; e) TsOH/MeOH (64%).

Fig.11. Synthesis of (2S,6R,8S)-2,8-dimethyl-1,7-dioxaspiro[5.5]undecane.

(iv) <u>Synthesis of grandisol</u>. Grandisol (**55**) is a component of the male-produced pheromone of the boll weevil <u>Anthonomus grandis</u> (ref.76). Starting from ethyl (<u>R</u>)-3-hydroxybutanoate (**41**), both the enantiomers of grandisol (**55**) were synthesized as shown in Fig.12 (ref.77). The key-step was the intramolecular cycloaddition of olefinic ketene **49** to give a mixture of **50** and **51**. Although this mixture could not be separated, the corresponding alcohols **52** and **53** were separable. Alcohols **52** and **53** gave back the ketones **50** and **51**, respectively, after Swern oxidation. The ketone **50** furnished (+)-grandisol (**55**), while **51** was converted to (-)-grandisol. Reflecting the 100% enantiomeric purity of (<u>R</u>)-**41**, grandisol enantiomers thus synthesized were of 100% e.e.

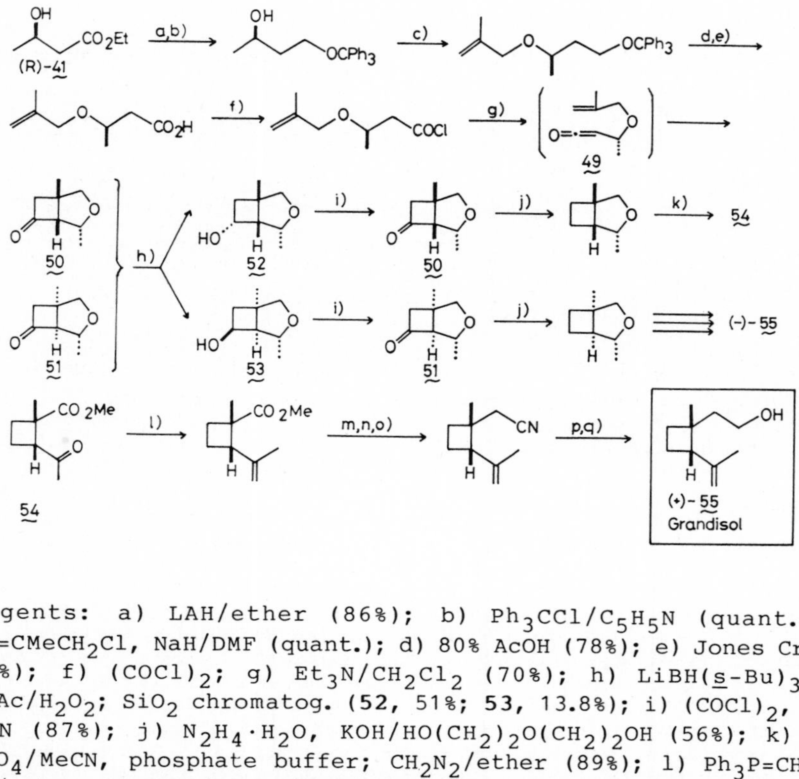

Reagents: a) LAH/ether (86%); b) Ph$_3$CCl/C$_5$H$_5$N (quant.); c) CH$_2$=CMeCH$_2$Cl, NaH/DMF (quant.); d) 80% AcOH (78%); e) Jones CrO$_3$/An (83%); f) (COCl)$_2$; g) Et$_3$N/CH$_2$Cl$_2$ (70%); h) LiBH(<u>s</u>-Bu)$_3$/THF; NaOAc/H$_2$O$_2$; SiO$_2$ chromatog. (**52**, 51%; **53**, 13.8%); i) (COCl)$_2$, DMSO, Et$_3$N (87%); j) N$_2$H$_4$·H$_2$O, KOH/HO(CH$_2$)$_2$O(CH$_2$)$_2$OH (56%); k) RuO$_2$, NaIO$_4$/MeCN, phosphate buffer; CH$_2$N$_2$/ether (89%); l) Ph$_3$P=CH$_2$; m) LAH/ether; n) TsCl/C$_5$H$_5$N (73% from **54**); o) NaCN/aq HMPA (77%); p) (<u>i</u>-Bu)$_2$AlH/<u>n</u>-pentane; q) LAH/ether (40%).

Fig.12. Synthesis of the enantiomers of grandisol.

4.3 Methyl 3-hydroxypentanoate as a building block

(i) <u>Preparation of the enantiomers of methyl 3-hydroxypentanoate</u>. β-Oxidation of pentanoic acid with a mutant of <u>Candida rugosa</u> is known

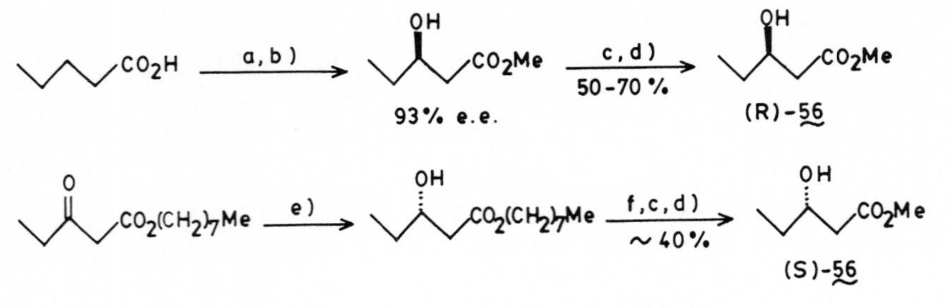

Reagents: a) <u>Candida rugosa</u>; b) MeOH-H_2SO_4 (80%); c) 3,5-
$(O_2N)_2C_6H_3CO_2H$, DMAP, DCC/CH_2Cl_2; recrystallization; d) KOH/THF-MeOH-
H_2O; e) <u>Saccharomyces cerevisiae</u> (70%); f) K_2CO_3/MeOH.

**Fig.13. Preparation of the enantiomers of
methyl 3-hydroxypentanoate.**

to give (<u>R</u>)-3-hydroxypentanoic acid of 93% e.e. (ref.78). The corresponding methyl ester **56** could be purified via its crystalline 3,5-
dinitrobenzoate to give (<u>R</u>)-**56** of 100% e.e. as shown in Fig.13 (ref.79).
Reduction of octyl 3-oxopentanoate with baker's yeast yielded octyl (<u>S</u>)-
3-hydroxypentanoate of 97% e.e. Conversion of this ester to the methyl
ester (<u>S</u>)-**56** was followed by its purification as the corresponding 3,5-
dinitrobenzoate to give (<u>S</u>)-**56** of 100% e.e. (ref.80).

Experimental. Preparation of (<u>R</u>)- and (<u>S</u>)-**56** (refs.79 and 80).
a) Purification of methyl (<u>R</u>)-3-hydroxypentanoate of 93 % e.e.:
      (1) Methyl (<u>R</u>)-3-(3'5'-dinitrobenzoyloxy)pentanoate. 3,5-Dinitrobenzoic acid (60.6 g) was
added to a stirred and ice-cooled soln of (<u>R</u>)-**56** (93% e.e., 25.0 g), DCC (43.7 g) and DMAP (1.89
g) in CH_2Cl_2 (380 ml). After stirring for 3 h at room temp, to this was added <u>n</u>-pentane (100 ml)
and the stirring was continued for 5 min at room temp. The mixture was filtered and the residue
was washed with CH_2Cl_2. The combined filtrate and washings were concentrated <u>in vacuo</u> to give a
crude product (100 g). This was chromatographed over SiO_2 followed by recrystallization from <u>n</u>-
hexane-ether (7:3, 1500 ml) to give the 3,5-dinitrobenzoate (53.4 g, 86.5 %). (This conversion
can also be effected with 3,5-dinitrobenzoyl chloride and pyridine followed by a usual work-up.)
The crystalline material was further recrystallized several times to give the 3,5-dinitrobenzoate
(38.6 g, 72% recovery) as pale yellow needles, m.p. 64-65°C, $[\alpha]_D^{22}$ -9.6° (c=2.64, $CHCl_3$); νmax
(nujol) 1740 (sh), 1738 (s), 1625 (m), 1555 (s), 1340 (s), 1295 (s), 1180 (s), 720 (s) cm^{-1}.
      (2) Methyl (<u>R</u>)-3-hydroxypentanoate (<u>R</u>)-**56**. To a stirred and ice-cooled soln of the 3,5-
dinitrobenzoate (35.0 g) in THF-MeOH (1:1, 440 ml) was added dropwise 1M KOH (113 ml) for a period
of 20 min. After stirring for 30 min at 0°C, the mixture was diluted with sat $NaHCO_3$ soln (330
ml) and extracted with CH_2Cl_2. The extract was washed with water and brine, dried ($MgSO_4$) and
concentrated <u>in vacuo</u>. The residue was distilled to give (<u>R</u>)-**56** (12.8 g, 90 %), b.p. 75-80.8°C/21
Torr, n_D^{23} 1.4224, $[\alpha]_D^{23}$ -37.8° (c=1.30, $CHCl_3$); νmax(film) 3470 (br.m), 1740 (s), 1175 (s) cm^{-1}.
(ref. 133).
      (3) HPLC analysis of the MTPA ester of (<u>R</u>)-**56**. HPLC (under the same condition for the MTPA
ester of **41**): Rt (before purification) 48.5 min (96.5%), 55.2 min (3.5%): Rt (after purification)
48.5 min (single peak). Therefore, the optical purities of these materials were 93% and 100%,
respectively.
b) Preparation of methyl (<u>S</u>)-3-hydroxypentanoate (<u>S</u>)-**56**:
      (1) Reduction with baker's yeast. Dry baker's yeast (Oriental Yeast Co., 800 g) was added to
a soln of sucrose (2 kg) in tap water (13000 ml) at 30°C. The mixture was stirred for 10 min at
30°C. Then octyl 3-oxopentanoate (81 g) was added to the stirred yeast suspension. After 2 h,

sucrose (2 kg) and yeast (100 g) was added to it and the fermentation was further continued for 5 h. The culture broth was mixed with ether and benzene, then filtered through Celite and the filter-cake was extracted with acetone. The extract was concentrated in vacuo. The filtrate was saturated with NaCl and extracted with EtOAc. This was concentrated in vacuo and the residue was combined with the extract from cells of yeast to give crude product (67.4 g). This was chromatographed over SiO$_2$ followed by distillation to give octyl (S)-3-hydroxypentanoate (49.2 g, 47.4 %), b.p. 112-122°C/0.66 Torr, n$_D^{23}$ 1.4371, [α]$_D^{21}$ +22.1° (c=1.08, CHCl$_3$); νmax (film) 3470 (br.s), 1740 (s), 1175 (s) cm^{-1}.

(2) Transesterification to give (S)-56 of 97% e.e. To a soln of octyl (S)-3-hydroxypentanoate (13.8 g) in MeOH (150 ml) was added portionwise K$_2$CO$_3$ (1.5 g) and the mixture was stirred at room temp for 2 h. To this was added AcOH to pH 7 at 0°C and concentrated in vacuo until precipitate appeared. The mixture was then filtered and the precipitate on the filter was washed with ether. The combined filtrate and washings were concentrated in vacuo to give a mixture of (S)-56 and 1-octanol (15.4 g). This was chromatographed over SiO$_2$ (300 g). Elution with n-hexane-EtOAc (25:1-10:1) followed by distillation gave (S)-56 (5.00 g, 63%), b.p. 75°C/21 Torr, [α]$_D^{23}$ +36.3° (c=0.99, CHCl$_3$).

(3) Purification via the corresponding 3,5-dinitrobenzoate. This was executed as described for (R)-56. The purified (S)-56 was obtained in 53 % yield from crude (S)-56. The pure (S)-56 showed the following properties: b.p. 84-88°C/25 Torr, n$_D^{25}$ 1.4211, [α]$_D^{24}$ +38.4° (c=1.32, CHCl$_3$).

(ii) Synthesis of serricornin. Serricornin (61) is the sex pheromone produced by the female cigarette beetle, Lasioderma serricorne, which is a serious pest of cured tobacco leaves (ref.81). Starting from (R)-56, we synthesized serricornin [(4S,6S,7S)-61] and its 4R-isomer as shown in Fig.14. (ref.79). Dianion alkylation of (R)-56 with methyl iodide was follwed by several steps to give 57. This was submitted to the Mitsunobu inversion using 3,5-dinitrobenzoic acid to give 58 of 100% e.e. in 42% yield after repeated recrystallization. Iodide 59 was derived from 58. Alkylation of diethyl ketone with 59 yielded 60, whose

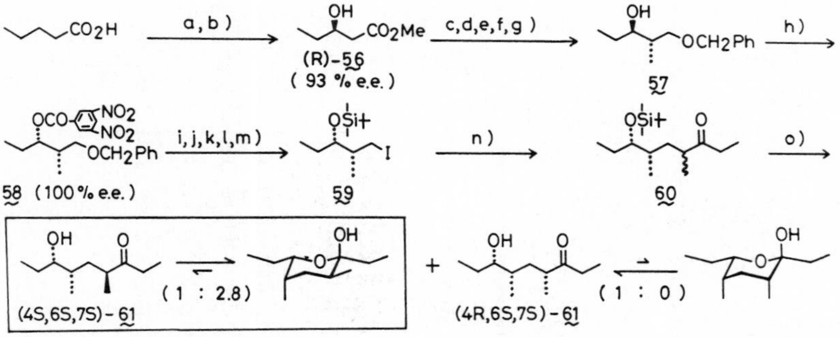

Reagents: a) Candida rugosa; b) methylation; c) LDA, MeI; d) DHP, PPTS (61% from 56); e) LAH (99%); f) NaH, PhCH$_2$Cl (94%); g) TsOH, MeOH (quant.); h), 3,5-(O$_2$N)$_2$C$_6$H$_3$CO$_2$H, Ph$_3$P, EtO$_2$CN=NCO$_2$Et, recrystallization (40%); i) KOH (95%); j) t-BuMe$_2$SiCl, imidazole (quant.); k) H$_2$/Pd-C (97%), l) TsCl/C$_5$H$_5$N; m) NaI (99.6%, 2 steps); n) Et$_2$CO, LDA (80%); o) AcOH-THF-H$_2$O (43%).

Fig.14. Synthesis of (4S,6S,7S)-serricornin.

deprotection gave a mixture of serricornin [(4S,6S,6S)-61] and its 4R-isomer. Separation of these two isomers was readily achieved by silica gel chromatography owing to the large difference in their ease of hemi-acetal formation as shown in Fig.14. The overall yield of the pure pheromone from (R)-56 was 7.6%.

(iii) <u>Synthesis of lardolure</u>. In 1982 lardolure (68) was isolated by Kuwahara <u>et al</u>. as the aggregation pheromone of the acarid mite, <u>Lardoglyphus konoi</u>, which is the primary pest for stored products such as dried meat and fish meal (ref.82). After clarifying its stereochem-istry as (1R,3R,5R,7R)-68 (ref.83), we carried out the synthesis of lardolure and its antipode employing the enantiomers of methyl 3-hydroxypentanoate (56) as shown in Fig.15 (ref.84). Three out of the

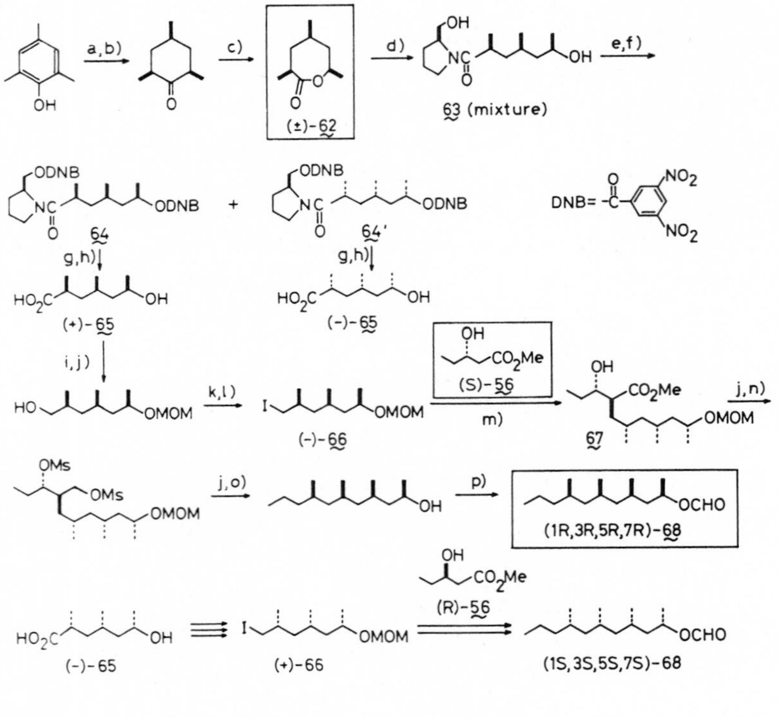

Reagents: a) H$_2$, Raney-Ni/EtOH; b) Jones CrO$_3$; c) MCPBA/CH$_2$Cl$_2$ (65%); d) (S)-Prolinol/toluene, Δ (91%); e) 3,5-(O$_2$N)$_2$C$_6$H$_3$COCl/C$_5$H$_5$N; f) chromatographic separation (SiO$_2$); g) K$_2$CO$_3$/MeOH; h) dil HCl (28% from 63); i) MOMCl, (i-Pr)$_2$NEt/CH$_2$Cl$_2$; j) LAH; k) TsCl/C$_5$H$_5$N; l) NaI/An-DMF (92% from 65); m) LDA/THF-HMPA (79%); n) MsCl, Et$_3$N; o) HCl/MeOH (62% from 67); p) HCO$_2$H

Fig.15. Synthesis of lardolure.

four chiral centers of **68** were derived from the lactone (±)-**62** after
resolving it by separating the diastereomeric mixture of **64** and **64'**.
The mixture of prolinol amide **63** could not be separated. Hydrolysis of
64 and **64'** gave crystalline acids (+)-**65** and (−)-**65**, respectively. The
acid (+)-**65** was converted to (−)-**66**, which was used for the dianion
alkylation of (<u>S</u>)-**56** to give **67** stereoselectively. The remaining steps
to (1<u>R</u>,3<u>R</u>,5<u>R</u>,7<u>R</u>)-(−)-lardolure (**68**) were straightforward. The overall
yield of (−)-**68** from 2,4,6-trimethylphenol was 6.7% in 18 steps. Simi-
larly, (−)-**65** furnished (1<u>S</u>,3<u>S</u>,5<u>S</u>,7<u>S</u>)-(+)-**68**. Only the naturally occurring (1<u>R</u>,3<u>R</u>,5<u>R</u>,7<u>R</u>)-(−)-**68** was bioactive against the acarid mite.

4.4 Ethyl (1R,2S)-5,5-ethylenedioxy-2-hydroxycyclohexane-1-carboxylate as a building block

(i)   Preparation of ethyl (1R,2S)-5,5-ethylenedioxy-2-hydroxycyclo-
hexane-1-carboxylate. Reduction of ethyl 2-oxocyclohexane-1-carboxylate
with baker's yeast was first studied by Ridley in 1976 (ref.59). It
occurred to us that the yeast reduction of ethyl 5,5-ethylenedioxy-2-
oxocyclohexane-1-carboxylate (**69**) might give a useful chiral building
block in terpene synthesis. The keto ester **69** was an intermediate in
the Sarett synthesis of corticosteroids (ref.85), and it was also used
by us in our synthesis of gibberellin-related compounds (ref.86). When
69 was treated with briskly fermenting baker's yeast, smooth reduction
took place to give ethyl (1<u>R</u>,2<u>S</u>)-5,5-ethylenedioxy-2-hydroxycyclohexane-
1-carboxylate (**70**) of 98.4% e.e. in 74% yield (ref.87).

Experimental.   Preparation of (1<u>R</u>,2<u>S</u>)-**70** (ref.87).

Reduction of **69** (15 g) with dry baker's yeast (Oriental Yeast Co., 200 g) in sucrose soln
(300 g in 2000 ml of tap water) for 2 days at 30°C gave 10.1 g (67% or 74% considering the
recovery of 1.55 g of **69**) of **70** after extraction with EtOAc, SiO_2 chromatography, and distillation, b.p. 117–118°C/0.35 Torr, n_D^{25} 1.4695; $[\alpha]_D^{23}$ +51.1° (c=1.02, $CHCl_3$); νmax (film) 3500(m),
1725(s) cm^{-1}.

(ii)   Synthesis of sporogen-AO 1.   In 1984 Marumo and his co-workers
isolated a sporogenic sesquiterpene from the culture broth of
<u>Aspergillus</u> <u>oryzae</u>, an important fungus in Japanese fermentation indus-
try (ref.88). They named it sporogen-AO 1, and clarified its structure
as depicted in **78** (Fig.16). After the completion of the synthesis of
both (+)- and (−)-**78**, it became clear that only the natural (+)-**78** was
bioactive as the inducer of sporulation (ref.89). We then achieved
another synthesis of (+)-**78** starting from **70** as shown in Fig.16
(ref.90). To generate the correct <u>R</u>-configuration of the hydroxyl group
of sporogen-AO 1, it was necessary to invert the <u>S</u>-configuration of the
hydroxyl group of **70**. This task was carried out by inversion at the
stage of **72**, which in turn was prepared from **71**. Mitsunobu inversion of

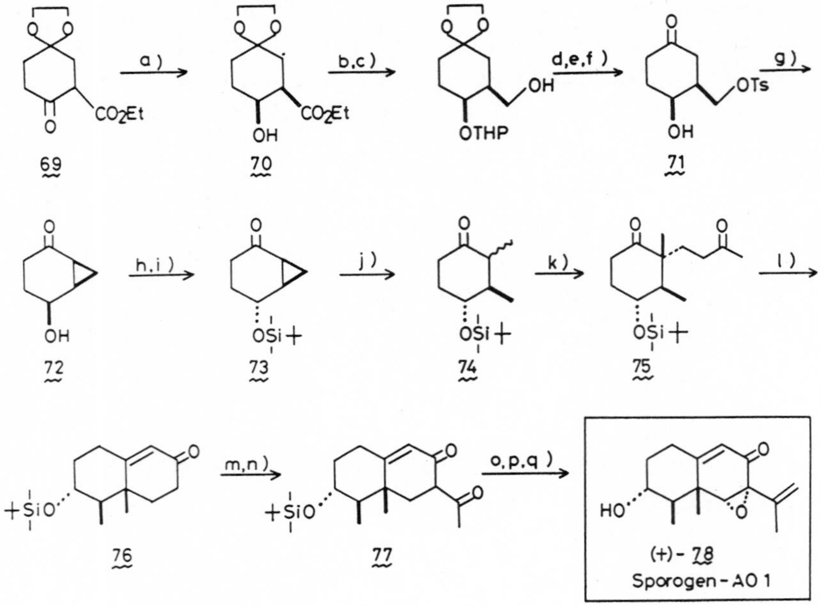

Reagents: a) baker's yeast (74%); b) DHP, PPTS (quant.); c) LAH/ether (98%); d) TsCl/C$_5$H$_5$N (95%); e) PPTS/MeOH; f) aq HClO$_4$-ether (85%); g) t-BuOK/t-BuOH (78.5%); h) Ph$_3$P, EtO$_2$CN=NCO$_2$Et, PhCO$_2$H/THF; LiOH/MeOH (80%); i) t-BuMe$_2$SiCl, imidazole-DMF (quant.); j) Li/NH$_3$-t-BuOH, MeI (83%); k) Me$_3$SiI, (Me$_3$Si)$_2$NH; CH$_2$=CHCOMe, BF$_3$·OEt$_2$/i-PrOH-MeNO$_2$ (68%); l) pyrrolidine/C$_6$H$_6$ (78.6%); m) LDA, MeCHO (92%); n) (COCl)$_2$, DMSO, Et$_3$N (78.5%); o) DDQ/ether (83%); p) t-BuOOH-Triton B (70%); q) Me$_3$SiCH$_2$MgCl; H$_2$SO$_4$-THF; HF-MeCN (88%).

Fig.16. Synthesis of (+)-sporogen-AO 1.

72 was followed by silylation to give **73**. A methyl group was introduced to **73** by reductive methylation to give **74**. Annulation of **74** to **76** was achieved via **75**. The bicyclic intermediate **76** gave (+)-sporogen-AO 1 (**78**) via **77**. The overall yield of **78** from **70** was 6.7% in 20 steps.

4.5 Other examples of yeast reduction of cyclic β-keto esters

(i) Synthesis of 6a-carbaprostaglandin I$_2$ and pentalenolactone E methyl ester. Reduction of (±)-2-ethoxycarbonyl-7,7-ethylenedioxybicyclo[3.3.0]octan-3-one (**79**) with baker's yeast or Saccharomyces bailii KI 0116 gave a mixture of **80** and **81**, which could be separated by silica gel chromatography (ref.91). In this case, the yeast reduction was useful in achieving the kinetic resolution. The keto ester **81** was our

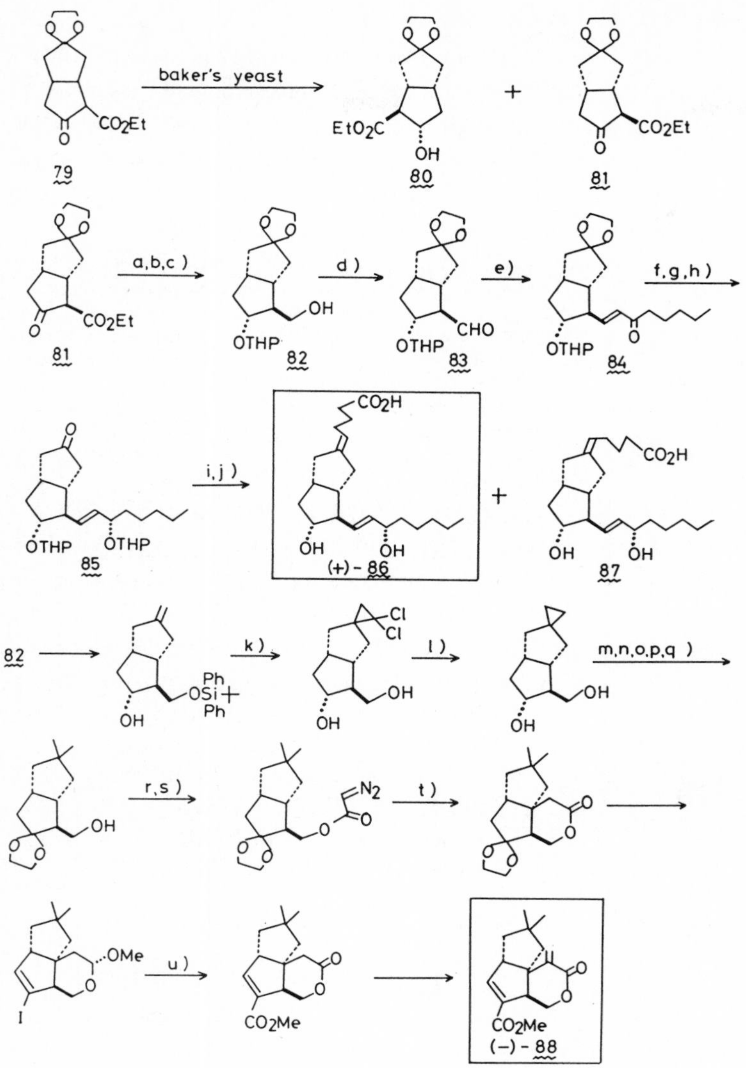

Reagents: a) NaBH$_4$/EtOH (76%); b) DHP, PPTS/CH$_2$Cl$_2$ (quant.);
c)LAH/ether (98%); d) PCC; e) (MeO)$_2$P(O)CH$_2$CO(CH$_2$)$_4$Me, NaH/THF (66%
from 82); f) NaBH$_4$/MeOH; g) AcOH-aq THF; h) DHP, PPTS/CH$_2$Cl$_2$ (44% from
84); i) Ph$_3$P=CH(CH$_2$)$_3$CO$_2$Na/DMSO; j) aq AcOH/THF (38% from 85); k)
CHCl$_3$, NaOH, Et$_3$PhCH$_2$NCl; Bu$_4$NF/THF (87%) l) Li/$\underline{t}$-BuOH-THF; m)
H$_2$/Pt/AcOH; n) $\underline{t}$-BuPh$_2$SiCl; o) PCC/CH$_2$Cl$_2$; p) HOCH$_2$CH$_2$OH, TsOH; q)
Bu$_4$NF/THF; r) ClCOCH=NNHTs, AgCN; s) Et$_3$N/CHCl$_2$ (65%); t) Rh$_2$(OAc)$_4$;
u) Ni(CO)$_4$, NaOMe, MeOH.

Fig.17. Synthesis of (+)-6a-Carbaprostaglandin I$_2$ and
(-)-pentalenolactone E methyl ester.

starting material for the synthesis of (+)-6a-carbaprostaglandin I$_2$ (**86**) (ref.91), and that of (−)-pentalenolactone E methyl ester (**88**) (ref.92) as shown in Fig.17. The synthesis of (+)-**86** involved aldehyde **83** as the key-intermediate, which was prepared from **81** via **82**. Two side-chains were attached to **83** by two Wittig reactions (**83** → **84** and **85** → **86**+**87**).

In the synthesis of (−)-**88**, attachment of two methyl groups onto the bicyclo[3.3.0]octane ring system was the major problem.

Experimental. Yeast reduction of 79 to give 80 and 81 (ref.91).

Wet cells of S. bailii (108 g) was dispersed to 0.1M phosphate buffer (pH 7, 1000 ml) containing glucose (200 g) at 37°C. After the flask was shaken at 37°C for 40 min, an emulsion of **79** (5.00 g) in 0.2% Triton X-100 soln (150 ml) was added to the fermentation mixture and the mixture was shaken at 37°C for 5 h. Then the yeast-cells were removed by centrifugation. The supernatant was saturated with NaCl and extracted with EtOAc (500 ml x 3). The combined EtOAc soln was washed with brine, dried (Na$_2$SO$_4$), and concentrated in vacuo. The residue was chromatographed over SiO$_2$ (75 g). The fraction earlier eluted with n-hexane-ether (4:1) gave (+)-**81** (2.01 g, 40%), n$_D^{21}$ 1.4927; [α]$_D^{21}$ +23.9° (c=2.55, CHCl$_3$). The fraction later eluted with n-hexane-ether (4:1) gave (+)-**80** (1.90 g, 38%), n$_D^{21}$ 1.4850; [α]$_D^{21}$ +4.1° (c=1.87, CHCl$_3$).

(ii) Synthesis of talaromycins A and B. (−)-Talaromycins A (**93**) and B (**94**) are fungal toxins isolated by Lynn et al. in 1982 from Talaromyces stipitatus (ref.93). We recently synthesized them employing the chiral building blocks of microbial origin as shown in Fig.18 (ref.94). The key-step was the reduction of **89** to **90** (64% e.e.) with baker's yeast. A Wittig reagent **91** prepared from **90** was coupled with

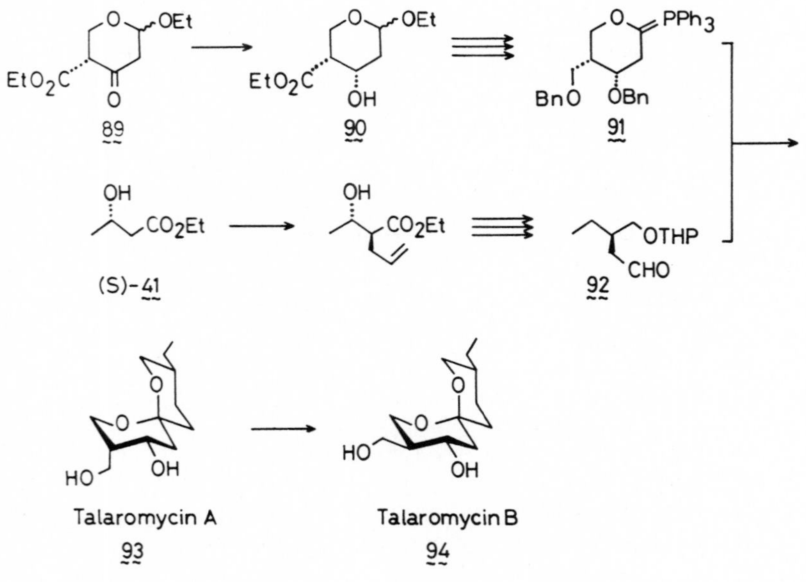

Fig.18. Synthesis of (−)-talaromycins A and B.

aldehyde **92,** which was synthesized from (<u>S</u>)-**41** to give talaromycin A (**93**) after deprotection and spiroacetalization. Acid-catalyzed isomerization of **93** yielded the more stable talaromycin B (**94**).

5 β-HYDROXY KETONES AS CHIRAL BUILDING BLOCKS

5.1 (S)-3-Hydroxy-2,2-dimethylcyclohexanone as a building block

(i) <u>Preparation of (S)-3-hydroxy-2,2-dimethylcyclohexanone.</u> Microbial reduction of prochiral cyclopentane- and cyclohexane-1,3-diones was extensively studied in 1960's in relation to steroid total synthesis (pp.28-31 of ref.5). Reduction of 2,2-dimethylcyclohexane-1,3-dione (**95**) to (<u>S</u>)-3-hydroxy-2,2-dimethylcyclohexanone (**96**) was first reported by Kieslich, Djerassi and their co-workers (ref.95). They reduced **95** with <u>Kloeckera magna</u> ATCC 20109, and obtained (<u>S</u>)-**96** of >95% e.e. We found that the reduction of **95** can also be effected with conventional baker's yeast, and secured (<u>S</u>)-**96** of 98-99% e.e. as determined by the HPLC analysis of the corresponding (<u>S</u>)-α-methoxy-α-trifluoromethyl-phenylacetate (MTPA ester) (refs.96 and 97). (<u>S</u>)-3-Hydroxy-2,2-dimethylcyclohexanone **96** was proved to be a versatile chiral building block in terpene synthesis.

Experimental. <u>Yeast reduction of</u> **95** <u>to</u> (<u>S</u>)-**96** (ref.96).
A soln of **95** (15.0 g) in 95% EtOH (30 ml) and 0.2% Triton X-100 soln (150 ml) was added to a suspension of dry yeast (Oriental Yeast Co., 200 g) in a 15% sucrose soln in water (3000 ml) at 30°C with stirring. The mixture was stirred at 30°C with aeration (16000 ml/min) for 48 h. Ether (200 ml) was then added to the mixture, and the mixture was left to stand overnight. After the flocculated cells of the yeast had precipitated, the mixture was filtered through Celite. The filtrate was saturated with NaCl and extracted with EtOAc, and the filter-cake was washed with EtOAc. The extract and washings were combined, washed with brine, dried (MgSO$_4$) and concentrated in vacuo. The residual oil (27 g) was chromatographed over SiO$_2$ (220 g). Elution with n-hexane-EtOAc (5:1) recovered **95** (4.74 g). Elution with EtOAc gave (<u>S</u>)-**96** (8.8 g, 79%, based on the consumed **95**). A small portion was distilled to give an analytical sample, b.p. 85-87°C/3.7 Torr, n$_D^{21}$ 1.4747, [α]$_D^{21}$ +24.1° (c=1.12, CHCl$_3$); νmax (film) 3470 (s), 1705 (s), 1055 (s), 985 (s) cm^{-1}.

(ii) <u>Synthesis of polygodial.</u> Polygodial (**100**) is a hot-tasting sesquiterpene first isolated from <u>Polygonum hydropiper</u> (refs.98 and 99). It was also isolated from <u>Warburgia stuhlmanni</u> and shown to possess antifeedant activity against some pest insects (ref.100). Racemic polygodial [(±)-**100**] has been synthesized several times to date (ref.101). The first chiral synthesis of the enantiomers of polygodial (**100**) was achieved by us starting from (<u>S</u>)-**96** as shown in Fig.19 (ref.102). The driving force to carry out the synthesis was Kubo's claim that the specific absolute configuration of the antifeedant appeared to govern the hotness of its taste (ref.103). Our synthetic strategy was to use the hydroxyl group of **96** as the handle to facilitate the chiral synthesis. The Diels-Alder reaction between **97** and dimethyl acetylenedicarboxylate was the key-reaction. The product was a mixture

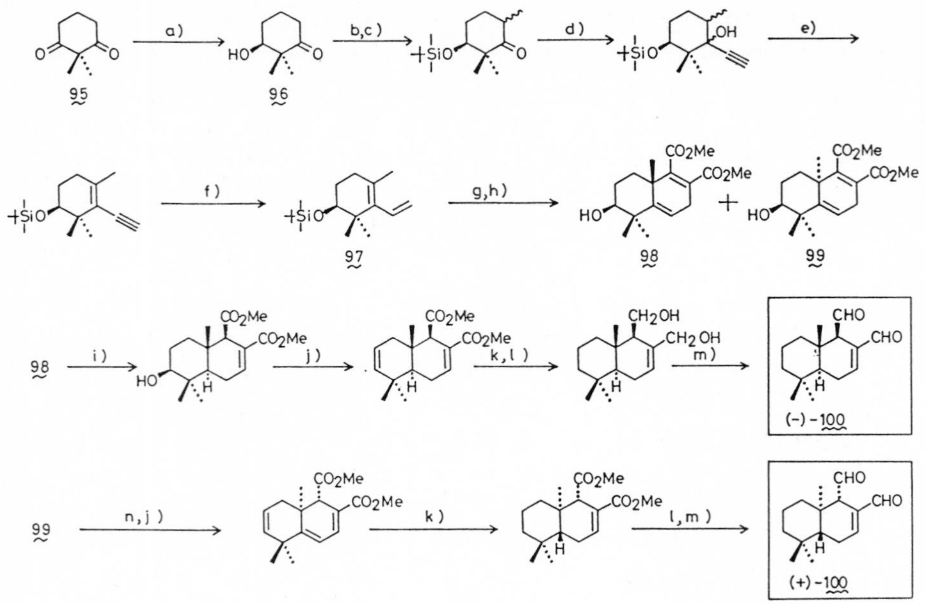

Reagents: a) baker's yeast (63-79%); b) t-BuMe$_2$SiCl (81%); c) LDA/THF-HMPA, MeI (89%); d) NaC≡CH/liq NH$_3$ (99%); e) CuSO$_4$/xylene, Δ (51%); f) H$_2$/Pd-CaCO$_3$, quinoline in n-pentane (quant.); g) MeO$_2$CC≡CCO$_2$Me, 110°C, 30 h (97%); h) aq HF-MeCN (27% of 98 and 27% of 99 after MPLC); i) DBU/THF, Δ; H$_2$/Pd-C (80% from 98); j) TfCl-DMAP/CH$_2$Cl$_2$ (86%); k) H$_2$/Pd (89%); l) LAH/ether (82%); m) (COCl)$_2$, DMSO, Et$_3$N/CH$_2$Cl$_2$ (63%); n) DBU/THF

Fig.19.   Synthesis of the enantiomers of polygodial.

of two isomers.   Fortunately, after desilylation, 98 and 99 were readily separable by chromatography.   (-)-Polygodial (100), the natural enantiomer, was obtained in 3.0% overall yield from 96.   Likewise, (+)-100 was also synthesized.   Both the enantiomers of polygodial (100) showed identical biological properties such as hot taste to human tongue, antifeedant activity against insects, and pisicidal activity against fishes.

   (iii) Synthesis of O-methylpisiferic acid.   (+)-O-Methylpisiferic acid (106) was first isolated by Yatagai and Takahashi in 1980 from the leaves of Chamaecyparis pisifera (ref.104).   In 1984 Marumo and his co-workers demonstrated that (+)-106 inhibits the hatching of the two-spotted spider mite, Tetranychus urticae (ref.105).   Since the two-spotted spider mite is a serious pest to many crops, we became inter-

ested in synthesizing this diterpene **106.** In Fig.20 is shown our synthesis of O-methylpisiferic acid starting from (S)-**96** (ref.106). Here
again we used the hydroxyl group of **96** as the handle to facilitate the
chiral synthesis. β-Keto ester **101** was converted into a stereoisomeric
mixture of bicyclic keto esters **102.** Treatment of **102** with p-toluene-

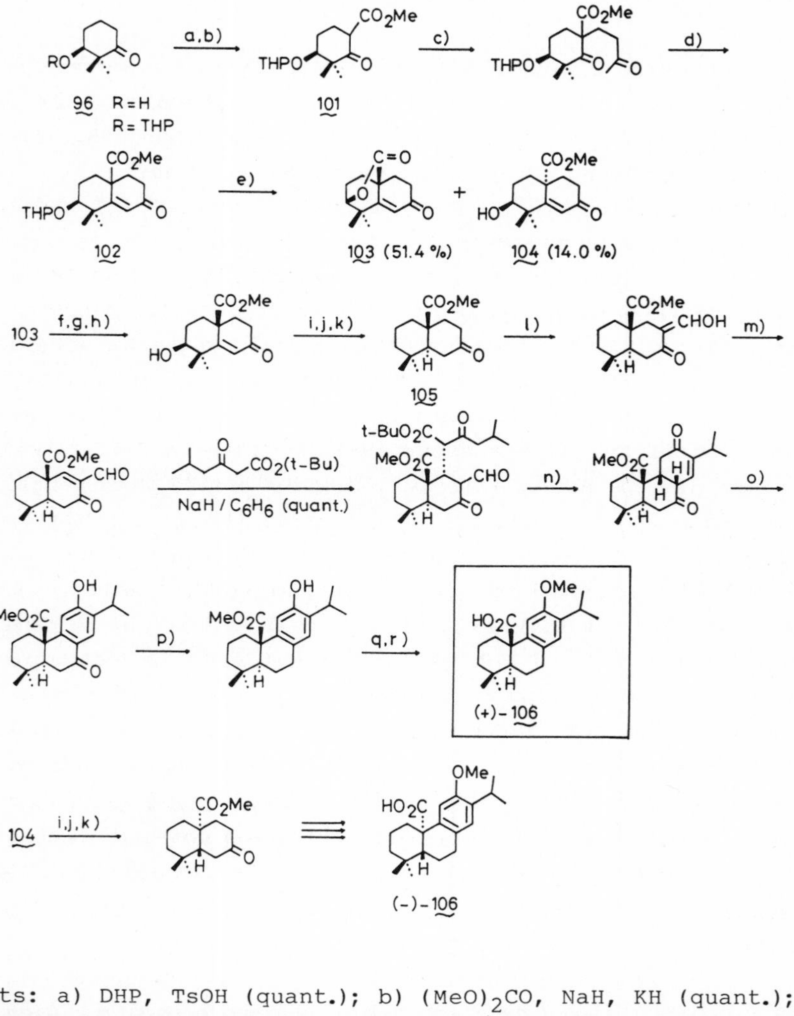

Reagnts: a) DHP, TsOH (quant.); b) (MeO)$_2$CO, NaH, KH (quant.); c),
MeCOCH=CH$_2$, MeONa (93%); d) pyrrolidine/C$_6$H$_6$ (85%); e) TsOH; f) aq
KOH; g) H$^+$; h) CH$_2$N$_2$ (70%); i) TfCl, DMAP (80%); j) H$_2$, PtO$_2$ (quant.);
k) CrO$_3$ (80%); l) NaH, HCO$_2$Me (quant.); m) DDQ (94%); n) TsOH/AcOH
(84%); o) C$_5$H$_5$NHBr$_3$/AcOH (quant.); p) H$_2$, Pd-C, H$_2$SO$_4$/EtOAc (89%); q)
Me$_2$SO$_4$, K$_2$CO$_3$ (82%); r) t-BuOK/DMSO (71%).

Fig.20. Synthesis of the enantiomers of O-methylpisiferic acid.

sulfonic acid in methanol gave a separable mixture of **103** and **104.** The major product **103** yielded keto ester **105.** Conversion of **105** to the final product **106** was carried out essentially in the same manner as reported by Meyer et al. in their synthesis of related racemic diterpenes (ref.107). The overall yield of (+)-**106** from **96** was 9% in 18 steps. Similarly, (-)-**106** was prepared in 1.8% overall yield from **96** in 16 steps. Only the natural (+)-**106** was shown to be bioactive on the mite.

(iv) Synthesis of juvenile hormone III. Juvenile hormone III [JH III, (+)-**117**], which was first isolated from organ cultures of corpora allata of the tobacco hornworm moth (Manduca sexta) (ref.108), was later found in at least one stage of development in nearly all insects surveyed to date (ref.109). Very recently, (±)-JH III and methyl farnesoate were detected even in a crustacean, adult spider crabs (Libinia emarginata) (ref.110). A number of syntheses of JH III were published (ref.111). The chiral synthesis of JH III, however, was a difficult task, and the existing three chiral syntheses afforded the enantiomers of JH III of obscure enantiomeric purity (refs. 112-114).

Our recent synthesis of the enantiomers of JH III started from (S)-3-hydroxy-2,2-dimethylcyclohexanone (**96**) as shown in Fig.21 (ref.115). The key-step was the Baeyer-Villiger oxidation of **107** with m-chloroperbenzoic acid to give ε-lactone **108.** Because **108** was crystalline, it could be purified by recrystallization to give **108** of 100% e.e. The lactone **108** was converted to acyclic acetylene **109.** Methoxycarbonylation of **109** to **110** was followed by the Michael addition of thiophenol to give a mixture of **111** and **112.** After chromatographic separation, **111** was treated with methylmagnesium bromide and cuprous iodide to give **113.** Chain-elongation of **113** to **114** was followed by methylation with lithium dimethylcuprate to give **115.** Deprotection of **115** gave dihydroxy ester **116a,** which was enantiomerically pure as checked by the HPLC analysis of **116b** and **116c.** The natural (+)-JH III (**117**) was obtained by the ring-closure of the monomesylate of **116a,** while the unnatural (-)-JH III was prepared from **116a** via an acetoxy bromide. The JH activity of the enantiomers of JH III is currently under investigation.

5.2 (2S,3S)-2-Ethyl-3-hydroxy-2-methylcyclohexanone as a building block

(i) Preparation of (2S,3S)-2-ethyl-3-hydroxy-2-methylcyclohexanone. After the completion of our synthesis of JH III, we decided to synthesize all of the other JH's, especially JH I (**124**) and JH II (**125**). These two JH's were isolated from the adult male giant silk moth

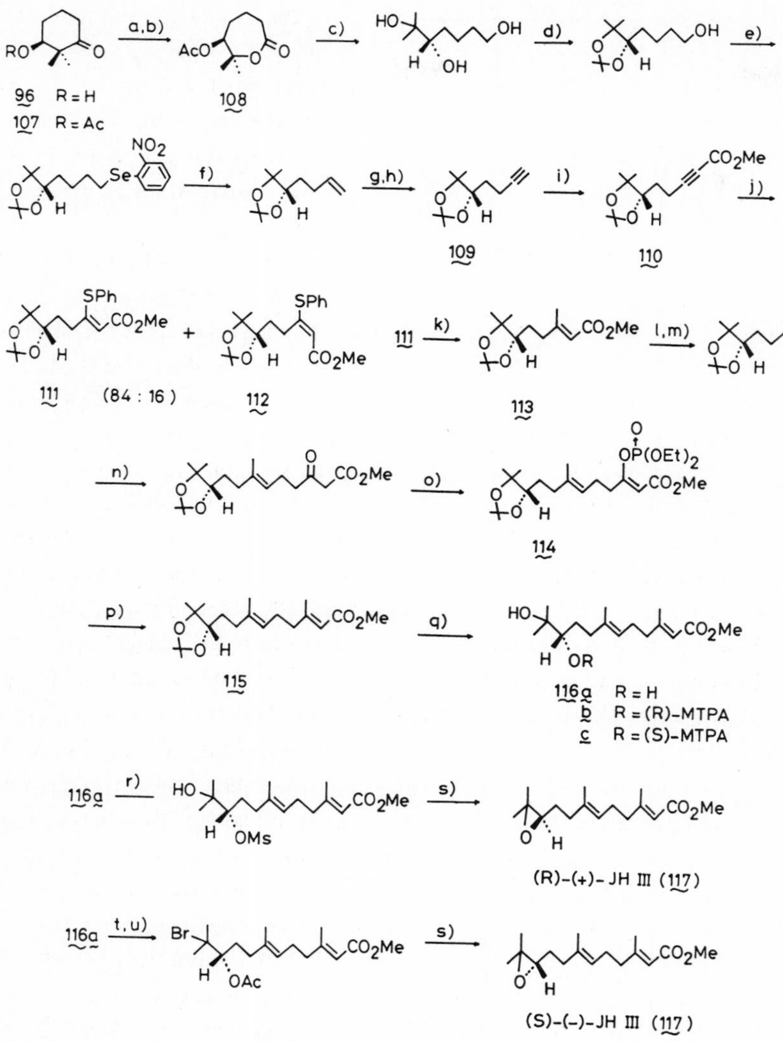

Reagents: a) Ac$_2$O (quant.); b) MCPBA (92%); c) LAH (quant.); d) Me$_2$C(OMe)$_2$, TsOH/Me$_2$CO (99%); e) $\underline{o}$-(O$_2$N)C$_6$H$_4$SeCN, ($\underline{n}$-Bu)$_3$P (quant.); f) H$_2$O$_2$ (80%); g) C$_5$H$_5$NHBr$_3$ (87%); h) NaNH$_2$/liq NH$_3$ (72%); i) $\underline{n}$-BuLi, ClCO$_2$Me (85%); j) PhSH, NaOH/MeOH (91%), k) MeMgBr, CuI (71%); l) LAH (90%); m) $\underline{n}$-BuLi, TsCl, LiBr; n) MeCOCH$_2$CO$_2$Me, NaH, $\underline{n}$-BuLi; o) NaH, (EtO)$_2$POCl; p) Me$_2$CuLi (40%, 4 steps); q) 75% AcOH (98%); r) Ms$_2$O, Et$_3$N (quant.); s) MeONa/MeOH [79% for ($\underline{R}$)-**117**; 83% for ($\underline{S}$)-**117**]; t) Ac$_2$O; u) PBr$_3$ (quant.).

Fig.21. Synthesis of the enantiomers of juvenile hormone III.

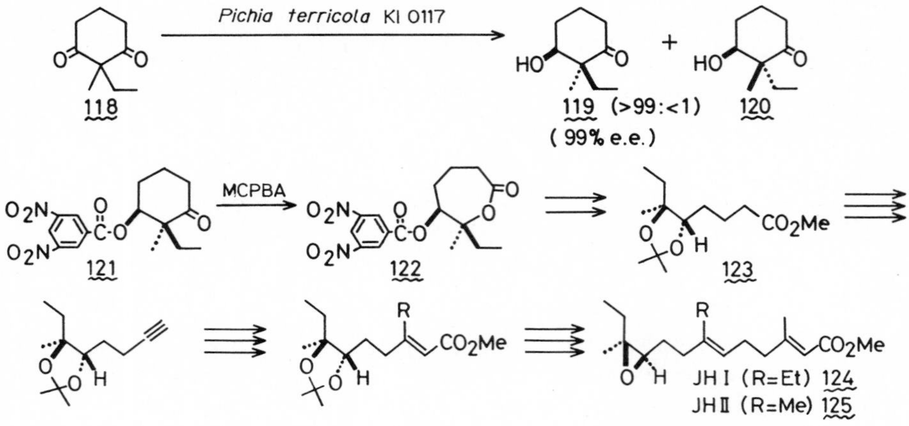

Fig.22. Synthesis of juvenile hormones I and II.

(Hyalophora cecropia) by Roller et al. in 1967 (ref.116) and by Meyer et al. in 1968 (ref.117). The structures of JH I and JH II demanded that we should prepare hydroxy ketone **119**. Yeast reduction of a prochiral 1,3-diketone such as **118** with two different alkyl groups at C-2 was known to yield a diastereomeric mixture of two hydroxy ketones like **119** and **120** (ref.118). After some screening experiments, we found that Pichia terricola KI 0117 reduces **118** diastereo- and enantioselectively to give the desired **119** of 99% e.e. (ref.119). It is therefore impor-tant to find a microorganism which is most appropriate for the specific biotransformation required in that project.

Experimental. Yeast reduction of **118** to **119** (ref.119).
2-Ethyl-2-methylcyclohexane-1,3-dione **118** (8.17 g) in Triton X-100 (0.2%, 80 ml) was treated with the cells of Pichia terricola KI 0117 (87 g) in glucose soln (total 175 g/1000 ml) at 37°C for 24 h. A usual workup and purification gave 7.08 g (86%) of ketol **119**, b.p. 109–114°C/5 Torr, n_D^{24} 1.4743, $[\alpha]_D^{24}$ +65.6° (c=1.07, CHCl$_3$); vmax (film) 3450 (s), 1695 (s), 1060 (s), 990 (s) cm^{-1}.

(ii) Synthesis of juvenile hormones I and II. Although four chiral syntheses of JH I and JH II were reported, none of them afforded the pure natural enantiomers of JH I and JH II (refs.120-123). Starting from **119**, we accomplished the synthesis of (+)-JH I and (+)-JH II as shown in Fig.22 (ref.119). Thanks to the crystalline nature of **121** and **122**, we were able to obtain the final products of 100% e.e. The syn-thetic routes to JH I and JH II were almost the same as that employed for the synthesis of JH III (Fig.21; ref.115) except that an ethyl group was attached to C-7 in the case of JH I.

6   CONCLUSION

I reviewed our works on the preparation and use of the chiral building blocks of microbial origin. In conclusion, versatility of our
building blocks enabled us to achieve numerous chiral syntheses of
bioactive natural products so as to help the cooperation between
chemists and biologists. Ethyl 3-hydroxybutanoate, methyl 3-hydroxy-
pentanoate and (S)-3-hydroxy-2,2-dimethylcyclohexanone were most useful
in our synthetic endeavor. Natural products derived from these three
chiral building blocks are shown in Figs.23-25 with refs. below the
formulae. Finally I must add that recent developments in analytical
methods to determine the absolute configuration and enantiomeric purity
of chiral organic compounds (refs.143 and 144) were fully taken
advantage of in executing our chiral syntheses.

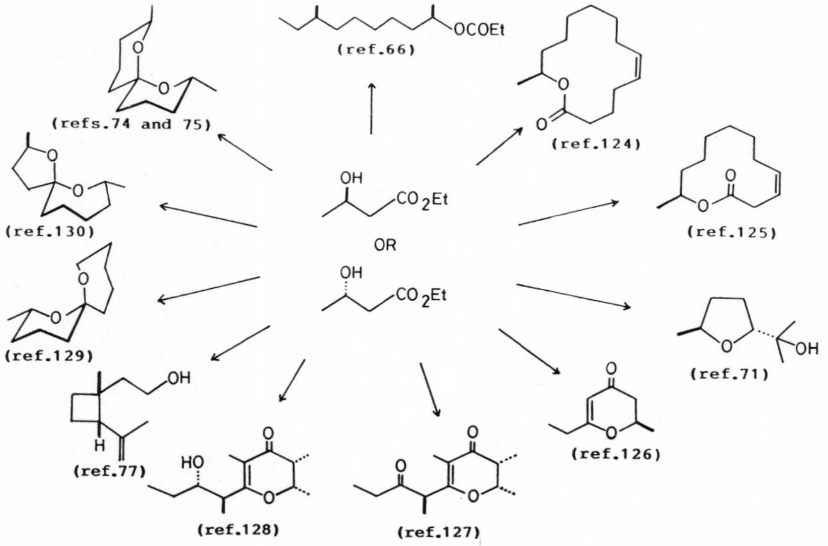

Fig.23. Pheromones synthesized from ethyl 3-hydroxybutanoate.

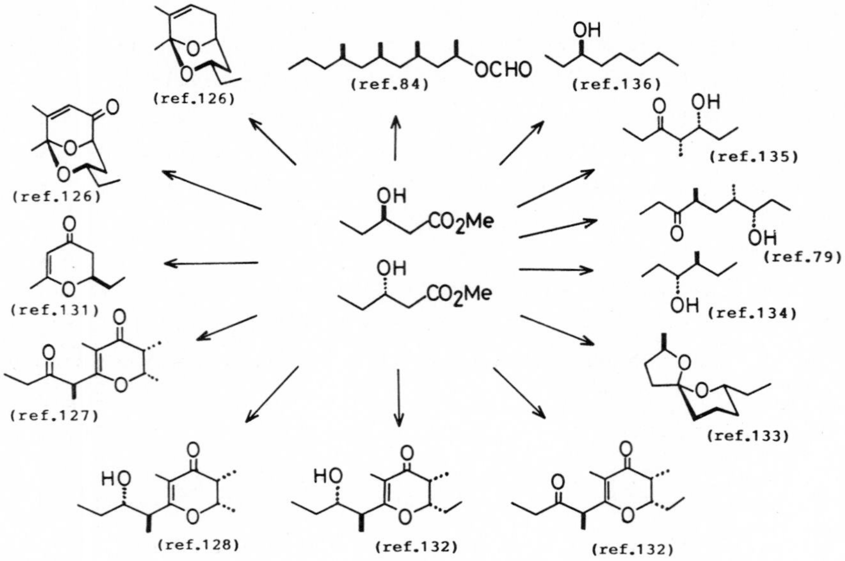

Fig.24. Pheromones synthesized from methyl 3-hydroxypentanoate.

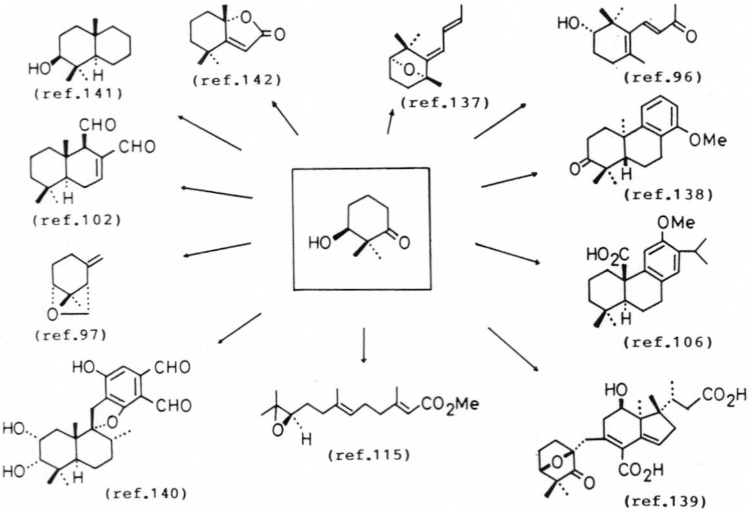

Fig.25. Natural products synthesized from (S)-3-hydroxy-
2,2-dimethylcyclohexanone.

ACKNOWLEDGEMENT

I thank Profs. Y. Minoda and T. Beppu of this Department for their advice on microbiological experiments. The skill and enthusiasm of my co-workers whose names were cited in refs. were the key to our success-ful syntheses. My thanks are due to Drs. T. Sugai and S. Kuwahara and my students for the preparation of the camera-ready typescript.

REFERENCES

1 L. Pasteur, Compt. rend., 46 (1858) 615.
2 J. Jacques, C. Glos and S. Bourcier, Absolute configuration of 6000 selected compounds with one asymmetric carbon atom, in: H. B. Kagan (Ed.), Stereochemistry, Vol. 4, Georg Thieme, Stuttgart, 1977, pp. 1-602.
3 J. W. Scott, Readily available chiral carbon fragments and their use in synthesis, in: J. D. Morrison and J. W. Scott (Eds.), Asym-metric Synthesis, Vol. 4, Academic Press, Orlando, 1984, pp. 1-226.
4 J. B. Jones, C. J. Sih and D. Perlman (Eds.), Application of Bio-chemical Systems in Organic Chemistry, Part 1, John Wiley, New York, 1976.
5 K. Kieslich, Microbial Transformations of Non-Steroid Cyclic Com-pounds, Georg Thieme, Stuttgart, 1976.
6 A. Fischili, Chiral building blocks in enantiomer synthesis using enzymatic transformations, in: R. Schefford (Ed.), Modern Synthetic Methods, Vol. 2, Salle/Sauerlander, Frankfurt a. M., 1980, pp. 269-350.
7 J. B. Jones, Enzymes as chiral catalysts, in: J. D. Morrison (Ed.), Asymmetric Synthesis, Vol. 5, Academic Press, Orlando, 1985, pp. 309-344.
8 J. Tramper, H. C. van der Plas and P. Linko (Eds.), Biocatalysts in Organic Synthesis, Elsevier, Amsterdam, 1985.
9 R. Porter and S. Clark (Eds.), Enzymes in Organic Synthesis (CIBA Foundation Symposium 111), Pitman, London, 1985.
10 W. Bartmann and K. B. Sharpless (Eds.), Stereochemistry of Organic and Bioorganic Transformations, VCH, Weinheim, 1987.
11 J. B. Jones, Tetrahedron, 42, (1986) 3351.
12 S. Butt and S. M. Roberts, Nat. Prod. Rep., 3 (1986) 489.
13 H. Simon, J. Bader, H. Gunther, S. Neumann and J. Thanos, Angew. Chem. Int. Ed., 24 (1985) 539.
14 G. M. Whitesides and C.-H. Wong, Angew. Chem. Int. Ed., 24 (1985) 617.
15 S. Butt and S. M. Roberts, Chem. Br., (1987) 127.
16 A. Akiyama, M. Bednarski, M.-J. Kim, E. S. Simon, H. Waldmann and G. M. Whitesides, Chem. Br., (1987) 645.
17 P. J. Fodor, V. E. Price and J. P. Greenstein, J. Biol. Chem., 178 (1949) 503.
18 I. Chibata, Applications of immobilized enzymes for asymmetric reactions, in: E. L. Eliel and S. Otsuka (Eds.), Asymmetric Reac-tions and Processes in Chemistry, Am. Chem. Soc., Washington, D. C., 1982.
19 T. Okazaki, T. Noguchi, K. Igarashi, Y. Sakagami, H. Seto, K. Mori, H. Naito, T. Masumura and M. Sugahara, Agric. Biol. Chem., 47 (1983) 2949.
20 K. Mori, T. Okazaki, T. Noguchi and H. Naito, Agric. Biol. Chem., 47 (1983) 2131.
21 K. Mori, T. Sugai, Y. Maeda, T. Okazaki, T. Noguchi and H. Naito, Tetrahedron, 41 (1985) 5307.

710

22   Y. Ito, T. Noguchi and H. Naito, Anal. Biochem., 151 (1985) 28.
23   G. Kawai and Y. Ikeda, Biochim. Biophys. Acta, 719 (1982) 612.
24   G. Kawai and Y. Ikeda, Biochim. Biophys. Acta, 754 (1983) 243.
25   K. Mori and Y. Funaki, Tetrahedron, 41 (1985) 2369.
26   K. Mori and Y. Funaki, Tetrahedron, 41 (1985) 2379.
27   K. Mori, Tetrahedron, 33 (1977) 289 and refs. therein.
28   K. Mori and H. Iwasawa, Tetrahedron, 36 (1980) 2209.
29   S. Senda and K. Mori, Agric. Biol. Chem., 51 (1987) 1379.
30   H. Yamane, Y. Sato, N. Takahashi, K. Takano and M. Furuya, Agric.
      Biol. Chem., 44 (1980) 1697.
31   Y. Masaoka, M. Sakakibara and K. Mori, Agric. Biol. Chem., 46
      (1982) 2319.
32   T. Sugai and K. Mori, Agric. Biol. Chem., 48 (1984) 2155.
33   T. Sugai and K. Mori, Agric. Biol. Chem., 48 (1984) 2497.
34   K. Mori and T. Otsuka, Tetrahedron, 41 (1985) 547.
35   B. Cambou and A. M. Klibanov, J. Am. Chem. Soc., 106 (1984) 2687.
36   P. Mohr, L. Rösslein and C. Tamm, Helv. Chim. Acta, 70 (1987) 142.
37   V. Kerscher and W. Kreiser, Tetrahedron Lett., 28 (1987) 531.
38   R. Riva, L. Banfi, B. Danieli, G. Guanti, G. Lesma and G.
      Palmisano, J. Chem. Soc., Chem. Commun., (1987) 299.
39   T. M. Stokes and A. C. Oehlschlager, Tetrahedron Lett, 28 (1987)
      2091.
40   E. Guibé-Jampel, G. Rousseau and J. Salaün, J. Chem. Soc., Chem.
      Commun., (1987) 1080.
41   Z.-F. Xie, H. Suemune and K. Sakai, J. Chem. Soc., Chem. Commun.,
      (1987) 838.
42   A. J. Pearson, H. S. Bansal and Y.-S. Lai, J. Chem. Soc., Chem.
      Commun., (1987) 519.
43   S. Takano, K. Tanigawa and K. Ogasawara, J. Chem. Soc., Chem.
      Commun., (1976) 189.
44   Y.-F. Wong, C.-S. Chen, G. Girdaukas and C. J. Sih, J. Am. Chem.
      Soc., 106 (1984) 3695.
45   K. Laumen and M. P. Schneider, Tetrahedron Lett., 25 (1984) 5875.
46   K. Laumen, E. H. Reimerdes and M. P. Schneider, Tetrahedron Lett.,
      26 (1985) 407.
47   K. Laumen and M. P. Schneider, J. Chem. Soc., Chem. Commun., (1986)
      1298.
48   D. R. Deardorff, A. J. Matthews, D. S. McMeekin and C. L. Craney,
      Tetrahedron Lett., 26 (1985) 5615.
49   T. Sugai and K. Mori, Synthesis, in press.
50   B. J. Baker, R. K. Okuda, P. T. K. Yu and P. J. Scheuer, J. Am.
      Chem. Soc., 107 (1985) 2976.
51   K. Mori and T. Takeuchi, Tetrahedron, in press.
52   E. M. Kleiner, S. A. Pliner, V. S. Soifer, V. V. Onoprienko, T. A.
      Balasheva, B. V. Rozynov and A. S. Khokhlov, Bioorg. Khim., 2
      (1976) 1142.
53   K. Mori and K. Yamane, Tetrahedron, 38 (1982) 2919.
54   K. Mori, Tetrahedron, 39 (1983) 3107.
55   K. Mori and N. Chiba, unpublished results.
56   G. M. R. Tombo, H.-P. Schär, X. F. Busquets and O. Ghisalba,
      Tetrahedron Lett., 27 (1986) 5707.
57   Y.-F. Wang and C. J. Sih, Tetrahedron Lett., 25 (1984) 4999.
58   D. Breitgoff, K. Laumen and M. P. Schneider, J. Chem. Soc., Chem.
      Commun., (1986) 1523.
59   B. S. Deol, D. D. Ridley and G. W. Simpson, Aust. J. Chem., 29
      (1976) 2459.
60   E. Hungerbühler, D. Seebach and D. S. Wasmuth, Helv. Chim. Acta, 64
      (1981) 1467.
61   K. Mori, Tetrahedron, 37 (1981) 1341.
62   B. Wipf, E. Kupfer, R. Bertazzi and H. G. W. Leuenberger, Helv.
      Chim. Acta, 66 (1983) 485.

63   M. Hirama, M. Shimizu and M. Iwashita, J. Chem. Soc., Chem. Commun., (1983) 599.
64   T. Sugai, M. Fujita and K. Mori, Nippon Kagaku Kaishi, (1983) 1315.
65   D. Seebach and M. F. Züger, Helv. Chim. Acta, 65 (1982) 495.
66   K. Mori and H. Watanabe, Tetrahedron, 40 (1984) 299.
67   T. Fukui, A. Yoshimoto, M. Matsumoto, S. Hosokawa, T. Saito, H. Nishikawa and K. Tomita, Arch. Microbiol., 110 (1976) 149.
68   K. J. Byrne, A. A. Swigar, R. M. Silverstein, J. H. Borden and E. Stokkink, J. Insect Physiol., 20 (1974) 1895.
69   K. Mori, Tetrahedron, 31 (1975) 3011.
70   J. H. Borden, L. Chong, J. A. McLean, K. N. Slessor and K. Mori, Science, 192 (1976) 894.
71   K. Mori and P. Puapoomchareon, Liebigs Ann. Chem., (1987) 271.
72   W. Francke, W. Reith, G. Bergström and J. Tengö, Naturwiss., 67 (1980) 149.
73   K. Mori and K. Tanida, Heterocycles, 15 (1981) 1171.
74   K. Mori and K. Tanida, Tetrahedron, 37 (1981) 3221.
75   K. Mori and H. Watanabe, Tetrahedron, 42 (1986) 295.
76   J. H. Tumlinson, D. D. Hardee, R. C. Gueldner, A. C. Thompson, P. A. Hedin and J. P. Minyard, Science, 166 (1969) 1010.
77   K. Mori and M. Miyake, Tetrahedron, 43 (1987) 2229.
78   J. Hasegawa, S. Hamaguchi, M. Ogura and K. Watanabe, J. Ferment. Technol., 59 (1981) 257.
79   K. Mori and H. Watanabe, Tetrahedron, 41 (1985) 3423.
80   K. Mori, H. Mori and T. Sugai, Tetrahedron, 41 (1985) 919.
81   T. Chuman, M. Kohno, K. Kato and M. Noguchi, Tetrahedron Lett., (1979) 2361.
82   Y. Kuwahara, L. T. M. Yen, Y. Tominaga, K. Matsumoto and Y. Wada, Agric. Biol. Chem., 46 (1982) 2283.
83   K. Mori and S. Kuwahara, Tetrahedron, 42 (1986) 5545.
84   K. Mori and S. Kuwahara, Tetrahedron, 42 (1986) 5539.
85   R. M. Lukes, G. I. Poos and L. H. Sarett, J. Am. Chem. Soc., 74 (1952) 1401.
86   K. Mori, M. Matsui and Y. Sumiki, Agric. Biol. Chem., 25 (1961) 205.
87   T. Kitahara and K. Mori, Tetrahedron Lett., 26 (1985) 451.
88   S. Tanaka, K. Wada, S. Marumo and H. Hattori, Tetrahedron Lett., 25 (1984) 5907.
89   K. Mori and H. Tamura, Liebigs Ann. Chem., in press.
90   T. Kitahara, H. Kurata and K. Mori, Tetrahedron, to be submitted.
91   K. Mori and M. Tsuji, Tetrahedron, 42 (1986) 435.
92   K. Mori and M. Tsuji, Tetrahedron, to be submitted.
93   D. G. Lynn, N. J. Phillips, W. C. Hutton, J. Shabanowitz, D. J. Fennell and R. J. Cole, J. Am. Chem. Soc., 104 (1982) 7319.
94   K. Mori and M. Ikunaka, Tetrahedron, 43 (1987) 45.
95   Y. Lu, G. Barth, K. Kieslich, P. D. Strong, W. L. Duax and C. Djerassi, J. Org. Chem., 48 (1983) 4549.
96   M. Yanai, T. Sugai and K. Mori, Agric. Biol. Chem., 49 (1985) 2373.
97   K. Mori and H. Mori, Tetrahedron, 41 (1985) 5487.
98   A. Ohsuka, Nippon Kagaku Zasshi, 83 (1962) 757.
99   C. S. Barnes and J. W. Loder, Aust. J. Chem., 15 (1962) 322.
100   I. Kubo, Y.-W. Lee, M. Pettei, F. Pilkiewicz and K. Nakanishi, J. Chem. Soc., Chem. Commun., (1976) 1013.
101   D. M. Hollingshead, S. C. Howell, S. V. Ley, M. Mahon, N. M. Ratcliffe and P. A. Worthington, J. Chem. Soc., Perkin Trans. I (1983) 1579 and refs. therein.
102   K. Mori and H. Watanabe, Tetrahedron, 42 (1986) 273.
103   I. Kubo and J. Ganjian, Experientia, 37 (1981) 1063.
104   M. Yatagai and T. Takahashi, Phytochem., 19 (1980) 1149.
105   J.-W. Ahn, K. Wada, S. Marumo, H. Tanaka and Y. Osaka, Agric. Biol. Chem., 48 (1984) 2167.

712

106 K. Mori and H. Mori, Tetrahedron, 42 (1986) 5531.
107 W. L. Meyer, R. A. Manning, E. Schindler, R. S. Schroeder and D. C. Shew, J. Org. Chem., 41 (1976) 1005.
108 K. J. Judy, D. A. Schooley, L. L. Dunham, M. S. Hall, B. J. Bergot and J. B. Siddall, Proc. Natl. Acad. Sci. USA, 70 (1973) 1509.
109 D. A. Schooley, Analysis of the naturally occurring juvenile hormones, in: R. B. Turner (Ed.), Analytical Biochemistry of Insects, Elsevier, Amsterdam, 1977, pp. 241-287.
110 H. Laufer, D. Borst, F. C. Baker, C. Carrasco, M. Sinkus, C. C. Reuter, L. W. Tsai and D. A. Schooley, Science, 253 (1987) 202.
111 K. Mori, Synthetic chemistry of insect pheromones and juvenile hormones, in: R. Bognár, V. Bruckner and Cs. Szántay (Eds.), Recent Developments in the Chemistry of Natural Carbon Compounds, Akadémiai Kiadó, Budapest, 1979, pp. 9-209.
112 Y. Suzuki, K. Imai, S. Marumo and T. Mitsui, Agric. Biol. Chem., 36 (1972) 1849.
113 K. Imai, S. Marumo and T. Ohtaki, Tetrahedron Lett., (1976) 1211.
114 D. A. Schooley, M. J. Bergot, W. Goodman and L. J. Gilbert, Biochem. Biophys. Res. Commun., 81 (1978) 743.
115 K. Mori and H. Mori, Tetrahedron, in press.
116 H. Röller, K. H. Dahm, C. C. Sweeley and B. M. Trost, Angew. Chem. Int. Ed., 6 (1967) 179.
117 A. S. Meyer, H. A. Schneiderman, E. Hanzmann and J. H. Ko, Proc. Natl. Acad. Sci. USA, 60 (1968) 853.
118 D. W. Brooks, H. Mazdiyasni and P. G. Grothaus, J. Org. Chem., 52 (1987) 3223.
119 K. Mori and M. Fujiwhara, Tetrahedron, in press.
120 P. Loew and W. S. Johnson, J. Am. Chem. Soc., 93 (1971) 3765.
121 D. J. Faulkner and M. R. Petersen, J. Am. Chem. Soc., 93 (1971) 3766.
122 K. Imai, S. Marumo and K. Mori, J. Am. Chem. Soc., 96 (1974) 5925.
123 G. D. Prestwich and C. Wawrzeńczyk, Proc. Natl. Acad. Sci. USA, 82 (1985) 5290.
124 T. Sakai, H. Hamamoto and K. Mori, Agric. Biol. Chem., 50 (1986) 1621.
125 T. Sakai and K. Mori, Agric. Biol. Chem., 50 (1986) 177.
126 K. Mori and H. Kisida, Tetrahedron, 42 (1986) 5281.
127 K. Mori and T. Ebata, Tetrahedron, 42 (1986) 4413.
128 K. Mori and T. Ebata, Tetrahedron, 42 (1986) 4685.
129 K. Mori and M. Katsurada, Liebigs Ann. Chem., (1984) 157.
130 K. Mori, H. Soga and M. Ikunaka, Liebigs Ann. Chem., (1985) 2194.
131 S. Hayashi and K. Mori, Agric. Biol. Chem., 50 (1986) 3209.
132 T. Ebata and K. Mori, Agric. Biol. Chem., in press.
133 K. Mori and M. Ikunaka, Tetrahedron, 40 (1984) 3471.
134 M. Kato and K. Mori, Agric. Biol. Chem., 49 (1985) 3073.
135 K. Mori and T. Ebata, Tetrahedron, 42 (1986) 4421.
136 M. Fujiwhara and K. Mori, Agric. Biol. Chem., 50 (1986) 2925.
137 K. Mori and H. Tamura, Tetrahedron, 42 (1986) 2643.
138 T. Sugai, H. Tojo and K. Mori, Agric. Biol. Chem., 50 (1986) 3127.
139 K. Mori and H. Watanabe, unpublished work.
140 K. Mori and M. Komatsu, Liebigs Ann. Chem., in press.
141 K. Mori, H. Mori and M. Yanai, Tetrahedron, 42 (1986) 291.
142 K. Mori and Y. Nakazono, Tetrahedron, 42 (1986) 283.
143 J. D. Morrison (Ed.), Asymmetric Synthesis, Vol.1, Analytical Methods, Academic Press, New York, 1983.
144 K. Mori, The significance of chirality: methods for determining absolute configuration and optical purity of pheromones and related compounds, in: H. E. Hummel and T. A. Miller (Eds.), Techniques in Pheromone Research, Springer Verlag, New York, 1984, pp. 323-370.

SUBJECT INDEX

714